2020 International Conference on Simulation of Semiconductor Processes and Devices (SISPAD 2020)

Kobe, Japan
23 September – 6 October 2020

IEEE Catalog Number: CFP20SSD-POD
ISBN: 978-1-7281-7354-2

Copyright © 2020, The Japan Society of Applied Physics (JSAP)
All Rights Reserved

*** *This is a print representation of what appears in the IEEE Digital Library. Some format issues inherent in the e-media version may also appear in this print version.*

IEEE Catalog Number: CFP20SSD-POD
ISBN (Print-On-Demand):   978-1-7281-7354-2
ISBN (Online):   978-4-86348-763-5
ISSN: 1946-1569

Additional Copies of This Publication Are Available From:

Curran Associates, Inc
57 Morehouse Lane
Red Hook, NY 12571 USA
Phone: (845) 758-0400
Fax: (845) 758-2633
E-mail: curran@proceedings.com
Web: www.proceedings.com

2020 International Conference on Simulation of Semiconductor Processes and Devices (SISPAD 2020)

Kobe, Japan
23 September – 6 October 2020

IEEE Catalog Number: CFP20SSD-POD
ISBN: 978-1-7281-7354-2

2020 International Conference on Simulation of Semiconductor Processes and Devices (SISPAD)

SISPAD 2020

September 23-October 6, 2020
ALL-VIRTUAL Conference

Co-sponsored by
  The Japan Society of Applied Physics
  The Murata Science Foundation
Technical co-sponsored by
  The IEEE Electron Devices Society
Supported by
  The IEEE Electron Devices Society Kansai Chapter
Sponsors
  KIOXIA Corporation
  Sony Semiconductor Solutions Corporation

2020 International Conference on Simulation of Semiconductor Processes and Devices (SISPAD)

For reprint or republication permission, email to JSAP office at secretariat@jsap.or.jp.

FOREWORD

On behalf of the conference committee, we would like to welcome you to the 2020 International Conference on Simulation of Semiconductor Processes and Devices (SISPAD), which is held as an All-Virtual conference on September 23-October 6, 2020.

The SISPAD is a leading forum for the presentation and discussion of recent advances and developments in the field of semiconductor process, device, and circuit simulations, and their applications to advanced devices. The main themes of interest for SISPAD 2020 are: 1) Band Structure, 2) Computational Methodology, 3) Nanowire, 4) Material and Geometry Impact, 5) Reliability, 6) Power and Optoelectronic Devices, 7) Non-volatile Memory, 8) Transport, 9) High Speed Switching Devices, 10) Emerging Devices, 11) 2D and Nano System, 12) FET Devices and Design Technology Co-Optimization, and 13) Machine Learning.

This year's conference program includes 3 plenary talks, 6 invited talks, 59 fifteen-minutes contributed talks, 27 ten-minutes contributed talks, and 1 late news. We are happy having a total of 125 abstracts submitted from 16 countries and regions. The contributed talks and late news were carefully selected by the members of the Technical Program Committee. The excellent program owes much to the dedicated efforts of the Technical Program Committee chaired by Dr. Tatsuya Kunikiyo and Prof. Satofumi Souma.

Given the global health concerns surrounding COVID-19, the organization of SISPAD 2020 has decided to hold the SISPAD 2020 as ALL-VIRTUAL conference. Despite the difficult times brought by the pandemic, we have received high-quality submissions from the authors, and we would like to appreciate their efforts. We also wish to express our sincere appreciation to all the members of the organizing committee and steering committee for planning and organizing this conference. It is with great pleasure that we extend a warm welcome to all of the speakers and the attendants of SISPAD 2020.

Nobuya Mori
Steering Committee Chair
2020 International Conference on Simulation of
Semiconductor Processes and Devices

Yoshinari Kamakura
Conference Chair
2020 International Conference on Simulation of
Semiconductor Processes and Devices

COMMITTEE MEMBERS

Organization:

Steering Chair	N. Mori	(Osaka Univ.)
Conference Chair	K. Kamakura	(Osaka Inst. Technol.)
Program Chair	T. Kunikiyo	(Renesas Electronics)
Program Vice-Chair	S. Souma	(Kobe Univ.)
Audit	J. Hattori	(AIST)
Local Arrangement	S. Sato	(Kansai Univ.)
	S. Souma	(Kobe Univ.)
Publication	K. Matsuzawa	(Kioxia)
General Affairs	T. Iizuka	(Hiroshima Univ.)
	H. Minari	(Sony Semiconductor Solutions)
	N. Nakano	(Keio Univ.)
	K. Sonoda	(Renesas Electronics)

International Steering Committee:

D. Esseni	(Univ. of Udine, Italy)
F. Gamiz	(Univ. of Granada, Spain)
N. Goldsman	(Univ. of Maryland, USA)
Y. Kamakura	(Osaka Inst. Technol., Japan)
J. Lorenz	(Fraunhofer IISB, Germany)
N. Mori	(Osaka Univ., Japan)
L. F. Register	(Univ. of Texas at Austin, USA)
K. Sonoda	(Renesas Electronics, Japan)
W. Vandenberghe	(Univ. of Texas at Dallas, USA)

Technical Program Committee:

Chair	T. Kunikiyo	(Renesas Electronics, Japan)
Vice-Chair	S. Souma	(Kobe Univ., Japan)

Members

M. Bazizi	(Applied Materials, USA)	Y. Li	(National Chiao Tung Univ., Taiwan)
L. Filipovic	(TU Wien, Austria)	B. M.-Kope	(TSMC USA, USA)
A. Godoy	(Univ. of Granada, Spain)	S. Martinie	(CEA-LETI, France)
J. Hattori	(AIST, Japan)	H. Minari	(Sony Semiconductor Solutions, Japan)
A. Hiroki	(Kyoto Inst. Technol., Japan)	V. Moroz	(Synopsys, USA)
S.-M. Hong	(Gwangju Inst. Sci. and Technol., Korea)	S. Reggiani	(Univ. of Bologna, Italy)
C. Jungemann	(Univ. of Aachen, Germany)	F. Register	(Univ. of Texas at Austin, USA)
C. Kaneta	(Tohoku Univ., Japan)	H. Tanaka	(Kyoto Univ., Japan)
M. Karner	(Global TCAD Solutions, Austria)	W. Vandenberghe	(Univ. of Texas at Dallas, USA)
K. Kukita	(Kioxia, Japan)	H. Y. Wong	(San Jose State Univ., USA)
U. Kwon	(Samsung Electronics, Korea)	J. Wu	(TSMC, Taiwan)

SISPAD 2020 Conference Schedule (1/3)

1st group

Presentation	Video viewing start time: 9:00am on Sep. 23, JST Video viewing end time: 11:59pm on Sep. 28, JST **Opening and Welcome Remarks:** Session 1: Plenary Session 2: Band Structure Session 3: Computational Methodology Session 4: Nanowire Session 5: Material and Geometry Impact
Q&A Session	Q&A on bulletin board system (from Sep. 24 to Sep. 25, JST)

SISPAD 2020 Conference Schedule (2/3)

2nd group

Presentation	**Video viewing start time: 9:00am on Sep. 28, JST** **Video viewing end time: 11:59pm on Oct. 3, JST** **Session 6: Reliability** **Session 7: Power and Optoelectronic Devices** **Session 8: Non-Volatile Memory I Flash and Phase Change** **Memory** **Session 9: Transport** **Session 10: Non-Volatile Memory II ReRAM and MRAM** **Session 11: High Speed Switching Devices**
Q&A Session	**Q&A on bulletin board system (from Sep. 29 to Sep. 30, JST)**

SISPAD 2020 Conference Schedule (3/3)

3rd group

Presentation	Video viewing start time: 9:00am on Oct. 1, JST Video viewing end time: 11:59pm on Oct. 6, JST Session 12: Emerging Devices Session 13: 2D and Nano System I Session 14: FET devices and design technology co-optimization Session 15: Machine Learning Session 16: 2D and Nano System II Late News
Q&A Session	Q&A on bulletin board system (from Oct. 2 to Oct. 3, JST)

SISPAD 2020 Timetable (JST) (1/2)

	Sep. 23 Wed	Sep. 24 Thu	Sep. 25 Fri	Sep. 26 Sat	Sep. 27 Sun	Sep. 28 Mon	Sep. 29 Tue
1st group (Session 1 to 5)	Video viewing period	Q&A Session					
2nd group (Session 6 to 11)						Video viewing period	Q&A Session
Workshop 1	Video viewing period						
Company Exhibition	Video viewing period						
	Q&A Session						
Special Event	Opening and Welcome Remarks						

SISPAD 2020 Timetable (JST) (2/2)

	Sep. 30 Wed	Oct. 1 Thu	Oct. 2 Fri	Oct. 3 Sat	Oct. 4 Sun	Oct. 5 Mon	Oct. 6 Tue
2nd group (Session 6 to 11)	Video viewing period →						
	Q&A Session						
3rd group (Session 12 to 16, Late News)		Video viewing period →					
			Q&A Session				
Workshop 2	Video viewing period →						
						Q&A Session	
Company Exhibition	Video viewing period →						
			Q&A Session				
Special Event	SISPAD2021 Presentation						Closing

x

SISPAD 2020 Technical Program

1st group (from 9:00 am on Sep. 23 to 11:59 pm on Sep. 28, JST)

20min.　**Opening and Welcome Remarks**
Yoshinari Kamakura (Osaka Inst. Tech., Japan)

Session 1: Plenary

Chairperson:　Tatsuya Kunikiyo (Renesas, Japan)

1-1　40 min.　**invited talk**
"Forefront of Silicon Quantum Computing"
Kohei M. Itoh (Keio Univ., Japan)..1

1-2　40 min.　**invited talk**
"Ab-initio quantum transport with a basis of unit-cell restricted Bloch functions and the NEGF formalism"
Marco Pala (CNRS, Univ. Paris-Sud, France)..3

1-3　40 min.　**invited talk**
"Future of Power Electronics from TCAD Perspective"
Terry Ma (Synopsys, U.S.A.) 7

Session 2: Band Structure

Chairperson: Chioko Kaneta (Tohoku Univ., Japan)

2-1　25 min.　**invited talk**
"Computics Approach toward Clarification of Atomic Reactions during Epitaxial Growth of GaN"
Atsushi Oshiyama (Nagoya Univ., Japan) 11

2-2　15 min.
"Estimation of Phonon Mean Free Path in Small-Scaled Si Wire by Monte Carlo Simulation"
Y. Suzuki[1], Y. Fujita[1], K. Fauziah[2], T. Nogita[2] H. Ikeda[2], T. Watanabe[3], Y. Kamakura[1]
([1]Osaka Inst. Tech., [2]Shizuoka Univ., [3]Waseda Univ., Japan).............................15

2-3　15 min.
"First-principles study of dopant trap level and concentration in $Si(110)/a\text{-}SiO_2$ interface"
G.Kang, J. Jeon, J. Kim, H. Ahn, I. Jang, D. Kim (Samsung Electronics, Korea) 19

2-4　15 min.
"Energy Band Calculation of $Si/Si_{0.7}Ge_{0.3}$ Nanopillars in k Space"
M-H Chuang, Y. Li (National Chiao Tung Univ., Taiwan).......................................23

2-5 15 min.

"Full Band Monte Carlo simulation of phonon transfer at interfaces"

N. D. Le[1], B. Davier[1,2,] P. Dollfus[1], M. Pala[1], A. Bournel[1], J. Saint-Martin[1]

([1]Universite Paris-Saclay, CNRS, France, [2]Univ. Tokyo, Japan) 27

2-6 10 min.

"First Principle Simulations of Electronic and Optical Properties of a Hydrogen Terminated Diamond Doped by a Molybdenum Oxide Molecule"

J.McGhee, V. P. Georgiev (Univ. Glasgow, U.K.)...31

Session 3: Computational Methodology

Chairpersons: Yiming Li (National Chao Tung Univ., Taiwan)
Victor Moroz (Synopsys, U.S.A.)

3-1 10 min.

"High-sigma analysis of DRAM write and retention performance: a TCAD-to-SPICE approach"

S. M. Amoroso[1], J. Lee[1], A. R. Brown[1], P. Asenov[1], X. W. Lin[2], T. Yang[3], V. Moroz[2]

([1]Synopsys Europe, U.K., [2]Synopsys, U.S.A., [3]Synopsys Taiwan, Taiwan)..........................35

3-2 15 min.

"Generative Model Based Adaptive Importance Sampling for Flux Calculations in Process TCAD"

A. Scharinger[1], P. Manstetten[1], A. Hossinger[2], J. Weinbub[1]

([1]TU Wien, Austria, [2]Silvaco Europe, U.K.) 39

3-3 15 min.

"Implant heating contribution to amorphous layer: a KMC approach"

P. L. Julliard[1,2], P. Dumas[1], F. Monsieur[1], F. Hilario[1], D. Rideau[1], A. Hemeryck[2],

F.Cristiano[2] ([1]STMicroelectronics, France, [2]LAAS-CNRS, Univ. Toulouse, France)............43

3-4 15 min.

"Automatic Modeling of Logic Device Performance Based on Machine Learning Utilizing Feature Engineering"

S. Kim, K. Lee, Y. Shin, K. Chang, J. Jeong, S. Baek, M. Kang, K. Cho, D. Kim

(Samsung Electronics, Korea) 47

3-5 15 min.

"Gummel-cycle Algebraic Multigrid Preconditioning for Large-scale Device Simulations"

H. Koshimoto[1], H. Ishimabuchi[2], J. Yoo[2], Y. Kayama[1], S. Yamada[1], U. Kwon[2], D. S. Kim[2]

([1]Samsung R&D Inst. Japan, Japan, [2]Samsung Electronics, Korea).....................................51

3-6 15 min.

"A continuous cellular automaton method with flux interpolation for two-dimensional electron gas electron transport analysis"

K.Fukuda[1], J. Hattori[1], H. Asai[1], J. Yaita[2], J. Kotani[2] ([1]AIST, Japan, [2]Fujitsu, Japan)..........55

3-7 15 min.
"Geometric Advection Algorithm for Process Emulation"
X.Klemenschits, S. Selberherr, L. Filipovic (TU Wien, Austria)............................59

Session 4: Nanowire

Chairperson: Susanna Reggiani (Univ. Bologna, Italy)

4-1 15 min.
"Performance and Leakage Analysis of Si and Ge NWFETs Using a Combined Subband BTE and WKB Approach"
Z. Stanojevic, K. Steiner, G. Strof, O. Baumgartner, G. Rzepa, M. Karner
(Global TCAD Solutions, Austria)..63

4-2 15 min.
"Molecular Dynamics Modeling of the Radial Heat Transfer from Silicon Nanowires"
I. Bejenari[1], A. Burenkov[1], P. Pichler[1], I. Deretzis[2], A. La Magna[2]
([1]Fraunhofer IISB, Germany, [2]CNR-IMM, Italy)..67

4-3 15 min.
"Advanced simulations on laser annealing: explosive crystallization and phonon transport corrections"
A. Sciuto[1,2], I. Deretzis[1], S. F. Lombardo[1], M. G. Grimaldi[2], K. Huet[3], B. Curvers[3],
B. Lespinasse[3], A. Verstraete[3], I. Bejenari[4], A. Burenkov[4], P. Pichler[4], A. La Magna[1]
([1]CNR-IMM, Italy, [2]Univ. Catania, Italy, [3]LASSE laser systems and solutions of Europe, France, [4]Fraunhofer IISB, Germany) 71

4-4 10 min.
"Effect of Unit-Cell Arrangement on performance of Multi-Stage-planar Cavity-free Unileg Thermoelectric Generator Using Silicon Nanowires"
K. Abe[1], K. Oda[1], M. Tomita[1], T. Matsukawa[2], T. Matsuki[1,2], T. Watanabe[1]
([1]Waseda Univ., Japan, [2]AIST, Japan) 75

4-5 10 min.
"Characteristics of Gate-All-Around Silicon Nanowire and Nanosheet MOSFETs with Various Spacers"
S. R. Kola, Y. Li, N. Thoti (National Chiao Tung Univ., Taiwan).........................79

Session 5: Material and Geometry Impact

Chairpersons: Jun'ichi Hattori (AIST, Japan)
William Vandenberghe (Univ. Texas at Dallas, U.S.A.)

5-1 25 min. **invited talk**
"On the Physical Mechanism of Negative Capacitance Effect in Ferroelectric FET"
Masaharu Kobayashi (Univ. Tokyo, Japan)...83

5-2 15 min.

"Undoped SiGe material calibration for numerical laser annealing simulations"
A-S. Royet[1], L. Dagault[1,2], S. Kerdiles[1], P. Acosta-Alba[1], J. P. Barnes[1], F. Cristiano[2], H. Huet[3]
([1]Univ. Grenoble Alpes, France, [2]LAAS, CNRS Univ. Toulouse, France,
[3]Laser Systems & Solutions of Europe, France)..89

5-3 15 min.

"TCAD simulation for transition metal dichalcogenide channel Tunnel FETs consistent with
ab-initio based NEGF calculation"
H. Asai[1], T. Kuroda[2], K. Fukuda[1], J. Hattori[1], T. Ikegami[1], N. Mori[2]
([1]AIST, Japan, [2]Osaka Univ., Japan)...93

5-4 15 min.

"Ab Initio Study of Magnetically Intercalated Tungsten Diselenide"
P. D. Reyntjens[1,2,3], S. Tiwari[1,2,3], M. L. Van de Put[1], B. Sor´ee[2,3,4], W. G. Vandenberghe[1]
([1]Univ. Texas at Dallas, U.S.A., [2]Imec, Belgium, [3]KU Leuven, Belgium,
[4]Univ. Antwerp, Belgium) 97

5-5 15 min.

"A Study of Wiggling AA modeling and Its Impact on Device Performance in Advanced
DRAM"
Q.Wang, Y. D. Chen, J. Huang, W. Liu, E. Joseph (Lam Research, China)..........................101

5-6 10 min.

"Reactive Force-Field Molecular Dynamics Study of the Silicon-Germanium Deposition
Processes by Plasma Enhanced Chemical Vapor Deposition"
N. Uene[1], T. Mabuchi[1], M. Zaitsu[2], S. Yasuhara[2], T. Tokumasu[1]
([1]Tohoku Univ., Japan, [2]Japan Advanced Chemicals, Japan) 105

2nd group (from 9:00 am on Sep. 28 to 11:59 pm on Oct. 3, JST)

Session 6: Reliability

Chairpersons: Markus Karner (Global TCAD Solutions, Austria)
Hajime Tanaka (Kyoto Univ., Japan)

6-1 15 min.

"Universal Feature of Trap-Density Increase in Aged MOSFET and Its Compact Modeling"
F. Avila Herrera[1], M. Miura-Mattausch[1], T. Iizuka[1], H. Kikuchihara[1], H. J. Mattausch[1],
H.Takatsuka[2] ([1]Hiroshima Univ., Japan, [2]USJC, Japan) 109

6-2 10 min.

"TCAD Incorporation of Physical Framework to Model N and P BTI in MOSFETs"
R. Tiwari, N. Chowdhury, T. Samadder, S. Mukhopadhyay, N. Parihar, S. Mahapatra
(Indian Inst. Tech., India)..113

6-3 10 min.
"Benchmarking Charge Trapping Models with NBTI, TDDS and RTN Experiments"
S.Bhagdikar, S. Mahapatra (Indian Inst. Tech., India).......................................117

6-4 10 min.
"A TCAD Framework for Assessing NBTI Impact Under Drain Bias and Self-Heating Effects in Replacement Metal Gate (RMG) p-FinFETs"
U.Sharma, S. Mahapatra (Indian Inst. Tech., India) 121

6-5 10 min.
"Model analysis for effects of spatial and energy profiles of plasma process-induced defects in Si substrate on MOS device performance"
T.Hamano, K. Urabe, K. Eriguchi (Kyoto Univ., Japan) 125

Session 7: Power and Optoelectronic Devices

Chairpersons: Blanka Magyari-Kope (TSMC at U.S.A., U.S.A.)
Hideki Minari (Sony Semiconductor Solutions, Japan)

7-1 25 min. **invited talk**
"Modeling and Simulation of Si IGBTs"
Naoyuki Shigyo (Tokyo Inst. Tech., Japan) 129

7-2 15 min.
"Full Band Monte Carlo simulations of GaAs p-i-n Avalanche PhotoDiodes: What are the Limits of Nonlocal Impact Ionization Models?"
A. Pilotto[1], F. Driussi[1], D. Esseni[1], L. Selmi[2], M. Antonelli[3], F. Arfelli[3,4], G. Biasiol[5], S. Carrato[3], G. Cautero[6,4], D. De Angelis[6], R. H. Menk[6,4], C. Nichetti[6,3], T. Steinhartova[5], P. Palestri[1] ([1]Univ. Udine, Italy, [2]Univ. Modena and Reggio Emilia, Italy, [3]Univ. Trieste, Italy, [4]INFN, Italy, [5]IOM CNR, Italy, [6]Elettra-Sincrotrone, Italy) 133

7-3 15 min.
"A technique for phase-detection auto focus under near-infrared-ray incidence in a back-side illuminated CMOS image sensor pixel with selectively grown germanium on silicon"
T. Kunikiyo, H. Sato, T. Kamino, K. Iizuka, K. Sonoda, T. Yamashita
(Renesas Electronics, Japan)...137

7-4 15 min.
"Investigation of the relationship between current filament movement and local heat generation in IGBTs by using modified avalanche model of TCAD"
T.Suwa (Toshiba Electronic Devices & Storage, Japan) 141

7-5 15 min.
"Verilog-A model for avalanche dynamics and quenching in Single-Photon Avalanche Diodes"
Y. Oussaiti[1,2], D. Rideau[1], J. R. Manouvrier[1], V. Quenette[1], H. Wehbe-Alause[1], B. Mamdy[1], A. Lopez[1], G. Mugny[1], M. Agnew[1], E. Lacombe[1], J. Grebot[1], P. Dollfus[2], M. Pala[2] ([1]STMicroelectronics, France, [2]Centre de Nanosciences et de Nanotechnologies, France)...145

7-6 15 min.

"A Novel Full-Band Monte Carlo Device Simulator with Real-Space Treatment of the Short-Range Coulomb Interactions for Modeling 4H-SiC Power Devices"

C-Y. Cheng, D. Vasileska (Arizona State Univ., U.S.A.) 149

7-7 10 min.

"Tight-binding simulation of optical gain in h-BCN for laser application"

D.Maki, M. Ogawa, S. Souma (Kobe Univ., Japan) 153

7-8 10 min.

"Predictive Compact Modeling of Abnormal LDMOS Characteristics Due to Overlap Length Modification"

T. Iizuka[1], D. Navarro[1], M. Miura-Mattausch[1], H. Kikuchihara[1], H. J. Mattausch[1], D.R. Nestor[2] ([1]Hiroshima Univ., Japan, [2]Allegro MicroSystems, U.S.A.) 157

Session 8: Non-Volatile Memory I   Flash and Phase Change Memory

Chairperson: Kentaro Kukita (Kioxia, Japan)

8-1 15 min.

"A TCAD Study on Mechanism and Countermeasure for Program Characteristics Degradation of 3D Semicircular Charge Trap Flash Memory"

N. Kariya, M. Tsuda, T. Kurusu, M. Kondo, K. Nishitani, H. Tokuhira, J. Shimokawa, Y. Yokota, H. Tanimoto, S. Onoue, Y. Shimada, T. Kato, K. Hosotani, F. Arai, M. Fujiwara, Y.Uchiyama, K. Ohuchi (Kioxia, Japan) 161

8-2 15 min.

"Impact of Random Phase Distribution in 3D Vertical NAND Architecture of Ferroelectric Transistors on In-Memory Computing"

G.Choe, W. Shim, J. Hur, A. I. Khan, S. Yu (Georgia Inst. Tech., U.S.A.) 165

8-3 15 min.

"TCAD Modeling and Optimization of 28nm HKMG ESF3 Flash Memory"

A. Zaka, T. Herrmann, R. Richter, S. Duenkel, R. Jain (GLOBALFOUNDRIES, Germany)...169

8-4 15 min.

"Coupling the Multi Phase-Field Method with an Electro-Thermal Solver to Simulate Phase Change Mechanisms in Ge-rich GST based PCM"

R. Bayle[1,2,3], O. Cueto[1], S. Blonkowski[1], T. Philippe[3], H. Henry[3], M. Plappa[3] ([1]CEA-LETI, France, [2]STMicroelectronics, France, [3]Ecole Polytechnique, France) 173

Session 9: Transport

Chairperson: Christoph Jungemann (Univ. Aachen, Germany)

9-1 15 min.
"Efficient partitioning of surface Green's function: toward ab initio contact resistance study"
G. Gandus[1,2], Y. Lee[2], D. Passerone[1], M. Luisier[2]
([1]nanotech@surfaces (EMPA), Switzerland, [2]Integrated Systems Laboratory (ETH Zurich),
Switzerland) 177

9-2 15 min.
"Quantum transport in Si: P δ-layer wires"
J. P. Mendez, D. Mamaluy, X. Gao, L. Tracy, E. Anderson, D. Campbell, J. Ivie, T.-M. Lu,
S.Schmucker, S. Misra (Sandia National Laboratories, U.S.A.) 181

9-3 15 min.
"Analytical Formulae for the Surface Green's Functions of Graphene and 1T' MoS2
Nanoribbons"
H.Kosina, V. Sverdlov (TU Wien, Austria)..185

9-4 15 min.
"Numerical Solution of the Constrained Wigner Equation"
R.Kosik, J. Cervenka, H. Kosina (TU Wien, Austria) 189

9-5 10 min.
"Calibrated Si Mobility and Incomplete Ionization Models with Field Dependent Ionization
Energy for Cryogenic Simulations"
H. Y. Wong (San Jose State Univ., U.S.A.)...193

Session 10: Non-Volatile Memory II   ReRAM and MRAM

Chairperson: Uihui Kwon (Samsung, Korea)

10-1 15 min.
"Monte Carlo Simulation of a Three-Terminal RRAM with Applications to Neuromorphic
Computing"
A.Balasingam, A. Levy, H. Li, P. Raina (Stanford Univ., U.S.A.) 197

10-2 15 min.
"Fully Analog ReRAM Neuromorphic Circuit Optimization using DTCO Simulation
Framework"
A. Nguyen[1], H. Nguyen[1], S. Venimadhavan[1], A. Venkattraman[2], D. Parent[1], H. Y. Wong[1]
([1]San Jose State Univ., U.S.A., [2]Univ. California Merced, U.S.A.) 201

10-3 10 min.
"Effect of Shape Deformation due to Edge Roughness in Spin-Orbit Torque Magnetoresistive
Random-Access Memory"
J. Byun, D. H. Kang, M. Shin (KAIST, Korea)...205

10-4 10 min.

"Computation of Torques in Magnetic Tunnel Junctions through Spin and Charge Transport Modeling"

S. Fiorentini[1], J. Ender[1], M. Mohamedou[1], R. Orio[1], S. Selberherr[1], W. Goes[2], V. Sverdlov[1] ([1]TU Wien, Austria, [2]Silvaco Europe, U.K.) 209

10-5 10 min.

"Efficient Demagnetizing Field Calculation for Disconnected Complex Geometries in STT-MRAM Cells"

J. Ender[1], M. Mohamedou[1], S. Fiorentini[1], R. Orio[1], S. Selberherr[1], W. Goes[2], V. Sverdlov[1] ([1]TU Wien, Austria, [2]Silvaco Europe, U.K.) 213

10-6 10 min.

"Properties of Conductive Oxygen Vacancies and Compact Modeling of IV Characteristics in HfO_2 Resistive Random-Access-Memories"

J.Park, M.-J. Kim, J.-H. Jang, S.-M. Hong (Gwangju Inst. Sci. Tech., Korea) 217

Session 11: High Speed Switching Devices

Chairpersons: Akira Hiroki (Kyoto Inst. Tech., Japan)
Sebastien Martinie (CEA-LETI, France)

11-1 15 min.

"MOS-like approach for compact modeling of HEMT transistor"

A. Vaysset, S. Martinie, F. Triozon, O. Rozeau, M.-A. Jaud, R. Escoffier, T. Poiroux (CEA, LETI, Univ. Grenoble Alpes, France) 221

11-2 15 min.

"Compact modeling of gate leakage phenomenon in GaN HEMTs"

K. Li[1,3], E. Yagyu[2], H. Saito[2], K. H. Teo[1] ([1]Mitsubishi Electric Research Labs, U.S.A., [2]Mitsubishi Electric Corp., Japan, [3]Univ. Illinois at Urbana-Champaign, U.S.A.)................225

11-3 15 min.

"Effect of Atomic Interface on Tunnel Barrier in Ferroelectric HfO_2 Tunnel Junctions"

J.Seo, M. Shin (KAIST, Korea)...229

11-4 15 min.

"Surge Current Capability in lateral AlGaN/GaN Hybrid Anode Diodes with p-GaN/Schottky Anode"

G. Atmaca[1], M.-A. Jaud[1], J. Buckley[1], R. Gwoziecki[1], A. Yvon[2,] E. Collard[2], M. Plissonnier[1], T.Poiroux[1] ([1]CEA, LETI, Univ. Grenoble Alpes, France, [2]STMicroelectronics, France).....233

11-5 15 min.

"Dynamic Simulation of Write '1' Operation in the Bi-stable 1-Transistor SRAM Cell"

T. Dutta[1], F. Adamu-Lema[1], A. Asenov[1], Y. Widjaja[2], V. Nebesnyi[3] ([1]Semiwise, U.K., [2]Zeno Semiconductor, [3]MCPG) 237

11-6 10 min.

"Simulation of gated GaAs-AlGaAs resonant tunneling diodes for tunable terahertz communication applications"

V. Georgiev, A. Sengupta, P. Maciazek, O. Badami, C. Medina-Bailon, T. Dutta, F.Adamu-Lema, A. Asenov (Univ. Glasgow, U. K.) 241

11-7 10 min.

"Theoretical Study of Double-Heterojunction AlGaN/GaN/InGaN/δ-doped HEMTs for Improved Transconductance Linearity"

T.-H. Yu (Inforsight Computing, Taiwan) 245

11-8 10 min.

"Nanoscale FET: How To Make Atomistic Simulation Versatile, Predictive, and Fast at 5nm Node and Below"

P. Blaise[1], U. Kapoor[1], M. Townsend[1], E. Guichard[1], J. Charles[2], D. A. Lemus[2], T. Kubis[2] ([1]Silvaco, U.S.A., [2]Purdue Univ., U.S.A.) 249

3rd group (from 9:00 am on Oct. 1 to 11:59 pm on Oct. 6, JST)

Session 12: Emerging Devices

Chairpersons: Andres Godoy (Univ. Granada, Spain)
Sung-Ming Hong (Gwangju Inst. Sci. Tech., Korea)

12-1 25 min. **invited talk**

"TCAD-Assisted MultiPhysics Modeling & Simulation for Accelerating Silicon Quantum Dot Qubit Design"

Fahd Ayyalil Mohiyaddin (imec, Belgium) 253

12-2 10 min.

"Physics-augmented Neural Compact Model for Emerging Device Technologies"

Y.Kim, S. Myung, J. Ryu, C. Jeong, D. S. Kim (Samsung Electronics, Korea)...................257

12-3 15 min.

"A Modeling Study on Performance of a CNOT Gate Devices based on Electrode-driven Si DQD Structures"

H.Ryu, J.-H. Kang (Korea Inst. Sci. Tech. Info., Korea).....................................261

12-4 15 min.

"Simulation and Evaluation of Plasmonic Circuits"

M.Fukuda, Y. Ishikawa (Toyohashi, Univ. Tech., Japan)....................................265

12-5   15 min.

"Numerical study of surface chemical reactions in 2D-FET based pH sensors"
A. Toral-Lopez[1], E. G. Marin[1], J. Cuesta[1], F. G. Ruiz[1], F. Pasadas[2], A. Mediana-Rull[1],
A.Godoy[1] ([1]Univ. Granada, Spain, [2]Univ. Autonoma Barcelona, Spain) 269

12-6   10 min.

"A Combined First Principle and Kinetic Monte Carlo Study of Polyoxometalates Based
Molecular Memory Devices"
P. Lapham, O. Badami, C. Medina-Bailon, F. Adamu-Lema, T. Dutta, V. Georgiev,
A.Asenov (Univ. Glasgow, U.K.) 273

12-7   10 min.

"Modeling Assisted Room Temperature Operation of Atomic Precision Advanced
Manufacturing (APAM) Devices"
X. Gao, L. Tracy, E. Anderson, DeAnna Campbell, J. Ivie, T.-M. Lu, D. Mamaluy,
S.Schmucker, S. Misra (Sandia National Lab., U.S.A.) 277

Session 13: 2D and Nano System I

Chairperson: Jeff Wu (TSMC, Taiwan)

13-1   15 min.

"Effects of the Dielectric Environment on Electronic Transport in Monolayer MoS2: Screening
and Remote Phonon Scattering"
M. L. Van de Put, G. Gaddemane, S. Gopalan, M. V. Fischetti
(Univ. Texas at Dallas, U.S.A.) 281

13-2   15 min.

"Impact of Schottky Barrier on the Performance of Two-Dimensional Material Transistors"
S.-K. Su, J. Cai, E. Chen, L.-J. Li, H.-S. Philip Wong (TSMC, Taiwan) 285

13-3   15 min.

"AC NEGF Simulation of Nanosheet MOSFETs"
S.-M. Hong, P.-H. Ahn (Gwangju Inst. Sci. Tech., Korea) 289

13-4   10 min.

"Enhanced Capabilities of the Nano-Electronic Simulation Software (NESS)"
C. Medina-Bailon, O. Badami, H. Carrillo-Nunez, T. Dutta, D. Nagy, F. Adamu-Lema,
V.Georgiev, A. Asenov (Univ. Glasgow, U.K.) 293

13-5   10 min.

"Electrostatic Potential Profile Generator for Two-Dimensional Semiconductor Devices"
S.-C. Han, J. Choi, S.-M. Hong (Gwangju Inst. Sci. Tech., Korea) 297

Session 14: FET Devices and Design Technology Co-Optimization

Chairpersons: Mehdi Bazizi, (Applied Materials, U.S.A.)
Lado Filipovic (TU Wien, Austria)

14-1 25 min. **invited talk**
"Agile Pathfinding Technology Prototyping: the Hunt for Directional Correctness"
Daniel Chanemougame (TEL at Albany, U.S.A.) 301

14-2 15 min.
"Self-Aligned Single Diffusion Break Technology Optimization Through Material Engineering for Advanced CMOS Nodes"
A. Pal, E. M. Bazizi, L. Jiang, M. Saremi, B. Alexander, B. Ayyagari-Sangamalli
(Applied Materials, U.S.A.) 307

14-3 15 min.
"L-UTSOI: A compact model for low-power analog and digital applications in FDSOI technology"
S. Martinie[1], O. Rozeau[1], T. Poiroux[1], P. Scheer[1], S. E. Ghouli[2], M. Kang[3], A. Juge[2], H. Lee[3]
([1]CEA, LETI, Univ. Grenoble Alpes, France, [2]STMicroelectronics, France,
[3]Samsung, Korea)...311

14-4 15 min.
"Electromigration Model for Platinum Hotplates"
L.Filipovic (TU Wien, Austria) 315

14-5 15 min.
"Compact Modeling of Radiation Effects in Thin-Layer SOI-MOSFETs"
M. Miura-Mattausch, H. Kikuchihara, D. Navarro, T. Iizuka, H. J. Mattausch
(Hiroshima Univ., Japan) 319

14-6 15 min.
"Complementary FET Device and Circuit Level Evaluation Using Fin-Based and Sheet-Based Configurations Targeting 3nm Node and Beyond"
L. Jiang, A. Pal, E. M. Bazizi, M. Saremi, R. He, B. Alexander, B. Ayyagari-Sangamalli
(Applied Materials, U.S.A.) 323

14-7 10 min.
"Via Size Optimization for Optimum Circuit Performance at 3 nm node"
S. Mittal[1], A. Pal[2], M. Saremi[2], E. M. Bazizi[2], B. Alexander[2], B. Ayyagari-Sangamalli[2]
([1]Applied Materials, India, [2]Applied Materials, U.S.A.) 327

14-8 10 min.
"Time-Resolved Mode Space based Quantum-Liouville type Equations applied onto DGFETs"
L.Schulz, D. Schulz (TU Dortmund, Germany) 331

Session 15: Machine Learning

Chairpersons: Satofumi Souma (Kobe Univ., Japan)
Hiuyung Wong (San Jose State Univ., U.S.A.)

15-1 25 min.   **invited talk**
"Power Device Degradation Estimation by Machine Learning of Gate Waveforms"
Makoto Takamiya (Univ. Tokyo, Japan)...335

15-2 15 min.
"Machine Learning Prediction of Formation Energies in a-SiO_2"
D.Milardovich, M. Jech, D. Waldhoer, M. Waltl, T. Grasser (TU Wien, Austria)..............339

15-3 15 min.
"Novel Optimization Method using Machine-learning for Device and Process Competitiveness of BCD Process"
J. Kim, J.-H. Yoo, J. Jung, K. Kim, J. Bae, Y.-S. Kim, O.-K. Kwon, U.-H. Kwon, D.-S. Kim
(Samsung Electronics, Korea) 343

15-4 15 min.
"Real-Time TCAD: a new paradigm for TCAD in the artificial intelligence era"
S. Myung, J. Kim, Y. Jeon, W. Jang, J. Kim, S Han, K.-H. Baek, J. Ryu, Y.-S. Kim, J. Doh,
C.Jeong, D. -S. Kim (Samsung Electronics, Korea) 347

15-5 15 min.
"Application of Noise to Avoid Overfitting in TCAD Augmented Machine Learning"
S. S. Raju[1], B. Wang[2], K. Mehta[1], M. Xiao[2], Y. Zhang[2], H.-Y. Wong[1]
([1]San Jose Univ., U.S.A., [2]Virginia Polytech. Inst. State Univ., U.S.A.)...................351

15-6 15 min.
"Automatic Device Model Parameter Extractions via Hybrid Intelligent Methodology"
C.-C. Liu, Y. Li, Y.-S. Yang, C.-Y. Chen, M.-H. Chuang
(National Chiao Tung Univ., Taiwan)..355

15-7 10 min.
"Physics-Informed Graph Neural Network for Circuit Compact Model Development"
X. Gao, A. Huang, N. Trask, S. Reza
(Sandia National Lab., U.S.A.)..359

Session 16: 2D and Nano System II

Chairperson: Frank Register (Univ. Texas at Austin, U.S.A.)

16-1 15 min.
"Theoretical study of electronic transport in monolayer SnSe"
S. Gopalan[1], G. Gaddemane[2], M. L. Van de Put[1], M. V. Fischetti[1]
([1]Univ. Texas at Dallas, U.S.A., [2]imec, Belgium)..363

16-2 15 min.
"Transient simulation of graphene FET gated by electrolyte medium"
K. Arihori[1], M. Ogawa[1], S. Souma[1], J. Sato-Iwanaga[2], M. Suzuki[2]
([1]Kobe Univ., Japan, [2]Panasonic, Japan) 367

16-3 15 min.
"Quantum Transport Simulations of Phosphorene Nanoribbon MOSFETs: Effects of Metal Contacts, Ballisticity and Series Resistance"
M.Poljak, Mislav Matic (Univ. Zagreb, Croatia) 371

16-4 10 min.
"High-Performance Metal-Ferroelectric-Semiconductor Nanosheet Line Tunneling Field Effect Transistors with Strained SiGe"
N. Thoti[1], Y. Li[1], S. R. Kola[1], S. Samukawa[2]
([1]National Chaio Tung Univ., Taiwan, [2]Tohoku Univ., Japan) 375

16-5 10 min.
"A First-Principles Study on the Strain-induced Localized Electronic Properties of Dumbbell-shape Graphene Nanoribbon for Highly Sensitive Strain Sensors"
Q.Zhang, K. Suzuki, H. Miura (Tohoku Univ., Japan).........................379

Late News

Chairperson: Tatsuya Kunikiyo (Renesas Electronics, Japan)

LN 10 min
"Multiband Phase Space Operator for Narrow Bandgap Semiconductor Devices"
L.Schulz, D. Schulz (TU Dortmund, Germany) 383

Forefront of Silicon Quantum Computing

Kohei M. Itoh
Dept. Applied Physics
Keio University
Yokohama, Japan
kitoh@appi.keio.ac.jp

Abstract—**Forefront of the silicon quantum computer development is described.**

Keywords—quantum computing, silicon

I. CONTENTS

A. Why quantum computing

Quantum computing will complement a number of advanced tasks that will be performed by the extension of today's computing frameworks. Its application is expected to span many fields such as optimization, machine learning, security, data search, etc. Here, as examples, the current states of quantum computer algorithm and software developments in the area of finance [1] and chemical engineering [2] sectors are introduced.

B. Why silicon (See Ref. 3 for an excellent review on silicon quantum computing)

A quantum computer is going to operate in collaboration with silicon CMOS electronics and silicon AI chips. Therefore, integration of the classical and quantum circuits together in one silicon chip is strongly preferred. Use of the state-of-the-art silicon processing technology will allow to shrink the size of silicon quantum bits (qubits), interconnects, electrodes, etc. for larger and more reproducible integration. Moreover, it has been shown recently that the silicon spin qubits can operate at "high-enough temperature" of ~5K [4-6] at which classical CMOS can also function together. The temperature of more than 4K can be achieved easily by mechanical cooling, allowing to have a very large sample space for hosting necessary electronics. Such high temperature operation possibility is a sharp contrast to the case of superconducting qubits that can only work at temperatures much less than 0.1 K requiring dilution refrigerators, whose sample space is severely limited for scaling.

C. Variations in silicon qubits)

Silicon qubit forefronts are categorized into the following five groups.

i) Gate-defined MOS spin qubit [7, 8]
One electron spin captured under a positively biased Si MOS gate is employed as a qubit. The two levels, spin up and down states, of the single electron are used to represent qubit states.

ii) Gate-defined Si/SiGe spin qubit [9, 10]
One electron spin captured in the layer of strained Si grown on the lattice mismatched SiGe virtual substrate in the region under a positively biased MOS gate is utilized as a qubit.

iii) Gate-defined Si/SiGe singlet-triplet qubit [11-13]
Multiple electrons confined in a pair of gate-defined Si/SiGe quantum dots form a single qubit state.

iv) MOSFET structure converted to spin qubit [4, 14, 15]
A MOSFET structure is employed to confine a single electron in a certain position of the channel to form a qubit.

v) Donor qubit [16-18]
A phosphorus donor electron spin qubit is placed in silicon by ion-implantation or scanning tunneling microscope lithography.

D. Substrates [See Ref. 19 for a review on the isotope engineering of Si for quantum information processing]

Naturally available silicon (^{nat}Si) is composed of ^{28}Si, ^{29}Si, and ^{30}Si stable isotopes with the fixed compositions of 92.2%, 4.7%, and 3.1%, respectively. Among them, only ^{29}Si has nuclear spin 1/2 and the magnetic field fluctuation arising from the ^{29}Si nuclear spins turn out to be the major source of qubit decoherence. For this reason, isotopically enriched ^{28}Si epilayers grown on the standard 300 mm Si wafers [20, 21] and isotopically purified ^{28}Si strain layers grown on SiGe virtual substates [10, 14, 22] are being developed.

E. Interconnects

Transfer of quantum information between two distant qubits is a challenging task. It is possible to perform sequential nearest-neighbor information swapping to move quantum information from one qubit to another within a small number (~20) of quantum bit array. However, a longer distance quantum information transfer requires conversion of the electron spin qubit information to a different kind of a mobile qubit such a microwave photon [23, 24]. The microwave photon qubit can travel through a waveguide to connect two distance qubits.

F. Architecture and integration with classical CMOS circuits

There have been many proposals on integration of silicon quantum and classical computers [1, 25-27]. The circuit layout design including essential elements (qubits, readout devices, microwave waveguides, microwave antennas, interconnects, etc.) plays a key role.

G. Conclusion and outlook

Advancements are accelerating towards realization of silicon quantum computers. Based on the current state-of-the-art, outlook will be given.

ACKNOWLEDGMENT

This work was supported by Spintronics Research Network in Japan.

REFERENCES

[1] Y. Suzuki, S. Uno, R. Raymond, T. Tanaka, T. Onodera, and N. Yamamoto, "Amplitude Estimation without Phase Estimation," *Quantum Information Processing* **19**, 75 (2020).

[2] Q. Gao, H. Nakamura, T. P. Gujarati, G. O. Jones, J. E. Rice, S. P. Wood, M. Pistoia, J. M. Garcia, and N. Yamamoto,

"Computational Investigations of the Lithium Superoxide Dimer Rearrangement on Noisy Quantum Devices," arXiv:1906.10675

[3] L.M.K. Vandersypen, and M.A. Eriksson, "Quantum Computing with Semiconductor Spins," *Physics Today* **72**, 38 (2019)

[4] K. Ono, T. Mori, and S. Moriyama, "High-temperature operation of a silicon qubit," *Scientific Reports* **9**, 469 (2019).

[5] L. Petit, J.M. Boter, H.G.J. Eenink, G. Droulers, M.L.V. Tagliaferri, R. Li, D.P. Franke, K.J. Singh, J.S. Clarke, R.N. Schouten, V.V. Dobrovitski, L.M.K. Vandersypen, and M. Veldhorst, "Spin Lifetime and Charge Noise in Hot Silicon Quantum Dot Qubits," *Physical Review Letters* **121**, 076801 (2018).

[6] C. H. Yang, R. C. C. Leon, J. C. C. Hwang, A. Saraiva, T. Tanttu, W. Huang, J. C. Lemyre, K. W. Chan, K. Y. Tan, F. E. Hudson, K. M. Itoh, A. Morello, M. Pioro-Ladriere, A. Laucht, A. S. Dzurak, "Operation of a silicon quantum processor unit cell above one kelvin," *Nature* **7803**, 350 (2020).

[7] M. Veldhorst, C. H. Yang, J. C. C. Hwang, W. Huang, J. P. Dehollain, J. T. Muhonen, S. Simmons, A. Laucht, F. E. Hudson, K. M. Itoh, A. Morello, and A. S. Dzurak, "A two-qubit logic gate in silicon," *Nature* **526**, 410 (2015).

[8] W. Huang, C. H. Yang, K. W. Chan, T. Tanttu, B. Hensen, R. C. C. Leon, M. A. Fogarty, J. C. C. Hwang, F. E. Hudson, K. M. Itoh, A. Morello, A. Laucht, and A. S. Dzurak, "Fidelity benchmarks for two-qubit gates in silicon," *Nature* **569**, 532 (2019).

[9] A. Pateras, J. Park, Y. Ahn, J.A. Tilka, M.V. Holt, C. Reichl, W. Wegscheider, T.A. Baart, J.P. Dehollain, U. Mukhopadhyay, L.M.K. Vandersypen, and P.G. Evans, "A programmable two-qubit quantum processor in silicon," *Nature* **555**, 633 (2018).

[10] J. Yoneda, K. Takeda, T. Otsuka, T. Nakajima, M. R. Delbecq, G. Allison, T. Honda, T. Kodera, S. Oda, Y. Hoshi, N. Usami, K. M. Itoh, and S. Tarucha,"A quantum-dot spin qubit with coherence limited by charge noise and fidelity higher than 99.9%," *Nature Nanotechnology* **13**, 102 (2018).

[11] K. Takeda, A. Noiri, J. Yoneda, T. Nakajima, and S. Tarucha, "Resonantly Driven Singlet-Triplet Spin Qubit in Silicon," *Physical Review Letters* **124**, 117701 (2020)

[12] R. W. Andrews, C. Jones, M. D. Reed, A. M. Jones, S. D. Ha, M. P. Jura, J. Kerckhoff, M. Levendorf,S. Meenehan, S. T. Merkel, A. Smith, B. Sun, A. J. Weinstein, M. T. Rakher, T. D. Ladd, and M. G.Borselli, "Quantifying error and leakage in an encoded Si/SiGe triple-dot qubit," *Nature Nanotechnology* **14**, 747 (2019).

[13] D. Kim, Z. Shi, C. B. Simmons, D. R. Ward, J. R. Prance, T. S. Koh, J. K. Gamble, D. E. Savage, M. G. Lagally, M. Friesen, S. N. Coppersmith, M. A. Eriksson, "Quantum control and process tomography of a semiconductor quantum dot hybrid qubit," *Nature* **511**, 70 (2014).

[14] M. Urdampilleta, D. J. Niegemann, E. Chanrion, B. Jadot, C. Spence, P.-A. Mortemousque, C. Bauerle,L. Hutin, B. Bertrand, S. Barraud, R. Maurand, M. Sanquer, X. Jehl, S. De Franceschi, M. Vinet, and T. Meunier, "Gate-based high fidelity spin readout in a CMOS device," *Nature Nanotechnology* **14**, 737 (2019).

[15] R. Maurand, X. Jehl, D. Kotekar-Patil, A. Corna, H. Bohuslavskyi, R. Lavieville, L. Hutin, S. Barraud,M. Vinet, M.

Sanquer, and S. De Franceschi, "A CMOS silicon spin qubit," *Nature Communications* **7**, 13575 (2016).

[16] S. B. Tenberg, S. Asaad, M. T. Madzik, M. A. Johnson, I, B. Joecker, A. Laucht, F. E. Hudson, K. M. Itoh, A. M. Jakob, B. C. Johnson, D. N. Jamieson, J. C. McCallum, A. S. Dzurak, R. Joynt, and A. Morello, "Electron spin relaxation of single phosphorus donors in metal-oxide-semiconductor nanoscale devices," *Physical Review B* **99**, 205306 (2019).

[17] Y. He, S. K. Gorman, D. Keith, L. Kranz, J. G. Keizer, and M. Y. Simmons, "A two-qubit gate between phosphorus donor electrons in silicon," *Nature* **571**, 371 (2019).

[18] J. T. Muhonen, A. Laucht, S. Simmons, J. P. Dehollain, R. Kalra, F. E. Hudson, S. Freer, K. M. Itoh, D. N. Jamieson, J. C. McCallum, A. S. Dzurak, and A. Morello, "Quantifying the quantum gate fidelity of single-atom spin qubits in silicon by randomized benchmarking," *Journal of Physics Condensed Matter* **27**, 154205 (2015).

[19] K. M. Itoh and H. Watanabe, "Isotope engineering of silicon and diamond for quantum computing and sensing applications," *MRS Communications* **4**, 143 (2014).

[20] D. Sabbagh, N. Thomas, J. Torres, R. Pillarisetty, P. Amin, H.C. George, K. Singh, A. Budrevich, M. Robinson, D. Merrill, L. Ross, J. Roberts, L. Lampert, L. Massa, S. Amitonov, J. Boter, G. Droulers, H.G.J. Eenink, M. van Hezel, D. Donelson, M. Veldhorst, L.M.K. Vandersypen, J.S. Clarke, and G. Scappucci, Quantum transport properties of industrial 28Si/28SiO2," *Physical Review Applied* **12**, 014013 (2019).

[21] V. Mazzocchi, P. G. Sennikov, A. D. Bulanov, M. F. Churbanov, B. Bertrand, L. Hutin, J. P. Barnes,M. N. Drozdov, J. M. Hartmann, and M. Sanquer, "99.992% Si-28 CVD-grown epilayer on 300 mm substrates for large scale integration of silicon spin qubits," *Journal of Crystal Growth* **509**, 1 (2019).

[22] A. J. Sigillito, J. C. Loy, D. M. Zajac, M. J. Gullans, L. F. Edge, and J. R. Petta, "Site-Selective Quantum Control in an Isotopically Enriched Si-28/Si0.7Ge0.3 Quadruple Quantum Dot," *Physical Review Applied* **11**, 061006 (2019).

[23] X. Mi, M. Benito, S. Putz, D. M. Zajac, J. M. Taylor, G. Burkard, and J. R. Petta, "A coherent spin-photon interface in silicon," *Nature* **555**, 599 (2018).

[24] N. Samkharadze, G. Zheng, N. Kalhor, D. Brousse, A. Sammak, U.C. Mendes, A. Blais, G. Scappucci, and L.M.K. Vandersypen, "Strong spin-photon coupling in silicon," *Science* **359**, 1123 (2018).

[25] R. Li, L. Petit, D.P. Franke, J.P. Dehollain, J. Helsen, M. Steudtner, N.K. Thomas, Z.R. Yoscovits, K.J. Singh, S. Wehner, L.M.K. Vandersypen, J.S. Clarke, and M. Veldhorst, "A Crossbar Network for Silicon Quantum Dot Qubits," *Science Advances* **4**, e3960 (2018).

[26] L. M. K. Vandersypen, H. Bluhm, J. S. Clarke, A. S. Dsurak, R. Ishihara, A. Morello, D. J. Reilly, L. R. Schreiber, and M. Veldhorst, "Interfacing spin qubits in quantum dots and donors – hot, dense and coherent," *npj Quantum Information* **3**, 34 (2017).

[27] M. Veldhorst, H. G. J. Eenink, C. H. Yang, A. S. Dzurak, "Silicon CMOS architecture for a spin-based quantum computer," *Nature Communications* **8**, 1766 (2017).

Ab-initio quantum transport with a basis of unit-cell restricted Bloch functions and the NEGF formalism

Marco Pala
Université Paris-Saclay, CNRS
Centre de Nanosciences et de Nanotechnologies
91120 Palaiseau, France
marco.pala@c2n.upsaclay.fr

David Esseni
DPIA
University of Udine
33100 Udine, Italy
david.esseni@uniud.it

Abstract—**This invited contribution illustrates the theory and application of a first-principle transport methodology employing a basis set obtained directly from the Bloch functions computed with a plane wave (PW) *ab-initio* solver. We start from a PW density functional theory (DFT) Hamiltonian, use a unitary transformation to real space in the transport direction, and then discuss a basis of Bloch functions enabling a huge reduction of the size of the Hamiltonian blocks and an effective suppression of possible unphysical states. Our methodology enables *ab-initio* transport simulations with a good computational efficiency, and we here present results for self-consistent simulations of a single-gate monolayer PtSe$_2$ field effect transistor.**

Index Terms—**Density functional theory, quantum transport, NEGF, 2D materials**

I. INTRODUCTION

In recent years many innovations in nanoelectronics devices have been driven by the exploitation of new materials, such as atomically thin 2D materials and their heterostructures [1]–[5]. *Ab-initio* methods based on the DFT have been shown to be quite effective to investigate the electronic and optical properties of these new materials, but their direct use for quantum transport calculations has been hindered by the size and complexity of their Hamiltonians. Remarkable exceptions are the contributions employing Hamiltonians based on the linear combination of local orbitals (LCAO) [6], or on the maximally localized Wannnier functions (MLWF) [7], [8].

Here we discuss an alternative approach to *ab-initio* quantum transport [9], that leverages directly on the Hamiltonian matrix and Bloch functions obtained with plane-wave DFT solvers [10], [11]. We believe that such a methodology, differently from the widely used MLWF approach, can be naturally integrated into the workflow of an *ab-initio* solver, and thus provide a fairly automatic pathway to *ab-initio* quantum transport simulations.

II. METHODOLOGY

In order to obtain the DFT Hamiltonian, we start by self-consistently solving the Kohn-Sham (KS) equations

$$H_{KS}\,\Psi_n = E_n\,\Psi_n, \qquad H_{KS} = T + V_{scf} \qquad (1)$$

where the KS Hamiltonian H_{KS} is the sum of the operators corresponding to the kinetic energy T and to the self-consistent

potential $V_{scf} = V_{eI} + V_H + V_{xc}$, where V_{eI} is the electron-ion interaction potential, V_H the Hartree electrostatic potential, V_{xc} the exchange-correlation potential. To this goal we use a PW basis set and pseudopotentials to represent the valence electron-nuclei interactions. The KS Hamiltonian is then written as

$$
\begin{aligned}
H(\mathbf{k}+\mathbf{G}, \mathbf{k}+\mathbf{G}') &= \frac{\hbar^2}{2m}(\mathbf{k}+\mathbf{G})^2 \delta_{\mathbf{G},\mathbf{G}'} + V_L(\mathbf{G}-\mathbf{G}') \\
&+ V_{NL}(\mathbf{k}+\mathbf{G}, \mathbf{k}+\mathbf{G}') \qquad (2)
\end{aligned}
$$

where $V_L(\mathbf{G}-\mathbf{G}')$ is the Fourier transform of the local part of the total potential and $V_{NL}(\mathbf{k}+\mathbf{G}, \mathbf{k}+\mathbf{G}')$ is the spatially non-local contribution of the pseudopotential, that for most atoms is short ranged and vanishes for $|\mathbf{r}-\mathbf{r}'| > r_c$, with r_c being the radius beyond which pseudo and true KS orbitals coincide [12].

For transport calculations it is convenient to use an orthorombic unit cell [9], where the real space unit vectors can be written as $\mathbf{a_1}=(a_x, 0, 0)$, $\mathbf{a_2}=(0, a_y, 0)$, $\mathbf{a_3}=(0, 0, a_z)$, with x being the transport direction. Hence the unit vectors of the reciprocal lattice are $\mathbf{b_1}=(2\pi/a_x, 0, 0)$, $\mathbf{b_2}=(0, 2\pi/a_y, 0)$, $\mathbf{b_3}=(0, 0, 2\pi/a_z)$. The Brillouin zone (BZ) is therefore given by the conditions $-\pi/a_s < k_s \le \pi/a_s$ (with $s = x, y, z$).

A. Discussion of the reduced basis set

The DFT Hamiltonian in Eq. (2) is a dense matrix with a rank equal to the number, N_G, of reciprocal lattice vectors, and thus it cannot be directly employed in transport calculations. Thus our first step is a unitary transformation from PW to real space along the transport direction x, that provides the Hamiltonian in the hybrid $x\mathbf{K}_{yz}$ basis given by real-space along x and plane waves in the (y, z) plane [13]. In this representation the Hamiltonian matrix is a block tri-diagonal matrix, which is crucially important in order to exploit recursive algorithms for non-equilibrium Green's function (NEGF) calculations [14]. In this hybrid $x\mathbf{K}_{yz}$ basis each block describes an a_x long spatial region consisting of N_{dx} discretization points, so that the matrix blocks have a rank $N_G=N_{dx}N_{Gy}N_{Gz}$ (with $N_{dx}=N_{Gx}$). Even if the $x\mathbf{K}_{yz}$ basis has been used to perform quantum transport calculations as in [13], [15], the size N_G of the blocks is large and it increases by enlarging the cutoff energy used in DFT calculations. Huge computational advantages can thus be obtained by moving to a basis of unit-cell restricted Bloch functions, that preserves the block tridiagonal structure of the Hamiltonian matrix [9], [16].

To this purpose we first argue that, due to the block tridiagonal form of the Hamiltonian, the unit cell restricted Bloch functions $\{\Psi_{\mathbf{k}}^0\}$ corresponding to the wave vector $\mathbf{k}=(k_x,\mathbf{k}_{yz})$ satisfy the secular equation

$$\left[\mathbf{H}_{0,1}^\dagger e^{-ik_x a_x} + \mathbf{H}_{0,0} + \mathbf{H}_{0,1} e^{ik_x a_x} \right] \{\Psi_{\mathbf{k}}^0\} = E(\mathbf{k}) \{\Psi_{\mathbf{k}}^0\} \tag{3}$$

with $\mathbf{H}_{0,0}$, $\mathbf{H}_{0,1}$ being the Hamiltonian blocks describing respectively one unit cell and its interaction with the nearest neighbor cell. Our reduced basis set consists of a subset of the $\{\Psi_{n\mathbf{k}}^0\}$ corresponding to a few k_x values in the reduced zone $-\pi/a_x < k_x \le \pi/a_x$ along the transport direction and, for each k_x, to some tens of bands $E_n(\mathbf{k})$ up to the energies of interest for the problem at study.

It is noteworthy that the $\{\Psi_{n\mathbf{k}}^0\}$ are obtained directly from the PW representation provided by the DFT solver, namely *we do not need to solve Eq. (3) to find the basis functions.*

The size of the basis can be written as

$$N_B = \sum_{n=n_i}^{n_f} N_{kB,n} \tag{4}$$

where $N_{kB,n}$ denotes the number of k_x values included in the basis for the n-th band, with n ranging between n_i and n_f. The $\{\Psi_{n\mathbf{k}}^0\}$ functions are not orthogonal for different k_x, but they can be orthogonalized and we denote by Φ the basis set orthonormalized over a unit cell. The Hamiltonian blocks in the Φ basis have a size N_B that is typically much smaller than the size N_G in the hybrid $x\mathbf{K}_{yz}$ basis.

A first validation of the reduced basis set requires that it allows us to calculate the band structure for any k_x in the reduced zone, which is discussed below in Sec.II-B.

the high-symmetry points of the primitive BZ obtained from the DFT Hamiltonian in the PW basis. The DFT calculation was performed with the Quantum ESPRESSO code [11], by using the Perdew-Burke-Ernzerhof [18] approximation to the exchange-correlation functional and a norm-conserving pseudopotential [19]. We employed a $12 \times 12 \times 1$ Monkhorst-Pack k-points grid and a cutoff energy of $E_w = 60$ Ry.

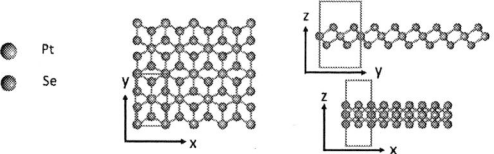

Fig. 2. Orthorombic unit cell of the 1T phase monolayer PtSe$_2$.

For transport calculations we first define the orthorombic cell composed of 6 atoms and sketched in Fig. 2. The unit vectors are $\mathbf{a_1}=(a_x,0,0)$, $\mathbf{a_2}=(0,a_y,0)$, $\mathbf{a_3}=(0,0,a_z)$ with $a_x=3.712$ Å, $a_y=6.429372$ Å and $a_z=32$ Å, while the vertical distance between the Pt and Se atoms is 1.312 Å. In order to validate the reduced Bloch state basis, we compared the bands computed with the Hamiltonian blocks in the reduced basis with the results of the plane-wave DFT solver. The accuracy of the reduced basis depends on the included number of k_x Bloch functions [9], hence we optimized the basis by using a larger $N_{kB,n}$ for bands with energies in the range of interest, and smaller $N_{kB,n}$ for bands at much lower or higher energies.

In Fig. 3 we show an excellent band-structure reconstruction obtained by using $N_{kB,n}=4$ for the bands close to the energygap and $N_{kB,n}=2$ for smaller energy bands, and for an overall number $N_B=74$ of Bloch functions. Such a definition of the reduced basis results in no observable unphysical states [20], [21].

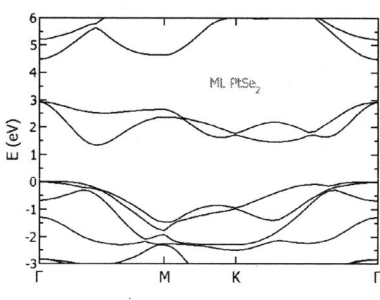

Fig. 1. Bandstructure of the monolayer PtSe$_2$.

B. Material system and validation of the reduced basis set

In this paper we exemplify our methodology by focusing on the monolayer PtSe$_2$, which is a noble metal dichalcogenure identified as a promising channel material for MOSFETs due to the small effective mass and large density of states (DOS) [17]. Figure 1 reports the corresponding band structure along

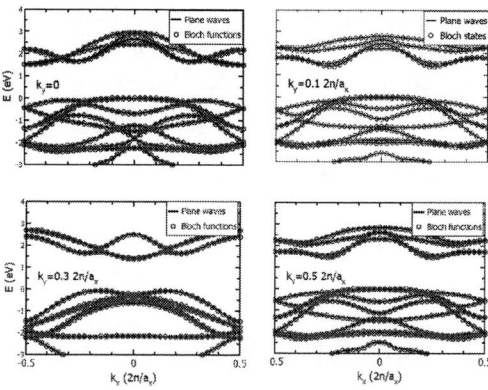

Fig. 3. Bandstructure of the monolayer PtSe$_2$ for various k_y values computed with (circles) the Bloch function basis and (lines) the PW basis.

C. Carrier densities and electron current

All relevant physical quantities are defined in terms of the retarded(advanced), $[\mathbf{G}^{r(a)}]_\Phi$, and the lesser(greater)-than Green's functions, $[\mathbf{G}^{<(>)}]_\Phi$, computed in the reduced Bloch functions basis. In order to compute the carrier concentrations in the real space, we first compute $[\mathbf{G}^{<(>)}]_\Phi$, then we reconstruct the Green's function in the hybrid basis according to the expansion

$$
\begin{aligned}
\mathbf{G}^{<(>)}(x_j\mathbf{K}_{yz}, x_j\mathbf{K}'_{yz}; E) &= \sum_{n,m=1}^{N_B} \mathbf{G}^{<(>)}(n,m;E) \\
&\times \Phi_n(x_j, \mathbf{G}_{yz}) \Phi_m^*(x_j, \mathbf{G}'_{yz})
\end{aligned}
\tag{5}
$$

and finally evaluate electron and hole densities as described in Ref. [9]. Thanks to this procedure we are able to evaluate the 3D electron and hole concentrations, $n(\mathbf{r})$ and $p(\mathbf{r})$, with atomistic accuracy as shown in the example of Fig. 4.

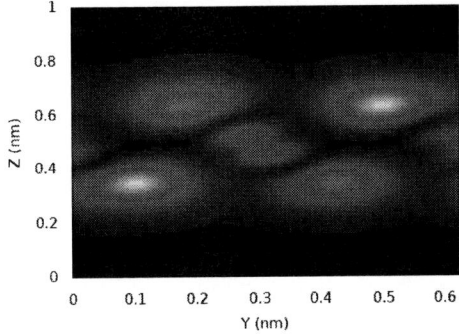

Fig. 4. Contourplot of the cross section of the electron density at an abscissa x corresponding to the center of the orthorombic unit cell. Colors are in arbitrary units.

However, by leveraging the fact that the electrostatic potential varies over larger spatial scales compared to the discretization used in Eq. (5), in the self-consistent solution of the NEGF transport equations and the Poisson equation we evaluate $n(\mathbf{r})$ and $p(\mathbf{r})$ on a coarser mesh. For the 3D Poisson problem we assume Diriclect conditions at the metal boundaries and Neumann conditions elsewhere.

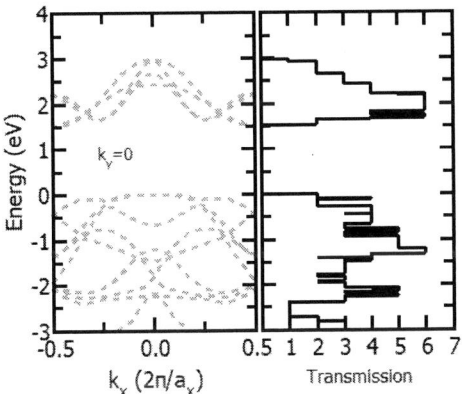

Fig. 5. Bandstructure of an infinitely-long monolayer PtSe$_2$ and the corresponding transmission probability obtained with the Hamiltonian in the reduced basis consisting of unit-cell-restricted Bloch states. $k_y = 0$.

The electron current is obtained within the elastic approximation as

$$
I_D = \frac{ge}{h} \int T(E)\left[f_D(E) - f_S(E)\right] dE \tag{6}
$$

with g being the degeneracy and $f_{S(D)}$ the Fermi-Dirac function at the source (drain) contact. Figure 5 illustrates the bandstructure and transmission of an infinitely long monolayer PtSe$_2$ at $k_y = 0$ calculated by using the reduced basis and showing that, as expected, the transmission at a given energy E equals the number of available bands.

The device under investigation is sketched in Fig. 6. It is a single-gate FET with doped source and drain access regions having a length $L_{S,D} \approx 11.1$ nm and an undoped channel region with $L_G \approx 18.5$ nm. The donor concentration in the source and drain region is $N_D = 10^{20}$ cm^{-3}. The equivalent oxide thickness (EOT) of the high-k dielectric layer is 0.65 nm. In the device width direction, y, we assume periodic boundary conditions, described by sampling the BZ along the k_y direction with a step $\Delta k_y = 0.1 \times 2\pi/a_y$.

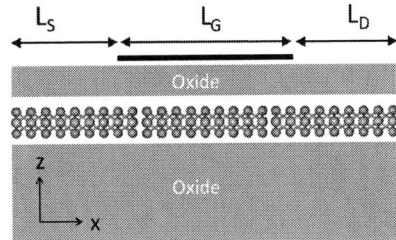

Fig. 6. Sketch of the single-gate monolayer PtSe$_2$ FET.

Figure 7 shows the transfer characteristics for different V_{DS} values. In the sub-threshold voltage regime we observe an ideal sub-threshold swing of about 60 mV/dec, which stems from the excellent electrostatic integrity of the device due, in turn, to the ultra-thin channel and the aggressive EOT. The device exhibits also a high transconductance resulting in a high on-state current, that confirms the promising properties of this material for high-performance MOSFETs. Of course the transconductance and on-state current in Figure 7 should be considered as upper limit figures, that can be significantly degraded by series resistance and scattering [22].

Quite interestingly, in Figure 7 we can also observe a non-monotonic I_D dependence on V_{DS}, in fact the I_D at large V_{TG} increases with V_{DS} for V_{DS} up to 0.3 V, but then it decreases for larger V_{DS}. This behavior has been already observed in simulated FETs with a monolayer MoS$_2$ channel [23], and it has been ascribed to peculiar features of the DOS. We verified that a similar explanation applies to our results, in fact Fig. 8 reports the local density of states (LDOS) at $V_{DS}=0.3$ V and 0.6 V in the on-state regime (i.e. at $V_{TG}=0.8$ V), showing that in the drain region the LDOS for energies close to the top of the barrier is larger at $V_{DS}=0.3$ V than it is at $V_{DS}=0.6$ V.

Such a non-monotonic I_D dependence on V_{DS}, however, is expected to disappear in the presence of significant inelastic scattering mechanisms, such as electron-phonon scattering.

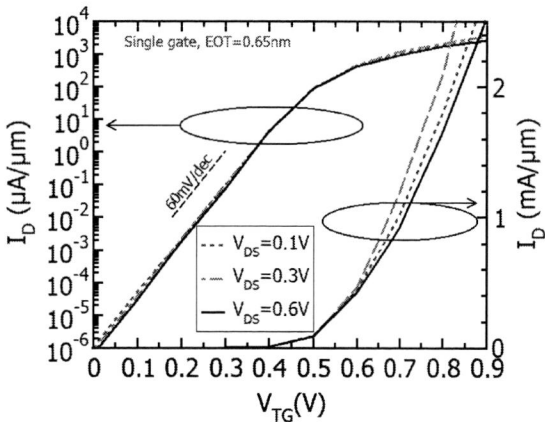

Fig. 7. Transfer characteristics of the monolayer PtSe$_2$ FET for V$_{DS}$=0.1, 0.3 and 0.6 V.

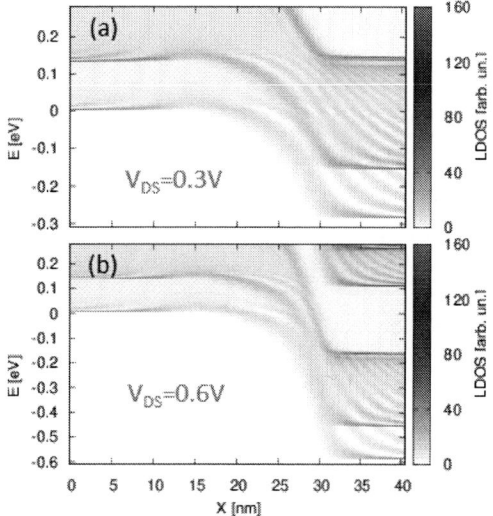

Fig. 8. LDOS computed for (a) V$_{DS}$=0.3 V and (b) V$_{DS}$=0.6 V and for $V_{TG} = 0.8$ V. The transverse wave-vector is $k_y = 0.1 \times 2\pi/a_y$.

III. CONCLUSIONS AND OUTLOOK

We have shown that a unit-cell restricted Bloch functions basis enables band structure and transport calculations with a significant reduction of the computational burden compared to the original plane-wave DFT formulation. This methodology is a viable approach for *ab initio* and semi-empirical quantum transport simulations, and it can be seen as an alternative to the MLWF approach. In fact the Bloch functions can be obtained directly from the DFT solvers, thus bypassing the extraction of a MLWF basis and of the corresponding tight-binding Hamiltonian. We foresee that further developments of the methodology will enable us to deal with non homogenous systems (i.e. including hetero-jnctions), and to include inelastic scattering mechanisms.

REFERENCES

[1] L. Mingda, *et al.*, "Single Particle Transport in Two-dimensional Hetero-junction Interlayer Tunneling Field Effect Transistor," *Journal of Applied Physics*, vol. 115, p. 074508, 2014.

[2] D. Sarkar, *et al.*, "A subthermionic tunnel field-effect transistor with an atomically thin channel," *Nature*, vol. 526, no. 7571, pp. 91–95, 2015.

[3] J. Cao, *et al.*, "Operation and Design of van der Waals Tunnel Transistors: A 3-D Quantum Transport Study," *IEEE Transactions on Electron Devices*, vol. 63, no. 11, pp. 4388–4394, 2016.

[4] D. Logoteta, *et al.*, "A steep-slope MOS$_2$-nanoribbon MOSFET based on an intrinsic cold-contact effect," *IEEE Electron Device Letters*, vol. 40, no. 9, pp. 1550–1553, 2019.

[5] E. G. Marin, *et al.*, "Lateral heterostructure field-effect transistors based on two-dimensional material stacks with varying thickness and energy filtering source," *ACS Nano*, vol. 14, no. 2, pp. 1982–1989, 2020.

[6] J. M. Soler, *et al.*, "The SIESTA method for ab initio order-n materials simulation," *Journal of Physics: Condensed Matter*, vol. 14, no. 11, pp. 2745–2779, mar 2002.

[7] N. Marzari and D. Vanderbilt, "Maximally localized generalized Wannier functions for composite energy bands," *Phys. Rev. B*, vol. 56, p. 12847, Nov 1997.

[8] A. Szabó, R. Rhyner, and M. Luisier, "Ab initio simulation of single- and few-layer MoS$_2$ transistors: Effect of electron-phonon scattering," *Phys. Rev. B*, vol. 92, p. 035435, Jul 2015.

[9] M. G. Pala, P. Giannozzi, and D. Esseni, "Unit cell restricted Bloch functions basis for first-principle transport models: Theory and application," *Phys. Rev. B*, vol. 102, p. 045410, Jul 2020.

[10] G. Kresse and J. Furthmüller, "Efficiency of ab-initio total energy calculations for metals and semiconductors using a plane-wave basis set," *Computational Materials Science*, vol. 6, no. 1, pp. 15 – 50, 1996.

[11] P. Giannozzi *et al.*, "Quantum Espresso: a modular and open-source software project for quantum simulations of materials," *Journal of Physics: Condensed Matter*, vol. 21, no. 39, p. 395502 (19pp), 2009.

[12] P. Giannozzi *et al.*, "Advanced capabilities for materials modelling with Quantum Espresso," *Journal of Physics: Condensed Matter*, vol. 29, no. 46, p. 465901, 2017.

[13] M. G. Pala and D. Esseni, "Full-band quantum simulation of electron devices with the pseudopotential method: Theory, implementation, and applications," *Phys. Rev. B*, vol. 97, p. 125310, 2018.

[14] M. P. Anantram, M. S. Lundstrom, and D. E. Nikonov, "Modeling of Nanoscale Devices," *Proceedings of the IEEE*, vol. 96, no. 9, pp. 1511–1550, Sept 2008.

[15] M. Pala, O. Badami, and D. Esseni, "NEGF based transport modeling with a full-band, pseudopotential Hamiltonian: Theory, Implementation and Full Device Simulations," *IEEE International Electron Devices Meeting*, pp. 35.1.1–35.1.4, 2017.

[16] M. L. Van de Put, M. V. Fischetti, and W. G. Vandenberghe, "Scalable atomistic simulations of quantum electron transport using empirical pseudopotentials," *Computer Physics Communications*, vol. 244, pp. 156–169, 2019.

[17] A. AlMutairi, D. Yin, and Y. Yoon, "PtSe$_2$ field-effect transistors: New opportunities for electronic devices," *IEEE Electron Device Letters*, vol. 39, no. 1, pp. 151–154, 2018.

[18] J. P. Perdew, K. Burke, and M. Ernzerhof, "Generalized gradient approximation made simple," *Phys. Rev. Lett.*, vol. 77, pp. 3865–3868, Oct 1996.

[19] D. R. Hamann, "Optimized norm-conserving Vanderbilt pseudopotentials," *Phys. Rev. B*, vol. 88, p. 085117, Aug 2013.

[20] G. Mil'nikov, N. Mori, and Y. Kamakura, "Equivalent transport models in atomistic quantum wires," *Phys. Rev. B*, vol. 85, p. 035317, Jan 2012.

[21] M. Shin, W. J. Jeong, and J. Lee, "Density functional theory based simulations of silicon nanowire field effect transistors," *Journal of Applied Physics*, vol. 119, no. 15, p. 154505, 2016.

[22] D. Esseni, P. Palestri, and L. Selmi, *"Nanoscale MOS Transistors"*, 1st ed. Cambridge University Press., 2011.

[23] J. Chang, L. F. Register, and S. K. Banerjee, "Atomistic full-band simulations of monolayer MoS$_2$ transistors," *Applied Physics Letters*, vol. 103, no. 22, p. 223509, 2013.

SISPAD 2020

Future of Power Electronics from TCAD Perspective

Terry Ma
Synopsys, Inc., Mountain View, CA, USA
terryma@synopsys.com

Power electronics are an integral part of our daily life. The applications of power electronics are widespread, supporting multiple industries such as automotive, telecommunication, transportation, utility systems, aerospace, etc. According to a new market research report including the analysis of the COVID-19 impact [1], the global power electronics market is expected to grow at a Compounded Annual Growth Rate (CAGR) of 4.7% from $35.1 billion in 2020 to $44.2 billion by 2025. As depicted in Figure 1, the key drivers of growth are 1) increasing integration of power electronics, and 2) increasing use of wide bandgap (WBG) materials. In terms of regional growth, Asia Pacific (APAC) including China, Japan, South Korea, and India is expected to grow the fastest compared to North America, Europe, and Rest of the World (Figure 2).

Power electronics can be classified as discrete such as IGBT, which can also be configured in modules for EV/HEV automotive applications, Si-based ICs for power management in mobile applications, and Wide Bandgap (WBG) devices for medium (up to 500V) and high (>1000V) voltage applications. While the demand for Si-based discrete and power IC continues to grow, higher efficiency power electronics drive the development of WBG devices based on GaN and SiC materials. Figure 3 shows the applications of both Si-based and WBG power electronics with respect to power and switching (frequency) levels.

One of the areas that fuels the growth of Si-based power electronics is the increasing electronics content in automotive as depicted in Figure 4. Power electronics provides efficient conversion, control, and conditioning of electric power in automotive (Figure 5). By 2030, the electronics system cost can reach as high as 50% of the cost of a car [2]. In addition to the growth in automotive industry, power electronics, particularly, power management IC used in consumer and industrial electronics such as smartphones, tablets, TV, electrical appliances, etc. will continue to drive the overall market.

Unlike advanced CMOS logic development that follows the Moore's Law, development of power electronics is based on demand in a wide range of applications that serve multiple industries. Figure 6 contrasts the requirements between mobile and automotive chip designs. In terms of development direction, IGBT and power management IC will continue to drive the Si-based development to improve energy efficiency and push to reach their theoretical limits. For instance, IGBT development is in its 7th generation to reduce its size, lower Ron, and operate at higher junction temperature as shown in Figure 7. As mentioned in the first paragraph, increasing integration of power electronics is one of the growth drivers. Thus, the development focus is not only on individual power devices, but it must also include co-optimizing the components and packaging in the power module or system. Figure 8 shows a TCAD simulation example of co-optimizing a power module over multiple design parameters including electrical, thermal and mechanical effects.

For WBG applications, SiC IGBT significantly reduces switching loss and size (form factor) compared to Si-based IGBT. GaN-on-Si is gaining momentum to become the main technology for medium power applications as it offers lower cost and compatibility with Si fab lines. The device performance of both SiC and GaN is still far from the theoretical limit, motivating continued exploration and innovation to drive higher performance and efficiency (see Figure 9), as well as increasing reliability [3]. In terms of growth, the WBG power electronics market is expected to grow at a CAGR of 30% between 2020 and 2025. The largest market opportunity for WBG devices is in power converters and inverters for EV/HEV vehicles.

To support the increasing demand for power electronics, top manufacturers have been shifting from 200mm to 300mm fab lines. Figure 10 shows the timeline and investment that major players made and are expected to make. In addition, WBG technology has recently achieved a critical milestone for high volume production with the availability of 200mm wafer.

From the TCAD perspective, the future of power electronics is bright as the development of new Si-based and WBG applications for power electronics is unabated. However, the boom in the power electronics market represents both challenges and opportunities for TCAD. Over the past decade, the nature of TCAD simulation of power electronics has evolved rapidly in four areas: 1) 2D to 3D, 2) small to large scale, 3) Silicon to WBG materials [4], and 4) single device to integrated modules. Despite these challenges, the recent improvements and enhancements in commercial TCAD tools such as Synopsys' Sentaurus line of products enable power electronics suppliers to explore and innovate new capabilities and functionalities in the form of single discrete device and multiple devices as power ICs, as well as new materials (e.g. GaN, SiC) and device architectures (e.g. super-junction). The types of analysis range from the traditional breakdown characteristics (Figure 11) to electro-thermal interactions (Figure 12) of power electronics modules [5] and reliability impact [6]. TCAD is poised to help innovate and develop a new wave of power electronics to support the growth in adoption in the automotive, consumer, industrial, and renewable industries in the next decade.

References

[1] Power Electronics Market with COVID-19 Impact Analysis by Device Type (Power Discrete, Power Module and Power ICs), Material (Silicon, Silicon Carbide and Gallium Carbide), Voltage (Low Voltage, Medium Voltage and High Voltage), Vertical (ICT, Consumer Electronics, Industrial, Automotive & Transportation, Aerospace & Defense), and Geography - Global Forecast to 2025, Power Electronics Market, Published Date: June 2020 | Report Code: SE 2434.

[2] Semiconductors – The Next Wave Opportunities and Winning Strategies for Semiconductor Company, April 2019, Deloitte.

[3] Improving Reliability For GaN And SiC: Why these chips are gaining ground, and what still needs to be addressed, June 18, 2020, M. Lapedus, https://semiengineering.com/improving-reliability-for-gan-and-sic/

[4] S. Stoffels, et al., "From TCAD Device Simulation to Scalable Compact Model Development for GaN HEMT Powerbar Designs," 1st IEEE Workshop on Wide Bandgap Power Devices and Applications, October 2013.

[5] A. Chvála, et al., "Advanced 3-D Device and Circuit Electrothermal Simulations of Power Integrated Circuit," 46th European Solid-State Device Research Conference (ESSDERC), September 2016.

[6] T. Cilento, et al., "Investigation of Layout Effects in Diode-triggered SCRs under very-fast TLP Stress through Full-size, Calibrated 3D TCAD Simulation," Microelectronics Reliability 88–90 (2018) 1103–1107.

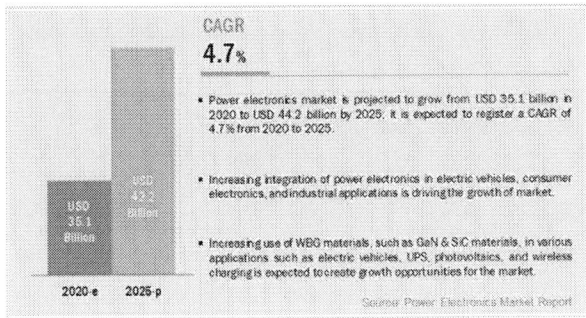

Figure 1. Power electronics market is expected to grow to $42.2 billion by 2025.

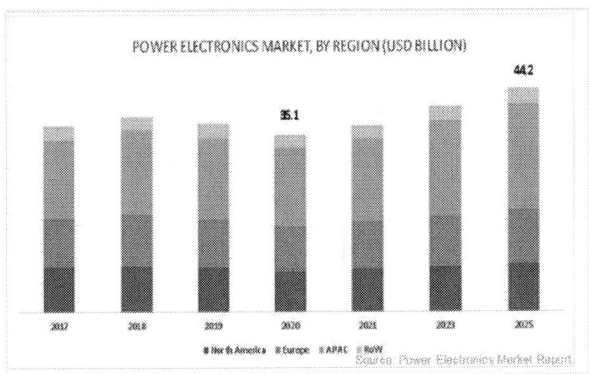

Figure 2. Asia Pacific grows the fastest compared to North America, Europe and Rest of the World.

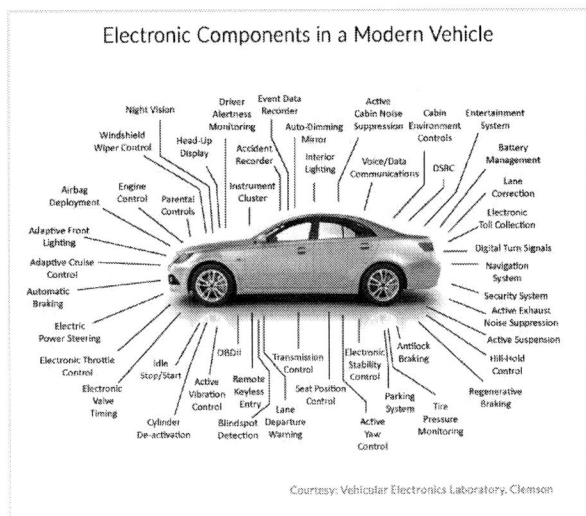

Figure 4. Cost of electronics in modern vehicles can reach as high as 50% of the car cost.

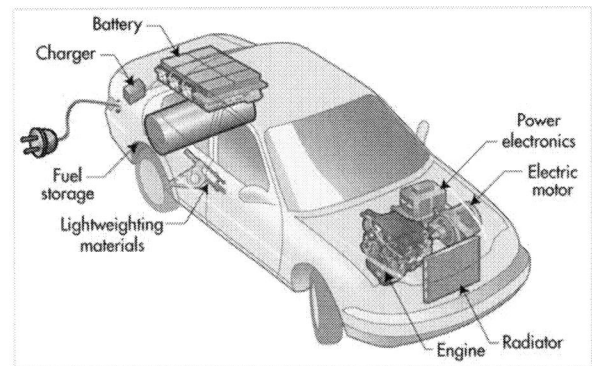

Figure 5. Power electronics play an important role of controlling electrical systems in modern vehicles.

Figure 3. Power electronics and their applications.

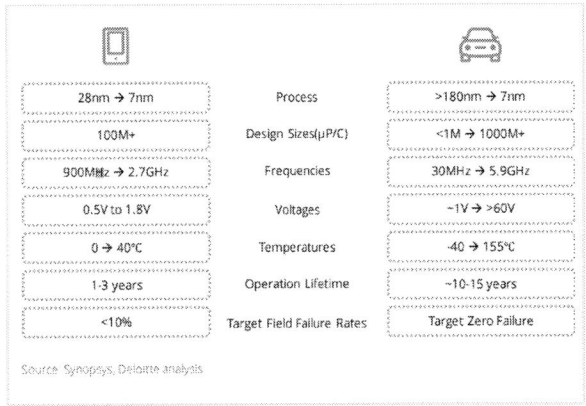

Figure 6. Comparison of IC requirements between mobile and automotive.

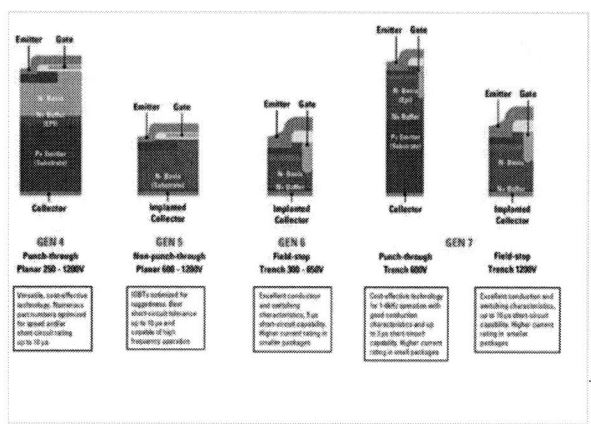

Figure 7. Evolution of IGBT development from 4th to 7th generation.

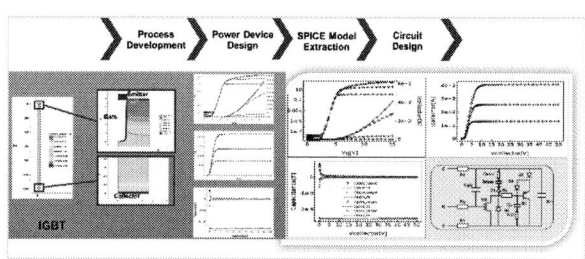

Figure 8. An example of using TCAD simulation for co-optimization of a power module,

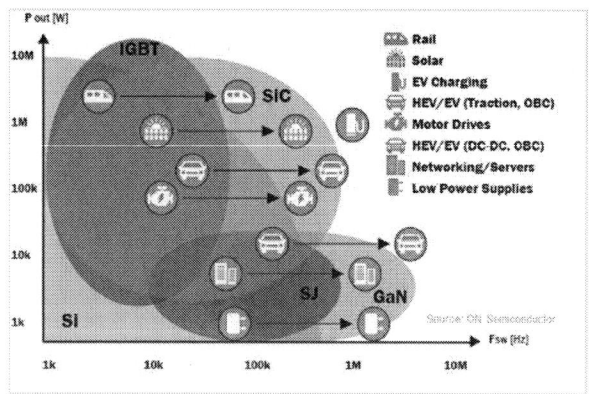

Figure 9. Development of SiC and GaN to support the growth in applications requiring higher energy efficiency.

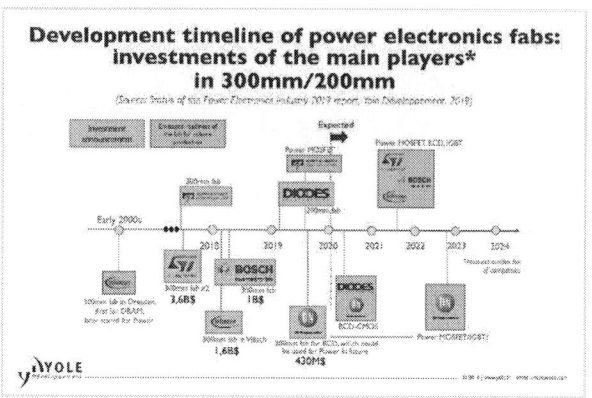

Figure 10. Key power electronics manufacturers are preparing to switch from 200mm to 300mm production.

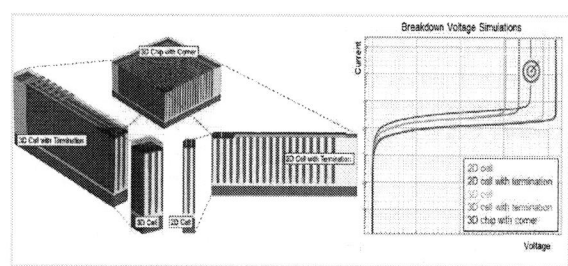

Figure 11. TCAD simulation expands from 2D to 3D and increased scale from a small portion of the device to multiple junctions of the device.

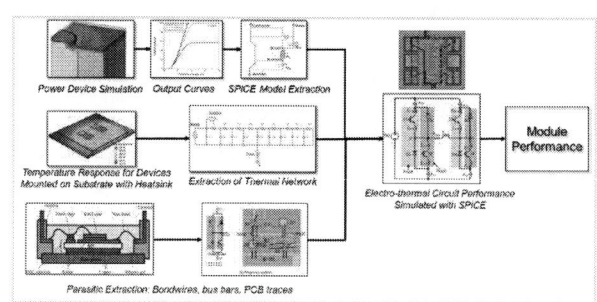

Figure 12. An example of using TCAD for Design Technology Co-Optimization (DTCO) for electro-thermal analysis of device-circuit interaction of power electronics modules.

Computics Approach toward Clarification of Atomic Reactions during Epitaxial Growth of GaN

Atsushi Oshiyama[1], Kieu My Bui[1], Mauro Boero[2,1], Yoshihiro Kangawa[3,1], and Kenji Shiraishi[1,4]

[1]Institute of Materials and Systems for Sustainability, Nagoya University, Nagoya 464-8601, Japan
{oshiyama, bui.my}@imass.nagoya-u.ac.jp
[2]University of Strasbourg, Institut de Physique et Chimie des Matériaux de Strasbourg, CNRS, UMR 7504, F67034, Strasbourg, France
mauro.boero@ipcms.unistra.fr
[3]Research Institute for Applied Mechanics, Kyushu University, Fukuoka 816-8580, Japan
kangawa@riam.kyushu-u.ac.jp
[4]Graduate School of Engineering, Nagoya University, Nagoya 464-8603, Japan
shiraishi@cse.nagoya-u.ac.jp

Abstract—We report first-principles calculations based on the density-functional theory that clarify atomic reactions of ammonia decomposition and subsequent nitrogen incorporation during GaN epitaxial growth. We find that Ga-Ga weak bonds are ubiquitous on Ga-rich growing surface and responsible for the growth reactions. Furthermore, Car-Parrinello Molecular Dynamics simulations predict the existence of 2-dimensional Ga liquid phase, providing new insight into the epitaxial growth. The obtained results are expected to become basics for multi-scale growth simulations in future.

Keywords—density functional theory, Car-Parrinello molecular dynamics, GaN, epitaxial growth

I. INTRODUCTION

Density-functional theory (DFT) [1] combined with the Kohn-Sham (KS) scheme [2] was innovated more than a half-century ago and then provided a new way to study physical and chemical properties of real materials quantum mechanically. Subsequent progress in computational methods allows us to apply this KSDFT scheme to a numerous number of materials and unprecedented success has been achieved [3]. In addition to reveal atomic and electronic properties of stable and metastable structures, dynamical properties have been also explored by introducing molecular dynamics (MD) scheme. In particular, the MD combined with DFT proposed by Car and Parrinello (CPMD) is proved to be particularly powerful [4-6].

Performance of supercomputers increases exponentially and now it reaches 410 PFLOPS (Fugaku in Japan) as its theoretical peak performance [7]. This situation of the computers seems to reinforce the computational approach. However, it is not necessarily true. The performance of a single compute node is already saturated. Hence the only way to make a high-performance computer is to gather a huge number of compute nodes and connect them with high-speed network (massively parallel architecture). Moreover hardware accelerators have been introduced to perform a part of the scientific calculations. Hence, without the cross-disciplinary collaboration among the field of physical science or device simulation and the field of computer science or applied mathematics, it would be impossible to exploit the tremendous performance of the supercomputers. We call this new stage of the collaboration *computics* [8].

In this article, we present our efforts to clarify microscopic mechanisms of Metal-Organic Vapor-Phase Epitaxy (MOVPE) of gallium nitride (GaN) [9] based on the DFT calculations. These efforts just get started in *computics* viewpoints but the obtained results are already interesting [10-12]. We find that the existence of Ga-Ga weak bonds on the growing surface is essential for the epitaxial growth.

GaN known as a premier material in optoelectronics is now emerging in power electronics due to its wider band-gap, higher mobility and robustness under harsh environment [13]. In GaN MOVPE, trimethylgallium (TMG) and ammonia (NH_3) are used as gas sources. Many experimental and theoretical efforts have revealed that TMG is decomposed in the gas phase and the growing surface is generally Ga rich [14-16]. On the other hand, the mechanism of the decomposition of NH_3 is unclear though it is postulated without any evidence to be decomposed in the gas phase and forms several NH_x unit [17]. On the contrary, recent high-resolution time-of-flight (TOF) measurements [16] show that ammonia is rarely decomposed in the gas phase: Only small amount (0.1 %) of NH_2 is detected in the gas phase except for NH_3. Hence clarification of the atomistic behavior of NH_3 on the growing GaN surface is the first priority to be attacked by the DFT calculations.

II. COMPUTATIONAL METHODS

All the calculations have been performed in DFT with the generalized gradient approximation [18] to the many-body exchange-correlation energy functional. Nuclei and core electrons are described by norm-conserving pseudo-potentials [19]. The Ga 3d states are treated as core-electron states with or without the partial core correction [20]. The geometry-optimized electronic-structure computations have been done using our real-space-scheme code called RSDFT [21] which has been developed so as to achieve the state-of-the-art high-performance computing [22]. Dynamical simulations have been performed by Car-Parrinello molecular dynamics (CPMD) as is implemented in the plane-wave basis-set scheme [23]. The GaN (0001) surface is simulated by a repeating slab model in which each slab consisting of six GaN double layers are separated from its images by the 15-Å thick vacuum regions. The bottom N surface is passivated by fractionally charged (0.75 *e*) hydrogen to annihilate the dangling bonds and mimic the semi-infinite GaN substrate [24]. Other calculational details and their validity are found in References 10 and 12. Reaction pathways and corresponding energy barriers are calculated by using the hyperplane constraint method (HPC) [25].

III. DECOMPOSITION OF NH₃ AND INCORPORATION OF N

Thermodynamical analyses combined with DFT calculations [15] have shown that the growing surface in GaN MOVPE at the typical range of temperature and gas-source pressures is Ga rich. We have calculated structures and binding energies of a Ga adatom (Ga$_{ad}$) on Ga-face GaN (0001) surface and found that the Ga adatom is adsorbed on top of N atoms (T4 site [10]). Relying on the electron counting rule (ECR) [26,27], the coverage of Ga adatoms is expected to be 0.25 where all the Ga dangling bond become unoccupied. Then we consider 2x2 lateral periodicity with one additional Ga atom at T4 site is a representative of the Ga-rich GaN(0001). On this Ga-rich surface, we explore stable adsorption sites for an NH₃ molecule. We have found two possibilities (Fig. 1): One is on the top Ga atom (T1 site) and the other is on the Ga$_{ad}$ (T$_{ad}$ site hereafter) [10]. As shown in Fig. 1, the strong bond formation between the N atom in NH₃ and the top Ga atom is evident. This is the reason for the stability of the two configurations. It is noteworthy that both the T$_{ad}$ and the T1 configurations satisfy the ECR with the Ga dangling bond unoccupied. Furthermore, we have found that the Ga-Ga bonds in the structures are rather weak, as is shown by the smaller isovalue surfaces. This weak spots may be a candidate place for the N incorporation (see below).

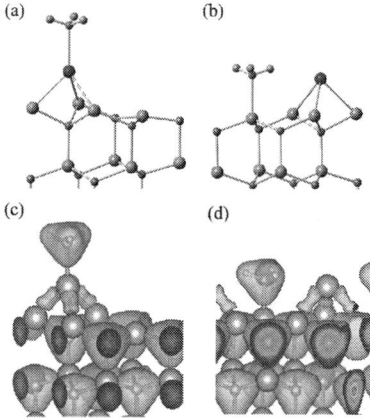

Fig. 1. Stable adsorption structure and corresponding electron density of NH₃ at T$_{ad}$ site [(a) and (c)] and at T1 site [(b) and (d)] on the Ga-rich 2x2 GaN (0001) surface. Green (large), blue (small), salmon pink (small) and burgandy (large) balls depict Ga, N, H atoms and Ga adatoms, respectively. Electron density is represented by yellow isovalue surfaces and by contour color plots at the cross-sections. (after Bui *et al.* [10])

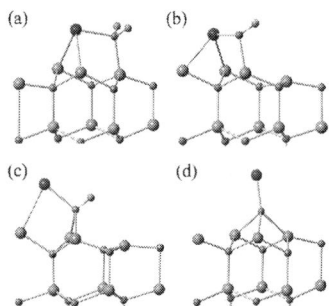

Fig. 2. Stable adsorption structures of NH$_x$ units on the Ga-rich GaN (0001). (a) NH₂ on the T1 site, (b) NH on the T1 site, (c) NH on the BR site, and (d) substitutional adsorption of N. The color code is the same as in Fig. 1. (after Bui *et al.* [10])

Behaviors of other species, NH₂, NH and N, are also important since they may be generated by the reactions of NH₃ on the surface. We have thus explored the most stable sites for those species. Fig. 2 shows the obtained stable structures. For NH₂, we have found that the most stable site is the T1 site with the binding energy being 3.96 eV [Fig. 2(a)]. In the final structure Ga$_{ad}$ is dislodged to the center of the hexagon with the NH₂ intervening in the Ga-Ga bond. Surprisingly, this geometry is reached without any energy barrier in spite that one of the three Ga$_{ad}$-Ga bonds is cleaved. We have also found that the NH unit is adsorbed at several sites including T1 and the bridge sites (BR site) and then intervenes in the Ga-Ga bond again with an undetected energy barrier [Fig. 2 (b) and (c)]. Another surprise for the N atom is its substitutional adsorption. The N atom approaches the Ga adatom, then destruct the bonds between the Ga adatom and the top Ga atoms spontaneously, and forms a Ga-N-Ga3- configuration [Fig. 2 (d)].

The calculations for NH₂ and NH reported above show that the weak Ga-Ga bond is likely to be the reaction site. Hence we next explore a pathway and its corresponding energy barrier for the reaction,

$$[\text{NH}_3 \text{ on } T_{ad}] \rightarrow [\text{NH at } T1] + \text{H}_2 , \qquad (1)$$

using the HPC method. Fig. 3 shows the obtained energy profile of the reaction from NH₃ at the T$_{ad}$ [Fig. 1 (a)] toward NH near the T1 [Fig. 2(b)] with the H₂ molecule being desorbed. It is noteworthy that the final configuration satisfies ECR with the N dangling bond being occupied by 2 electrons and the two Ga dangling bonds being unoccupied. The energy barrier for this reaction of this NH₃ decomposition and the subsequent NH incorporation in the GaN bond network is calculated to be 0.63 eV. This value is surprisingly small in spite that the reaction involves the cleavage of a Ga-Ga bond. The distance from the incorporated N of the NH unit to the two Ga atoms are 1.90 Å and 1.92 Å, indicating that new Ga-N bonds are already formed.

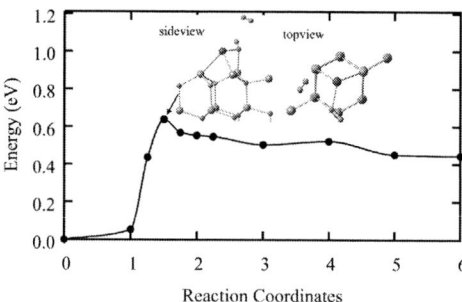

Fig. 3. Calculated energy profile for the reaction of the NH₃ decomposition and the NH incorporation (see text) on the Ga-rich GaN(0001) growing surface. The transition-state geometry is also shown. (after Bui *et al.* [10])

The next issue is how the hydrogen atoms are desorbed in a form of H₂ and leaves the N atom surrounded by four neighboring Ga atoms. We have explored several possibilities [10] and found a plausible reaction in which the -(Ga adatom)-(NH₂)-Ga- structure [Fig. 2 (a)] becomes the incorporated N being four-fold coordinated with Ga atoms [Fig. 2 (d)]. We have identified the transition state geometry and the calculated energy barrier is 2.0 eV. Examining the energy profile, we have found that this is the energy cost to make the final state with H_2 being desorbed. Hence, when the energy of the final state is lowered, the reaction is likely to occur. This indeed occurs in the typical growth conditions: the hydrogen chemical potential at 1300 K in the gas phase is evaluated to be -2.1 eV, considering translation, vibration and rotation motions of H_2. This value of the chemical potential compensates for the zero-temperature energy barrier computed above. Then the whole reaction of the formation of GaN unit and the H_2 desorption that we have found is possible in GaN MOVPE.

IV. GA-GA WEAK BONDS AT SURFACE STEPS

The epitaxial growth is usually conducted on GaN(0001) surface inclined by less than a few degrees toward either [11-20] or [1-100] direction. Such vicinal surfaces consist of terraces and steps and the reactions of the decomposition and the incorporation discussed above occur at the step edges especially in the step-flow growth. We have examined all the possible mono-bilayer atomic steps by the RSDFT calculations and obtained the stable edge structures and the formation energies. Fig. 4 shows the most stable atomic steps toward [11-20] direction. The most prominent feature is the existence of weak bonds between upper- and the lower-terrace Ga atoms. Ga adatom also participate in the formation of the weak bonds. These weak Ga-Ga bonds are expected to be stages of the epitaxial growth.

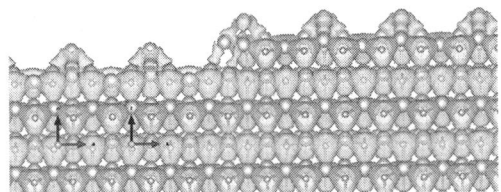

Fig. 4. Side view of a stable step structure of the [11-20]-inclined GaN (0001) vicinal surface under the Ga-rich condition. Green(large) and violet (small) balls depict Ga and N atoms, respectively. Electron density is repreented by the yellow isovalue surface.

V. SURFACE GA ADATOMS: TWO-DIMENSIONAL LIQUIDS

The growth temperature of GaN MOVPE is about 1300 K. The behavior of surface atoms at atomistic level is totally unknown. To respond to this challenge, we have performed CPMD simulations for Ga-rich GaN(0001) surface. To allow the migration of Ga adatoms with long distance, we use 4x4 lateral periodicity for our simulation cell. (The 4x4 is obviously insufficient to perform more realistic simulations which are now in progress.)

A stable structure was first obtained via standard geometry optimization. Then this system was gradually heated up from room temperature to 1300 K. During this stage, we monitored the behavior of the system at each temperature plateau by computing the pair correlation functions (PCFs):

$$g(r) = \frac{V}{4\pi r^2 N^2} \langle \sum_i \sum_{j \neq i} \delta(r - r_{ij}) \rangle , \qquad (2)$$

being N the number of atoms contained in a volume V and r_{ij} the interatomic distance. The average is computed on the trajectory produced by the dynamical simulation and, since both positions and velocities become simultaneously available, another quantity describing the dynamics, the diffusion

coefficient can be directly obtained upon integration of the velocity-velocity autocorrelation function [28,29]:

$$D = \frac{1}{3} \int_0^{t \to \infty} \langle \frac{1}{N} \sum_{I=1}^{N} v_I(t) \, v_I(0) \rangle \, dt \qquad (3)$$

At room temperature, as shown in the bottom panel of Fig. 5, a sharp peak at 6.4 Å in the PCF is the signature of the relative high stability of Ga adatoms (Ga_{ad}) on the sites where they have been placed. However, at increasing temperature, this peak broadens significantly from 500 to 1100 K. At 1300 K, the Ga_{ad} undergoes the large fluctuation and the disruption of the weak bond between the Ga_{ad} and the top surface Ga on the (0001) takes place. Then, Ga_{ad} is dislodged and starts migrating on the surface. This result is well represented by the Ga_{ad}-Ga_{ad} PCF at 1300 K in the range from 2.0 to 6.5 Å with a peak centered at 2.6-2.7 Å. This peak is a clear evidence of the formation of chemical bonds between two Ga_{ad} atoms and its corresponding spread accounts rather well for all the possible Ga_{ad}-Ga_{ad} chemical bonds, from single (~2.59 Å) to double (~2.41-2.25 Å) and triple (~2.45-2.21 Å) bonds. The long tail extending rather uniformly up to the borders of the simulation cell, visible in the upper panel of Fig.5 resemble indeed what can be observed in non-polar liquids [30].

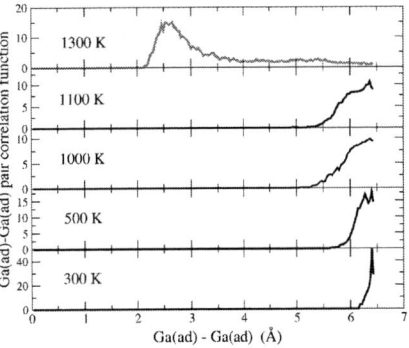

Fig. 5. Evolution of the Ga(ad)-Ga(ad) pair correlation function (equation (2)) during the heating process. The initial distance of 6.4 Å corresponds to the relaxed system from which dynamical simulations were started. (after Bui et al. [12])

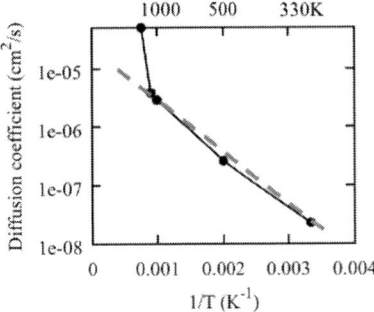

Fig. 6. Diffusion coefficient of a Ga adatom on Ga-rich GaN (0001) surface as a function of the inverse temperature 1/T for the temperature range (300-1300 K) investigated in the present work. The dashed red line shows the Arrhenius plot obtained by excluding the point at 1300 K, where the system deviates from the standard trend. (after Bui et al. [12])

What can be inferred from the analysis of the PCFs is that Ga_{ad} is destabilized by the high temperature, starts wandering on the GaN (0001) surface and, due to their relatively high density, start clustering. Such an enhanced mobility is confirmed by the diffusion coefficient D computed at the various temperatures explored (Fig. 6). A feature that can be immediately remarked is that, at variance with lower temperatures, D deviates from the Arrhenius behavior, when T=1300 K. This increased diffusivity corroborates the evidence about the enhanced mobility of Ga_{ad}, which behaves like a liquid state, thus providing further support to the picture offered by the PCFs. Indeed, if we neglect the value of the diffusivity at 1300 K, we can accommodate all the points of Fig. 6 in the Arrhenius plot with the diffusion barrier of 0.28 eV and the pre-exponential factor of 7.94×10^{-5} cm^2/s. Instead, at 1300 K the Ga_{ad} is destabilized and the system shows a liquid-like two-dimensional character with a sudden increase of D arising and reaching a value of 0.492×10^{-4} cm^2/s which amounts to one order of magnitude higher than the value of D (0.382×10^{-5}) at 1100 K.

The epitaxial growth has been so far believed to occur by cleavage of individual chemical bonds on the solid surface and the subsequent formation of new chemical bonds with arriving atomic and molecular species. However, the finding of the two-dimensional liquid-like phase leads to a new picture: At the growth temperature, the solid Ga-covered surface transforms to a two-dimensional liquid surface; the N species then easily intervenes into the subsurface region and bridges the Ga adatom in the liquid phase and the Ga atom in the sub surface layer. This is the formation of a new Ga-N network. This picture is consistent with a well-known fact that the temperature range in which the thin films grow is very narrow: The lower limit of this temperature range is the temperature at which the liquid-phase is formed, whereas the upper limit is the temperature at which the liquid evaporates into the gas phase.

VI. SUMMARY

In summary, we have performed the DFT-based total-energy electronic-structure calculations and CPMD simulations to clarify atomic reactions during the GaN MOVPE. We have found that the formation of the Ga-Ga weak bond in the stable structures and also in the novel liquid phase causes the NH$_3$ decomposition and the N incorporation, thus being essential for the epitaxial growth.

ACKNOWLEDGMENT

The work was supported by MEXT(Japan) programs on "Post-K Supercomputer", "Supercomputer Fugaku" and "Next Generation Semiconductors for Energy-Saving Society (Contract No = JPJ005357)", and by Grants-in-aid under contract No. 18H03873.

REFERENCES

[1] P. Hohenberg and W. Kohn, "Inhomogeneous electron gas" Phys. Rev. **136**, B864-B871 (1964).

[2] W. Kohn and L. J. Sham, "Self-consistent equations including exchange and correlation effects" Phys. Rev. **140**, A1133-A1138 (1965).

[3] e.g., R. O. Jones, "Density functional theory: Its origins, rise to prominence, and future" Rev. Mod. Phys. **87**, 897-923 (2015).

[4] R. Car and M. Parrinello, "Unified approach for molecular dynamics and density-functional theory" Phys. Rev. Lett. **55**, 2471-2474 (1985).

[5] D. Marx and J. Hutter, "Ab Initio Molecular Dynamics (Cambridge University Press, Cambridge, 2009).

[6] M. Boero and A. Oshiyama, Encycl. Nanotechnol (Springer, Netherlands, Dordrecht, 2015), pp. 1–10.

[7] See http://www.top500.org/ for TOP 500 supercomputer site.

[8] See http://computics-material.jp/ for Materials Design through Computics.

[9] H. Amano, "Nobel Lecture: Growth of GaN on sapphire via low-temperature deposited buffer layer and realization of p-type GaN by Mg doping followed by low-energy electron beam irradiation" Rev. Mod. Phys. **87**, 1133-1138 (2015).

[10] K. M. Bui, J.-I. Iwata, Y. Kangawa, K. Shiraishi, Y. Shigeta, and A. Oshiyama, "A reaction pathway of surface-catalyzed ammonia decomposition and nitrogen incorporation in epitaxial growth of gallium nitride" J. Phys. Chem. C **122**, 24665-24671 (2018).

[11] K. M. Bui, J.-I. Iwata, Y. Kangawa, K. Shiraishi, Y. Shigeta, and A. Oshiyama, "First-principle study of ammonia decomposition and nitrogen incorporation on the GaN surface in metal organic vapor phase epitaxy" J. Cryst. Growth **507**, 421-424 (2019).

[12] K. M. Bui, M. Boero, K. Shiraishi, and A. Oshiyama, "A two-dimensional liquid-like phase on Ga-rich GaN (0001) surfaces evidenced by first principles molecular dynamics" Jpn. J. Appl. Phys. **59**, SGGK04 (2020).

[13] W. Saito, Y. Takada, M. Kuraguchi, K.Tsuda, I. Omura, T. Ogura, and H. Ohashi, "High breakdown voltage AlGaN-GaN Power FEMT Design and high current density switching behavior" IEEE Transactions on Electron Devices **50**, 2528-2531 (2003).

[14] K. Shiraishi, K. Sekiguchi, H. Shirakawa, K. Chokawa, M. Araidai, Y. Kangawa, and K. Kakimoto, "First principles and thermodynamics studies on metal organic vapor phase epitaxy of GaN" ECS Transactions, **80**, 295-301 (2017).

[15] A. Kusaba *et al.*, "Thermodynamic analysis of (0001) and (000-1) GaN MetalOrganic Vapor Phase Epitaxy" Jpn. J. Appl. Phys. **56**, 070304 (2017).

[16] K. Nagamatsu, S. Nitta, Z. Ye, H. Nagao, S. Miki, Y. Honda, and H. Amano, "Decomposition of trimethylgallium and adduct formation in a metalorganic vapor phase epitaxy reactor analyzed by high-resolution gas monitoring system" **254**, 1600737 (2017).

[17] L. C. Grabow, J. J. Uhlrich, T. F. Kuech, and M. Mavrikakis, "Effectiveness of in situ NH3 annealing treatments for the removal of oxygen from GaN surfaces" Surf. Sci. **603**, 387-399 (2008).

[18] J. P. Perdew, K. Burke, and M. Ernzerhof, "Generalized Gradient Approximation Made Simple" Phys. Rev. Lett. **77**, 3865-3868 (1996).

[19] N. Troullier and J. L. Martins, "Efficient pseudopotentials for plane-wave calculations" Phys. Rev. B **43**, 1993-2006 (2010).

[20] S. G. Louie, S. Froyen, and M. L. Cohen, "Nonlinear ionic pseudopotentials in spin-density-functional calculations" Phys. Rev. B **26**, 1738-1742 (1982).

[21] J.-I. Iwata, D. Takahashi, A. Oshiyama, T. Boku, K. Shiraishi, S. Okada, and K. Yabana, "A massively-parallel electronic-structure calculations based on real-space density functional theory" J. Comp. Phys. **229**, 2339-2363 (2010).

[22] Y. Hasegawa *et al.*, "Performance evaluation of ultra-largescale first-principles electronic structure calculation code on the K computer" Int. J. High Performance Computing Applications **28**, 335-355 (2014).

[23] CPMD, http://www.cpmd.org/, Copyright IBM Corp 1990–2019, Copyright MPI fur Festkoerperforschung Stuttgart 1997–2001.

[24] K. Shiraishi, "A new slab model approach for electronic structure calculation of polar semiconductor surface" J. Phys. Soc. Jpn. **59**, 3455-3458 (1990).

[25] S. Jeong and A. Oshiyama, "Adsorption and diffusion of Si adatom on hydrogenated Si(100) surfaces" Phys. Rev. Lett. **79**, 4425-4428 (1997).

[26] D. J. Chadi, "Atomic Structure of GaAs(100)-(2x1) and (2x4) Reconstructed Surfaces" J. Vac. Sci. Tecn. A **5**, 834-837 (1987).

[27] M. D. Pashley, "Electron counting model and its application to island structures on molecular-beam epitaxy grown GaAs(001) and ZnSe(001)" Phys. Rev. B **40**, 10481-10487 (1989).

[28] M. S. Green, "Markoff random processes and the statistical mechanics of time-dependent phenomena. II. Irreversible processes in fluids" J. Chem. Phys. **22**, 398-413 (1954).

[29] R. Kubo, "Statistical-mechanical theory of irreversible processes. I. General theory and simple applications to magnetic and conduction problems"J. Phys. Soc. Jpn **12**, 570-586 (1957).

[30] Y. Zhao, Z. Wu, and W. Liu, "Theoretical and analytical radial distribution function for non-polar mixtures" Physica A **390**, 2812-2818 (2011).

Estimation of Phonon Mean Free Path in Small-Scaled Si Wire by Monte Carlo Simulation

Yuhei Suzuki
Faculty of Information Science and Technology
Osaka Institute of Technology
Osaka, Japan
yuhei.suzuki@rsh.oit.ac.jp

Yuma Fujita
Faculty of Information Science and Technology
Osaka Institute of Technology
Osaka, Japan
e1q17075@st.oit.ac.jp

Khotimatul Fauziah
Research Institute of Electronics
Shizuoka University
Shizuoka, Japan
khotimatul.fauziah@gmail.com

Takuto Nogita
Research Institute of Electronics
Shizuoka University
Shizuoka, Japan
nogita.taketo.15@shizuoka.ac.jp

Hiroya Ikeda
Research Institute of Electronics
Shizuoka University
Shizuoka, Japan
ikeda.hiroya@shizuoka.ac.jp

Takanobu Watanabe
Graduate School of Engineering
Waseda University
Tokyo, Japan
watanabe-t@waseda.jp

Yoshinari Kamakura
Faculty of Information Science and Technology
Osaka Institute of Technology
Osaka, Japan
yoshinari.kamakura@oit.ac.jp

Abstract—**A phonon transport in Si wire structures were simulated based on a Monte Carlo method to clarify the influence of the wire geometry and the surface roughness on thermal conductivity and the phonon-drag component of Seebeck coefficient. The mean free path (MFP) spectrum was estimated by tracing the simulated phonons. The MFPs of 1 THz phonons which mainly contribute to Seebeck coefficient become shorter with a decrease of the wire width for rough surfaces. This agrees with experimental observation of Seebeck coefficient. The MFPs of 3 THz phonons which mainly contribute to thermal conductivity were influenced even by small-roughness surfaces.**

Index Terms—**Si, phonon, Monte Carlo simulation**

I. INTRODUCTION

In recent years, thermoelectric devices attract much attention as one of the applications of nanostructured Si CMOS process technology [1]. Since the conversion efficiency monotonically increases with the figure of merit $ZT(\propto \sigma S^2/\kappa$, where σ, S, and κ are the electrical conductivity, the Seebeck coefficient, and the thermal conductivity, respectively), the efficient thermoelectric devices require the higher σ and S, and the lower κ. Especially, the nanostructures, such as nanowires [2], [3], and nano-porous structure [4], are greatly expected to enhance ZT by decreasing κ. On the other hand, the decrease of S has been experimentally observed in narrow Si wires with the width of < 1 μm [5], [6], whose mechanism has been supposed to be due to the vanishment of the phonon drag contribution to S [7]. However, the detailed mechanisms have not been clarified yet, and the theoretical analyses are needed.

So, in this study, we carry out a phonon transport simulation in Si wire structures based on a Monte Carlo (MC) method to solve the phonon's Boltzmann transport equation [8]. Then, the effect of the wire geometry and the surface roughness on κ and the phonon-drag component of S (S_{pd}) is discussed in terms of the phonon mean free path (MFP) spectrum.

II. SIMULATION METHOD

The phonon density of states(DOS) and the group velocity averaged over constant-energy surfaces were calculated from the realistic dispersion relation (Fig. 1) [8]. For the phonon scattering models, the phonon-phonon, the phonon-impurity, and the phonon-boundary scattering processes were taken into account. The formula given in [9] was adopted for the Umklapp phonon scattering processes, but the parameters were readjusted to reproduce the measured bulk κ (Fig. 2) [10]. For phonon-impurity scattering processes, the formula is given in [11] which considers the difference in the mass, the radius, and the compressibility between Si and P atoms. The parameters for the phonon-impurity scattering were also readjusted to reproduce the measured bulk κ (Fig. 3) [12]–[16].

The specularity of crystal boundary p was employed to calculate the phonon-boundary scattering processes given in the following formula [17]:

$$p = \exp(-16\pi^2\eta^2\cos^2(\theta)/\lambda^2), \qquad (1)$$

where η is the root-mean-square (RMS) roughness of crystal surface, λ is the phonon wave length, θ is the incident angle of phonons onto the surface.

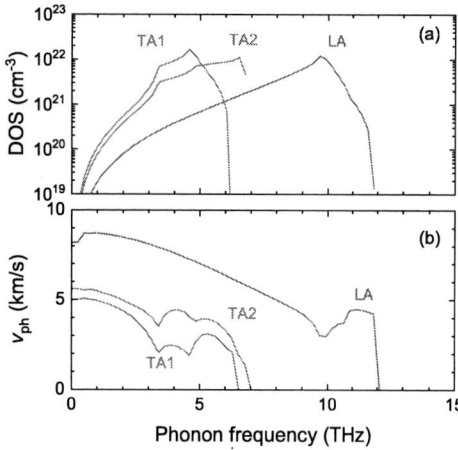

Fig. 1: Phonon frequency dependence of (a) density of states and (b) group velocity (v_{ph}) of transverse acoustic (TA1, TA2) and longitudinal acoustic (LA) phonons calculated from the approximated dispersion curves [8].

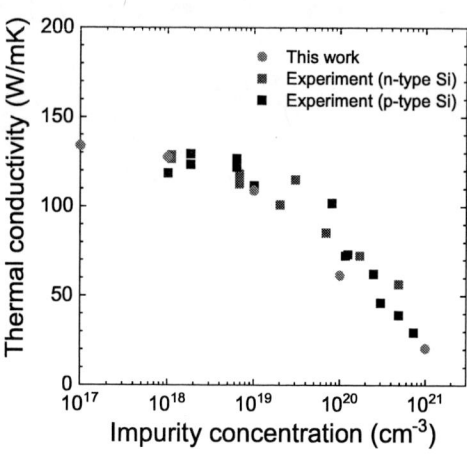

Fig. 3: Comparison of impurity concentration-dependency on thermal conductivity of bulk Si between the simulation (red) and experiment for n-type (blue) and p-type (black) [12]–[16].

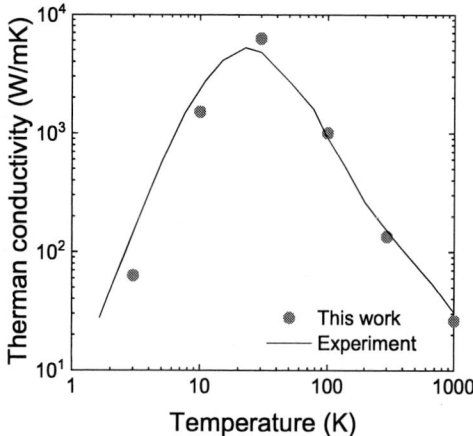

Fig. 2: Comparison of temperature-dependency of thermal conductivity of bulk Si between the simulation (dots) and experiment (line) [10].

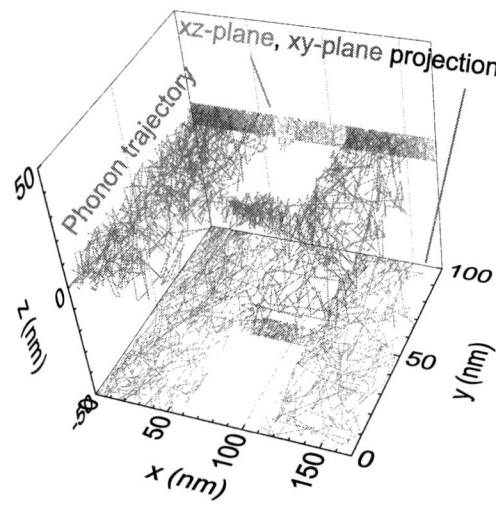

Fig. 4: Phonon trajectory in an actual Si wire structure with Si pads formed in SOI substrate.

III. SIMULATION RESULTS AND DISCUSSION

Figure 4 shows an example of a simulated phonon trajectory in a typical structure of Si wire together with Si contact pads which were formed on an silicon-on-insulator (SOI) layer. In this work, to reduce the computational time, we simulated the phonon transport only in the Si wire region as shown in Fig. 5. The wire thickness was assumed to be 30 nm. The RMS roughness at the top and the bottom surfaces were fixed to 0.2 nm, while the sidewall roughness η_{side} was varied from 0.2 to 2.0 nm. The doping (P) concentration was set to 10^{19} cm^{-3}, which is normally used to achieve higher ZT with Si [18].

Figure 6 shows the simulated phonon trajectories in the Si

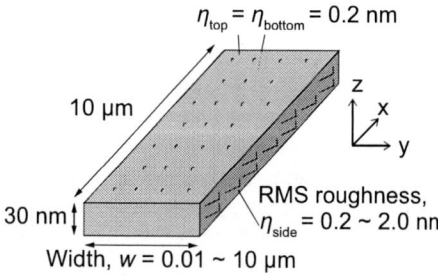

Fig. 5: Schematic of test device structure to simulate phonon mean free path.

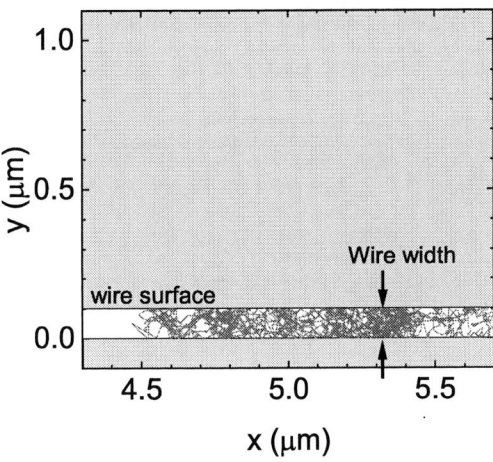

(a) $W = 0.1 \ \mu m$

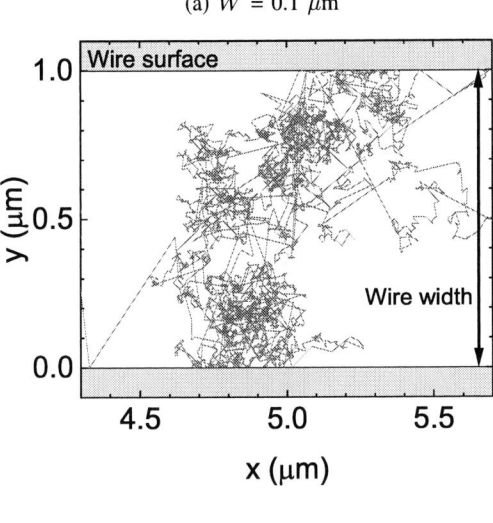

(b) $W = 1 \ \mu m$

Fig. 6: Phonon trajectories in the device structure shown in Fig. 5 for the width of (a) 0.1 μm and (b) 1 μm. η_{side} is 1.4 nm.

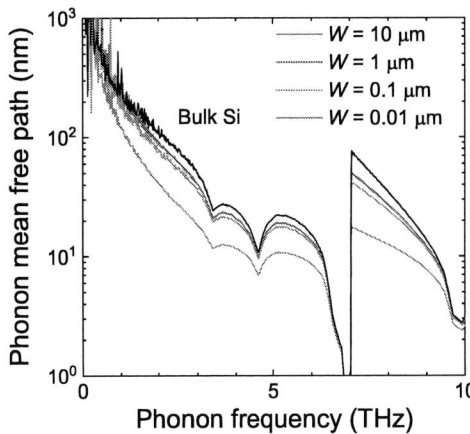

Fig. 7: Phonon mean free path for Si wire and bulk Si with respect to phonon frequency. η_{side} is 1.4 nm.

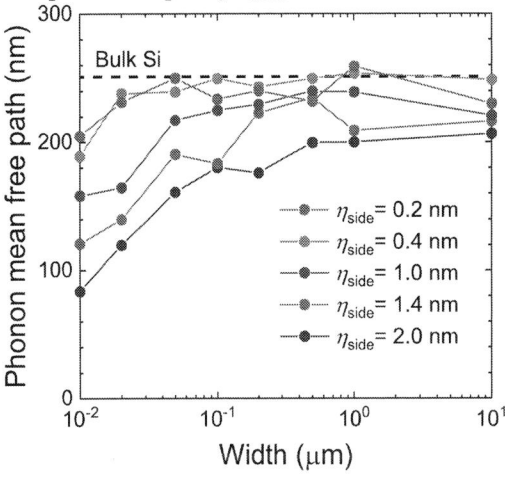

(a) $f = 1$ THz

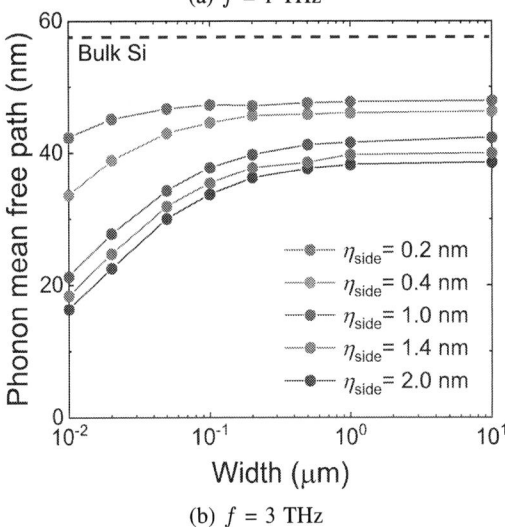

(b) $f = 3$ THz

Fig. 8: Comparison of phonon mean free path among RMS roughness with respect to wire width for phonon frequencies f of (a) 1 THz and (b) 3 THz.

wires with the wire width W of 0.1 and 1 μm at the $\eta_{\text{side}} = 1.4$ nm. Phonons in the case of $W = 0.1 \ \mu$m reach the surfaces of the sidewalls per unit time more frequently than the case of $W = 1 \ \mu$m, that is, the probability of the phonon-boundary scattering in a narrow-wire structure is higher.

The phonon MFP spectrum were estimated by tracing the phonon trajectrories during the simulation as plotted in Fig. 7. The spectrum of the phonon MFP is affected by the shape of DOS shown in Fig. 1 (a). At the phonon frequencies from 2 to 5 THz, the MFPs of the Si wires are shorter than that of the bulk Si and become progressively lower with decreasing the Si wire width. In contrast, at the low phonon frequencies, the MFPs of wide Si wires ($W \geq 1 \ \mu$m) are closed to the MFPs of the bulk Si. In Fig. 8, the phonon MFPs as a function of W for the phonon frequencies of 1 THz and 3 THz are shown. As suggested in [7], S_{pd} and κ at 300 K are mainly contributed by the different frequency phonons with $\sim$ 1 THz and $\sim$ 3 THz, respectively. The MFPs of 1 THz phonons are dependent on η_{side} and become shorter with decreasing W below 1 μm for $\eta_{side} \geq 1.0$ nm, because the low-frequency, i.e., long-λ, phonons are specularly reflected at the boundaries especially with smaller η as shown in Eq. (1).

In the experiment of [6], the η_{side} is estimated to be $>\sim 1.0$ nm. The result shown in Fig. 8a indicates that S_{pd} ($\propto$ MFP [19]) decreases with W (< 1 μm) for large η_{side}, which is consistent with the observation in [6]. On the other hand, the MFPs of 3 THz phonons also dependent on η_{side} and converge on $\sim$ 48 nm at $\eta_{side} = 0.2$ nm, but are not asymptotic to that of bulk Si (Fig. 8b). The reduction of the MFPs at 3 THz phonons from the bulk Si is caused by the phonon-boundary scattering at the top and bottom surface even though they are smooth as $\eta_{top} = \eta_{bottom} = 0.2$ nm. Furthermore, it is suggested that if the smaller η_{side} could be achieved, then the decrease of κ is expected without the penalty of the decrease of S_{pd}, which is desirable for ZT.

IV. Conclusions

The simulation using a MC method for a nanoscaled Si wire to estimate the phonon MFP has been carried out. The dependency of simulated phonon MFP on the wire width was consistent with the experimental result. The Si wire with smooth surface roughness is expected to enhance the conversion efficiency due to a preserved S_{pd} and a decrease in κ.

Acknowledgment

This work was supported by JST-CREST under Grant JPMJCR19Q5.

References

[1] M. Tomita, S. Oba, Y. Himeda, R. Yamato, K. Shima, T. Kumada, M. Xu, H. Takezawa, K. Mesaki, K. Tsuda et al., "Modeling, simulation, fabrication, and characterization of a 10-μw/cm^2 class si-nanowire thermoelectric generator for iot applications," *IEEE Transactions on Electron Devices*, vol. 65, no. 11, pp. 5180–5188, 2018.

[2] A. I. Boukai, Y. Bunimovich, J. Tahir-Kheli, J. K. Yu, W. A. Goddard, and J. R. Heath, "Silicon nanowires as efficient thermoelectric materials," *Nature*, vol. 451, no. 7175, pp. 168–171, 2008.

[3] A. I. Hochbaum, R. Chen, R. D. Delgado, W. Liang, E. C. Garnett, M. Najarian, A. Majumdar, and P. Yang, "Enhanced thermoelectric performance of rough silicon nanowires," *Nature*, vol. 451, no. 7175, pp. 163–167, 2008.

[4] R. Anufriev and M. Nomura, "Reduction of thermal conductance by coherent phonon scattering in two-dimensional phononic crystals of different lattice types," *Physical Review B*, vol. 93, no. 4, p. 045410, 2016.

[5] F. Salleh, T. Oda, Y. Suzuki, Y. Kamakura, and H. Ikeda, "Phonon drag effect on seebeck coefficient of ultrathin p-doped si-on-insulator layers," *Applied Physics Letters*, vol. 105, no. 10, p. 102104, 2014.

[6] K. Fauziah, Y. Suzuki, Y. Narita, Y. Kamakura, T. Watanabe, F. Salleh, and H. Ikeda, "Effect of phonon-drag contributed seebeck coefficient on si-wire thermopile voltage output," *IEICE Transactions on Electronics*, vol. 102, no. 6, pp. 475–478, 2019.

[7] J. Zhou, B. Liao, B. Qiu, S. Huberman, K. Esfarjani, M. S. Dresselhaus, and G. Chen, "Ab initio optimization of phonon drag effect for lower-temperature thermoelectric energy conversion," *Proceedings of the National Academy of Sciences*, vol. 112, no. 48, pp. 14 777–14 782, 2015.

[8] K. Kukita and Y. Kamakura, "Monte carlo simulation of phonon transport in silicon including a realistic dispersion relation," *Journal of Applied Physics*, vol. 114, no. 15, p. 154312, 2013.

[9] P. Chantrenne, J. Barrat, X. Blase, and J. Gale, "An analytical model for the thermal conductivity of silicon nanostructures," *Journal of applied physics*, vol. 97, no. 10, p. 104318, 2005.

[10] M. Holland, "Analysis of lattice thermal conductivity," *Physical review*, vol. 132, no. 6, p. 2461, 1963.

[11] P. Klemens, "The scattering of low-frequency lattice waves by static imperfections," *Proceedings of the Physical Society. Section A*, vol. 68, no. 12, p. 1113, 1955.

[12] G. A. Slack, "Thermal conductivity of pure and impure silicon, silicon carbide, and diamond," *Journal of Applied physics*, vol. 35, no. 12, pp. 3460–3466, 1964.

[13] Y. Lee and G. S. Hwang, "Mechanism of thermal conductivity suppression in doped silicon studied with nonequilibrium molecular dynamics," *Physical Review B*, vol. 86, no. 7, p. 075202, 2012.

[14] M. Asheghi, K. Kurabayashi, R. Kasnavi, and K. Goodson, "Thermal conduction in doped single-crystal silicon films," *Journal of applied physics*, vol. 91, no. 8, pp. 5079–5088, 2002.

[15] A. Stranz, J. Kähler, A. Waag, and E. Peiner, "Thermoelectric properties of high-doped silicon from room temperature to 900 k," *Journal of electronic materials*, vol. 42, no. 7, pp. 2381–2387, 2013.

[16] Y. Ohishi, J. Xie, Y. Miyazaki, Y. Aikebaier, H. Muta, K. Kurosaki, S. Yamanaka, N. Uchida, and T. Tada, "Thermoelectric properties of heavily boron-and phosphorus-doped silicon," *Japanese Journal of Applied Physics*, vol. 54, no. 7, p. 071301, 2015.

[17] J. M. Ziman, *Electrons and phonons: the theory of transport phenomena in solids.* Oxford university press, 2001.

[18] M. H. Elsheikh, D. A. Shnawah, M. F. M. Sabri, S. B. M. Said, M. H. Hassan, M. B. A. Bashir, and M. Mohamad, "A review on thermoelectric renewable energy: Principle parameters that affect their performance," *Renewable and sustainable energy reviews*, vol. 30, pp. 337–355, 2014.

[19] E. Behnen, "Quantitative examination of the thermoelectric power of n-type si in the phonon drag regime," *Journal of applied physics*, vol. 67, no. 1, pp. 287–292, 1990.

First-principles study of dopant trap level and concentration in Si(110)/a-SiO$_2$ interface

Gijae Kang
DIT center
Samsung Electronics
Hwasungsi, Gyeonggi-do, South korea
gijae.kang@samsung.com

Joohyun Jeon
DIT center
Samsung Electronics
Hwasungsi, Gyeonggi-do, South korea
jooh.jeon@samsung.com

Junsoo Kim
Semiconductor R&D center
Samsung Electronics
Hwasungsi, Gyeonggi-do, South korea
junsooya.kim@samsung.com

Hyoshin Ahn
DIT center
Samsung Electronics
Hwasungsi, Gyeonggi-do, South korea
hyoshin.ahn@samsung.com

Inkook Jang
DIT center
Samsung Electronics
Hwasungsi, Gyeonggi-do, South korea
inkook.jang@samsung.com

Daesin Kim
DIT center
Samsung Electronics
Hwasungsi, Gyeonggi-do, South korea
daesin.kim@samsung.com

Abstract—We investigate the dopant trap level and equilibrium concentration of Si(110)/a-SiO$_2$ interface with a wide variety of dopants (B, C, N, Br, Cl, F and H). The electronic and atomic properties of intrinsic and extrinsic defects are analyzed using First-principles calculation. It is shown that the average trap levels for hole and electron deepen as the electronegativity of the dopant increases. Also, we applied a simple thermodynamic model to evaluate the equilibrium concentration of active trap as a function of dopant concentration at the interface. From the model it turns out that H and F completely passivate the intrinsic Pb center of Si and reduce the trap concentration, while other elements, especially N, Br and Cl, induces new trap states which amounts to several times more than the pre-existing Pb center.

Keywords—Interface, semiconductor, trap level, doping concentration

I. INTRODUCTION

Si/SiO$_2$ interface is one of the most investigated system among the semiconductor/oxide interface due to its success in the semiconductor industry. The high degree of perfection achieved in the interface leads the low density of the interface trap, which is directly related to the device reliability.

It is known that the dangling bond of Si atoms at the interface, so called Pb center, acts as a major trap in the Si/SiO$_2$ system, and the Si/SiO$_2$ interface shows trap density less than 10^{11} cm^{-2} without external dopants [1]. However, various types of elements can diffuse to the Si/SiO$_2$ interface during or after the deposition process, which will play a critical role in a variety of physical and chemical properties of the interface. It is noteworthy that the effect of doping can be significantly different even among the elements which belong to the same group. For example, large dose of Cl atoms in the interface is known to worsen the interface integrity and increase the trap density, while F atoms in the interface monotonically decreases the interface trap [2,3]. Even though electronic and atomistic properties of the Si/SiO$_2$ interface have been extensively studied both experimentally and theoretically [4-6], the effect of extrinsic dopants on the electronic properties of the interface are not explored thoroughly.

Here we calculate the trap level and the concentration of such various defects (B, C, N, Br, Cl, F and H) to address the effect of the dopants in the Si(110)/a-SiO$_2$ interface. The average trap level of each dopant is calculated using First-

principles methods. Thermodynamic model which uses formation energy to calculate the equilibrium concentration of dopants based on the reaction energy of trapping and passivation is established and applied to the aforementioned dopants.

II. METHODOLOGY

A. Computational setup

We performed first-principles calculations using the Vienna Ab-initio Simulation package [7]. The generalized-gradient approximation (GGA) was used for the geometry optimization and first-principles molecular dynamics, while the electronic structures were calculated within HSE06 functional [8] to give a better description for the trap level position. The energy cutoff of 400 eV and gamma point sampling were employed. The cell geometry of each structure is fully relaxed with the atomic force less than 0.02 eV/Å.

B. Atomic structure preperation

To generate the crystal-amorphous interface structure, we firstly sprayed 40 formula units of Si and O atoms on the 12 layers of Si(110) surfaces, and applied melt-quench molecular dynamics (MD) simulations to equilibrate the SiO$_2$ amorphous layers. Firstly, the amorphous portion of interface structure is pre-melted in 5000 K with time step of 4 fs to erase the randomness of initial sprayed structure. After that, melting of 3000 K was applied for 6 ps was carried out to equilibrate the amorphous structure. Finally, the structure was quenched to 300 K within 10 ps. The melting and quenching process were performed with the time step of 3 fs. The melt-quenched atomic structure is fully relaxed before energy and electronic structure calculation.

C. Interface structure selecton

Because of the periodic boundary condition, the unit cell of the simulated structure should contain two interfaces. If trap states in those two interfaces have different density or charges, the potential banding which can be problematic in estimating the trap level can occur. In the case of crystalline interface, this can be avoided by creating same trap states in the two interfaces symmetrically, but in the amorphous interface inducing the same trap states in the different interfaces can only be achieved by making a lot of samples. To select the samples in which the internal potential band banding is small, we plotted local-projected density of states (local p-DOS) of

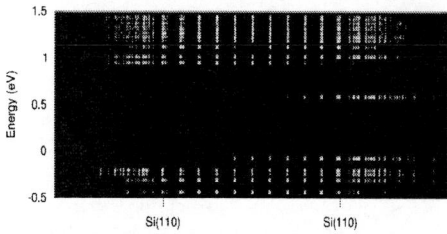

Fig. 1. Local-projected density of states of Si/SiO₂ interface. The energy is plotted with respect to the valence band maximum of the system.

Si/SiO₂ interfaces (Fig. 1). The example structure in Fig. 1 has one deep trap states (around 0.6 eV from valence band maximum) on the right interface, which leads slight band banding in the Si bulk region. By checking the local p-DOS of each structure, we analyzed only the cases where the same number of traps exist at each interfaces and the potential banding is not too significant.

III. RESULT AND DISCUSSION

We generated 20 Si(110)/a-SiO₂ interface structures using the aforementioned melt-quench procedure and select 10 samples which show small band banding by plotting local p-DOS. We added a dopant at each interface region and performed atomic relaxation. Five random positions at the interface region were chosen for each dopant, which means 360 structures (doped structures with B, C, N, Br, Cl, F and H with 10 samples and 5 random positions, plus 10 pristine structures) were generated. A representative structure for this ensemble is shown in Fig. 2(a). Most of the Si atoms positioned in the interface are almost passivated with O atoms, and the change of the density of the a-SiO₂ layer after the cell optimization was less than 5 %.

The density of states (DOS) for the interface structure with and without dopants are presented in Fig. 2(b) and Fig. 2(c), respectively. The inset of Fig. 2(c) shows the representative atomic configuration of Cl-doped case. It should be mentioned that in the most of the cases (except N) the trap states are mostly originated from Si dangling bond, so called Pb center. This means that the dopant generates a trap state by breaking Si-Si bond and passivate one of the Si atom, rather than by forming the localized orbital of itself.

In Fig. 3, the calculated trap levels of intrinsic Pb center and each dopant are plotted with respect to the electronegativity of the dopant elements. It is clear that both

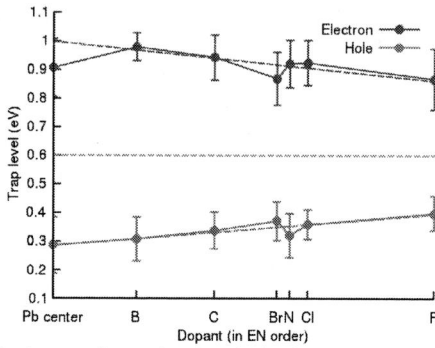

Fig. 3. Averaged trap levels of Si Pb center and extrinsic dopants.

electron and hole trap levels deepen as the electronegativity of the dopant increases. The dependency of trap levels on the electronegativity can be explained by the orbital hybridization between dopants and the Si dangling bond, and the local potential from dopant atoms exerted to the Pb center.

The trap concentration, which means the concentration of defect states that can act as a carrier trap, can differ from the concentration of the dopants at the interface. Here we applied simple thermodynamic model which relates the trap concentration and dopant concentration at the interface by using calculated formation energy and counting the active sites for the dopant:

$$p_{n+1} = p_n + C_{add} * \frac{D - p_n}{(N - t_n + D - p_n)} * \exp(-\frac{E_p}{kT})$$
$$t_{n+1} = t_n + C_{add} * \frac{N - t_n}{(N - t_n + D - p_n)} * \exp(-\frac{E_t}{kT})$$

(1)

where p_n and t_n are the number of passivated and induced trap site at nth step, respectively, and C_{add} is the amount of added dopant. D is the number of Pb center in the pristine interface and N is the number of of Si-Si bonds. E_p and E_t are the reaction energy for passivation and trapping. By iterating (1) with small value of C_{add}, the concentration of dopant at the interface (nC_{add}), the number of induced traps and passivated sites, and hence the number of non-passivated traps can be calculated. The total trap concentration of interface is the sum of induced traps and the non-passivated traps. from dopant atoms exerted to the Pb center.

In Fig. 4 is the calculated equilibrium trap concentration with respect to the dopant concentration at the interface. It is shown that H and F completely passivate the intrinsic Pb centers and decrease the trap concentration, while B and C have small variation with the dopant concentration. For the other elements, N, Br and Cl, the concentration of newly induced trap states amounts to several times the intrinsic Pb center concentration with the considered doping range. In the

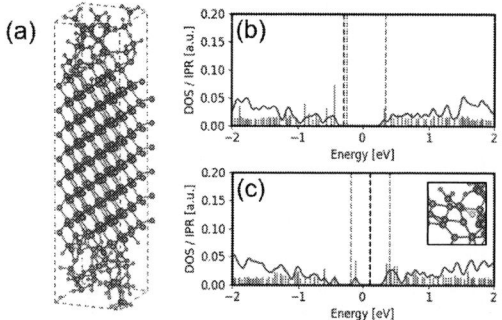

Fig. 2. (a) Atomistic structure of Si(110)/a-SiO₂ interface. Blue and red circles represent Si and O atoms, respectively. The DOS of (b) clean interface without traps and (c) interface with trap states.

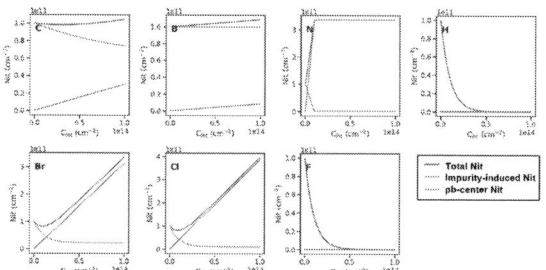

Fig. 4. Interface trap concentration as a function of dopant concentration at the interface.

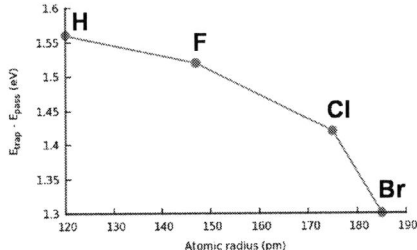

Fig. 5. The average coordination number and the difference in the reaction energy of trapping and passivation.

case of halogen atoms or H, this tendency can mostly be explained with the size of the dopants, as the bigger dopants induces structural strain which ease the formation of the trap states, while small dopants such as F and H induces no such strain and rather prefer the passivation of the existing trap sites. To illustrate this point more clearly, we plotted the difference between reaction energy of trapping and passivation versus atomic radius of halogen atoms and H in Fig. 5. In the figure, the difference in the energy decreases as the dopant size increases, which means the formation of trap site is more dominant in the case of big dopants, while passivation prevails in the interfaces with small dopants. This size effects can also exist in the case of non-halogen atoms such as C, B, and N, but we find that the relation between the atomic characteristic and the reaction energy are more complex due to the covalent character of those atoms.

IV. CONCLUSION

In conclusion, we calculated the dopant trap level and equilibrium concentration for wide range of dopants in Si(110)/a-SiO$_2$. It is found that the average trap level of the interface approaches to midgap as the electronegativity of the dopant atom increases. We also calculated the active trap concentration as a function of dopant concentration, and it is shown that H and F completely passivate the intrinsic trap center of Si/SiO$_2$ interface, while N, Br and Cl induces new trap states which can worsen the device performance..

REFERENCES

[1] S. C. Witczak, J. S. Suehle, M. Gaitan, "An experimental comparison of measurement techniques to extract Si-SiO$_2$ interface trap density," Solid-State Electron. 35, 345–355 (1992).

[2] Y.□B. Park, X. Li, S-.W. Rhee, "Characterization of the Si/SiO$_2$ interface formed by remote plasma enhanced chemical vapor deposition from SiH$_4$/N$_2$O with or without chlorine addition," J. Vac. Sci. Technol. B 14, 2660 (1996).

[3] A. Balasiński, M.H.Tsai, L.Vishnubhotla, T.P.Ma, H.H.Tseng, P.J.Tobin, "Interface properties in fluorinated (100) and (111)Si/SiO$_2$ MOSFETs," Microelectron. Eng., 22, 97 (1993)

[4] A. Pasquarello, M. S. Hybertsen, R. Car, "Interface structure between silicon and its oxide by first-principles molecular dynamics," Nature 396, 58–60 (1998).

[5] H. Kageshima, K. Shiraishi, "First-Principles Study of Oxide Growth on Si(100) Surfaces and at SiO$_2$/Si(100) Interfaces," Phys. Rev. Lett. 81, 5936–5939 (1998).

[6] A. Pasquarello, M. S. Hybertsen, R. Car, "Structurally relaxed models of the Si(001)–SiO$_2$ interface," Appl. Phys. Lett. 68, 625–627 (1996).

[7] G. Kresse, J. Furthmüller, "Efficient iterative schemes for ab initio total-energy calculations using a plane-wave basis set," Phys. Rev. B 54, 11169 (1996).

[8] J. Heyd, G. E. Scuseria, M. Ernzerhof, "Hybrid functionals based on a screened Coulomb potential," J. Chem. Phys. 118, 8207 (2003).

22

Energy Band Calculation of Si/Si$_{0.7}$Ge$_{0.3}$ Nanopillars in $\vec{k}$ Space

Min-Hui Chuang
Parallel and Scientific Computing Laboratory
Institute of Communications Engineering
National Chiao Tung University
Hsinchu City, Hsinchu 300, Taiwan

Yiming Li
Parallel and Scientific Computing Laboratory
Institute of Communications Engineering
National Chiao Tung University
Hsinchu City, Hsinchu 300, Taiwan
ymli@faculty.nctu.edu.tw

Abstract—**In this work, we explore the energy band of the well-aligned silicon (Si) nanopillars (NPs) embedded in Si$_{0.7}$Ge$_{0.3}$ matrix fabricated by neutral beam etching. Instead of real-space modeling, we formulate and solve the Schrödinger equation with an effective mass approach using 3D finite-element simulation in $\vec{k}$ space. This approach enables us to calculate the electronic structure in a computationally effective manner. The effects of the height, radius, separation, and shape of Si NPs on the energy band and density of states are calculated and discussed. The effect of the radius on the electron energy band control is significant while that of the shape is marginal owing to high geometry aspect ratio. In contrast with the results of electrons, both the radius and separation play crucial role in tuning the energy band of holes; consequently, they govern the variation of energy band gap of Si/Si$_{0.7}$Ge$_{0.3}$ NPs.**

Index Terms—**energy band, density of state, electrons, light holes, Si/Si$_{0.7}$Ge$_{0.3}$ nanopillar, Schrödinger equation, $\vec{k}$ space**

I. INTRODUCTION

Silicon nanopillars (Si NPs) are promising for more than Moore applications due to the tunable energy-band profiles compared with the materials of bulk Si [1]. Highly periodical Si NPs have been fabricated by the neutral-beam (NB) etching techniques [2], [3]. A thin SiO$_2$ layer is formed via an oxygen NB at 573 K under 0.14 Pa. A solution of PEG-ferritin (0.05 mg/ml) in 30 mM ammonium acetate is used to form a uniform coating. The wafer is annealed in oxygen to remove the ferritin protein shell. The surface oxide layer is etched using NF$_3$/H radicals at 300 K under 40 Pa. The iron oxide and surface oxide are used as masks for etching. Via Cl$_2$ NB etching, the depth of 90 nm can be formed. We remove the masks and then grow Si$_{0.7}$Ge$_{0.3}$ by thermal CVD. The fabricated sample before depositing the matrix material has good uniformity and alignment, as shown in Fig. 1(a). In our earlier work, modeling and simulation of the electron and hole energy band in real space were reported [4], [5]. However, finite-element simulation of the 3D Schrödinger equation in real space is a very time-consuming task. In this work, we calculate the energy

This work was supported in part by the Ministry of Science and Technology, Taiwan, under Grant MOST 109-2221-E-009-033, Grant MOST 108-2221-E-009-008, and Grant MOST 108-3017-F-009-001, and in part by the "Center for mmWave Smart Radar Systems and Technologies" under the Featured Areas Research Center Program within the framework of the Higher Education Sprout Project by the Ministry of Education in Taiwan.

band by solving the Schrödinger equation in $\vec{k}$ space [6], [7]. We model the structure as a superlattice with periodical boundary conditions on the sidewalls. The conduction- and valence-band edges, cut from Fig. 1(b), are in Fig. 1(c) and the band structure is discussed. To model nonideal etching processes, three different shapes [8], as shown in Figs. 1(d)–(f), are also simulated. The simulation flow for the electronic structure calculation and a schematic plot of the irreducible Brillouin zone (IBZ) are plotted in Fig. 1(g).

II. COMPUTATIONAL MODEL

To calculate the energy states, the 3D Schrödinger equation in real space $r = (x, y, z)$ is given by

$$\nabla[\frac{-\hbar^2}{2m^*}\nabla\vec{\psi}_n(\vec{r})] + V(\vec{r})\vec{\psi}_n(\vec{r}) = E_{n,k}\vec{\psi}_n(\vec{r}), \quad (1)$$

where $\hbar$, m^*, $V(\vec{r})$, $E_{n,k}$, and $\vec{\psi}_n(\vec{r})$ are the reduced Plank's constant, the effective mass, the position-dependent potential energy, quantum energy levels, and the corresponding wave function, respectively. The effective masses of electrons along the transverse and longitudinal direction are setting as $0.19m_0$ ($0.12m_0$) and $0.98m_0$ ($1.14m_0$) for Si (Si$_{0.7}$Ge$_{0.3}$) [10], respectively. The conduction band edges of Si and Si$_{0.7}$Ge$_{0.3}$ are 0.56 and 0.575 eV, respectively. Assuming that the structure is periodical, by introducing the Bloch theorem, Eq. (1) can be formulated in $\vec{k}$ space as

$$\nabla[\frac{-\hbar^2}{2m^*}\nabla\vec{u}_{n,k}] - \frac{i\hbar^2}{m^*}\vec{k}\cdot\nabla\vec{u}_{n,k}(\vec{r}) + [V(\vec{r}) + \frac{\hbar^2\vec{k}^2}{2m^*}]\vec{u}_k(\vec{r})$$
$$= E_{n,k}\vec{u}_{n,k}(\vec{r}). \quad (2)$$

Furthermore, the density of states per IBZ (DOS) is [9]

$$DOS(E) = \frac{2}{(2\pi)^2}\sum_{n,m}\frac{l_n(E, k_m)}{|\nabla_k E_n(k_m)|}$$
$$= \frac{2}{(2\pi)^2}\sum_{n,m} DOS_{n,m}(E), \quad (3)$$

where the factor 2 is the twofold electron spin degeneracy, $1/(2\pi)^2$ is from the conversion of $\vec{k}$ space summation into integration in energy, and l_n is the length of the individual

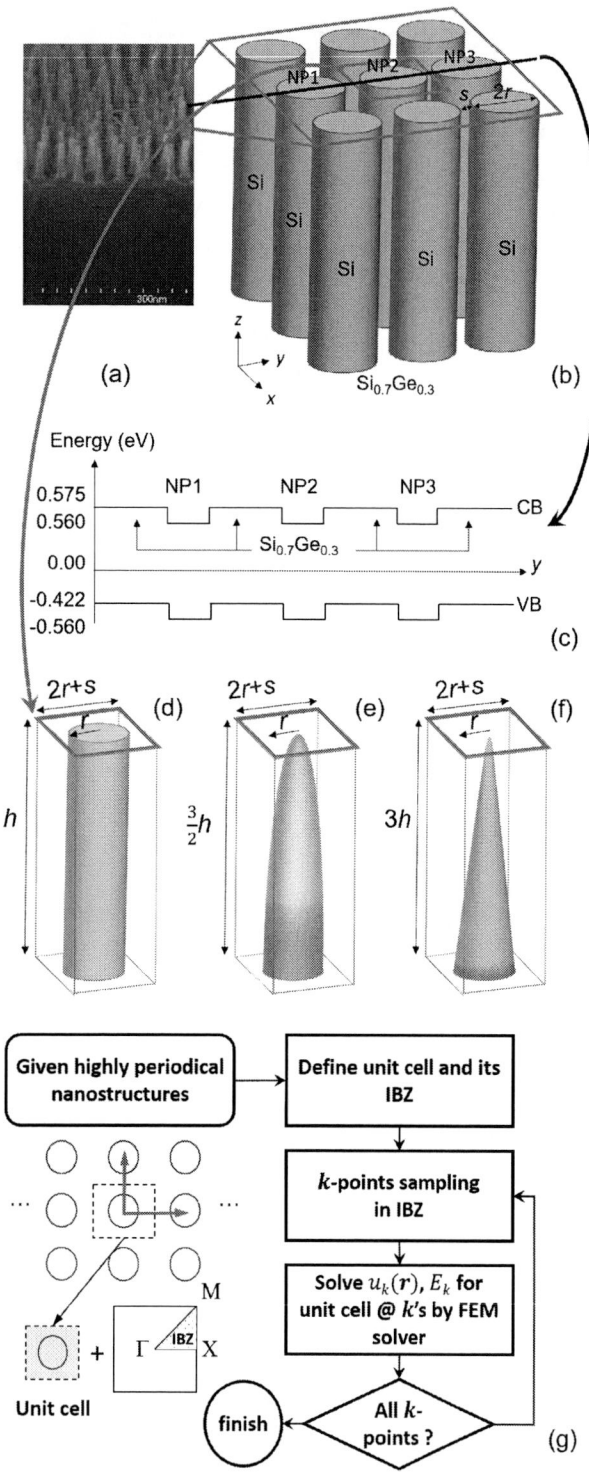

Fig. 1. (a) The side view of the fabricated Si NPs [2], [3]. Schematics of (b) the simulated structure, and (c) the conduction band (CB) and valence band (VB) of Si NPs embedded in $Si_{0.7}Ge_{0.3}$ matrix along y-axis. Note that electrons are confined in the CB of Si NPs while holes are confined in $Si_{0.7}Ge_{0.3}$. To model the nonideal etching processes, three different Si NPs with (d) cylindrical, (e) elliptical, and (f) conical shapes are constructed and simulated by solving Eq. (2) in $\vec{k}$ space, respectively. (g) The simulation flow for the electronic structure calculation for a given highly periodical nanostructure by introducing the concept of IBZ in $\vec{k}$ space. For each sampling point in IBZ, the eigenvalue problem of Eq. (2) is solved.

segments with a constant energy E in the 2D $\vec{k}$ space at sampling point m.

III. RESULTS AND DISCUSSION

To verify the computational efficiency of the simulation in $\vec{k}$ space, we first compute the electron ground-state energy in real and $\vec{k}$ spaces by solving Eqs. (1) and (2). As shown in Fig. 2(a), the electron ground-state energy converges to the same result when the number of Si NPs in real space is increased. However, the larger problem size, the longer computational time. More specifically, the calculation of the energy states of a single point in $\vec{k}$ space only takes 25 s; however, to reach the same accuracy, at least 49 Si NPs should be simulated in real space which requires more than 104 s. Figs. 2(b), (c) and (d) show the s-orbit and p-orbit of the Si NP with $r = 5$ nm and $s = 4$ nm for the electrons with an isotropic effective mass along z-, x- and y-direction, respectively. Thus, the Schrödinger equation in $\vec{k}$ space is adopted for the following simulations. The height, radius, separation, and shape effect on the energy band of the Si NPs–$Si_{0.7}Ge_{0.3}$ composite for electrons.

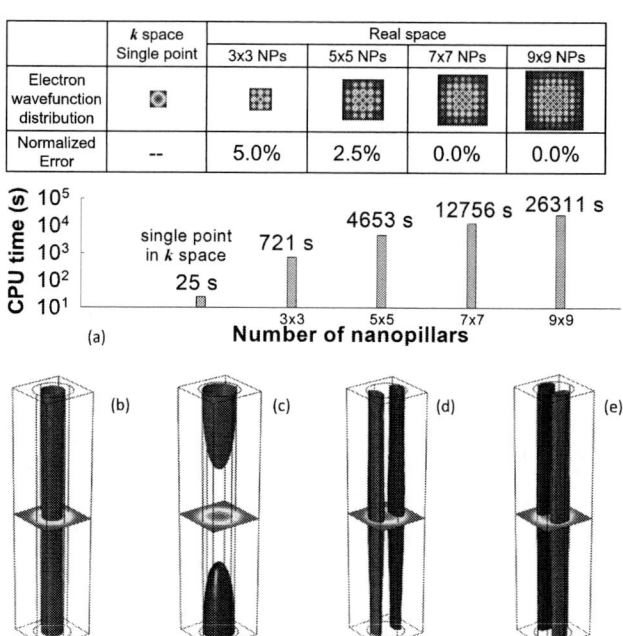

Fig. 2. (a) The computational time of Eq. (1) and (2) with respect to the number of Si NPs, where the case of $r = 5$ nm and $s = 2$ nm is simulated. For the $\vec{k}$ space simulation by Eq. (2), only 25 s are needed at a $\vec{k}$ point; however, to reach the same accuracy of the electron ground-state energy level, at least $7 \times 7 = 49$ Si NPs are needed in real space and the computational time is more than 10000 sec. The (b) s-orbit and p-orbit of the Si NP along (c) z-, (d) x-, and (e) y-direction for electrons with an isotropic effective mass.

A. Height Effect

To understand the height effect of Si NPs on the energy band for electrons, we split three conditions. Fig. 3 shows the electron energy states and DOS calculated from cylindrical Si NPs with the radius of 2 nm, the separation of 5 nm, and the

24

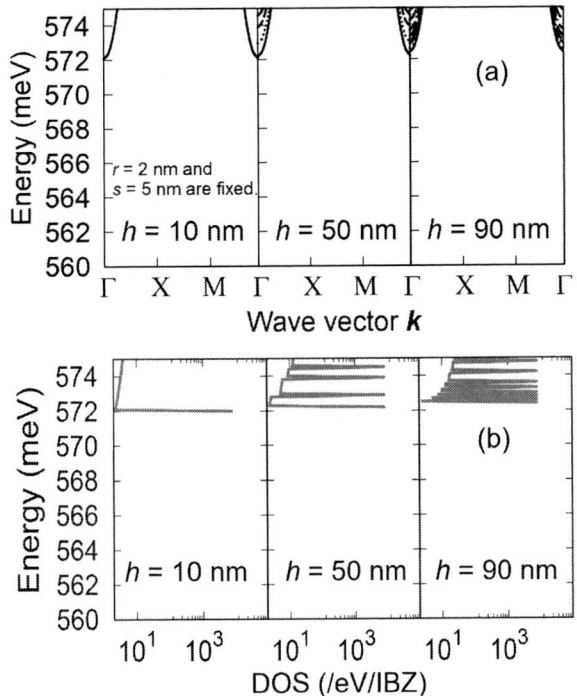

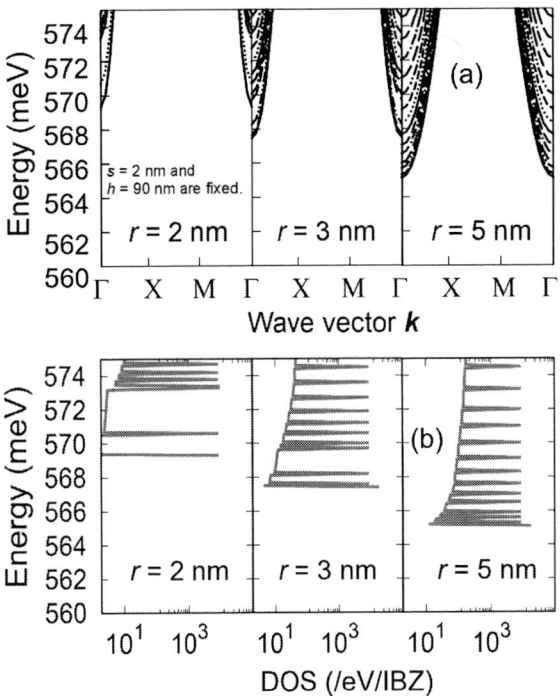

Fig. 3. The (a) electron energy states and (b) density of DOS of the cylindrical Si NPs with the radius of 2 nm and the separation of 5 nm with respect to different heights. The height effect on the ground-state energy is with a little increase as the height of Si NPs are increased; however, it is non-significant. The number of bound states is increased as the height of Si NPs is increased.

Fig. 4. The (a) electron energy states and (b) DOS of the cylindrical Si NPs with the separation of 2 nm and the height of 90 nm with respect to different radii. As the radius is increased, the wave function is well confined in the Si NPs and lead to a lower ground-state energy.

height of 10, 50, and 90 nm, respectively. The ground-state energy is with a little increase as the height of Si NPs are increased; however, it is insignificant. The number of bound states is increased as the height of Si NPs is increased.

B. Radius Effect

Because the Si NPs is regarded as the potential well to confine the wave function, we fixed the height of Si NPs at 90 nm consistent with our fabrication result and do the energy-band simulation for electrons. Figs. 4(a) and (b) show the electron energy states and DOS of the cylindrical Si NPs with the separation of 2 nm with respect to different radii. As the radius is increased, the wave function is well confined in the Si NPs and lead to a lower ground-state energy.

C. Separation Effect

The separation effect on the electron energy states and DOS of the cylindrical Si NPs with the radius of 2 nm and the height of 90 nm are plotted in Figs. 5(a) and (b). When the separation is small, the potential barrier is thin for electrons; thus, the wave function could penetrate the barrier and lead to a low ground state energy.

D. Shape Effect

Fig. 6(a) shows the influence of shape on energy is negligible; nevertheless, the cylindrical NPs possess the lowest energy among three shapes owing to the best wave function

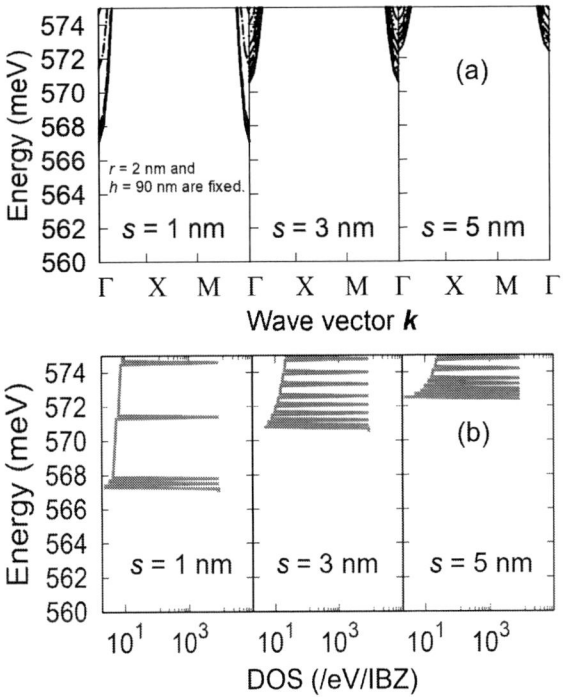

Fig. 5. The (a) electron energy states and (b) DOS of the cylindrical Si NPs with the radius of 2 nm and the height of 90 nm with respect to different separations. As the separation is small, the potential barrier is thin for electrons; thus, the wave function could penetrate the barrier and lead to a low ground state energy.

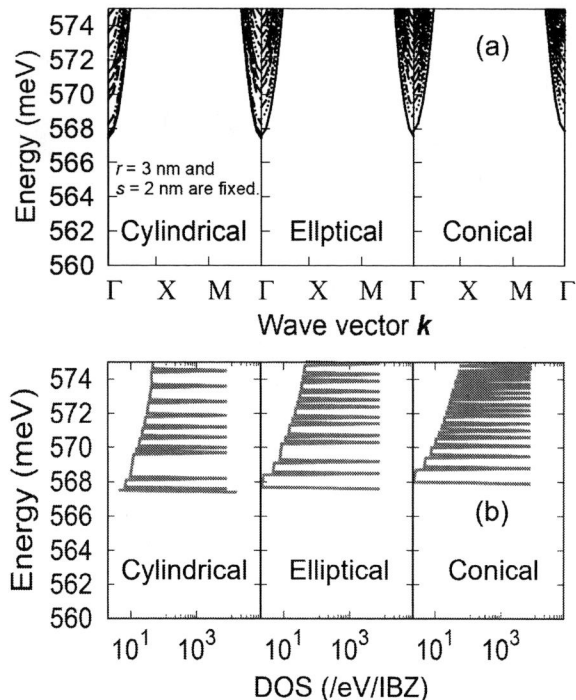

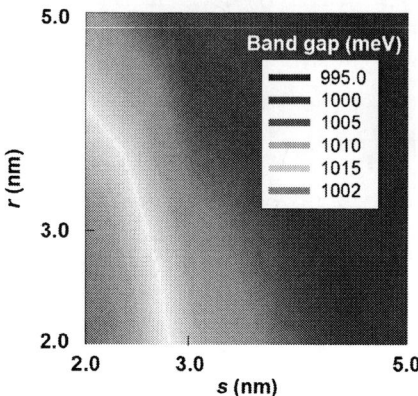

Fig. 8. The band gap of Si/Si$_{0.7}$Ge$_{0.3}$ NPs as a function of the radius and the separation. As the separation is decreased, the band gap becomes large and closed to that of bulk Si because of Si NPs rich in the Si$_{0.7}$Ge$_{0.3}$ matrix.

NPs with respect to different radii and separations. More than 20 meV variation can be altered by changing the separation of Si NPs which is mainly contributed by heavy holes.

IV. CONCLUSIONS

In summary, we have performed 3D finite element simulations on highly uniform Si/Si$_{0.7}$Ge$_{0.3}$ NPs for electronic structure calculation. The dependences of energy band and DOS on the heights, radii, separation distances, and shapes of Si NPs have been discussed. The separation of Si NPs plays a key role in tuning the band gap among four factors.

Fig. 6. The (a) electron energy states and (b) DOS of the cylindrical, elliptical, and conical Si NPs with $s = 2$ nm and $r = 3$ nm. The volume is setting as 2544.6 nm^3. As the shape of Si NPs changes from cylinder to cone, the ground-state energy is increased; however, the effect of the shape of Si NPs on the energy band and DOS is marginal owing to high geometry aspect ratio.

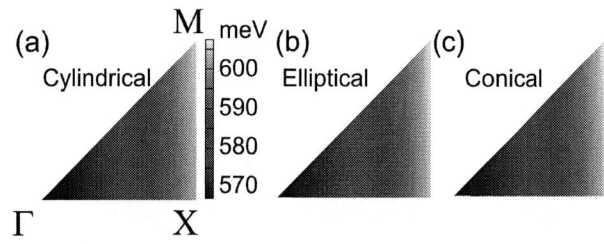

Fig. 7. Plot of the electron ground-state energy on the IBZ for (a) cylindrical, (b) elliptical, and (c) conical Si NPs. All shapes of Si NPs have similar energy distribution owing to similar accommodation structures. At the Γ-valley, the ground-state energy is the lowest.

confinement. As shown in Fig. 6(b), the DOS of the conical NPs is different from others. Figs. 7(a)–(c) show their ground-state-energy distributions on the IBZ. The energy distributions on the IBZ are corresponding to the results in Fig. 6. All shapes of Si NPs have similar energy distribution owing to similar accommodation structures.

We conduct similar simulation for heavy holes and light holes. In contrast with the results of electrons, the radius and separation play crucial role in tuning the energy band of light holes. Especially, not shown here, the separation of Si NPs provides even more significant tunable range on the energy band of light holes. Fig. 8 shows the band gap of Si/Si$_{0.7}$Ge$_{0.3}$

REFERENCES

[1] N. Neophytou, A. Paul, and G. Klimeck, "Bandstructure effects in silicon nanowire hole transport," IEEE Trans. Nanotechnol., vol. 7, pp. 710–719, Nov. 2008.

[2] A. Kikuchi, A. Yao, I. Mori, T. Ono, and S. Samukawa, "Composite films of highly ordered Si nanowires embedded in SiGe$_{0.3}$ for thermoelectric applications," J. Appl. Phys., vol. 122, p. 165302, Oct. 2017.

[3] M.-H. Chuang, D. Ohori, Y. Li, K.-R. Chou, and S. Samukawa, "Fabrication and simulation of neutral-beam-etched silicon nanopillars," Vacuum, vol. 181, p. 109577, Nov. 2020.

[4] W. Hu, M. Igarashi, M.-Y. Lee, Y. Li, and S. Samukawa, "50% Efficiency Intermediate Band Solar Cell Design Using Highly Periodical Silicon Nanodisk Array," in Tech. Dig. IEDM, San Francisco, CA, Dec. 10–12, 2012, pp. 6.1.1–6.1.4.

[5] W. Hu, M. M. Rahman, M.-Y. Lee, Y. Li, and S. Samukawa, "Simulation study of type-II Ge/Si quantum dot for solar cell applications," J.Appl. Phys., vol. 114, p. 124509, Sep. 2013.

[6] M.-Y. Lee, Y. Li, and S. Samukawa, "Miniband calculation of 3-D nanostructure array for solar cell applications," IEEE Trans. Electron Dev., vol. 62, pp. 3709–3714, Nov. 2015.

[7] M.-Y. Lee, Y. Li, M.-H. Chuang, D. Ohori, and S. Samukawa, "Numerical Simulation of Thermal Conductivity of SiNW–SiGe$_{0.3}$ Composite for Thermoelectric Applications," IEEE Trans. Electron Devices, vol. 67, pp. 2088–2092, May 2020.

[8] Y.-C. Tsai, M.-Y. Lee, Y. Li, and S. Samukawa, "Miniband formulation in Ge/Si quantum dot array," Jpn. J. Appl. Phys., vol. 55, p. 04EJ14, Mar. 2016.

[9] J.-H. Lee, T. Shishidou, and A. J. Freeman, "Improved triangle method for two-dimensional Brillouin zone integrations to determine physical properties," Phys. Rev. B, vol. 66, p. 233102, Dec. 2002.

[10] F. Schaffler, "Silicon-Germanium," in Properties of Advanced Semiconductor Materials: GaN, AlN, InN, BN, SiC, SiGe, M. E. Levinshtein, S. L. Rumyantsev, and M. S. Shur, Eds., Wiley: New York, 2001.

Full Band Monte Carlo simulation of phonon transfer at interfaces

N. D. Le[1], B. Davier[1,2], P. Dollfus[1], M. Pala[1], A. Bournel[1], J. Saint-Martin[1]

[1]Université Paris-Saclay, CNRS, Centre de Nanosciences et de Nanotechnologies, 91120, Palaiseau, France

[2]Department of Mechanical Engineering, The University of Tokyo, Tokyo 113-856, Japan

Abstract: Our home made Full-Band particle Monte Carlo is used to investigate the thermal interface conductance at Silicon/Germanium heterojunctions.

1. Introduction

As self-heating effect could degrade electronic device performances [16], understanding the nature of heat transport at the nanoscale is important for the CMOS industry [1]. In particular, the contribution of out-of-equilibrium phonons in heat transfer at interfaces remains an important issue.

Molecular Dynamics (MD) [2,3] and Non Equilibrium Green's function (NEGF) [4,5] simulations have been used to investigate this topics. The method of MD provided important results to the study of heat transport as the complex structures of interfaces can be described at the atomic level in terms of roughness, strain and anharmonic phonon/phonon scattering mechanisms [6–8]. However, the size of the system that can be studied is limited to few thousands of atoms. Whereas MD only takes into account the classical nature of phonon transport, NEGF, by contrast, considers its quantum nature as interference and Bose-Einstein statistics can be captured. It is a very efficient method to calculate the transmission coefficient through interface, but is preferred for small and low-dimensional systems rather than micrometer long systems [9–12].

The particle Monte Carlo (MC) algorithm [13–15] is a powerful and complementary tool to investigate the heat transfer as the simulated system size can vary from the nanoscale to the microscale. MC solves the Boltzmann Transport Equation (BTE) for phonons without assumptions, thus the out-of-equilibrium transport regime is naturally taken into account. Besides, with convenient modeling, particle scattering mechanisms can be accurately described in the full phase space especially at the interfaces.

In this work, our Full-Band (FB) particle Monte Carlo [17] approach has been extended to treat the phonon transport at semi-transparent interfaces and used to extract and analyze the thermal interface conductance at Silicon/Germanium heterojunctions.

2. Full Band Monte Carlo

In the particle Monte Carlo method dedicated to phonon transport, the trajectories of a large number of phonons are randomly selected according to relevant scattering rates and phonon dispersion that is in our case a Full-Band description. "Full Band Monte Carlo" means that all the phonon states in the first 3D Brillouin zone are considered as detailed in Ref. [17]. Here, the phonon dispersion, phonon velocity in cubic Si and cubic Ge have been calculated by using the Adiabatic Bond Charge Model (ABCM) [18] and employed as an input for the Monte Carlo simulations. The first Brillouin zone was discretized into a sampling of 29791 (31×31×31) wave vectors.

For Silicon, the phonon scattering rates have been derived from DFT as presented in Ref. [19]. For Germanium which is a heavier atom, the DFT approach is more complex. Consequently, a set of Holland parameters specific to a FB approach [20] has been adjusted to fit the bulk thermal conductivity in the full temperature range. The considered relaxation times are those of the three-phonon scattering and phonon-impurity scattering processes.

According to the definition used in Ref. [21], three out-of-equilibrium temperatures T, T^+ and T^- are defined. The pseudo temperature T is the temperature of an equilibrium distribution of phonons having the same energy density and considering the total population of phonon. T^+ and T^- are called "hemispherical temperatures" and are defined by considering only the phonons with positive or only those with negative velocity, respectively.

3. Semi-transparent interface modeling

When a particle collides with a semi-transparent interface, both elastic and diffusive mechanisms are considered. Besides, the probability of transmission of a phonon colliding the interface is obtained within the Full-Band Diffusive Mismatch Model (DMM) [22]. In this model, the interface is assumed to have many defects so that each phonon colliding the interface loses totally the memory of its initial state. The wave vector and the mode of the phonon in the final state are independent of those in the initial state. The only conserved quantity is the phonon energy. Thus, the transmission coefficient depends only on the phonon energy, and neither on the wave vector, nor on the state of the phonon.

As the reflection and transmission processes are assumed to be diffusive and elastic, the angular distribution of the wave vector of reflected or transmitted phonon is chosen randomly over the states belonging to the same isoenergy surface to satisfy the energy conservation. Moreover, the appearance of a net flux along the interface must be prohibited and a Full-Band version of the Lambert's emission law [23] must be applied to avoid a non-physical accumulation of phonons near the interface. Thus, among the states belonging to the same isoenergy surface, the final state j and thus the final wave vector $\vec{q}_j$ (and mode) are selected according to this following statistic weight:

$$P_j \propto \vec{v}_j . \vec{n} \qquad (1)$$

where $\vec{v}_j$ is the group velocity of the state j and $\vec{n}$ is the vector normal to the interface.

4. Simulated Devices

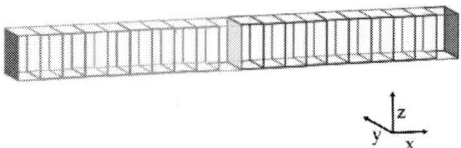

Fig. 1. *Si/Ge heterojunction. Orange and violet cells are made of Si and Ge, respectively. Hot and cold thermostats are in red and blue, respectively. DMM interface is in Yellow. White faces stand for periodic boundaries.*

Fig. 1 shows the simulated Si/Ge heterostructure of length L. One device end is in contact with a hot thermostat at temperature T_H and the other end is in contact with a cold thermostat at temperature T_C. The Si/Ge junction is located at the middle of the device ($x = L/2$). The interface between Si and Ge is assumed to be abrupt. The width w and height h of the heterostructure (along y and z directions, respectively) are equal to 100 nm. As the cross-plane transport configuration is simulated, the structures are supposed to be infinite along the y and z directions. In that purpose, external interfaces that are normal to y and z directions follow periodic boundary conditions.

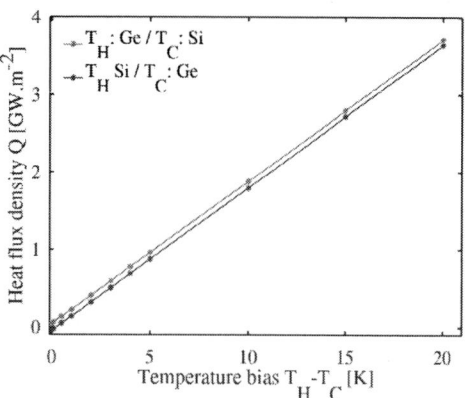

Fig. 2. *Heat flux density in a Si/Ge heterostructure of length*

$L=20$ nm as a function of the applied temperature bias T_H-T_C. $(T_H+T_C)/2=300K$.

The system is primarily uniformly meshed along the x-axis by using 20 cells. Then, this mesh could be refined at 5 nm around the heterojunction and near the thermostats to keep the cell length equal to or lower than 1 nm. It should be noticed that the size of a cell should be significantly larger than the atomic radii of Ge (125 pm) and Si (110 pm) [24] i.e. it should remain in the order of magnitude of nanometer.

5. Results

Fig. 2 shows the simulated heat flux density Q as a function of the temperature difference between the two thermostats for a 20 nm long Si/Ge heterostructure. For all studied temperature biases, the cold and hot thermostats have been swapped. The evolution of the flux is clearly a linear function of the temperature difference. The linear coefficient gives a total thermal conductance of the heterostructure (for both thermostat configurations) of G_{tot} = 183 MW/m²/K.

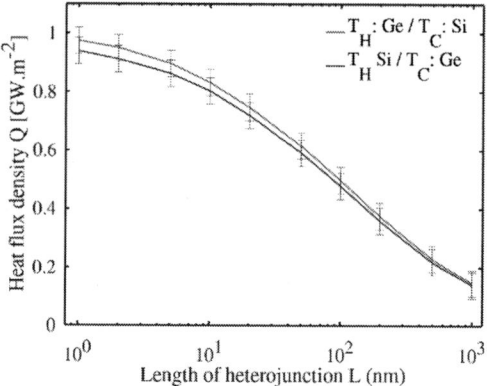

Fig. 3. *Heat flux density in a Si/Ge heterostructure as a function of the device length L. T_H=302 K and T_C=298 K.*

However for T_H - T_C = 0 K, a non-zero flux of δQ = 45.3 MW/m² is extracted from the simulation. This value is assumed to be a good estimate of the MC simulation resolution. The statistical error for the calculation of a physical quantity f with the Monte Carlo method is [25]:

$$|f - \langle f \rangle| < \alpha \sqrt{\frac{\langle f^2 \rangle - \langle f \rangle^2}{N}}, \qquad (2)$$

where N is the number of random observations of the quantity f. Assuming that f follows the normal distribution, the interval of 95% confidence corresponds to α = 1.96. By increasing N, the statistical error can be reduced as needed. Nevertheless, the resolution of our MC simulation is the lower limit of the error related to the heat flux density estimation and will be used in the following to define the error bars.

In Fig. 3, the heat flux density Q is plotted as a

function of the device length L for a temperature bias T_H-T_C of 4.0 K. In ultra-short devices (L<100 nm), the fluxes tend to saturate and to become independent of the length as the phonon transport becomes ballistic. In long devices (L > 100 nm) in which the transport is diffusive, the heat flux density tends to decrease linearly when L increases. Moreover, the thermostat swap induces a shift of the flux smaller than the error bar of the MC simulation, thus no significant thermal rectification can be observed.

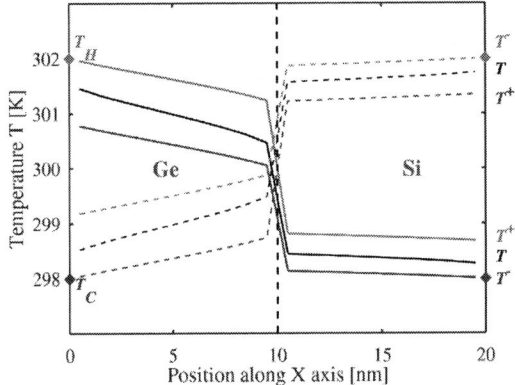

Fig. 4. *Temperature profiles in a Si/Ge heterostructure of length L=20 nm in contact with T_H = 302 K (298 K), T_C = 298 K (302 K) in continuous lines (in dashed lines, respectively).*

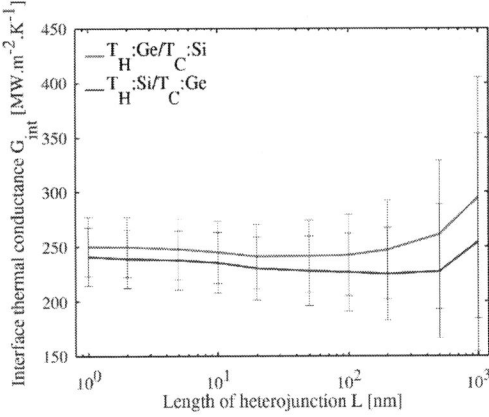

Fig. 5. *Thermal interface conductance in a Si/Ge heterostructure as a function of the device length L at 300 K and with a temperature bias of T_H-T_C=4 K*

Fig. 4 presents the temperature profiles: T, T^+ and T^- in a 20 nm long device for both positive and negative temperature biases of 4 K. In the first configuration, the Ge bar is in contact with the hot thermostat and the Si bar is in contact with the cold thermostat (solid line) and, as the flux is positive along the x axis, T^+ is the highest temperature. By swapping the heterostructure bias

(dashed line), the thermal flux is reversed and the phonons having positive x-velocity, i.e. propagating from the cold to the hot thermostat, are minority. Thus, in this configuration T^+ is the lowest temperature.

The temperature drop at the interface is sharp while the temperature gradient in the Si and Ge bars are weak. Besides, all the temperatures gradients (of T, T^+ and T^-) are higher in Ge than in Si as Ge exhibits lower thermal conductivity. Moreover, the gradient of T is slightly higher than those of T^+ and T^-. It should be noted that, the net thermal flux is governed by the gradient of T within the Fourier's formalism or the difference between T^+ and T^- at a given position within the formalism of Ref. [21].

The interface thermal conductance is finally computed from dividing the heat flux density Q by the difference of the hemispherical temperatures T^+ and T^- on each side of the interface [21]:

$$G_{int} = \frac{Q}{T^+(L/2-\delta x)-T^-(L/2+\delta x)} \qquad (3)$$

where δx stands for an infinitely small distance.

From the previous results of the resolution of the heat flux density and the total thermal conductance, the resolution of the temperature bias is estimated by using the following ratio:

$$\delta(\Delta T) = \frac{\delta Q}{G_{tot}} = 0.25\ K \qquad (2)$$

For the device of length L, the lower bound for the statistical error of the interface thermal conductance $\delta G_{int}(L)$ is deduced from the resolutions of the heat flux density and the temperature bias according to:

$$\frac{\delta G_{int}(L)}{G_{int}(L)} = \frac{\delta Q(L)}{Q(L)} + \frac{\delta(\Delta T)}{T_H - T_C} \qquad (3)$$

The resulting G_{int} are plotted in Fig. 5 for several device lengths. The width of error bar increases when the length L increases as the temperature difference (T_H − T_C) near the interface also decreases leading to higher uncertainty in terms of extracted temperature.

However, the variation of G_{int} as a function of the device length remains smaller than the error bars. Thus, G_{int} appears to be equal to 243.3±41.9 MW/m² independently of the device length. This is consistent with the DMM theory which predicts a unique value for the Si/Ge interface conductance of 225 MW/m²/K [22]. The modification of the transport regime of phonon in the studied length range does not affect G_{int}.

It should be noticed that the standard definition of the interface thermal conductance $G_{int}^T = Q/[T(L/2 - \delta x) - T(L/2 + \delta x)]$ using only the pseudo temperature T leads to different results. Indeed, the values of G_{int}^T that are 360 MW/m² for L=1nm and 350 MW/m² for $L = 100$ nm. are significantly higher than the DMM results.

Assuming by assumptions spectral energy balance, the DMM does not provide by itself any physical ingredient that could induced thermal rectification. However, the presence of out of equilibrium phonons could have generated such asymmetry in the phonon transport. In Fig. 5, the difference between the values of G_{int} when heat flux is reversed is still smaller than the error bars and no thermal rectification due to "hot" phonons is observed.

6. Conclusion

In this work, the heat transfer through Si/Ge interfaces was investigated by using Full Band Monte Carlo simulation for phonon using DMM for modeling the phonon transmission at semi-transparent interfaces. Our MC simulator is able to calculate all the relevant temperature profiles, the heat flux density and finally the interface thermal conductance at heterojunctions. It can simulate short systems working in ballistic regime as well as long devices working in diffuse transport regime.

A linear relationship was obtained between the heat flux density and the applied temperature difference across a short heterostructure of 20 nm. Thus, even in the case of a strongly ballistic phonon transport, the concept of thermal conductance remains relevant. Moreover, the value of the interface thermal conductance obtained from this MC approach are consistent with the semi-analytical DMM model by using the concept of hemispherical temperature T^+ and T^-. Besides, no thermal rectification is observed in our model at the Si/Ge interface.

Finally, this work demonstrates the relevance of this MC simulator to investigate the heat transport at the nanoscale as it provides very detailed insight into thermal transfer at interfaces.

References

[1] C. Prasad, IEEE Transactions on Electron Devices **66**, 4546 (2019).

[2] S. Merabia and K. Termentzidis, Physical Review B - Condensed Matter and Materials Physics **86**, 094303 (2012).

[3] E. S. Landry and A. J. H. McGaughey, Physical Review B - Condensed Matter and Materials Physics **80**, 165304 (2009).

[4] N. Mingo, Physical Review B - Condensed Matter and Materials Physics **74**, 125402 (2006).

[5] S. Sadasivam, U. V. Waghmare, and T. S. Fisher, Physical Review B **96**, 174302 (2017).

[6] P. Chantrenne and J. L. Barrat, Journal of Heat Transfer **126**, 577 (2004).

[7] R. Rurali, X. Cartoixà, and L. Colombo, Physical Review B - Condensed Matter and Materials Physics **90**, 041408 (2014).

[8] K. R. Hahn, M. Puligheddu, and L. Colombo, Physical Review B - Condensed Matter and Materials Physics **91**, 195313 (2015).

[9] N. Mingo and L. Yang, Physical Review B - Condensed Matter and Materials Physics **68**, 245406 (2003).

[10] N. Mingo and L. Yang, Physical Review B -

[11] W. Zhang, T. S. Fisher, and N. Mingo, Journal of Heat Transfer **129**, 483 (2007).

[12] C. Monachon, L. Weber, and C. Dames, Annual Review of Materials Research **46**, 433 (2016).

[13] S. Mazumder and A. Majumdar, Journal of Heat Transfer **123**, 749 (2001).

[14] D. Lacroix, K. Joulain, and D. Lemonnier, Physical Review B - Condensed Matter and Materials Physics **72**, 064305 (2005).

[15] J. P. M. Péraud and N. G. Hadjiconstantinou, Physical Review B - Condensed Matter and Materials Physics **84**, 205531 (2011).

[16] T. Thu Trang Nghiêm, J. Saint-Martin, and P. Dollfus, Journal of Applied Physics **116**, 074514 (2014).

[17] B. Davier, J. Larroque, P. Dollfus, L. Chaput, S. Volz, D. Lacroix, and J. Saint-Martin, Journal of Physics Condensed Matter **30**, 495902 (2018).

[18] W. Weber, Physical Review B **15**, 4789 (1977).

[19] L. Chaput, J. Larroque, P. Dollfus, J. Saint-Martin, and D. Lacroix, Appl. Phys. Lett. **112**, 033104 (2018).

[20] M. G. Holland, Physical Review **132**, 2461 (1963).

[21] B. Davier, P. Dollfus, S. Volz, J. Shiomi, U. Paris-saclay, C. De Nanosciences, and D. Nanotechnologies, ArXiv (2020).

[22] J. Larroque, P. Dollfus, and J. Saint-Martin, Journal of Applied Physics **123**, 025702 (2018).

[23] W. J. Smith, *The Design of Optical Systems: General*, 4th ed. (McGrawHill, 2008).

[24] J. C. Slater, The Journal of Chemical Physics **41**, 3199 (1964).

[25] S. Volz, R. Carminati, P. Chantrenne, S. Dilhaire, S. Gomez, N. Trannoy, and G. Tessier, *Microscale and Nanoscale Heat Transfer* (Springer, 2007).

Condensed Matter and Materials Physics **70**, 249901(E) (2004).

2-6

First Principle Simulations of Electronic and Optical Properties of a Hydrogen Terminated Diamond Doped by a Molybdenum Oxide Molecule

Joseph McGhee, Vihar P. Georgiev,

Device Modelling Group

University of Glasgow, Glasgow G12 8QQ, United Kingdom
E-mail: vihar.georgiev@glasgow.ac.uk

ABSTRACT

In this work we investigate the surface transfer doping process induced between a hydrogen-terminated (100) diamond and a metal oxide MoO_3, using the Density Functional Theory (DFT) method. DFT allows us to calculate the electronic and optical properties of the hydrogen-terminated diamond (H-diamond) and establish a link between the underlying electronic structure and the charge transfer between the oxide materials and the H-diamond. Our results show that the metal oxide molecule can be described as an electron acceptor and extracts the electrons from the diamond creating 2D hole gas in the diamond surface. Hence, this metal oxide molecule acts as a p-type doping material for the diamond.

INTRODUCTION

Diamond is a compound semiconductor material with many electronic applications such as microwave electronic devices [1], bipolar junction transistors [2] and Schottky diodes [3]. However, one of the most promising areas for diamond industrial application is in high-performance field effect transistors (FETs) used in the production of high frequency and high-power electronic devices [4]. Its properties potentially enable devices that are beyond the scope of current systems in terms of operating frequency, power handling capacity, operating voltage, thermal robustness and operating environment. This is due to the fact that diamond has a wide band-gap of 5.5eV, a thermal conductivity 5 times greater than 4H-SiC of 24 W/cm*K (for CVD diamond), a high breakdown field of $20Wcm^{-1}$ and high hole and electron carrier velocities of $0.8x10^7cm/s$ and $2.0x10^7cm/s$ respectively. As a result, diamond is a superior new candidate for high frequency and high-power device applications [5].

However, the primary issue that has inhibited the application of diamond is the lack of a suitably efficient and stable doping mechanism. The most promising way to dope diamond is by surface transfer doping (STD), which is achieved by depositing different materials on the surface of diamond. For example, organic molecules, such as C_{60} [6] and $C_{60}F_{48}$ [7], can be used

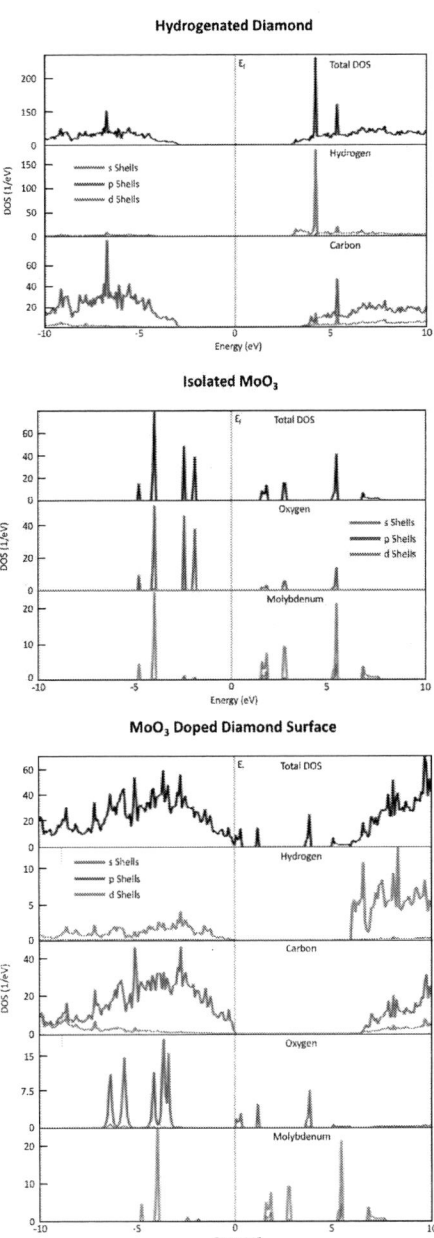

Figure 1. Top) Hydrogen terminated (100) diamond Density of States (DOS); Middle) MoO_3 molecules DOS; Bottom) MoO_3 molecule on the top of an H-Diamond.

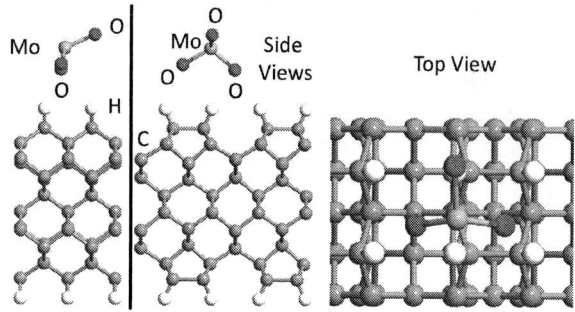

Figure 2. Simulated systems. The (left hand side) side view and (right hand side) top view of charge density different for the most stable MoO_3-doped diamond surface.

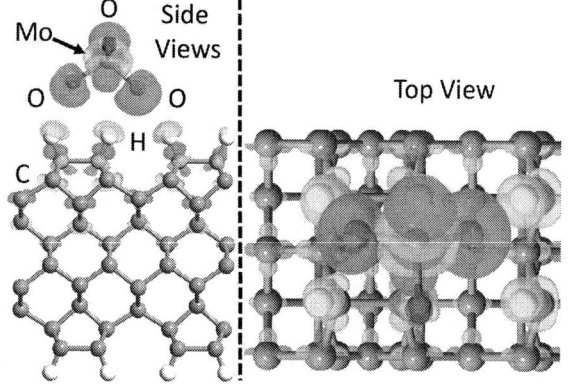

Figure 3. The (left hand side) side view and (right hand side) top view of charge density different for the most stable MoO_3-doped diamond surface. The purple regions represent electron accumulation and the green regions represent electron depletion (hole accumulation). The isosurface values are ±0.001 Bohr^{-3}.

to deposit on the diamond surface and they will act as surface acceptors to induce holes in hydrogen terminated diamond (H-diamond). However, the organic molecules are not very stable in high temperatures and they are not compatible with the fabrication process. Hence, an alternative approach is to use inorganic components for STD, such as chromium oxide (CrO_3) [8], molybdenum oxide (MoO_3) [9, 10] and vanadium oxide (V_2O_5) [11, 12].

In this paper, we would like to explore further than the previous research and explain in detail the process of STD of metal oxide materials, such as molybdenum oxide (MoO_3), as a material for achieving such STD in diamond. In order to achieve our aim, we have performed numerical first principle simulations based on the Density Functional Theory (DFT).

SIMULATION METHODOLOGY

All calculations were carried out using the Quantumwise Atomistix ToolKit (ATK) software using the DFT method [13]. The Generalised Gradient Approximation (GGA) exchange correlation was used for the geometry optimisations of the system and to obtain the total energies of the interfaced systems and the individual component parts, i.e. H-diamond and the oxide layer in question. For all geometry optimisations a force

tolerance of 0.01eV/Å was used. GGA-1/2 exchange correlation was used for all electronic structure calculations. To obtain a more accurate electronic description of the systems, DFT-1/2 method was used [14]. DFT-1/2 method is a semi-empirical approach that can overcome the error that local and semi-local exchange correlation density functionals inherently have when working with semiconductors and insulators. It works by correcting the self-interaction error of DFT by cancelling out the electron-hole self-interaction energy by defining an atomic self-energy potential. However, DFT-1/2 method is not suitable for calculating total energies, hence, we used GGA for the geometry optimisations and calculating adsorption energies. The Perdew, Burke and Ernzerhof (PBE) functional was chosen for all calculations because of the good match with experimental data (around 5.5 eV) regarding the value on the band gap in bulk diamond [11].

A Monkhorst-Pack scheme with an 8 x 8 x 1 k-point density (Å) mesh was used for the Brillouin zone integration. An iteration control tolerance of 0.0001 with a density cut off of 1×10^{-6} was used for all calculations with a medium basis set. The number of pseudo-atomic orbitals in a medium basis set is typically comparable to that of a double-zeta polarized (DZP) basis set [15]. The pseudopotential is SG15 and the density mesh cut-off is 185Ha, which gives a high accuracy with a medium computational efficiency.

RESULTS AND DISCUSSION

Fig. 1 shows the total Density of States (DOS) and the partial DOS (PDOS) of a hydrogen terminated (100) diamond, the DOS of a single MoO_3 molecule and the DOS of a MoO_3-doped H-diamond obtained from the DFT simulations. The PDOS gives us the possibility to separate the contribution of each atom in the overall profile of the full DOS of the materials. Also, the PDOS provides information about the different electronic shells of each element involved in the simulations. For example, when a diamond is hydrogen terminated the band gap is 5.9 eV (please see the top plot in Fig. 1). This is very close to the experimental value of 5.5 eV. Also, PDOS shows that the conduction band is mainly determined by the carbon p-shells. This is exactly what is expected because the carbon is a p-type element with valence electrons occupying p_x, p_y and p_z atomic orbitals. The same plot reveals that the conduction band is mainly formed by the hydrogen atoms with electrons in the s-shell. Indeed, each H atom has $1s^1$ electronic configuration that makes H an s-type element.

Following the discussion above, it can be concluded that the MoO_3 molecule has a band gap of around 2.5 eV. This again is comparable to the reported experimental values of 3.2 eV and 2.8 eV for bulk and polycrystalline MoO_3, respectively,

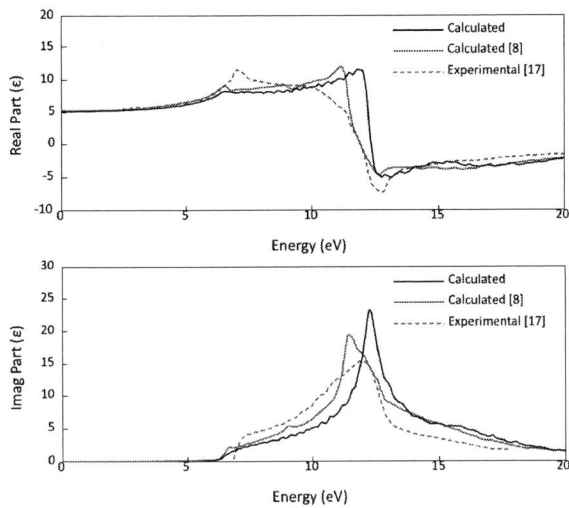

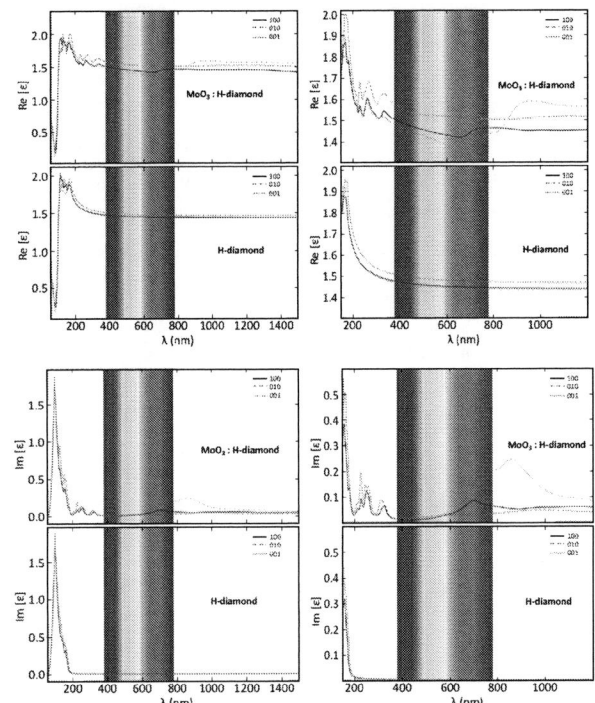

Figure 4. Dielectric constant plots for a bulk diamond comparing our calculated results with calculated results by Xiang et al. [8] and experimental results by Philipp and Taft [17].

obtained by the absorption spectra measurements [16]. The PDOS plots in Fig. 1 show that the oxygen p-shells are dominant in the valence bands while the d-shell (states) of the metal atoms are localised both in the conduction and the valence band. The Mo atom is a d-element and hence only the d-shells are visible in the PDOS and active in the chemical reaction.

The bottom part of Fig. 1 shows the DOS and the PDOS diagram for MoO_3 adsorbed on the H-diamond surface. There is a shift of the DOS to higher energies and the Fermi Level (E_F) is in contact with the valence states. Moreover, the band gap of the diamond remains constant but states of the Valance Band Maximum (VBM) have crossed the Fermi Level, which in turn means that charge transfer has occurred as previously occupied states within the H-diamond are now vacant. This demonstrates that electrons have transferred from the diamond surface to the MoO_3 molecule leaving the hole accumulation layers. Moreover, this transfer of electrons can be directly linked with exchange of charge between the oxygen atoms from the MoO_3 molecule and the carbon hydrogen atoms of the H-diamond. The PDOS for the Mo atoms is virtually not modified in comparison to the isolated molecule presented in Fig. 1.

Fig. 2 shows the atomic position of the simulated MoO_3:H-diamond interface. This configuration has the lowest energy in comparison to the other twelve systems of varying MoO_3 positions and orientations that we simulated. Hence, we used this system for further analysis.

Fig. 3 shows the electron density change for the most energetically stable MoO_3:H-diamond interface. The blue regions show where there is a loss of electron density and thus the accumulations of holes, and the purple regions show where there is an increase in electron density. Hence, there is a hole

Figure 5. The dielectric function of the real (Re [ε]) and imaginary (Im [ε]) parts for the clean hydrogenated diamond surface and the most preferred MoO_3-doped diamond surface. The results are superimposed of the visible spectrum.

accumulation that occurs near the surface of the diamond while most of the electron density gained by the MoO_3 clearly migrates to the O atoms. Therefore, the MoO_3 acts as a p-type doping material. This indeed is in agreement with the discussions above and the data presented in Fig. 1.

To further validate the reliability of our calculation parameters, we calculated the optical spectrum of a bulk diamond and plotted the dielectric constant to compare our results to a recent theoretical study by Xiang *et.al.* [8] and experimental data published by Philipp and Taft [17].

The comparison plots for the real (Re [ε]) and imaginary (Im [ε]) parts of the dielectric constant in Fig. 4 show that our DFT method produces results in agreement with other published theoretical work and experimental data. For the bulk diamond, the spectrums of [100, 010, 001] crystallography orientation are degenerate and hence only one result for [100] orientation is presented. The small discrepancies between the calculated and the experimental curves are due to the inter-band transitions and neglected exitonic effects in the simulations. It should be pointed out that the absorption edge corresponding to the bulk gap of diamond of 5.5 eV is suppressed in the experimental spectre due to its indirect transition nature.

Fig. 5 shows the calculated real (Re [ε]) and the imaginary (Im [ε]) parts of the dielectric function of the type H-diamond and the MoO_3:H-diamond system. For the pure H-diamond systems the spectra of [100, 010, 001] diamond crystallography

33

orientations are degenerate due to inherent structural anisotropy. However, this is not the case for the MoO_3:H-diamond [100, 010, 001] surfaces. Depending on the substrate orientation, there is a significant difference in the visible part of spectrum. Also, the most significant difference between the pure diamond and the MoO_3:H-diamond systems is at below 400 nm, where, due to the MoO_3 molecule, there are additional absorption features. Hence, this shows that the adsorption of the MoO_3 molecule on the H-diamond surface introduces new empty states within the diamond band gap that enhances the optical absorption of the MoO_3-doped H-diamond near the infrared (IR) region.

CONCLUSION

In this work we have performed the DFT simulations of a single molecule absorbed on a hydrogen terminated diamond substate. The PDOS data shows that there is charge transfer between the MoO_3 molecule and the H-diamond. This is also confirmed by the charge density difference data. Hence, our results suggest that the MoO_3 molecule acts as a p-type doping material to the H-diamond surface.

Moreover, our optical simulation suggests that the MoO_3-doped H-diamond may have great potential in application of opto-electronic devices for IR and near IR light detection. This is due to the fact that the MoO_3-doped H-diamond systems have well pronounced peaks below the wavelength of the visible spectrum in the near IR wave lengths.

REFERENCES

1. R.J. Trew, J. Yan and P.M. Mock, "The potential of diamond and SiC electronic devices for microwave and millimeter-wave power applications," *Proceedings of the IEEE*, vol. 79, no. 5, 1991, pp. 598-620; DOI 10.1109/5.90128.

2. M. Willander, M. Friesel, Q.-u. Wahab and B. Straumal, "Silicon carbide and diamond for high temperature device applications," *Journal of Materials Science: Materials in Electronics*, vol. 17, no. 1, 2006, pp. 1; DOI 10.1007/s10854-005-5137-4.

3. C. Wort and R. Balmer, "Diamond as an Electronic material," *Materials Today*, vol. 11, no. 1, 2008.

4. J. Liu, H. Ohsato, X. Wang, M. Liao and Y. Koide, "Design and fabrication of high-performance diamond triple-gate field-effect transistors," *Scientific Reports*, vol. 6, no. 1, 2016, pp. 34757; DOI 10.1038/srep34757.

5. S. Russell, S. Sharabi, A. Tallaire and D.A.J. Moran, "RF Operation of Hydrogen-Terminated Diamond Field Effect Transistors: A Comparative Study," *IEEE Transactions on Electron Devices*, vol. 62, no. 3, 2015, pp. 751-756; DOI 10.1109/TED.2015.2392798.

6. P. Strobel, M. Riedel, J. Ristein, L. Ley and O. Boltalina, "Surface transfer doping of diamond by fullerene," *Diamond and Related Materials*, vol. 14, no. 3, 2005, pp. 451-458; DOI https://doi.org/10.1016/j.diamond.2004.12.051.

7. M.T. Edmonds, M. Wanke, A. Tadich, H.M. Vulling, K.J. Rietwyk, P.L. Sharp, C.B. Stark, Y. Smets, A. Schenk, Q.H. Wu, L. Ley and C.I. Pakes, "Surface transfer doping of hydrogen-terminated diamond by C60F48: Energy level scheme and doping efficiency," *The Journal of Chemical Physics*, vol. 136, no. 12, 2012, pp. 124701; DOI 10.1063/1.3695643.

8. Y. Xiang, M. Jiang, H. Xiao, K. Xing, X. Peng, S. Zhang and D.-C. Qi, "A DFT study of the surface charge transfer doping of diamond by chromium trioxide," *Applied Surface Science*, vol. 496, 2019, pp. 143604; DOI https://doi.org/10.1016/j.apsusc.2019.143604.

9. K. Xing, Y. Xiang, M. Jiang, D.L. Creedon, G. Akhgar, S.A. Yianni, H. Xiao, L. Ley, A. Stacey, J.C. McCallum, S. Zhuiykov, C.I. Pakes and D.-C. Qi, "MoO3 induces p-type surface conductivity by surface transfer doping in diamond," *Applied Surface Science*, vol. 509, 2020, pp. 144890; DOI https://doi.org/10.1016/j.apsusc.2019.144890.

10. S.A.O. Russell, L. Cao, D. Qi, A. Tallaire, K.G. Crawford, A.T.S. Wee and D.A.J. Moran, "Surface transfer doping of diamond by MoO3: A combined spectroscopic and Hall measurement study," *Applied Physics Letters*, vol. 103, no. 20, 2013, pp. 202112; DOI 10.1063/1.4832455.

11. J. McGhee and V.P. Georgiev, "Simulation Study of Surface Transfer Doping of Hydrogenated Diamond by MoO3 and V2O5 Metal Oxides," *Micromachines*, vol. 11, no. 4, 2020, pp. 433.

12. C. Verona, F. Arciprete, M. Foffi, E. Limiti, M. Marinelli, E. Placidi, G. Prestopino and G. Verona Rinati, "Influence of surface crystal-orientation on transfer doping of V2O5/H-terminated diamond," *Applied Physics Letters*, vol. 112, no. 18, 2018, pp. 181602; DOI 10.1063/1.5027198.

13. A.T. (2017.2), "Atomistix ToolKit (2017.2)," *Book Atomistix ToolKit (2017.2)*, Series Atomistix ToolKit (2017.2), ed., Editor ed.^eds., pp.

14. L.G. Ferreira, M. Marques and L.K. Teles, "Slater half-occupation technique revisited: the LDA-1/2 and GGA-1/2 approaches for atomic ionization energies and band gaps in semiconductors," *AIP Advances*, vol. 1, no. 3, 2011, pp. 032119; DOI 10.1063/1.3624562.

15. S. Smidstrup, T. Markussen, P. Vancraeyveld, J. Wellendorff, J. Schneider, T. Gunst, B. Verstichel, D. Stradi, P.A. Khomyakov, U.G. Vej-Hansen, M.E. Lee, S.T. Chill, F. Rasmussen, G. Penazzi, F. Corsetti, A. Ojanpera, K. Jensen, M.L.N. Palsgaard, U. Martinez, A. Blom, M. Brandbyge and K. Stokbro, "QuantumATK: an integrated platform of electronic and atomic-scale modelling tools," *J Phys Condens Matter*, vol. 32, no. 1, 2020, pp. 015901; DOI 10.1088/1361-648X/ab4007.

16. S. K. C, R.C. Longo, R. Addou, R.M. Wallace and K. Cho, "Electronic properties of MoS2/MoOx interfaces: Implications in Tunnel Field Effect Transistors and Hole Contacts," *Scientific Reports*, vol. 6, no. 1, 2016, pp. 33562; DOI 10.1038/srep33562.

17. H.R. Philipp and E.A. Taft, "Optical Properties of Diamond in the Vacuum Ultraviolet," *Physical Review*, vol. 127, no. 1, 1962, pp. 159-161; DOI 10.1103/PhysRev.127.159.

High-sigma analysis of DRAM write and retention performance: a TCAD-to-SPICE approach

Salvatore Maria Amoroso,
Jaehyun Lee, Andrew R. Brown,
Plamen Asenov

Synopsys Northern Europe Ltd.,
Glasgow, G3 8HB, Scotland, UK

Xi-Wei Lin, Victor Moroz

Synopsys Inc, Mountain View, CA
94043, USA

Thomas Yang

Synopsys Taiwan Ltd., Hsinchu, TW

Abstract—**This paper presents a TCAD-to-SPICE high-sigma analysis of DRAM write and retention performance. Both statistical and process-induced variability are taken into-account. We highlight that the interplay between discrete traps and discrete dopants is ruling the leakage statistical tails and therefore can play a fundamental role in determining yield and reliability of ultra-scaled DRAMs.**

Keywords—DRAM, leakage, retention time, DTCO, TCAD.

I. INTRODUCTION

Continuing efforts in advancing DRAM process technology have enabled dramatic feature-size reduction and unprecedented levels of integration [1-4]. However, the resulting increase in memory density per chip comes at the expense of increased severity of parasitic effects [5]. In particular, during the design cycle, attention has to be put on the DRAM cell transistor leakage current, which dictates DRAM refresh time (tREF) and, in turn, affects manufacturing yields. It is of utmost importance to highlight that the ultimate failure in the tREF performance is not governed by the average cell behaviour, but by the leakage current of extreme-tail cells ($<10^{-6}$ probability). These cells may exhibit a few orders of magnitude higher leakage than the nominal cell, with a statistical distribution that is influenced by both process (e.g. geometry, doping profiles) and intrinsic statistical variability (e.g. random discrete dopants, random traps). Innovative characterization techniques have been proposed to experimentally evaluate the DRAM cell transistor leakage current distributions [6]. Of equal importance becomes the availability and timely deployment of modelling platforms that enable the Design-Technology Co-Optimization (DTCO) of DRAM circuits to evaluate and optimize tREF in the presence of process and statistical variability with reduced requirements on costly and slow silicon manufacturing cycles.

II. SIMULATION METHODOLOGY

In this paper we present a TCAD-to-SPICE modelling approach enabling the early injection of statistical metrics into the design/optimization cycle. The flow, which allows accurate and extensive exploration of the design space by taking into account both leakage (tREF) and writing (tWR) statistical behaviour and their correlation, consists of the following steps (Fig.1): (i) accurate process structure generation by means of Process Explorer (layout to 3D structure) and Sentaurus Process (doping profiles) [7-8] to capture process and doping profile variations, (ii) accurate

device simulation of the nominal transistors by means of Sentaurus Device [9], (iii) Garand VE [10] for the physics-based variability simulation of trap-assisted leakage current in presence of random discrete dopants (RDD), (iv) Mystic [11] to extract statistical compact models, and (v) RandomSpice [12] for generation of models at arbitrary trap densities in statistical SPICE simulations.

It has been previously shown that discrete doping can play a fundamental role in determining the stochastic dispersion of band-to-band tunnelling in transistors [13]. Here, we also take into account the trap-assisted band-to-band tunnelling (TAT), as the measurements clearly show that the transistor leakage current is a function of the number of defects in silicon, their energy level in the bandgap, and the electric field [6]. The trap-assisted contribution is modelled through an enhancement of the trap capture cross-section in the conventional Shockley-Read-Hall (SRH) generation term. The enhancement can either be computed by Hurkx-like local models or by non-local tunnelling path approaches. Garand VE simulates hundreds of statistical RDD instances for each process condition under consideration. For each RDD configuration, thousands of single-trap positions are evaluated to gather the TAT leakage statistics. Arbitrary trap density statistics can then be obtained later in RandomSpice by convolution of the single-trap cumulative distribution functions (this assumes single traps to act as independent entities).

III. TCAD SIMULATION RESULTS

The DRAM test structure used for this study (Fig.2) is representative of a $6F^2$ cell with features reported in the side table.

The 3D structures are generated, for different process conditions, by using Process Explorer, which allows for fast, yet accurate, structure generation. The parameters *WLetch* and *Dose* (roll-off) are varied to generate a range of structures corresponding to different process conditions, or process variations.

A. RDD simulation results

Initial Garand VE analysis was performed to evaluate the impact of RDD on the on-current for the DRAM cell, across the *WLetch* and *Dose* process variations space, showing a 10% variation in the mean on-current (Fig.3). Furthermore, the on-current standard deviation also shows a consistent process-dependence, varying from 3% to 6% of the nominal on-current value and, hence, highlighting the interplay

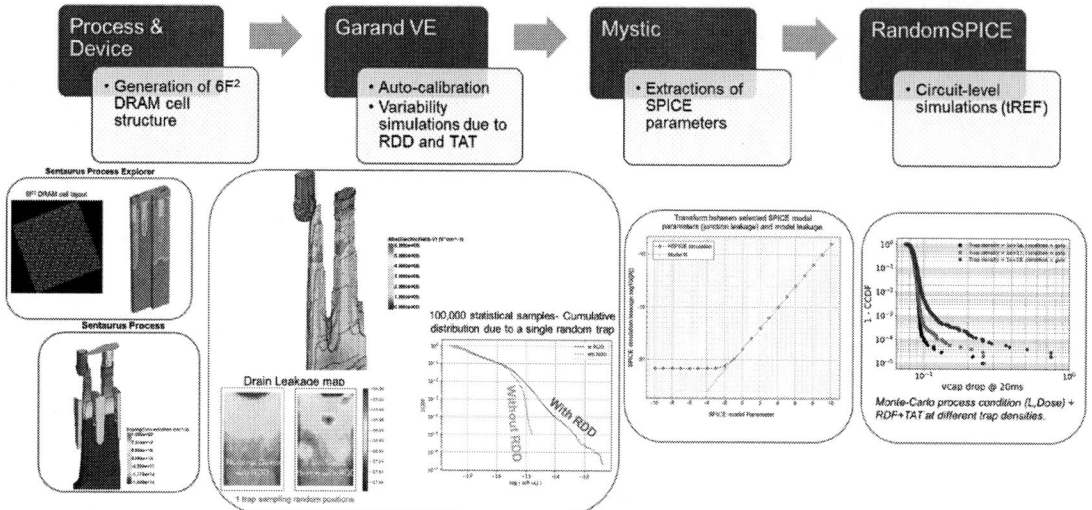

Figure 1:TCAD-to-SPICE flow for the high-sigma analysis of DRAM write and retention performance, consisting of: (i) layout-based structure generation and process emulation, (ii) device simulation of the nominal transistors, (iii) physics-based variability simulation of trap-assisted leakage current in the presence of random discrete dopants, (iv) statistical compact model extraction, and (v) model card generation at arbitrary trap densities in statistical SPICE simulation.

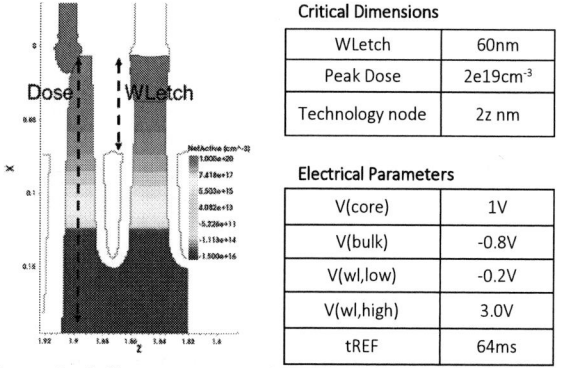

Critical Dimensions

WLetch	60nm
Peak Dose	$2e19cm^{-3}$
Technology node	2z nm

Electrical Parameters

V(core)	1V
V(bulk)	-0.8V
V(wl,low)	-0.2V
V(wl,high)	3.0V
tREF	64ms

Figure 2: (left) cross-section of generated DRAM cell structure, corresponding to a $6F^2$ technology. (right) table of critical dimensions and electrical parameters for the DRAM structure under test.

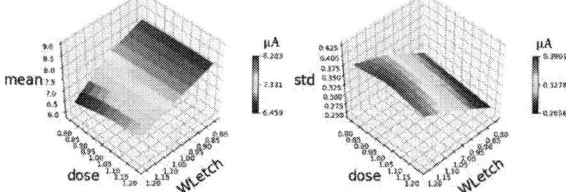

Figure 3: Mean ON-current (left) and its standard deviation (right) across the space of process variation. The plot on the right highlights the clear interplay between statistical and process variability.

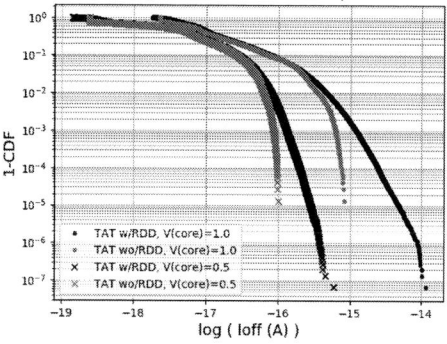

Figure 4: Complementary CDF of single trap leakage. The "w/RDD" plot represents 200 RDD configurations with ~70,000 trap evaluations per RDD configuration. The "wo/RDD" is instead obtained for a device featuring a continuous doping profile.

between process and statistical variability. The results of the analysis can be understood by considering that the combination of *WLetch* and *Dose* define the gate to source/drain overlap. With a high *WLetch*, there is significant underlap, leading to low on-current and high variability.

B. Single trap analysis

We consider 200 RDD configurations. For each of them we sweep a single trap position across the silicon area under investigation (the drain pillar region) with a 0.5nm spacing, leading to ~70,000 trap evaluations per each RDD configuration (14,000,000 trap configurations for each simulated process condition). Results in Fig.4 highlight that the interaction between traps and random dopants can enhance the leakage. In this case we see a greater than 10x increase in leakage at 10^{-5} probability level and, moreover, the tail of the distribution is now unbounded. Fig.5 shows the map of leakage as a function of the trap vertical position, highlighting that the peak of the leakage is determined by the

p-n junction and, therefore, by the process conditions (Fig.6). The leakage tail is also strongly influenced by process conditions, with *WLetch* and *Dose* having a reverse impact on leakage compared to I_{on}, with worst case leakages encountered when *WLetch* is small (i.e. the metal gate gets closer to the storage capacitor).

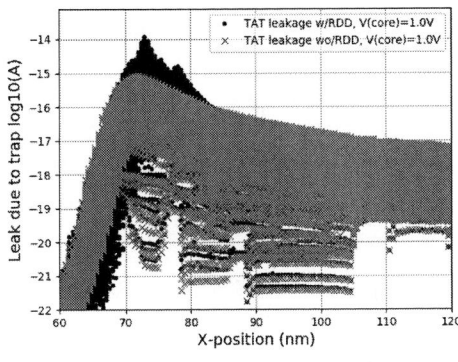

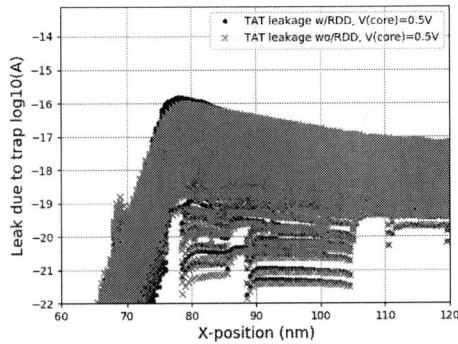

Figure 5: (top) Map of trap-related leakage against vertical position along the source/drain pillar connected to the storage capacitor, at V(core)=1.0V. (bottom) The same picture, at V(core)=0.5V.

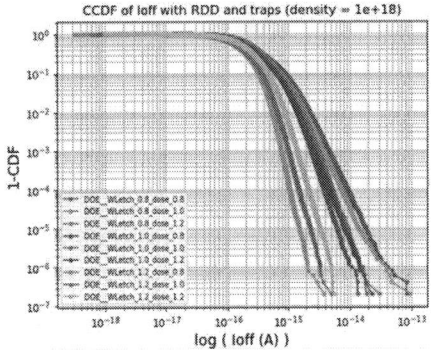

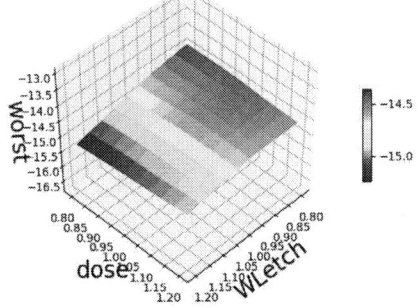

Figure 6: (top) Leakage distributions for each process condition showing ~2 orders of magnitude difference in tail leakage for this set of experiments. (bottom) tail leakage values across different process conditions, highlighting a clear interplay between process and statistical variability also for the leakage behaviour.

C. Arbitrary trap density

The results for many traps can be obtained analytically by self-convolution of the single-trap statistics. Fig.7 confirms the validity of the approach, by comparing analytical results with full TCAD simulation.

IV. SPICE SIMULATION RESULTS

The simulated TCAD data is then propagated into a SPICE model. For each Monte-Carlo instance of the DRAM cell, a unique leakage current is generated using the fitted TCAD data distributions. These randomised leakage values are then converted to BSIM4 junction leakage parameters. To verify the SPICE model is reproducing the TCAD data, Fig.8 shows the leakage complementary cumulative distributions from the TCAD fitting and leakage values extracted from the SPICE model. Because leakage values are directly generated, it becomes simple to skip any SPICE simulation where leakage current is not high enough to cause cell failures. In Fig.8, SPICE simulations are only enabled for circuits where the DRAM cell leakage current at nominal bias >1 fA. As a result, only ~400k of 10M (representing roughly 10Mbit) SPICE tWR+tREF simulations were run – enabling high-sigma analysis. Finally, SPICE simulations were run combining on-current and leakage variability where a write-then-hold operation was performed. The results of these simulations are shown in Fig.9, where the left figure shows (target - final storage voltage), and the right figure shows storage voltage after a 64ms hold. The simulations are performed at 3 process conditions, which are described in Table 1.

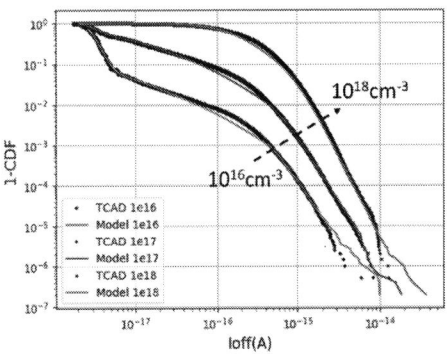

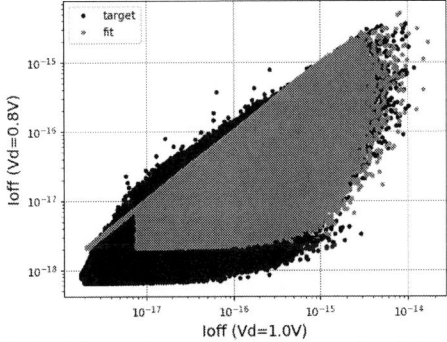

Figure 7: (top) Comparing analytically generated statistical leakage (obtained by convolution of the single trap distribution) vs TCAD-generated data for different trap densities. (bottom) The analytical generation is also able to maintain good agreement with TCAD data with respect to drain bias dependence (here shown for the case of single trap).

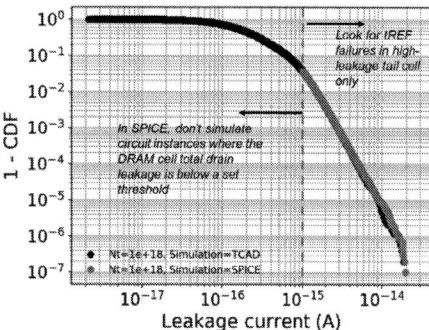

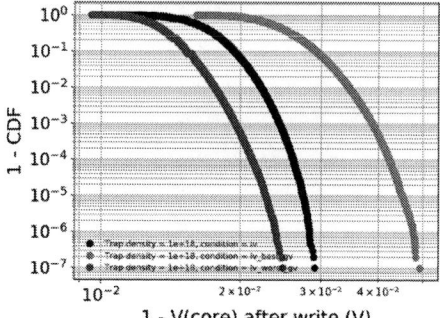

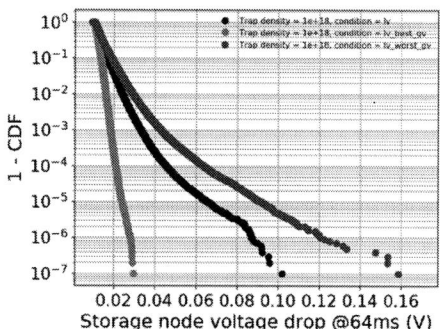

Figure 8: CCDF of leakage current in the DRAM cell: TCAD vs SPICE simulation. The SPICE tail methodology allows fast and reliable high-sigma analysis.

Process Conditions Definitions			
	lv	lv + worst_gv	lv + best_gv
WLetch	1.0	0.8	1.2
Dose	1.0	0.8	0.8
4.5σ voltage underwrite	29mV	22mV	50mV
4.5σ hold voltage drop	70mV	110mV	24mV

Table 1 Process conditions for the statistical circuit simulation analysis reported in Figure 10.

Finally, the results shown in Fig.10 provide a quantitative analysis of the trade-offs between better write behaviour (tWR), and better leakage (tREF).

V. CONCLUSIONS

We have presented a TCAD-to-SPICE methodology for the high-sigma analysis of DRAM write and retention performance. This modelling platform can enable the Design-Technology Co-Optimization (DTCO) of DRAM circuits to evaluate and optimize tREF in the presence of process and statistical variability.

REFERENCES

[1] S-W Park et al., "Highly Scalable Saddle-Fin Transistor for Sub-50nm DRAM Technology", 2006 VLSI Symposium.
[2] S-W Ryu et al; "Overcoming the reliability limitation in the ultimately scaled DRAM using silicon migration technique by hydrogen annealing", IEDM 2017.
[3] C-M Yang et al; "Suppression of Row Hammer Effect by Doping Profile Modification in Saddle-Fin Array Devices for Sub-30-nm DRAM Technology", IEEE TDMR, 2016.
[4] S. H. Jang et al., "A Fully Integrated Low Voltage DRAM with Thermally Stable Gate-first High-k Metal Gate Process", IEDM 2019.
[5] S-K Park, "Technology Scaling Challenge and Future Prospects of DRAM and NAND Flash Memory", IMW 2015.
[6] M.H. Cho et al., "An Innovative Indicator to Evaluate DRAM Cell Transistor Leakage Current Distribution", IEEE J-EDS, 2018.
[7] Process Explorer User Guide, Synopsys, CA, USA, 2019.
[8] Sentaurus Process User Guide, Synopsys, CA, USA, 2019.
[9] Sentaurus Device User Guide, Synopsys, CA, USA, 2019.
[10] Garand VE User Guide, Synopsys, CA, USA, 2019.
[11] Mystic User Guide, Synopsys, CA, USA, 2019.
[12] RandomSpice User Guide, Synopsys, CA, USA, 2019.
[13] A. Ghetti et al., "Evidence for an atomistic-doping induced variability of the band-to-band leakage current of nanoscale device junctions", IEDM-12

Figure 9: (top) Delta V(core) distribution after a write operation, representing variability in on-current due to random dopant fluctuations. (bottom) Distribution of storage capacitor voltage drop (relative to final storage node voltage value), after 65ms hold time.

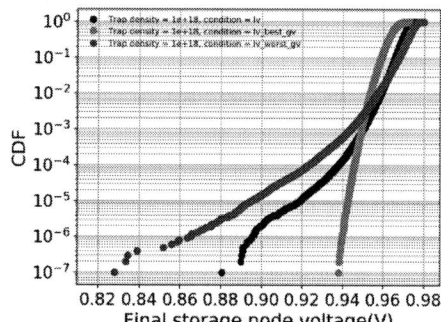

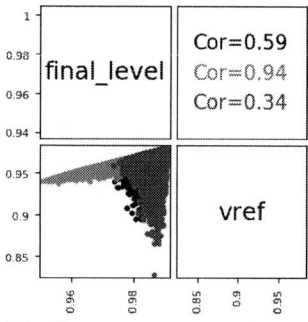

Figure 10: (top) Final storage capacitor voltage after a write and 64ms hold operation. (bottom) scatter plot of final storage node voltage after write (final level) vs. final storage node voltage after write and 64ms hold. The red data points represent worst tWR/best tREF process conditions, the blue represents best tWR/worst tREF conditions, and black the "typical" case.

Generative Model Based Adaptive Importance Sampling for Flux Calculations in Process TCAD

Alexander Scharinger*, Paul Manstetten†, Andreas Hössinger‡, and Josef Weinbub*

*Christian Doppler Laboratory for High Performance TCAD at the
†Institute for Microelectronics, TU Wien, Gußhausstraße 27-29/E360, 1040 Wien, Austria
‡Silvaco Europe Ltd., Compass Point, St Ives, Cambridge, PE27 5JL, United Kingdom
Email: scharinger@iue.tuwien.ac.at

Abstract—A key part of advanced three-dimensional feature-scale etching and deposition simulations is calculating the particle flux distributions. The most commonly applied flux calculation approach is top-down Monte Carlo which, however, introduces numerical noise. In principal, this noise can be reduced by increasing the number of simulated particles but doing so also increases the overall running time. For complex geometries, especially high aspect ratio structures, which are very prominent in state of the art three-dimensional electronic device designs, increasing the number of samples is not a viable approach: Only a very small subset of simulated particles contributes to reducing the noise in remote and obscured surface regions. We thus propose an adaptive importance sampling approach based on a generative model to more efficiently focus the sampling on those surface regions with high noise levels. We show that, for a constant number of simulated particles, our approach reduces the noise levels in the calculated flux by about 33% for a representative high aspect ratio test structure.

I. INTRODUCTION

Three-dimensional feature-scale simulations of etching and deposition processes require an accurate simulation of the particle transport to obtain realistic particle flux distributions on the structure surface. A common approach to compute the particle flux distributions is to use a Monte Carlo simulation [1]. In particular, *top-down* Monte Carlo particle tracing is a popular and versatile approach because the integration of adsorption and reflection characteristics of any detail is straightforward and flexible [2]. A top-down Monte Carlo flux simulation consists of sampling (pseudo) random particles (origin and direction) from a given source distribution. Each particle carries a weight or energy value. The trajectories of the particles are traced through the simulation domain assuming ballistic transport. When a particle hits a material surface the local adsorption and reflection characteristics determine the effect of the particle on the local flux and its further treatment, e.g., re-emission of subsequent particles.

A major downside in conventional Monte Carlo particle tracing algorithms is the inherent numerical noise present in the results. This is particularly the case for (very) high aspect ratio (HAR) structures which are prominent in many semiconductor devices, e.g., NAND flash cells [3]. Even worse, the noise in the calculated flux distribution can be highly heterogeneous across the simulation domain [4], [5]. A straightforward but computationally very expensive way to deal with the noise is to excessively increase the number of simulated particles. However, this typically leads to massive oversampling in many parts of the surface while regions remote from the particle source and obscured from direct flux contributions still reveal high noise levels.

We present an adaptive Monte Carlo importance sampling algorithm applied to the source of the particles to reduce oversampling and to drastically decrease the noise levels in remote and obscured regions without increasing the total number of particles. In our approach, we start by distributing a limited number of particles in the conventional way to identify regions with high noise levels. We use this information to construct a Gaussian mixture model (GMM) [6], [7], which is an established generative model from the field of machine learning. The GMM's probability distribution is used to generate the remaining (major) share of particles, ultimately allowing to increase the accuracy in regions with high noise levels. In Section II we introduce the method and in Section III the method is applied and analyzed: Our approach reduces the overall noise in the calculated flux distribution on a representative HAR structure by 33% compared to the conventional Monte Carlo technique when maintaining the number of simulated particles.

II. METHOD

The proposed adaptive sampling algorithm for computing the flux distribution is illustrated in Fig. 1. It is inspired by the cross-entropy method for rare event simulation and combinatorial optimization [8].

The first stage consists of performing a conventional Monte Carlo flux calculation by sampling from the original uniform distribution at the source of the particles (Fig. 1a). This first stage yields for each surface element a flux estimate and a relative error of the estimate (Fig. 1b). The relative error is defined by $\sqrt{\mathrm{Var}\,(\hat{f})}/\mathbb{E}\,(\hat{f})$ ($\mathrm{Var}\,(\hat{f})$ and $\mathbb{E}\,(\hat{f})$ denote the variance and the expected value of the flux estimate $\hat{f}$, respectively) and quantifies the convergence of the Monte Carlo estimation [1].

The second stage uses a small number of particles to track which locations at the particle source hit surface elements with a flux estimate with high relative error (Fig. 1c). We call this stage *importance mapping* and it produces a set of relevant locations at the particle source.

Next, a GMM [6], [7] is fitted to the output of the importance mapping (Fig. 1d).

A GMM is a parametric probability density function characterized by the weighted sum of a finite number of Gaussian components: $p(\boldsymbol{x}) = \sum_{k=1}^{K} \pi_k \mathcal{N}(\boldsymbol{x}|\boldsymbol{\mu}_k, \boldsymbol{\Sigma}_k)$ where K is the number of components, π_k the mixing weights, and $\boldsymbol{\mu}_k$ and $\boldsymbol{\Sigma}_k$ are the the mean vectors and covariance matrices of the Gaussian distributions $\mathcal{N}(\boldsymbol{x}|\boldsymbol{\mu}_k, \boldsymbol{\Sigma}_k)$, respectively. The mixing weights π_k, the mean vectors $\boldsymbol{\mu}_k$, and the covariance matrices $\boldsymbol{\Sigma}_k$ are GMM parameters. A GMM is a generative probabilistic model, that is, it can be used to generate new random data points reproducing its probability distribution. The particle source in the three-dimensional simulation domain is a plain surface, hence the Gaussian distributions are two-dimensional. Given the set of relevant samples from the importance mapping step, the parameters of the GMM are computed using maximum likelihood estimation and the expectation maximization algorithm [7]. The optimal value for K (the number of components in the GMM) is automatically selected via the the integrated completed likelihood (ICL) criterion [9]. Random sampling from a GMM requires only minor computational overhead compared to sampling from a uniform distribution using, e.g., the Box-Muller approach for Gaussian sampling [1].

The final stage of the algorithm is to use the probability distribution of the obtained GMM for Monte Carlo importance sampling [1]: Particles are generated according to the probability distribution of the GMM and each particle is weighted with the inverse sampling probability such that the Monte Carlo estimates remain unbiased (Fig. 1e), resulting in the final surface flux distribution and corresponding relative errors (Fig. 1f).

III. RESULTS

We assess our algorithm with a representative HAR test structure shown in Fig. 2a. We compare conventional sampling using 32 M (million) particles with the proposed adaptive importance sampling algorithm using about 8 M particles for the first stage ($S_{\text{conv.}} \approx 8$ M in Fig. 1a), about 100 k (thousand) particles for the importance mapping ($S_{\text{imp.}} \approx 100$ k in Fig. 1c), and about 24 M particles for the importance sampling ($S_{\text{GMM}} \approx 24$ M in Fig. 1e), totaling again 32 M particles. For the importance mapping (Fig. 1c) we consider all surface elements with a relative error greater than 0.1 (i.e. 10%).

Fig. 3a shows the set of relevant samples obtained from the importance mapping step when running the proposed algorithm on the test structure. About 100 k samples for the importance mapping result in 634 relevant samples. It is apparent that the contributions from particles sampled right above the circular holes and the trench dominate the set of relevant samples.

From the set of relevant samples the fitting step (cf. Fig. 1d) results in a GMM with three components ($K = 3$). Fig. 3b visualizes the probability density function of this GMM (i.e. the superposition of three two-dimensional Gaussian distributions). This distribution is used for Monte Carlo importance sampling in the importance sampling step (cf. Fig. 1e). Although the geometry is axis-symmetric, the distribution of

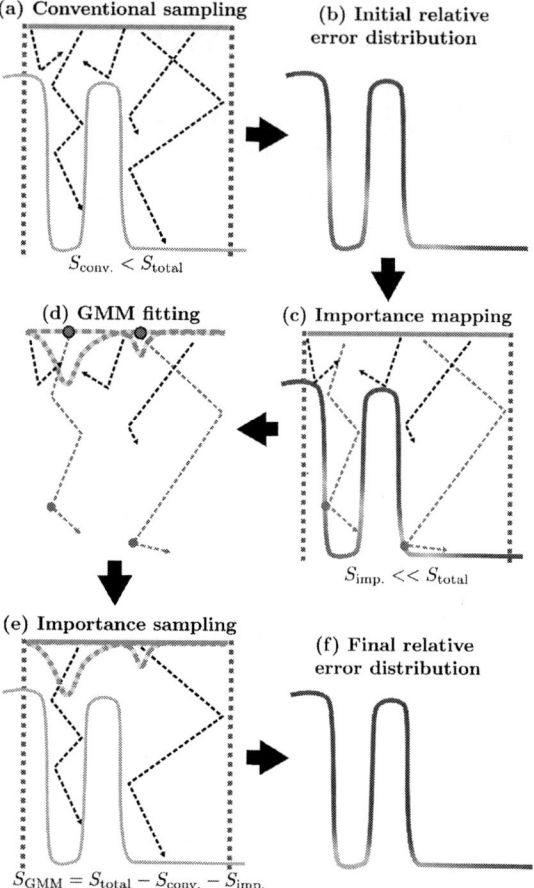

Fig. 1: Schematic sequence of the proposed adaptive importance sampling algorithm: Instead of a full conventional sampling of S_{total} particles, only a portion of particles $S_{\text{conv.}}$ is sampled conventionally to generate an initial error distribution. This initial distribution is used to identify important samples from a small subset $S_{\text{imp.}}$, which are used to fit a GMM. Finally, the majority of particles S_{GMM} is generated using importance sampling based on the distribution of the fitted GMM.

the GMM in Fig. 3b is not strictly symmetric. As the set of relevant samples is produced using a Monte Carlo approach these deviations from symmetry are expected.

Fig. 2b shows the relative error of the flux estimates using conventional sampling. The bottom of the two circular holes in the test structure shows high levels of relative error of the calculated flux. In comparison to this, Fig. 2c shows the relative error using the proposed algorithm when applying adaptive importance sampling. The error distributions are visualized in Fig. 4 which compares the distributions of the relative errors using conventional sampling and the proposed adaptive importance sampling: One can clearly see that our approach significantly reduces the number of surface elements with high relative error.

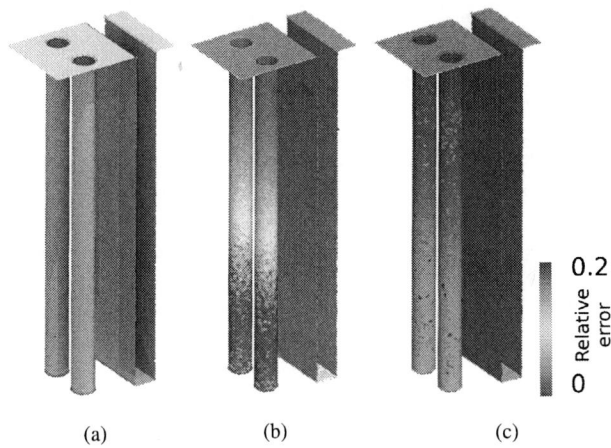

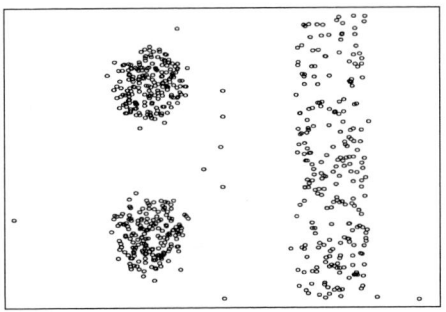

(a) Set of relevant samples.

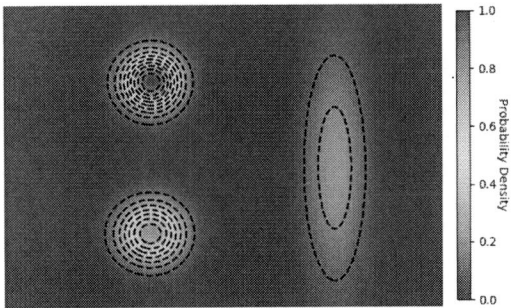

(b) Probability density function of the GMM ($K = 3$).

Fig. 2: HAR test structure consisting of about 40 k surface primitives: (a) The plain geometry, (b) the relative error of the surface flux distribution using 32 M conventionally sampled particles, and (c) the relative error of the surface flux distribution using 32 M particles sampled with the proposed adaptive importance sampling algorithm.

The noise in the calculated flux distribution is reduced by about 33% (from 0.3 to below 0.2 in Fig. 4) for surface regions with high noise in remote and obscured areas of the geometry.

To provide a statistically unbiased sampling of the particle source distribution, the weights of the particles sampled from the GMM exhibit a broad range of weights (energies). A very small number of these particles carry a very high weight value. These particles actually have a negative effect on the computed flux and its relative error: When they hit the surface they increase the relative error of the flux estimate at the surface element, especially in regions with low particle flux, e.g., towards the bottom of HAR structures. This effect is apparent in Fig. 2c: A few surface elements in the otherwise continuous gradient exhibit greater relative errors than others. A straightforward way to deal with high-energy particles would be to introduce a particle splitting scheme which can be found, for example, in [1].

IV. Summary and Outlook

We propose an adaptive importance sampling approach based on a generative model to more effectively sample particles in flux calculations. We assess our algorithm on a representative HAR test structure and show that it reduces the noise by about 33% – without increasing the number of Monte Carlo samples. A promising idea to improve the applicability of this approach to a wide range of geometries is to apply the adaptive importance sampling steps iteratively in order to progressively improve the focus of the importance sampling until a desired noise level is obtained for all surface elements.

Fig. 3: Top view onto the HAR test structure shown in Fig. 2: (a) Visualization of the set of relevant samples (origins on source plain) computed in the importance mapping step of the proposed algorithm (Fig. 1c), (b) Visualization of the probability density function of the GMM for the set of relevant samples shown in (a) computed in the fitting step of the proposed algorithm (Fig. 1d).

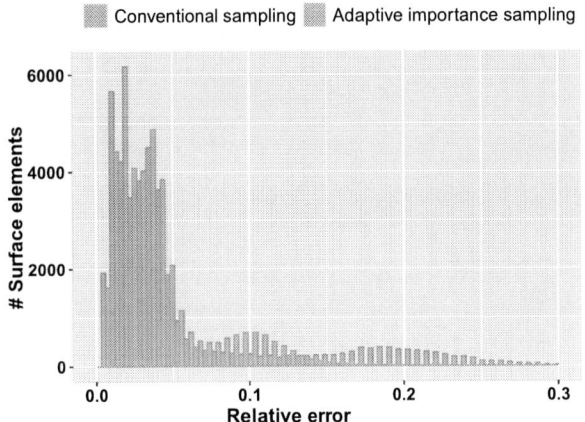

Fig. 4: Histogram of the relative errors of the flux estimates for the test structure with 32 M conventionally sampled particles as in Fig. 2b and 32 M particles with the proposed adaptive importance sampling algorithm as in Fig. 2c.

ACKNOWLEDGMENT

The financial support by the Austrian Federal Ministry for Digital and Economic Affairs, the National Foundation for Research, Technology and Development, and the Christian Doppler Research Association is gratefully acknowledged.

REFERENCES

[1] R. Rubinstein and D. Kroese, *Simulation and the Monte Carlo Method.* Wiley & Sons, 2016.

[2] X. Klemenschits, S. Selberherr, and L. Filipovic, "Modeling of Gate Stack Patterning for Advanced Technology Nodes: A Review," *Micromachines*, vol. 9, no. 12, p. 631, 2018.

[3] P. Dimitrakis, *Charge-Trapping Non-Volatile Memories: Volume 1 – Basic and Advanced Devices.* Springer International Publishing, 2015.

[4] O. Ertl and S. Selberherr, "Three-Dimensional Level Set Based Bosch Process Simulations using Ray Tracing for Flux Calculation," *Microelectronic Engineering*, vol. 87, no. 1, pp. 20–29, 2010.

[5] P. Manstetten, L. Filipovic, A. Hössinger, J. Weinbub, and S. Selberherr, "Framework to Model Neutral Particle Flux in Convex High Aspect Ratio Structures Using One-Dimensional Radiosity," *Solid-State Electronics*, vol. 128, pp. 141–147, 2017.

[6] G. J. McLachlan and D. Peel, *Finite Mixture Models.* Wiley & Sons, 2004.

[7] K. P. Murphy, *Machine Learning: A Probabilistic Perspective.* MIT press, 2012.

[8] P. de Boer, D. P. Kroese, S. Mannor, and R. Y. Rubinstein, "A Tutorial on the Cross-Entropy Method," *Annals of Operations Research*, vol. 134, no. 1, pp. 19–67, 2005.

[9] C. Biernacki, G. Celeux, and G. Govaert, "Assessing a Mixture Model for Clustering with the Integrated Completed Likelihood," *IEEE Transactions on Pattern Analysis and Machine Intelligence*, vol. 22, no. 7, pp. 719–725, 2000.

3-3
Implant heating contribution to amorphous layer formation: a KMC approach

P.L. Julliard*[1,2], P. Dumas[1], F. Monsieur[1], F. Hilario[1], D. Rideau[1], A. Hemeryck[2], F. Cristiano[2]

[1]ST Microelectronics, Crolles, France

[2]LAAS-CNRS, Université de Toulouse, CNRS, Toulouse, France

Email: pierre-louis.julliard@st.com

Abstract—The present work investigates the influence of implantation induced heating on the amorphization profile in silicon wafer. A simulation approach based on a Kinetic Monte Carlo method is compared to experimental implantations and characterizations. We demonstrate that a backside pressure cooling can be used to tune amorphous layer thickness.

Keywords—*amorphization, Kinetic Monte Carlo, implantation induced heating, machine parameters*

I. INTRODUCTION

Ion implantation is a widely used process in the semiconductor industry to introduce dopants into silicon substrates and leads to amorphization under specific conditions. While amorphous silicon recrystallizes upon annealing, excess self-interstitials lying in crystalline areas evolve in extended defects such as {311} or dislocation loops [1]. Extended defects are known to induce several issues in semiconductor devices. The first one is the degradation of electronic performances as dislocation loops can induced leakage current in transistor [2]. Extended defects also act as recombination or emission center for electron-hole pair and are involved in the degradation of optical properties [3]. Dark current in images sensors has been associated in some cases with the presence of dislocations [4]. Furthermore, extended defects can dissolve during the annealing. The dissolution of a dislocation or a {311} is followed by an increase of the supersaturation of interstitials. This increase of interstitials produced a well-known phenomenon: the transient enhanced diffusion (TED). Dopant diffusion can increase by a factor 100 or more during the first instant of the annealing and lead to a spreading of the dopants distribution [5].

A knowledge of the formation kinetic is necessary to control the presence of extended defects. The growth kinetics of such defects is not only driven by the annealing temperature but also by their position relative to the surface [6]. Therefore, the spatial distribution of both amorphous/crystalline areas after ion implantation is a key parameter for process optimization. Process optimization can be improved using Technology Computer-Aided Design (TCAD). Reliable TCAD simulations therefore need complete and accurate models for implantation induced amorphization [7].

Amorphization is a complex phenomenon involving several parameters such as the ion dose, the wafer temperature, the mass of the implanted ion and the frequency between ions collisions (dose rate) [8]. Certain parameters driving amorphization are themselves dependent on the machine parameters used for implantation. The dopants are implanted by the mean of a beam sending ions on a delimited area on the wafer. The part of the wafer exposed to the beam changes with time to obtain a uniformly implanted wafer. A given area on the wafer is therefore alternatively heated and cooled down, depending if it is exposed to the beam or not.

Industrial implanters include a cooling system that, notably, prevents the resist from being damaged during implantation [9]. However, despite the cooling system, the ion beam heats up the wafer, which could impact the amorphization dynamic. In most cases, the temperature increase can be relatively modest, which has no impact on amorphization [10]. However, this work focuses on implantations with higher beam energy than in [10] which could lead to a greater heating. Machine parameters are essential in the heating phenomenon and therefore for the amorphization phenomenon. This work investigates the impact of backside cooling pressure on amorphization comparing experimental data and Kinetic Monte Carlo (KMC) simulations.

II. EXPERIMENTAL AND SIMULATION METHOD

The implanter used here is an industrial Viista HCS implanter supplied by Applied Material. Its end-station configuration is presented in Fig. 1. A fixed ribbon beam sends ions with a constant flux on the wafer which moves back and forth face to the ribbon beam. An implantation is divided into N_{passes} back and forth on the wafer to reach the total dose. Incidents ions hit the wafer with an energy E_{ions}. The ions flux is given by the current I_{beam}. The first step of this work is to simulate the temperature profile during implantation depending on machine parameters. The temperature is calculated for one given wafer sub-area at the wafer center. For this given area, the heat equation is separated in two parts: a heating phase when the beam irradiates the area and a cooling phase when the beam is located elsewhere on the wafer. In equation (1), p_{beam} is the power induced per unit area ($\frac{E_{ions}*I_{beam}}{qA}$ with q the Coulomb constant and A the surface of the implanted area)

$$\rho c_p \frac{dT}{dt} = -\alpha p_{cooling}(T - T_f) + p_{beam}(t) \qquad (1)$$

ρ and c_p are the silicon mass density and thermal capacitance. T_f is the water temperature (see Fig. 1). α is an surface roughness constant and $p_{cooling}$ is the cooling gas pressure.

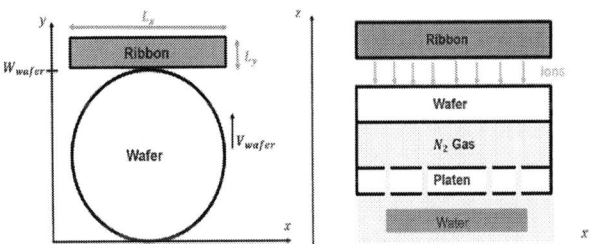

Fig. 1: Scheme of the implanter end-station used in this work (left) and of its cooling system (right).

This equation assumes an instantaneous diffusion of the heat along the depth of the wafer. The temperature profile is composed of N_{passes} cycles of heating-cooling, each pass starting at the temperature reached at the previous one. The beam exposition time and the time between two successive beam expositions for the considered sub area are directly function of the machine parameters used during the implantation. These parameters can be evaluated during the implantation by machine sensors and are extracted to obtain the real dynamic of the implantation. The beam exposition time $t_{exposed}$ and the time between two passes on the center of the wafer $t_{cooling}$ are calculated with the following formulas:

$$t_{exposed} = \frac{L_x}{v_{wafer}} \quad (2) \quad \text{and} \quad t_{cooling} = \frac{W - L_x}{v_{wafer}} \quad (3)$$

Parameters such as the currents used during implantation or the number of passes are also extracted from the equipment sensors.

The temperature trend with time is a saw-tooth like increase. The temperature profile is coupled to the input files of Sentaurus KMC software [11] to model implantation at the calculated temperatures. The total implantation is subdivided in N_{passes} implantations and N_{passes} cooling times between implantations in the input files of Sentaurus. Each sub implantation is simulated at the corresponding temperature. Time between implantations allows diffusion and recombination of defects.

Sentaurus KMC calculates each point defect but also enables to plot average 1-D profiles along the depth of the wafer. When the concentration of point defects overcomes $1.5 \, 10^{22} \, cm^{-3}$ in a region, it is considered as amorphous. This enables amorphization 1-D profiles along the depth of the wafer.

In the presented method to calculate wafer heating, different cooling gas pressures lead to different local temperature increases of the wafer. A solution to highlight temperature contribution is to simulate the same implantation with different cooling pressures. Fig. 2 shows the simulation of temperature for a 60 keV carbon ion implantation with a dose of $1.5 \, 10^{15} cm^{-2}$ at three different backside cooling pressures (4, 9 and 15 Torr). Fig. 2 highlights that the smaller the cooling pressure, the higher the wafer heating, in agreement with equation (1).

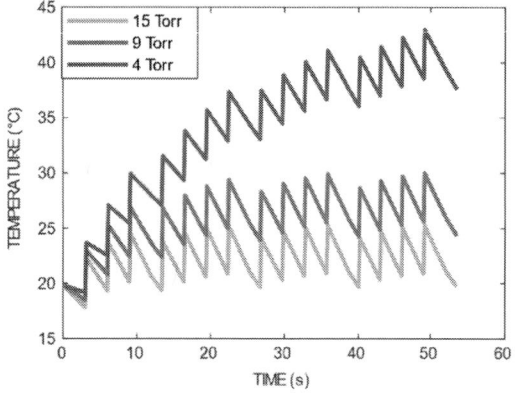

Fig. 2: Simulation of temperature evolution during a 60 keV carbon implant with a dose of $1.5 \, 10^{15} cm^{-2}$ for three different cooling pressures (4, 9 and 15 Torr).

III. RESULTS AND DISCUSSION

The amorphization generated by the previously mentioned 60 keV carbon implantations is simulated with Sentaurus KMC coupled with the generated time dependent temperature profile. The amorphous layer is defined by the depth where the concentration of point defects is above the amorphization threshold value. The results demonstrate that the amorphous layer thickness increases with the cooling pressure used during implantation (Fig. 3). The amorphous layer thickness for a 4 Torr cooling pressure only reaches 90 nm as compared to the 120 nm amorphous thickness obtained at 15 Torr. This result can be compared to the amorphous layer thickness measured by transmission electron microscopy (TEM) images given on Fig. 3. Estimation for amorphous thickness could be extracted by plotting the laterally integrated contrast versus depth of the TEM images. Silicon is considered amorphous when its brightness is at 85 percent of the brightest point. Simulation and experimental results for these implantations are summarized in Table I.

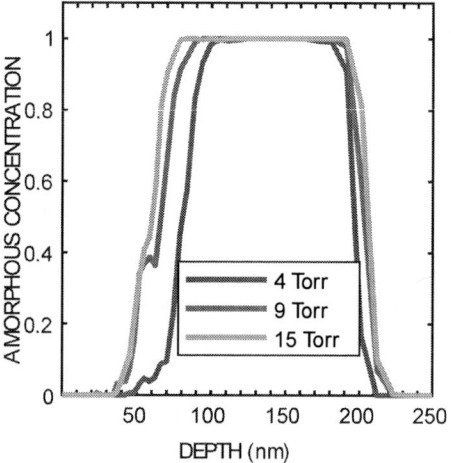

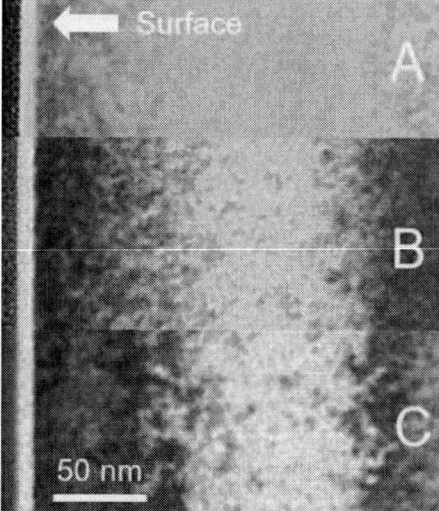

Fig. 3: (Top) Simulation of amorphous concentration as function of the depth for a 60 keV carbon implantation with a dose of $1.5 \, 10^{15} \, cm^{-2}$ and for cooling pressures of 15, 9 and 4 Torr. (Bottom) A, B and C are the TEM images of 60 keV carbon implanted wafers with a dose of $1.5 \, 10^{15} \, cm^{-2}$ and a cooling pressure of 15, 9 and 4 Torr respectively.

The effect of the cooling pressure on amorphization for a 30 keV carbon implantation at a dose of $2 \cdot 10^{15} cm^{-2}$ and cooling pressures of 4 and 9 Torr is also simulated in this work. The difference in temperature for the two implantation conditions can be observed in Fig. 4. The effect of the pressure is the same as for the 60 keV implantation. A smaller cooling pressure induced higher temperature during implantation. Amorphization profile for these two conditions is simulated in Fig. 5. The difference of simulated amorphization is much smaller but follows the same trend as the 60 keV carbon implantation. The higher the cooling pressure, the thicker the amorphous layer. TEM images of these 30 keV implantations show more crystalline areas near the surface for the high-pressure case (Fig. 4). This confirms that amorphization tends to increase with cooling pressure.

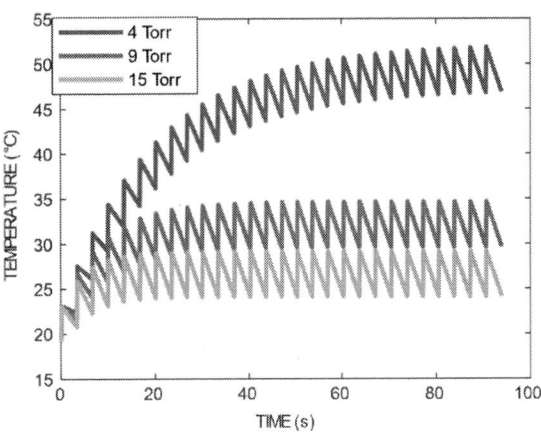

Fig. 6: Simulation of temperature evolution during a 30 keV carbon implant with a dose of $2 \cdot 10^{15} cm^{-2}$ for two different cooling pressures (4, 9 Torr).

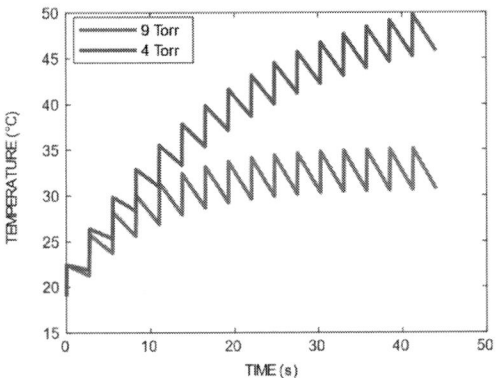

Fig. 4: Simulation of temperature evolution during a 30 keV carbon implant with a dose of $2 \cdot 10^{15} cm^{-2}$ for two different cooling pressures (4, 9 Torr).

This trend is particularly highlighted by the ion and the dose investigated in the two previous experiments. The dose of $1.5 \cdot 10^{15} cm^{-2}$ is close to the amorphization threshold dose $D_{critical}$ for carbon at room temperature. The more the implanted dose is superior to the threshold dose value, the less the effect of cooling pressure on amorphization is observed. For an implanted dose of $4 \cdot 10^{15} cm^{-2}$ and for cooling pressures of 4, 9 and 15 Torr, the difference in temperature is still observed and even more present than in the previous $2 \cdot 10^{15} cm^{-2}$ implantation (see Fig. 6). The difference in amorphization is not observed in TEM images. The KMC simulation does not predict difference in amorphization for these temperature differences (Fig. 7). Simulation and experiment are consistent and show a diminution of the effect when the dose increases.

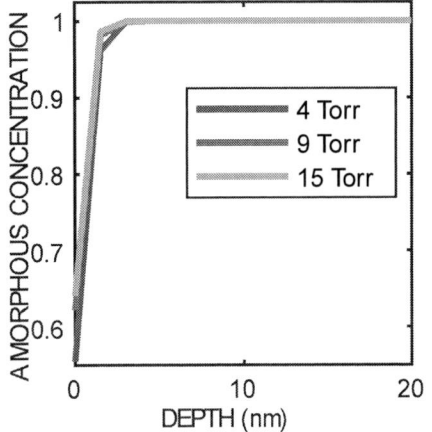

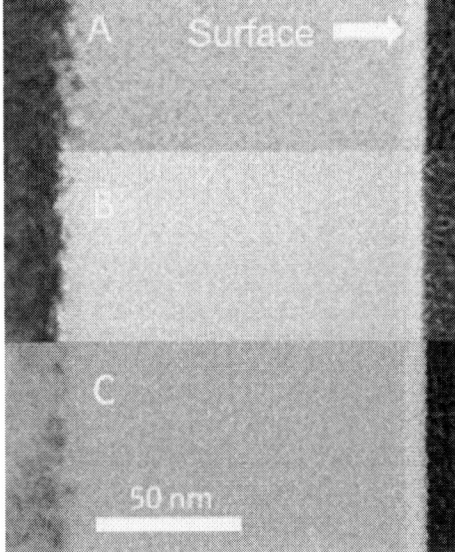

Fig. 7: (Top) Simulation of amorphous concentration as function of the depth for a 30 keV carbon implantation with a dose of $4 \cdot 10^{15} cm^{-2}$ and for cooling pressures of 4, 9 and 15 Torr. (Bottom) A, B and C are the TEM images of 30 keV carbon implanted wafers with a dose of $4 \cdot 10^{15} cm^{-2}$ and a cooling pressure of 15, 9 and 4 Torr respectively.

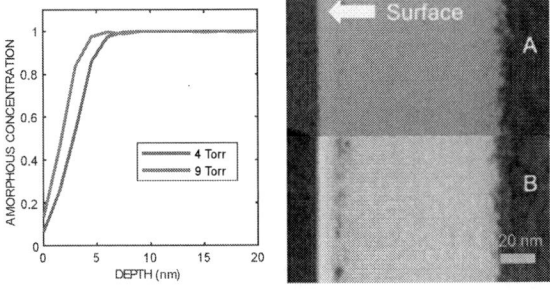

Fig. 5: (Left) Simulation of amorphous concentration as function of the depth for a 30 keV carbon implantation with a dose of $2 \cdot 10^{15} cm^{-2}$ and for cooling pressures of 9 and 4 Torr. (Right) A and B are the TEM images of 30 keV carbon implanted wafers with a dose of $2 \cdot 10^{15} cm^{-2}$ and a cooling pressure of 9 and 4 Torr respectively.

This phenomenon is linked with the dynamic of growth of an amorphous layer. The growth of an amorphous layer is the consequence of damages accumulation in silicon. The damages accumulation regarding the implanted dose is composed in three distinct regimes for a light ion implantation (see Fig. 8). The first regime is a sublinear growth of the damages in silicon until small amorphous areas are formed. For higher implanted doses the defects growth follows a superlinear behavior where silicon is composed of both crystalline and amorphous areas. The third regime is a platen corresponding to the creation of an amorphous layer and is reached when the dose overcomes a value $D_{critical}$. The damages profile is not uniform along the depth of the implanted wafer. When the critical dose is reached, the amorphous layer is formed at the depth where damages are maximum. If the implanted dose overcomes $D_{critical}$, damages are still increasing in non-amorphous areas. The amorphous layer expends in the neighboring regions where the damages accumulated are close to the amorphization threshold for a $D_{critical}$ implanted dose. The expansion of the amorphous layer with dose stops when all damaged regions are amorphous. This explains the constant amorphous layer thickness in Fig. 7. The dose is high enough to amorphize along all the damages profile for all temperature conditions. For $1.5\ 10^{15}\ cm^{-2}$ implantation, region between the surface and the amorphous layer are in the superlinear behavior and a small change in implantation parameters induces a visible difference in amorphization profile.

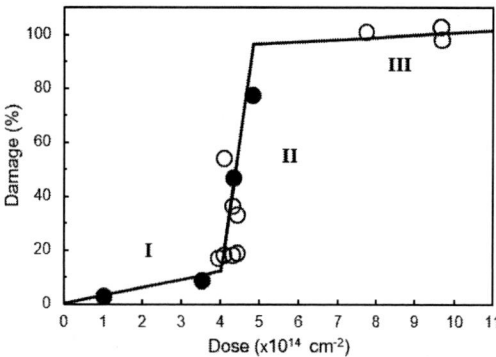

Fig. 8: Superlinear behavior of the damage vs dose for a 230 keV Si implant into S (100) at room temperature reproduced from [8]. Symbols represent the maximum defect concentration as extracted from channeling spectra.

The superlinear behavior is also an effect correlated with the mass of the implanted ion. The lighter the implanted ion specie, the sharper the superlinear slope [12]. Heavy species ions do not demonstrate strong super-linear behavior compared to carbon. Heating during implantation is therefore expected to be more relevant for light ion species.

TABLE I. COMPARAISON BETWEEN SIMULATION AND EXPERIMENTAL DATA FOR AMORPHOUS LAYER THICKNESS

Amorphous layer thickness (nm)		
Wafer	*KMC*	*Experimental*
Carbon 60 keV $1.5\ 10^{15} cm^{-2}$ 4 Torr	80	80
Carbon 60 keV $1.5\ 10^{15} cm^{-2}$ 9 Torr	103	100
Carbon 60 keV $1.5\ 10^{15} cm^{-2}$ 15 Torr	120	110

IV. CONCLUSION

KMC simulations and experimental data are consistent and showed that amorphous thickness can be tuned from dozen of nm to less than 5 nm by changing backside pressure cooling on a standard industrial implanter. The impact of the backside cooling pressure is more relevant when the implanted dose approaches the threshold value of amorphous layer formation. This behavior is also predicted by KMC simulations.

V. REFERENCES

[1] C. Bonafos, D. Mathiot and A. Claverie. "Ostwald ripening of end-of-range defects in silicon." Journal of Applied Physics, vol. 83, no. 6, pp. 3008-3017. (1998).

[2] C. Nyamhere, A. Scheinemann, A. Schenk, A. Olivie, F. Cristiano . "A comprehensive study of the impact of dislocation loops on leakage currents in Si shallow junction devices. " Journal of Applied Physics, vol. 118, no 18, pp. 184501.(2015)

[3] Y. Wei et al. "Analysis of dark current dependent upon threading dislocations in Ge/Si heterojunction photodetectors." Microelectronics international (2012).

[4] T. Hoang, J. Holleman, P. Leminh *et al.* "Influence of dislocation loops on the near-infrared light emission from silicon diodes." IEEE transactions on electron devices, vol. 54, no 8, pp. 1860-1866.(2007).

[5] S.C. Jain, W. Schoenmaker, R. Lindsay *et al.* "Transient enhanced diffusion of boron in Si. " Journal of applied physics, vol. 91, no 11, p. 8919-8941.(2002).

[6] B. Colombeau, N.E.B Cowern, F. Cristiano, P. Calvo, N. Cherkashin, Y. Lamrani and A. Claverie. "Time evolution of the depth profile of {113} defects during transient enhanced diffusion in silicon." Applied physics letters, vol. 83, no. 10,pp. 1953-1955. (2003).

[7] N. Zographos, and I. Martin-Bragado "A Comprehensive Atomistic Kinetic Monte Carlo Model for Amorphization/Recrystallization and its Effects on Dopants" Mater. Res. Soc. Symp. Proc. vol. 1070, (2008).

[8] L. Pelaz, LA. Marqués, and J. Barbolla. "Ion-beam-induced amorphization and recrystallization in silicon." Journal of applied physics,vol. 96, no.11, pp. 5947-5976.(2004).

[9] T. C.Smith, "Wafer cooling and photoresist masking problems." Ion Implantation: Equipment and Techniques: Proceedings of the Fourth International Conference Berchtesgaden, Fed. Rep. of Germany, September 13–17, 1982. vol. 11. Springer Science and Business Media, (2012).

[10] HJ. Gossmann, N. Zographos et al. "Predictive process simulation of cryogenic implants for leading edge transistor design." AIP Conference Proceedings. vol. 1496., no. 1. American Institute of Physics, (2012).

[11] Sentaurus Process User Guide, 2019.03, Synopsys Inc.

[12] E. C. Baranova., V. M. Gusev, Y.V. Martynenko *et al.* "On Silicon Amorphisation During Different Mass Ions Implantation. " Ion implantation in semiconductors and other materials. Springer, Boston, MA, p. 59-71.(1973).

3-4 Automatic Modeling of Logic Device Performance Based on Machine Learning and Explainable AI

Seungju Kim[1], Kwangseok Lee[1], Hyeon-Kyun Noh[1], Youngkyu Shin[1], Kyu-Baik Chang[1], Jaehoon Jeong[1],
Sangwon Baek[2], Myunggil Kang[2], Keunhwi Cho[2], Dong-Won Kim[2], Daesin Kim[1]

[1]Data & Information Technology Center, [2]R&D Center / Device Solution Business
Samsung Electronics Co., Ltd., Gyeonggi-do 18448, Republic of Korea.
* Email Address: seungjju.kim@samsung.com (S. Kim).

Abstract— **In this paper, we propose a machine learning framework for predicting performances of semiconductor devices that can automatically reflect modifications in process conditions. While standard TCAD simulators require intensive modeling and calibration works to capture new process conditions, our proposed framework can learn these conditions from data efficiently and directly. Furthermore, by applying recently attention-getting explainable AI techniques, important factors that affecting device performances can be discovered automatically from the proposed model. Specifically, our model quantifies significance of each process step as the game-theoretic Shapley value, that cannot be achieved by TCAD simulators.**

Keywords— **Logic device, electrical test, machine learning, explainable AI, SHAP (Shapley Additive explanation)**

I. INTRODUCTION

In the sub-10-nm technology era, various process techniques have been introduced to achieve the optimal performance of devices [1-2]. As a result, process conditions that should be estimated have increased explosively. Traditionally, TCAD simulations have been used to predict device performances under numerous process conditions. However, TCAD simulations require time-consuming procedures such as generating device structures, converting process steps as simulation models, and calibrating various parameters. Hence, they suffer from reflecting new device scheme and various modifications in process conditions within practical time scale. To overcome such limitations, in this paper, we propose a machine learning (ML) framework that can predict device performances with automatic reflections of changes in complex process conditions. In addition, the proposed frameworks can deduce important factors that determine device performances, by combining the ML models and recently-developed explainable AI (XAI) technique [3].

Fig. 1 illustratively summarizes the overall procedure of the proposed framework. The proposed framework consists of three principled stage: the data preparation (blue one), pre-processing (orange one) and modeling (green one). In the data preparation stage, raw manufacturing data is collected daily and engineered to extract some useful information for ML model, i.e., features. Then, in the pre-processing stage, the extracted features are further refined. Finally, in the modeling stage, the ML models are (re-)trained and updated with incoming the pre-processed features. Since whole steps mentioned above progress automatically, device engineers can use the model that learns the up-to-date process condition. Details on each stage will be explained in the following Section. Table I shows major differences between traditional TCAD simulations and the proposed framework. There are several merits of the proposed framework over the TCAD: automatic data collection, improved input factors based on process step ID that inaccessible with standard TCADs, e.g., new process scheme or dependency of equipment, and significantly reduced TAT.

TABLE I. MAJOR DIFFERENCES OF TCAD AND THIS WORK

	TCAD	This work
Data collection	Fab data & device engineer	Database (automatic extraction)
X Factor	Device structures, process conditions, simulation parameters, etc.	Process step ID based engineering data (device structures, process conditions, equipment, etc.)
Y Factor	Device performances (AC/DC), drain-induced barrier lowering (DIBL), subthreshold swing (SS), threshold voltage (V_t), channel/external resistances ($R_{ch/ext}$), source/drain contact resistances ($R_{cnt-s/d}$), etc.	
Model	TCAD model	ML model (ensemble methods, e.g., random forest [10], gradient boosting [11], XGBoost [12], Catboost [13])
Tool (S/W)	Commercial TCAD simulator	SQL, Python
TAT	1~2 weeks (include data collection)	~10 min
Explanation	Physical formula	XAI

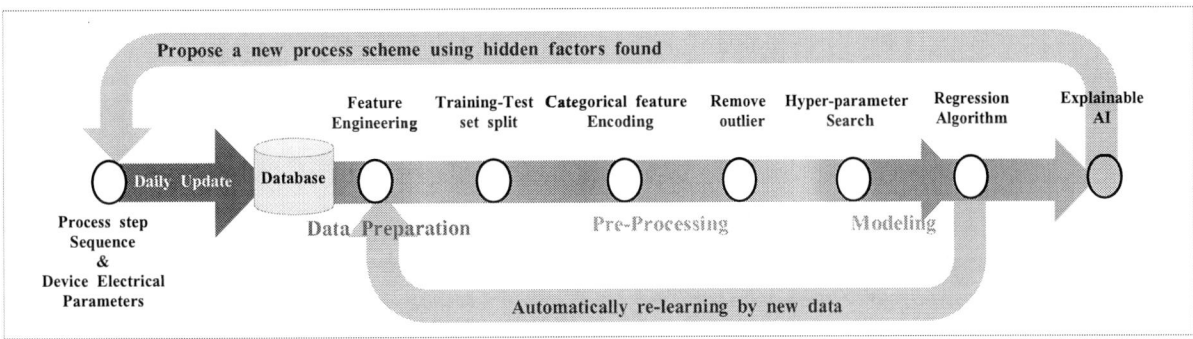

Fig. 1. Main process of automatically analyzing device performance from the process information database

II. DATA PREPARATION

A. Collecting rawdata

Millions of process progress information is generated from thousands of equipment every day. We collect information such as execution time, facility type, facility location, process type, and process conditions for each wafer in a table format and store it in a database once a day. Process conditions are mostly simplified or encrypted for security by engineers, so additional data (feature) engineering work is required to change them to a form for use in learning.

B. Feature Engineering

Logic process information contains hundreds of process progress sequences per wafer, includes various unit process factors, facility information, and structure information. One cannot directly use this raw process information for ML models because of the following reasons: (1) when hundreds of process steps are used as a feature of the model, the model becomes excessively complicated and sometimes the predictability of the model is rather degraded. (2) The process information is encoded as strings, thus it is impossible to recognize numerical correlations between features, such as temperature and thickness changes.

We solve these problems through the feature engineering based on domain knowledge. First, we select 56 of important processes and structure information that should be learned essentially. In addition, we develop a parser that detect numeric process information encoded as string then automatically convert it into numeric form. We compare the correlation matrix to check the effect of this parsing module in Fig. 2. Because the categorical as well as numerical features are exist, we use the Theil's Uncertainty [4] and Pearson's correlation coefficient for former and latter, respectively [5]. Fig. 2 clearly shows that the correlation matrix after the parsing process is more pronounced than the unprocessed one.

III. DATA PRE-PROCESSING

While the extracted features can be used in its form, they still have several problems in practical. It is mainly due to characteristics of the semiconductor process data. First, there are outliers in electrical measurement result. They distort the test model performance generally, thus should be detected and removed. Next, semiconductor wafers are grouped by lot, and process characteristics in the same lot are almost identical. In this case, dataset should be split into training and test sets appropriately to prevent the overfitting issue. Last, some data contain categorical features. To fully exploit their

information, the proper encoding method should be found. For these reasons, we conduct additional data pre-processing with sufficient consideration of such data characteristics. We use the random forest regressor [10] as a baseline model for data pre-processing in this section.

A. Outlier detection method

In this paper, we evaluate following three outlier detection methods: Tukey method [6], Z-score (standard score) method, and modified Z-score method [7]. Table II shows the equation of each outlier detection method and the corresponding thumb-rule threshold. We summarize the evaluation result in Table III. As a result of the evaluation, the IQR-based Turkey method shows stable outlier detection and removal performances.

B. Dataset splitting method

Semiconductor wafers are grouped into lots. Generally, wafers in the same lot have almost identical process conditions. On the other hand, the process conditions can vary abruptly for wafers from different lots. Thus, if the process conditions for all lots are not regularly distributed in the training set, the predictive performance of the trained model can be degraded significantly. To deal with this problem, we include one or more labels of all types of categorical features in the training set, to ensure stable model performance. In this paper, this method is called a categorical feature-wise splitting.

Fig. 3 compares the performance distributions of models trained with random and proposed splitting methods. To extract the performance distribution, we repeat the experiment 100 times. From the figure, one can find that splitting method used in our paper shows clear improvement over the random one, by checking the mean values and variances of two distributions.

TABLE II.   OUTLIER DETECTION METHODS

Outlier detection methods	Equation	Thumb-rule threshold
Tukey method	IQR(InterQuartile Range) = 3rd Quartile – 1st Quartile	Upper Bound = 3rd Quartile + (IQR * 1.5) Lower Bound = 1st Quartile – (IQR * 1.5)
Z-Score	Z-score = $(x - \mu)/\sigma$	|Z-score| > 3
Modified Z-Score	$M_i = 0.6745(x_i - \tilde{x})/MAD$ $MAD = median(\|x_i - \tilde{x}\|)$	|Modified Z-score| > 3.5

TABLE III.   COMPARISON OF OUTLIER DETECTION METHODS

Outlier detection methods	Model performance [a]		Standard deviation
	Train R^2	Test R^2	
Original data	0.435	-11.41	147.712
Z-Score	0.811	0.333	27.773
Modified Z-score	0.949	0.722	19.577
Tukey method	**0.954**	**0.786**	**19.312**

[a.] Model: Random forest regressor, Target Electrical Test: NMOS Contact Resistant

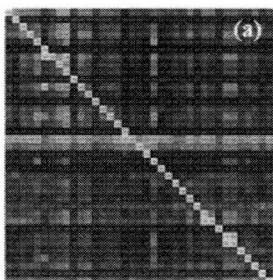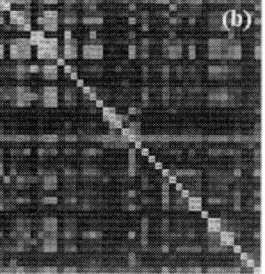

Fig. 2. Comparison of correlation matrix before and after feature engineering. (a) Before feature engineering (only string data) (b) After feature engineering (string data and numeric data)

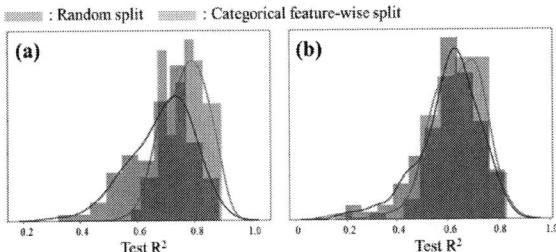

Fig. 3. Comparison of model performance (test R^2) histogram

C. Categorical data encoding

Semiconductor data often contains various categorical features such as equipment or material information. It is obvious that for such data the model performance highly depends on the used categorical encoding technique. In ML field, there are six representative encoding methods used frequently [8]. We compare these encoding methods to search the proper one for logic process data.

Table IV summarize the performance of random forest models with respect to the encoding method. We find that the target encoding with the categorical feature-wise data splitting shows the best performance across the various combinations. One notable thing is that the target encoding shows degraded performance than one-hot encoding when the dataset is randomly split. It is because the target encoding converts the categorical labels as a function of the target value y. Thus, in the random splitting, which contains unseen categorical features in the test dataset, the target encoding cannot properly work. However, since the categorical feature-wise splitting can regularly distribute the features, it does a better job. It reveals that the data splitting and categorical encoding methods should be considered simultaneously to find the optimal data preprocessing rule.

IV. MODELING AND RESULTS

It is noteworthy that our proposed models not only predict the device performance, but also deduce important factors that critically affect the device performance, with the help of the recently attention-getting XAI technique.

A. Prediction model

In this paper, we evaluate the tree-based ensemble models including random forest, gradient boosting [11], XGBoost [12], and Catboost [13]. They have various desirable properties for our target task: first, they show much reliable performance under a small amount of training data. It should be noted that the number of accessible data for semiconductor manufacturing does not exceed ~1,000 generally. In this case, complicated models such as deep neural networks show rather deteriorated performances. Second, they outperform other ML models, e.g., linear model and kernel based model, when the categorical features are included in the input [9]. Last but not least, their prediction results can be easily explained. To help the decision of device engineers, the explainability of the model, i.e., *why the model arrived at a specific prediction*, is highly preferred. The tree-based ensemble models are naturally XAI considering the interpretability of decision trees. Furthermore, some advanced techniques can be directly and efficiently applied to the tree-based ensemble models.

TABLE IV. EVALUATION RESULTS OF REGRESSION MODEL USING 6 TYPES OF ENCODING METHODS

Split type + Encoding Method	Train R^2	Test R^2
Random split + One-hot	0.791	0.724
Random split + Sum	0.852	0.603
Random split + Helmert	0.910	0.472
Random split + Target	0.831	0.640
Random split + Leave-One-Out	0.992	-0.139
Random split + Catboost	0.918	0.566
Categorical feature-wise split + One-hot	0.864	0.742
Categorical feature-wise split + target	**0.921**	**0.826**

The experimental conditions are as follows: the number of training data (wafer) and selected features (process steps and measured device structure information) are 323 and 56, respectively. For hyper-parameter tuning, we use the grid search, whose searching space are listed in Table V. We summarize the searched best model's prediction scores for AC, DC, and other electrical test (ET) parameters, in the case of n-type and p-type MOSFETs, in Table VI. It shows that ML model can achieve ~0.7 R^2 for unseen test data.

B. Explainable AI

XAI is one of the most active research area in recent ML fields. Frequently used XAI models include decision tree, a simple model-specific method, LIME (Local Interpretable Model-agnostic Explanations) [14], SHAP (Shapley Additive exPlanations) [3], PDP (Partial Dependence Plot), and LRP (Layer-wise Relevance Propagation).

In this paper, we use the Tree SHAP [15], that utilizing the standard kernel-based SHAP and tree-based models.

TABLE V. HYPER-PARAMETER SET

Hyper-parameter	Parameter set
n_estimators	[10, 50, 100, 250]
max_depth	[5, 10, 20, 40]
max_features	[10, 20, 'auto', 'log2']
min_samples_leaf	[1, 0.5, 0.3, 0.1]
loss	['ls', 'lad', 'huber', 'quantile']
subsample	[1.0, 0.8, 0.6]
learning_rate	[0.01, 0.05, 0.1, 0.2]

TABLE VI. PREDICTION PERFORMANCE OF REGRESSION MODEL

ET	NMOS		PMOS	
	Train R^2	Test R^2	Train R^2	Test R^2
AC performance	0.915	0.702	0.924	0.605
DC performance	0.941	0.758	0.956	0.803
R_{cnt-d}	0.939	0.752	0.971	0.836
R_{cnt-s}	0.943	0.793	0.968	0.741
R_{ch}	0.934	0.645	0.984	0.783
R_{ext}	0.973	0.683	0.962	0.871
DIBL	0.921	0.854	0.903	0.826
SS	0.931	0.597	0.947	0.649
V_t	0.967	0.831	0.908	0.672

Tree SHAP can calculate the Shapley value [16] of features accurately and efficiently. The Shapley value represent a responsibility of a feature, i.e., a process step, for a change in the model output, i.e., the predicted ET value. More specifically, it is equal to the difference between the predicted output with and without a certain feature, for a given data point. It should be noted that the Shapley value is a function of the feature value, not just a feature itself.

As an example, Fig. 4 (a) shows the Shapley value on ET value with varying the feature value of 10 representative process steps. For TREATMENT #1, as its feature value increases (from blue dot to red dot), its corresponding Shapley value also increases significantly (from -4 to 4). It clearly shows that the predicted ET value is highly depend on the TREATMENT #1. On the other hand, there is no significant variation in the Shapley value with respect to the feature value of ETCH #5. Based on this, one can find that the condition of TREATMENT #1 is relatively important for controlling ET value than that of ETCH #5. Fig. 4 (b) sorts process steps according to the averaged absolute values of the Shapley value on ET value. It plays a similar role with the feature importance, but shows game-theoretically meaningful rank than conventional impurity-based one. We find that the importance ranking derived through the Shapley value matches well with the domain knowledge of device engineers.

Furthermore, with the Shapley value, one can find the optimal process condition that cannot searched well due to its non-linear nature. For example, Fig. 5(a) shows the effect of a certain structure width on Δ Shapley value. It clearly shows that this structure condition shows non-linear effects on ET value. For this process, human engineers predicted that the

ET value would decrease as the structure width increased from 12 to 14, assuming it would have linear relationship. However, using the Shapley value plot, they could find optimal ET value could be achieved around 13. Similar analysis can be performed in the case of the categorical feature, as shown in Fig. 5 (b). It shows among 9 different (categorical) conditions for certain etch process, #5, #6, #7, and #8 conditions and the others show respectively positive and negative effects on ET value.

V. CONCLUSION

In this paper, we propose the automatic performance modeling framework based on feature engineering, ML and XAI techniques. Our ML model utilizing information from process step ID can achieve prediction score ~0.7 test R^2. Furthermore, by applying the tree SHAP to ML model, hidden key factors that have significant impact on performances can be derived easily.

REFERENCES

[1] Ha, Daewon, et al. "Highly manufacturable 7nm FinFET technology featuring EUV lithography for low power and high performance applications." 2017 Symposium on VLSI Technology. IEEE, 2017.

[2] Bae, Geumjong, et al. "3nm GAA technology featuring multi-bridge-channel FET for low power and high performance applications." 2018 IEEE International Electron Devices Meeting (IEDM). IEEE, 2018..

[3] Lundberg, Scott M., and Su-In Lee. "A unified approach to interpreting model predictions." Advances in neural information processing systems. 2017.

[4] Press, William H., et al. Numerical recipes in Fortran 77: volume 1, volume 1 of Fortran numerical recipes: the art of scientific computing. Cambridge university press, 1992.

[5] Pearson, Karl. "VII. Note on regression and inheritance in the case of two parents." proceedings of the royal society of London 58.347-352 (1895): 240-242.

[6] Hoaglin, David C., Boris Iglewicz, and John W. Tukey. "Performance of some resistant rules for outlier labeling." Journal of the American Statistical Association 81.396 (1986): 991-999.

[7] Rousseeuw, Peter J., and Mia Hubert. "Robust statistics for outlier detection." Wiley Interdisciplinary Reviews: Data Mining and Knowledge Discovery 1.1 (2011): 73-79.

[8] McGinnis, William D., et al. "Category encoders: a scikit-learn-contrib package of transformers for encoding categorical data." Journal of Open Source Software 3.21 (2018): 501.

[9] Razi, Muhammad A., and Kuriakose Athappilly. "A comparative predictive analysis of neural networks (NNs), nonlinear regression and classification and regression tree (CART) models." Expert Systems with Applications 29.1 (2005): 65-74.

[10] Liaw, Andy, and Matthew Wiener. "Classification and regression by randomForest." R news 2.3 (2002): 18-22.

[11] Friedman, Jerome H. "Greedy function approximation: a gradient boosting machine." Annals of statistics (2001): 1189-1232.

[12] Chen, Tianqi, and Carlos Guestrin. "Xgboost: A scalable tree boosting system." Proceedings of the 22nd acm sigkdd international conference on knowledge discovery and data mining. 2016.

[13] Prokhorenkova, Liudmila, et al. "CatBoost: unbiased boosting with categorical features." Advances in neural information processing systems. 2018.

[14] Ribeiro, Marco Tulio, Sameer Singh, and Carlos Guestrin. "'Why should I trust you?' Explaining the predictions of any classifier." Proceedings of the 22nd ACM SIGKDD international conference on knowledge discovery and data mining. 2016.

[15] Lundberg, Scott M., Gabriel G. Erion, and Su-In Lee. "Consistent individualized feature attribution for tree ensembles." arXiv preprint arXiv:1802.03888 (2018).

[16] Shapley, Lloyd S. "A value for n-person games." Contributions to the Theory of Games 2.28 (1953): 307-317.

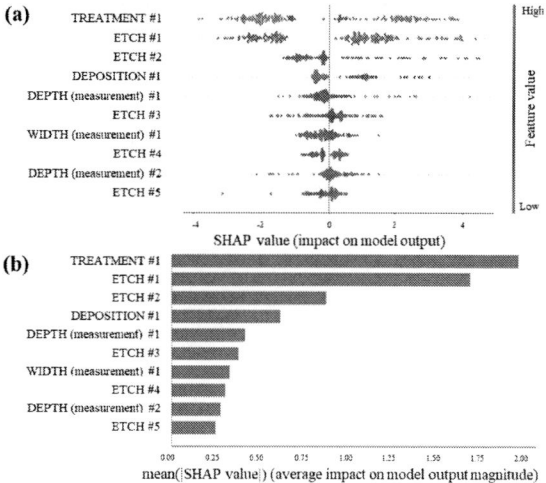

Fig. 4. Rank of process steps and structure measurement values calculated by SHAP value

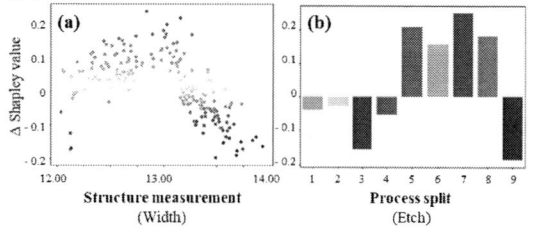

Fig. 5. Analysis of the impact on the ET according to the value of the process step and structure measurement information

Gummel-cycle Algebraic Multigrid Preconditioning for Large-scale Device Simulations

Hiroo Koshimoto*, Hisashi Ishimabushi†, Jaehyun Yoo†,
Yasuyuki Kayama*, Satoru Yamada*, Uihui Kwon† and Dae Sin Kim†

Samsung R&D Institute Japan (SRJ)
Yokohama 230-0027, Japan
†*Data and Information Technology Center, Samsung Electronics Co., Ltd.*
Gyeonggi-do 18448, Korea

Abstract—It has been proven that the multigrid method is promising on large-scale scientific simulations. However there still remains some difficulties on applying the multigrid method, which is the system of systems such as FEM on stress analysis or coupled PDEs. Above all, the drift-diffusion model widely used in the device modeling is a typical case belonging to the problems. Because the model has a tight coupling between the electrostatic field and the carrier movements and this property prevents the multigrid method from working effectively. In this paper, we propose a technique to apply the multigrid method to the drift-diffusion model. The technique consists of reflection process between systems coupled in the equation. Consequently the technique helps to solve large-scale device simulations. We show the case of power devices.

Index Terms—power device, device simulation, drift–diffusion, multigrid, large-scale

I. INTRODUCTION

The drift–diffusion (DD) model has been widely used in the device modeling and is still the fundamental tool on devices not focusing on downsizing such as power devices. And many methods have been developed to solve the DD model. However, turning to power devices, the difficulty of large-scale device structures is remaining. In particular, it is difficult for 2D simulations to predicate carrier behaviors caused by dynamic avalanche capability or electrostatic discharge. In such a situation, 3D simulations with large-scale device structures are required recently.

On power devices high biased electric fields shift carries a lot, so a fine discrete mesh is required to find a good approximation of carrier profiles, and besides the coefficient matrices in simulations tend to be ill conditioned and need a lot of fill-ins for robust and fast convergence. Consequently, it is crucial how to handle large memory requirements for power devices. There are two known approacehes, one is the domain decomposition method [5] [6], and another is the multigrid method [2].

We focus on the multigrid method and propose a technique to apply the method to the Newton method on device simulations. The Newton method on device simulations treats the coupled DD model and the strong coupling factors make the equations ill conditioned. The conventional multigrid methods can not handle this difficulties arising from a system of

systems like FEM on stress analysis. The Gummel iteration [1] plays an important role in our technique.

II. DRIFT–DIFFUSION MODEL

The DD model in the steady analysis is:

$$\begin{cases} E(\psi,n,p) := -\nabla \cdot (\varepsilon \nabla \psi) - q\,(p - n + C) = 0 \\ F(\psi,n,p) := -\nabla \cdot j_n + qR = 0 \\ G(\psi,n,p) := \nabla \cdot j_p + qR = 0 \end{cases}$$

where ψ is the electrostatic potential, n and p are the electron density and the hole density, ε is the dielectric permittivity depending on materials, C is the doping concentration, R is the total recombination rate and q is the elementary charge. $j_{n,p}$ are the carrier current densities as following:

$$\begin{cases} j_n = q\,(D_n \nabla n - \mu_n n \nabla \psi) \\ j_p = -q\,(D_p \nabla p + \mu_p p \nabla \psi) \end{cases}$$

Here $\mu_{n,p}$ are the carrier mobilities and $D_{n,p}$ are the carrier diffusivities. R and $\mu_{n,p}$ are nonnlinear models. Besides we apply the Scharfetter–Gummel scheme to discretize F, G. Thus they consist of nonlinear PDEs. Therefore we employ the Newton method to find ψ, n, p defined on a given discrete mesh Ω.

As disscussed in O. Schenk *et al.* [3], solving the DD model by the Newton method introduces highly ill-conditioned and significantly more demanding unsymmetric sparse matrices. We describe its coefficient matrix J and the vectrized function values f formaly here:

$$J = \begin{pmatrix} E_\psi & E_n & E_p \\ F_\psi & F_n & F_p \\ G_\psi & G_n & G_p \end{pmatrix}, f(\psi,n,p) = \begin{pmatrix} E(\psi,n,p) \\ F(\psi,n,p) \\ G(\psi,n,p) \end{pmatrix}$$

$E_\square, F_\square$ and $G_\square$ mean the derivatives of each equation of the variable $\square$, for instance $E_\psi = \frac{\partial E}{\partial \psi}$. The linear equation $J\delta = -f(\psi,n,p)$ arises to update nonlinear iterative solutions in the Newton method. The direct methods could solve these ill-conditioned matrices robustly. However its huge memory consumption is an obstacle for large-scale device simulations. So far, many practical device simulations use the sparse iterative solvers with preconditioners based on the incomplete LU-factorization. Unfortunately these preconditioners need pretty much fill-ins to solve ill-conditioned matrices. The preconditioned iterative solvers also have the large-scale difficulty.

In scientific simulations dominated by diffusion or the Poisson equation, the multigrid method [4] could overcome the difficulty of solving large-scale matrices without computing fill-ins. Instead of computing fill-ins, the multigrid method computes errors defined on a mesh coarsen from the original (fine) mesh to retrieve the factors converging slowly, called smooth errors. However, the matrix J is composed of 3 different systems and it is hard to construct a coarse mesh corresponding to smooth errors. We propose a technique similar to the Gummel iteration to overcome this difficulty.

III. GUMMEL-CYCLE ALGEBRAIC MULTIGRID METHOD

We employ the GMRES method to solve the linear equation $J\delta = -f(\psi, n, p)$ and use the algebraic multigrid method as its preconditioning. For simplicity, we denote the equation as $Ax = b$. The preconditioning K is applied with to compute $w_{k+1} = AK^{-1}v_k$ where w_{k+1} is a new vector of the Hessenberg matrix in the k-iteration and v_k is the computed vectors of the Hessenber matrix. We describe the procedure of applying the algebraic multigrid method in a way limited to preconditioning. So we rewrite $Kw_{k+1} = v_k$ as $Ax = b$ after this.

A. Algebraic Multigrid Method (AMG)

A typical algebraic multigrid iteration solving $Ax = b$ is: (1) Relax a iterative solution x_k^h on the fine mesh Ω^h, (2) Restrict a residual vector $r_k^h = b - Ax_k^h$ to $r_k^H = Rr_k^h$ defined on the coarse mesh Ω^H, (3) Relax an error e_k^H on Ω^H by solving $A^H e_k^H = r_k^H$, (4) Prolongate the error to $e_k^h = Pe_k^H$ defined on the fine mesh Ω^h, (5) Update the iterative solution by $x_{k+1}^h = x_k^h + e_k^h$. Here R is a restriction map: $\Omega^h \to \Omega^H$ and P is a prolongation map: $\Omega^H \to \Omega^h$. For better convergence we expect e_k^H would be an approximation of the smooth error defined on the coarse mesh Ω^H. However it is difficult for systems equation like the DD model because we do not know how to coarse coupled multiple variables.

B. Co-cycle

The complemental procedure: co-cycle is introduced. We suppose that we solve a system composed of systems $Mx = b$:

$$M = \begin{pmatrix} A & A_B \\ B_A & B \end{pmatrix}, b = \begin{pmatrix} b_A \\ b_B \end{pmatrix}$$

Here, the sub systems A and B are defined on the same mesh Ω^h, however, on applying the algebraic multigrid methods, they have two different coarse meshes: Ω_A^H and Ω_B^H. The co-cycle $A^H \Rightarrow B^H$ resolves this discrepancy by following steps(Fig. 1): (1) Prolongate the error e_A^H accompanied with A^H to $e_A^h = P_A e_A^H$ on the fine mesh Ω^h, (2) Map the error e_A^h to the right hand side of the B system, $r_B^h = b_B^h - B_A e_A^h$, (3) Restrict the residual r_B^h to $r_B^H = R_B r_B^h$ on the coarse mesh Ω_B^H, (4) Relax the error e_B^H on Ω_B^H by solving $B^H e_B^H = r_B^H$. On deeper coarse levels than 2, prolongation and restriction steps are applied recursively until an error defined on each coarse mesh could be mapped by the off diagonal sub matrix A_B or B_A. This co-cycle is associative, and we could get

the cyclic errors by repeated applications. For instance, the recursive error e_A^H could be given by $A^H \Rightarrow B^H \Rightarrow A^H$. The co-cycle could make convergence of the algebraic multigrid method stabilized however it is a time consuming process because it requires several matrix–vector multiplications depend on the depth. We show the trade off later.

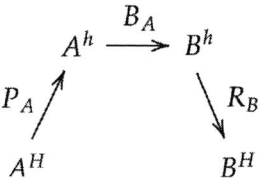

Fig. 1. Multigrid Co-cycle

C. Gummel-cycle AMG

Our approach is, defining an independent coarse mesh $\Omega_\square^H$ for each variable ψ, n, p respectively, and employing the co-cycle to get self-consistent smooth errors. The former trick enables the multigrid method to work well for each system: $E_\psi x_\psi = b_\psi, F_n x_n = b_n$ and $G_p x_p = b_p$. The latter trick recovers the components of smooth errors missed by each system. The overview of the algorithm is Algorithm1.

Algorithm 1 Gummel cycle Algebraic Multigrid

Input: A, b

Output: x

    Initialisation :

1: Build coarsen coefficient submatrices: E_ψ^H, F_n^H, G_p^H

2: Prepare retrictions $R_\square$, prolongations $P_\square$ for ψ, n, p
    LOOP until : $\frac{\|Ax-b\|}{\|b\|} \le \varepsilon$

3:   *Pre-smoothing*: Relax a fine error e_ψ^h

4:   Restrict a residual $r_\psi^H = R_\psi \left(b_\psi^h - E_\psi^h e_\psi^h \right)$

5:   *Pre-smoothing*: Relax a coarse error e_ψ^H

6:   **for** $i = 1$ to N **do**

7:     Resolve the co-cycle: $E_\psi^H \Rightarrow F_n^H \Rightarrow E_\psi^H$

8:     Resolve the co-cycle: $E_\psi^H \Rightarrow G_p^H \Rightarrow E_\psi^H$

9:   **end for**

10:   *Post-smoothing*: Relax a coarse error e_ψ^H

11:   Prolongate and update an error $e_\psi^h \leftarrow e_\psi^h + P_\psi e_\psi^H$

12:   *Post-smoothing*: Relax a fine error e_ψ^h

13:   **for** $i = 1$ to N **do**

14:     Resolve the co-cycle: $E_\psi^h \Rightarrow F_n^h \Rightarrow E_\psi^h$

15:     Resolve the co-cycle: $E_\psi^h \Rightarrow G_p^h \Rightarrow E_\psi^h$

16: **end for**

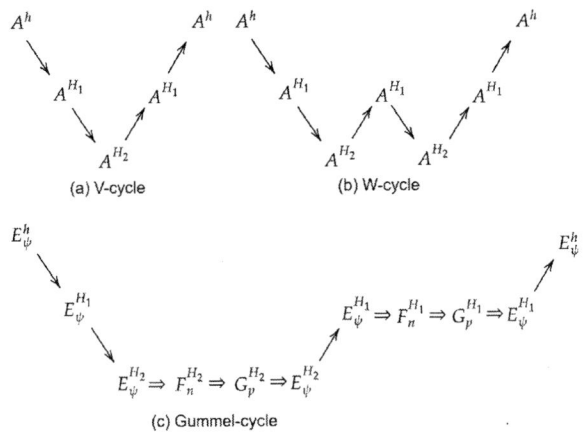

Fig. 2. Multigrid cycles: V-cycle, W-cycle and Gummel-cycle

We call the intermediate cycles: $E_\psi \Rightarrow F_n \Rightarrow E_\psi$ and $E_\psi \Rightarrow G_p \Rightarrow E_\psi$ the Gummel-cycle because this process traces the traditional Gummel iteration [1]. The conventional multigrid methods employ V-cycle or W-cycle to reduce errors but our Gummel-cycle is used to resolve the discrepancy between each system: E, F and G (Fig. 2).

IV. EVALUATION

A. Trade-off

As mentioned before, the co-cycle could make convergence of the multigrid method stabilized, but on the other hand, it is a time consuming process consisting of several matrix–vector multiplications. There is a trade off between the number of co-cycles per. a multigrid cycle and the number of iterations required to converge (FIG. 3). To reduce the time of co-cycle, pre-calculated matrices could be provided. However, generating such a matrix requires sparse matrix–matrix multiplications, which are also time consuming. Besides the memory consumption for the pre-calculated matrices is not small. So this issue is still opened.

B. Scaling

To evaluate capability of the algebraic multigrid method with the Gummel-cycle, the strong and weak scaling benchmarks are performed on the matrices generated by the terminal structure of the power device (FIG. 4) under the near breakdown voltage (FIG. 5.) In the both benchmarks, the GMRES method is used with preconditioning. For comparison, the ILUT preconditioning is chosen as the conventional one. As shown in the strong scaling benchmark (Fig. 6), the Gummel-cycle algebraic multigrid method shows almost linear scaling property. Even though the cost of the co-cycle is not cheap, having enough parallel units could defeat the conventional methods. This trend is remarkable on the weak scaling benchmark which loads a constant task per a same compute unit. As using a finer mesh (ie. increasing a number of nodes), the total numbers of fill-ins for solving by the conventional methods are getting larger. Some 15,360 variables are filled for a cocmpute unit and the flat MPI is used for parallelization. As shown in FIG. 7, our method exceeds the other conventional methods at the larger cases.

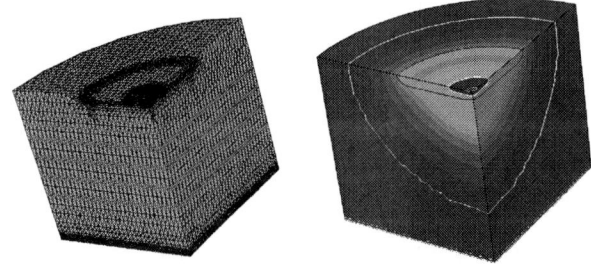

Fig. 4. Device structure for bench- Fig. 5. Electric Field under the near
mark   breakdown

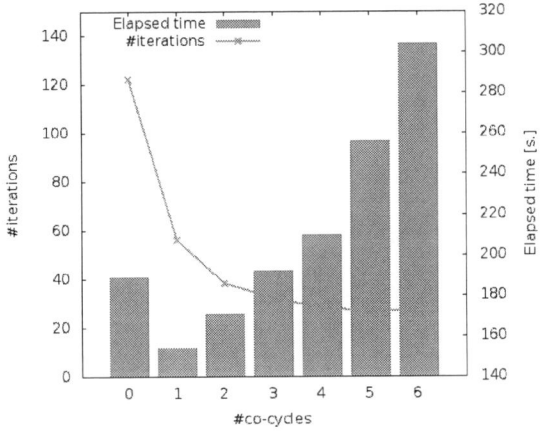

Fig. 3. Trade off btw. #of co-cycles and #of iterations

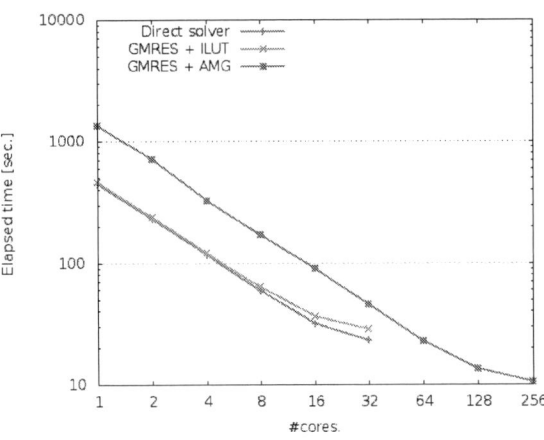

Fig. 6. Comparison on strong scaling with each method

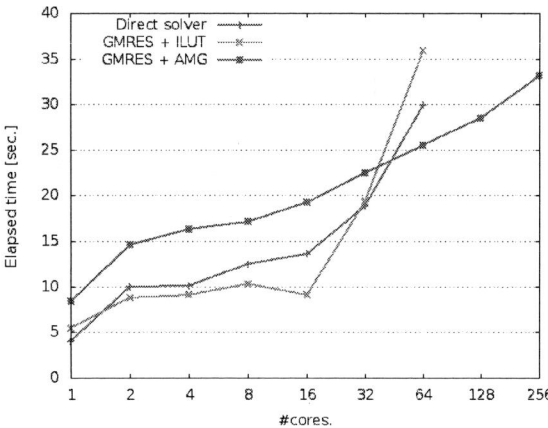

Fig. 7. Comparison on weak scaling with each method

V. CONCULUSION

In this work, we have proposed a new preconditioning technique based on the algebraic multigrid with Gummel-cycle, which can scale up without having fill-ins. The proposed technique will also meet a situation such as electric fields spread into the bulk owing to the algebraic multigrid manner. And so this approach helps simulations of deep depletion layers or wide bandgap devices like GaN, SiC and Ga2O3. However, there requires a cafeful treatment on applying our method because it has the trade off between its stability and time consumption. The optimal balance of the number of co-cycles or an appropriate relaxation approach should be introduced in future.

REFERENCES

[1] Gummel, Hermann K., "A self-consistent iterative scheme for one-dimensional state transistor calculations," IEEE Transactions on electron devices, vol. 11, no. 10, pp. 455–465, 1964.

[2] M. Brezina, R. Falgout, S. MacLachlan, T. Manteuffel, S. McCormick, and J. Ruge, "Adaptive smoothed aggregation (α sa) multigrid," SIAM review, vol. 47, no. 2, pp. 317–346, 2005.

[3] O. Schenk, S.Rollin, and A. Gupta, "The effects of unsymmetric matrix permutations and scalings in semiconductor device and circuit simulation," IEEE Transactions on Computer-Aided Design of Integrated Circuits and Systems, vol. 23, no. 3, pp. 400–411, 2004.

[4] P. T. Lin, J. N. Shadid, R. S. Tuminaro and M. Sala, "Performance of a Petrov–Galerkin algebraic multilevel preconditioner for finite element modeling of the semiconductor device drift–diffusion equations," International journal for numerical methods in engineering, vol. 74, no. 4, pp. 448–469, 2010.

[5] A. Toselli and O. Widlund "Domain decomposition methods - algorithms and thery," Springer Science & Business Media, vol. 34, 2006.

[6] S. Sho and S. Odanaka, "A hybrid MPI/OpenMP parallelization method for a quantum drift-diffusion model," International Conference on Simulation of Semiconductor Process and Devices (SISPAD), pp. 33-36, 2017.

3-6 A continuous cellular automaton method with flux interpolation for two-dimensional electron gas electron transport analysis

Koichi Fukuda
AIST
Tsukuba, Japan
https://orcid.org/0000-0002-3148-6010

Junichi Hattori
AIST
Tsukuba, Japan

Hidehiro Asai
AIST
Tsukuba, Japan

Junya Yaita
Fujitsu Limited
Atsugi, Japan

Junji Kotani
Fujitsu Limited
Atsugi, Japan

Abstract— **Due to the innovation of microwave communication using GaN-based HEMT, further improvement of HEMT device performance is expected. Prediction of transport properties of 2D electron gas is indispensable for designing HEMT devices. Since electron energy becomes high in HEMT channel because of its high electric field, a simulation method which covers the effects of band nonparabolicity, subband, and upper valley is required. By combining the Poisson-Schrodinger solver with the continuous cellular automaton method, a new simulation method is realized which stably obtains the electron distribution function over a wide range including the high-energy tail. It is reported that selfconsistent simulation is realized for the case where electron concentration redistribution by intersubband transitions affects subband energies through the Poisson-Schrodinger method.**

Keywords—GaN, HEMT, Poisson-Schrodinger, Cellular Automaton, device simulation, two-dimensional electron gas

I. INTRODUCTION

High electron mobility transistor (HEMT) is widely used as a high-power and high-speed device for microwave communication and is expected to be improved to higher performance [1][2]. Due to the innovation of microwave communication using GaN-based HEMT [3], further improvement of HEMT device performance is expected. The electron transport property of the two-dimensional electron gas formed by the energy barrier due to the heterolayer structure is the key to the high performance of HEMT. Therefore, it is important to predict the electron transport characterization in the two-dimensional electron gas 2DEG. Since the carrier energy increases due to the high acceleration electric field in the channel, the nonparabolicity of the energy band and the upper valley effect cannot be ignored.

A combination of the Poisson-Schrodinger method [4][5] and the Monte Carlo method [6][7] is often used to predict the electron transport properties of 2DEG [8][9]10]. In such approaches, the energy level and electron concentration profile of the subband are obtained by the Poisson-Schrodinger method, and the electron transport in the subband is obtained by the Monte Carlo method [11][12]. However, it is difficult for the method to consider the redistribution of electrons between subband and between valleys stably. Distribution function of electron over a wide energy range changes over many orders of magnitude cannot be efficiently obtained by the Monte Carlo method.

We propose the continuous cellular automaton method with the flux interpolation method [13] for this problem. This method calculates a continuous distribution function for orders of magnitude by introducing distribution function tables. The method reveals how the valley and the subband electron distribution changes over low to high electric fields and how the resulting carrier redistribution affects the subband energies in turn. The method clarified how the redistribution of carrier concentration affects the subband of HEMT. In the following chapter, the simulation method is de-scribed, and obtained 2DEG physics is discussed for a simple HEMT channel case.

II. SIMULATION METHOD

As shown in Fig. 1, the proposed method uses the Poisson-Schrodinger method to determine the subband and electron concentration distribution for the quantum confinement direction as z. Numerical tables for a two-dimensional wave vector space in the x and y directions are used for the distribution function representation for each subband. As an initial condition, the total amount of electron in each subband is obtained by the Poisson-Schrodinger method, and the energy distribution is set as thermal equilibrium.

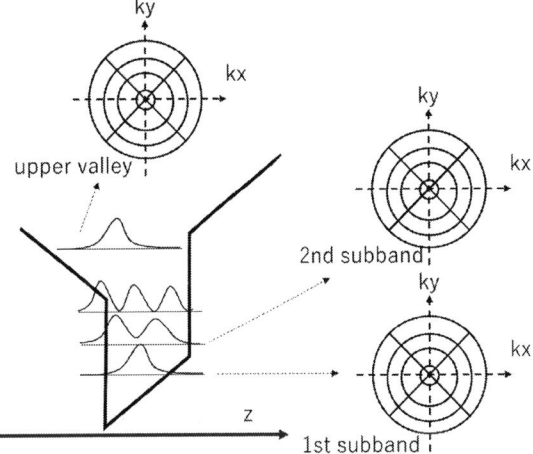

Fig. 1. Concepts of the proposed method. Distribution function tables for 2D momentum spaces are prepared for each subband and each valley obtained by Poisson-Schrodinger method.

As shown in Fig. 2, the cellular automaton method [13] is applied to update the distribution function at each time step. Since electrons are exchanged between valleys and subbands due to inter-valley and inter-subband scattering, the amount of carriers in each subband changes. Since electron depth profiles are different between each subband, carrier exchanges between subbands cause electron profile redistribution. This affects the potential distribution through the Poisson equation, and it is necessary to solve the Poisson-Schrodinger method self-consistently. Because of the stable carrier distribution function obtained by the cellular automaton method, such effects can be accurately captured.

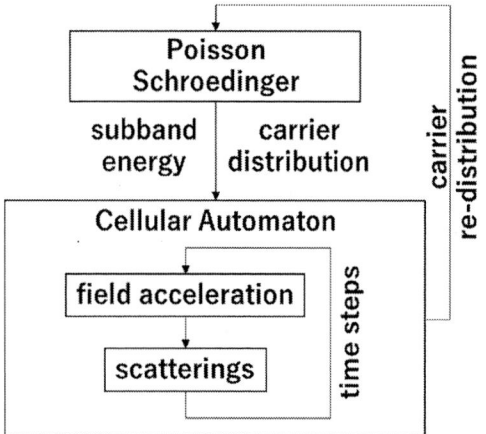

Fig. 2. Calculation flow of the proposed method. The distribution function tables are updated for each time step, according to the acceleration by the electric field and the scattering of various mechanisms. The distribution function updated by the cellular automaton method is used to update the carrier profiles, and is feed backed to Poisson-Schrodinger solver to consider the carrier redistribution effects.

Since the distribution function has wide orders of magnitude, a fine k-space mesh is required to suppress carrier pseudo-diffusion in the momentum space due to the large difference of values of the adjacent k-meshes when calculating fluxes at each time step. Fine k-meshes are not desirable because the calculation time of the method is significantly increased. For the k-space distribution function tables, logarithmic interpolation is assumed to calculate the flux which suppresses the pseudo diffusion problem with a reasonable number of k-meshes as conceptually explained in Fig. 3. The details are described in [13], which deals with the silicon as the bulk semiconductor case.

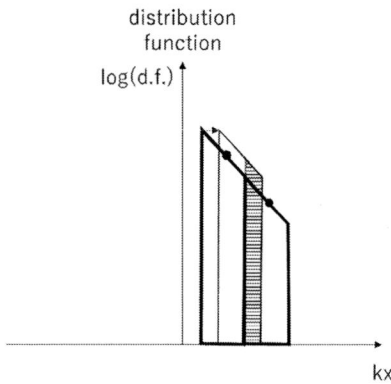

Fig. 3. One dimensional concept of the interpolation in the distribution function cell table. Each table covers a certain wave vector range shown by thick line box represented by a center point shown by black dot. When the representation point moved to the arrow direction, the distribution function flux is calculated as the hatched trapezoid. In the present method this scheme is applied to the 2D wave vector space.

For physical models of GaN-based HEMTs, electron scattering models from [14][15] and the model parameters from [16][17] are implemented into the present simulator. The heavy part of the simulation code is the cellular automaton loop and the loop is parallelized for distribution function tables by the OpenMP technology. For the top and the bottom boundary condition, a simple Schottky model is applied. The strain related interface charge is considered in the Poisson equation, which determine the electron charge area density for the equilibrium condition.

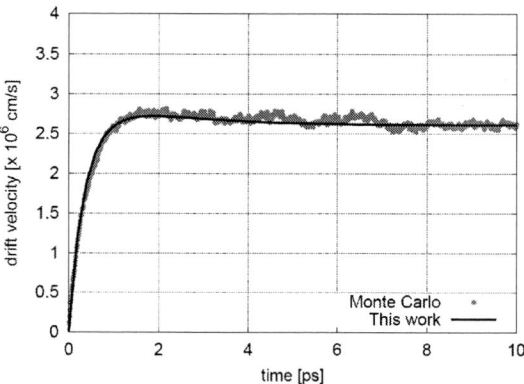

Fig. 4. Time dependent electron drift velocity after applying acceleration electric field of 1 kV/cm calculated by the present approach compared with the Monte Carlo results. In the Monte Carlo, 100,000 particles are used which is still not enough to reduce the fluctuation of the drift velocity.

Fig. 4 shows the transient electron drift velocity behavior after applying the low acceleration electric field of 1 kV/cm and is compared to the Monte Carlo method result using the same physical model. Assumed HEMT structure is AlN / GaN / AlN with 10 nm thickness for all three layers. Although for the Monte Carlo calculation, 100000 particles are used, the Monte Carlo results suffers such small fluctuations in the results since the random numbers are used to select scattering mechanisms. In the cellular automaton method, the amount of distribution function in each cell is divided and delivered to the final scattering condition in proportional to the each scattering mechanisms as schematically shown in Fig. 5.

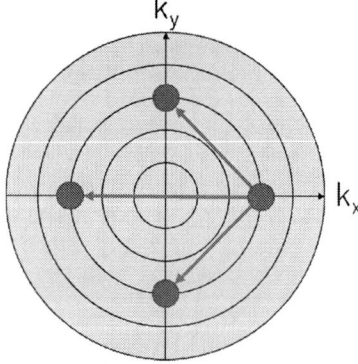

Fig. 5. The concept of the distribution function redistribution for each time step. The amount of the redistribution is calculated in proportional to the each scattering probability. This is the intra-subband and elastic scattering case, but same for inter-valley, inter-subband cases and for inelastic cases.

The final distribution function obtained for the same condition of Fig.4 is shown in Fig. 6. Thus, the distribution function obtained smoothly for wide, several tens orders of magnitudes by the present method, which enables seamless analysis including high energy subbands and upper valleys. This feature plays an indispensable role especially for the high power III-V HEMTs with high voltage or hot carrier conditions.

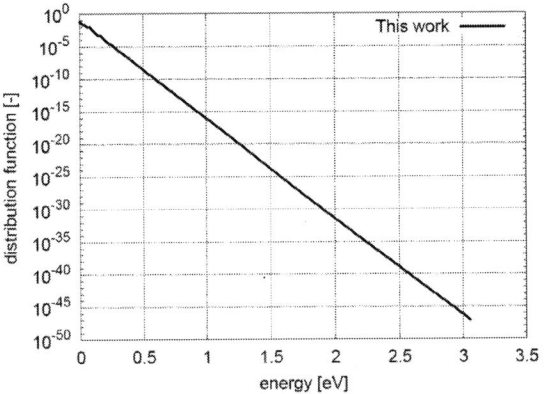

Fig. 6. The final distribution function obtained for the same low field condition of Fig. 4 calculation. The distribution function over wide orders of magnitude is smoothly obtained which enables seamless wide energy range analyses.

III. SIMULATION RESULTS AND DISCUSSIONS

Fig. 7 shows the simple HEMT channel structure assumed as a test case for the proposed method. Both sides of the GaN channel layer are sandwiched by the barrier AlN layers. The z-axis in the figure is in the substrate depth direction which is the carrier confinement direction. The x-axis is assumed to be the horizontal direction to which the electron accelerating electric field is applied. When such an HEMT structure is analyzed, it is common to incorporate the interface charge due to the strain between different material layers caused by the lattice mismatch. The values of the interface charge assumed in this work are shown in the figure, which are rather high value corresponding to the large mismatch between pure AlN and GaN. It is worth mentioning that the program supports multiple layer structures in ordered to be used as the HEMT design tool.

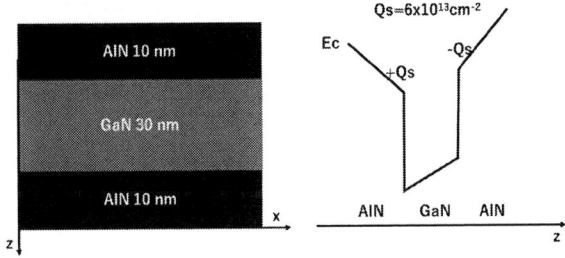

Fig. 7. Assumed device structure of GaN-based HEMT with strain induced charge assumed at semiconductor interfaces.

In GaN HEMT power applications, the acceleration electric field is large, so the carrier energy becomes high and it is necessary to consider the nonparabolicity of the electron energy band. The nonparabolicity affects not only the carrier transport but also the energy subband levels, and both are considered in the present simulation. Fig. 8 shows the band energy and the electron profiles for with and without the nonparabolicity consideration. There are clear differences in the subband energy levels and electron concentration

distributions between the nonparabolic (red) and the parabolic energy band (blue). The nonparabolicity effectively increase the effective mass of higher energy subbands, which reduces the energy of the higher subband energy levels. As a results the carrier concentration is confined stronger by considering the nonparabolicity as shown in Fig. 8.

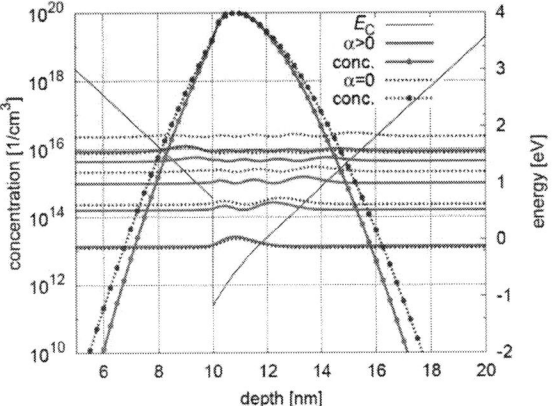

Fig. 8. Difference of subband (with marks) and electron concentration (lines) with (red) and without (blue) the nonparabolicity α consideration. The effective mass becomes heavier in higher energy, and the higher subbands energies reduced. This causes the stronger confinement which is considered by the present simulator.

Such carrier redistribution is also caused by applying large electric field. With high field, the electron energy becomes higher and the role of upper subbands or even upper valleys becomes significant. Electron depth profile differs depending on the subband levels which are determined by the confinement law of the Schrodinger equation. This means that the carrier profile is changed by applied acceleration fields. Fig. 9 shows the change of electron profile in the case of the large driving electric field of 500 kV/cm. Electron redistribution from the first to the upper subbands changes the electron profile wider, and the subband energies are affected by the redistribution in turn. It should be mentioned that these effects depend on the HEMT structures and band parameters.

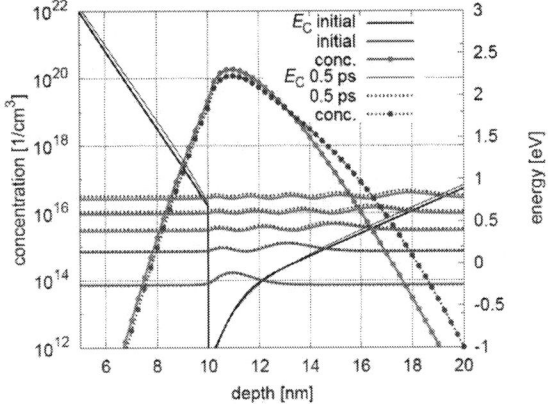

Fig. 9. Changes of subband energies and electron profile in the case of applied horizontal (x-direction) field of 500 kV/cm. Electron redistribution from the first to the upper subbands changes the electron profile with affects the subband energies in turn.

Fig. 10 shows the electron velocity overshoot characteristics for the condition of Fig. 9. Comparing to the overshoot in the low electric field in Fig. 4, the overshoot under strong field is clearer and its time scale is shorter, which is often observed in the Monte Carlo analyses. It should be mentioned that the result contains the carrier redistribution effects especially after the overshoot. This overshoot result ensures that the obtained results incorporate the nonequilibrium effects of the Boltzmann transport equation as obtained by Monte Carlo method. The proposed method has additional merits in obtaining the distribution function stably for wide orders of magnitude regardless of the analysis conditions. This feature of the present method ensures the method to be a strong tool to design HEMTs, including the hot carrier effects important for power devices. Finally, it should be commented that the present approach incorporates the scattering mechanisms in proportional to their probabilities, and therefore does not cover the random nature of the carrier scattering which causes the noises by a finite number of electrons.

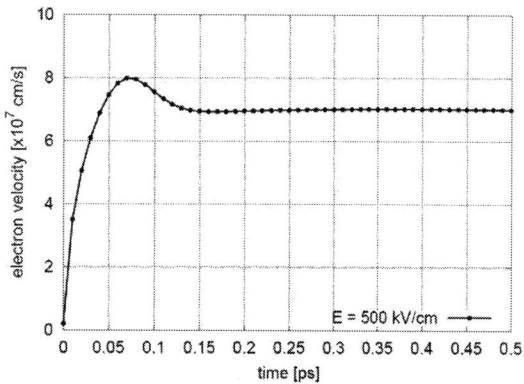

Fig. 10. Electron velocity overshoot characteristics for the same electric field condition of Fig.9. No comparison with the Monte carlo, because of the lack of the corresponding Monte Carlo code, coupled with Poisson-Schrodinger, with subbands and upper valley considerations. The overshoot is clearer than the low field condition in Fig. 4 and its time scale is shorter which are commonly observed in general Monte Carlo results.

IV. CONCLUSIONS

The continuous cellular automaton method coupled with the Poisson-Schrodinger method is proposed for the analysis of two-dimensional electron gas transport under high electric field. Since the carrier distribution function over a wide range is obtained stably and seamlessly, the method enables to investigate the effect of upper subbands and upper valleys on the electron transport, including the effect of carrier redistributions. It was confirmed through the test case of AlN-GaN HEMT structure, that the present method is a strong tool for carrier transport analysis of two-dimensional electron gas for wide range of electric field such as in HEMTs for power applications. In addition, the method is also a strong tool to design and optimize the HEMT layer structures based on the fundamental physics.

Acknowledgment

This work was partially supported by Innovative Sci-ence and Technology Initiative for Security Grant Number JPJ004596, ATLA, Japan.

REFERENCES

[1] T. Mimura, "The early history of the high electron mobility transistor (HEMT)," IEEE Trans. Microw. Theory Tech., vol. 50, pp. 780-782, 2002.

[2] P. D. Ye, B. Yang, K. K. Ng, J. Bude, G. D. Wilk, S. Halder, and J. C. M. Hwang, "GaN metal-oxide-semiconductor high-electron-mobility-transistor with atomic layer deposited Al2O3 as gate dielectric," Appl. Phys. Lett., vol. 86, no. 6, p. 063501, 2005.

[3] T. Ohki, A. Yamada, Y. Minoura, K. Makiyama, J. Kotani, S. Ozaki, and N. Nakamura, "An over 20-W/mm S-band InAlGaN/GaN HEMT with SiC/diamond-bonded heat spreader," IEEE Electron Device Lett., vol. 40, no. 2, pp. 287-290, 2019.

[4] T. Ando, A. B. Fowler, and F. Stern, "Electronic properties of two-dimensional systems," Rev. Mod. Phys., vol. 54, no. 2, pp. 437-672, 1982.

[5] C. Jacoboni and L. Reggiani, "The Monte Carlo method for the solution of charge transport in semiconductors with applications to covalent materials," Rev. Mod. Phys., vol. 55, no. 3, pp. 645-705, 1983.

[6] W. Fawcett, A. D. Boardman, and S. Swain, "Monte Carlo determination of electron transport properties in gallium arsenide," J. Phys. Chem. Solids, vol. 31, no. 9, pp. 1963-1990, 1970.

[7] S. E. Laux and F. Stern, "Electron states in narrow gate-induced channels in Si," Appl. Phys. Lett., vol. 49, no. 2, pp. 91-93, 1986.

[8] P. Gorre, R. Vignesh, R. Arya, and S. Kumar, "A review of mm-wave power amplifiers for net-generation 5G communication," Soft Computing: Theories and

[9] M.V. Fischetti and S. E. Laux, "Monte Carlo study of electron transport in silicon inversion layers," Phys. Rev. B, vol. 48, no. 4, pp. 2244-2274, 1993.

[10] C. Jngemann, A. Emunds, and W. L. Engl, "Simulation of linear and nonlinear electron transport in homogeneous silicon inversion layers," Solid-State Electron., vol. 36, no. 11, pp. 1529-1540, 1993..

[11] J. L. Thobel, L. Baudry, P. Bourel, F. Dessenne, and M. Charef, "Monte Carlo modeling of high-field transport in III-V heterostructures," J. Appl. Phys., vol. 74, no. 10, pp. 6274-6280, 1993.

[12] J. Fang, M. V. Fischetti, R. D. Schrimpf, R. A. Reed, E. Bellotti, and S. T. Pantelides, "Electron transport properties of $Al_xGa_{1-x}N$/GaN transistors based on first-principles calculations and Boltzmann-equation Monte Carlo simulations," Phys. Rev. Appl., vol. 11, no. 4, p. 044045, 2019.

[13] K. Fukuda and K. Nishi, "An interpolated flux scheme for cellular automaton device simulation," IEEE Trans. CAD, vol. 17, no. 7, pp. 553-560, 1998.

[14] H. Tanimoto, N. Yasuda, K. Taniguchi, and C. Hamaguchi, "Monte Carlo study of hot electron transport in quantum wells," Jpn. J. Appl. Phys., vol. 27, no. 4R, pp. 563-571, 1988.

[15] S. Yamakawa, E. Ueno, K. Taniguchi, C. Hamaguchi, K. Miyatsuji, K. Masaki, and U. Ravaioli, "Study of interface roughness dependence of electron mobility in Si inversion layers using the Monte Carlo method," J. Appl. Phys., vol. 79, no. 2, pp. 911-916, 1996.

[16] B. E. Foutz, S. K. O'Leary, M. S. Shur, and L. F. Eastman, "Transient electron transport in wurtzite GaN, InN, and AlN," J. Appl. Phys., vol. 85, no. 11, pp. 7727-7734, 1999.

[17] M. Farahmand, C. Garetto, E. Bellotti, K.F. Brennan, M. Goano, E. Ghillino, and. P. P. Ruda, "Monte Carlo simulation of electron transport in the III-nitride wurtzite phase materials system: binaries and ternaries," IEEE Trans. Electron Devices, vol. 48, no. 3, pp. 535-542, 2001.

Geometric Advection Algorithm
for Process Emulation

Xaver Klemenschits
Institute for Microelectronics
TU Wien
Vienna, Austria
Email: klemenschits@iue.tuwien.ac.at

Siegfried Selberherr
Institute for Microelectronics
TU Wien
Vienna, Austria
Email: selberherr@iue.tuwien.ac.at

Lado Filipovic
Institute for Microelectronics
TU Wien
Vienna, Austria
Email: filipovic@iue.tuwien.ac.at

Abstract—**An algorithm is developed, which advects a material interface analytically, according to purely geometric considerations. This algorithm is implemented in ViennaLS, a sparse level set library and its applicability to common microelectronic fabrication processes is demonstrated. A pinch-off plasma CVD process is emulated using the presented algorithm. This algorithm is compared to common advection algorithms, showing a significant improvement in accuracy, with a performance penalty of a factor of about 2 when compared to simple advection schemes and a performance benefit of a factor of 6 when compared to more sophisticated schemes.**

I. INTRODUCTION

In topography simulations, implicit surface representations such as the level set (LS), offer significant advantages over explicit representations. These include robust merger and separation of surfaces, and simple extraction of geometrical properties like surface normals and curvature [1], which is especially important for the simulation of processes such as air-void creation with chemical vapour deposition (CVD). The complex merger of surfaces at the pinch-off point required for the accurate simulation of this process is critical. It is therefore important for the movement of surfaces to be as robust as possible, without losing information about the precise location of the surface. Implicit surface representations offer intuitive handling of such complex deformations and no additional correction steps are required, while they are essential when advecting explicit surfaces. An algorithm providing these important features was developed and implemented in ViennaLS [2], an open-source, high performance sparse level set library tailored towards the simulation of semiconductor fabrication processes.

II. LEVEL SET ADVECTION

Advancing the surface described by a LS is deemed advection and is achieved by solving the equation [3]

$$\frac{\partial \phi_{(\vec{x})}}{\partial t} + V_{(\vec{x})} |\nabla \phi_{(\vec{x})}| = 0 \quad , \tag{1}$$

where $\phi_{(\vec{x})}$ is the signed distance function represented by the LS, and $V_{(\vec{x})}$ is a scalar velocity field describing the movement of the surface. Since this is a Hamilton-Jacobi equation, several numerical schemes are available for its solution. However, they all have some restrictions on how much the values of $\phi_{(\vec{x})}$ are permitted to change in one advection step in order to achieve a desired accuracy. The largest step which is permitted, regardless of the numerical scheme applied to solve Eq. (1), is given by the Courant-Friedrichs-Lewy (CFL) condition [4]. Therefore, advection must be repeated several times in order to advance the surface corresponding to a single fabrication step. This leads to the propagation of numerical errors introduced during the first advection step, resulting in problems such as flattened corner geometries during deposition [5]. Therefore, in recent years, the concept of process emulation has grown in interest, where the surface is effectively re-drawn to its new state in a single geometric step, corresponding to a complete processing step. We present a novel geometric advection algorithm and its implementation in ViennaLS, allowing for fast and accurate surface advancement.

III. GEOMETRIC ADVECTION ALGORITHM

If the exact geometry of an advected surface after a fabrication step is known in advance, there is no need to directly solve Eq. (1) repeatedly while adhering to the CFL condition. The geometric relationship between the initial surface and the known final surface can simply be applied in a single step. For example, for a highly conformal deposition process, we know from the onset, that the surface will be advanced outwards isotropically by a distance calculated by multiplying the deposition rate by the process time. Therefore, the new surface can simply be deduced from the initial surface and our knowledge of the applied process. Here, we propose a geometric advection algorithm which is able to advance the surface in a single step, given that the geometric properties of the fabrication process are known.

A. Geometric Distributions

The proposed algorithm is applicable, if the geometric properties of a process can be described by a distribution. This distribution defines how each point on the surface influences the final geometry. Conceptually, the final geometry is created by performing a union between the initial surface and the geometric distribution centred at each point on the surface. Creating these unions explicitly is not necessary, since the new level set values can be calculated directly. Referring to Fig. 1, first, the distance vector $\vec{v}$ between a point on the initial surface $\vec{P}_{contribute}$ and a point close to the new surface $\vec{P}_{candidate}$ is computed. $\vec{v}$ is then used as the position vector of the geometric distribution to calculate the new LS value at $\vec{P}_{candidate}$. As can be seen in Fig. 1, for a spherical distribution with radius r, which corresponds to a perfectly conformal deposition/etching process, the signed distance d_s for a candidate point is computed using

$$d_s = |\vec{P}_{candidate} - \vec{P}_{contribute}| - r \quad . \tag{2}$$

The signed distance value d_s can directly be applied to construct the LS describing the surface after the processing step.

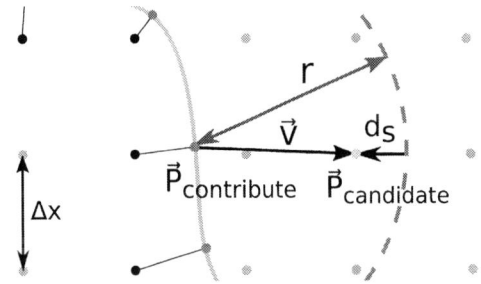

Fig. 1: Signed distance calculation for candidate point $\vec{P}_{candidate}$ (green) from an explicit point $\vec{P}_{contribute}$ (red) on the initial surface. The spherical distribution is depicted by the dashed red circle. d_s is then used as the new LS value for the grid point $\vec{P}_{candidate}$. The LS grid spacing is shown as Δx.

B. The Algorithm

The advection algorithm consists of the following steps, as shown in Fig. 2, where point colours refer to those given in Fig. 1:

1. Extract the explicit point cloud of the initial surface by shifting grid points by their LS value in the direction of the surface normal at that grid point [6] (blue points).

2. Identify candidate grid points which may be close to the new surface (orange points). In the simplest case, the bounding box of the initial surface is extended by the size of the geometric distribution in each dimension and all points in this new bounding box are checked. This simple approach was used in the presented implementation.

3. Iterate over candidate grid points while applying the following procedure:

   a) Identify initial surface points which are close enough to the current candidate point (green) to contribute to the signed distance calculation (contribute points).

      i) Iterate over contribute points: Find the signed distance using Eq. (2). If d_s is negative (inside the surface) and its absolute value is larger than Δx, discard the candidate point because it is too far away to improve the description of the surface.

      ii) Apply the d_s with the lowest value found in i) as the final signed distance value for this candidate point.

4. Construct the new LS from the final values of each candidate point which was not discarded.

After the algorithm has completed, all sparse LS points will be set with the correct signed distance, thus resulting in a valid sparse field LS. This algorithm combines the advantages of the explicit representation of the initial surface and the implicit representation of the new surface. The explicit initial surface can be accessed and reordered quickly, while the new surface, described by a sparse field LS, automatically discards self-intersecting or overlapping points.

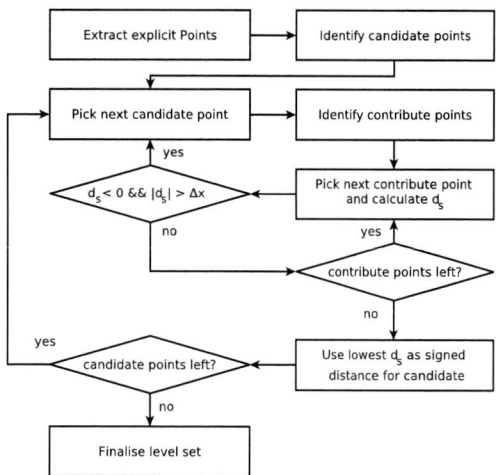

Fig. 2: Flowchart of the geometric advection algorithm.

C. Limitations

The algorithm, as presented above, is used if the surface is moved outwards (deposition), while inwards movement (etching) requires some changes to work correctly. During deposition, all level set values must become smaller, since the surface moves outwards. Therefore, the lowest level set value is used in Item 3.(a)ii. However, the opposite is the case for etching,

as all level set values must become larger. Identification of candidate and contribute points remains the same, but due to the changed signs of the advection, Item 3.(a)i must be changed to discard level set values which are larger than Δx and positive, rather than negative. Similarly, in Item 3.(a)ii the d_s with the largest value is used as the final distance. These changes in the algorithm are applied automatically depending on the sign of the distribution. All other parts, as well as the robustness concerning self-intersections or overlapping phenomena remain unchanged.

The change in logic depending on the direction of movement of the surface is the only limitation of this algorithm compared to iterative advection schemes. It is not possible to advect the surface outwards at some points and inwards at others, as it is not clear whether the smallest or largest level set value should be chosen. In our implementation, this change in logic is applied automatically, based on the geometric distribution passed to the algorithm. If the distribution has a negative extent, it is treated as etching, while a positive extent means deposition. It is therefore not possible to define a distribution with a positive extent in one axis direction and a negative extent in another. Therefore, this technique may not be applicable for certain processes, where deposition and etching occurs at the same time, such as ion-enhanced plasma etching [7]. However, many of these processes could likely be approximated by splitting them each into a separate etching emulation followed by a deposition emulation.

D. Performance and Parallelisation

Perfectly isotropic deposition or etching results in a very simple signed distance calculation, as shown in Eq. (2). Since signed distance calculations are conducted for all combinations of candidate and contribute points, great care should be taken to make sure that the calculation is executed as efficiently as possible. However, since this is more or less given by the chosen geometric distribution, the possibility for runtime reduction is limited.

A far bigger performance benefit can be achieved by identifying candidate grid points efficiently, as many signed distances must be calculated for every candidate point. For the ideal performance case, no candidate points would be discarded in Item 3.(a)i, as only candidate points which are part of the final level set will have been examined. The simplest approach is setting a bounding box for the anticipated result of the advection and using all grid points in this bounding box as candidate points. This is performed in the current implementation, as it already provides reasonable performance and is the simplest solution.

The efficient identification of contribute points presents a very similar problem to picking candidate points. Since contribute points are represented explicitly, standard meshing algorithms could be applied to sort them quickly based on their proximity to the current candidate point. In the current implementation, a linear search is used to identify contribute points, which does impact performance negatively for large structures.

Since the calculation of the signed distance for each candidate point is completely independent, parallelisation is straight forward. However, the choice of suitable candidate points is important for proper load balancing, if course-grained parallelisation is used, i.e. the domain is split into a subdomain for each used thread. A more general and robust method is distributing single candidate points to threads as they become available, thus resulting in fine-grained parallelisation and efficient load balancing. However, this would require more instances of random access into the initial level set structure, which is comparably slow in the used data structure.

E. Future Improvements

Currently, the identification of candidate points follows a simple bounding box approach, as described in Section III-D. However, a more complex approach which involves reordering the points based on proximity to the old surface could provide significant performance improvements and should be the topic of future work. This way, the new level set values would only be calculated for points a certain distance away from the initial surface, which is likely to contain the final surface.

Contribute points are identified by a simple linear search. Sorting them by proximity to the current candidate point would allow for the quick discarding of all remaining contribute points as soon as the first is identified to not contribute to the level set value. Using a more complex sorting procedure would therefore likely lead to considerable performance benefits and is intended be studied in future work.

Calculation of signed distances for candidate points could be parallelised by distributing single signed distance calculations to threads as they become available. This would result in optimal load balancing and would greatly improve the performance of the parallelised algorithm.

IV. PINCH OFF PLASMA CVD PROCESS

In order to demonstrate the capabilities of the geometric advection algorithm, a chemical deposition process was emulated using ViennaLS. A pinch-off plasma CVD process [8] was chosen, because its geometric properties can be deduced from scanning transmission electron microscope (STEM) measurements to generate a predictive process emulation model solely reliant on geometric considerations. The simplest approximation for this anisotropic deposition process is based on the view factors of the points in the trench to the opening plane. This approach has already been used to describe deposition processes in trenches and vias [9]. For an infinite trench, an ideal solution exists and is given by [10]

$$ F(d) = \frac{1}{2} + \frac{cos(\theta) - \frac{d}{D}}{2\sqrt{1 - 2\frac{d}{D}cos(\theta) + \left(\frac{d}{D}\right)^2}} \quad , \qquad (3) $$

where $F(z)$ is the view factor of a point a distance d down the sidewall, which is tapered by the angle θ. This is the view factor with respect to an opening plane of width D. Since the deposition at each point is isotropic a spherical advection kernel was chosen to represent the process. However, since the deposition rate changes with depth down the trench, the radius of the different advection distributions differs, as shown in Fig. 3. Since $F(d)$ only depends on the downward distance along the trench sidewall d, the radius of the spherical advection kernel $R(d)$ also only depends on d with

$$ R(d) = R_0 F(d) \quad , \qquad (4) $$

where R_0 is the deposition distance at the opening of the trench. For the modelled process R_0 was chosen as one half of the opening width of the trench D, meaning that the timing of the trench pinch-off will be exact during the process. The distance down the slanted sidewall d is not directly available in the simulator, but can be found from the vertical coordinate z of a point on the sidewall:

$$ d = \frac{z}{cos(\theta - \pi/2)} \qquad (5) $$

The angle of the sidewall slant of the copper lines in the chosen process is $\theta = 95°$ and the pitch of the copper lines is 48nm. The copper line width decreases linearly from a width of 25nm at the top to 18nm at the bottom. The height of copper lines is 54nm [8]. R_0 was chosen to be 12nm, slightly more than half the opening width of 23nm, to ensure a proper closing of the trench.

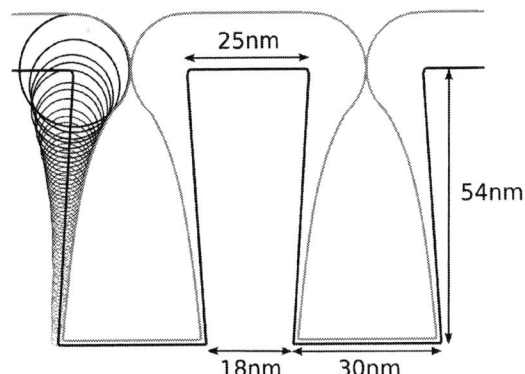

Fig. 3: Schematic showing how the described model creates the resulting geometry (green) using spherical geometric distributions (black circles). The outline of the circles thus creates the final geometry, based on the view factor calculated using Eq. (3).

V. RESULTS

The presented pinch-off plasma CVD process was emulated with high resolution using several advection schemes. The presented algorithm was the only one that could produce the expected results, matching experimental data. The result of a three-dimensional simulation with periodic boundary conditions is shown in Fig. 4

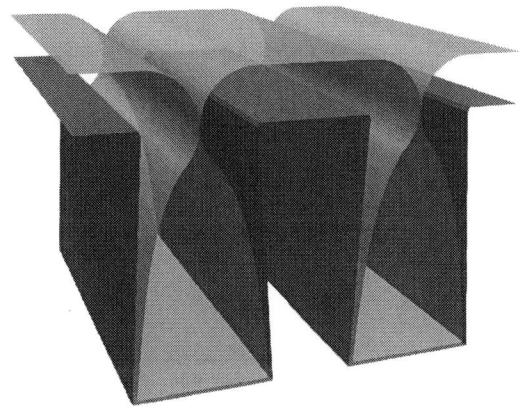

Fig. 4: Result of three-dimensional emulation of the pinch-off plasma CVD model. As expected, the resulting surface is pinched off exactly since the advection is purely geometric.

A comparison of the geometric advection and a sophisticated iterative scheme, the Stencil-Local Lax-Friedrichs (SLLF) scheme [11], is shown in Fig. 5. The exact timing of the closing of the trench is important, as the geometry inside the trench is fixed thereafter. Since the new surface can be calculated directly, the pinch-off follows the predicted time exactly, because it is produced analytically. As evidenced in Fig. 5, the SLLF scheme is unable to reproduce the pinch-off time, meaning that the air void will not have the desired shape at the end of the simulation. In addition to SLLF, the presented emulation algorithm was compared to several other iterative schemes, as summarized in Table I. As is evident from these results, iterative schemes are unable to predict this pinch-off correctly, as they include numerical errors (i.e., corner flattening discussed earlier), even though exactly the same model is applied. In this case, more complex schemes produce even larger errors, as they include damping terms, slowing the advection.

A runtime comparison with conventional advection schemes is also given in Table I. As can be seen, the geometric advection algorithm provides high accuracy with reasonable performance. Simple advection schemes, such as the Engquist-Osher and Lax-Friedrichs schemes provide smaller runtimes, but suffer from the introduction of numerical errors originating from the numerical derivatives. Furthermore, like all iterative schemes, they are susceptible to the propagation of errors, as discussed previously. More complex schemes, such as the Local-Local-Lax-Friedrichs and the Local-Lax-Friedrichs scheme, produce even larger errors, due to the larger damping terms they include, which makes them more stable in complex geometries, but leads to slower overall front propagation, ergo a larger pinch-off error. Even the complex SLLF scheme, which includes numerous corrections [11], cannot emulate the deposition process correctly. This comparative study further suggests that simply using a higher-order and higher-complexity scheme does not necessarily provide more accurate results. One must always be aware of the geometry which is being simulated and an advection scheme should be chosen accordingly. Here, we note that, for trench CVD and pinch-off, the Lax-Friedrichs scheme is most appropriate for pinch-off simulations; nevertheless, it cannot match the accuracy of the proposed geometric advection scheme.

Another problem of iterative schemes is that they may be accurate for certain applications and geometries, but not all. Therefore, the accuracy they provide is not easily predictable. The great advantage of the proposed geometric advection algorithm is that it always provides algebraic accuracy. This is possible, because it does not rely on the derivatives generated in the level set, which always include some numerical errors. Thus, the proposed advection algorithm is suitable for fast, accurate and predictive high-performance process emulation with predictable accuracy and performance.

Advection scheme	Runtime (s)	Pinch-off error (nm)
Engquist-Osher (1st)	8.5	3.7
Engquist-Osher (2nd)	18.3	2.8
Lax-Friedrichs (1st)	8.5	2.5
Lax-Friedrichs (2nd)	19.0	2.1
Local-Local-Lax-Friedrichs (1st)	8.2	4.0
Local-Local-Lax-Friedrichs (2nd)	18.4	3.0
Local Lax-Friedrichs (1st)	143.3	6.5
Local Lax-Friedrichs (2nd)	171.0	4.1
SLLF (1st) [11]	181.3	3.6
Geometric Advection	**30.6**	**0.0**

TABLE I: Runtimes of different advection schemes for the three-dimensional pinch-off plasma process with 62,662 LS points on the initial surface compared to the presented geometric advection (Fig. 4). The pinch-off error describes, how wide the opening of the trench is at the expected pinch-off time. The approximation order for derivatives is listed in the parenthesis. All simulations are performed on an AMD Ryzen 3950X CPU using 32 threads.

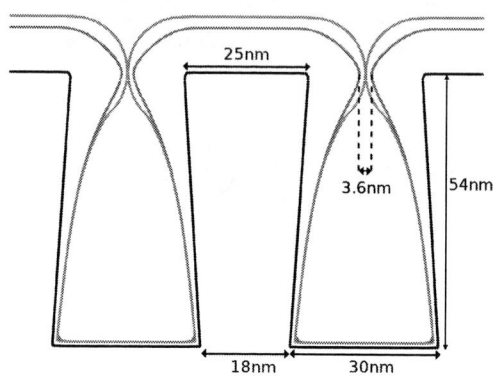

Fig. 5: Slice of a three-dimensional emulation of the pinch-off plasma CVD process inside tapered trenches (black surface). The green surface is the result of the geometric advection, while even the most robust iterative scheme (SLLF [11], red surface) is unable to predict the closing of the trench.

VI. CONCLUSION

A purely geometric advection scheme was developed and implemented in the level set framework, ViennaLS, allowing for the emulation of an entire deposition process in a single step with higher accuracy than conventional advection schemes. This is due to numerical errors introduced when calculating derivatives in the level set stemming from its limited resolution. Furthermore, conventional schemes must be applied iteratively as there is a limit on how much the values of the level set are permitted to change in a single advection step before the numerical scheme becomes unstable. This leads to the propagation – and possibly the amplification – of errors, such as the flattening of corner geometries.

In order to test the applicability of the proposed algorithm, a pinch-off CVD process was emulated and compared to conventional advection schemes. The proposed advection algorithm was shown to outperform even sophisticated advection schemes in both runtime and accuracy. Compared to the sophisticated SLLF schemes, up to 6 times faster runtimes have been achieved. Although the proposed algorithm was shown to be slower than simple iterative advection schemes, the provided accuracy still warrants the performance penalty of a factor of 2. It could therefore be shown, that the proposed geometric advection algorithm is suitable for efficient and highly accurate emulation of complex semiconductor fabrication processes.

REFERENCES

[1] O. Ertl, Ph.D. dissertation, TU Wien, 2010.

[2] X. Klemenschits, O. Ertl, P. Manstetten, J. Weinbub, and L. Filipovic. ViennaLS. [Online]. Available: https://github.com/ViennaTools/ViennaLS

[3] S. Osher and J. A. Sethian, *J. Comput. Phys.*, vol. 79, no. 1, pp. 12–49, 1988. doi: 10.1016/0021-9991(88)90002-2

[4] R. Courant, K. Friedrichs, and H. Lewy, *Math. Ann.*, vol. 100, no. 1, pp. 32–74, 1928. doi: 10.1007/BF01448839

[5] J. Sethian and D. Adalsteinsson, *IEEE Trans. Semicond. Manuf.*, vol. 10, no. 1, pp. 167–184, 1997. doi: 10.1109/66.554505

[6] O. Ertl and S. Selberherr, in *Proc. SISPAD*, 2008. pp. 325–328. doi: 10.1109/SISPAD.2008.4648303

[7] X. Klemenschits, S. Selberherr, and L. Filipovic, *Micromachines*, vol. 9, no. 12, p. 631, 2018. doi: 10.3390/mi9120631

[8] S. Nguyen, T. Haigh, K. Cheng, C. Penny, C. Park, J. Li, S. Mehta, T. Yamashita, L. Jiang, and D. Canaperi, *ECS J. Solid State Sc.*, vol. 7, no. 10, pp. P588–P594, 2018. doi: 10.1149/2.0021811jss

[9] P. Manstetten, L. Filipovic, A. Hössinger, J. Weinbub, and S. Selberherr, *Solid-State Electronics*, vol. 128, pp. 141–147, 2017. doi: 10.1016/j.sse.2016.10.029

[10] M. F. Modest, *Radiative Heat Transfer*, 3rd ed. Academic Press, 2013. ISBN 9780123869449

[11] A. Toifl, M. Quell, X. Klemenschits, P. Manstetten, A. Hössinger, S. Selberherr, and J. Weinbub, *IEEE Access*, vol. 8, pp. 115 406–115 422, 2020. doi: 10.1109/ACCESS.2020.3004136

Performance and Leakage Analysis of Si and Ge NWFETs Using a Combined Subband BTE and WKB Approach

Z. Stanojević, G. Strof, O. Baumgartner, G. Rzepa, and M. Karner

Global TCAD Solutions GmbH., Bösendorferstraße 1/12, 1010 Vienna, Austria

Email: {z.stanojevic|g.strof|o.baumgartner|g.rzepa|m.karner}@globaltcad.com

Abstract—We are the first to present a subband-BTE solver with a fully integrated source/drain-tunneling current calculation based on the WKB-approximation. The method is validated against ballistic NEGF calculations showing good agreement. An investigation of Si and Ge-based NWFETs is performed showing that intra-band source/drain-tunneling is not a concern for Si devices. For Ge-based PMOS devices however, tunneling leakage limits sensible L_G-scaling to around 20 nm.

I. INTRODUCTION

As technology development progresses to the 5 nm-node and beyond, source/drain-leakage is found to be the ultimate obstacle to CMOS-scaling. In silicon, even with printed gate-lengths as low as 20 nm, intra-band source/drain-tunneling is not seen as a concern. However, this does not necessarily hold for alternative channel materials such as SiGe and Ge; both are considered candidates for the replacement of Si for high-performance CMOS at 3 nm [1, 2]. Such high-mobility materials however come at the price of increased S/D-leakage [3].

II. METHODOLOGY

The Subband-Boltzmann transport equation (SBTE) solver underlying to this work has been presented in [4]. It is commercially available as the GTS Nano Device Simulator (NDS)[5], part of GTS Framework. The solver was shown to accurately reproduce the device characteristics of highly-scaled stacked-nanowire FETs [6]. In this work, the solver is extended to cover source/drain-tunneling by implementing a WKB-based scheme operating on the complex subband structure.

A. Complex subband-structure

The complex band structure is obtained by extending the $\mathbf{k}$-vector to the complex plane. For each subband $\mathbf{k}_0$ i.e. the minimum (for electrons, maximum for holes) is found and the Hamiltonian $H(\mathbf{k}_0 + i\kappa)$ is solved to obtain the $E_n(\mathbf{k}_0 + i\kappa)$-relations of the evanescent (i.e. tunneling) states (Fig. 1). For simulation of planar devices, this process has to be repeated for every $\mathbf{k}$-grid-line parallel to the transport direction as shown in the bottom part of Fig. 1.

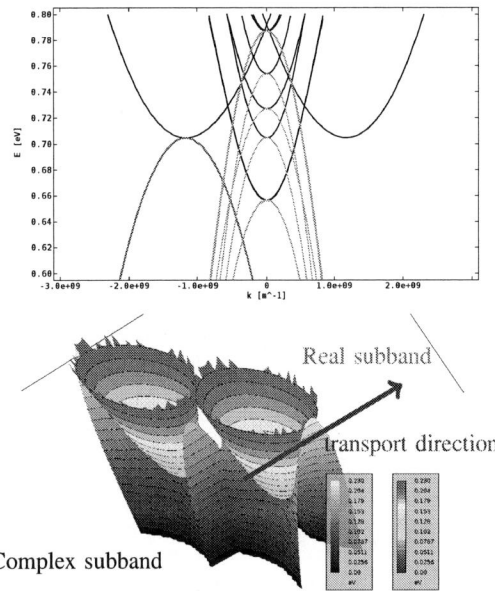

Fig. 1. Top: real (black) and complex (colored) 1D conduction band structure of a Si-NMOS-NWFET; complex subbands are expanded at the minimum of each of their real counterparts; bottom: real (colored) and complex (gray) 2D conduction band structure in a Si-NMOS-UTBFET

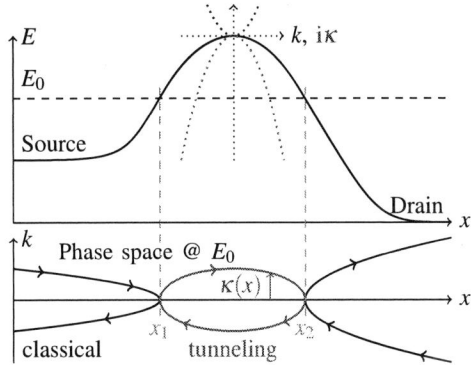

Fig. 2. Typical n-MOSFET potential in off-state along with the local real $E(x,k)$ and complex dispersion relation $E(x,i\kappa)$; a cut through the $E(k,x)$ landscape at energy E_0 reveals the classical and tunneling trajectories an electron can take; the classical turning points are marked as x_1 and x_2.

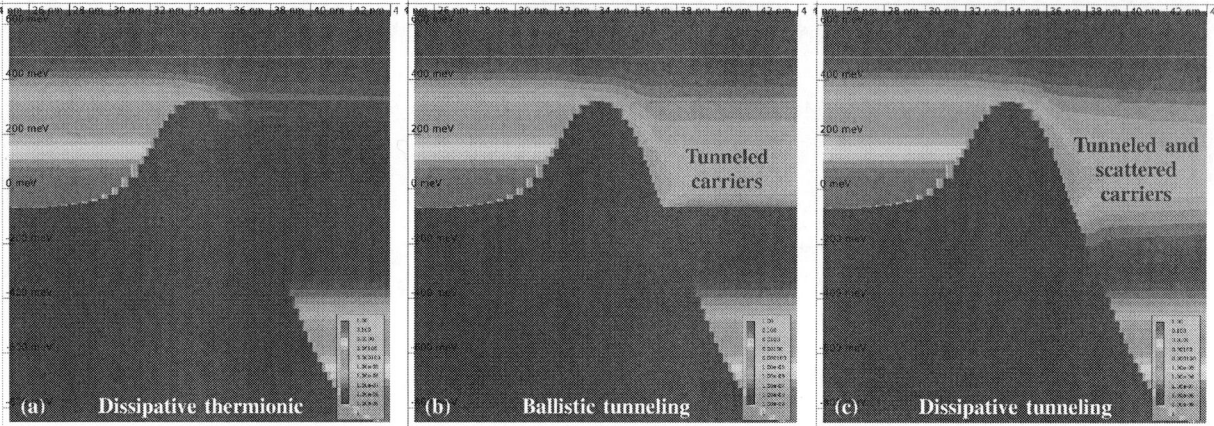

Fig. 3. Hole spectra in a Si-PMOS with $L_G = 5\,\text{nm}$ biased at $V_{GS} = 0\,\text{V}$ and $V_{DS} = -0.8\,\text{V}$; for the thermionic-only case (a) few holes cross the barrier and are being scattered to lower energies; however the majority of the current is due to tunneling transport, shown in (b) the ballistic and (c) the dissipative cases

B. WKB-implementation

Based on the complex subbands, tunneling paths are constructed that allow the evaluation of the WKB-tunneling-integral,

$$\theta(E_0) = \exp \left[\int_{x_1}^{x_2} \kappa(E_0, x)\,dx \right]. \tag{1}$$

computed for each energy E_0 between the turning points x_1 and x_2. κ is the imaginary wave number in the classically forbidden region, which for parabolic bands can be computed from

$$\kappa(E_0, x) = \frac{\sqrt{2m^*(V(x) - E_0)}}{\hbar}, \tag{2}$$

but can also be calculated for arbitrary numerical representations of the complex band structure by integrating along the E_0-contour of the complex phase-space dispersion relation, $E(x, i\kappa)$. This procedure is visually represented in Fig. 2 and allows us to use both effective mass and $\mathbf{k} \cdot \mathbf{p}$ band models in this work. From the integral the first and second order approximations of the transmission coefficients (TC) can be calculated,

$$\text{TC}^{1^{st}} = \frac{1}{\theta^2}, \quad \text{TC}^{2^{nd}} = \frac{1}{(\theta + \frac{1}{4\theta})^2} \tag{3}$$

The transmission coefficients are used to derive effective tunneling rates which are entered into the SBTE system of equations at the respective turning points of each energy. The net spectral current density per subband through the barrier of a NWFET is given by the Tsu-Esaki-formula,

$$j_n(E_0) = -\frac{g_v q}{\pi \hbar} \text{TC}_n(E_0) \left[f_n(E_0, x_1) - f_n(E_0, x_1) \right], \tag{4}$$

where g_v denotes valley degeneracy. In the case of a planar MOSFET with a 2D $\mathbf{k}$-space, an additional integration over the lateral $\mathbf{k}$-space component has to be performed. Since we are only concerned with steady-state simulations, it can be assumed that the the carrier tunneling happens instantaneously.

Furthermore, the tunneling process is assumed to not be affected by scattering.

The seamless integration of WKB-tunneling in SBTE becomes is apparent in Fig. 3 (b) and (c), especially the latter where tunneling and scattering are shown interacting. The WKB-integrals are simple repeated 1D integrations of $\kappa(E_0, x)$ and entering the effective tunneling rates into the SBTE system does not increase its rank; this means that there is barely any impact on simulation run-time when WKB is used.

III. RESULTS

A. Validation against NEGF-reference

To validate the presented approach, we compared the results of ballistic SBTE (with and without WKB tunneling enabled) with our in-house ballistic NEGF solver, which operates in full 3D-real-space as opposed to the mode-space approach employed for the SBTE. As test case, we selected the shortest of the Si NMOS devices described in Section III-B with a gate length of 5 nm in order to emphasize the source-drain tunneling effect. Both SBTE and NEGF simulations were performed self-consistently with electrostatics and using a parabolic effective mass band structure model.

Transmission coefficient and current spectrum in the off-state are shown in excellent agreement for the lowest subband in Fig. 4. A very good agreement is also achieved for the transfer characteristic at $V_{DS} = 0.7\,\text{V}$ across a wide range of gate voltages. This is surprising considering the vastly different approaches of SBTE + WKB and NEGF. No perceivable difference was noted between using the 1st and 2nd-order approximations for TC in the SBTE + WKB results.

It is worth noting that the self-consistent solutions of SBTE + WKB, which are typically calculated in a few minutes from scratch, were used to provide initial potentials for the NEGF simulations, which then took hours to complete on 20-core servers highlighting the efficiency of the presented approach.

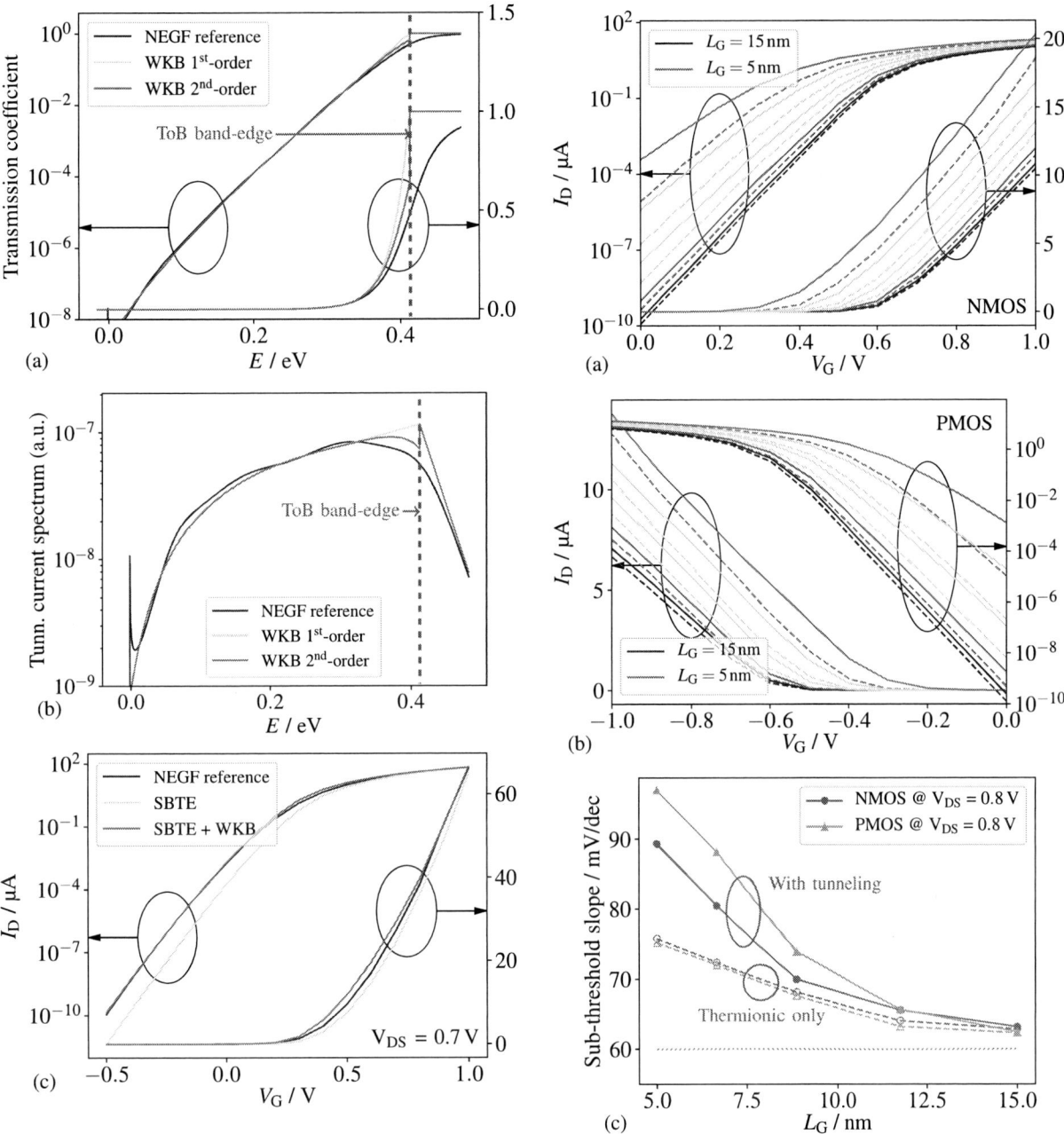

Fig. 4. Validation of ballistic SBTE (with and without WKB) against ballistic NEGF (eff. mass) using the $L_G = 5$ nm NMOS device; graphs (a) and (b) show the comparison of transmission coefficient and tunneling current spectrum, respectively; NEGF and WKB agree very well except near the band edge, where the WKB has unity value by definition; there is also excellent agreement in the transfer characteristic in graph (c).

Fig. 5. Saturated I_D/V_G-curves for Si-NMOS (a) and Si-PMOS (b) for varying L_G; dashed curves represent thermionic-only transport while solid curves represent transport with tunneling enabled; SS-roll-off is shown in graph (c) for both device types, where PMOS shows slightly higher SS-degradation

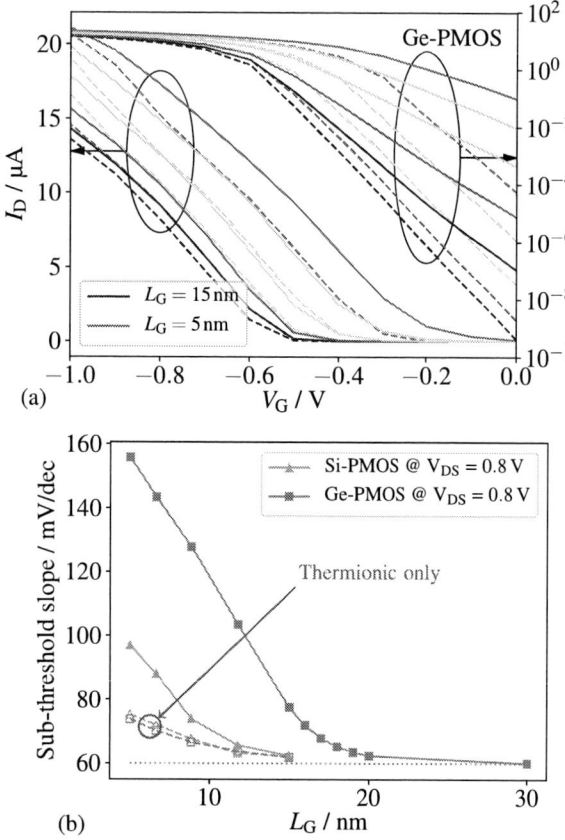

Fig. 6. Saturated I_D/V_G-curves for Ge-PMOS (a) – same as Fig. 5 (b); SS-roll-off is shown in graph (c) for Si & Ge-PMOS, where the Ge-PMOS shows severe SS-degradation for $L_G < 20$ nm

B. Scaling study of Si and Ge-based NWFETs

Due to their superior electrostatic integrity, GAA devices allow further shortening of the printed gate length without compromising their low leakage current – at least until the onset of source-drain tunneling. To investigate the limits of gate length scaling, Si NMOS and PMOS NWFETs with 6 nm diameter, 7 nm spacer length and following gate lengths were simulated: 15 nm, 11.8 nm, 8.9 nm, 6.7 nm, 5 nm (33 % decrements). Channel orientation was set to ⟨110⟩. Intrinsic channel and 10^{20} cm^{-3} source/drain-doping with abrupt junctions between gate and spacers were assumed. Two-band and six-band **k·p**-models were used for the conduction and valence band structure, respectively. Acoustic, optical, intervalley phonon scattering, surface roughness scattering, and Coulomb scattering off impurities were included. The sub-threshold slopes (SS) were recorded at $|I_D| = 0.1$ nA. The results shown in Fig. 5 for Si-based NMOS and PMOS show no effect at $L_G = 15$ nm, slowly progressing to just below 100 mV at $L_G = 5$ nm.

The situation is however different when germanium is used as PMOS channel material. Holes in Ge are known to have sig-

nificantly lower effective masses than in Si, which is beneficial for boosting on-current but also increases the tunneling TC, adversely affecting off current. For a quantitative assessment, we replaced Si with Ge in the PMOS devices mentioned above while performing the same scaling analysis. The results show that tunneling has a much more severe impact in Ge-base PMOS NWFETs, as shown in Fig. 6. Source-drain tunneling is so severe that it results in unacceptable SS below a printed gate lenght of 20 nm. For a more realistic device with diffused junctions and gate overlap, the minimum required L_G might be even higher.

IV. CONCLUSION

We presented a methodology for including intra-band source/drain-tunneling in semi-classical transport based on WKB. The methodology is shown to be of comparable accuracy to ballistic NEGF for MOSFET devices while leveraging the advanced transport modeling capabilities of the subband-BTE framework. We showed that while source/drain-tunneling is not a serious concern in Si-based devices, Ge-based PMOS NWFETs suffer from increased tunneling leakage prohibiting L_G-scaling below 20 nm.

REFERENCES

[1] L. Witters, F. Sebaai, A. Hikavyy, A. P. Milenin, R. Loo, A. De Keersgieter, G. Eneman, T. Schram, K. Wostyn, K. Devriendt, A. Schulze, R. Lieten, S. Bilodeau, E. Cooper, P. Storck, C. Vrancken et al., "Strained germanium gate-all-around pmos device demonstration using selective wire release etch prior to replacement metal gate deposition," in 2017 VLSIT, June 2017, pp. T194–T195.

[2] H. Arimura, L. Witters, D. Cott, H. Dekkers, R. Loo, J. Mitard, L. . Ragnarsson, K. Wostyn, G. Boccardi, E. Chiu, A. Subirats, P. Favia, E. Vancoille, V. De Heyn, D. Mocuta, and N. Collaert, "Performance and electrostatic improvement by high-pressure anneal on si-passivated strained ge pfinfet and gate all around devices with superior nbti reliability," in 2017 VLSIT, June 2017, pp. T196–T197.

[3] S. Jin, A. Pham, and Y. Nishizawa, "Performance evaluation of ingaas, si, and ge nfinfets based on coupled 3d drift-diffusion/multisubband boltzmann transport equations solver," in 2014 IEDM, Dec 2014, pp. 7.5.1–7.5.4.

[4] Z. Stanojević, M. Karner, O. Baumgartner, H. W. Karner, C. Kernstock, H. Demel, and F. Mitterbauer, "Phase-space solution of the subband Boltzmann transport equation for nano-scale TCAD," in 2016 SISPAD, Sept 2016, pp. 65–67.

[5] Global TCAD Solutions, "Nano Device Simulator," http://www.globaltcad.com/nds.

[6] M. Karner, O. Baumgartner, Z. Stanojević, F. Schanovsky, G. Strof, C. Kernstock, H. W. Karner, G. Rzepa, and T. Grasser, "Vertically stacked nanowire MOSFETs for sub-10nm nodes: Advanced topography, device, variability, and reliability simulations," in 2016 IEDM, Dec 2016, pp. 30.7.1–30.7.4.

Molecular Dynamics Modeling of the Radial Heat Transfer from Silicon Nanowires

Igor Bejenari[a,b], Alexander Burenkov[a], Peter Pichler[a,c], Ioannis Deretzis[d], Antonino La Magna[d]

[a]Fraunhofer Institute for Integrated Systems and Device Technology IISB, Erlangen, Germany
[b]E-Mail: igor.bejenari@iisb.fraunhofer.de
[c]Chair of Electron Devices, University of Erlangen-Nuremberg, Erlangen, Germany
[d]CNR-IMM, Catania, Italy

Abstract—Thermal transport in radial direction in Si nanowires embedded into amorphous silicon dioxide has been studied using nonequilibrium molecular dynamics simulations. For comparison, we also considered the axial heat transfer. For Si nanowires with a radius of 2.6 nm, both radial and axial thermal conductivities were found to be about independent of the SiO_2 thickness ranging from 1 nm to 3 nm. The radial thermal conductivity of the Si core and of the covering SiO_2 material are similar and nearly equal to 1 $W\cdot K^{-1}\cdot m^{-1}$. Thermal resistances for the heat transfer from uniformly heated nanowires in radial direction are by a factor of 3 to 4 lower than those for the heat transfer in axial direction.

Keywords—heat transport, thermal conductivity, nanowire, Molecular Dynamics (MD)

I. INTRODUCTION

Silicon nanowires build the basic structure for the next generations of CMOS transistors. In such structures, because of phonon confinement and boundary scattering, the effective thermal conduction is strongly reduced. This poses a limiting factor for many aspects of their application as devices but also for their fabrication via processing steps with low thermal budget like laser annealing. The transport of heat within and from the nanowires to their environment goes in two directions: First, along the wire to the metallic source and drain contacts and, second, in radial direction through the walls of the nanowire and the surrounding dielectric layers to the gate electrode or to other parts of the integrated circuit. The heat transport along the nanowire axis was studied in the literature both experimentally [1] and by simulations [2]. The radial heat transport, on the other hand, has been less addressed and is not well quantified [3]. However, in our earlier work [4] on SOI-FETs we found that it makes an important contribution to the total heat dissipation and we expect similar effects for nanowire FETs.

II. NONEQUILIBRIUM MOLECULAR DYNAMICS SIMULATIONS METHOD

In this work, we investigated nanoscale heat transfer by means of non-equilibrium molecular dynamics (NEMD) simulations exploiting free available software LAMMPS [5], stable version (3 Mar 2020). The Tersoff potential was used to define the structural properties of the Si and SiO_2 atomic systems [6]. In the simulations, the system was first equilibrated in the isobaric-isothermal (NPT) ensemble, then in the canonical (NVT) ensemble. Next, a heat flux was imposed on the system. Finally, when the system has reached a steady state, the temperature gradient between the heat source and sink was calculated as system's response. Thermally grown silicon dioxide as used in electronic technology is amorphous. In this work, the respective atomic structure was created by heating the SiO_2 beyond the melting temperature followed by rapid cooling-down [7].

For the study of the radial heat transport, we use atomic structures like the one shown in Fig. 1. The cylindrical Si nanowire with a length L_{Si} of 10 nm and a radius ρ_{Si} of 2.6 nm is surrounded by amorphous silicon dioxide with thicknesses t_{ox} of 1, 2, and 3 nm. Two different cylindrical heat sources, co-axial with the wire structure, with radii of 0.4 and 2.6 nm and powers of 21.7 and 62.5 $eV\cdot ps^{-1}$, respectively, were used for heat generation. In the first case, this corresponds to a heat source in the structure core, while in the second case the whole Si part of nanowire is heated uniformly. To establish a steady state, a heat sink of equal power was placed on the surface of the structure.

For the study of the heat transport in the direction along the nanowire, we used atomic structures like the one shown in Fig. 2. Here, the heat source with a power of 15 $eV\cdot ps^{-1}$ is uniformly distributed over the silicon part. The left and right lids of the silicon cylinder are covered by a thin layer of silicon dioxide to create a similar material topology for the longitudinal heat transport as it is for the radial one. Two heat sinks with thicknesses of 0.5 nm and a total power of -15 $eV\cdot ps^{-1}$ are located on the external surfaces of the silicon dioxide layers, which are perpendicular to the nanowire axis. The molecular dynamics simulation box indicated by the black lines in Fig. 1 and Fig. 2 defines the borders on which periodic boundary conditions were set-up.

To validate our method, we have simulated the heat conduction along a Si nanowire with a size of 27×27×81 Å. It resulted in a thermal conductivity of 1.75 $W\cdot K^{-1}\cdot m^{-1}$ which is in perfect agreement with the value of 1.8 $W\cdot K^{-1}\cdot m^{-1}$ calculated at 300 K in [2] under the same conditions.

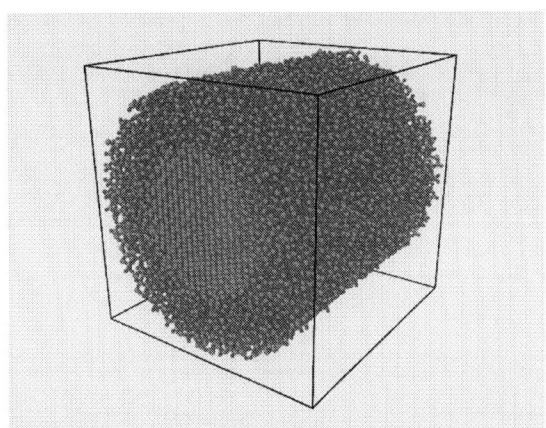

Fig. 1. Atomic arrangement for the study of the radial heat transport in a cylindrical silicon nanowire with a length of 10 nm and growth direction <100>. Radius of the silicon nanowires is 2.6 nm, the nanowire side wall is covered by silicon dioxide with a thickness 2 nm.

The research leading to these results has received funding from the European Union's Horizon 2020 research and innovation programme under grant agreement No. 871813 MUNDFAB.

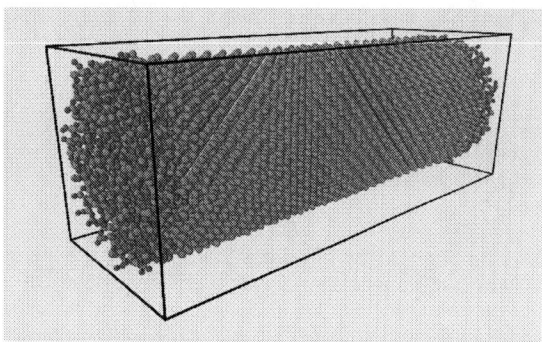

Fig. 2. Atomic arrangement for the study of the axial heat transport in a cylindrical silicon nanowire with a length of the silicon part of 14 nm. Again, the axis of the nanowire corresponds to the <100> crystal direction. The radius of the silicon nanowire is 2.6 nm, top and bottom of the silicon cylinder are covered by silicon dioxide slabs with a thickness 1 nm.

III. HEAT TRANSFER FROM LOCALLY HEATED SILICON NANOWIRES

As demonstrator, we address exemplarily the important case of silicon nanowires clad with silicon dioxide as shown in Fig. 1 and variants made entirely of silicon dioxide. For a numerical example, silicon nanowires with a diameter of 5.2 nm and a length of 10 nm surrounded by silicon dioxide with thicknesses of 1, 2, and 3 nm have been chosen.

Fig. 3 shows the stationary temperature distribution along the nanowire radius for the nanowires like the one shown in Fig. 1 but with different thicknesses of the oxide. The heat transport was calculated in radial direction for the cylindrical heat source with a radius of 0.4 nm and a power of 21.7 eV·ps^{-1}. Please note that the radius scale in all the temperature distribution figures is logarithmic, because the classical macroscopic theory of heat conduction in core-shell cylindrical structures of homogeneous materials predicts a linear dependence in this coordinate system outside the heat sources. The temperature profile has been obtained by averaging the data set of temperatures from 0.6 to 1 ns. As can be seen in Fig. 3, the calculated temperature distribution exhibits several about linear parts that have different slopes in different radii ranges. This is due to the dependence of the phonon properties on the material, on the radius of the wire, and on the oxide thickness.

Fig. 4 shows similar calculations for mono-material nanowires of crystalline silicon or amorphous SiO$_2$. In comparison to Fig. 3, the temperature distribution in the crystalline silicon nanowire has a higher maximum temperature, lower minimum temperature and about three regions with different slopes. This is due to the phonon boundary scattering and Umklapp phonon scattering at high temperature.

In the amorphous silicon dioxide nanowire, a more smooth temperature distribution is observed predominately because of a relatively small phonon mean free path and, therefore, negligible phonon size quantization effect.

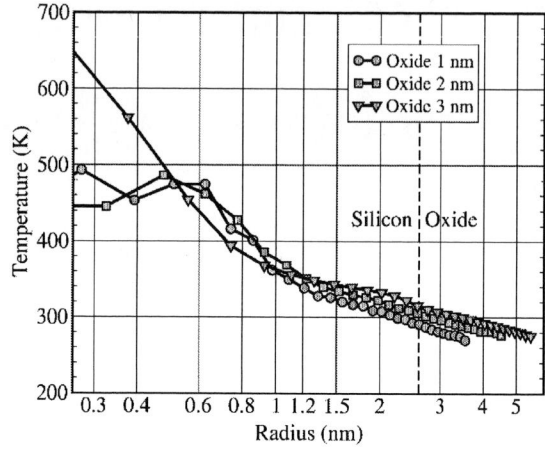

Fig. 3. Temperature distribution as a function of the radius from the axis of the cylindrical nanowire. The radius of the silicon nanowire is 2.6 nm, its cylindrical surface is covered by silicon dioxide with thicknesses of 1, 2, or 3 nm as indicated in the legend. A heat source with a power of 21.7 eV·ps^{-1} is distributed in the inner Si core with a diameter of 0.8 nm. NEMD simulations were performed for 1 ns.

The temperature difference between the maximum and minimum temperatures in these distributions is proportional to the radial thermal resistance. For different materials or material combinations, both the radial thermal conductivity and the thermal resistance were calculated using Fourier's law considering different regions with the constant temperature gradient, see Table I and Table II. We have estimated in Table I also the effective radial interface thermal resistance considering the temperature difference across the smooth Si-SiO$_2$ interface region with a thickness of $\Delta\rho \sim 0.3$ nm This value is small, because the thermal conductivities in Si and SiO$_2$ nanowire parts are similar. In addition, due to the atom relaxation, the atomic structure is gradually changing between the two adjacent materials. Therefore, the phonon modes are smoothly changing across the interface resulting in a relatively large phonon spectra overlap.

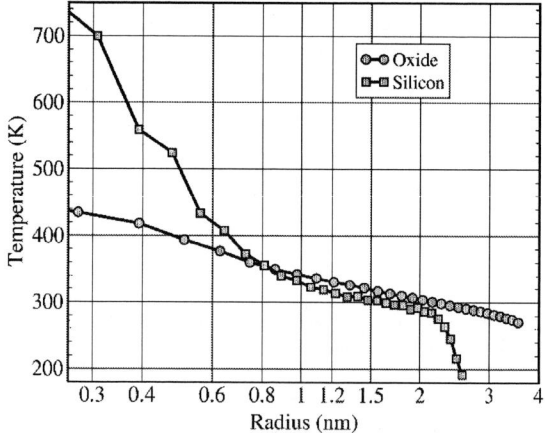

Fig. 4. Temperature distribution as a function of the radius from the axis of cylindrical c-Si and a-SiO$_2$ nanowires. The radius of the silicon nanowire is 2.6 nm, the one of the silicon dioxide nanowire is 3.6 nm. The heat source with a power of 21.7 eV·ps^{-1} is distributed in the inner nanowire core with a diameter of 0.8 nm. NEMD simulations were performed for 1 ns.

68

TABLE I. RADIAL THERMAL RESISTANCES
(10 NM LENGTH, 0.4 NM SOURCE RADIUS)

Radial Oxide thickness	Radius (nm)	$R*_{th}$ (10^{+7} K·W^{-1}) linear part / whole radius	$R_{th, interface}$ (10^{+7} K·W^{-1}) from Fourier Law
No oxide	ρ_{Si} = 2.6	0.785 / 17.7	-
1 nm	ρ_{Si} = 2.6	1.47 / 5.02	0.126
2 nm	ρ_{Si} = 2.6	1.82 / 5.24	0.193
3nm	ρ_{Si} = 2.6	2.09 / 12.7	0.346
Oxide only	ρ_{SiO2} = 3.6	1.60 / 5.39	-

TABLE II. RADIAL THERMAL CONDUCTIVITIES
(10 NM LENGTH, 0.4 NM SOURCE RADIUS)

Radial Oxide thickness	Radius (nm)	κ_{Si} (W·K^{-1}·m^{-1})	κ_{Si-ox} (W·K^{-1}·m^{-1})	κ_{ox} (W·K^{-1}·m^{-1})
No oxide	ρ_{Si} = 2.6	1.05	-	-
1 nm	ρ_{Si} = 2.6	1.10	1.03	0.839
2 nm	ρ_{Si} = 2.6	1.03	1.04	1.18
3nm	ρ_{Si} = 2.6	1.09	1.10	1.12
Oxide only	ρ_{SiO2} = 3.6	-	-	1.04

IV. HEAT TRANSFER FROM UNIFORMLY HEATED SILICON NANOWIRES

A uniform heating of a silicon nanowire is a typical case in many technical applications. For example, in case of laser processing, the light is absorbed mainly in the silicon while the silicon dioxide remains unheated as long as the silicon dioxide is transparent for the chosen wavelength of light. Also in nanowire CMOS transistors, heat is generated in the silicon only because the oxide is electrically not conducting. In this section, we consider the heat transfer from uniformly heated cylindrical silicon nanowires. We compare the heat transport in two directions, radial transport through the sidewalls of the nanowires and axial transport along the nanowires through the end lids.

In the case of radial heat transport, we varied the thickness of the silicon oxide layers that enwrap the silicon nanowire. In CMOS technology, this would mean a variation of the gate oxide thickness of the nanowire transistors. For the heat source power of 65.2 eV·ps^{-1}, distributed uniformly in the silicon part of the nanowire, the resulting temperature distribution along the radius from the nanowire axis is shown in Fig. 5. Because of the periodicity of the simulation setup in the axial direction and the uniformity of the atomic arrangement in this direction, the temperature depends only on the radial position ρ.

At small radius values, ρ < 1 nm, i.e. in the middle of the nanowire, the temperature is about constant. Then, at ρ > 2 nm, there is a steep decrease of the temperature towards the silicon-to-dioxide interface at ρ_{Si} = 2.6 nm. The decrease of the temperature continues also in the silicon dioxide layer. At the interface, a change in the slope of the curves is clearly visible.

To properly understand the role of radial heat transport from the nanowires, we will compare the absolute values of the thermal resistances for nanowires with the values typical

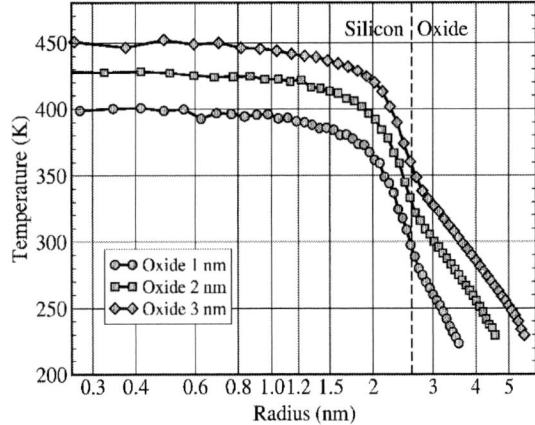

Fig. 5. Temperature distribution as a function of the radial position from the axis of the cylindrical nanowire. The radius of the silicon nanowire is 2.6 nm, the nanowire side wall is covered by silicon dioxide with a thickness of 1, 2, or 3 nm as indicated in the legend. The heat source power is 65.2 eV·ps^{-1}.

for applications in the CMOS technology. Accordingly, we also calculated the temperature distributions for the case of axial thermal transport. We varied the length of the nanowires and the thickness of the oxide on the silicon surface through which the heat transfer is established.

Fig. 6 shows axial temperature distributions for a silicon nanowire with a length of 11 nm covered on the end lids with an oxide layer of different thickness. The temperature profiles are qualitatively similar to the ones shown in the previous Fig. 5. The thicker the oxide is, the higher the maximum temperature and the lower the temperature at the boundary points of the curves. From the temperature distributions like the ones presented in Figs 5 and 6, thermal resistances for the heat transfer from the nanowires to external environment were calculated and the results are presented below.

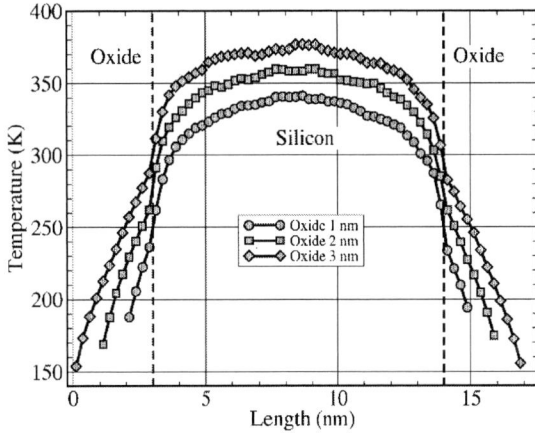

Fig. 6. Temperature distribution along the axis of the cylindrical nanowire with a length of 11 nm. The radius of the silicon nanowire is 2.6 nm, the nanowire top and bottom sides are covered by silicon dioxide with a thickness of 1, 2, or 3 nm as indicated in the legend. The heat source power is 15 eV·ps^{-1}.

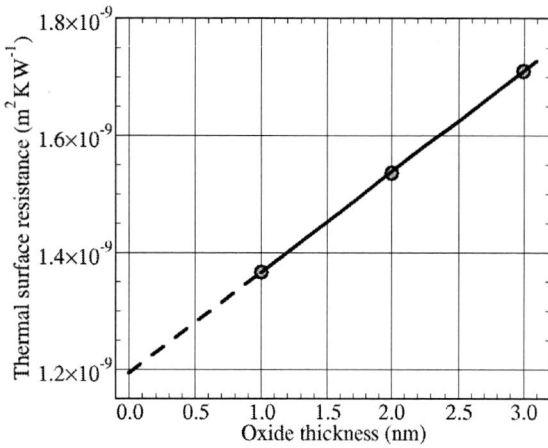

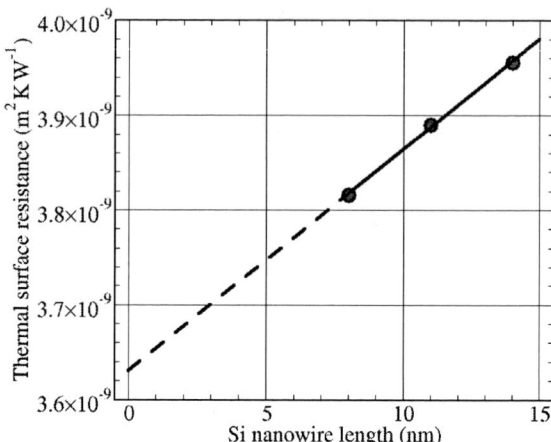

Fig. 7. Thermal surface resistance between the axis of the cylindrical silicon nanowire and the surface of the nanowire stack as a function of the silicon dioxide thickness covering the cylindrical side wall of the silicon nanowire.

Fig. 8. Thermal resistance between the middle of the cylindrical silicon nanowires and the external top and bottom surfaces of the nanowire sample as a function of the silicon nanowire length. Thickness of the oxide covering the top and bottom sides of the silicon cylinders is 3 nm.

Fig 7 shows the thermal surface resistance R_{ths} as a function of oxide thickness for the radial heat transfer from the silicon nanowire to the outside of the oxide shell surrounding the nanowire as shown in Fig. 1. R_{ths} corresponds to the total thermal resistance multiplied by the area of the silicon surface through which the heat is transferred.

As demonstrated in Fig. 7, the radial thermal resistance increases linearly with the thickness of the oxide layer surrounding the silicon. This is qualitatively in accordance with the macroscopic thermal transport theory, but there are peculiarities for nanowires. First, in macroscopic theory the linearity holds for thin isolation layers $t_{ox} \ll \rho_{Si}$ only. Here, we have $t_{ox} = 3$ nm, and $\rho_{Si} = 2.6$ nm, but the linearity still holds. Then, if we extrapolate the value of R_{ths} to $t_{ox} = 0$, the R_{ths} for $t_{ox} = 0$ still has a considerable residual value of about $1.2 \cdot 10^{-9}$ $m^2 KW^{-1}$. This value is well comparable with the values of R_{ths} for oxide thicknesses in technologically relevant cases of $t_{ox} = 1$ to 3 nm. This residual value has two important contributions, the silicon-to-oxide interface resistance and the distributed resistance during the heat transport in silicon. Due to a strong impact of nano-size effects, which we demonstrated in Section 3, we could not separate these two contributions.

Fig.8 shows the axial thermal surface resistance R_{ths} in dependence on the nanowire length for the set-up in Fig.2. As for the radial heat transport, a linear dependence is observed. For a cylindrical silicon nanowire with $\rho_{Si} = 2.6$ nm and $L_{Si} = 10$ nm, a direct comparison of the values of the thermal resistances for radial and for axial transport in Table III shows that the resistances for the radial transport are about factor 3 to 4 lower than the resistances for axial heat transport along the <100>-oriented nanowire axis. This means that for the nanoscale structures investigated here, the radial heat transfer through the gate oxides can be more efficient than the axial heat transfer through the ends of the nanowires.

V. CONCLUSIONS

In conclusion, we have studied both radial and axial heat transfer in silicon nanowires surrounded by silicon dioxide using NEMD simulations. The results indicate a more intensive heat transfer in radial direction than along the axis.

TABLE III. THERMAL RESISTANCE TO HEAT FLOW FROM THE UNIFORMLY HEATED CYLINDRICAL SILICON NANOWIRES THROUGH THIN OXIDE LAYERS AT THE LIDS TO THE EXTERIOR

Direction of Heat transfer	Oxide Thickness (nm)	Silicon Nanowire Length (nm)	R_{th} (KW^{-1})
Radial	1	10	$1.67 \cdot 10^7$
Radial	2	10	$1.87 \cdot 10^7$
Radial	3	10	$2.09 \cdot 10^7$
Axial	1	10	$6.14 \cdot 10^7$
Axial	2	10	$7.75 \cdot 10^7$
Axial	3	10	$9.10 \cdot 10^7$

The observed deviation of the temperature distribution near the heat source in the silicon part of the locally heated nanowire from Fourier's law requires a further analysis.

[1] A. Boukai, Y. Bunimovich, J. Tahir-Kheli et al., "Silicon nanowires as efficient thermoelectric materials," Nature vol. 451, pp. 168–171, January 2008; A. Hochbaum, R. Chen, R. Delgado et al., "Enhanced thermoelectric performance of rough silicon nanowires," Nature vol. 451, pp. 163–167, January 2008.

[2] S. G. Volz, G. Chen, "Molecular dynamics simulation of thermal conductivity of silicon nanowires," Applied Physics Letters vol. 75 (14), pp. 2056-2058, October 1999.

[3] M. Verdier, Y. Han, D. Lacroix, P-O. Chapuis. and K. Termentzidis, "Radial dependence of thermal transport in silicon nanowires," J. Phys. Mater., vol. 2, pp. 015002(1-8), November 2019.

[4] A. Burenkov, V. Belko and J. Lorenz, "Self-heating of nano-scale SOI MOSFETs: TCAD and molecular dynamics simulations," 19th International Workshop on Thermal Investigations of ICs and Systems (THERMINIC), Berlin, 2013, pp. 305-308.

[5] S. Plimpton, "Fast Parallel Algorithms for Short-Range Molecular Dynamics," J. Comp. Phys., vol. 117, pp. 1-19, March 1995; http://lammps.sandia.gov

[6] S Munetoh, T. Motooka, K. Moriguchi, and A. Shintani, "Interatomic potential for Si-O systems using Tersoff parameterization," Comput. Materials Sci., vol. 39, pp. 334-339, 2007.

[7] W. Zhu, G. Zheng, S. Cao, H. He, "Thermal conductivity of amorphous SiO2 thin film: A molecular dynamics study," Sci. Rep., vol. 8, pp. 10537(1-9), 2018.

Advanced simulations on laser annealing: explosive crystallization and phonon transport corrections

Alberto Sciuto

Dipartimento di Fisica e Astronomia
Ettore Majorana
Università di Catania
Via S. Sofia 64, 95123 Catania, Italy
and CNR IMM
Z.I. VIII Strada 5, 95121 Catania, Italy
Email: alberto.sciuto@imm.cnr.it

Ioannis Deretzis,
Giuseppe Fisicaro,
Salvatore F. Lombardo,
Antonino La Magna

CNR IMM
Z.I. VIII Strada 5, 95121 Catania, Italy

Maria Grazia Grimaldi

Dipartimento di Fisica e Astronomia
Ettore Majorana
Università di Catania
Via S. Sofia 64, 95123 Catania, Italy

Karim Huet,
Bobby Lespinasse,
Armand Verstraete,
Benoit Curvers

LASSE
laser systems and solutions of Europe
14-38 Rue Alexandre, 92230 Gennevilleres, France

Igor Bejenari,
Alexander Burenkov,
Peter Pichler

Fraunhofer IISB
Schottkystrasse 10, 91058 Erlangen, Germany

Abstract—Current semiconductor device manufacturing often needs the integration of annealing process steps with a low thermal budget; and, among them, pulsed laser annealing (LA) is a reliable option. Consequently, the use of LA specialized Technology Computer Aided Design (TCAD) models is emerging as a support for the development of this particular heating methods. Anyway, models already implemented in academic or commercial packages usually consider some approximations which can lead to inaccurate predictions if they are applied in rather common configurations of nano-device: i.e. structures with *nm* wide elements where amorphous pockets are also present. In particular, in these cases non-diffusive thermal transport and explosive crystallization could take place. Here we present upgrades of the LA TCAD models allowing the simulation of these phenomena. We will demonstrate that these models can be reliably integrated in the current TCAD packages discussing the key features of the numerical solutions features in some particular cases.

I. INTRODUCTION

Thermal annealing with low thermal budget is nowadays a common processing step in the semiconductor device manufacturing. Laser annealing (LA), in particular, can be integrated in the process flow for micro-and nano-electronics, yielding versatile and powerful solutions in extremely constrained space and time scales. However, optimal control is a key

The research leading to these results has received funding from the European Union's Horizon 2020 research and innovation programme under grant agreement No. 871813 MUNDFAB.

issue for the successful application of LA [1], [2] and, within this context, reliable simulations of LA processes are required for optimizing the process parameters while reducing the number of experimental tests. Despite LA models have been already implemented in academic or commercial packages, a further evolution of the methods is necessary for the general application in complex structures with *nm* wide elements, where the thermal transport is strongly limited by phonon effects. Moreover, due to the frequent presence of amorphous pockets in processed device structures, ultra-fast (explosive) amorphous-to-crystal phase transitions occurring during the irradiation of the devices could present a further challenging objective for LA modelling and simulation.

In this context, we integrate phonon transport corrections [3] and explosive crystallization models [4] to our existing simulation tool, LIAB (LASSE Innovation Application Booster) [5], which allow for reproducing experimental data while predicting the behavior of semiconductor structures upon laser annealing. We have simulated thermal transport induced by either conventional and laser annealing in various structures, comparing the results in which standard conditions and corrections of the Fourier law were applied. We demonstrate that corrections due to the finite phonon mean free paths can be suitably included in annealing process simulations of three-dimensional nanosystems. Moreover, the reliability of the corrections has been verified comparing continuous solutions and Molecular Dynamic simulations in equivalent systems and heating conditions [6]. Finally we will demonstrate that the implementation in LIAB of a three phase mode 1 [4] allows accurate numerical solutions of the explosive crystallization phenomena in structured samples, therefore the impact of this phenomenon in the processing of nano-device can be potentially predicted.

II. PHONONIC CORRECTION TO THERMAL TRANSPORT

A possible approach for the TCAD of a laser annealing process in systems with very small structures is the preliminary investigation based on Molecular Dynamics or Boltzmann Transport Equation of the analysed structure and the use of the estimated dependence of $k_{app}(T)$ in the Time dependent Fourier Law.

$$C(T)\frac{\partial T}{\partial t} = \nabla \cdot [k_{bulk}(T)\nabla T] + S_T(t) \qquad (1)$$

where $C(T)$ is the thermal capacitance, $k_{bulk}(T)$ the bulk thermal conductivity and $S_T(t)$ is the time dependent internal heat source which implicitly depends on the T field.

However, this procedure is computationally expensive and practically unfeasible for general applications. An alternative method has been treated in the case of LA simulation in [3]. This method is consistent with the general result that the temperature field is ruled by Eq. 1 with the bulk value of the conductivity $k_{bulk}(T)$ whilst corrections due to the finite phonon mean free path have to be considered only for the boundary conditions. In particular, using jump-lick BCs, exact solutions in the diffusive $\langle Kn \rangle << 1$ and ballistic $\langle Kn \rangle >> 1$ limit can be obtained, while the solutions deviate only by a few percent with respect to the BTE solution in the intermediate $\langle Kn \rangle \sim 1$ region. In this formalism, Dirichelet boundary conditions for a system in contact with a thermostat at a T_0 is substituted by

$$\hat{n} \cdot \vec{F}_Q = k_{bulk}\lambda^{-1}(T - T_0) \qquad (2)$$

Where $\vec{F}_Q = \vec{F}_Q^+ - \vec{F}_Q^-$ is the combined phonon flux coming from the left and right size of the junction between the nano-system and the thermostat, k_{bulk} is the bulk thermal conductivity of the material and λ is the average phonon scattering length, which can be related to known and measurable material properties as

$$\lambda(T) = 4k_{bulk}(T)/C(T)v_s(T) \qquad (3)$$

Moreover, the material boundary is ruled by the jump condition

$$-k_{bulk}^1 \nabla T_1 = \hat{\Pi}(T_1|_{\Gamma}^+ \hat{n}_1 + T_2|_{\Gamma}^- \hat{n}_2) \qquad (4)$$

for the temperatures T_1 and T_2 at the interface Γ between two regions made of different materials. In Eq. 4 $\hat{n}_1 = -\hat{n}_2$ are the local outward normal unit vectors to the two sides of the boundary and $\hat{\Pi} = (k_{bulk}^1 \lambda_1^{-1} + k_{bulk}^2 \lambda_2^{-1})|_{\Gamma}$ is the local phonon scattering functional. Moreover, the energy conservation condition at the boundary has to be imposed as:

$$-k_{bulk}^1 \nabla T_1|_{\Gamma} = -k_{bulk}^2 \nabla T_2|_{\Gamma} \qquad (5)$$

Eq. 5 acts as a closure of the model at the interface location. We note that an external boundary simulating a virtual interface with the same material is still ruled by the Neumann relation $\nabla T = 0$.

The reliability of the approach can be demonstrated by calculating the apparent conductivity k_{app} which can be directly compared with MD estimates or experimental data [3]. According to Ref. [7] k_{app} can be calculated by:

$$k_{app}(T) = k_{bulk}(T)\frac{\langle T_{standard} - T_0 \rangle}{\langle T_{corrected} - T_0 \rangle} \qquad (6)$$

where $T_{correct}$ and $T_{standard}$ are the corrected and diffusive temperature fields evaluated numerically, the symbol T_0 is the thermostat temperature while $\langle \rangle$ indicates the average of the field expression over the space region occupied by the structures.

As an example of the corrections' impact, we consider a Si nanowire (NW) having an almost regular polygon with 12 edges, in contact with one or two thermostats along one or two perimeter edges, while internally heated by a uniform source. In Fig. 1(a) we report a cross section perpendicular to the Si NW axis, of the thermal field obtained for the standard Dirichlet BC(s) (left side) and the corrected BC(s) expression (right side) for the edge(s) in contact with the thermostat. The temperature obtained in the two cases along the NW diameter connecting the center of the thermostat edge and the opposite edge is shown in Fig. 1(b). The difference between the standard and the corrected solution is relevant for this structure and it tends to enlarge (decrease) for smaller (larger) NWs.

The $k_{app}(T)$ value estimated with the formula 6 for the NW with a diameter of 56nm is $k_{app}(300K) = 21.5[W/mK]$, for the thermal contact configuration of fig. 1, which is in nice agreement with the experimental measurement of ref. [8]. $k_{app}(T)$ reduces monotonically with the NW diameter; e.g. its value for T=300K and a NW of a 5.2 nm is in the range $k_{app} \sim 1 - 2[W/mK]$, in dependence of the contact realization. These values agree with MD estimates for NW of same size.

After the integration of the corrected boundary conditions in the LIAB package we can simulate the full LA process. Our computational code [5] uses the FEniCS (https://fenicsproject.org/) computing platform for solving the evolution of continuum fields and the Gmsh (http://gmsh.info/) library as a 3D finite element mesh generator.

As an example, we have investigated the laser thermal treatment of a 56 nm wide infinite Ge nanowire on a Si substrate. The presence in the device structures of interface between crystalline and amorphous materials needs a proper "ad-hoc" calibration for the interface boundary expression since the average phonon scattering length cannot be properly estimated by means of material properties in the amorphous phase. A ubiquitous example is the interface between the structure and the air. In the modeling of the process we introduced a scattering length λ_{air} to represent the air side of the interface between the studied system and the air portion itself. Fig. 2(a) shows the heat source distribution deriving

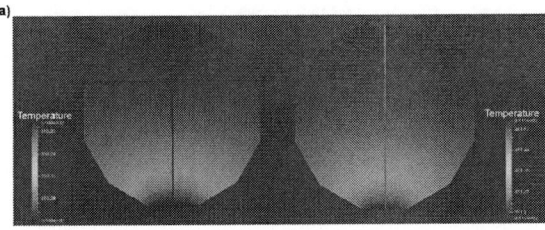

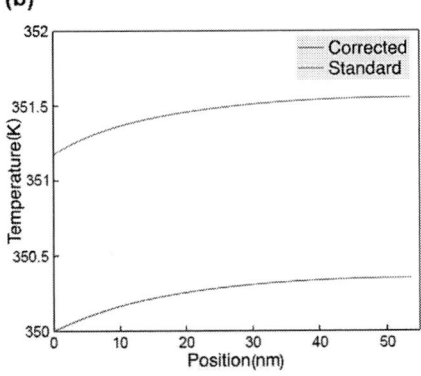

(b)

Fig. 1. (a) Cross section of a heated Si nanowire with a diameter of 56 nm: standard (left) and corrected (right) solutions. (b) Standard and corrected T values along the nanowire diameter line shown in (a). The correction is implemented as jump-like (i.e. Robin type) boundary condition in the Finite Elements Methods used to solve the heat equation. The discontinuity at the device/thermostat interface reproduces the incomplete thermalization of the internal phonon modes due to the nanosize of the system.

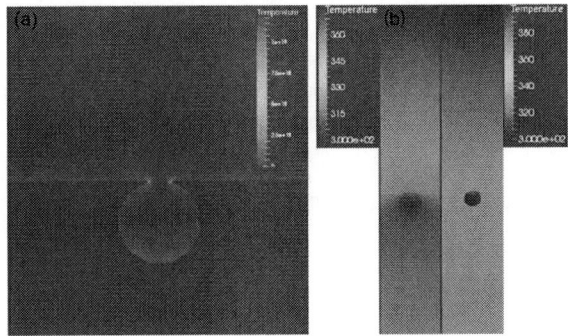

Fig. 2. (a) Source distribution simulated by LIAB after 5 ns heating with λ_{air} = 3 nm,
(b) Comparison between the results of the pulsed LA heating process after 30 ns from pulse starting instant with λ_{air} = 3 nm with (right) and without (left) the corrections

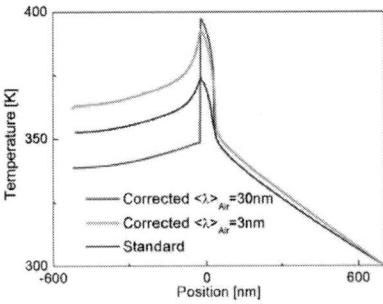

Fig. 3. Temperature profiles of 30 ns heating with different values of λ_{air}.

from the absorption of a laser pulse of $\sim 160ns$ width and wave length of $308nm$ after 5 ns heating for a value of the scattering length of λ_{air} = 3 nm. Using these computed source distribution we simulated the heating using corrected and standard boundary conditions. The role of λ_{air} parameter is shown in Fig. 3 where the temperature profiles for different values of λ_{air} are plotted. A strong confinement is simulated for values of the scattering length similar to the ones derived in crystal material (red line) while the $\lambda_{air} = 3nm$ one gives a smoother and more correct behavior. A further analysis of corrected versus standard solutions can be seen in Fig. 2(b), where the effect of heat confinement is more evident. We notice that the setting of the interface parameters for crystal-amorphous boundary is a critical issue of the method and deserves of further studies also with the aid of MD simulations.

III. EXPLOSIVE CRYSTALLIZATION

Annealing processes of amorphous materials are rather usual in device technology and in most of the cases the crystal phase is recovered at the end of the process. In the case of LA application, the process can easily lead to a complex ultra fast phenomenon known as explosive crystallization. The Explosive Crystallization (EC) mechanism in covalent elemental semiconductors (Si and Ge) is the result of a significant larger (negative) latent heat for the liquid to crystal phase transition

with respect to the one (positive) for the amorphous to liquid transition. The explosive kinetics of the process has to be naturally recovered by the model since there is the unbalance of the thermodynamic parameters of the three simultaneous phases.

The quantitative modeling of the EC phenomenon should allow the computation of the kinetics of both the crystal-liquid and liquid-amorphous front on the basis of the thermodynamic parameters of the materials in study. In ref. [4] a new method for the tracking of the interface is proposed. The formalism is a natural extension of the phase field one that we normally use to study melting and ensures an efficient computational way to solve a multiple phase problem. The new formulation relies on the presence of three local minima of the potential energy density at $-1, 0, 1$ (in contrast with the two minima of the classical phase field Wheeler's formulation [9] that represent, respectively, the amorphous, liquid and crystal phase. All the physical parameters (including the ones ruling the dopant evolution) in this approach must interpolate the calibrated values in the three phases. As an example, heat capacity, one of the parameters of the model that clearly depends on temperature and phase, behaves in this particular way:

$$C(\phi,T) = \theta(\phi)\,C_c(T) + [1 - \theta(\phi)]\,C_a(T) + C_l(T)\,[1 - |\phi|] \tag{7}$$

where $\theta(\phi)$ is the Heaviside step function while the c, a, l subscripts indicate crystal, amorphous and liquid phases respectively.

We have moreover successfully simulated this challenging behavior that includes secondary melting during the process. Fig. 4 shows a 2D melting front driven by an ultrafast re-crystallization in a simple Si slab with a 50 nm narrowing of the structure, in order to test the robustness of the solution with hard edges.

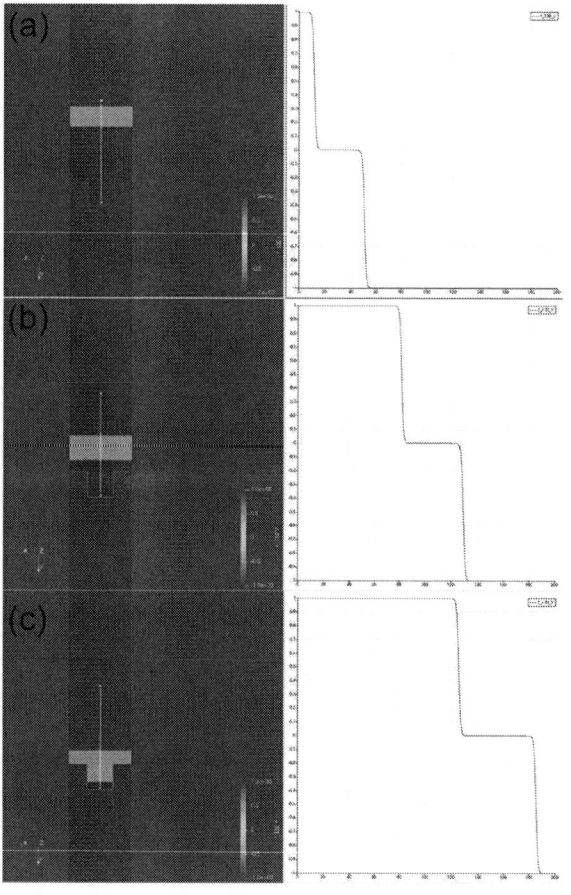

Fig. 4. Snapshots of 2D explosive crystallization at (a) 78 ns (b) 84 ns and (c) 89 ns. The red, blue and grey regions are crystal, amorphous and liquid, respectively. The initial thickness of the amorphous region is indicated by the vertical line, whilst the top region in red is filled by air. The lateral dimension of the structure is 50 nm and periodic boundary conditions are applied along the x-axis.

CONCLUSIONS

This work focuses on two crucial, industry standard driven, aspects of the current advancements in LA simulations. A more refined control on thermal transport is key to correctly simulate the behaviour of Laser Annealing, especially in smaller structures, where geometry and size plays an important role. The failure of classical heat transport in this structures opens the possibility to explore new ways of accounting for phonon driven effects without the need of solving directly transport equations for the particles, the method that we proposed, as we have seen is, in fact, less resource demanding and can be easier to implement with the current state of the art TCAD tools. Another crucial aspect that we implemented in our simulation code is explosive crystallization. The possibility to investigate this ultrafast phenomenon is key to a better understanding and control of advanced processes especially when accounting for dopant diffusion and alloy behaviour. For this purpose, we have implemented a multi-well phase-field model specifically suited for the simulation of explosive crystallization induced by pulsed laser irradiation in the nanosecond time scale. The numerical implementation of the model is robust despite the discontinuous jumps of the interface speed induced by the phenomenon.

REFERENCES

[1] S. F. Lombardo, S. Boninelli, F. Cristiano, G. Fisicaro, G. Fortunato, M. G. Grimaldi, G. Impellizzeri, M. Italia, A. Marino, R. Milazzo, E. Napolitani, V. Privitera, and A. La Magna, "Laser annealing in Si and Ge: Anomalous physical aspects and modeling approaches," *MATERIALS SCIENCE IN SEMICONDUCTOR PROCESSING*, vol. 62, no. SI, pp. 80–91, MAY 2017.

[2] K. Huet, J. Aubin, P. E. Raynal, B. Curvers, A. Verstraete, B. Lespinasse, F. Mazzamuto, A. Sciuto, S. F. Lombardo, A. La Magna, P. Acosta-Alba, L. Dagault, C. Licitra, J. M. Hartmann, and S. Kerdiles, "Pulsed laser annealing for advanced technology nodes: Modeling and calibration," *APPLIED SURFACE SCIENCE*, vol. 505, MAR 1 2020.

[3] A. Sciuto, I. Deretzis, G. Fisicaro, S. F. Lombardo, M. G. Grimaldi, K. Huet, B. Curvers, B. Lespinasse, A. Verstraete, and A. La Magna, "Phononic transport and simulations of annealing processes in nanometric complex structures," *Physical Review Materials*, vol. 4, no. 5, p. 056007, 2020.

[4] S. Lombardo, S. Boninelli, F. Cristiano, I. Deretzis, M. Grimaldi, K. Huet, E. Napolitani, and A. La Magna, "Phase field model of the nanoscale evolution during the explosive crystallization phenomenon," *Journal of Applied Physics*, vol. 123, no. 10, p. 105105, 2018.

[5] S. Lombardo, G. Fisicaro, I. Deretzis, A. La Magna, B. Curver, B. Lespinasse, and K. Huet, "Theoretical study of the laser annealing process in finfet structures," *Applied Surface Science*, vol. 467, pp. 666–672, 2019.

[6] A. Burenkov, V. Belko, and J. Lorenz, "Self-heating of nano-scale SOI MOSFETs: TCAD and molecular dynamics simulations," in *19th International Workshop on Thermal Investigations of ICs and Systems (THERMINIC)*. IEEE, 2013, pp. 305–308. [Online]. Available: 10.1109/THERMINIC.2013.6675230

[7] J. Kaiser, T. Feng, J. Maassen, X. Wang, X. Ruan, and M. Lundstrom, "Thermal transport at the nanoscale: A fourier's law vs. phonon boltzmann equation study," *Journal of Applied Physics*, vol. 121, no. 4, p. 044302, 2017. [Online]. Available: https://doi.org/10.1063/1.4974872

[8] J. Anaya, J. Jimenez, and T. Rodriguez, "Thermal transport in semiconductor nanowires," in *Nanowires*, X. Peng, Ed. Rijeka: IntechOpen, 2012, ch. 11. [Online]. Available: https://doi.org/10.5772/52588

[9] A. A. Wheeler, W. J. Boettinger, and G. B. McFadden, "Phase-field model for isothermal phase transitions in binary alloys," *Phys. Rev. A*, vol. 45, pp. 7424–7439, May 1992. [Online]. Available: http://link.aps.org/doi/10.1103/PhysRevA.45.7424

4-4 Effect of Unit-cell Arrangement on Performance of Multi-stage-planar Cavity-free Unileg Thermoelectric Generator Using Silicon Nanowires

Katsuki Abe
Waseda University
Tokyo, Japan
katsube.aki@akane.waseda.jp

Kaito Oda
Waseda University
Tokyo, Japan
oda-fringe@moegi.waseda.jp

Motohiro Tomita
Waseda University
Tokyo, Japan
m.tomita3@aoni.waseda.jp

Takeo Matsuki
Waseda University, AIST
Ibaragi, Japan
takeo-matsuki@aist.go.jp

Takashi Matsukawa
AIST
Ibaragi, Japan
t-matsu@aist.jp

Takanobu Watanabe
Waseda University
Tokyo, Japan
watanabe-t@waseda.jp

Abstract— We compare the thermoelectric (TE) performances between two types of unit cell arrangements of planar unileg TE generators of silicon-nanowire (Si-NW). The planar TE generators are driven by the temperature gradient along the Si-NW. The TE performance highly depends on how the hot and cold sides of the unit cell is oriented with respect to the neighboring cells. If the hot sides of neighboring cells are placed next to each other, TE power is improved compared to the case where the hot and cold side of TE generator is arranged alternately. Optimal conditions of Si-NW length and metal wiring structure are also discussed.

Keywords—Thermoelectric generator, Si, Nanowire, Semiconductor

I. INTRODUCTION

Energy harvesting such as TE power generation is expected as a new power source technology for the autonomous IoT system. Recently, it has been revealed that nano-sized Si shows excellent TE conversion ability [1]. Our research group proposed a cavity-free planar TE generator using Si-NW [2]. Generally, the output power of single TE generator is small to drive a DC/DC boost converter [3]. Thus, a large number integration of the TE generator is required to develop the output power.

We have investigated two types of unit cell arrangements to achieve the optimum structure which maximizes the power generation of the large-scale integrated cavity-free planar TE generators by means of FEM simulation. TE generators are categorized into two groups: unileg and bileg TE generator. An unileg TE generator is composed of only n-type or p-type semiconductor, and a bileg TE generator is composed of both n- and p-type types semiconductors. Bileg

generators are advantageous to achieve high conversion efficiency between heat and electricity, but the unileg TE generators are also beneficial to reduce the manufacturing cost, and hence, to reduce the cost per power generation from ambient heat energy [4].

In cavity-free planar TE generators, the key to an efficient operation is how the temperature difference across the Si-NW is maintained. To do so, it is important to control the heat flow into the TE generator. In our previous studies [5,6,7], we reported the effects of the substrate structure and electrode area on the TE performance. In the present work, we focus on how to arrange the element of the unileg TE generator into an integrated TE module.

One of the arrangement of unileg TE generator is H-C-H-C arrangement where the hot and cold side of TE generator is arranged alternately as Hot-Cold-Hot-Cold…. However, the distance between the hot and cold side of neighboring TE generators is close to each other. As the result, the temperature difference will be diminished due to the heat leakage from the hot side to the cold side.

We propose C-H-H-C arrangement where the hot side of TE generator is arranged next to each other [7]. This arrangement is realized by a meander wiring [8]. In this study, we compare the TE performance of H-C-H-C and C-H-H-C arrangement of the cavity-free planar TE generators by FEM simulation.

II. SIMULATION METHOD

The model used in FEM calculation are shown in Fig. 1 and 2. Figure 1 and 2 show a cross-sectional view of the unit cell and the bird's-eye view of the 1st Al wiring layer and underlying layer of the unileg TE generator with H-C-H-C and C-H-H-C type arrangement, respectively. Figure 3 shows the top views. C-H-H-C arrangement requires a meander electric wiring to connect the hot and the cold side electrodes. To minimize the circuit footprint, the Al-bridge connecting hot and cold side is passed over the TE generators, but the structure has a slight drawback that the heat flow from the heat guide toward TE generator is impeded at the Al bridge, which will be discussed later.

COMSOL Multiphysics® 5.3 was used for the FEM simulation. The thermal and electrical conductivities of Si-pads, Si-NWs and SiO₂ layer are summarized in Table I. The employed equations in the simulation are summarized in Table II.

Table I Thermal conductivities and electrical conductivities used in FEM simulation

	Si-pad	Si-NW	SiO₂
Thermal conductivity [W/m·K]	131	1.4	1.4
Electrical conductivity [S/m]	232	232	

Table II The equations employed in this study

	Formula	
Heat transfer equation	$\nabla \cdot (\boldsymbol{q} + \boldsymbol{q_e}) = 0$	(1)
Fourier's law	$\boldsymbol{q} = -k \cdot \nabla T$	(2)
Second Thomson relation	$P = ST$	(3)
Seebeck equation	$\boldsymbol{J_e} = -\sigma S \nabla T$	(4)
Peltier equation	$\boldsymbol{q_e} = P\boldsymbol{J}$	(5)
Current continuity equation	$\nabla \cdot \boldsymbol{J} = Q_{j,v}$	(6)
Ohrm's law	$\boldsymbol{J} = \sigma(-\nabla V) + \boldsymbol{J_e}$	(7)

In Table II, $\boldsymbol{q}$ is the heat flux caused by the temperature gradient, $\boldsymbol{q_e}$ is the Peltier heat flux, T is the temperature, k is the thermal conductivity, P is the Peltier coefficient, S is the Seebeck coefficient, $\boldsymbol{J_e}$ is the TE current, σ is the electrical conductivity, $\boldsymbol{J}$ is the total current density, $Q_{j,v}$ is the change of charge density per unit of time and V is the electric potential. The heat current distribution was simulated by simultaneously solving (1) - (5). The electric current was obtained by solving (3) - (7). The above calculations are repeated until both heat current distribution and electric current reach a steady state. A two-dimensional periodic boundary condition is adopted in x and y-axis (Fig. 1(b) and 2(b)).

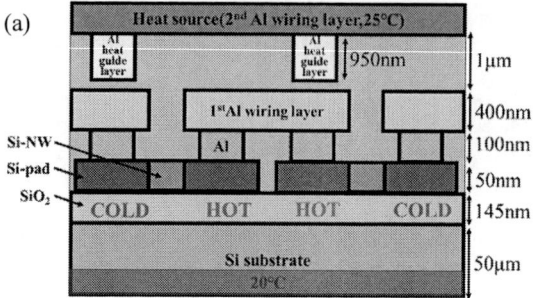

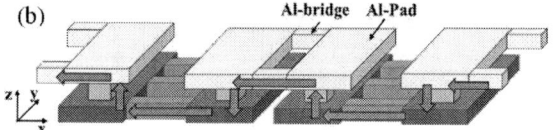

Fig. 1 Schematic of H-C-H-C device (a) cross section view, (b) Bird's-eye view. The green arrows show the current through the device.

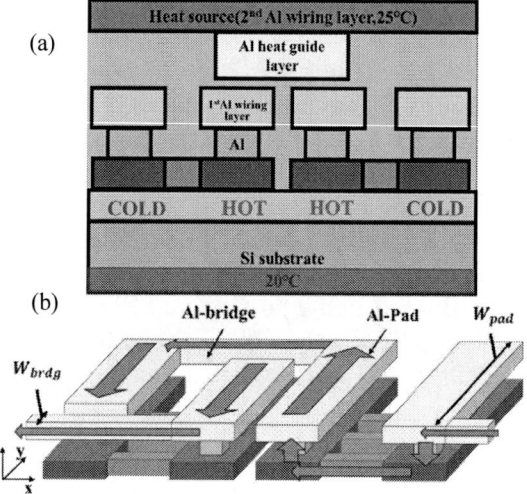

Fig. 2 Schematics of the unit cell of C-H-H-C device. (a) Cross-sectional view. (b) Bird's-eye view.

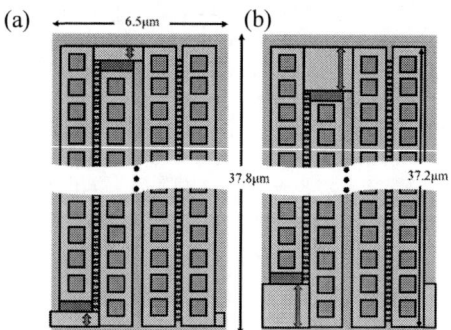

Fig. 3 Schematics of the unit cell of C-H-H-C device. Top view of the 1st Al wiring layer and underlying layers. Red arrow shows the W_{brdg}. (a) The $W_{brdg} \leq$ 1.8 µm. (b) The $W_{brdg} > 1.8$ µm. In case of (b), the Al-pad width (W_{pad}) on the hot side becomes narrower than the W_{pad} when the $W_{brdg} = 1.8$ µm.

In this paper, the interconnecting Al-bridge width (W_{brdg}) of 1.8 μm and the Si-NW length (L_{NW}) of 250 nm are used as reference values. Then, the W_{brdg} was varied from 0.1 μm to 37.2 μm to find the optimum W_{brdg}. Furthermore, the L_{NW} was varied from 250 nm to 1000 nm to find the optimum L_{NW}. The TE power per unit area of the two types of unit cell arrangements were compared.

III. RESULTS&DISCUSSION

The highest TE power and power density of the C-H-H-C type were achieved to 1.61 nW and 655 μW/cm², respectively, when L_{NW} = 250 nm and W_{brdg} = 1.8 μm. The TE power and power density of H-C-H-C type were 1.13 nW and 459 μW/cm², respectively. Figure 4 and 5 show the cross-sectional temperature distributions. Figure 6 shows the temperature profile at the Si-NW surface along its longitudinal direction which is extracted from Fig. 4 and 5. Figure 6 shows the temperature profiles in the Si-NW for H-C-H-C and C-H-H-C type devices.

The temperature difference between both ends of Si-NW in C-H-H-C type is larger than that of H-C-H-C type. The temperature at the hot side of C-H-H-C type is maintained higher than that of H-C-H-C type due to decreasing the heat leakage from the hot side to the cold side. The injection pitch of the heat current has a greater impact on the TE performance, even if the wiring part partially impedes the heat current as shown in Fig. 7. The heat current leakage through the Al-bridge will have an unignorable impact on the performance of the TE generator if we shrink the device width in y-axis (see Fig.7(a)).

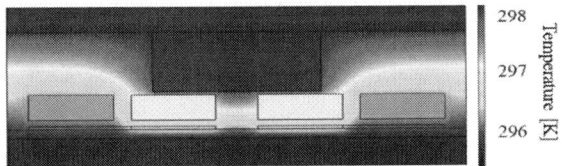

Fig. 4 Cross-sectional temperature distribution in C-H-H-C device (L_{NW}=250 nm, W_{brdg} = 1.8 μm)

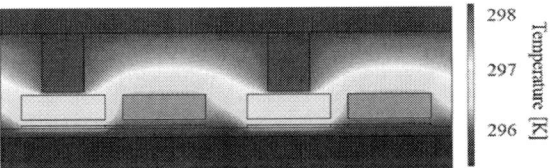

Fig. 5 Cross-sectional temperature distribution in H-C-H-C device (L_{NW} = 250 nm, W_{brdg} = 1.8 μm)

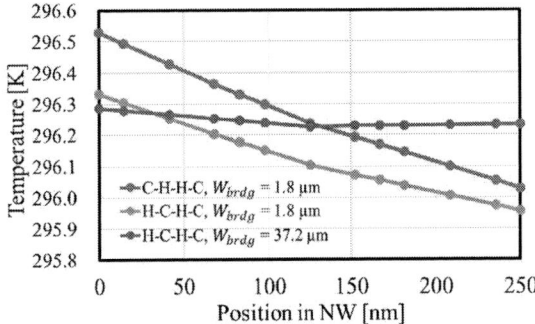

Fig. 6 One dimensional temperature profiles along the Si-NW

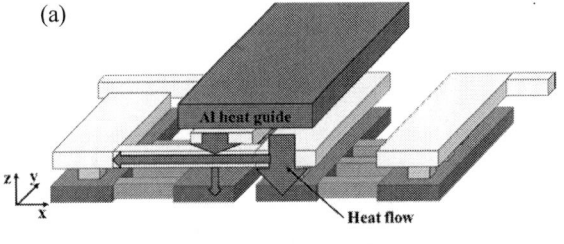

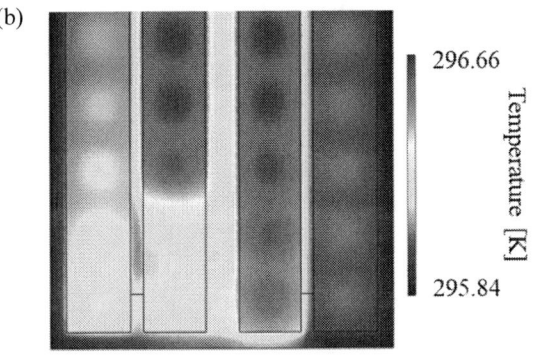

Fig. 7 Disadvantage of the Al-bridge structure passed over the TE generators. (a) Schematic illustration of the heat flow from the heat guide impeded at the bridge. (b) Planar temperature distribution in SOI layer in C-H-H-C device (L_{NW} = 250nm, W_{brdg} = 1.8 μm)

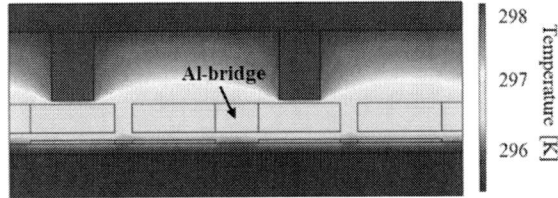

Fig. 8 Cross-sectional temperature distribution in H-C-H-C device (W_{brdg}= 37.2 μm)

Figure 8 shows the cross-sectional temperature distribution in H-C-H-C type when the W_{brdg} = 37.2 μm. The temperature profile along the Si-NW surface is shown in Fig. 6. The 37.2 μm-wide W_{brdg} has low electrical resistance, but also low thermal resistance. Thus, the temperature difference in the TE generator is diminished due to the large heat leakage through the Al-bridge from the hot side to the cold side.

The TE power of C-H-H-C and H-C-H-C type arrangements was calculated with varying the W_{brdg} from 0.1 μm to 9 μm, to find the optimum W_{brdg}. The results are shown in Fig. 9. In C-H-H-C type, the highest power is achieved when the W_{brdg} is 1.8 μm. On the other hand, in H-C-H-C type, the highest power is achieved when the W_{brdg} is 0.3 μm. The TE power of C-H-H-C type surpasses that of H-C-H-C type for all values of W_{brdg}.

Figure 10 show the TE power as a function of L_{NW} for C-H-H-C and H-C-H-C type arrangements. The TE power monotonically increases with L_{NW}. This is because the thermal resistance of Si-NW increases with L_{NW}, thereby the temperature difference across the Si-NW increases.

Figure 11 shows the areal power density vs L_{NW} for C-H-H-C and H-C-H-C type arrangements. From the view point of the areal power density, there exists optimum length L_{NW} in both types of TEGs. In both cases, the power density has a maximum at around L_{NW} = 750 nm. As, the thickness of Si substrate is reduced, the optimum NW length may shorten [6].

From these results, it is concluded that the TE power of C-H-H-C type surpasses that of H-C-H-C type arrangement even though the meander electric wiring partially hinder the injecting heat current.

IV. CONCLUSION

We investigated the effect of unit cell arrangements on the multi-stage-planar cavity-free unileg TE generator performance. We compared two different unit cell arrangements by FEM simulation. C-H-H-C type structure, in which the hot side of the adjacent TE generators are placed opposite to each other, exhibits higher power generation performance than H-C-H-C type. The result indicates that it is quite effective to increase the spatial interval of heat injection pitch to increase the temperature difference across Si-NW legs.

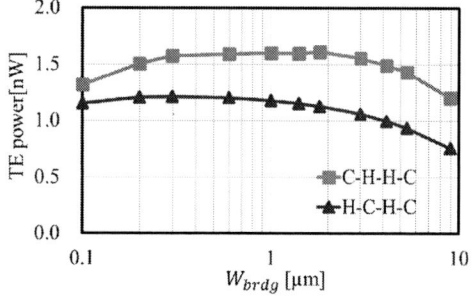

Fig. 9 TE power of H-C-H-C device and C-H-H-C device when the W_{brdg} was varied from 0.1 μm to 9 μm.

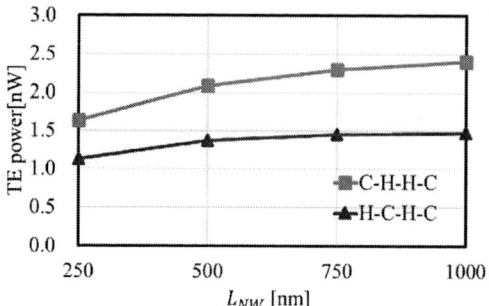

Fig. 10 TE power of H-C-H-C and C-H-H-C type devices.

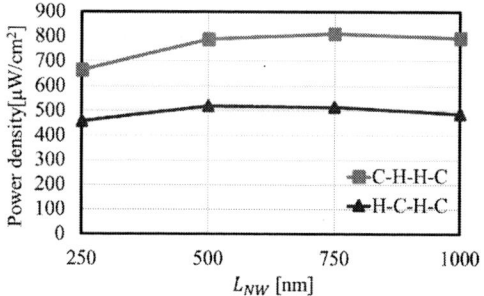

Fig. 11 Power density of H-C-H-C and C-H-H-C type devices.

V. ACKNOWLEDGMENT

This research is supported by JST-CREST (JPLJCR19Q5).

VI. REFERENCES

[1] A. Boukai, Y. Bunimovichet, et al., "Silicon nanowires as efficient thermoelectric materials", Nature, 451.7175 (2008) 168.

[2] H. Zhang, T. Xu, S. Hashimoto, and T. Watanabe, "The Possibility of mW/cm2-Class On-Chip Power Generation Using Ultrasmall Si Nanowire-Based Thermoelectric Generators", IEEE TED 65 (2018) 2016.

[3] M. Tomita, H. Takezawa, et al., "Evaluation of Multi-stage, Unileg, Si-nanowire Thermoelectric Generator with A Cavity-free Planar Device Architecture", Ext. Abstr. SSDM (2019).

[4] Gao Min, "Chapter 11: Thermoelectric Module Design Theories", Thermoelectrics Handbook, ed. By D. M. Rowe, CRC Press, (2006).

[5] K. Oda, K. Shima, M. Tomita, T. Matsuki, and, T. Watanabe, "Optimum Design of Multi-stage Planar Unileg Thermoelectric Generator Using Si Nanowire", Abstr. JSAP Spring, (2019).

[6] K. Shima et al., "Optimum substrate design of planar type Si nanowire thermoelectric generator", Ext. Abstr. SSDM, (2018).

[7] K. Oda, M. Tomita, T. Matsukawa, T. Matsuki, and, T. Watanabe, "Backend Engineering of Cavity-free Planar Si-nanowire Thermoelectric Generator", Ext. Abstr. SSDM, (2019).

[8] T. Huesgen ,N. Kockmann, and P. Woias, "DESIGN AND FABRICATION OF A MEMS THERMOELECTRIC GENERATOR FOR ENERGY HARVESTING", Sensors and Actuators A: Physical, vol.145-146, July-August, pp.423-429 (2008).

Characteristics of Gate-All-Around Silicon Nanowire and Nanosheet MOSFETs with Various Spacers

Sekhar Reddy Kola[1,2], Yiming Li[1-4,*], and Narasimhulu Thoti[1,2]

[1]Parallel and Scientific Computing Laboratory; [2]EECS International Graduate Program; [3]Institute of Communications Engineering; [4]Department of Electrical and Computer Engineering, National Chiao Tung University, 1001 Ta-Hsueh Rd., Hsinchu 300, Taiwan, *Tel: +886-3-5712121 ext. 52974; Fax: +886-3-5726639; Email: ymli@faculty.nctu.edu.tw

Abstract—We estimate DC characteristics and single-charge trap (SCT) induced random telegraph noise (RTN) of gate-all-around (GAA) silicon nanowire (NW) and nanosheet (NS) metal-oxide-semiconductor field effect transistor (MOSFETs) for sub-5-nm nodes. Devices with various dielectric spacers from low- to high-κ including asymmetric dual spacers (ADS) are considered. More than 31% boost on the normalized on-state currents is observed for the explored devices with high-κ and ADS spacers. Similarly, for the normalized off-state currents, more than 50% reduction is achieved. The largest magnitude of the RTN ($\Delta I_D / I_D \times 100\%$) is 6.7% for the nominal GAA Si NS MOSFET with an effective channel width of 40-nm.

Keywords—DC characteristics; SCT; RTN; Gate-all-around; Nanowire; Nonosheet; MOSFETs.

I. INTRODUCTION

Gate-all-around nanowire (NW) and nanosheet (NS) MOSFETs have been of great interest in futuristic technological nodes because of their ultimate channel controllability and high immune to short channel effects (SCEs) [1]-[5]. Ultimately, ultra-small devices suffer from various fluctuations including work function fluctuation, random dopant fluctuation and process variations, interface trap fluctuations, line edge roughness [3],[6]. Furthermore, random telegraph noise (RTN) is a low-frequency noise phenomenon observed in emerging CMOS devices; it changes as discrete jumps in the magnitude of drain current of MOSFETs, due to capture and emission of conduction carriers by single charge trap (SCT) specifically at the interface of channel/oxide [7]. In short-channel devices, the existance of SCT increases the device variability of random telegraph noise (RTN) [8]. Electrical characteristics of silicon GAA NW and NS MOSFETs including FinFETs have recently been reported [4]-[5], [9]-[11]. In GAA NW and NS MOSFETs, the wider cross-section simultaneously affects drive current, electrostatics, and parasitic capacitances. Notably, electrical characteristics MOSFETs can be boosted by using differnt topologies of spacers [3]. The structural and electrical characteristics differences among FinFETs, GAA NW and NS MOSFETs were reported [11], but the impacts of RTN and spacer technology on these devices have not been clearly discussed yet.

In this study, we explore DC characteristics and RTN induced by the acceptor-type SCT appearing in the middle of the interface between channel and oxide of the explored devices: GAA NW MOSFETs and GAA NS MOSFETs with various spacers. Compared with the conventional SiO_2 spacer ($\varepsilon = 3.7\varepsilon_0$), the studied device with asymmetric dual spacer (ADS) consisting of 50% SiO_2 and 50% HfO_2 ($\varepsilon = 22\varepsilon_0$) exhibit interesting DC characteristics and parameters of the short-channel effect (SCE).

TABLE I. THE DEVICE SETTINGS, ACHIEVED DC CHARACTERISTIC AND THE SCE PARAMETERS OF THE 16-NM-GATE GAA NS MOSFET WITH A 40-NM ECW CORRESPONDING TO SUB-5-NM NODES.

Parameters	Settings and Achieved Specification
Gate Length (L_G) (nm)	16
Channel Doping (cm^{-3})	5×10^{17}
Channel Width (W_{Ch}) (nm)	5, 6, 10, 15
Channel Height (H_{Ch}) (nm)	4, 5, 10
S_{ext}/D_{ext} Length (nm)	5
S/D Doping (cm^{-3})	1×10^{20}
S_{ext}/D_{ext} Doping (cm^{-3})	4.8×10^{18}
EOT (nm) ($T_{sio2}+T_{hfo2}\times\varepsilon_{sio2}/\varepsilon_{hfo2}$) $(0.6+2\times4/22) = 0.9636$	0.963
I_{ON} (A)	7.21×10^{-6}
I_{OFF} (A)	5.62×10^{-16}
Threshold Voltage (mV)	260
I_{ON}/I_{OFF} Ratio (A/A)	1.282×10^{10}
SS (mV/dec)	62.8
DIBL (mV/V)	25.9

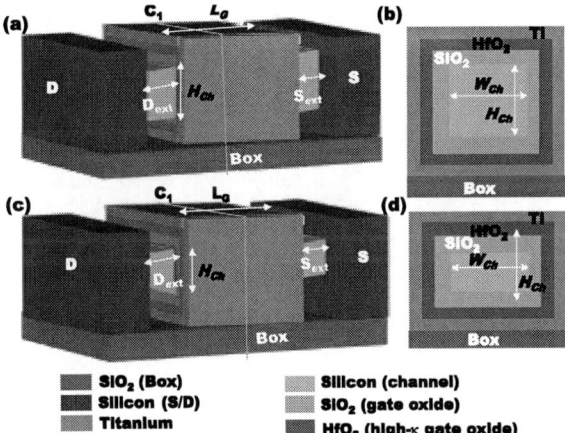

Fig. 1. (a) and (c) are simulated GAA NW MOSFET and GAA NS MOSFET. (b) and (d) are the cross sections of the channel which are 2D cuts (cut-C_1) from (a) and (c), respectively.

II. COMPUTATIONAL DEVICES

A 3D quantum-mechanically correction device simulation is intensively performed by solving a set of density-gradient drift-diffusion equations. Physical parameters of the device simulation are firstly validated through the result of the nonequilibrium green function simulation [7],[12]. The thin layer mobility model engaged with mobility degradation at the interface of semiconductor-insulator [13] is also adopted. To provide the best accuracy of device simulation, calibration with measurement has been reported in our recent study [7]. Devices with 16-nm channel length are with a rectangular channel of 20- and 40-nm effective channel widths (*ECWs*)

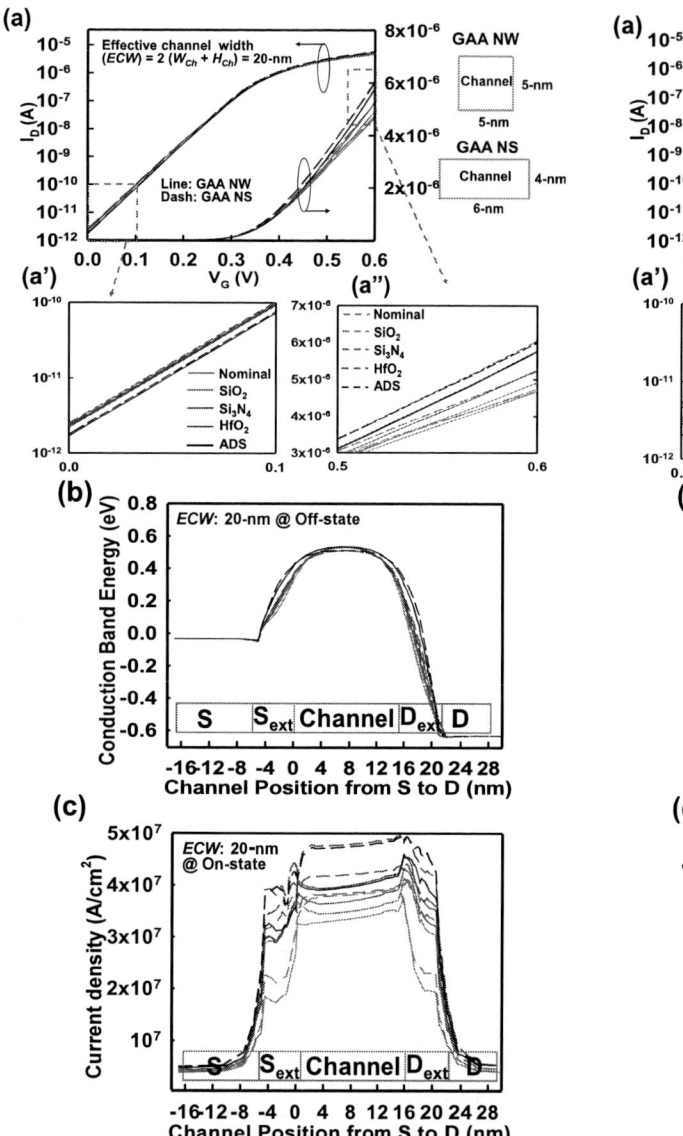

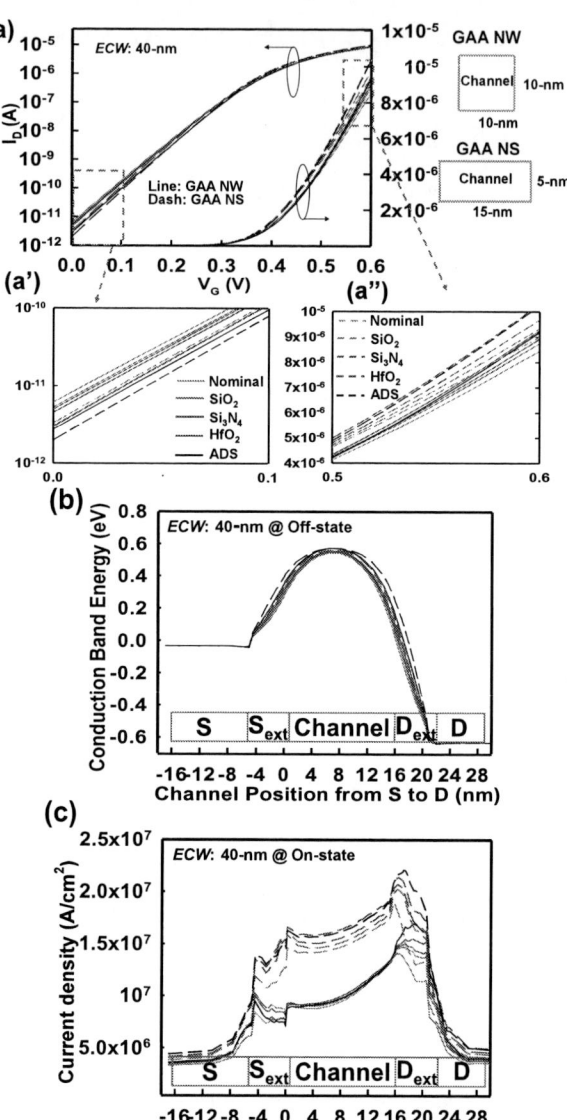

Fig. 2. For the same threshold voltage 260 mV without SCT, (a) simulated I_D-V_G curves of the GAA NW and GAA NS devices with various spacers and a 20-nm ECW. (a') and (a'') Zoom-in plots near the off- and on-state regions. (b) Without SCT, 1D cut views of the off-state ($V_D = 0.6$ V and $V_G = 0$ V) conduction band energy from S to D along the channel for the GAA NW and NS devices with various spacers and the 20-nm ECW. (c) The on-state ($V_D = 0.6$ V and $V_G = 0.6$ V) current density of the GAA NW and NS devices with various spaces and the 20-nm ECW.

are examined, by using the expression as stated in Eq. (1).

$$ECW = 2(W_{ch} + H_{ch}), \quad (1)$$

where W_{ch} and H_{ch} are the width and height of the channel of the explored GAA NW and NS devices. Notably, the 16-nm channel length is correspond to sub-5-nm technological nodes, according to projections of the International Technology Roadmap for Semiconductors [14]. Figs. 1(a)-(d) show the studied devices and cross-sectional views of the GAA NW and NS devices. The simulated structure consists of a titanium metal gate with low-κ (SiO₂) and high-κ (HfO₂) stack, and the effective oxide thickness (EOT) of 0.963 nm, calculated by

Fig. 3. (a) I_D-V_G curves of the GAA NW and NS devices with various spacers and a 40-nm ECW. (a') and (a'') Zoom-in plots near the off- and on-state regions. (b) Without SCT, 1D cut views of the off-state ($V_D = 0.6$ V and $V_G = 0$ V) conduction band energy from S to D along the channel for the GAA NW and NS devices with various spacers and the 40-nm ECW. (c) The on-state ($V_D = 0.6$ V and $V_G = 0.6$ V) current density of the GAA NW and NS devices with various spacers and the 40-nm ECW.

$$EOT = T_{sio2} + T_{hfo2} \times \frac{\varepsilon_{sio2}}{\varepsilon_{hfo2}}. \quad (2)$$

To assess the effect of spacers on device characteristics, we consider the nominal case (i.e., devices without any spacer), low-κ (SiO₂), Si₃N₄, high-κ (HfO₂), and 50% SiO₂ + 50% HfO₂ for the case of ADS to cover the source (S) and drain (D) extensions [3].

The adopted device parameters and the achieved characteristics including SCE parameters are listed in Table I, for the nominal n-type device. According to our recent study [7], we assume that SCT has a rectangular shape with (2 nm)³ at the middle of the interface between channel and oxide.

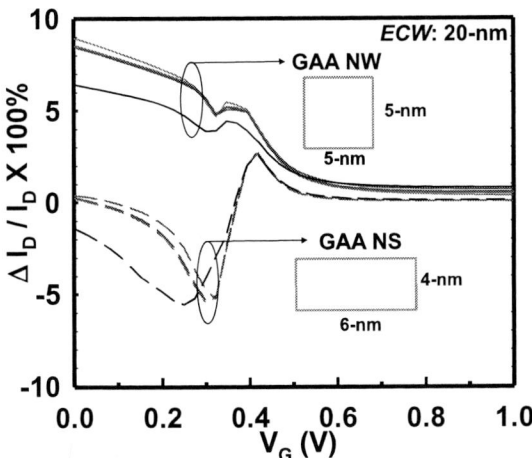

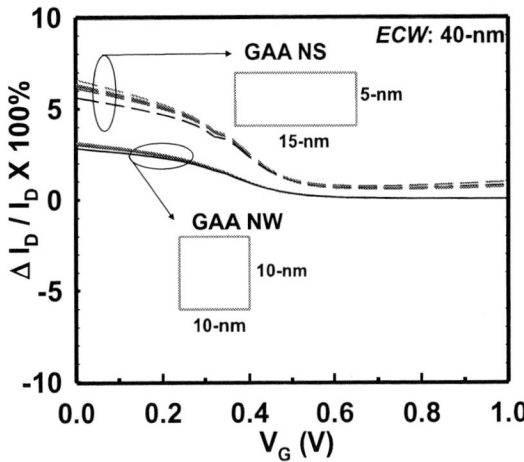

Fig. 4. (a) and (b) The statistically calculated magnitude of the RTN induced by the SCT locating at middle of the surface of the channel, under the biasing of $V_D = 0.01$ V and $V_G = 1$ V for both the GAA NW and NS MOSFETs with various spacers of the 20-nm ECW.

Fig. 5. (a) and (b) The statistically calculated magnitude of the RTN induced by the SCT locating at middle of the surface of the channel, under the biasing of $V_D = 0.01$ V and $V_G = 1$ V for both the GAA NW and NS MOSFETs with various spacers of the 40-nm ECW.

To explore the most severe impact of SCT on RTN, the density of interface trap (D_{it}) of 5.6×10^{11} cm^{-2}eV^{-1} is assumed for all stated devices [7].

III. RESULTS AND DISCUSSION

In order to investigate the difference stemming only from the different geometries of GAA NW and NS MOSFETs, it is necessary to impose similar threshold voltage for all devices. The transfer characteristics for all nominal devices are tuned to the same threshold voltage (V_{th}) at 260 mV by adjusting the metal work function. The I_D-V_G curves of the devices with the 20-nm ECW, as shown in Fig. 2(a), show the comparison among the aforementioned devices with various spacers. The comparison reveals that the impact of spacers on the enhancement of the on-state current and the reduction of the off-state current which are further examined via the zoom-in plots, as shown in Figs. 2(a')-(a''). It is not shown here, first we cut from the simulated 4D results to a 3D plot of the conduction band energies for the devices with the 20-nm ECW. Then, the 3D plot is further sliced into a 2D off-state conduction band energy. As shown in Fig. 2(b), the devices with all spacers are plotted together. Similarly, the on-state current densities of the 20-nm ECW are shown in Fig. 2(c). Furthermore, the I_D-V_G curves of the devices with the 40-nm ECW, as shown in Fig. 3(a), compare the devices with aforementioned various spacers. The comparison reveals that the impact of spacer on the enhancement of the on-state current and the reduction of the off-state current are further examined as zoom-in results shown in Figs. 3(a')-(a''). The off-state conduction band energy profiles of the devices with the 40 nm ECW are shown in Fig. 3(b). In addition, the on-state current densities of the devices with the 40-nm ECW are shown in Fig. 3(c). The majority of electron transport is controlled through large overlapped gate along the channel direction; therefore, the scattering of electrons can be effectively controlled in GAA NS MOSFETs that resemble low off-state current; thus, the conduction band energy is higher during the off-state in the GAA NS MOSFET compared to the NW one. For the GAA NS MOSFET with the ADS, the highest normalized on-state current is increased by 31.4%; similarly, a 65.4% reduction on the normalized off-state current is observed, as shown in Figs. 2(a) and 3(a). At high gate bias, the fringing electric field will penetrate from

the S/D extension regions owing to heavy dopant; thus, the gate fringing electric field cannot change the resistance of S/D extensions. This leads to a relatively high on-state current and low off-state currents. We do study the relationship of the dielectric constant and characteristics are further examined by observing the off-state conduction band energy profiles of the 20- and 40-nm $ECWs$, as shown in Figs. 2(b) and 3(b), respectively. The higher conduction band energy at off-state leads to lower off-state current; therefore, the ADS spacer records high conduction band compared to other spacers. Similarly, for the on-state, due to the extension in the spacer region constitutes for a high fringing electric-field area that leads to high current density, thereby improving the on-state current, as shown in Figs. 2(c) and 3(c) for the 20- and 40-nm $ECWs$, respectively. Both the GAA NW and NS MOSFETs with the ADS spacer reports high electric field compared to other spacers; thus, the ADS records high on-state current for both devices in Figs. 2(c) and 3(c). The effect of spacers on SCEs for both the GAA NW and NS MOSFETs with the 20- and 40-nm $ECWs$ are compared and listed in Table II. The comparison results reveal that the I_{ON}/I_{OFF} current ratio is increased for the GAA NS and NW MOSFETs from low- to high-κ spacers. The extracted parameters of SCE are suppressed for the GAA NS MOSFETs compared to the GAA NW MOSFET from low- to high-κ spacers.

It has been explored that the magnitudes of RTNs induced by SCT are significant in planar MOSFETs and bulk FinFETs [7]; however, the explored GAA NW and NS MOSFETs have improved immunity to the impact of SCT. Fig. 4 shows the magnitude of RTN induced by SCT of the devices with the 20-nm ECW. For the 20-nm-ECW, the channel height of the nominal GAA NS MOSFET is similar to the GAA NW device, owing to strong gate controllability which leads to higher immunity to SCT, as shown in Fig. 4. The magnitude of RTN induced by SCT of the devices with the 40-nm ECW is shown in Fig. 5. However, the largest magnitude of RTN is 6.7% (about 3% for the case of GAA NW device) in Fig. 5 for the GAA Si NS MOSFET with the 40-nm ECW. The off-state current of the GAA NS MOSFET is lower than that of the GAA NW one at small V_G. When V_G increases, the device starts to induce channel and conduct, vast conduction carriers are generated in the channel which enhances the SCT capture

TABLE II. WITHOUT SCT, COMPARISON OF THE SIMULATED AND EXTRACTED SCE PARAMETERS, STATISTICALLY CALCULATED MAGNITUDES OF RTN FOR THE GAA NW AND NS MOSFETs WITH A 20 AND 40-NM *ECWs* AND VARIOUS SPACERS. THE COMPARISON IS WITH THE SAME THRESHOLD VOLTAGE OF 260 MV. THE CASES OF THE NOMINAL SPACER ARE THE DEVICES WITHOUT SPACERS. NOTABLY, FOR THE CASES FROM LOW- TO HIGH-κ, THE SIMULATED SCE OF ALL DEVICES ARE IMPROVED.

Device		20-nm *ECW*		40-nm *ECW*	
Spacer	Parameter	GAA NW MOSFET	GAA NS MOSFET	GAA NW MOSFET	GAA NS MOSFET
Nominal	I_{ON}/I_{OFF} Ratio	3.5×10^9	3.4×10^9	1.32×10^9	3.7×10^9
	SS (mV/dec)	61.3	61	66.8	63.4
	DIBL (mV/V)	11	10.2	44	25
	Max. RTN ($\Delta I_D/I_D \times 100\%$)	8.9	0.3	3.1	6.7
SiO$_2$	I_{ON}/I_{OFF} Ratio	4×10^9	3.6×10^9	1.7×10^9	4.4×10^9
	SS	61.4	61.2	66.2	63.2
	DIBL	10.8	9.6	41	22
	Max. RTN	8.8	0.26	2.7	6.12
Si$_3$N$_4$	I_{ON}/I_{OFF} Ratio	4.2×10^9	3.95×10^9	1.9×10^9	4.6×10^9
	SS	61.2	61.3	66.1	63.3
	DIBL	10.8	9.5	40	22
	Max. RTN	8.4	-0.8	2.8	6.05
HfO$_2$	I_{ON}/I_{OFF} Ratio	5.7×10^9	4.22×10^9	2.24×10^9	4.9×10^9
	SS	61.2	61	65.4	63.2
	DIBL	12	9.2	39.6	23
	Max. RTN	8.2	-1.2	2.72	6
ADS	I_{ON}/I_{OFF} Ratio	6.5×10^9	6.2×10^9	4.23×10^9	8.3×10^9
	SS	60.9	60.9	64.3	62
	DIBL	9.4	9	32.7	19
	Max. RTN	6.4	-1.9	2.48	5.5

and emission of carriers; consequently, it leads to large magnitude of RTN. It is because of that the magnitude of RTN is essentially dependent on the conducting carriers in the channel. It is also infers that for all spacer devices, the increase in spacer dielectric further reduces the tunneling gate leakage there by a reduction in the magnitude of RTN.

IV. CONCLUSIONS

In summary, the reported rectangular shape channel GAA NW and NS MOSFETs has been studied for the DC and RTN induced by SCT with various spacers materials. The ADS has wonderful characteristics compared to other spacer materials; in particular, for the off-state current, it is about 65.4% reduction and the boost of on-state current is up to 31.4% for the GAA NS MOSFET of 40-nm ECW. We have explored the impacts of the SCT on the RTN for the GAA NW and NS MOSFETs. Thus, the influence of RTN (at low gate bias) is more significant in GAA NW MOSFETs compared to GAA NW MOSFETs at scaled ECWs.

ACKNOWLEDGMENT

This work was supported in part by the Ministry of Science and Technology, Taiwan, under Grant MOST 109-2221-E-009-033, Grant MOST 108-2221-E-009-008 and Grant MOST 108-3017-F-009-001, and in part by the "Center for mm Wave Smart Radar Systems and Technologies" under the Featured Areas Research Center Program within the framework of the Higher Education Sprout Project by the Ministry of Education in Taiwan.

REFERENCES

[1] F. M. Bufler, R. Ritzenthaler, H. Mertens, G. Eneman, A. Mocuta, and N. Horiguchi, "Performance Comparison of n-Type Si Nanowires, Nanosheets, and FinFETs by MC Device Simulation," *IEEE Electron Device Lett.*, vol. 39, no. 11, pp. 1628–1631, 2018.

[2] A. Dasgupta, S. S. Parihar, H. Agarwal, P. Kushwaha, Y. S. Chauhan, and C. Hu, "Compact Model for Geometry Dependent Mobility in Nanosheet FETs," *IEEE Electron Device Lett.*, vol. 41, no. 3, pp. 313–316, 2020.

[3] S. R. Kola, Y. Li, and N. Thoti, "Effects of a Dual Spacer on Electrical Characteristics and Random Telegraph Noise of Gate-All-Around Silicon Nanowire p-type Metal-Oxide-Semiconductor Field-Effect Transistors," *Jpn. J. Appl. Phys.*, vol. 59, no. SGGA02, pp. 1-5, 2020.

[4] P. Harsha Vardhan, Amita, S. Ganguly, and U. Ganguly, "Threshold Voltage Variability in Nanosheet GAA Transistors," *IEEE Trans. Electron Devices*, vol. 66, no. 10, pp. 4433–4438, 2019.

[5] Y. Lee *et al.*, "Design Study of the Gate-All-Around Silicon Nanosheet MOSFETs," *Semicond. Sci. Technol.*, vol. 35, no. 3, pp. 1–6, 2020.

[6] W. L. Sung and Y. Li, "DC/AC/RF Characteristic Fluctuations Induced by Various Random Discrete Dopants of Gate-All-Around Silicon Nanowire n-MOSFETs," *IEEE Trans. Electron Devices*, vol. 65, no. 6, pp. 2638–2646, 2018.

[7] S. R. Kola, Y. Li, and N. Thoti, "Random Telegraph Noise in Gate-All-Around Silicon Nanowire MOSFETs Induced by a Single Charge Trap or Random Interface Traps," *J. Comput. Electron.*, vol. 19, pp. 253-262, 2020.

[8] E. Caruso, F. Bettetti, L. Del Linz, D. Pin, M. Segatto, and P. Palestri, "Modeling 1/f and Lorenzian Noise in III-V MOSFETs," *Int. Conf. Simul. Semicond. Process. Devices, SISPAD*, pp. 4–7, 2019.

[9] J. S. Yoon, J. Jeong, S. Lee, and R. H. Baek, "Multi-V$_{th}$ Strategies of 7-nm node Nanosheet FETs with Limited Nanosheet Spacing," *IEEE J. Electron Devices Soc.*, vol. 6, pp. 861–865, 2018.

[10] N. Loubet *et al.*, "T17-5 Stacked Nanosheet Gate-All-Around Transistor to Enable Scaling Beyond FinFET T230 T231," *VLSI Technol. 2017*, vol. 5, no. 1, pp. 14–15, 2017.

[11] D. Jang *et al.*, "Device Exploration of NanoSheet Transistors for Sub-7-nm Technology Node," *IEEE Trans. Electron Devices*, vol. 64, no. 6, pp. 2707–2713, 2017.

[12] S. R. Kola, Y. Li, N. Thoti, W. L. Sung and C. Y. chen, "Spacer Effects on Electrical Characteristic and Random Telegraph Noise of Gate-All-Around Silicon Nanowire MOSFETs Induced by Single Charge Trap," *Solid State Devices Mater.*, SSDM, pp. 633–634, 2019.

[13] C. Lombardi, S. Manzini, and A. Saporito, "A Physically Based Mobility Model for Numerical Simulation of Nonplanar Devices," vol. 7, no. 11, 1988.

[14] S. Natarajan *et al.*, "A 14nm Logic Technology Featuring 2nd-Generation FinFET, Air-Gapped Interconnects, Self-Aligned Double Patterning and a 0.0588 μm² SRAM Cell Size," *Tech. Dig. - Int. Electron Devices Meet. IEDM*, pp. 3.7.1-3.7.3, 2015.

5-1

On the Physical Mechanism of Negative Capacitance Effect in Ferroelectric FET

Masaharu Kobayashi

System Design Lab (d.lab), School of Engineering, The University of Tokyo, Tokyo, Japan
Masa-kobayashi@nano.iis.u-tokyo.ac.jp

Abstract—**Negative capacitance FET is a promising CMOS technology booster which may break the limit of 60mV/dec in subthreshold swing (SS) without degrading performance. We investigated the physical mechanism of negative capacitance in ferroelectric FET (FeFET) by considering the dynamics of the polarization in ferroelectric gate insulator: transient negative capacitance (TNC). Polarization switching and depolarization effect are essential to cause negative capacitance effect, that is, apparent surface potential amplification in deep subthreshold region with small depletion layer capacitance. Moreover, unique features of reverse DIBL and negative differential resistance (NDR) are also reproduced by the transient negative capacitance theory. Modeling charged defect in FeFET, hysteresis-free sub-60mV/dec SS can be realized. TNC theory is regarded as a comprehensive framework to model subthreshold characteristics of FeFET.**

Keywords—*negative capacitance, ferroelectric FET, polarization, subthreshold swing, DIBL, NDR, charged defect.*

I. INTRODUCTION

For future energy-efficient computing, ferroelectric FET (FeFET) with sub-60 mV/dec subthreshold slope (SS) caused by negative capacitance (NC) effect has been proposed [1]. NC effect was originally proposed based on a static model (Quasi static NC: QSNC). The metastable NC region in double well-shaped free energy landscape and S-shaped polarization-voltage (P-V) curve of ferroelectric (FE) based on the phenomenological Landau theory can be stabilized and accessible with an appropriate positive capacitor connected in series [1, 2]. According to this theory, the capacitance matching for steep SS should be achieved near inversion region instead of subthreshold region with standard channel design [2]. However, many experimental results of long channel transistors show steep SS in deep subthreshold region with negligible hysteresis in quasi-static condition, which may not be fully explained only within the framework above [3-5]. Moreover, the original theory assumes single-domain configuration in which all domains flip simultaneously in response to electric field as a large single-domain. This is not consistent with the classical FE physics. It is natural that multi-domain switching occurs via anti-parallel configuration from the perspective of thermal dynamics [6, 7]. In addition, it was reported that polarization switching plays an important role in steep SS phenomenon. While the stabilized NC in S-shaped P-V curve can be accessed without polarization switching in unipolar sweep according to the original theory, in Ref. [5], sub-60 mV/dec SS happens only when the gate voltage sweep range is large enough to initialize polarization and trigger large polarization switching in bipolar sweep. Furthermore, NC is static and steep SS behavior should be time independent according to the original theory. However, in Ref. [8], sub-60 mV/dec SS can be only realized within certain measurement time window determined by switching

dynamics of polarization, which indicates that NC effect has a transient aspect. Therefore, several groups have been exploring alternative interpretations for the NC effect as well as steep SS phenomenon [6, 9-13]. what is the physical mechanism of such transient NC (TNC) and how it affects subthreshold behavior of FeFET are not fully clarified yet and need to be investigated.

FeFET with sub-60mV/dec SS also exhibits unique device characteristics such as reverse drain-induced barrier lowering (R-DIBL) and negative differential resistance (NDR) [14-17]. Previously, these special behaviors were predicted by the QSNC theory and regarded as the indication of the stabilized NC in ferroelectric [18-20]. The theory of NC needs to give reasonable explanation not only for steep SS but also R-DIBL and NDR consistently. It should be also investigated that TNC theory can also explain R-DIBL and NDR based on transient device simulation.

One another important aspect is an impact of defects such as fixed charge and charge trapping in FeFET on hysteresis behavior. There are quite a number of reports that shows almost hysteresis-free and sub-60mV/dec subthreshold characteristics. Charge trapping appears to compensate the ferroelectric hysteresis, resulting in hysteresis-free. But this has not been fully investigated yet.

In this paper, we overview our research progresses and explain the current understanding of negative capacitance in FeFET from TNC perspective.

II. MODEL DESCRIPTION & SIMULATION METHOD

The static saturation polarization-voltage (P-V) loop of ferroelectric (Fig. 1 (a)) is described by Miller model which is an analytical version of multi-domain Preisach model [21]. Here, P_s and P_r are saturation and remnant polarization, respectively. V_{fe} and V_c are voltage across ferroelectric and coercive voltage of ferroelectric, respectively. By considering a paraelectric component, the static saturation charge-voltage (Q-V) loop (Fig. 1 (b)) is calculated by the sum of spontaneous polarization and the paraelectric component. Here, ε_r and t_{fe} are the relative dielectric constant and the thickness of ferroelectric, respectively.

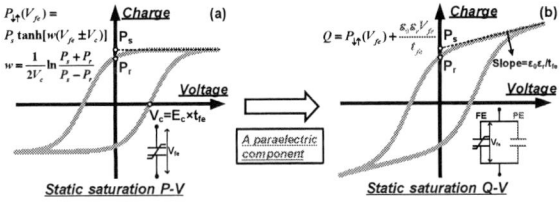

Fig. 1. The static saturation (a) P-V and (b) Q-V of a ferroelectric capacitor.

In order to capture minor loops due to partial polarization switching and ferroelectric history, the turning point method

83

is applied [22, 23]. Fig. 2 (b) is an example of minor P-V loops staring from the saturation loop and passing through points a, b, c, d, and e in sequence. The minor loop is calculated according to the last two turning points ((V_{i-1}, P_{i-1}) and (V_i, P_i)), as shown by the equations in Fig. 2 (b). Then, the minor Q-V loop can be obtained by considering the paraelectric component, which is the same as the method in Fig. 1.

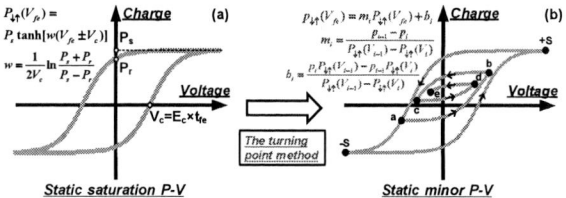

Fig. 2. (a) The static saturation P-V loop and (b) the static minor P-V loop of a ferroelectric capacitor.

Finally, polarization switching delay is introduced to simulate the dynamic behavior of ferroelectric [12,13]. The delay is determined by a first order differential equation (the equation in Fig. 3) where the switching delay (τ) is a constant. The equivalent circuits of a ferroelectric capacitor in both static and transient conditions are illustrated in Fig. 3. In transient conditions, the actual driving force of spontaneous polarization is the auxiliary voltage (V_{aux}). There is certain delay between V_{aux} and the actual voltage across ferroelectric (V_{fe}).

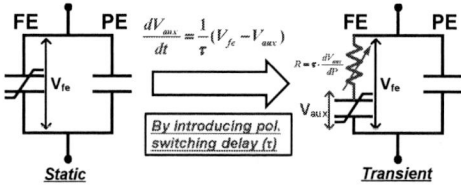

Fig. 3. Equivalent circuits of a ferroelectric capacitor in both static and transient conditions.

Fig. 4 (a) illustrates the simulated transistor structures and parameters. Both FeFET and reference MOSFET are simulated. The only difference between them is the additional HfZrO$_2$ (HZO) thin film in the gate stack of FeFET.

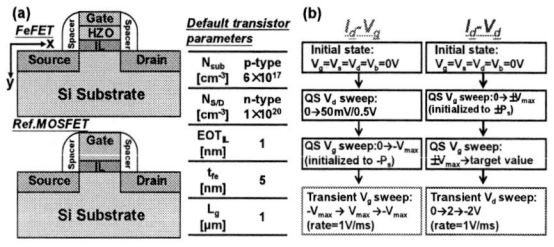

Fig. 4. (a) The simulated device structures and parameters. (b) The flow chart for both I_d-V_g and I_d-V_d simulation.

Fig. 4 (b) shows the flow chart for transistor simulation. Before transient I_d-V_g and I_d-V_d simulation, quasi-static (QS) sweep is applied to initialize certain polarization states and bias conditions [24]. After that, double V_g or V_d sweep is applied in transient condition. Note that the value of V_{max} in Fig. 4 (b) is large enough to initialize saturation polarization (P_s) in this study.

To extract the parameters and verify the abovementioned ferroelectric model, a ferroelectric capacitor in response to triangular waveforms is simulated and fitted to the measurement result [25].

III. RESULTS AND DISCUSSIONS

A. TNC in capacitors

To verify this model and extract parameters of FE, quasi-static Q-V curve of a 10 nm FE capacitor is simulated and fitted to our previous experimental result of a FE-HfZrO (HZO) capacitor as shown in Fig. 5 (a) [25]. The extracted P_r, P_s, ε_r, and E_c (quasi-static parameters) are 20.1 µC/cm^2, 23 µC/cm^2, 35, and 1.16 MV/cm, respectively. Meanwhile, the experimental displacement current of the FE capacitor in response to a triangular waveform can be reproduced with the same fitting parameters shown in Fig. 5 (b).

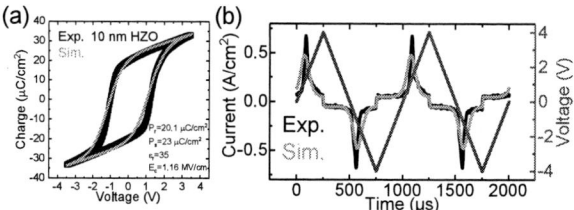

Fig. 5. (a) Q-V curve (100 Hz) with parameter fitting for 10 nm HZO. (b) Displacement current of FE capacitor in response to a triangular waveform with same parameter fitting [19].

In order to extract dynamic parameter, τ, resistor-metal/FE/metal (R-MFM) netlist which is the same as the experimental set-up in Ref. [25] is reconstructed in simulation. Fig. 6 (a)-(c) show transient responses of R-MFM netlist where R=20 kΩ (Fig. 6 (e)) and the input voltage pulse is from −4 V to 4 V. For the best fitting, τ=4 µs is obtained. Fig. 6 (d) shows reconstructed Q-V_{fe} curves for both measurement and simulation where TNC is directly observed as a voltage snapback [26].

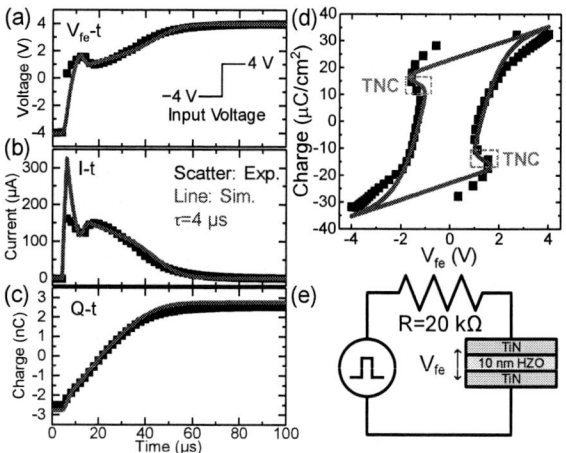

Fig. 6. (a)-(c) Measured and simulated transient responses of R-MFM. (d) Reconstructed Q-V_{fe} curves for both measurement and simulation. (e) Experiment and simulation set-up [19, 20]

By using the dynamic FE model and FE parameters extracted above, transient characteristics of a FE-DE series capacitor which is a simplified model of a FeFET gate capacitor is simulated in response to a triangular waveform (Fig. 7 (a) inset). Fig. 7 (a) and (b) plot the simulated internal voltage amplification (A_v) as function of gate voltage (V_g) and corresponding Q-V_{fe} curves, respectively. A_v is larger than 1 in certain V_g region and TNC is observed as a voltage snapback for τ=4 µs, while A_v is always smaller than 1 and no

84

TNC occurs for $\tau=0$ (quasi-static condition). This means that finite polarization switching delay is responsible for TNC.

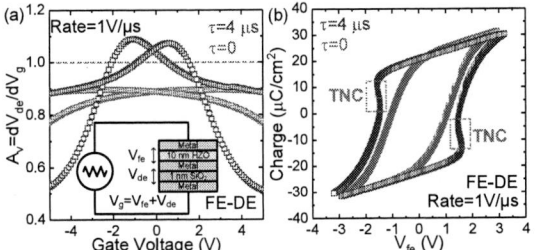

Fig. 7. (a) Simulated internal voltage amplification (A_v) as function of gate voltage (V_g) in FE-DE series capacitor. (b) Corresponding Q-V_{fe} curves. A_v is larger than 1 and TNC is observed as a voltage snapback for $\tau=4$ μs, while A_v is always smaller than 1 and no TNC is observed for $\tau=0$ (quasi-static condition).

To better understand the physical mechanism of TNC, Fig. 8 (a) shows the simulated polarization switching current (dP/dt), total free charge current (dQ/dt), and voltage drop across FE (V_{fe}) as function of time for $\tau=4$ μs. TNC can be understood as the consequence of incomplete screening of spontaneous polarization charge (depolarization effect) [8, 23-25]. Fig. 8 (b) shows the schematic illustration of the physical mechanism for TNC in FE-DE series capacitor. Initially, as V_g is swept in forward direction from the maximum negative value, V_{fe} increases by charging the PE component. After a certain time period, dP/dt increases due to the response of spontaneous polarization. When the increased dP/dt is larger than dQ/dt, in order to satisfy the charge balance condition: $Q=\varepsilon_0\varepsilon_r E_{fe}+P$, V_{fe} has to drop, which is regarded as TNC. Meanwhile, there is a voltage gain in the internal node because of the reduced V_{fe} and more increased V_{de}. It should be noted that, in quasi-static condition, dP/dt is never larger than dQ/dt and TNC is not observed.

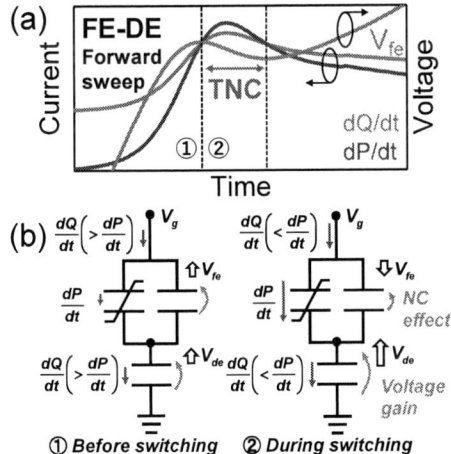

Fig. 8. (a) Simulated polarization switching current (dP/dt), total free charge current (dQ/dt), and voltage drop across FE (V_{fe}) as function of time for $\tau=4$ μs in FE-DE series capacitor. (b) Schematic illustration of physical mechanism for TNC.

B. Steep SS, Reverse DIBL and NDR in FeFET

Fig. 9 (a)-(c) plot simulated I_d-V_g curves and SS in forward and reverse sweep, respectively. Sub-60 mV/dec SS is achieved more prominently in reverse sweep for the FeFET and counter-clockwise hysteresis is caused by polarization switching. FE-type hysteresis can be compensated by

threshold voltage shift caused by charge trapping and detrapping [10, 17, 27], which can be a reasonable explanation for sub-60 mV/dec SS with negligible hysteresis observed in some experiments and revisited later. Note that SS is also lower than 60 at very low I_d level in forward sweep for both FeFET and reference MOSFET is due to the displacement current at the sweep rate.

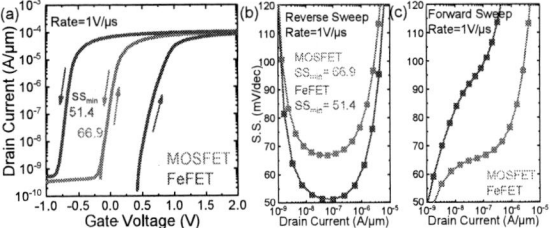

Fig. 9. (a) Simulated I_d-V_g curve for both FeFET and reference MOSFET with $\tau=4$ μs and sweep rate of 1 V/μs. (b) SS in reverse sweep. (c) SS in forward sweep.

Similarly to the discussion for the FE-DE series capacitor, dP/dt and dQ/dt versus time are plotted in Fig. 10 (a) and (b) for further understanding the relationship between TNC and steep SS phenomenon. During forward/reverse sweep, two TNC region (|dP/dt|>|dQ/dt|) are observed. The first one is in accumulation/inversion region and the second one is in transition between accumulation and inversion. These two TNC region can be also observed as two negative slope regions with different slopes in Fig. 10 (b). In subthreshold region, small depletion layer capacitance suppresses both dQ/dt and dP/dt and causes strong depolarization effect with dP/dt > dQ/dt, thus steep SS. Sub-60 mV/dec is observed in wide I_d range only for reverse sweep, since the second TNC covers most of the weak inversion and depletion regions for reverse sweep (Fig. 10 (b)) but not forward sweep (Fig. 10 (a)). Fig. 10 (c) plots zoomed-in Q-V_{fe} curve around 0 charge corresponding to the subthreshold region in Fig 10 (a) and (b), which also indicates that NC is more prominent in wide I_d range in reverse sweep. With different FE parameters, it is possible to make TNC cover most of the weak inversion region in forward sweep as well, thus leading to sub-60 mV/dec SS near V_{th} bidirectionally (not shown, this time).

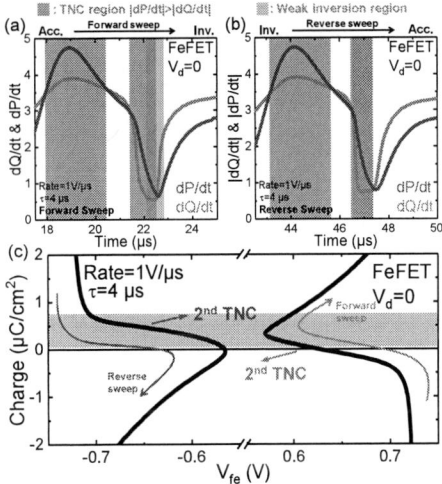

Fig. 10. Simulated polarization switching current (dP/dt) and total free charge current (dQ/dt) as a function of time for (a) forward and (b) reverse sweep. (c) Zoomed-in Q-V_{fe} curve around 0 charge.

Fig. 11 plots the simulated transient I_d-V_g characteristics and the calculated SS-I_d for FeFET. Prominent sub-60mV/dec SS and R-DIBL are observed in reverse sweep. R-DIBL can be also judged by the increased barrier height as shown in Fig. 12. To explore the mechanism of R-DIBL, we investigated the charge density in the channel (N_{ch}) and the voltage across the ferroelectric layer (V_{fe}) at different V_d (Fig. 13). For fixed V_g, higher V_d will lead to higher channel potential and thus lower N_{ch} and more depletion near the drain. This happens even for conventional MOSFET (Fig. 13 (a)). However, for FeFET, higher V_d (lower N_{ch} and more depletion) will result in higher V_{fe} (reduced $|V_{fe}|$) due to TNC (Fig. 13 (b)). The increased V_{fe} will further decrease N_{ch} (Fig. 13 (c)), raise the channel electron potential shown in Fig. 12, and thus I_d decreases [24]. This way, R-DIBL happens.

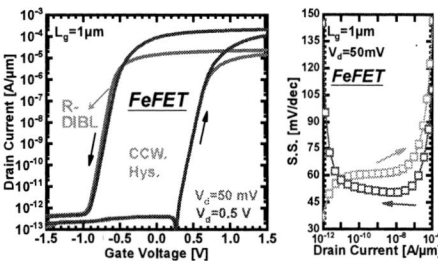

Fig. 11. The simulated transient I_d-V_g characteristics and the calculated SS-I_d for FeFET.

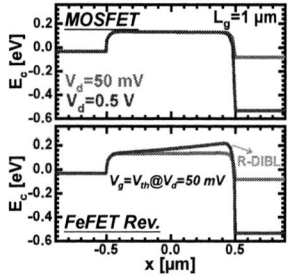

Fig. 12. Extracted conduction band-edge energies (E_c). R-DIBL can be judged by the increased barrier height at higher V_d.

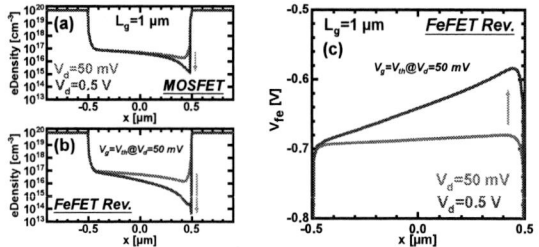

Fig. 13. Simulated charge density in the channel (N_{ch}) at different V_d for (a) reference MOSFET and (b) FeFET. (c) Voltage across ferroelectric (V_{fe}) at different V_d.

Fig. 14 (a) and (b) plot the simulated transient I_d-V_d characteristics for FeFET with +P_s (low V_{th}) and −P_s (high V_{th}) initialization, respectively. In the case of +P_s initialization, I_d shows large hysteresis. Moreover, I_d decreases as V_d increases in forward V_d sweep, which is regarded as NDR. However, in the case of −P_s initialization, I_d is nearly hysteresis-free and no NDR is observed. These results are consistent with experiments demonstrated in Ref. [17].

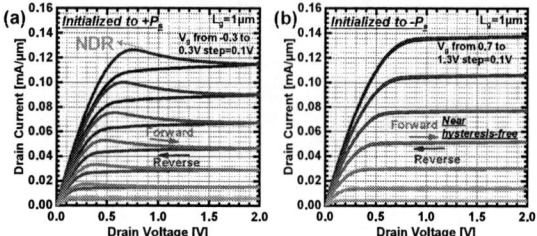

Fig. 14. The simulated transient I_d-V_d characteristics for FeFET with (a) +P_s (low V_{th}) and (b) −P_s (high V_{th}) initialization. NDR is observed only in forward V_d sweep with +P_s initialization.

According to the simulated I_d-V_d characteristics (Fig. 14), ferroelectric history should play an important role in NDR. We investigated the Q-V trajectory of ferroelectric (Q_{fe}-V_{fe}) near the drain during the V_d sweep (Fig. 15). Note that TNC can be induced during V_d sweep instead of V_g sweep. In the case of +P_s initialization, Q_{fe}-V_{fe} shows large hysteresis and TNC is observed only in forward sweep because there is a pathway where Q_{fe} decreases while V_{fe} increases as a part of major loop. Whereas, in the case of −P_s initialization, Q_{fe}-V_{fe} is nearly hysteresis-free and no TNC is observed because there is no pathway where Q_{fe} increases while V_{fe} decreases in minor loop. The behaviors of Q_{fe}-V_{fe} are consistent with I_d-V_d (Fig. 14), which means TNC is responsible for NDR observed in forward sweep, if +P_s is initialized.

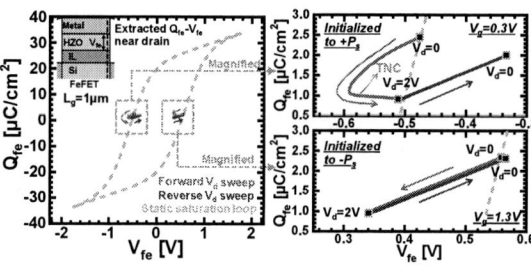

Fig. 15. Extracted Q-V trajectory of ferroelectric (Q_{fe}-V_{fe}) during the transient V_d sweep. The magnified plots are shown in the right side.

The mechanism of NDR is similar to R-DIBL and it can be illustrated as follows [24]. For fixed V_g, as V_d increases, N_{ch} decreases and the channel near the drain depletes more due to the increased channel potential near the drain. Then, in TNC region, V_{fe} increases according to the charge balance condition. This increased V_{fe} further reduces N_{ch} near the drain, the electron potential is raised, and I_d decreases. Therefore, NDR happens. This can be also interpreted as R-DIBL in that V_{th} increases and I_d decreases as V_d increases.

C. Charged defects in FeFET

As already mentioned in the earlier sections, TNC describes sub-60mV/dec steep SS but with hysteresis. This hysteresis is often not seen in experimental reports. Such hysteresis-free steep SS behavior can be due to the existence of charged defect such as charge trapping and fixed charge. We also incorporate charged defects in FeFET model as shown in Fig. 16 (a). Trap band of FE-HfO$_2$ such as HZO has been recently explored and identified [28]. We applied the similar defect profile for charge trap and fixed charge in our work. Fig. 16 (b) shows the example of FeFET with HZO gate insulator, where almost hysteresis-free operation with sub-60mV/dec has been obtained [29]. The interaction between

86

ferroelectric property and such charged defects are important to consider for understanding the device operation.

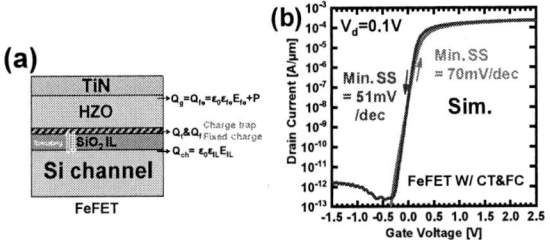

Fig. 16 (a) Schematic of FeFET with charged defects. (b) Simulated I_d-V_g characteristics with the defects.

IV. SUMMARY

In this paper, we overviewed the framework of TNC in FeFET referring to our researches and other groups. TNC provides us physical understanding of steep subthreshold behavior, R-DIBL, NDR, hysteresis-free operation in FeFET. This framework was intended for understanding NCFET but now can be practically applied to FeFET memory application.

ACKNOWLEDGMENT

The author thanks Dr. Chengji Jin, who is the main contributor in this paper. The author thanks Takuya Saraya and Toshiro Hiramoto for technical discussions and support. This work is supported by JST PRESTO (Grant Number 15656058) and MEXT/JSPS Grant-in-Aid and 18H01489.

REFERENCES

[1] S. Salahuddin and S. Datta, "Use of negative capacitance to provide voltage amplification for low power nanoscale devices," Nano Lett., vol. 8, no. 2, pp. 405–410, 2008, doi: 10.1021/nl071804g.

[2] A. I. Khan, C. W. Yeung, C. Hu, and S. Salahuddin, "Ferroelectric negative capacitance MOSFET: Capacitance tuning & antiferroelectric operation," in IEDM Tech. Dig., Dec. 2011, pp. 3–11, doi: 10.1109/IEDM.2011.6131532.

[3] M. H. Lee et al., "Physical thickness 1.x nm ferroelectric HfZrO$_x$ negative capacitance FETs," in IEDM Tech. Dig., Dec. 2016, pp. 12.1.1–12.1.4. doi: 10.1109/IEDM.2016.7838400.

[4] C.-C. Fan, C.-H. Cheng, Y.-R. Chen, C. Liu, and C.-Y. Chang, "Energy efficient HfAlO$_x$ NCFET: using gate strain and defect passivation to realize nearly hysteresisfree sub-25mV/dec switch with ultralow leakage," in IEDM Tech. Dig., Dec. 2017, pp. 561–564. doi: 10.1109/IEDM.2017.8268444.

[5] P. Sharma et al., "Impact of total and partial dipole switching on the switching slope of gate-last negative capacitance FETs with ferroelectric hafnium zirconium oxide gate stack," in Proc. IEEE VLSI Technol., Jun. 2017, pp. T154–T155, doi: 10.23919/VLSIT.2017.7998160.

[6] S. J. Song et al., "Alternative interpretations for decreasing voltage with increasing charge in ferroelectric capacitors," Sci. Rep., vol. 6, p. 20825, Feb. 2016, doi: 10.1038/srep20825.

[7] M. Kobayashi, "A perspective on steep-subthreshold-slope negative-capacitance field-effect transistor," Appl. Phys. Exp., vol. 11, no. 11, pp. 1–20, Oct. 2018, doi: 10.7567/APEX.11.110101.

[8] P. Sharma, J. Zhang, K. Ni, and S. Datta, "Time-resolved measurement of negative capacitance," IEEE Electron Device Lett., vol. 39, no. 2, pp. 272–275, Feb. 2018, doi: 10.1109/LED.2017.2782261.

[9] B. Obradovic, T. Rakshit, R. Hatcher, J. Kittl, and M. S. Rodder, "Modeling of negative capacitance of ferroelectric capacitors as a non-quasi static effect," (2018), arXiv:1801.01842.

[10] B. Obradovic, T. Rakshit, R. Hatcher, J. A. Kittl, and M. S. Rodder, "Ferroelectric Switching Delay as Cause of Negative Capacitance and the Implications to NCFETs," in Proc. IEEE VLSI Technol., Jun. 2018, pp. T51–T52.

[11] B. Obradovic, T. Rakshit, R. Hatcher, J. A. Kittl and M. S. Rodder, "Modeling Transient Negative Capacitance in Steep-Slope FeFETs," in IEEE Transactions on Electron Devices, vol. 65, no. 11, pp. 5157-5164, Nov. 2018, doi: 10.1109/TED.2018.2868479

[12] A. K. Saha, S. Datta, and S. K. Gupta, ""Negative capacitance" in resistor-ferroelectric and ferroelectric-dielectric networks: Apparent or intrinsic?," J. Appl. Phys., vol. 123, no. 10, p. 105102, 2018. doi: 10.1063/1.5016152.

[13] C. Jin, T. Saraya, T. Hiramoto, and M. Kobayashi, "On the physical mechanism of transient negative capacitance effect in deep subthreshold region," IEEE J. Electron Devices Soc., vol. 7, pp. 368–374, 2019. doi: 10.1109/JEDS.2019.2899727.

[14] M. H. Lee et al., "Extremely Steep Switch of Negative-Capacitance Nanosheet GAA-FETs and FinFETs," in IEDM Tech. Dig., Dec. 2018, pp. 31.8.1-31.8.4. doi: 10.1109/IEDM.2018.8614510

[15] H. Zhou et al., "Negative Capacitance, n-Channel, Si FinFETs: Bi-directional Sub-60 mV/dec, Negative DIBL, Negative Differential Resistance and Improved Short Channel Effect," in Proc. IEEE VLSI Technol., Jun. 2018, pp. 53-54. doi: 10.1109/VLSIT.2018.8510691.

[16] J. Zhou et al., "Frequency dependence of performance in Ge negative capacitance PFETs achieving sub-30 mV/decade swing and 110 mV hysteresis at MHz," in IEDM Tech. Dig., Dec. 2017, pp. 15.5.1-15.5.4. doi: 10.1109/IEDM.2017.8268397.

[17] M. Jerry, J. A. Smith, K. Ni, A. Saha, S. Gupta and S. Datta, "Insinhts on the DC Characterization of Ferroelectric Field-Effect-Transistors," 2018 76th Device Research Conference (DRC), Santa Barbara, CA, 2018, pp. 1-2. doi: 10.1109/DRC.2018.8442191.

[18] H. Ota, T. Ikegami, J. Hattori, K. Fukuda, S. Migita and A. Toriumi, "Fully coupled 3-D device simulation of negative capacitance FinFETs for sub 10 nm integration," in IEDM Tech. Dig., Dec. 2016, pp. 12.4.1-12.4.4. doi: 10.1109/IEDM.2016.7838403.

[19] A. K. Saha, P. Sharma, I. Dabo, S. Datta, and S. K. Gupta, "Ferroelectric transistor model based on self-consistent solution of 2D Poisson's, nonequilibrium Green's function and multi-domain Landau Khalatnikov equations," in IEDM Tech. Dig., Dec. 2017, pp. 13.5.1–13.5.4, doi: 10.1109/IEDM.2017.8268385.

[20] G. Pahwa, A. Agarwal and Y. S. Chauhan, "Numerical Investigation of Short-Channel Effects in Negative Capacitance MFIS and MFMIS Transistors: Subthreshold Behavior," in IEEE Transactions on Electron Devices, vol. 65, no. 11, pp. 5130-5136, Nov. 2018. doi: 10.1109/TED.2018.2870519.

[21] S. L. Miller, J. R. Schwank, R. D. Nasby, and M. S. Rodger, "Modeling of ferroelectric capacitor switching with asymmetry nonperiodic input signals and arbitrary initial conditions," J. Appl. Phys., vol. 70, no. 5, pp. 2849–2860, 1991. doi: 10.1063/1.349348.

[22] B. Jiang, P. Zurcher, R. E. Jones, S. J. Gillespie, and J. C. Lee, "Computationally efficient ferroelectric capacitor model for circuit simulation," in Proc. IEEE VLSI Technol., Jun. 1997, pp. 141–142, doi: 10.1109/VLSIT.1997.623738.

[23] K. Ni, M. Jerry, J. A. Smith, and S. Datta, "A Circuit Compatible Accurate Compact Model for Ferroelectric-FETs," in Proc. IEEE VLSI Technol., Jun. 2018, pp. T131–T132.

[24] C. Jin, T. Saraya, T. Hiramoto and M. Kobayashi, "Physical Mechanisms of Reverse DIBL and NDR in FeFETs With Steep Subthreshold Swing", IEEE J. Electron Devices Soc., vol. 8, pp. 429–434, 2020. doi: 10.1109/JEDS.2020.2986345.

[25] M. Kobayashi, N. Ueyama, K. Jang, and T. Hiramoto, "Experimental study on polarization-limited operation speed of negative capacitance FET with ferroelectric HfO2," in IEDM Tech. Dig., Dec. 2016, pp. 12.3.1–12.3.4, doi: 10.1109/IEDM.2016.7838402.

[26] A. I. Khan et al., "Negative capacitance in a ferroelectric capacitor," Nature Mater., vol. 14, no. 2, pp. 182–186, 2015, doi: 10.1038/nmat4148.

[27] E. Yurchuk, J. Müller, S. Müller, J. Paul, M. Peší´c, R. van Bentum, U. Schroeder, and T. Mikolajick, "Charge-trapping phenomena in HfO2-based FeFET-type nonvolatile memories," IEEE Trans. Electron Devices, vol. 63, no. 9, pp. 3501–3507, Sep. 2016, doi: 10.1109/TED.2016.2588439.

[28] Y. Xiang, M. Garcia Bardon, Md Nur K. Alam, M. Thesberg, B. Kaczer, P. Roussel, M. I. Popovici, L. –A. Ragnarsson, B. Truijen, A. S. Verhulst, B. Parvais, N. Horiguchi, G. Groeseneken, and J. Van Houdt, " Physical Insights on Steep Slope FEFETs including Nucleation-Propagation and Charge Trapping", in IEDM Tech. Dig., pp. 510-513 (2019). https://doi.org/10.1109/IEDM19573.2019.8993492.

[29] C. Jin, S. Takuya, T. Hiramoto, and M. Kobayashi, to be published.

5-2

Undoped SiGe material calibration for numerical nanosecond laser annealing simulations

A-S. Royet
CEA-LETI, MINATEC
Campus
Université Grenoble Alpes
Grenoble, France
anne-sophie.royet@cea.fr

L. Dagault
CEA-LETI, MINATEC
Campus
Université Grenoble Alpes
Grenoble, France
sebastien.kerdiles@cea.fr

S. Kerdilès
CEA-LETI, MINATEC
Campus
Université Grenoble Alpes
Grenoble, France
sebastien.kerdiles@cea.fr

P. Acosta Alba
CEA-LETI, MINATEC
Campus
Université Grenoble Alpes
Grenoble, France
pablo.acostaalba@cea.fr

J.P Barnes
CEA-LETI, MINATEC Campus
Université Grenoble Alpes
Grenoble, France
jean-paul.barnes@cea.fr

F.Cristiano
LAAS, CNRS
Université de Toulouse
Toulouse, France
cfuccio@laas.fr

K. Huet
LASSE
SCREEN SPE
Gennevilliers, France
karim.huet@screen-lasse.com

Abstract— **Physical parameters calibration (dielectric and alloy properties) of $Si_{1-x}Ge_x$ alloys is presented in order to simulate the Ultra Violet-Nanosecond Laser Annealing (UV-NLA) of this material for Si/ $Si_{1-x}Ge_x$ based MOS devices. Optical and physical parameters are extracted and modeled from experimental characterizations for several Ge concentrations and then fitted to match experimental laser annealing results. A good prediction, in terms of melt depth and melting duration, is achieved for different Ge concentrations between 20 and 40%, usually encountered in $Si_{1-x}Ge_x$ CMOS integration process.**

Keywords— Laser annealing, melt depth, SiGe alloy characterization.

I. INTRODUCTION

Strained SiGe epi-layers are very interesting for MOSFET technology thanks to their higher mobility compared to Si as well as interesting optical properties [1-2]. However, strain relaxation and defect formation may occur during the growth process and the strain is difficult to maintain especially for high Ge concentrations. In addition, UV-NLA is a very promising technique for 3D sequential integration due to the combination of a very short pulse duration (hundred of nanoseconds) and the limited UV light penetration in most of semiconductors used for CMOS integration [3-4].These unique conditions make UV-NLA a good candidate for suppressing the propagation and growth of defects like dislocations. Laser Annealing of $Si_{1-x}Ge_x$ layers has been investigated for several years and has demonstrated many benefits such as a high electrical activation efficiency and a low contact resistance due to the formation of a Ge-rich layer during UV-NLA [5-7]. In this context, UV-NLA numerical simulation is a powerful tool to predict thermal budget in the structure and the process window in terms of laser beam energy density (ED, J/cm²). However, in order to obtain an accurate prediction, such numerical tool needs a calibrated material properties database as well. In addition to the thermal properties, the main material parameters concerned are optical (permittivity variations with temperature and Ge concentration) and alloy parameters (segregation coefficient). This work shows numerical simulation combined with a large number of experimental data used to calibrate SiGe material behavior upon UV-NLA. It is also shown that once the material is calibrated, the simulations are able to accurately describe molten material thicknesses, Ge concentration distributions and laser annealing kinetics.

II. SIMULATION TOOL AND EXPERIMENTAL DESCRIPTION AND STRATEGY

Numerical simulations were performed using LIAB simulation software described in [8]. This tool solves self-consistently the heat equation coupled to the time harmonic solution of Maxwell equation (UV laser light coupling), including temperature dependency of materials parameters and phase change (here using a phase field based approach [9]).

The experimental procedure is based on the stack, illustrated in Fig.1, consisting in a pseudomorphic 30 to 35 nm-thick undoped $Si_{1-x}Ge_X$ layer with Ge concentrations from 0 up to 40 %. Those layers were deposited on n-type Si (100) substrates using experimental conditions as described in [6]. Such $Si_{1-x}Ge_X$ layers have been widely characterized: optically by spectroscopic ellipsometry and physically thanks to a Secondary Ion Mass spectrometry (SIMS) system.

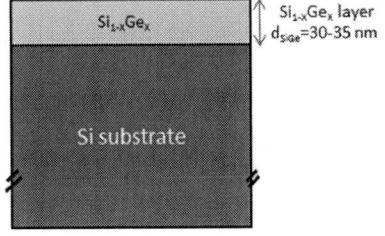

Fig. 1: Experimental stack used for $Si_{1-x}Ge_X$ parameter calibration.

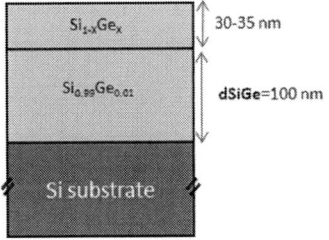

Fig. 2: Corresponding modeling for 1D numerical simulations with LIAB tool.

89

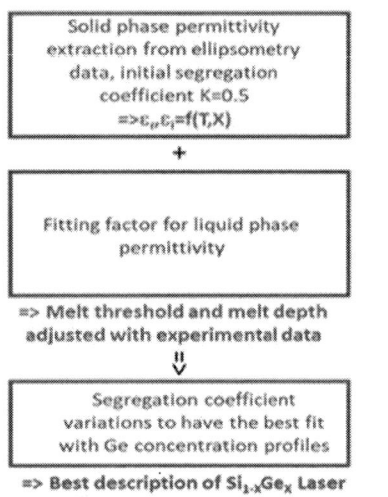

Fig. 3: Simulation strategy for $Si_{1-x}Ge_x$ parameters calibration.

The samples were submitted to pulsed laser annealing in a SCREEN-LT3100 system based on a XeCl excimer laser (308 nm wavelength and 145 ns pulse duration) at various ED. The resulting composition of $Si_{1-x}Ge_x$ layers has been extracted from (SIMS) measurements detailed elsewhere [6-8].

1D Numerical simulations were carried out using the stack shown in Fig.2 where the layer $Si_{0.99}Ge_{0.01}$ was introduced to take into account Ge diffusion into the Si substrate. These simulations were performed at various ED ranging from 1.2 to 2.4 J/cm². We follow the calibration methodology summarized in Fig.3. The first step is to calculate permittivity values in the solid phase from the experimental data and which can account for the real reflectivity and absorption in the material. After that, the liquid phase permittivity values will be calibrated in order to fit the experimental molten thickness and melting threshold values. Finally, the segregation coefficient can be adjusted to find germanium concentration profiles close to those given by the experimental measurements.

Table I: Description of material characteristics used in the LIAB database for numerical simulations and $Si_{1-x}Ge_x$ calibration. ε_r and ε_i are real and imaginary permittivity in solid phase respectively. ε_{Lr}, ε_{Li} are real and imaginary permittivity in liquid phase.

Material	Permittivity in solid phase	Permittivity in liquid phase	Segregation coefficient K
REF	$\varepsilon_r,\varepsilon_i$=f(T,X) with a simple Si and Ge parameter linearization	$\varepsilon_{Lr},\varepsilon_{Li}$=α*(1-X)-β*X	
1	$\varepsilon_r,\varepsilon_i$=f(T,X) with a independent constants between 20 and 40%		0.5
2		$\varepsilon_{Lr},\varepsilon_{Li}$=[α*(1-X)-β*X]*fc	
Calibrated			0.45

III. PARAMETERS CALIBRATION WITH Ge CONCENTRATION AND TEMPERATURE

The optical parameters of $Si_{1-x}Ge_x$ are not always well known and are often derived from the linearization of Si and Ge parameters from [10], considered here as the reference material, called REF, as indicated in Table I. In a first approach, we used ellipsometry measurements for different Ge concentrations, ranging from 10 to 40 % and temperatures from 25°C up to 600°C. The refractive index and the absorption coefficient were used to calculate real and imaginary part of the permittivity, ε_r and ε_I respectively. These values were fitted with polynomial functions depending on the temperature and the Ge content X. Fig.4 shows the permittivity values as a function of the temperature for various Ge contents and the corresponding fits. After that, a single polynomial function is deduced for ε_r and ε_I as indicated in eq.(1):

$$\varepsilon_r, \varepsilon_i = A(T-300) + B_n X^n + \ldots B_1 X + B_0 \qquad (1)$$

where A, B_0, …B_n are independent constants. This function is given as an input into the LIAB material properties database.

For the REF material, the variations of the solid phase permittivity as a function of the Ge content, X, are taken into account by linearization while the segregation coefficient is fixed at the value k = 0.5. Material 1, also described in Table I, takes into account the permittivity variation in the solid phase with no variations on the segregation coefficient while material 2 takes into account both in the liquid and solid phase variations. Finally, the segregation coefficient is slightly tuned in order to better fit segregation profiles. Fig.5 shows the experimental and simulated melt thicknesses versus ED in the cases of $Si_{0.80}Ge_{0.20}$ and $Si_{0.60}Ge_{0.40}$. After several iterations with the different materials tested on the simulation database, a set of optimal parameters values is chosen in order to best fit experimental data thus resulting in a well calibrated $Si_{1-x}Ge_x$ material.

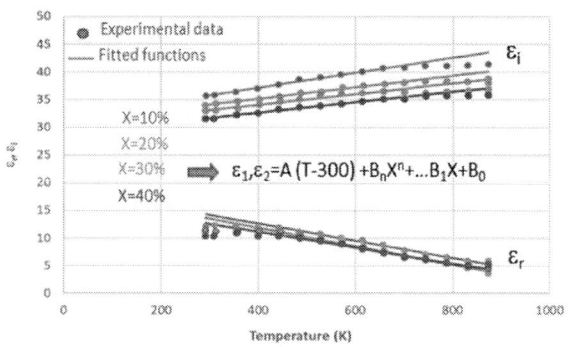

Fig. 4: Experimental permittivity variations (symbols) and fits (lines) versus temperature and Ge concentration. The polynomial function extracted from these results is given on the graph. All the constants A, B0, … Bn, are T and X-independent.

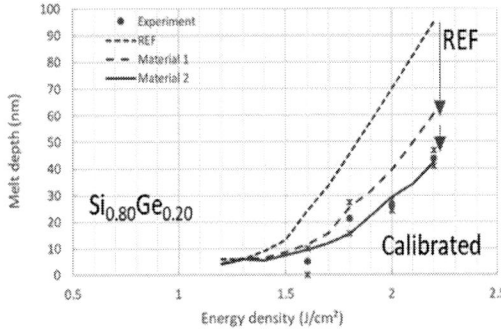

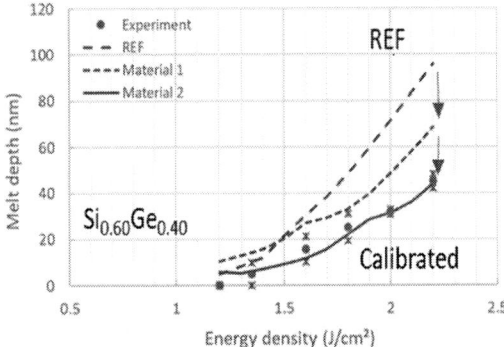

Fig. 5: Comparison of experimental melt depth (nm) and LTA simulation results for $Si_{0.80}Ge_{0.20}$ and $Si_{0.60}Ge_{0.40}$ as a function of the laser energy density .For $Si_{0.80}Ge_{0.20}$ the experimental point at 1.6 J/cm² can not be accurately measured. The molten thickness at this point strongly depends on the surface roughness and is evaluated at the average 5 nm.

IV. ALLOYS PROFILES AND LA KINETICS ASPECTS

Among the set of parameters, the segregation coefficient k = 0.45 is found to provide the best fit with Ge concentration depth profiles at various ED. In Fig.6, the experimental (full lines and points) and simulated (dotted lines) profiles are shown for $Si_{0.80}Ge_{0.20}$ and $Si_{0.60}Ge_{0.40}$ and for ED = around 1.8, 2.0, and 2.2 J/cm². Profiles are well described by the simulations in terms of melt depths and Ge contents at the surface especially at high ED (> 1.8 J/cm²).

In order to investigate kinetics aspects, the melting duration is extracted from experimental and simulated data. Fig.7 shows the laser annealing melting duration versus the melt depth. Even if the fit is not yet perfect, the calibrated simulations enable an acceptable estimation of the annealed material kinetic behavior for each Ge concentration. For example, at high Ge concentration (> 20 %), it is experimentally observed that the melt duration decreases when the full melt of the SiGe layer is reached, which is well reproduced by the simulation. This decrease is attributed to the solidification rate variation when the Si substrate begins to melt. Fig.8 shows the simulated melt depth as a function of time for various ED values. For the highest ED, leading to the full SiGe layer melt, the solidification rate is found almost constant. In contrast, for lower ED values, leading to maximum melt depth within SiGe layer, the solidification exhibits a different behavior with 2 slopes. From maximum melt depth down to ~10 nm, the solidification rate is slightly lower than that obtained in the full melt conditions. Below 10 nm melt depth, the solidification rate decreases further,

likely related to the high Ge concentration in the near surface region. This kinetic aspect can be taken into account by simulation with the calibrated material.

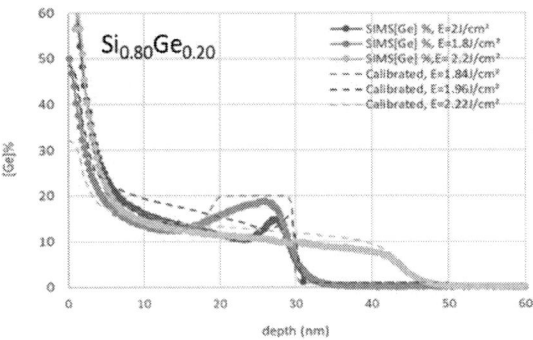

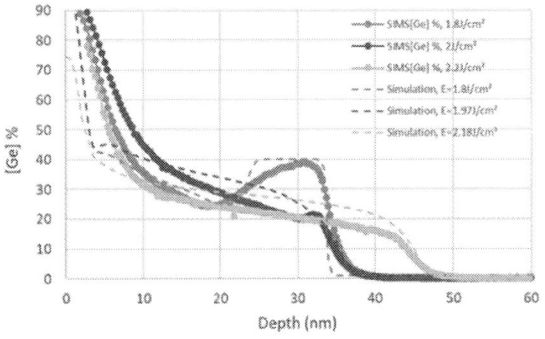

Fig. 6:: Experimental and simulated concentration depth profiles in $Si_{0.80}Ge_{0.20}$ and $Si_{0.60}Ge_{0.40}$ samples annealed at laser energy density from 1.8 to 2.2 J/cm².

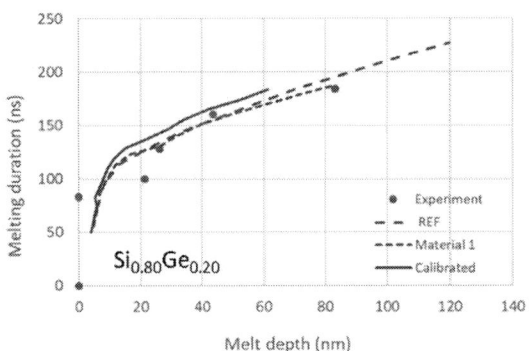

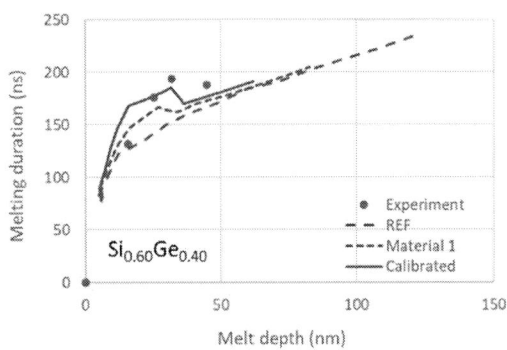

Fig. 7: Experimental and simulated melting duration (ns) as a function of melt depth (nm) for X = 20 and 40 %.

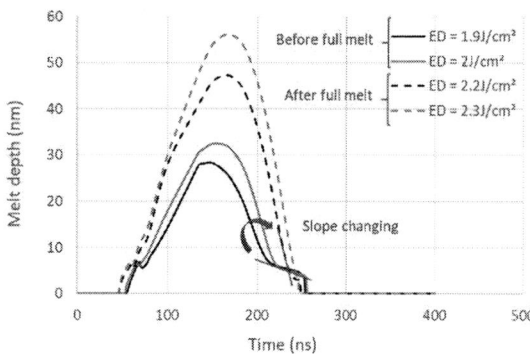

Fig. 8: Simulated melt depth (nm) as a function of time (ns). The slope variation at the end of the process represents the solidification rate variation after the full melt.

V. CONCLUSION

In this paper, we performed 1D numerical nanosecond laser annealing simulations of $Si_{1-x}Ge_X$ layers on Si substrate in order to calibrate this material by comparing simulation results with experimental measurements such as melt depth and duration of the material and Ge content segregation profiles. To have the best material calibration as possible, $Si_{1-x}Ge_X$ optical data measured from 25°C up to 600°C are fitted to obtained a polynomial function describing the permittivity variations with temperature in solid phase. Afterward, this function is introduced in the simulation tool database. The permittivity in liquid phase is also described by a linear X dependent equation and a correcting factor. The segregation coefficient in the material database is then fine tuned to have the best fit with experimental Ge segregation profiles for various Ge contents between 20 and 40 %. Kinetics aspects, such as melting duration are also investigated thanks to simulation results. The Ge concentration range of validity of these simulations is considered to be between 20 and 40 %.

Below 20 %, the dielectric permittivity loses in linearity. This concentration range practically corresponds to the needs of many applications.

REFERENCES

[1] G.E. Jellison, T.E Haynes, "Optical functions of silicon-germanium alloys determined using spectroscopic ellipsometry", Optical Materials 2, pp.105-117, 1993.

[2] R. Milazzo, M. Linser, D. Scarpa, A. Carnera, A. Andrighetto, E. Napolitani, "Indiffusion of oxygen in germanium induced by pulsed laser melting", Materials Science in Semiconductor Processing, vol. 88, pp.93-96, 2018.

[3] K. Huet, F. Mazzamuto, T. Tabata, I. Toqué-Tresonne, and Y. Mori, "Doping of semiconductor devices by laser thermal annealing", Materials Science in Semiconductor Processing, vol. 62, pp.92-102, 2017.

[4] F. Liu, H. Wong, K. Ang, M. Zhu, X. Wang, D.M. Lai, P. Lim and Y. Yeo, "Laser annealing of amorphous germanium on silicon-germanium source/drain for strain and performance enhancement in pMOSFETs", IEEE Electron Device Letters, vol. 29 (8), pp. 885-888, 2008.

[5] O. Gluschenkov, et al, "External resistance reduction by nanosecond laser anneal in Si/SiGe CMOS technology", in Proc. IEEE Int. Electron Devices Meeting (IEDM), 2018.

[6] L Dagault et al, "Impact of UV nanosecond laser annealing on composition and strain of undoped $Si_{0.8}Ge_{0.2}$ epitaxial layers", ECS Journal of Solid State Science and Technology Vol. 8, (3), pp202-208, 2019.

[7] L.Dagault et al, "Investigation of recrystallization and stress relaxation in nanosecond laser annealed Si1−xGex/Si epilayers" Applied Surface Science vol. 527, (2020) 146752.

[8] S.F. Lombardo et al, "Theoretical study of the laser annealing process in FINFET structures", Applied Surface Science vol. 467-468, pp. 666-672, 2019.

[9] G. Fisicaro, A. La Magna, G. Picitto and V. Privitera, "Laser annealing of SiGe and Ge based devices", Microelectronic Engineering vol. 88, pp. 488-491, 2011.

[10] K. Huet et al, "Pulsed laser annealing for advanced technology nodes: Modeling and calibration", Applied Surface Science vol. 505, 2020, 144470.

TCAD simulation for transition metal dichalcogenide channel Tunnel FETs consistent with *ab-initio* based NEGF calculation

Hidehiro Asai
Device Technology Research Institute
National Institute of Advanced
Industrial Science and Technology
Tsukuba, Japan
hd-asai@aist.go.jp

Tatsuya Kuroda
Graduate School of Engineering
Osaka University
Suita, Japan
kuroda@si.eei.eng.osaka-u.ac.jp

Koich Fukuda
Device Technology Research Institute
National Institute of Advanced
Industrial Science and Technology
Tsukuba, Japan
fukuda.koichi@aist.go.jp

Junichi Hattori
Device Technology Research Institute
National Institute of Advanced
Industrial Science and Technology
Tsukuba, Japan
j.hattori@aist.go.jp

Tsutomu Ikegami
National Institute of Advanced
Industrial Science and Technology
Tsukuba, Japan
t-ikegami@aist.go.jp

Nobuya Mori
Graduate School of Engineering
Osaka University
Suita, Japan
mori@si.eei.eng.osaka-u.ac.jp

Abstract—We perform TCAD simulation for TMDC channel TFETs with the material parameters considering ab initio band structure. By using the WKB-based non-local band-to-band tunneling (BTBT) model with the above parameters, we find that the current voltage characteristics of the TFETs are in good agreement with those obtained by microscopic NEGF calculation. Based on this approach, we also investigate the dependence of tunnel leakage current on the gate length. Our simulation method paves the way for reliable macroscopic device simulations for TMDC channel TFET.

Keywords—*TMDC channel TFET, TCAD simulation, NEGF simulation*

I. INTRODUCTION

Atomically thin transition metal dichalcogenides (TMDCs) such as MoS2 and WS2 are promising candidates as channel material for high performance transistors because of their atomically flat structure and excellent electrostatic controllability [1]. In particular, abrupt van der Waals interface and atomically thin structure will be strong advantage for designing tunnel transistor (TFET) having ultra short tunnel distance (i.e. high ON current). Moreover, wide variety of TMDC materials allows us to design high performance TFET by band engineering approach. Therefore, TMDC based TFETs have been extensively studied [2-6] so far. At the same time, for the design of such novel TFETs, modeling of TMDC devices for technology computer aided design (TCAD) simulation is strongly required. However, the band structures of TMDC materials drastically change with the number of stacking layers. Thus, the precise device modeling based on *ab initio* band structure becomes also crucial for the TCAD simulation. In this paper, we perform TCAD simulation for TMDC TFETs with parameter tuning considering *ab initio* band structure, and show the current-voltage

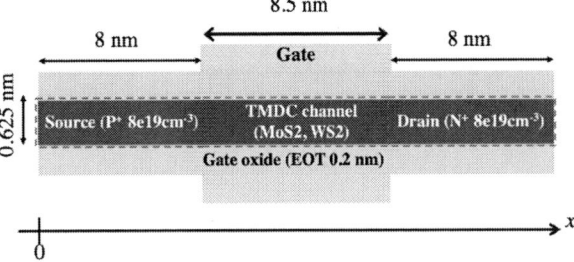

Fig. 1: Schematic figure of N-type TMDC channel TFET.

characteristics which reproduce well those of nonequilibrium Green function (NEGF) calculation.

II. MODEL & METHOD

In this paper, for the comparison between TCAD simulations and NEGF simulations, we considered simple monolayer N-type TFETs as sown in Fig. 1. The dimensions and the dopant concentrations of the TFETs are set to the same values of our previous NEGF calculation [7]. We considered MoS2 and WS2 as TMDC channel materials. In our previous NEGF simulation, we employed the three-band nearest-neighbor TB approximation of which considers three atomic d-orbitals of the transition metal atoms [8]. The TB parameters are extracted from the first-principles calculation using the Vienna *ab initio* simulation package VASP [9,10]. In the present paper, we utilized same band parameters for adjusting the band profile and the ON current of the TFET.

We used our homemade TCAD simulator named Impulse TCAD [11]. In order to calculate the tunneling current of the TFETs, we implemented non-local band-to-band tunnel (BTBT) model [12-14] into Impulse

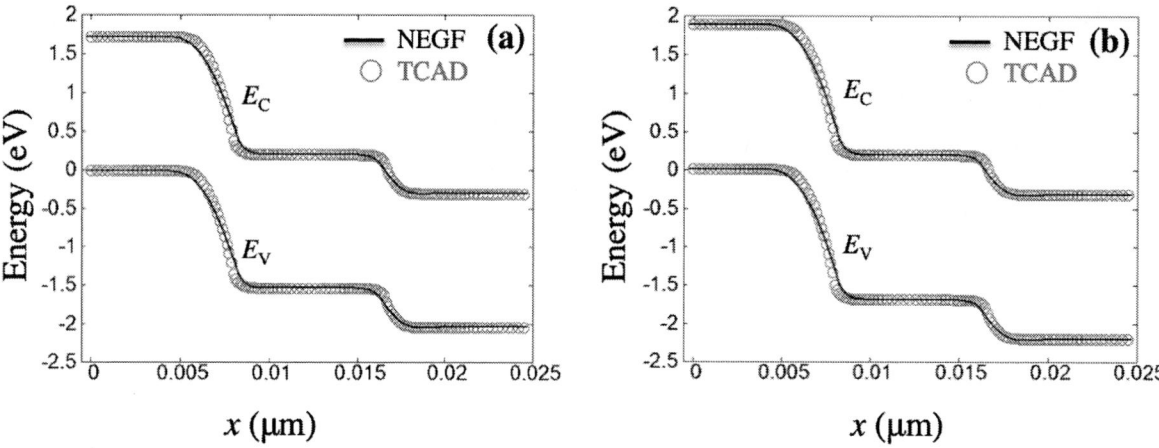

Fig. 2: Conduction and valence band profiles along the source-to-drain direction for (a) MoS2 channel and (b) WS2 channel TFET. The solid lines and the open circles indicate the results of NEGF and TCAD simulations, respectively.

TCAD. The tunneling paths are searched by tracing the energy surface along the steepest gradient direction. The generation rate due to the tunneling G_{BTBT} is calculated along this one-dimensional path. In order to consider the effect of non-uniform electric field along the path (from $x = 0$ to $x = l$), we calculate G_{BTBT} based on WKB method with Kane's relation [15,16],

$$G_{BTBT} = \frac{e}{36\hbar} E_t \left(\int_0^l \frac{dx}{\kappa} \right)^{-1} \left[1 - \exp\left(-k_m^2 \int_0^l \frac{dx}{\kappa} \right) \right]$$
$$\times \exp\left(-2 \int_0^l \kappa dx \right) \left[f_p(\epsilon) - f_n(\epsilon) \right],$$

$$f_n(\epsilon) = \frac{1}{1 + \exp\left(\frac{\epsilon - E_{fn}}{k_B T} \right)}, \quad f_p(\epsilon) = \frac{1}{1 + \exp\left(\frac{\epsilon - E_{fp}}{k_B T} \right)}.$$

Here, κ is the imaginary wave vector along the tunneling path, k_m is the maximum wave vector transverse to the path. E_t is the electric field for tunneling area, and we take the average of the field at the start and the end of the tunneling path. f_n and f_p are Fermi-Dirac distribution for tunneling energy ε. E_{fn} and E_{fp} are the quasi Fermi level for the conduction band side and the valence band side, respectively.

III. RESULTS

Firstly, in order to reproduce the potential profile in our previous NEGF calculation, we set the effective density of states for conduction band N_c and valence band N_v of the TMDC material. Figs. 2 (a) and (b) show the calculated conduction and valence band profiles along the source-to-drain direction for MoS2 and WS2 TFETs at $V_{GS} = 0$V. Here, we adjusted N_c and N_v so that these parameters are consistent with effective masses calculated by VASP. In both figures, TCAD

simulations reproduce well the band profiles of the NEGF simulations.

Then, we set tunneling mass parameters for WKB model and calculated I_D-V_{GS} curve of the TMDC TFETs. Here, we adjusted the tunneling mass by multiplying constant value to the effective masses calculated by VASP, and we chose 0.925 for both MoS2 and WS2 channel TFETs. Figure 3 shows the I_D-V_{GS} curves for MoS2 and WS2 channel TFETs, respectively. The calculated I_D-V_{GS} curves for both TFETs agree well with those of NEGF calculation. Figs. 4 (a) and (b) show the distribution of the BTBT generation rate in the MoS2 TFET for $V_{GS} = 0.15$ and 0.3 V. The generation rate in the source region indicates the generation of electrons, while the rate in the channel and the drain regions indicates the generation of holes. As shown in this figure, source to channel tunneling, which contributes to the ON current, occurs in high voltage condition (V_{GS}

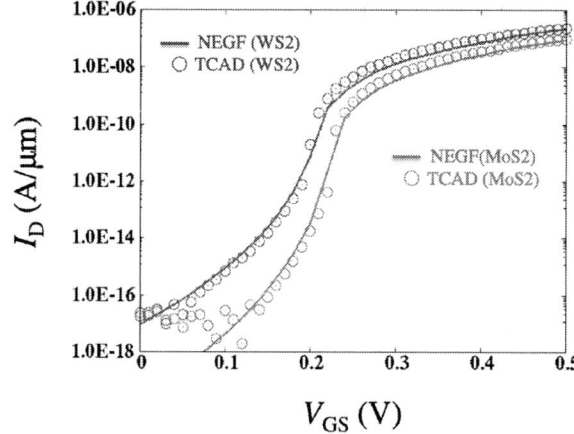

Fig. 3: I_D-V_{GS} curve of TMDC channel TFETs. The solid lines and the open circles indicate the results of NEGF and TCAD simulations, respectively.

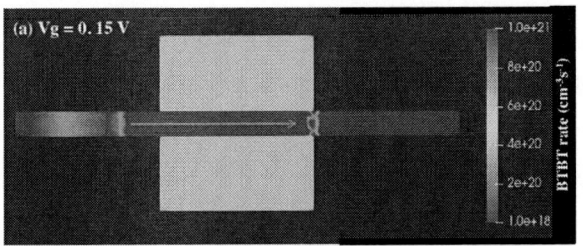

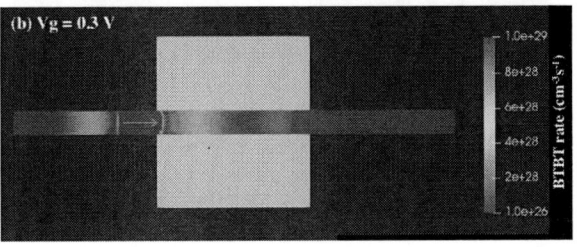

Fig. 4: The distribution of BTBT generation rate in MoS2 channel TFETs for (a) $V_{GS} = 0.15$ V, and (b) $V_{GS} = 0.3$ V.

$= 0.3$ V). Meanwhile source-to-drain tunneling, which contributes to the OFF-leak current, occurs in low voltage condition ($V_{GS} = 0.15$ V). Our TCAD simulation reproduces well both tunneling processes. The key parameters for both MoS2 and WS2 channel TFETs optimized in this study are listed in Table 1.

Finally, using the above parameters, we performed TCAD simulation for the TFET whose gate length L_G is much larger than that in the NEGF simulation. Figs. 5 (a) and (b) show I_D-V_{GS} curve of MoS2 and WS2 channel TFETs for $L_G = 8.5$-40 nm, respectively. The tunneling mass of WS2 is smaller than that of MoS2, because of which both the ON and OFF currents of the WS2 TFETs are larger than those of the MoS2 TFETs. However, as the gate length increases, only the OFF

	MoS2	WS2
N_c	6.73 e19 cm^{-3}	4 e19 cm^{-3}
N_v	9.51 e19 cm^{-3}	5.94 e19 cm^{-3}
m_c	0.446 m_e	0.301 m_e
m_v	0.561 m_e	0.391 m_e

m_e: electron mass

Table: The optimized key parameters for both MoS2 and WS2 channel TFETs.

leak current decreases, and it becomes negligibly small for large L_G. Fig. 6 shows I_{ON}/I_{OFF} ratio at an operating voltage $V_D = 0.15$ V as a function of L_G. Here, I_{ON} is defined as I_D at $V_{GS} - V_{th} = 0.15$ V, and the threshold voltages V_{th} is the voltage at which I_D reaches 10^{-13} A/μm. The degradation of the I_{ON}/I_{OFF} ratio due to the OFF-leak current is considerably large in the WS2 TFETs compared with that in the MoS2 TFETs. However, this degradation is suppressed for the devices having large gate length ($L_G > 15$ nm).

IV. CONCLUSION

We performed TCAD simulations for TMDC channel TFETs. By adjusting the parameters for effective density of states and the tunnel masses in a way that is consistent with *ab initio* band structure, we succeeded to reproduce well the current voltage characteristics of the TFETs by NEGF calculation. We also investigated the dependence of *I-V* characteristics of the TFETs on the gate length, and discussed the short channel effect in the TMDC channel TFETs. The simulation method presented in this paper is simple, but is a useful approach to achieve reliable design of novel TFETs based on *ab initio* calculations.

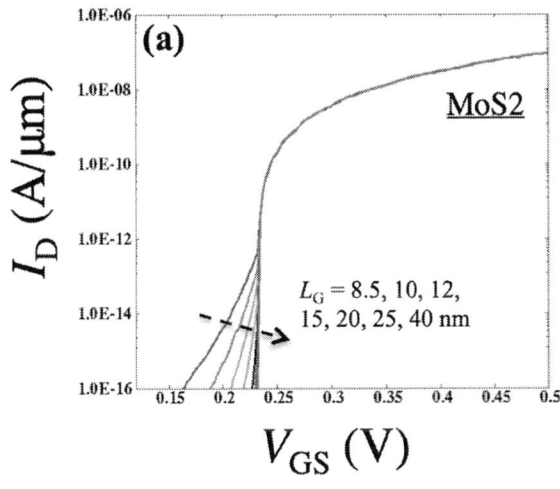

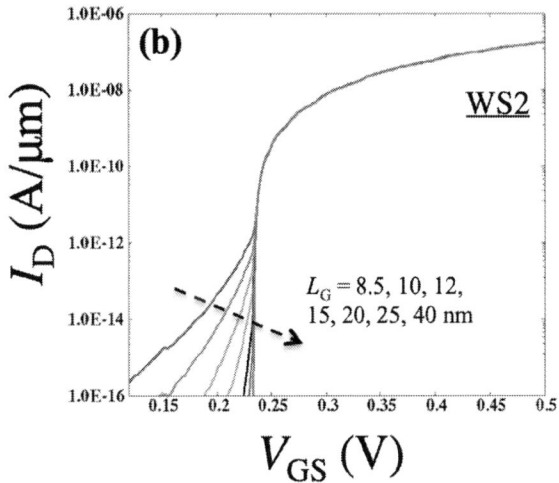

Fig. 5: I_D-V_{GS} curve of (a) MoS2, and (b) WS2 channel TFETs for $L_G = 8.5$-40 nm.

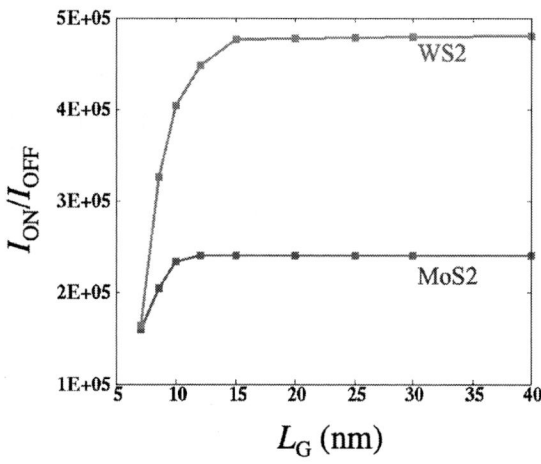

Fig. 6: I_{ON}/I_{OFF} ratio of the MoS2 and WS2 channel TFETs as a function of L_G.

ACKNOWLEDGEMENT

This work was supported by JST CREST (JPMJCR16F3).

REFERENCES

[1] A. Rai *et al.*, Crystals **8,** 316 (2018).
[2] D. Sarkar *et al.*, Nature **526,** 91 (2015).
[3] W. Cao *et al.*, IEDM Tech. Dig. **305** (2015).
[4] A. Nourbakhsh *et al.*, Nano Lett. **16,** 1359 (2016).
[5] J. He *et al.*, Adv. Electron. Mater. **4,** 1800207 (2018).
[6] T. Irisawa *et al.*, Jpn. J. Appl. Phys. **59,** SGGH05 (2020)
[7] T. Kuroda and N. Mori, Jpn. J. Appl. Phys. **57,** 04FP03 (2018).
[8] G. B. Liu, *et al.*, Phys. Rev. B **88,** 085433 (2013).
[9] G. Kresse and J. Hafner, Phys. Rev. B **47,** 558(R) (1993).
[10] G. Kresse and J. Furthmüller, Phys. Rev. B **54,** 11169 (1996).
[11] T. Ikegami *et al*, J. Comput. Electron. **18,** 534 (2019).
[12] K. Fukuda *et al.*, Jpn. J. Appl. Phys.**114,** 144512 (2013).
[13] K. Fukuda *et al.*, Jpn. J. Appl. Phys. **56,** 04CD04 (2017)
[14] H. Asai *et al.*, AIP Advances **8,** 095103 (2018).
[15] E. O. Kane, J. Appl. Phys. **32,** 83 (1961).
[16] Sentaurus Device User Guide. Version K-2015.06, June 2015, Synopsys.

Ab-initio Study of Magnetically Intercalated Tungsten Diselenide

Peter D. Reyntjens[1,2,3], Sabyasachi Tiwari[1,2,3], Maarten L. Van de Put[1],
Bart Sorée[2,4,5] and William G. Vandenberghe[1]

[1] Department of Materials Science and Engineering, The University of Texas at Dallas,
800 W Campbell Rd., Richardson, Texas 75080, USA.
[2] Imec, Kapeldreef 75, 3001 Heverlee, Belgium.
[3] Department of Materials Engineering, KU Leuven,
Kasteelpark Arenberg 44, 3001 Leuven, Belgium.
[4] Department of Electrical Engineering, KU Leuven,
Kasteelpark Arenberg 10, 3001 Leuven, Belgium.
[5] Department of Physics, Universiteit Antwerpen, Groenenborgerlaan 171, 2020 Antwerp, Belgium.

Abstract—We theoretically investigate the effect of intercalation of third row transition metals (Co, Cr, Fe, Mn, Ti and V) in the layers of WSe_2. Using density functional theory (DFT), we investigate the structural stability. We also compute the DFT energies of various magnetic spin configurations. Using these energies, we construct a Heisenberg Hamiltonian and perform a Monte Carlo study on each WSe_2 + intercalant system to estimate the Curie or Néel temperature. We find ferromagnetic ground states for Ti and Cr intercalation, with Curie temperatures of 31K and 225K, respectively. In Fe-intercalated WSe_2, we predict that antiferromagnetic ordering is present up to 564K. For V intercalation, we find that the system exhibits a double phase transition.

I. INTRODUCTION

Transition metal dichalcogenides (TMDs) which contain heavy elements such as tungsten (W) have attracted significant attention for their exceptional electronic properties [1], and their potential applications in spintronics [2]. The properties of TMDs can be enhanced or modified using dopants, further increasing their scope of potential applications.

An interesting field is that of spintronics, where magnetic materials are used to control the flow of information [2]. There are several ways of obtaining magnetism in TMDs. Some TMDs exhibit intrinsic magnetism in their monolayer form, for example VSe_2 [3]. Doping TMDs with magnetic transition metals as substitutional dopants [4] or as intercalants [5], is the most promising method of realizing TMDs which retain magnetic order up to room temperature.

In this work, we calculate the Curie or Néel temperatures of WSe_2, intercalated with the third-row transition metals Ti, V, Cr, Mn, Fe and Co. We use density functional theory (DFT) to find the optimal structures and the total energies of a set of spin-polarized states for each material. Next, we simulate the magnetic behavior for finite temperature using Monte Carlo simulations. We report the Curie (Néel) temperature for each (anti)ferromagnetic material. To assess the thermodynamic stability of the materials we study, we compute the formation energy for each host-dopant combination.

II. COMPUTATIONAL METHODS

Figure 1 shows a flowchart of the simulation process we use in this work. In all our DFT calculations, we employ an energetic plane-wave cut-off of 500 eV. We employ a Gamma-centered k-point grid, which we vary according to the supercell size.

The DFT simulations are performed using the Vienna *ab initio* Simulation Package (VASP) [6]. We use the Perdew-Burke-Ernzerhof (PBE) functional based on the generalized gradient approximation [7]. We consider the van der Waals interactions by using the DFT-D3 corrections of Grimme *et al.* [8]. We then calculate the Hubbard U parameter using the linear response method [9].

A. Structural relaxation

Figure 2 shows the structures we use in our work. We first use DFT to find the optimal atomic positions and lattice parameters. We intercalate into a $2 \times 2 \times 2$ supercell of WSe_2 where we place two transition metal dopant atoms above each other, but between different WSe_2 layers.

We perform the structural relaxation until the forces on the atoms are all below 5×10^{-3} eV/Å, using a $6 \times 6 \times 4$ k-grid.

B. Hubbard U

In $3d$ transition metals, the d-shell electrons are strongly correlated. To correctly calculate the properties of the system, we employ the Hubbard U-corrected DFT+U. We calculate the value of the Hubbard U parameter using the linear response technique of S. de Gironcoli and M. Coccocioni [9]. Once we have the value for the Hubbard U for each host-dopant combination, we proceed with the spin-polarized DFT+U calculations.

C. Magnetic states

We calculate the magnetic properties of the materials using collinear DFT within the collinear approximation. We simulate the unique magnetic states in the supercell by applying a positive or negative magnetic moment to each intercalant atom. To find the unique magnetic states, we first consider all possible magnetic states in the supercell and then remove symmetry-equivalent states. Later, in the Monte Carlo calculations, we take into account the multiplicity of the states.

In the spin-polarized calculations, we place the magnetic moments on each of the intercalant atoms, while leaving the host-TMD atoms unpolarized, with an initial magnetic moment of $0\mu_B$. For the initial magnetic moment on the intercalant atoms, we use the number of unpaired electrons in the d-shell of the dopant atom, times the Bohr magneton. This corresponds to an initial magnetic moment of $2\mu_B$ for Ti, $3\mu_B$ for Co and V, $4\mu_B$ for Fe and finally $5\mu_B$ for Cr and Mn. For the spin-polarized DFT calculations, we use a $4 \times 2 \times 2$ supercell, with four intercalant atoms. The k-point grid is a Gamma-centered $3 \times 6 \times 4$ grid.

To determine the temperature at which the system transitions between an ordered magnetic state and a disordered state, we calculate the magnetic susceptibility and specific heat from the results of our Monte Carlo runs. We take the peak of the specific heat as the phase transition temperature, namely the Curie temperature for ferromagnets and the Néel temperature for antiferromagnets.

D. Heisenberg Hamiltonian

To model the magnetic interactions, we employ a $J_1 - J_2$ model where J_1 is the in-plane interaction between in-plane nearest neighbors, while J_2 contains the out-of-plane nearest-neighbor interaction. We obtain the values of the exchange parameters using the parametrization scheme described in Refs. [10], [11]. After estimating the exchange parameters, we plug them into a classical Heisenberg Hamiltonian, where the positions of the intercalant atoms correspond to the magnetic sites in the model.

E. Monte Carlo simulations

For the Monte Carlo calculations, we employ a supercell consisting solely of the intercalant atoms, serving as magnetic sites in our model, with $9 \times 9 \times 8$ atoms. For each temperature step, we perform 2000 equilibration steps and 2000 Monte Carlo steps. Additionally, we run each simulation six times, each time starting from a different random initial magnetic configuration, and average the results over the six runs.

F. Formation energies

We calculate the formation energies of the various intercalated materials from the DFT total energies after relaxation. The formation energy of the material is calculated as

$$E_{\text{form}} = \frac{1}{N_1} \left(E_{\text{intercalated}} - N_1 E_{\text{pure}} - N_2 E_{\text{dopant}} \right) \quad (1)$$

where N_1 is the number of WSe$_2$ unit cells in the supercell, while N_2 is the number of dopant atoms per supercell. $E_{\text{intercalated}}$ is the DFT total energy of the intercalated WSe$_2$ supercell. E_{pure} is the DFT total energy of the WSe$_2$ unit cell, without intercalants. Finally, E_{dopant} is the DFT total energy of a dopant atom in its bulk metallic form.

III. RESULTS AND DISCUSSION

A. Crystal structure optimization

Figure 2 shows the structures we use in our work. We intercalated third-row transition metal atoms, namely Ti, V, Cr, Mn, Fe and Co, between the layers of WSe$_2$. In a first step, we start from the 2H AB stacked structure shown in Figs 2a and 2b for all intercalants. However, after the relaxations, we find that the distorted Bernal structure, shown in Figs. 2c and 2d, is more stable than the 2H AB structure for Ti, V, Cr and Mn intercalation. We use the distorted Bernal structure for all subsequent calculations involving X-intercalated WSe$_2$, where X = Ti, V, Cr, Mn. In the case of Fe and Co intercalation, we use the 2H AB structure for the subsequent calculations.

B. Magnetic states

Table I shows the values we obtain during the next steps in our workflow. We report the in-plane and out-of-plane exchange parameters, the Hubbard U parameter calculated using linear response, the final magnetic moment on the intercalant atoms and the estimated transition temperature, obtained from our Monte Carlo simulations. For Ti and Cr, DFT calculations predict a ferromagnetic ground state, while the most stable DFT state for V, Mn, Fe and Co intercalation is antiferromagnetic.

For each host-dopant combination we use the $J_1 - J_2$ model, with an in-plane and an out-of-plane interaction term. In the ferromagnetic materials, Ti and Cr-doped WSe$_2$, both exchange parameters are positive, indicating ferromagnetic in-plane and out-of-plane interactions. For Mn, the two exchange parameters are negative indicating purely antiferromagnetic interactions. Whereas V, Fe and Co-intercalated WSe$_2$ show ferromagnetic out-of-plane and antiferromagnetic in-plane interactions.

Figure 3 shows the susceptibility curves of the intercalated WSe$_2$ versus temperature. For each material, we divide the susceptibility by its maximum value, to normalize the curves for a better comparison of the different materials.

Figure 4 shows the specific heat curves for the intercalated WSe$_2$. We again normalize the curves for better comparison, by dividing each one by its maximum value. The transition temperatures reported in Table I are the locations of the specific heat peaks.

In Fig. 5, we plot the average magnetization and the specific heat of V-intercalated WSe$_2$ versus temperature. We see a cusp in the specific heat, at a somewhat lower temperature than the main peak. We see that the magnetization exhibits a rather complex structure. The peaks in the specific heat occur at the same temperature as abrupt changes in the magnetization, which are caused by distinct phase transitions. There are two such phase transitions, one at 118K and one at 36K. When cooling down from high temperature, the system undergoes a first phase transition at 118K, where some magnetic order appears. Upon further cooling, the system undergoes a second phase transition at 36K, where it transitions into its low-temperature antiferromagnetic ground state.

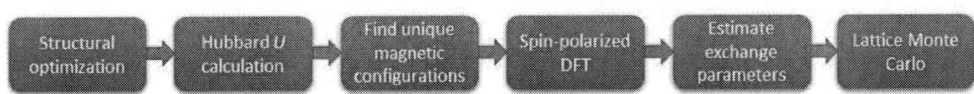

Fig. 1: Flowchart for our calculations. We perform a structural optimization, followed by a linear response calculation of the Hubbard U. After finding the unique spin configurations of the unit cell, we calculate the total energies which we use to calculate the magnetic exchange parameters. Finally, we calculate the Curie or Néel temperature by means of a lattice Monte Carlo simulation.

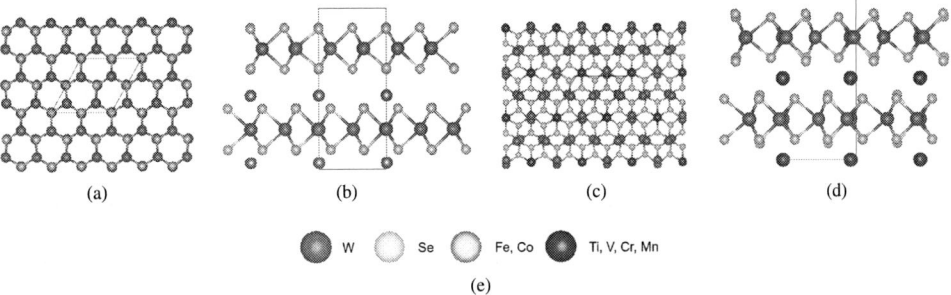

Fig. 2: (a) 2H AB structure for the intercalated WSe_2, viewed from the out-of-plane direction. (b) 2H AB structure for the intercalated WSe_2, viewed from the in-plane direction. For all intercalants, we start the structural optimization from this structure. For Fe and Co, we use the 2H AB structure for the spin-polarized DFT calculations. (c) Distorted Bernal structure for intercalated WSe_2, viewed from the out-of-plane direction and (d) viewed from the in-plane direction. For Ti, V, Cr and Mn intercalation, the distorted Bernal structure is more stable than the 2H AB structure, and we use it for the spin-polarized DFT calculations.

	WSe_2					
Intercalant	Ti	V	Cr	Mn	Fe	Co
J_1(meV)	0.255	-0.247	0.0658	-0.0307	-0.0362	-0.0482
J_2(meV)	0.0332	0.361	0.207	-0.0136	4.08	3.08
Hubbard U (eV)	4.64	3.87	4.82	7.08	5.51	5.05
Magnetic moment	$1.76\mu_B$	$2.75\mu_B$	$4.11\mu_B$	$4.48\mu_B$	$2.64\mu_B$	$1.15\mu_B$
T_C	34K	—	225K	—	—	—
T_N	—	118K	—	88K	564K	40K

TABLE I: For each host-dopant combination, the results of our DFT and Monte Carlo calculations. We calculate the Hubbard U using the linear response method. The exchange parameters in the $J_1 - J_2$ model are J_1, the interaction between in-plane nearest neighbors, and J_2, the interaction between out-of-plane nearest neighbors. Finally, we report the transition temperatures calculated using our Monte Carlo simulations.

C. Formation energies

Figure 6 shows the formation energies for the materials we study. The formation energies are positive for all cases, indicating poor stability. We see that the formation energy is highest for Cr intercalation, and lowest for Ti intercalation. The formation energy is proportional the number of unpaired valence electrons in the d-shell of the intercalants; Cr and Mn have 5 unpaired electrons, while Ti only has 2.

IV. CONCLUSIONS

We have calculated the magnetic properties of WSe_2 intercalated with the third-row transition metals Ti, V, Cr, Mn, Fe

and Co. After structural optimization, we have calculated the total energies of different magnetic states in each system. The calculation of the in-plane and out-of-plane nearest neighbor exchange parameters reveals a few important facts. For Ti and Cr intercalation, the combined system has ferromagnetic nearest neighbor interactions, both in and out of plane, which corresponds to a ferromagnetic ground state at low temperature. In the case of Mn intercalation, the nearest-neighbor interactions are purely antiferromagnetic. The other three cases, *i.e.*, V, Fe and Co intercalation, have in-plane antiferromagnetic interactions and out-of-plane ferromagnetic interactions.

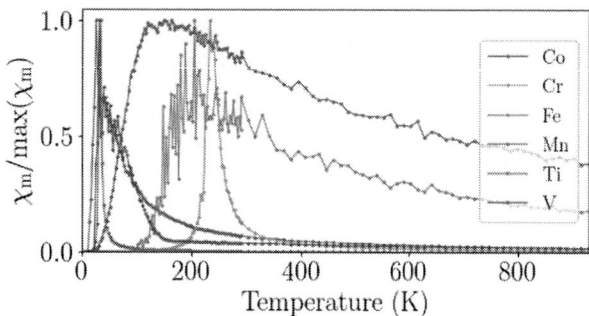

Fig. 3: The susceptibilities for WSe$_2$ intercalated with Ti, V, Cr, Mn, Fe and Co. We normalize the curves by dividing each one by its maximum value, in order to better compare the peaks.

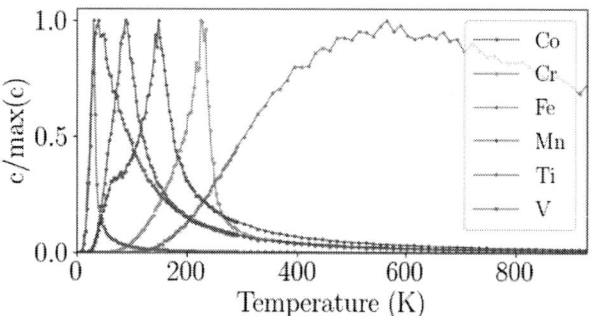

Fig. 4: The specific heat for WSe$_2$ intercalated with Ti, V, Cr, Mn, Fe and Co. For a better comparison between the different curves, we normalize them dividing them by their respective maximum values.

The specific heat curve of V-intercalated WSe$_2$ shows a double-peak structure, with a large peak at 118K and a cusp at 36K. The average magnetization of the system exhibits similar behavior, which we attribute to the existence of two phase transitions in the material.

While the formation energies of the materials studied here turn out to be positive, indicating poor stability, the results from the Monte Carlo simulations are encouraging. The magnetic behavior is very dependent on which atomic type is used to induce magnetism. In particular, the high transition temperature of Fe-intercalated WSe$_2$ and the double phase transition in V-intercalated WSe$_2$ are intriguing. More research into intercalated WSe$_2$ and other TMDs is necessary and may lead to extremely useful applications in the future.

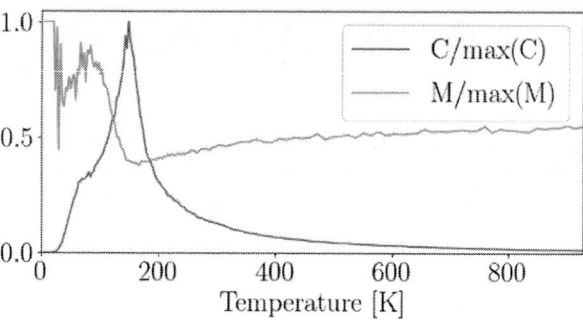

Fig. 5: The specific heat and magnetization for V-intercalated WSe$_2$. The specific heat exhibits a double-peak structure, with a large peak at 118K and a smaller one at 36K. Each peak corresponds to a distinct phase transtion. The two phase transitions are visible in the magnetization curve as well.

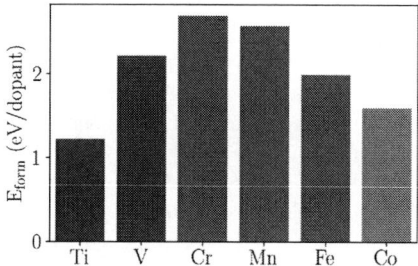

Fig. 6: The formation energies per intercalant for intercalated WSe$_2$. Cr-intercalated WSe$_2$ has the highest formation energy, while Ti-intercalated WSe$_2$ has the lowest formation energy.

REFERENCES

[1] Q. H. Wang, K. Kalantar-Zadeh, A. Kis, J. N. Coleman, and M. S. Strano, "Electronics and optoelectronics of two-dimensional transition metal dichalcogenides."

[2] I. Žutić, J. Fabian, and S. Das Sarma, "Spintronics: Fundamentals and applications," *Rev. Mod. Phys.*, vol. 76, pp. 323–410, Apr 2004.

[3] M. Bonilla, S. Kolekar, Y. Ma, H. C. Diaz, V. Kalappattil, R. Das, T. Eggers, H. R. Gutierrez, M.-H. Phan, and M. Batzill, "Strong room-temperature ferromagnetism in VSe$_2$ monolayers on van der waals substrates."

[4] B. Li, T. Xing, M. Zhong, L. Huang, N. Lei, J. Zhang, J. Li, and Z. Wei, "A two-dimensional Fe-doped SnS$_2$ magnetic semiconductor."

[5] Y. Jung, Y. Zhou, and J. J. Cha, "Intercalation in two-dimensional transition metal chalcogenides."

[6] G. Kresse and J. Furthmüller, "Efficient iterative schemes for ab initio total-energy calculations using a plane-wave basis set," *Phys. Rev. B*, vol. 54, pp. 11 169–11 186, Oct 1996.

[7] J. P. Perdew, K. Burke, and M. Ernzerhof, "Generalized gradient approximation made simple," *Phys. Rev. Lett.*, vol. 77, pp. 3865–3868, Oct 1996.

[8] S. Grimme, J. Antony, S. Ehrlich, and H. Krieg, "A consistent and accurate ab initio parametrization of density functional dispersion correction (DFT-D) for the 94 elements H-Pu."

[9] M. Cococcioni and S. de Gironcoli, "Linear response approach to the calculation of the effective interaction parameters in the LDA + U method," *Phys. Rev. B*, vol. 71, p. 035105, Jan 2005.

[10] S. Tiwari, M. L. van de Put, B. Sorée, and W. G. Vandenberghe, "Magnetism of 2D semiconductors, dichalcogenides and graphene," in *Bulletin of the American Physical Society*, vol. 65, no. 1, 2020.

[11] S. Tiwari, M. L. V. de Put, B. Soree, and W. G. Vandenberghe, "Critical behavior of ferromagnets CrI$_3$, CrBr$_3$, CrGeTe$_3$, and anti-ferromagnet FeCl$_2$: a detailed first-principles study," 2020.

A Study of Wiggling AA modeling and Its Impact on the Device Performance in Advanced DRAM

QingPeng Wang, Yu De Chen, Jacky Huang and Ervin Joseph

Coventor Inc., a Lam Research Company

Shanghai, China

Qingpeng.Wang@lamresearch.com

Abstract— **In this paper, a wiggling active area (fin) in an advanced 1x DRAM process was analyzed and modeled using the pattern-dependent etch simulation capabilities of the SEMulator3D® semiconductor modeling software. Nonuniformity in sidewall passivation caused by hard mask pattern density loading was identified as the root cause of the wiggling profile. The calibrated model mimicked these phenomena, giving nearly the same output AA shape as the real fabrication process. The wiggling profile's impact on device performance was assessed using the built-in drift-diffusion solver of SEMulator3D. Our analysis confirmed that the wiggling profile, induced by micro-loading during a pattern-dependent etch, has a large impact on overall electrical performance in the device. This was especially apparent with the off-state leakage, primarily due to a worse drain-induced barrier lowering effect in a fatter fin.**

Keywords— *DRAM; active area, sidewall passivation, micro-loading, modeling, virtual fabrication, device performance*

I. INTRODUCTION

Most leading semiconductor manufacturers strive to decrease transistor dimensions to lower cost and improve device performance [1]. However, with transistor size approaching the lower limits of what is achievable, variations occurring during the fabrication process become more and more important for determining final product performance and yield.

Process variations often fail a product in one of two ways. Large variations in transistor properties may lead directly to a hard fail, such as an open or short between two components. Small variations can introduce shifts in the final device in terms of threshold voltage, drive current, off-state leakage, resistance, capacitance, or other parameters, which can then lead to a soft fail or performance degradation due to mismatches between the circuit design and real silicon.

In a DRAM structure, the charging and discharging process of capacitor-based memory is directly controlled by the transistor [2]. The operation speed is modulated by the drive current of the transistor, while the data retention capabilities are limited by the off-state leakage current. Thus, variability in device properties, such as the transistor's AA (active area) dimension and profile, is one of the most important factors influencing the yield and performance of an advanced DRAM product.

A hard mask created by SAQP and the LELE process is widely used to form the AA (transistor fin) in an advanced 1x DRAM process. During the fin etch process, density loading from the hard mask can cause fin profile loading, which mainly occurs in two ways: macro-loading between the fin array's center and edge, and micro-loading within a single fin. The fin profile loading causes wiggling of the AA profile.

In this paper, the wiggling AA profile seen in advanced DRAM processes is investigated and analyzed. The loading profile during pattern-dependent etch is modelled using the SEMulator3D virtual fabrication platform developed by

Coventor Inc. [3]. In addition, the impact of the wiggling AA profile on device performance is simulated and analyzed based upon the modelled structures.

II. WIGGLING AA CONCEPTS AND MODEL

A. Wiggling AA and its Mechanism

In a 1x/y DRAM reverse engineering report from TechInsights, wiggling AA profiles were found in almost all commercialized DRAM products made by leading DRAM manufacturers. The wiggling AAs show not only wiggling center lines, but also CD differences in the regions neighboring the cut areas. In a specific case where a whole array was inspected, a much larger wiggle was found at the array edge [4,5,6]. (see Fig. 1)

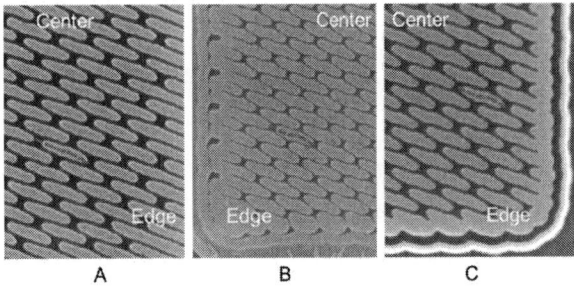

Fig.1. Planar view of the AA profiles of 1x/y DRAM from three different manufacturers [4,5,6] (Courtesy: TechInsights).

In the SAQP process, a spacer is usually deposited using an ALD process, which gives quite conformal and uniform results even in areas with different surface topography. Thus, the wiggling AA is unlikely to be due to a non-uniform spacer; rather, it more likely results from a phenomenon related to pattern dependent etching during the fin etch process, after the spacer is cut.

The main purpose of the fin etch process is to remove the exposed Si, which is not covered by the spacer hard mask, to form the AA and STI. Fig. 2 shows a brief illustration of the fin etch process. After the spacer cut process, the pattern density of the hard mask experiences loading between the cut area (see Fig. 2(a), site A) and the fin area (see Fig. 2(a), site B). During the fin dry etch process, the etch byproduct is more likely to stick to the sidewall of the structure than to be deposited on the bottom surface, because an anisotropic dry etch process provides a higher vertical etch rate on the bottom surface and a lower lateral etch rate on the sidewall. As a result, the sidewall will be passivated by the byproduct and exhibit a taper sidewall profile. Since more Si needs to be removed from the region around site A than the region around site B, more reagent will be consumed, and more byproduct will be generated around site A than around site B (see Fig. 2(b)). Finally, sidewall passivation near site A shows a much more tapered sidewall profile than the region near site B (see

Fig. 2(c)). A similar sidewall passivation etch profile is found in work reported by Yuri Granik et. al, which in that case is probably related to excessive polymer buildup [7].

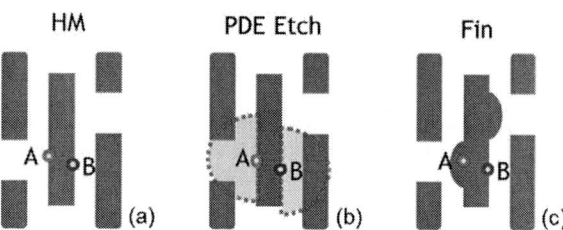

Fig. 2. Illustration of AA profile in Fin etch process (a) hard mask top view before etch, (b) pattern dependence etch amount comparison between site A and B, (c) top view after fin etch.

B. Wiggling AA Modeling

The SEMulator3D semiconductor platform provides a novel pseudo-3D approach to pattern dependence modeling based on 2D proximity functions. The proximity functions will be convoluted using a pattern-dependence mask, within a characteristic distance of a point of interest, and finally yield a 2D loading map. This 2D loading map modifies the behavior of a 3D behavioral etch algorithm in the software [8].

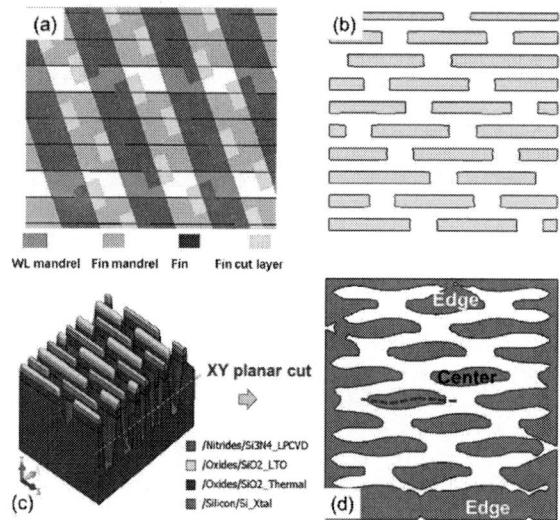

Fig. 3: (a) Layout design, (b) PDE mask generated from hard mask, (c) 3D structures after fin etch, (d) AA profile by planar cut on fin middle.

In this work, we create a pattern-dependence mask by cutting an XY plane from the 3D structure of the hard mask before fin etching occurs (see Fig. 3(b)). The related coefficients and characteristic distance in the proximity function can be calibrated from known Si data. Fig. 3(a) shows the layout design (rotated 20 degrees counterclockwise for easy viewing), 3(b) shows the generated pattern-dependent mask, and 3(c) and 3(d) show the 3D structure and planar view (polished to fin middle in the Z direction) from the simulation. These results were compared with hard silicon data shown in Fig. 1. Comparing 3(d) with Fig.1 (c), a similar wiggling AA profile is found at both the array center and the edge, demonstrating that the model properly reflects real Si. Fig. 4 displays the AA profile at different fin heights, providing 2D

and 3D modeled views of the AA profile evolution from fin bottom to fin top.

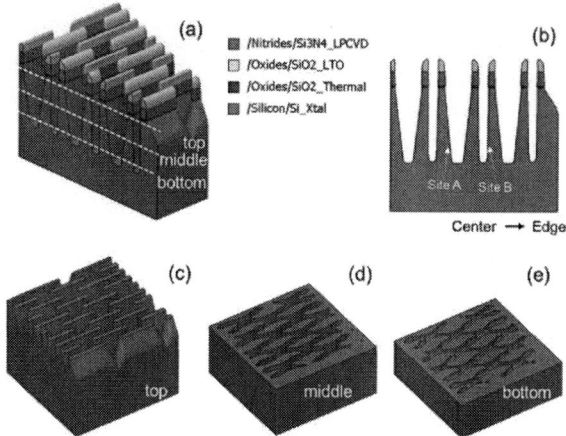

Fig. 4. AA profile at different fin heights (a)3d view cut on wordline, (b) cross-sectional view cut on wordline, (c) 3d view cut on fin top, (d) 3d view cut on fin middle, (e) 3d view cut on fin bottom

III. DEVICE PERFORMANCE SIMULATION AND ANALYSIS

A. Split Condition and Flow Build up

In the 1x/y DRAM cell with a buried wordline, the fin is recessed during the wordline formation process, and the final channel is a buried channel beneath the word line. Fig. 5(a) shows the top view of the structure after wordline formation (for easy observation, some metals and dielectric materials are made transparent). Fig. 5(b) shows a cross-sectional cut of the fin in the wordline direction while 5(c) shows the cross-sectional cut of the channel in the fin direction after wordline formation. The wiggling profile is more serious at the middle of the fin than at the fin top, due to the micro-loading effect occurring during the etch process (see Fig. 4 (c) and (d)). Since the channel is located near the middle of the fin, the final fin CD under the channel will be much larger due to micro-loading and sidewall passivation. Fin CD loading may introduce much different device performance, such as differences in the threshold voltage, off-state current, drive current, and so on.

To evaluate the impact of the wiggling AA profile on device performance, a full loop 1x DRAM process flow from the AA to the CC (capacitor contact) is built using SEMulator3D. In the flow, sidewall angle splits of 0.1, 2.5, and 5 degrees are used in the pattern-dependent etch process to understand the differences in final device performance. Fig. 6(a) shows the final 3D structure of the DRAM array, 6(b) shows a single device (cropped), and 6(c) shows the cross-section view cut along the fin, with ports defined for electrical analysis. Fig. 6(d), (e) and (f) show cross-sectional views of the fin profiles with different sidewall angle splits. A thicker and more asymmetrical fin is visible when the sidewall angle is larger.

102

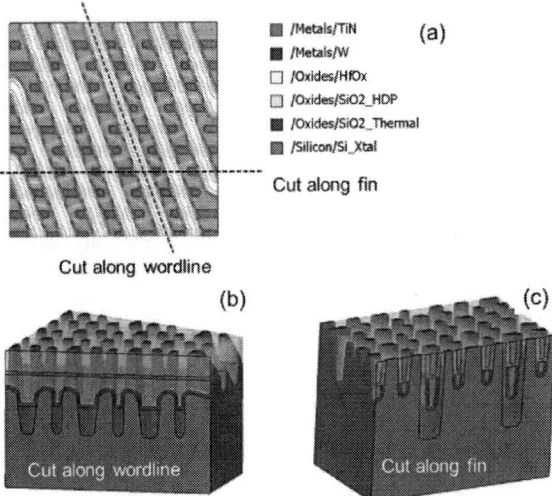

Fig. 5. DRAM structure after worldline formation (a) top view, (b) cross-sectional view along wordline, (c) cross-sectional view along fin.

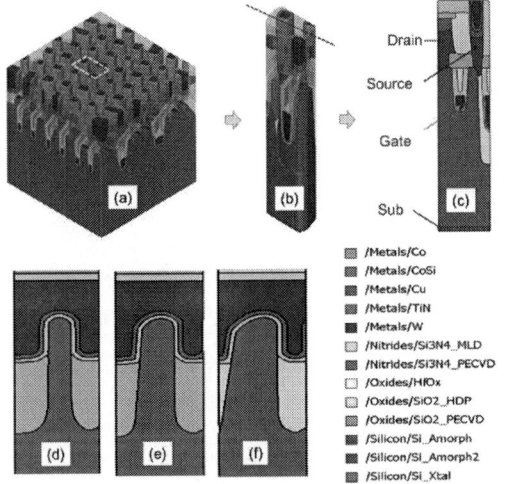

Fig. 6. DRAM structure after capacitance contact formation, (a) 3D view, (b) single device view, cropped, (c) View along fin cut with port definition, (d) SWA=0.1 deg (WL cut), (e) SWA=2.5 deg, (f) SWA=5 deg.

B. Device Simulation and Analysis

Based on the previously defined virtual structure, a single device is cropped out for electrical analysis (see Fig. 6(b)). The electrical ports for the source, drain, gate, and sub are assigned to the structure (see Fig. 6(c)). The built-in drift-diffusion solver of SEMulator3D is used to calculate electrical performance, including the Id-Vg curve, along with current distribution of these experimental sidewall angle splits. Fig. 7(a) shows the Id-Vg curve plotted on a log scale with V_d=3V, where V_g is swept from 0 to 1.2V. Fig. 7(b) shows the linear scale I_d-V_g curves of these splits.

From the linear I_d-V_g curves, it is noted that a slightly higher drive current can be obtained using a larger sidewall angle in the fin etch process. The on-state current difference is likely due to the differences in the effective channel width for different fin profiles. Usually, a fatter fin with a larger sidewall angle will have a wider effective channel width, which could be observed in Fig. 6(d, e, f).

From the logarithmic I_d-V_g curve, the off-state leakage and subthreshold swing of different splits can be observed. The fatter fin exhibits much higher off-state leakage and larger subthreshold swing, demonstrating that channel controllability is weak. Similar research with real Si data has been reported in a logic fin research paper [9]. Fig. 7(c) shows the drain-leakage current relationship for different drain voltages at V_g=0V. The leakage in a fatter fin will dramatically increase at higher drain voltages, while it is not so obviously changed in a thinner fin. Fig. 7(d) displays the drive current and leakage current (V_d=3V) with different splits. It shows that the drive current becomes slightly higher as the sidewall angle becomes tapered, but it saturates or even decreases with the fattest fin. On the other hand, the leakage current increases as the fin sidewall angle increases. The higher source-drain leakage current is due to the drain-induced barrier lowering effect (DIBL) in the short channel MOSEFT with weak gate controllability [10,11].

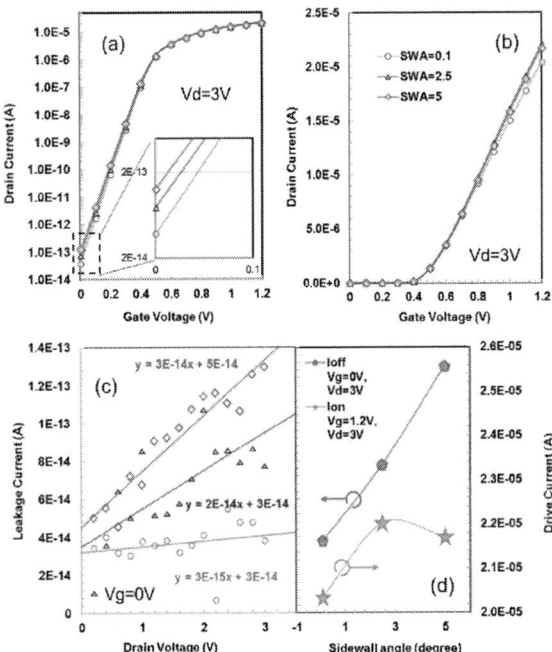

Fig. 7. Transfer characteristics (a) log I_d-V_g, (b) linear I_d-V_g, (c) off-state drain leakage at different V_d, (d) drain leakage and drive current at V_g=1.2V, V_d=3V.

Fig. 8 displays off-state leakage current distribution in the fin. It shows that most leakage current is concentrated in the fin center, far from the gate metal and not strongly controlled by the gate electrical field. Because of the diminished gate controllability in the fatter fin, the leakage current density is much higher than in a thinner fin.

Based on these results, it is easy to see that a narrower fin with a smaller sidewall angle and less micro loading during etching is preferred in an advanced DRAM process. The micro loading seems unavoidable due to the incoming hard mask density loading. However, if we change the integration scheme to first generate a uniform fin, and then cut the unnecessary fin after the fin etch process, the micro loading issue may be avoided. The tradeoff is that more difficult process control is required during the fin cut last etch process.

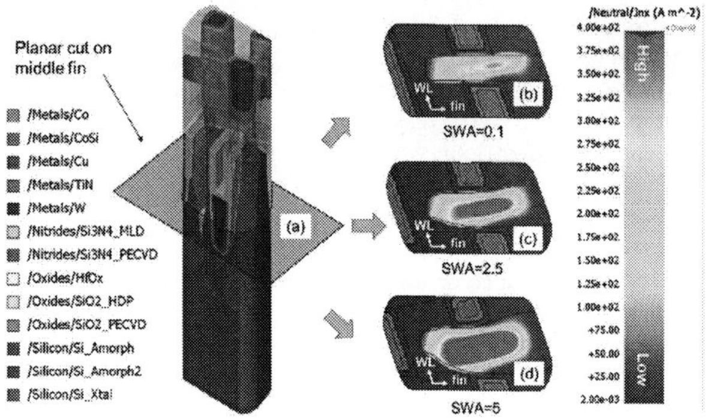

Fig. 8. Channel leakage profile from fin surface to fin center with different sidewall angle split.

IV. CONCLUSIONS

In this paper, wiggling AA in an advanced DRAM process was analyzed and modeled using the SEMulator3D platform. The impact of wiggling AA on device performance was reviewed using a drift-diffusion solver. The analysis shows that micro loading, induced during pattern-dependent etching, can cause a wiggling AA. This micro loading has a large influence on the device's electrical performance, especially the off-state leakage, which is a key factor in determining the data retention capabilities of DRAM cells. Effort should be devoted to minimizing this micro loading effect to further enhance device performance and yield. Potential solutions include optimizing the fin etch process, or changing the

integration scheme from fin cut first to fin cut last in the fin formation process.

REFERENCES

[1] G. E. Moore, "Cramming More Components onto Integrated Circuits", Electronics Magazine, vol. 38, no. 8, pp. 114-117, April 1965.

[2] D. Lee, Y. Kim, G. Pekhimenko, S. Khan, V. Seshadri, K. Chang, O. Mutlu, "Adaptive-latency DRAM: Optimizing DRAM timing for the common-case", In 2015 IEEE 21st International Symposium on High Performance Computer Architecture (HPCA) 2015 Feb 7, pp. 489-501. IEEE.

[3] http://www.coventor.com/products/semulator3d

[4] DDR4 SDRAM 1x nm tear down report from TechInsights

[5] LPDDR4X SDRAM 1y nm ear down report from TechInsights

[6] LPDDR4 SDRAM 1x nm tear down report from TechInsights

[7] Y. Granik, "Correction for etch proximity: new models and applications", In Optical Microlithography XIV 2001 Sep 14, Vol. 4346, pp. 98-112.

[8] D. Fried, K. Greiner, D. Faken, M. Kamon, A. Pap, R. Patz, M. Stock, J. Lehto, S. Breit, "Predictive modeling of pattern-dependent etch effects in large-area fully-integrated 3D virtual fabrication", In 2014 International Conference on Simulation of Semiconductor Processes and Devices (SISPAD) 2014 Sep 9 , pp. 209-212.

[9] Q. Wang, G. Mao, H. Zhao, C. Li, F. Xiao, R. Yang, S. Yu, "Fin critical dimension loading control by different fin formation approaches for FinFETs process", In 2017 China Semiconductor Technology International Conference (CSTIC) 2017 Mar 12, pp. 1-3.

[10] H.W. Su, Y. Li, Y.Y. Chen, C.Y. Chen, H.T. Chang, "Drain-induced-barrier lowering and subthreshold swing fluctuations in 16-nm-gate bulk FinFET devices induced by random discrete dopants", In 70th Device Research Conference 2012 Jun 18, pp. 109-110.

[11] R.H. Yan, A. Ourmazd, K.F. Lee, "Scaling the Si MOSFET: From bulk to SOI to bulk", IEEE Transactions on Electron Devices. 1992 Jul;39(7):1704-10.

5-6

Reactive Force-Field Molecular Dynamics Study of the Silicon-Germanium Deposition Processes by Plasma Enhanced Chemical Vapor Deposition

1st Naoya Uene
Graduate school of engineering
Tohoku university
Sendai, Japan
uene@nanoint.ifs.tohoku.ac.jp

2nd Takuya Mabuchi
Frontier research institute for
interdisciplinary sciences
Tohoku university
Sendai, Japan
mabuchi@tohoku.ac.jp

3rd Masaru Zaitsu
Rsearch & development
Japan advanced chemicals ltd.
Sagamihara, Japan
masaru.zaitsu@japanadvancedchemical
s.com

4th Shigeo Yasuhara
Rsearch & development
Japan advanced chemicals ltd.
Sagamihara, Japan
shigeo.yasuhara@japanadvancedchemi
cals.com

5th Takashi Tokumasu
Institute of fluid science
Tohoku university
Sendai, Japan
tokumasu@ifs.tohoku.ac.jp

Abstract— **In order to form a SiGe thin film by chemical vapor deposition (CVD) with a suitable quality for advanced devices, the relationships between materials/process and structure/composition are needed to be clarified at the atomic level. We simulated SiGe CVD by using reactive force-field (ReaxFF) molecular dynamics simulations, especially on binary systems of SiH_x + GeH_x, and derived the influence of the substrate temperature and these ratios of gaseous species on the crystallinity and compositions in the thin films. The crystallinity increases as the substrate temperature increases, and the lowest crystallinity is obtained at the ratios of gaseous species 0.5 and 0.7 for the SiH_3 and SiH_2, respectively. As the substrate temperature increases, the hydrogen content decreases while Si and Ge content tend to increase. These trends can be seen in relevant studies. Through this simulation we successfully observe that the reactivity of gaseous species greatly affects the crystallinity and compositions in the thin films.**

Keywords— Chemical Vapor Deposition, Reactive Force-Field Molecular Dynamics Simulation, Silicon-Germanium, Thin Film Deposition

I. INTRODUCTION

In the field of semiconductor manufacturing, chemical vapor deposition (CVD) and atomic layer deposition (ALD) are known as one of the common and powerful methods to deposit a high-quality thin film. In order to deposit promising thin films using CVD/ALD methods, the materials/process, structure/composition, and properties should be simultaneously optimized. For several decades, advances in semiconductor manufacturing have been increasing the number of materials and processes used for advanced device productions. As a result, an optimal combination should be selected from enormous number of options. This is hard issue and lead to increase the cost and development periods for device productions in the semiconductor manufacturing.

A new materials development tool called as "Materials integration" have appeared to deal with the hard issue. Materials integration aims to support the material development from an engineering viewpoint by combining all science and technology as shown in Fig. 1. In point of fact, some inorganic materials such as battery components have been successfully discovered for several years. Materials integration has also shown promising results in the semiconductor manufacturing. Chopra *et al.* developed a software tool for creating plasma etch recipes based on physical models and Bayesian inference [1] and applied it to prediction of experimental results [2]. They showed that the etching rate of SiO_2 with CF_4/Ar gases could be predicted. Suzuki *et al.* adopted machine learning approaches to optimize the plasma enhanced ALD process such as the uniformity [3]. They succeeded in achieving the target values and even outperforming the knowledgeable engineers. Chopra *et al.* and Suzuki *et al.* simply optimized only the process. The researches by Tanaka *et al.* are outstandingly noticeable from the viewpoint of simultaneous optimizing (co-optimization) materials and their processes [4]. They developed a predictive method to co-optimize by combining Bayesian optimization with data extracted from scientific papers, experimental data, and material databases. However, unfortunately, the results based on predictive model could not represent experimental results well.

Two approaches typically are known to improve the accuracy of materials integration; data-driven approach and physics-driven approach. The data-driven approach is based on correlations of material performance and structure extracted from systematically accumulated data. The physics-driven approach is based on the establishment of principles through elucidating controlling factors for material properties at each scale. Both approaches are necessary for all material development regardless of semiconductor manufacturing, and we selected latter. In order to proceed with the physics-driven

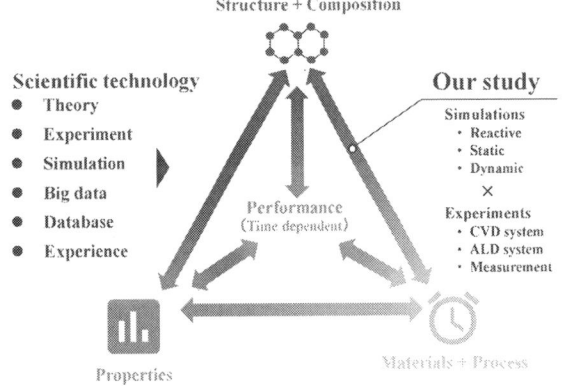

Fig. 1. A schematic diagram of Materials integration and our study position.

105

approach, the typical strategy is classifications into multi scale as shown in Fig. 2. At each scales, both physical dynamics and chemical reaction during CVD/ALD processes should be treated accurately. Reactive force-field (ReaxFF) MD simulations that is located between density functional theory (DFT), quantum mechanics (QM) calculations and empirical force-field (EFF) MD simulations can treat physical dynamics and chemical reactions [5].

Our final objective is to clarify the relationship between materials/process and structure/composition in thin films at the atomic level on CVD/ALD processes (see Fig. 1). As an example, we focused on a silicon-germanium (SiGe) thin film deposited by PECVD using SiH_4 and GeH_4, and H_2. The SiGe thin films, such as hydrogenated amorphous SiGe (a-SiGe:H) and hydrogenated microcrystalline SiGe (µc-SiGe:H), are known as a potential material for thin film solar cells [6], micro-electro-mechanical systems (MEMS) [7], and biomedical applications [8]. In the experiment, flow rates of SiH_4 and GeH_4 were changed to control the composition of Si and Ge in the thin film. At that time, various gaseous species that have a different reactivity are formed in the plasma. The different reactivity affects the structure/composition of the thin film, but the influence is not clear enough. To achieve a specific ratios of gaseous species experimentally is not easy, however this is an important aspect to optimize the deposition process. The SiH_3, SiH_2, GeH_3, and GeH_2 are suggested that they have a dominant effect on thin film deposition [9]. In this paper, we considered binary systems as shown in Fig. 3 and analyzed the influence of the substrate temperature and these ratios of gaseous species on the crystallinity and compositions in the thin films at the atomic level by ReaxFF MD simulations.

II. METHODS

A. Simulation details

ReaxFF MD simulations were performed to simulate the SiGe deposition processes with large-scale atomic/molecular massively parallel simulator (LAMMPS) [10]. An existing parameter set [11] was modified to match the dissociation energies in gaseous species because bond formation and breaking are particularly important in this study. The substrate was Si (100)-(2×1), and the size of the simulation box was 30.72 Å × 26.88 Å × 45.00 Å. A periodic boundary condition was applied in the x and y directions, and a fixed boundary condition was applied in the z direction. The first bottom layer of the substrate was fixed, and second and third layers were controlled constantly at desired temperature by the Berendsen

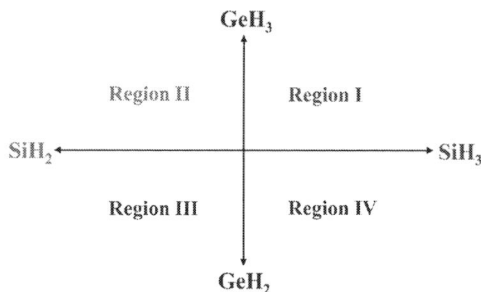

Fig. 3. A binary system of gaseous species ratios that dominates the deposition for SiGe thin film. Region I and II are described in this paper.

thermostat. The temperature was not controlled on the fourth and subsequent layers to prevent direct influences of changing the velocity of atom due to the temperature control on the surface reaction. The range of the substrate temperature was set at 800 K–1500 K. 2000 gaseous molecules with the velocity corresponding to 1300 K in the z direction were impinged one by one onto the substrate surface every 7.5 ps. Ratios of SiH_3 and GeH_3 were selected and then, GeH_3 flow rate X_{GeH3} was defined. The change of ratios of SiH_3 and GeH_3 corresponds to deposit films at various germanium contents by changing the gaseous composition (SiH_4, GeH_4 and, H_2). The initial position of the gaseous species was set at 10 Å from the top of the Si substrate in the z direction, and that in the in-plane (x-y) was randomly set. The initial position of gaseous molecules is changed from the surface with the constant velocity 1–2 Å /ns to prevent the interaction with the deposited atoms at initial condition. The Velocity-Verlet algorithm was used for time integration with a 0.25 fs time step. A pre-thermal annealing for 7.5 ps and a post-thermal annealing for 60 ps were performed before and after the deposition to equilibrate the substrate and deposited thin films.

B. Analysis details

Open visualization tool (OVITO) was used to analyze the structure and compositions in thin films [12]. In the case of SiGe thin films, the structure is mainly cubic diamond. The identify diamond structure algorithm was used to find atoms that is arranged in cubic diamond lattice. Using the algorithm, the local environment of each atom was analyzed up to the second neighbor shell to determine the four local structural type as a cubic diamond, 1st neighbor, 2nd neighbor, and others. These detail definition and algorithms for identifications are described in the reference [13]. As a structure, we evaluated the crystallinity (C_d) that was defined like following equation;

$$C_d = (\text{Cubic} + \text{1st neighbor} + \text{2nd neighbor}) / N_{depo} \quad [1]$$

Where, the cubic diamond, 1st neighbor, and 2nd neighbor represent the number of classified atoms as mentioned above. The atomic content as a composition were calculated for each atom by dividing the number of remained Si/Ge/H atoms by the number of total atoms (N_{depo}) in thin films as follows (x is H, Si, or Ge atom);

$$x \text{ content} = x \text{ atoms} / N_{depo} \quad [2]$$

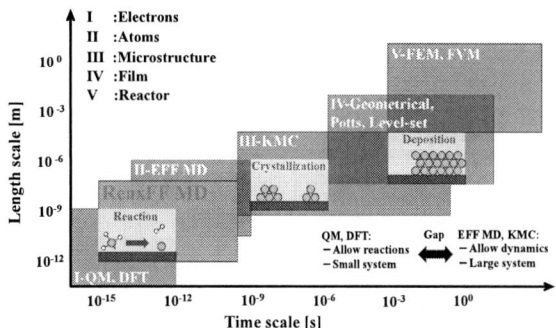

Fig. 2. Multiscale approach in CVD/ALD processes based on the physics driven approaches.

III. RESULTS AND DISCUSSION

In this paper, only the region I and II in fig. 3 are described (i.e., SiH_3 + GeH_3 or SiH_2 + GeH_3). Fig. 4 shows the deposited thin films with various ratios of gaseous species X_{GeH3} (= GeH_3 / SiH_x + GeH_3) and substrate temperatures. Here, x is 2 or 3. These snapshots were taken from the final configurations.

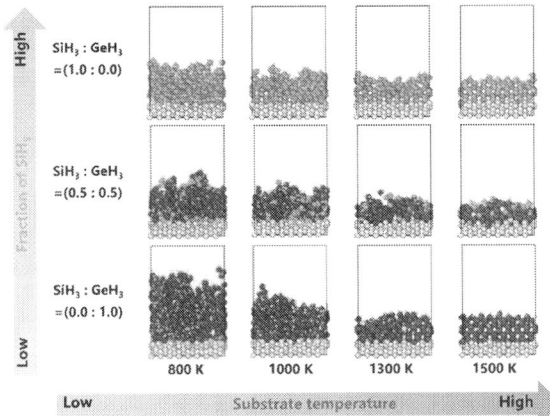

Fig. 4. Snapshots of the deposited thin films with various ratios of SiH_3 and GeH_3 X_{GeH3} and the substrate temperatures. The white, beige, yellow, and deep green represent added H atom, added Si atom, substrate Si atoms, and added Ge atoms, respectively.

A. Crystallinity

Fig. 5 shows the crystallinity as a function of the substrate temperature for ratios of SiH_3 and GeH_3 X_{GeH3}. The crystallinity C_d increases as the substrate temperature increases. The SiH_3 and GeH_3 more overcome the energy barrier from local minimum to global minimum (on lattice) with increasing the substrate temperature. As a result, higher crystal structure (more compact thin film) are obtained at high substrate temperature. These tendencies are related to experimental study for the deposition of a-SiGe:H [17]. Fig. 6 shows the crystallinity as a function of the X_{GeH3} in the case of SiH_2 and SiH_3. The insets represent the atomic configuration. The lowest crystallinity is obtained at the X_{GeH3} 0.5 and 0.7 for the SiH_3 and SiH_2, respectively. The crystallinity at the X_{GeH3} 1.0 is comparable to pure Si films (X_{GeH3} 0.0). This reflect that the uniform ratios of gaseous species are preferable for the highest crystallinity. It may lead

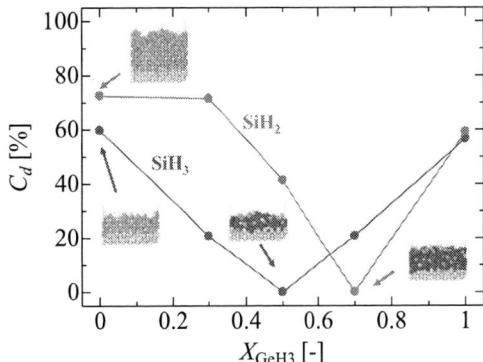

Fig. 6. Crystallinity as a function of the X_{GeH3} when the substrate temperature is 1500 K.

to the less difference of lattice constant in the films (Si: 5.43 Å, Ge: 5.68 Å). The crystallinity is related to the microstructure parameter R^* as measured with Fourier transform infrared (FTIR) spectroscopy. Please note that the microstructure parameter decreases at void-less films, thus it has opposite meanings with the crystallinity. Our trends can be quantitatively compared with the experimental ones reported in a literature [14].

B. Composition

Fig. 7 shows the hydrogen content as a function of the substrate temperature for various ratios of SiH_3 and GeH_3 X_{GeH3}. The hydrogen content decreases as the substrate temperature increases. While silicon and germanium content tend to increase with increasing substrate temperature (not shown here). The reason is more H_2 are formed and desorbed from the surface at higher substrate temperature, resulting in an increase in the silicon and germanium content. Fig. 8 shows Si and Ge content as a function of the X_{GeH3}. The increase of Ge content is remarkable with respect to X_{GeH3} in the case of SiH_3. In other words, Ge atoms were preferentially incorporated into the film compared with Si atoms. This Ge preferential incorporation from the gas phase into the film is typically observed during the deposition process for SiGe thin films [14]. Please note that the binding energy of Ge–H (2.97 eV/bond) is lower than that of Si–H (3.20 eV/bond) in gaseous species, therefore, GeH_3 is easier to dissociate than SiH_3 on the surface during surface diffusions and preferentially incorporated into the film.

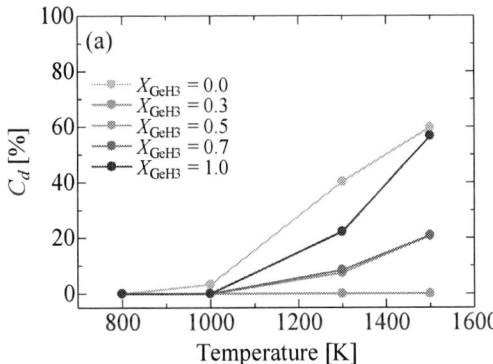

Fig. 5. Crystallinity as a function of the substrate temperature for various ratios of SiH_3 and GeH_3 X_{GeH3}.

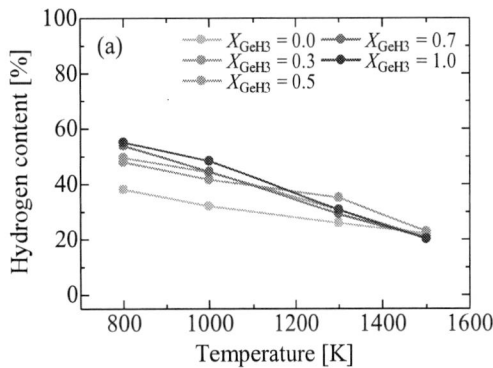

Fig. 7. Hydrogen content as a function of the substrate temperature for various ratios of SiH_3 and GeH_3 X_{GeH3}.

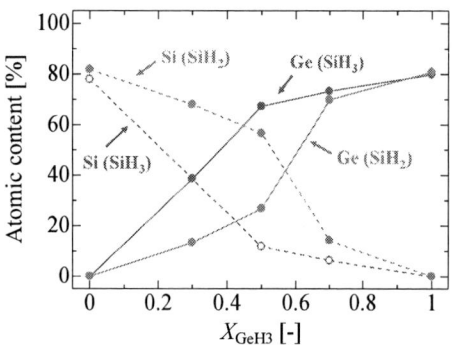

Fig. 8. Si and Ge content as a function of the ratios of X_{GeH3} when the substrate temperature is 1500 K.

On the other hands, in the case of SiH$_2$, the Ge content gradually goes up with respect to X_{GeH3} comparing to the case of SiH$_3$. This gradual increase may be originated from the difference of reactivity between SiH$_3$ and SiH$_2$. SiH$_2$ can be incorporated into the film without dissociation of Si–H bond. These results suggest that the Ge preferential incorporation from the gas phase into the film is also depend on the ratios of gaseous species in plasmas.

IV. CONCLUSIONS

We investigated SiGe thin film deposition with various ratios of SiH$_3$, SiH$_2$, and GeH$_3$ using ReaxFF MD simulations. The crystallinity and atomic contents in the deposited thin films were analyzed as a structure and compositions. The crystallinity C_d increased as the substrate temperature increases and could be explained by the higher mobility of SiH$_x$ and GeH$_x$ on the surface at elevated substrate temperature. The lowest crystallinity was obtained at the ratios of gaseous species X_{GeH3} 0.5 and 0.7 for the SiH$_3$ and SiH$_2$, respectively. The crystallinity at the X_{GeH3} 1.0 was comparable to pure Si films (X_{GeH3} 0.0). This reflect that the uniform ratios of gaseous species are preferable for the highest crystallinity. The hydrogen content decreased as the substrate temperature increases, while silicon and germanium content tended to increase with increasing substrate temperature. The reason is more H$_2$ molecules were formed and desorbed from the surface at higher substrate temperature, resulting in an increase in silicon and germanium content. In the case of SiH$_3$, Ge atoms were preferentially incorporated into the film compared with Si atoms. The reason is originated from lower binding energy of Ge–H (2.97 eV/bond) than Si–H (3.20 eV/bond) in gaseous species. These tendencies in this paper were in good agreement with relevant studies. On the other hands, in the case of SiH$_2$, the Ge content gradually went up with respect to the X_{GeH3} comparing to the SiH$_3$. It was originated from the difference of reactivity between SiH$_3$ and SiH$_2$. SiH$_2$ can be incorporated into the film without dissociation of Si–H bond. These results suggest that the Ge preferential incorporation from the gas phase into the film is also depend on the ratios of gaseous species in plasmas.

Finally, we observed that the substrate temperature and ratios of gaseous species affect the structure/composition of the thin film, obviously. Our simulation is promising to optimize the deposition process for high-quality thin films.

ACKNOWLEDGMENT

This work was supported by JSPS Grant-in-Aid for JSPS Research Fellow Grant Number JP20J20915. Numerical simulations were performed on the Supercomputer system "AFI-NITY" at the Advanced Fluid Information Research Center, Institute of Fluid Science, Tohoku University.

REFERENCES

[1] M. J. Chopra, R. Verma, A. Lane, C. G. Willson, and R. T. Bonnecaze, "A method to accelerate creation of plasma etch recipes using physics and Bayesian statistics," in *Proc.SPIE*, Mar. 2017, vol. 10149, doi: 10.1117/12.2263507.

[2] M. J. Chopra, X. Zhu, Z. Z. Zhang, S. Helpert, R. Verma, and R. Bonnecaze, "A model-based, Bayesian approach to the CF4/Ar etch of SiO2," in *SPIE*, 2018, no. March 2018, p. 15, doi: 10.1117/12.2297482.

[3] Y. Suzuki, S. Iwashita, T. Sato, H. Yonemichi, H. Moki, and T. Moriya, "Machine learning approaches for process optimization," in *IEEE International Symposium on Semiconductor Manufacturing Conference Proceedings*, 2019, vol. 2018-Decem, pp. 2–5, doi: 10.1109/ISSM.2018.8651142.

[4] F. Tanaka, H. Sato, N. Yoshii, and H. Matsui, "Materials informatics for process and material co-optimization," in *IEEE Transactions on Semiconductor Manufacturing*, 2019, vol. 32, no. 4, doi: 10.1109/TSM.2019.2943162.

[5] K. Chenoweth, A. C. T. van Duin, and W. A. Goddard, "ReaxFF Reactive Force Field for Molecular Dynamics Simulations of Hydrocarbon Oxidation," *J. Phys. Chem. A*, vol. 112, no. 5, pp. 1040–1053, 2008, doi: 10.1021/jp709896w.

[6] Z. Yu *et al.*, "Improved power conversion efficiency in radial junction thin film solar cells based on amorphous silicon germanium alloys," *J. Alloys Compd.*, vol. 803, no. 30, pp. 260–264, 2019, doi: 10.1016/j.jallcom.2019.06.276.

[7] Q. Wang and H. Vogt, "With PECVD Deposited Poly-SiGe and Poly-Ge Forming Contacts Between MEMS and Electronics," *J. Electron. Mater.*, vol. 48, no. 11, pp. 7360–7365, 2019, doi: 10.1007/s11664-019-07561-4.

[8] F. L. Huerta *et al.*, "Biocompatibility and surface properties of hydrogenated amorphous silicon-germanium thin films prepared by LF-PECVD," *IOP Conf. Ser. Mater. Sci. Eng.*, vol. 628, no. 1, p. 012003, 2019, doi: 10.1088/1757-899X/628/1/012003.

[9] L. Zhao, R. Hunsperger, and S. Hegedus, "Modeling and experimental study of SiH4/GeH4/H2 gas discharge for hydrogenated silicon germanium deposition by RF PECVD," in *Materials Research Society Symposium Proceedings*, 2012, vol. 1426, pp. 403–408, doi: 10.1557/opl.2012.841.

[10] S. Plimpton, "Fast Parallel Algorithms for Short-Range Molecular Dynamics," *J. Comput. Phys.*, vol. 117, pp. 1–19, 1995, doi: 10.1006/jcph.1995.1039.

[11] G. Psofogiannakis and A. C. T. Van Duin, "Development of a ReaxFF reactive force field for Si/Ge/H systems and application to atomic hydrogen bombardment of Si, Ge, and SiGe (100) surfaces," *Surf. Sci.*, vol. 646, pp. 253–260, 2016, doi: 10.1016/j.susc.2015.08.019.

[12] A. Stukowski, "Visualization and analysis of atomistic simulation data with OVITO-the Open Visualization Tool," *Model. Simul. Mater. Sci. Eng.*, vol. 18, no. 1, p. 015012, 2010, doi: 10.1088/0965-0393/18/1/015012.

[13] E. Maras, O. Trushin, A. Stukowski, T. Ala-Nissila, and H. Jónsson, "Global transition path search for dislocation formation in Ge on Si(001)," *Comput. Phys. Commun.*, vol. 205, pp. 13–21, 2016, doi: 10.1016/j.cpc.2016.04.001.

[14] L. W. Veldhuizen, C. H. M. Van Der Werf, Y. Kuang, N. J. Bakker, S. J. Yun, and R. E. I. Schropp, "Optimization of hydrogenated amorphous silicon germanium thin films and solar cells deposited by hot wire chemical vapor deposition," *Thin Solid Films*, vol. 595, pp. 226–230, 2015, doi: 10.1016/j.tsf.2015.05.055.

[15] M. K. Van Veen and R. E. I. Schropp, "Beneficial effect of a low deposition temperature of hot-wire deposited intrinsic amorphous silicon for solar cells," *J. Appl. Phys.*, vol. 93, no. 1, pp. 121–125, 2003, doi: 10.1063/1.1527208.

6-1 Universal Feature of Trap-Density Increase in Aged MOSFET and Its Compact Modeling

Fernando Avila Herrera
HiSIM Research
Hiroshima University
Higashihiroshima, Japan
herrera@hiroshima-u.ac.jp

Mitiko Miura-Mattausch
HiSIM Research
Hiroshima University
Higashihiroshima, Japan
mmm@hiroshima-u.ac.jp

Takahiro Iizuka
HiSIM Research
Hiroshima University
Higashihiroshima, Japan
iizuka@hiroshima-u.ac.jp

Hideyuki Kikuchihara
HiSIM Research
Hiroshima University
Higashihiroshima, Japan
kikuchihara532@hiroshima-u.ac.jp

Hans Jürgen Mattausch
HiSIM Research
Hiroshima University
Higashihiroshima, Japan
hjm@hiroshima-u.ac.jp

Hirotaka Takatsuka
Technology Developmen Division
United Semiconductor Japan Co.
Kanagawa, Japan
takatuka@usjpc.com

Abstract—Our investigation focuses on accurate circuit aging prediction for bulk MOSFETs. A self-consistent aging modeling is proposed, which considers the trap-density N_{trap} increase as the aging origin. This N_{trap} is considered in the Poisson equation together with other charges induced within MOSFET. It is demonstrated that a universal relationship of the N_{trap} increase as a function of integrated substrate current, caused by device stress, can describe the MOSFET aging in a simple way for any device-operating conditions. An exponential increase with constant and unitary slope of the N_{trap} is found to successfully predict the aging phenomena, reaching a saturation for high stress degradation. The model universality is verified additionally for any device size. Comparison with existing conventional aging modeling for circuit simulation is discussed for demonstrating the simplifications due to the developed modeling approach.

Keywords—MOSFET aging/reliability, Trap-density increase, channel-length dependence, compact model, aging simulation

I. INTRODUCTION

Circuit-reliability prediction requires several different simulation processes and tools to cover the whole range of chip development, from device-level to circuit-level aging. Such tools contain aging models, which are implemented in these tools separate from the MOSFET model [1]. Although the simulation tools provide advanced techniques, the accuracy of the results relies on the accuracy of the aging model itself, which is used to predict the device reliability during circuit operation. Our focus is therefore given on the development of an improved device-aging model for circuit-reliability prediction.

Conventional modeling for predicting transistor aging is derived on the basis of measurements. The accurate modeling of the device aging is normally based on the stress-degradation time and the bias stress applied in experiments with single devices, which covers one of the degradation issues. However, estimating the actually applied stress for each device during circuit operation is more complicated and therefore another issue. Consequently, important tasks for improving the reliability of the circuit simulation are to estimate each individual device stress accurately and also to model the resulting device aging accurately. Particularly, in N-MOSFETs the aging effect is caused by the hot-carrier effect, which is enhanced in new technology-node generations due to higher lateral electric field applied. Here, we investigate the aging effect in leading-edge technologies.

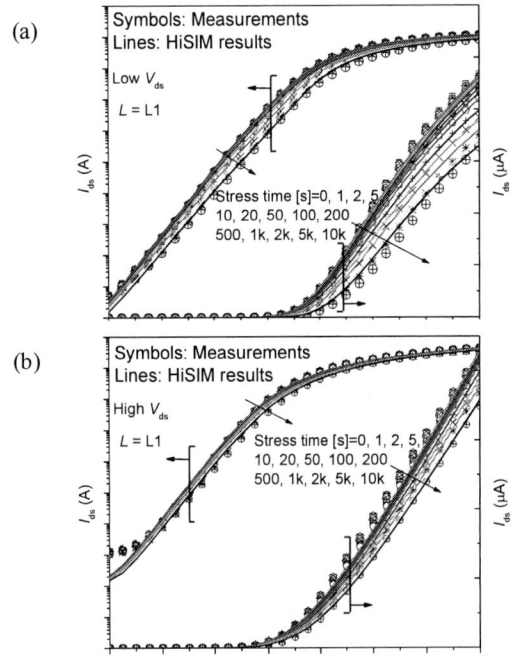

Fig. 1. Comparison of model calculation results with measurements for low (a) and high (b) V_{ds} values for the whole I-V characteristics.

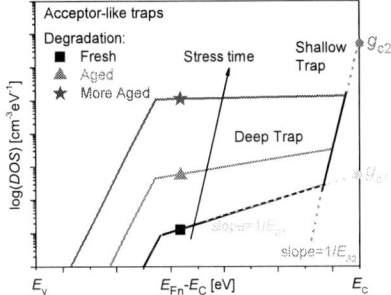

Fig. 2. Comparison of model calculation results with measurements for low (a) and high (b) V_{ds} values for the whole I-V characteristics.

109

Fig. 1 shows typically measured drain-current *vs.* gate voltage (I_{ds}-V_{gs}) aging characteristics for a 60nm MOSFET technology node at low (Fig. 1a) and high (Fig.1b) drain voltage V_{ds}. Usually, modeling is done separately for different operating conditions, considering individual macroscopic aging observables, such as threshold voltage (V_{th}) and maximum of I_{ds} ($I_{ds\ max}$), and adjusting the aging-model parameters associated with each of these observables [2] [3]. As a result, the burden for constructing an aging model becomes huge, with a complex model-parameter extraction. Such conventional models incorporate the model-equation deviations according to the stress for these operating conditions in order to describe the aging effect [4]. Here, our investigation aims at developing a new modeling approach, which considers the N_{trap} increase as the aging origin in a consistent way.

II. AGING MODELING: TRAP-DENSITY INCREASE BASED

Fig. 2 depicts a schematic of the density-of-state (DOS) model as a function of the state-energy difference from the conduction-band edge, where two parameters g_c and E_s are introduced to model the important features.

Measured $1/f$ noise I_{ds}-V_{gs} data is used for extracting the DOS parameters, because they allow N_{trap} extraction under exclusion of carrier-mobility effects [5]. The relation between N_{trap} and DOS is shown in Table I. From the analysis of measurements, it has been found that a universal relationship exists for DOS as a function of the integrated substrate current I_{sub} during stress time [6].

The obtained results for DOS aging (see Fig. 3a) exhibit an initial linear relationship with unity gradient, when plotted in double-log form (Fig. 3b). Only the extrapolated DOS-axis intercept is different, which is exactly a linear function of the I_{sub} for a fresh device (unstressed device), as depicted in Fig. 4. However, DOS does not increase unlimitedly, but starts to saturate around the value of 4×10^{18}cm^{-3}eV^{-1}. The N_{trap}-increase shows a universal relationship as a function of the substrate current I_{sub} and stress time t

$$N_{trap} = N_{trap0/max} + C \times I_{sub0}\left(I_{sub,t}\right)^n \quad (1)$$

where n=1 is obtained for relatively short t conditions; $I_{sub,t}$=I_{sub}·t. The n=1 relationship verifies the fact that aging is a linear function of $I_{sub,t}$, independently of the bias stress conditions. However, in Fig. 3b the universal relationship is shifted in parallel according to I_{sub0} (I_{sub} of fresh device) following a linear offset increase as Fig. 4 indicates. This result demonstrates that the stress-bias condition and the stress-duration time do not influence the aging equivalently, but that the stress-bias condition determines mainly the initial aging value. This initial aging is a linear function of I_{sub0} (see Fig. 4) and the fresh trapped charge N_{trap0}. Further, enhanced DC-stress reduces the N_{trap} rate (i.e. n<1) for longer stress times t, when reaching N_{trap} saturation at $N_{trap} = N_{max}$ of 4×10^{18}cm^{-3} (see Fig. 3b). Practically, $I_{sub}/W \cdot t$ has a value of the order of 100As/m and therefore the N_{trap} saturation will become hardly observable. N_{trap} is explicitly included in the Poisson equation to assure an accurate calculation of the surface-potential degradation and to determine the aged I-V characteristics [7] with the highest achievable physical consistency and correctness. Table I summarizes the main developed model equations and model parameters.

III. DISCUSSION

In order to validate the developed aging model, comparison of modeling and measurement results are performed for three different channel lengths: L1, L2=1.2·L1 and L3=1.7·L1. The studied devices have been stressed at high biases for both V_{ds} and V_{gs} and for several different stress times from 1s to 10ks. I-V characteristics are obtained for low and high V_{ds} cases.

TABLE I.   MAIN MODEL EQUATIONS AND MODEL PARAMETERS

Symbol	Quantity	Value
N_{trap}	Trap-density increase under stress	$N_{trap0} + C \times I_{sub0}\left(I_{sub,t}\right)^n$
N_{trap}	Trap-density increase under long-term stress	$N_{max} + C \times I_{sub0}\left(I_{sub,t}\right)^{n/2}$
N_{trap0}	Initial trap-density	
N_{max}	Maximum trap-density	4×10^{18} cm^{-3}
C	Coefficient for trap-density increase	$\approx 4\times10^{15}$ cm^{-3}
n	Initial slope of trap-density increase	1
I_{sub0}	Substrate current	
$I_{sub,t}$	Total stress condition with t duration	$I_{sub}\times t$
$d^2\phi_S/dx^2$	Poisson equation	$-q/\varepsilon_S \cdot (N_D - N_A + p - n - N_{trap})$
N_{trap} (g_c)	Trap-density equation based on DOS	$g_c E_s \cdot \dfrac{kT/E_s}{\sin\left(kT/E_s\right)} \times \exp\left(\dfrac{E_F - E_C}{E_s}\right)$

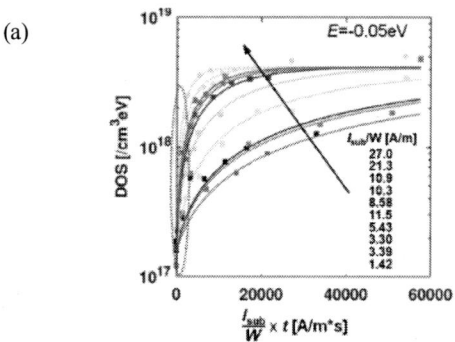

(a)

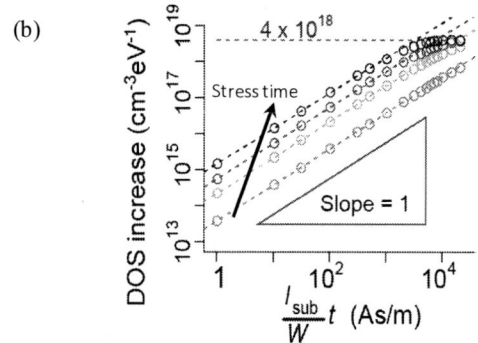

(b)

Fig. 3. Extracted DOS as a function of $I_{sub,t}$ for different stress conditions at the energy E of 0.05eV away from E_c in (a) semilog and (b) double-log scale, where N_{trap} is proportional to DOS.

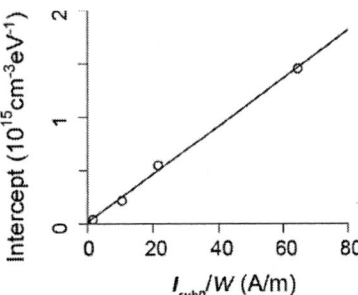

Fig. 4. Initial N_{trap} value at t=1 s as a function of I_{sub0}.

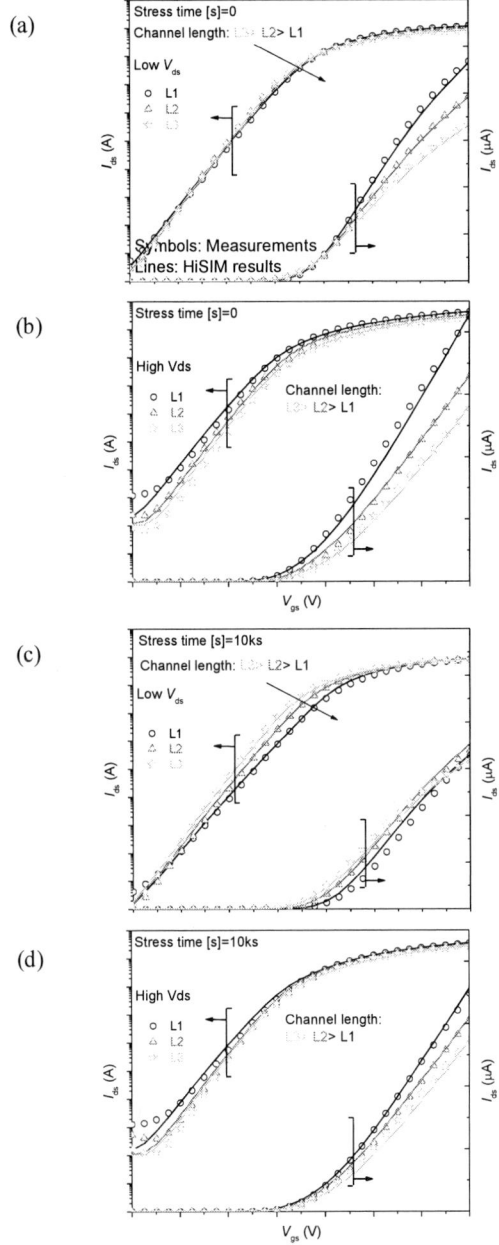

Fig. 5. Comparison of model calculation results with measurements at t=0 for low (a) and high (b) V_{ds} values and for an aged device with t=10ks at (c) low and (d) high V_{ds}. Three different channel lengths are compared by using one single model-parameter extraction. Channel lengths are L1, L2=1.2·L1 and L3=1.7·L1 of model Initial N_{trap} value at t=1 s as a function of I_{sub0}.

Fig. 1 shows the aged I_{ds}-V_{gs} characteristics for the transistor with the largest expected aging effects, namely for the case of the shortest channel length L = L1. An accurate prediction for the drain-current degradation is achieved for the whole operation range. Further drain-current comparison is drawn for the three different channel-lengths in Figs. 5a and b before stress at t=0 and in Figs. 5c and d after a stress time of t=10ks. In particular, the model is able to predict the subthreshold-slope degradation due to the trap-density increase.

To understand the degradation of the I-V characteristics, the N_{trap} extraction of the studied devices is shown Fig. 6a. As can be observed, the N_{trap} first increases linearly in this double-log-scale plot and tends to saturate for longer stress times. In Fig. 6b, the N_{trap} is plotted in log-log form as a function of the total stress for the maximum-stress case of 100 A/m·s, which can be a typical value reached for the devices in a circuit. The extracted slope is n=1 for all 3 channel lengths. Finally, Fig. 7 shows a comparison for typical aging parameters that are conventionally considered individually. Our modeling approach follows a cause-effect analysis, i.e., all aging observables are a consequence of the N_{trap} increase.

Therefore, these observables can be extracted without any further adjustment. Only one set of aging-model parameters can be used for all device structures and macroscopic aging observables.

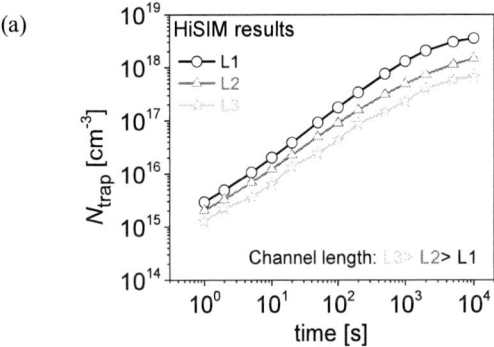

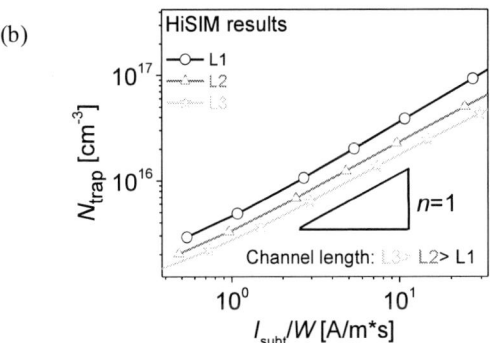

Fig. 6. Extracted N_{trap} as a function of (a) stress time t and (b) I_{subt} for three different channel lengths. A clear parallel offset is observed in Fig.6a for different channel lengths. Fig. 6b shows the range of interest for real applications. Aging saturation also occurs after very long degradation times with enhanced stress conditions. One single model-parameter set is used for the modeling of all structures.

111

IV. CONCLUSSION

The trap-density increase, the origin of aging phenomena, has been verified to have a universal relationship with the degradation process, i.e., an exponential increase with constant slope that reaches a saturation condition at high degradation times. Only the initial point of the aging process has been found to be substrate-current dependent. Verification is done for a family of devices, using a unique set of model parameters for describing all of their aging phenomena. All aging observables have been properly predicted along with the *I-V* characteristics.

REFERENCES

[1] T. Guo and J. Xie, "A Complete Reliability Solution: Reliability Modeling, Applications, and Integration in Analog Design Environment," 2018. [Online]. Available: http://www.mos-ak.org/beijing_2018/presentations/Tianlei_Guo_MOS-AK_Beijing_2018.pdf.

[2] C. Guerin, V. Huard and A. Bravaix, "The Energy-Driven Hot-Carrier Degradation Modes of nMOSFETs," in *IEEE Transactions on Device and Materials Reliability*, vol. 7, no. 2, pp. 225-235, June 2007, doi: 10.1109/TDMR.2007.901180.

[3] B. Tudor, J. Wang, Z. Chen, R. Tan, W. Liu and F. Lee, "An Accurate MOSFET aging model for 28 nm integrated circuit simulation," *Mielectronics Reliability*, vol. 52, pp. 1565-1570, Aug. 2012, doi: 10.1016/j.microrel.2011.12.008.

[4] C. Hu, S. C. Tam, F.-C. Hsu, P.-K. Ko, T.-Y. Chan and K. W. Terrill, "Hot-Electron-Induced MOSFET Degradation - Model, Monitor, and Improvement," in *IEEE Transactions on Electron Devices*, vol. 32, no. 2, pp. 375-385, Feb. 1985, doi: 10.1109/T-ED.1985.21952.

[5] M. Miura-Mattausch, H. Miyamoto, H. Kikuchihara, T. K. Maiti, N. Rohbani, D. Navarro and H. J. Mattausch, "Compact modeling of dynamic trap density for predicting circuit-performance aging," *Solid-State Electronics*, vol. 80, pp. 164-175, Jan. 2018, doi: 10.1016/j.microrel.2017.12.003.

[6] H. Tanoue, A. Tanaka, Y. Oodate, T. Nakahagi, D. Sugiyama, C. Ma, H. J. Mattausch and M. Miura-Mattausch, "Compact Modeling of Dynamic MOSFET Degradation Due to Hot-Electrons," in *IEEE Transactions on Device and Materials Reliability*, vol. 17, no. 1, pp. 52-58, March 2017, doi: 10.1109/TDMR.2017.2655519.

[7] M. Miura-Mattausch, H. J. Mattausch and T. Ezaki, *The Physics and Modeling of MOSFETs*, Singapore: World Scientific, 2008, doi: 10.1142/6159.

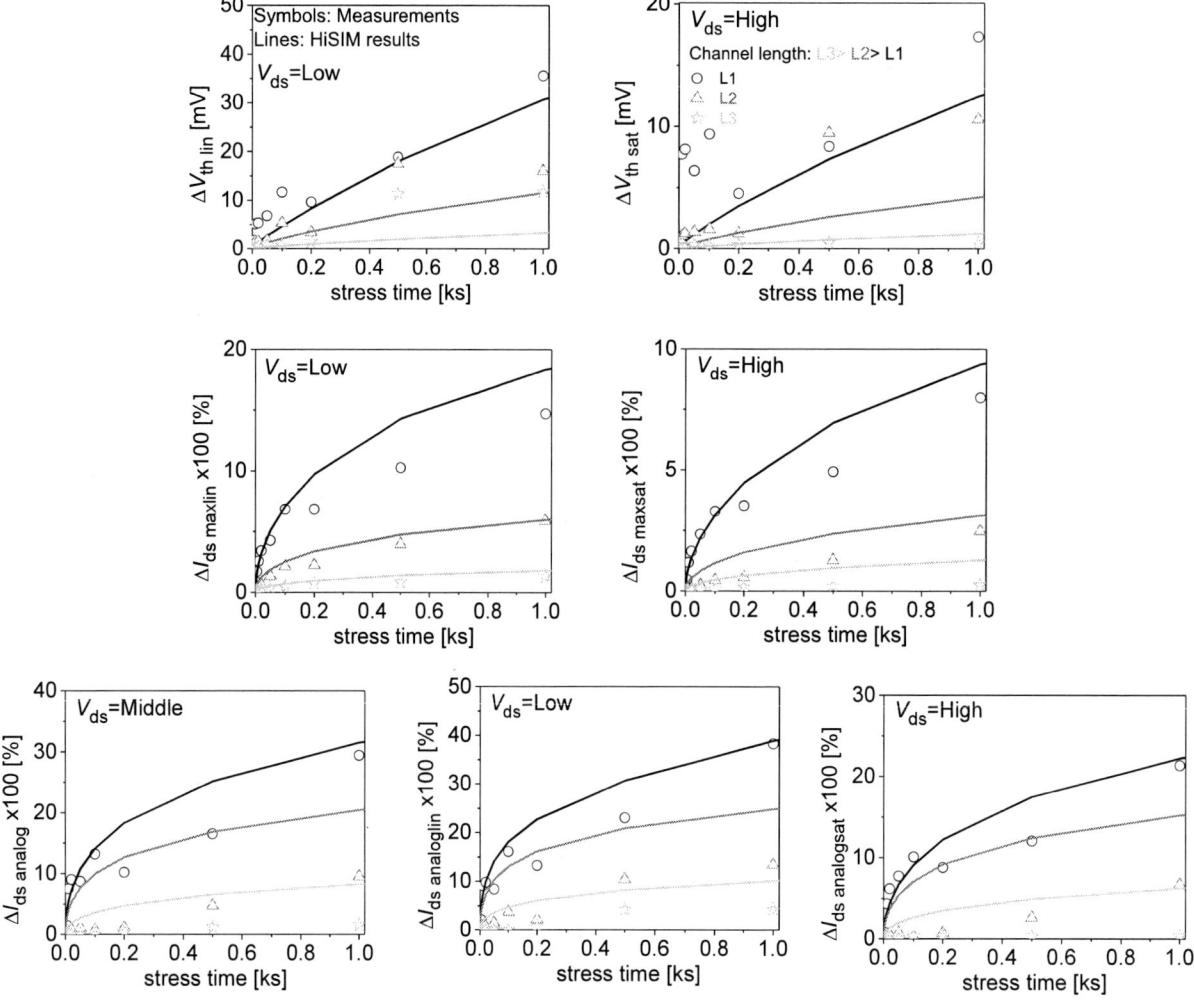

Fig. 7. Measured device characteristics as a function of stress duration for leading-edge MOSFET technology. Calculated results with the developed model are depicted together for the comparison. A single set of model parameters is used for all cases. Channel lengths are L1, L2=1.2·L1 and L3=1.7·L1. t=1ks corresponds to $I_{sub} \cdot t/W$=100As/m.

6-2
TCAD Incorporation of Physical Framework to Model N and P BTI in MOSFETs

Ravi Tiwari, Nilotpal Choudhury, Tarun Samadder, Subhadeep Mukhopadhyay, Narendra Parihar and Souvik Mahapatra

Department of Electrical Engineering, Indian Institute of Technology Bombay, Mumbai 400076, India

*Phone: +91-222-572-0408, Email: souvik@ee.iitb.ac.in

Abstract: Negative and Positive Bias Temperature Instabilities (NBTI, PBTI) respectively in P and N channel High-K Metal Gate (HKMG) MOSFETs are modeled by trap generation (TG) and charge trapping (CT) and validated against measured data. The mechanism of TG (interface) is incorporated into TCAD and is separately validated using independent experiments. BTI kinetics is modeled at different stress bias (V_G) and temperature (T). Impacts of Nitrogen (N%) and Equivalent Oxide Thickness (EOT) scaling on the magnitude of BTI and its time, V_G and T dependencies are modeled.

Keywords: threshold voltage shift, DCIV, NBTI, PBTI, EOT scaling, HKMG-MOSFETs, Nitrogen impact

I. INTRODUCTION

Bias Temperature Instability (BTI) is a crucial reliability issue in P and N channel HKMG MOSFETs [1]. It causes shift in several device parameters such as threshold voltage (ΔV_T), linear (ΔI_{DLIN}) and saturation (ΔI_{DSAT}) drain current, subthreshold slope (ΔSS) and transconductance (Δg_m) etc. over time, due to the buildup of positive charges in the gate insulator of the transistor. These parameters recover partially when the stress is lowered or removed, resulting in lower degradation during AC as compared to DC stress. The BTI mechanism is debated and various models are proposed [2]-[4]. It is now believed that ΔV_T during NBTI in P MOSFETs is due to TG at the channel/interlayer (IL) interface (ΔV_{IT-IL}) and IL bulk (ΔV_{OT-IL}), and CT (hole, ΔV_{HT-IL}) in IL bulk, and ΔV_T during PBTI in N MOSFETs is due to the TG at IL/HK interface (ΔV_{IT-HK}) and HK bulk (ΔV_{OT-HK}), and CT (electron, ΔV_{ET-HK}) in HK bulk, see Fig.1 [1]. The BTI Analysis Tool (BAT) utilizes the uncorrelated contributions from the above subcomponents of overall ΔV_T, Fig.2, and can model the ΔV_T kinetics during DC and AC stress and recovery, as V_G, T, pulse duty cycle and frequency are varied for PBTI [5] and NBTI [6], [7], and various process changes for NBTI [7]. BAT uses double interface Reaction-Diffusion Model (RDM) for interfacial TG (density ΔN_{IT-IL} or ΔN_{IT-HK}), their charged state and contribution (ΔV_{IT-IL} or ΔV_{IT-HK}) by the Transient Trap Occupancy Model (TTOM) and empirical models for CT and bulk TG. Recently, the Activated Barrier Double Well Thermionic Model (ABDWTM) is used for hole CT for NBTI [8]. The interfacial TG kinetics during NBTI has been modeled using TCAD and validated using Direct-Current I-V (DCIV) data [12], [13]. The TCAD-NBTI framework utilizes Capture-Emission Depassivation Model (CEDM), Fig.3, and Multi-State Configuration Model (MSCM), Fig.4 to calculate TG kinetics. The framework is validated using FinFETs with different channel materials, see [9] for further details.

II. SCOPE OF WORK

PBTI interfacial trap generation is incorporated in TCAD and validated using DCIV. Although experiments suggest TG at the channel/interlayer and interlayer/High-K respectively for N and P BTI [1], due to the absence of coupling between non-local tunneling and CEDM at present, TG is calculated at the channel/IL interface for both cases, but generated traps are assigned at the IL/High-K interface to obtain ΔV_T during PBTI. TG kinetics measured using DCIV method at various V_G and T, in N and P HKMG MOSFETs having different N% and IL thickness (data from [9]) is modeled by TCAD. ΔV_T time kinetics measured using ultra-fast I-V method for the same devices [5], [6] is modeled using BAT, with the underlying interfacial TG component validated using TCAD. Impact of N% induced IL scaling (leading to EOT scaling) is modeled.

III. BTI MECHANISM

Inversion layer holes tunnel to Hydrogen (H) passivated defect precursors at the channel/IL interface at $V_G<0V$ stress, react, break ($=\Delta N_{IT-IL}$) and release H, Fig.3. The released H atoms diffuse and react with other H passivated defects inside IL (HK) bulk to form H_2, which diffuses away, Fig.4. This is RDM for NBTI. Interfacial TG (from DCIV) shows similar power low time kinetics for both and N and P BTI [1]. A similar process as RDM for NBTI is used for PBTI at $V_G>0V$ stress, with inversion electrons induced dissociation of the defect precursors at the IL/HK interface ($=\Delta N_{IT-HK}$), H atoms diffusion and defect generation at the HK bulk, and diffusion of H_2 (Figs.3 and 4), which is RDM for PBTI. The chemical nature of defects is likely different for the interfacial TG in N and P BTI. ΔN_{IT-HK} can be created directly by electrons, or by the Anode Hole Injection (AHI [11]) process, *i.e.*, tunneling of electrons from cathode to anode, impact ionization and injection of energetic hole into gate oxide, although a direct link (the former) is used in TCAD at present. Moreover, the tunneling and trapping of holes (IL) and electrons (HK) result in CT, and AHI process is also responsible for bulk TG, and these effects are handled separately in BAT.

IV. TCAD FRAMEWORK

Two-dimensional P and N channel HKMG-MOSFET device structures are generated using process simulation [12]. In device simulation [13], defect dissociation by holes (NBTI) and electrons (PBTI) is handled using CEDM-MSC framework. Only four CEDM model parameters, *i.e.*, the pre-factor (K_{F10}), temperature activation (E_{AKF}), temperature (T) independent field acceleration (Γ_0) and bond polarization (α),

113

Fig.3, are related to bond dissociation, and are varied to model different IL based devices used in this work. MSCM is used for diffusion of released H after bond dissociation using CEDM, dimerization into H_2 molecule assisted by reaction with another H passivated defect (at IL/High-K interface) and H_2 diffusion. All MSCM parameters are consistent between studied devices for N & P BTI. Backend of $1\mu m$ is used to allow H_2 diffusion in both BTI (length depends on stress time and T that determine the distance of H_2 diffusion).

V. DEVICE AND MEASUREMENT DETAILS

Gate First (GF) N and P HKMG MOSFETs having ultra-thin thermal IL and HfO_2 HK are used [1]. IL scaling is done using thermal process tweak for D1 (5Å) and D2 (3Å), with N% in D2 for D3 (2.5Å) and N based IL, D4 (1.5Å), EOT of HK is 4.6Å. DCIV measured data in Figs.5-8 and Figs.12-15 are with delay correction [11], and in Figs.12-15 with (also) bandgap correction ($\Delta V_{IT}=K.\Delta N_{IT}$) [2]. ΔV_T in Figs.9-14 are measured with $10\mu s$ delay [5], [6].

VI. TCAD MODELING

Modeling of DCIV measured ΔN_{IT} time kinetics at various V_G and T are shown for N, P BTI, for devices having various N% and IL, in Figs.5 and 6. Only longer time (t>10s) data are available as DCIV is a slow method. Power law time kinetics is observed with $n\sim1/6$ for all cases (after delay correction), the slope is determined by H_2 diffusion and suggests similar process for the time kinetics in N, P BTI. Model accuracy is verified using data at different V_G x T, Fig.7. ΔN_{IT} increases for NBTI but reduces for PBTI as N% is increased, Figs.5, 6, 8, due to their different origin. The Voltage Acceleration Factor (VAF, slopes in Fig.8) reduces at higher N% for both N and P BTI.

VII. BTI MODELING

Measured and modeled ΔV_T kinetics are shown for N and P BTI, Figs.9, 10. The underlying subcomponents (ΔV_{IT} from RDM, ΔV_{HT} and ΔV_{ET} from ABDWTM) are shown at fixed V_G, T, Fig.9. Interface TG dominates ΔV_T for both N, P BTI at longer time (t>1s) and shows power-law time dependence with slope of $n\sim1/6$, due to H_2 diffusion in RDM as discussed before. DCIV calibrated TCAD with a fixed correction factor K is used for difference in scanned bandgap between DCIV and V_T measurements [2]. Bulk TG also has power-law time dependence with long-time slope $n\sim1/3$. CT saturates at long time ($n\sim0$ in a log-log plot). The model is validated across different V_G and T, Fig.10, 11 (only few examples are shown, the validation is done for other stress conditions as well).

VIII. EOT SCALING

Modeling of measured overall ΔV_T (from ultra-fast I-V) and ΔV_{IT} (from DCIV) versus EOT is shown in Fig.12. Time slope n, Fig.13, T activation energy (E_A), Fig.14 and VAF, Fig.15 versus EOT ($\sim$IL scaling) of ΔV_T and ΔV_{IT} are shown for N and P BTI. ΔV_{IT} dominates ΔV_T, both for NBTI and PBTI, in these experiments. There is a slight reduction in PBTI degradation at scaled EOT (due to Nitridation). This is consistent with higher N% resulting in higher ΔV_{IT-IL} and

lower ΔV_{IT-HK} (from calibrated TCAD), lower ΔV_{OT-IL} and ΔV_{OT-HK}, and higher pre-existing trap density (it is verified by 1/f noise measurements in [1]). Since CT saturates, n of ΔV_T reduces at lower EOT (more reduction than n of ΔV_{IT}), Figs.13. Since CT has lower E_A than TG [5]-[8], E_A of ΔV_T reduces (more reduction than ΔV_{IT}) at lower EOT due to Nitridation (D3). VAF for ΔV_{IT} hence ΔV_T reduces for both BTI as IL is scaled (as ΔV_{IT} dominates ΔV_T), Fig.15.

IX. CONCLUSION

TCAD is enabled for interface TG during BTI in N and P channel MOSFETs. BAT with RDM for interface TG (same as TCAD) and ABDWTM for CT is used to model N and P BTI in differently processed MOSFETs. TCAD and BAT are separately validated with measured BTI data from differently processed devices. Further work is underway to incorporate a Reaction Drift Diffusion Model (RDDM) for bulk TG and ABDWT for CT for complete BTI solution using TCAD.

REFERENCES

[1] K. Joshi et al., "HKMG process impact on N, P BTI: Role of thermal IL scaling, IL/HK integration and post HK nitridation," in Proc. IEEE Int. Rel. Phys. Symp. (IRPS), Apr. 2013, pp. 4C.2.1–4C.2.10. doi: 10.1109/IRPS.2013.653201.

[2] S. Mahapatra et al., "A comparative study of different physics based NBTI models," IEEE Trans. Electron Devices, vol. 60, no. 3. pp. 901–916, Mar. 2013.

[3] G. Rzepa et al, "Comphy — A compact-physics framework for unified modeling of BTI", Microelectronics Reliability, vol. 85, pp. 49-65, 2018. doi: 10.1016/j.microrel.2018.04.002

[4] J. H. Stathis et al, "Controversial issues in negative bias temperature instability," Microelectronics Reliability, vol. 81, pp. 244–251, Feb. 2018. doi: 10.1016/j.microrel.2017.12.035.

[5] S. Mukhopadhyay et al, "A Comprehensive DC and AC PBTI Modeling Framework for HKMG n-MOSFETs", IEEE Trans. Electron Devices, vol. 64, no. 4. pp. 1474–1484, April. 2017.

[6] N. Parihar, N. Goel, S. Mukhopadhyay, and S. Mahapatra, "BTI analysis tool-modeling of NBTI DC, AC stress and recovery time kinetics, nitrogen impact, and EOL estimation," IEEE Trans. Electron Devices, vol. 65, no. 2, pp. 392–403, Feb. 2018. doi: 10.1109/TED.2017.2780083.

[7] S. Mahapatra, N. Parihar, "Modeling of NBTI Using BAT Framework: DC-AC Stress-Recovery Kinetics, Material, and Process Dependence", IEEE Trans. Device and Material Reliability, vol. 20, pp. 4-23, March 2020, doi: 10.1109/TDMR.2020.2967696.

[8] N. Chaudhary et al, "Analysis of The Hole Trapping Detrapping Component of NBTI Over Extended Temperature Range" in Proc. IEEE Int. Rel. Phys. Symp. (IRPS), 2020, doi: 10.1109/IRPS45951.2020.9129245.

[9] R. Tiwari et al, "A 3-D TCAD Framework for NBTI— Part I: Implementation Details and FinFET Channel Material Impact" IEEE Trans. Electron Devices, vol. 66, TED May 2019, doi: 10.1109/TED.2019.2906339.

[10] S. Mukhopadhyay et al, "Trap Generation in IL and HK Layers During BTI /TDDB Stress in Scaled HKMG N and P MOSFETs", in Proc. IEEE Int. Rel. Phys. Symp. (IRPS), 2014, doi: 10.1109/IRPS.2014.6861146.

[11] M. A. Alam et al, "Field acceleration for oxide breakdown-can an accurate anode hole injection model resolve the E vs. 1/E controversy? in Proc. IEEE Int. Rel. Phys. Symp. (IRPS), 2000, doi: 10.1109/RELPHY.2000.843886.

[12] SentaurusTM Process User Guide, document N-2017.09, 2017.

[13] SentaurusTM Device user guide, document N-2017.09, 2017.

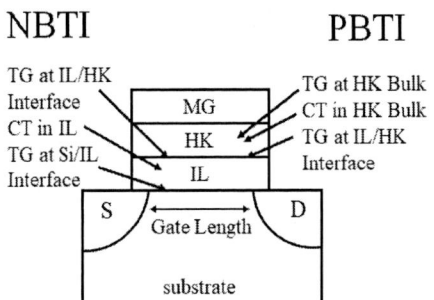

Fig.1. Schematic of the trap generation (TG) and trapping (CT) mechanism is shown under both NBTI and PBTI stress for a planar HKMG device.

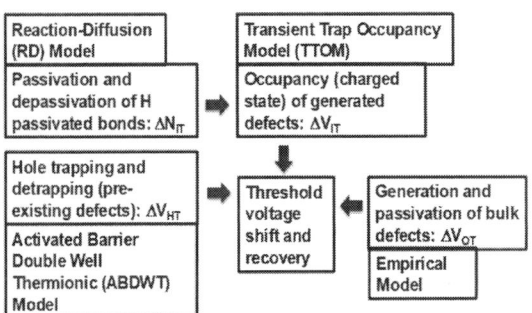

Fig.2. Schematic of BAT framework for both NBTI and PBTI stress-recovery for a planar HKMG device. ΔV_T subcomponents and corresponding models are shown.

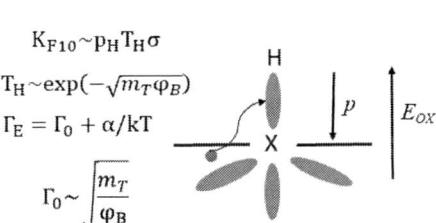

Fig.3. Schematic of H passivated bond dissociation process at the channel/IL interface used in CEDM [9].

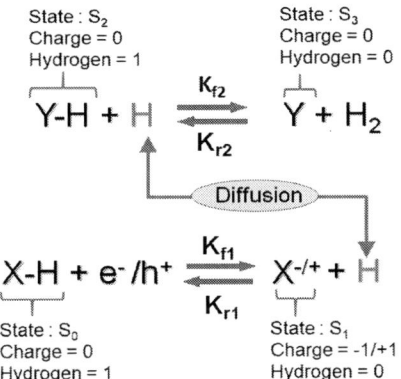

Fig.4. Multi State Configuration Model schematic, utilizing CEDM, showing Hydrogen Transport degradation mechanism at the two interfaces for N(h+) and P(e-) BTI.

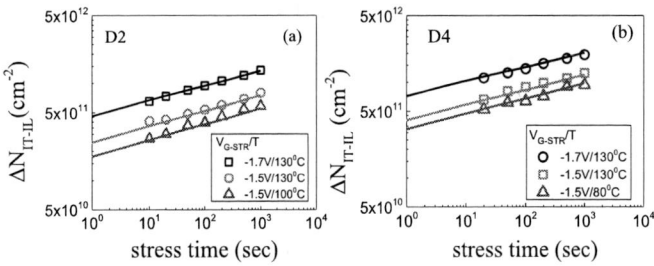

Fig.5. TCAD modeling of DCIV measured $\Delta N_{IT\text{-}IL}$ time kinetics showing increased NBTI degradation for (b) nitrided device (D4) as compared to (a) non-nitride device (D2). Lines: TCAD simulation, symbols: DCIV data.

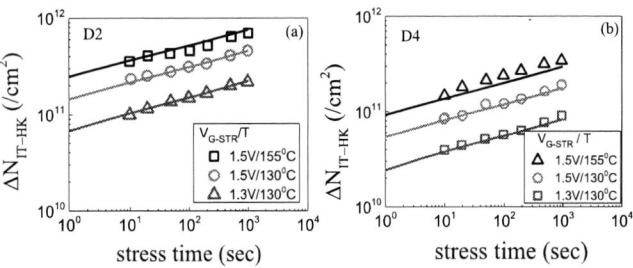

Fig.6. TCAD modeling of DCIV measured $\Delta N_{IT\text{-}HK}$ time kinetics showing lower PBTI degradation for (b) nitrided device (D4) as compared to (a) non-nitride device (D2). Lines: TCAD simulation, symbols: DCIV data.

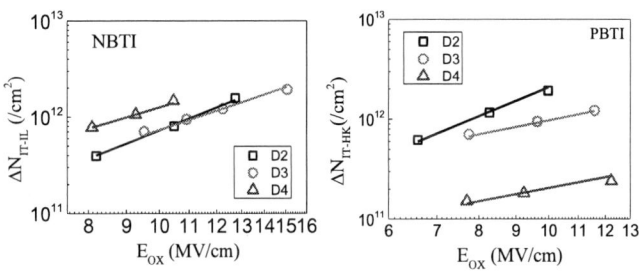

Fig.7. TCAD modeling of field dependence of DCIV measured $\Delta N_{IT\text{-}IL}$ for N (left), and $\Delta N_{IT\text{-}HK}$ P(right) BTI for different IL scaled devices. Lines: TCAD simulation, symbols: DCIV data.

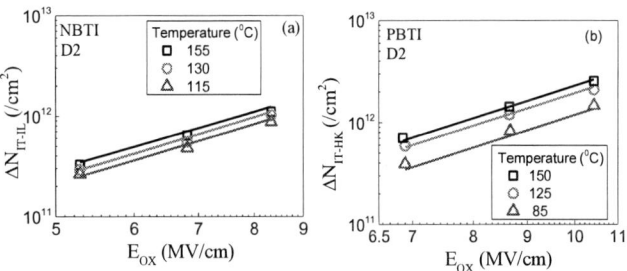

Fig.8. TCAD modeling of field dependence of DCIV measured TG, $\Delta N_{IT\text{-}IL}$ for (left) N and $\Delta N_{IT\text{-}HK}$ for (right) P BTI at different temperature. Lines: TCAD simulation, symbols: DCIV data.

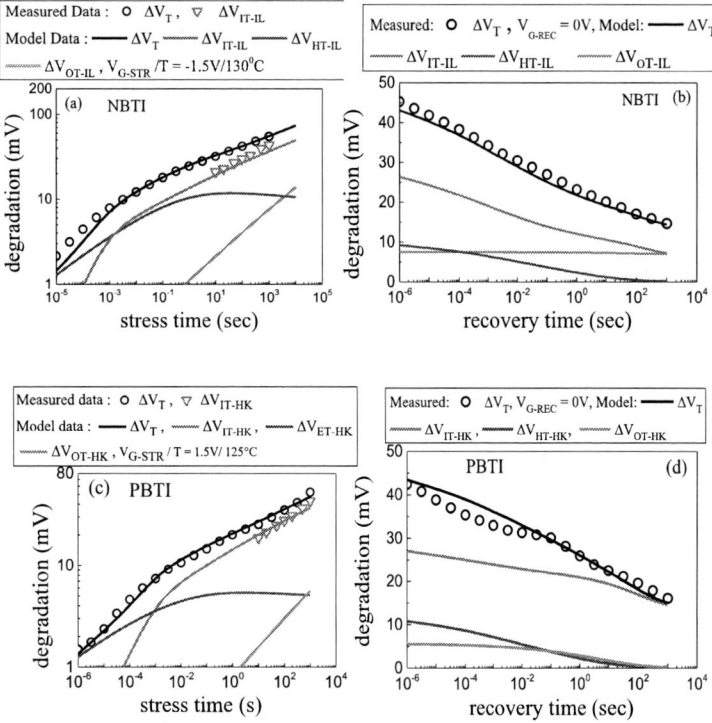

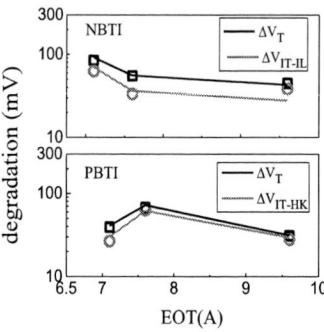

Fig.12. Ultra-fast measured fixed time ΔV_T and TG from DCIV, NBTI (top), and PBTI (bottom); for different IL-scaled devices are modeled using BAT. Measured data: symbols, model data: lines.

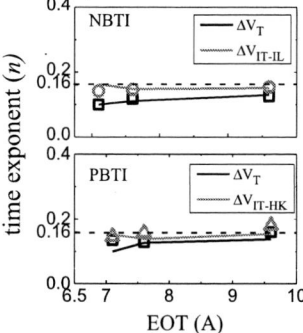

Fig.13. Time slope of ultra-fast measured ΔV_T and TG from DCIV, NBTI (top), and PBTI (bottom); for different IL-scaled devices are modeled using BAT. Measured data: symbols, model data: lines.

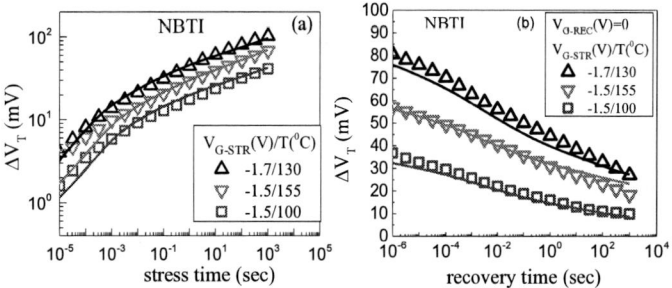

Fig.9. BAT modeling of ultra-fast measured ΔV_T and DCIV measured TG time kinetics for top: NBTI during (a) stress and (b) recovery; and bottom: PBTI during (c) stress and (d) recovery; at fixed $V_{G\text{-}STR}$-T and the sub-components are shown. Overall ΔV_T is dominated by ΔV_{IT} subcomponent.

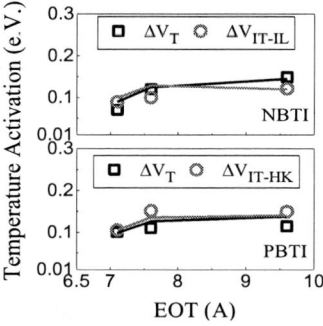

Fig.14. Temperature Activation Energy (E_A) of ultra-fast measured ΔV_T and TG from DCIV, NBTI (top), and PBTI (bottom); for different IL-scaled devices are modeled using BAT. Measured data: symbols, model data: lines.

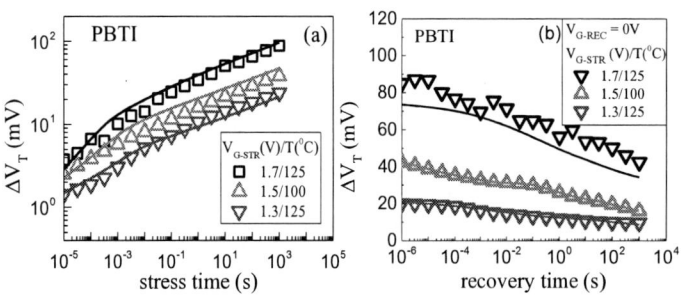

Fig.10. BAT modeling of ultra-fast measured NBTI ΔV_T time kinetics for (a) stress and (b) recovery at different $V_{G\text{-}STR}$-T are shown. Only few cases of $V_{G\text{-}STR}$-T are shown for better visibility. Measured data: symbols, model data: lines.

Fig.11. BAT modeling of ultra-fast measured NBTI ΔV_T time kinetics for (a) stress and (b) recovery at different $V_{G\text{-}STR}$-T are shown. Only few cases of $V_{G\text{-}STR}$-T are shown for better visibility. Measured data: symbols, model data: lines.

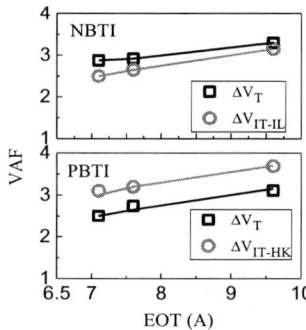

Fig.15. Voltage Acceleration Factor (VAF) of ultra-fast measured ΔV_T and TG from DCIV, NBTI (top), and PBTI (bottom); for different IL-scaled devices are modeled using BAT. Measured data: symbols, model data: lines.

⁶⁻³Benchmarking Charge Trapping Models with NBTI, TDDS and RTN Experiments

Sharang Bhagdikar and Souvik Mahapatra*
Department of Electrical Engineering
Indian Institute of Technology Bombay
Mumbai 400076, India
*Phone: +91-222-572-0408, Email: souvik@ee.iitb.ac.in

Abstract-- A systematic review and comparison of existing charge trapping models in literature is performed. A framework for simulating hole trapping/de-trapping kinetics is established to compute resultant threshold voltage degradation (ΔV_{HT}) and capture-emission time constants ($\tau_C - \tau_E$). The models are analyzed by using data from Negative Bias Temperature Instability (NBTI), Random Telegraph Noise (RTN) and Time Dependent Defect Spectroscopy (TDDS) experiments.

Keywords—NBTI, RTN, TDDS.

I. INTRODUCTION

Charge trapping and detrapping in MOS dielectric contributes to NBTI, RTN, stress-induced leakage currents (SILC) and time dependent dielectric breakdown (TDDB) [1]-[5]. It is established that charge (hole in pFETs) trapping constitutes the overall threshold voltage (ΔV_{th}) degradation along with the interface trap (ΔV_{it}) and bulk trap (ΔV_{ot}) generation. Although numerous models have been proposed in an attempt at modelling the charge trapping time kinetics, the exact physical mechanism governing the process stays uncertain [6][7]. The four-state Extended Nonradiative Multiphonon Model (eNMP) [8] is touted to provide the most complete description of hole trapping kinetics. Capture and emission time constants of individual defects are modelled using eNMP [2]. However, the large number of tuning parameters make eNMP intractable and limit its practical use. The Two Well Nonradiative Multiphonon Model (2WNMP) is an abstraction of the eNMP model that treats the neutral and charged states of a trap as two energy levels represented as intersecting parabolic potential wells [8]. The macroscopic implementation of 2WNMP is used to model hole trapping in pre-existing defects in large area devices [7][9], and the stochastic implementation is used for NBTI and TDDS kinetics in small area devices [10]. The double well thermionic (DWT) model represents the distinct defect states as energy levels separated by a thermionic barrier. The original model [11] is altered in [12] by introducing a temperature activated barrier for modelling BTI kinetics over a range of temperatures. This Activated Barrier Double Well Thermionic (ABDWT) model [12] is invoked to model NBTI stress-recovery transients over a range of biases and temperatures and across different technologies [13]. The bias and temperature couplings of capture (τ_C) and emission (τ_E) time constants measured from TDDS and RTN studies is modelled using ABDWT [14].

II. MODEL FRAMEWORK & SIMULATION SETUP

Fig.1 depicts a schematic of the NMP model. The neutral (E_1) and charged (E_2) defect energy levels are approximated as quadratic potential wells. Level E_2 is pinned to the energy of the reservoir which supplies the carriers i.e. either the substrate conduction band edge (for electrons) or the valence band edge (for holes). The point where the wells intersect provide the barrier heights ϵ_{12} and ϵ_{21} for hole capture (k_{12}) and emission (k_{21}) rates respectively. The bias dependence gets accounted in the fact that upon application of a gate voltage, level E_1 undergoes an electrostatic level shift relative to E_2 (substrate) which would revise the barrier heights and, hence, the reaction rates. The required energy for the transition is supplied/dissipated entirely via phonons. The expressions for the reaction rates are rigorously derived in [8] and are listed in Fig.1.

The ABDWT model provides transition rates for charge (hole in p-FET) capture and emission within a trap, Fig.2. A transition from a reference neutral state (E_1) to the charged state (E_2) via a thermally activated barrier (E_B) constitutes the hole capture reaction. A backward transition signals the hole emission reaction. The barrier E_B and state E_2 lowers when a gate bias (V_G) is applied to account for the bias dependence on the reaction rates [13]. The parameters for bias-dependent barrier lowering and thermal lowering are distinct and allows for decoupling of bias and T dependence on rate constants.

Defects are distributed uniformly spatially in the dielectric for performing macroscopic simulations using 2WNMP. Appropriate model parameters (mean + spread) are assigned to be consistent with experimental data. Only the defects (E_1) that transition above or below the Fermi level upon application of bias may take part in the capture-emission reaction and are otherwise assumed to remain in equilibrium . The setup for the ABDWT model is similar except that the defects are situated at the dielectric-substrate interface owing to the fact that spatial dependence is implicitly captured by its parameters. For replication of defect-centric data (RTN, TDDS), an individual defect is placed in the dielectric and assigned unique model parameters to generate k_{12} and k_{21}, which yield τ_C and τ_E respectively.

III. NBTI MODELLING

Fig.3-4 shows experimental ΔV_{HT} stress and recovery data from Gate First HKMG planar MOSFETs [15]. The hole trapping component ΔV_{HT} is isolated from the measured mean ΔV_{TH} data using the macroscopic BAT framework [1], which uses an empirical relation to compute ΔV_{HT}. The ΔV_{HT} is thus extracted over a range of temperatures and biases using the

BAT framework. The ABDWT model simulations are shown to map the T activation over the entire range (Fig.5). ABDWT is shown to accurately model ΔV_{HT} stress data over a range of V_{GSTR} (Fig.6). Appropriate parameters for 2WNMP are selected to reproduce the stress data over the entire temperature range in Fig.7. It is observed that when the same parameters are used to model ΔV_{HT} stress data over a range of V_{GSTR}, 2WNMP predicts a stronger bias (V_{GSTR}) activation and the dataset cannot be matched (Fig.8). In Fig.9, suitable 2WNMP parameters are chosen to model the bias activation. The same parameters predict a weaker T activation and cannot map the entire T dataset (Fig.10). Similar analysis is carried out for ΔV_{HT} extracted from RMG HKMG SOI FinFETs [16]. ABDWT is shown to model ΔV_{HT} data in Figs.11-12 whereas simultaneous realization of bias and T activation cannot be achieved using 2WNMP, Figs.13-14. Comparison of ΔV_{HT} recovery curves over different V_{GREC} (Figs.15-16) and different stress times (Figs.17-18) is performed. It is observed that, for concurrent stress curves, the 2WNMP model predicts consistently slower recovery than ABDWT.

IV. RTN AND TDDS VALIDATION

Fig.19(a)-(d) list the distinct types of bias couplings (types A-D) for capture and emission time constants obtained from RTN studies [17]. All the different V_G couplings can be reproduced by the ABDWT model upon selection of suitable parameters. The type-A and type-B bias couplings are reproduced by 2WNMP. The presence of a bias-dependent pre-factor (p) in the 2WNMP capture rate expression prevents τ_C from being bias agnostic. In Fig.19(c), 2WNMP predicts a weak negative coupling of τ_C as opposed to the zero coupling observed. A negative coupling of τ_E, as observed in Fig.19(d), is realized in the *weak* electron-phonon coupling regime [8] of 2WNMP. τ_C in this regime is also negatively coupled with bias and it is not possible to achieve type-D coupling using 2WNMP. The dependence of τ_C and τ_E on V_G at different temperatures is recorded from RTN experiments in [18] and reproduced in Fig. 20. In Fig.21, ABDWT is shown to capture similar temperature activation trends across all V_G, owing to the fact that the τ_C and τ_E are not strongly coupled and can be tuned independently unlike in 2WNMP.

Fig.22 illustrates the bias dependence of τ_C and τ_E acquired from TDDS measurements for a non-switching trap [2]. The bias dependence is reproduced by ABDWT across two different T using appropriate parameters. The time constants for the non-switching trap are also modelled using 2WNMP. Figs.23-24 show TDDS time constants for switching trap with weak [2] and strong [19] bias activation of τ_E in the subthreshold region respectively. 2WNMP cannot replicate the switching behavior i.e. positive coupling of τ_E below threshold voltage.

V. CONCLUSION

Macroscopic frameworks of ABDWT and 2WNMP are used to model experimental NBTI data over a range of stress biases, recovery biases and temperatures. The 2WNMP model in its present form cannot predict T and V_{GSTR} activation as ABDWT, and is unable to model extended stress dataset. 2WNMP predicts slower ΔV_{HT} recovery than ABDWT. 2WNMP cannot reproduce the available capture and emission time constant bias couplings observed in RTN experiments.

Switching trap time constants obtained using TDDS are not modelled by 2WNMP. τ_E and τ_C exhibit weaker correlation coming from ABDWT as compared to 2WNMP, which makes the former more versatile in modelling defect-centric data.

REFERENCES

[1] N. Parihar, N. Goel, S. Mukhopadhyay and S. Mahapatra, "BTI Analysis Tool-Modeling of NBTI DC, AC Stress and Recovery Time Kinetics, Nitrogen Impact, and EOL Estimation," in *IEEE Trans. Electron Devices*, vol. 65, no. 2, pp. 392-403, Feb.2018.

[2] T. Grasser, et al, "On the Microscopic Origin of the Frequency Dependence of Hole Trapping in pMOSFETs", *IEDM*, Dec. 2012, pp. 19.6.4.

[3] A. Kerber, A. Vayshenker, D. Lipp, T. Nigam and E. Cartier, "Impact of charge trapping on the voltage acceleration of TDDB in metal gate/high-k n-channel MOSFETs," 2010 *IEEE International Reliability Physics Symposium*, Anaheim, CA, 2010, pp. 369-372.

[4] W. Goes, M. Waltl, Y. Wimmer, G. Rzepa and T. Grasser, "Advanced modeling of charge trapping: RTN, 1/f noise, SILC, and BTI," 2014 *SISPAD*, Yokohama, 2014, pp. 77-80.

[5] N. Tega et al., "Increasing threshold voltage variation due to random telegraph noise in FETs as gate lengths scale to 20 nm," 2009 *Symposium on VLSI Technology*, Honolulu, HI, 2009, pp. 50-51.

[6] S. Mahapatra and Narendra Parihar, "A review of NBTI mechanisms and models," *Microelectron. Reliab*, Volume 81, 2018, Pages 127-135.

[7] Rzepa, G. et al. "Comphy - A compact-physics framework for unified modeling of BTI." *Microelectron. Reliab.* 85 (2018): 49-65.

[8] Grasser, T.. "Stochastic charge trapping in oxides: From random telegraph noise to bias temperature instabilities." *Microelectron. Reliab.* 52 (2012): 39-70.

[9] G. Rzepa et al., "Efficient physical defect model applied to PBTI in high-κ stacks," 2017 *IEEE International Reliability Physics Symposium (IRPS)*, Monterey, CA, 2017, pp. XT-11.1-XT-11.6.

[10] Anandkrishnan R, et al, "A Stochastic Modeling Framework for NBTI and TDDS in Small Area p-MOSFETs", *Simulation of Semiconductor Processes and Devices (SISPAD)* , Austin TX, Sep. 2018, pp. 181.

[11] D. Ielmini, et al, "A unified model for permanent and recoverable NBTI based on hole trapping and structure relaxation," *2009 IEEE International Reliability Physics Symposium*, Montreal, QC, 2009, pp. 26-32.

[12] S. Desai, et al, "A comprehensive AC / DC NBTI model: Stress, recovery, frequency, duty cycle and process dependence," *2013 IEEE International Reliability Physics Symposium (IRPS)*, Anaheim, CA, 2013, pp. XT.2.1-XT.2.11.

[13] N. Choudhury, N. Parihar, N. Goel, A. Thirunavukkarasu and S. Mahapatra, "A Model for Hole Trapping-Detrapping Kinetics During NBTI in p-Channel FETs: (Invited paper)," 2020 *4th IEEE Electron Devices Technology & Manufacturing Conference (EDTM)*, Penang, Malaysia, 2020, pp. 1-4.

[14] S. Bhagdikar and S. Mahapatra, "A Stochastic Hole Trapping-Detrapping Framework for NBTI, TDDS and RTN," *SISPAD*, Udine, Italy, 2019, pp. 1-4.

[15] N. Parihar, R. Anandkrishnan, A. Chaudhary and S. Mahapatra, "A Comparative Analysis of NBTI Variability and TDDS in GF HKMG Planar p-MOSFETs and RMG HKMG p-FinFETs," in *IEEE Transactions on Electron Devices*, vol. 66, no. 8, pp. 3273-3278, Aug. 2019.

[16] N. Parihar, U. Sharma, R. G. Southwick, M. Wang, J. H. Stathis and S. Mahapatra, "Ultrafast Measurements and Physical Modeling of NBTI Stress and Recovery in RMG FinFETs Under Diverse DC–AC Experimental Conditions," in *IEEE Transactions on Electron Devices*, vol. 65, no. 1, pp. 23-30, Jan. 2018.

[17] H. Miki, et al, "Understanding short-term BTI behavior through comprehensive observation of gate-voltage dependence of RTN in highly scaled high-κ / metal-gate pFETs", *Symposium on VLSI Technology - Digest of Technical Papers*, June 2011, pp. 149.

[18] H. Miki, et al, "Voltage and temperature dependence of random telegraph noise in highly scaled HKMG ETSOI nFETs and its impact on logic delay uncertainty", in *Symposium on VLSI Technology (VLSIT)*, pp. 138, June 2012.

[19] T. Grasser et al., "Advanced characterization of oxide traps: The dynamic time-dependent defect spectroscopy," 2013 *IEEE International Reliability Physics Symposium (IRPS)*, Anaheim, CA, 2013, pp. 2D.2.1-2D.2.7.

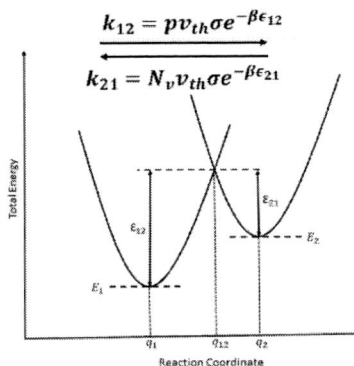

$$k_{12} = pv_{th}\sigma e^{-\beta\epsilon_{12}}$$
$$k_{21} = N_v v_{th}\sigma e^{-\beta\epsilon_{21}}$$

Fig.1. Schematic of 2WNMP model depicting potential wells of the neutral (q_1) and charged (q_2) states.

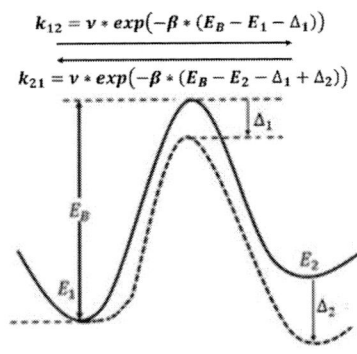

$$k_{12} = v * exp(-\beta * (E_B - E_1 - \Delta_1))$$
$$k_{21} = v * exp(-\beta * (E_B - E_2 - \Delta_1 + \Delta_2))$$

Fig.2. Schematic of ABDWT model. E_1, E_2 and E_b determine the energetic configuration of the trap.

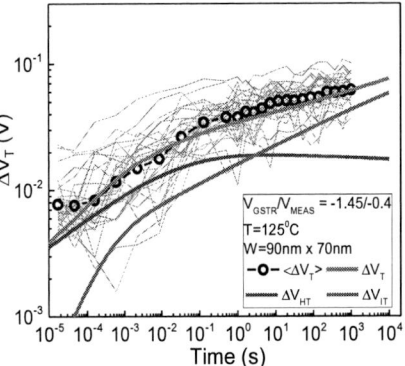

Fig.3. Individual (gray) and mean (black) measured ΔV_T traces during stress along with model calculated mean (red) and decomposition into subcomponents. Device is GF HKMG planar MOSFET.

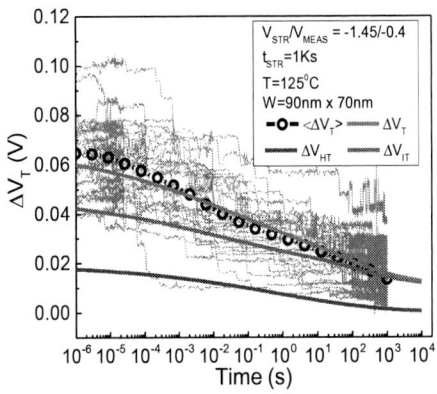

Fig.4. Individual (gray) and mean (black) measured ΔV_T traces during recovery along with model calculated mean (red) and decomposition into subcomponents

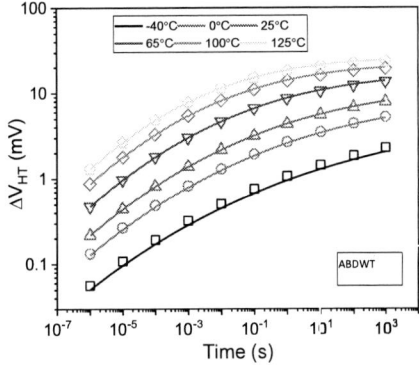

Fig.5. Modelling of ΔV_{HT} stress data (symbols) using ABDWT (solid lines) over a range of temperatures. GF HKMG planar MOSFETs

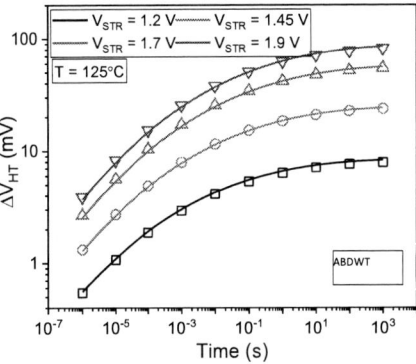

Fig.6. Modelling of ΔV_{HT} stress data (symbols) using ABDWT (solid lines) for a range of stress biases.

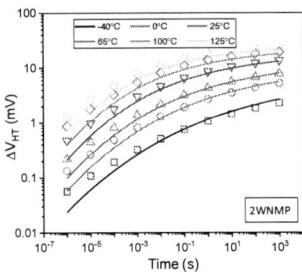

Fig.7. Modelling of ΔV_{HT} stress data (symbols) using 2WNMP (solid lines) over a range of temperatures.

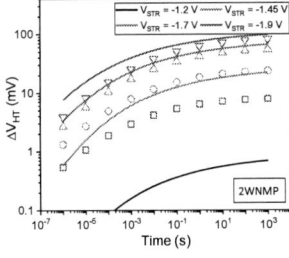

Fig.8. Modelling of ΔV_{HT} stress data using 2WNMP parameters of Fig.7. 2WNMP predicts stronger bias activation.

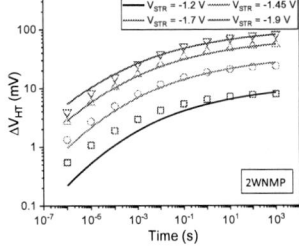

Fig.9. Modelling of ΔV_{HT} stress data (symbols) using 2WNMP (solid lines) for a range of biases.

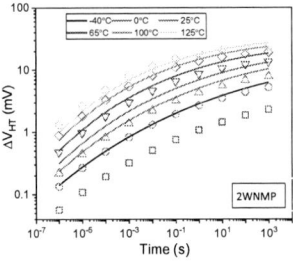

Fig.10. Modelling of ΔV_{HT} stress data using 2WNMP parameters of Fig.9. 2WNMP predicts weaker T activation.

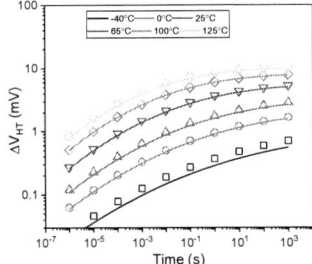

Fig.11. Modelling of ΔV_{HT} stress data (symbols) using ABDWT (solid lines) over a range of temperatures. Measurements are from RMG HKMG SOI FinFET.

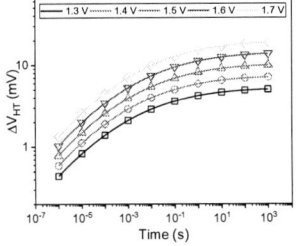

Fig.12. Modelling of ΔV_{HT} stress data (symbols) using ABDWT (solid lines) over a range of stress biases.

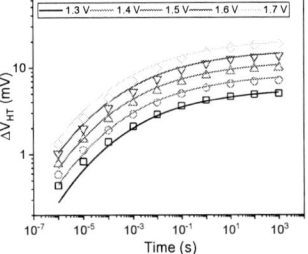

Fig.13. Modelling of ΔV_{HT} stress data (symbols) using 2WNMP (solid lines) for a range of biases.

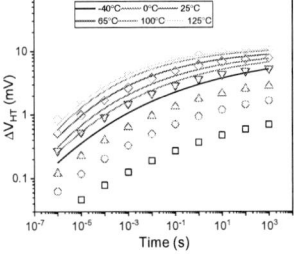

Fig.14. Modelling of ΔV_{HT} stress data using 2WNMP parameters used in Fig.13. 2WNMP predicts much weaker T activation.

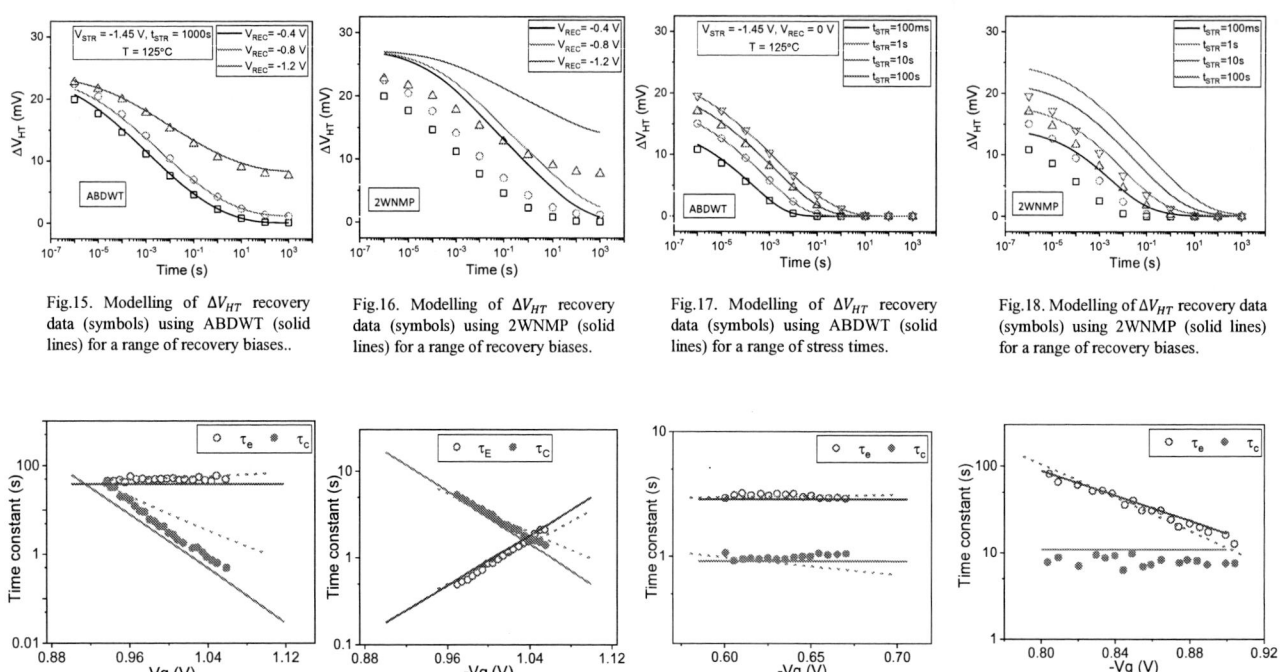

Fig.15. Modelling of ΔV_{HT} recovery data (symbols) using ABDWT (solid lines) for a range of recovery biases..

Fig.16. Modelling of ΔV_{HT} recovery data (symbols) using 2WNMP (solid lines) for a range of recovery biases.

Fig.17. Modelling of ΔV_{HT} recovery data (symbols) using ABDWT (solid lines) for a range of stress times.

Fig.18. Modelling of ΔV_{HT} recovery data (symbols) using 2WNMP (solid lines) for a range of recovery biases.

Fig.19(a)-(d). Panels depict various couplings of time constants to V_G extracted from RTN data (symbols) [12]. (A) $\tau_c < 0, \tau_e \sim 0$, (B) $\tau_c < 0, \tau_e > 0$, (C) $\tau_c \sim 0, \tau_e \sim 0$, (D) $\tau_c \sim 0, \tau_e < 0$. All the different couplings are reproduced using ABDWT model simulations (solid lines). 2WNMP simulations (dashed lines) can model type A and type B coupling upon selection of appropriate parameters. Zero coupling of τ_C with bias cannot be reproduced using 2WNMP and a negative coupling is observed in (c) and in (d) where τ_c is not shown (out of bounds, $\tau_C \gg \tau_E$).

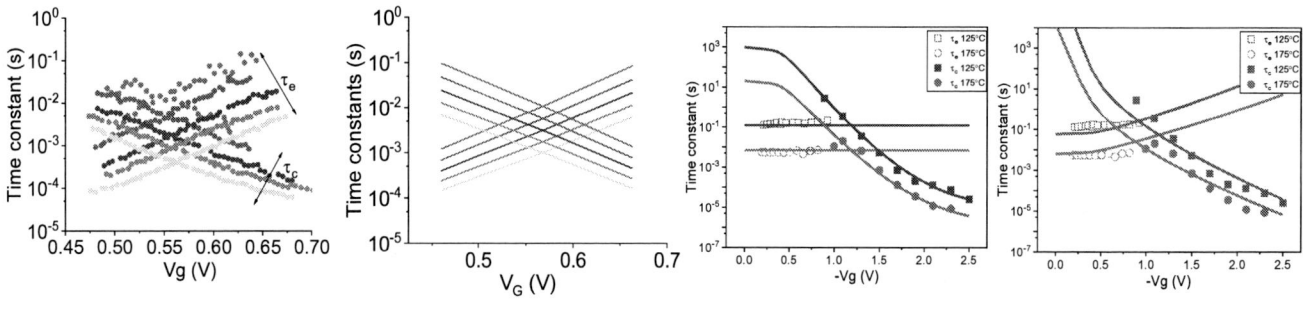

Fig.20. Time constants as a function of V_G at various temperatures from RTN experiments [18].

Fig.21. ABDWT model simulated time constants as a function of V_G at various temperatures.

Fig.21. Modelling of TDDS capture and emission time constants (symbols) for non-switching trap [2] using ABDWT (left) and 2WNMP (right). Non-switching nature is evident by the bias agnostic τ_E in the subthreshold regime.

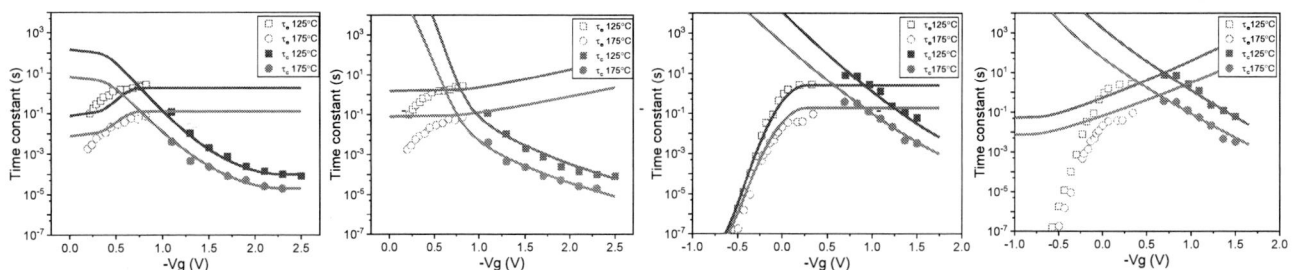

Fig.22. Modelling of TDDS capture and emission time constants (symbols) for non-switching trap [2] using ABDWT (left) and 2WNMP (right). Switching behavior is evident by the positive bias coupling of τ_e with V_G. 2WNMP in its present form is unable to replicate said coupling.

Fig.23. Modelling of TDDS capture and emission time constants (symbols) for switching trap B3 [19] using ABDWT (left) and 2WNMP (right).

6-4 A TCAD Framework for Assessing NBTI Impact Under Drain Bias and Self-Heating Effects in Replacement Metal Gate (RMG) p-FinFETs

Uma Sharma and Souvik Mahapatra

Department of Electrical Engineering, Indian Institute of Technology Bombay, Mumbai 400076, India
*Phone: +91-222-572-0408, Email: souvik@ee.iitb.ac.in

Abstract— Sentaurus TCAD is enabled to calculate interface trap generation (ΔN_{IT}) during Negative Bias Temperature Instability (NBTI) under drain bias (V_D) and self-heating (SH) effects. The setup is calibrated with pure NBTI (V_D=0V) experimental data, and is further used to determine the NBTI component during Hot Carrier Degradation (HCD) stress. Such decomposition of NBTI and HCD is demonstrated for multiple fin length (FL) p-FinFETs to model HCD experimental data at different V_G/V_D stress.

Keywords— FinFET, HKMG, BTI, SH, Fin length variation

I. INTRODUCTION

NBTI [1] and HCD [2], [3] are the key issues that impact reliability of advanced FinFETs [4]. Although a digital circuit (Fig.1) can suffer from BTI and HCD of both FETs, the BTI for NFET (Positive BTI) is negligible in modern FinFETs [4]. As per the standard practice, BTI and HCD are evaluated at device level during technology development, compact models are developed and are calibrated with measured data, and circuit simulations are done with calibrated compact models. The worst-case HCD mode for FinFETs is at $V_G=V_D$ stress. While NFETs are only impacted by HCD, PFETs are impacted by both NBTI and HCD in this stress condition. It is important to isolate their contribution to aid compact model development. The presence of SH effect in FinFETs, due to increased phonon confinement and less heat dissipation area [5], [6] further complicate this effort.

II. SCOPE OF WORK

Our earlier TCAD framework for pure NBTI for V_D=0V [7], [8], is enhanced to simulate non-zero V_D values. The pure NBTI framework is calibrated with experimental data, and the resulting setup is used to estimate NBTI contribution under non-uniform carrier density and electric field along the channel together with local T rise near drain due to SH effect. It is shown that ΔN_{IT} distribution is non-uniform along the channel (higher ΔN_{IT} near the source end), and transfer I-V sweeps are needed to calculate its impact on threshold voltage shift (ΔV_T). The usual electrostatic approach using average ΔN_{IT} is shown to be invalid for non-uniform ΔN_{IT} distribution under non-zero V_D. Careful decomposition of NBTI and pure HCD are demonstrated across multiple FL p-FinFETs.

III. FRAMEWORK

Sentaurus Process and Device TCAD [9] are used respectively for structure generation and NBTI simulation. The Reaction-Diffusion model (RDM) is incorporated in Sentaurus Device to calculate interface traps Capture-Emission Depassivation (CED) and Multi-State

Configuration (MSC) frameworks [7]. MSC model (Fig.2) deals with breaking of X-H bonds at the channel/oxide interface and Y-H bonds inside gate oxide bulk, (the exact chemical nature of electrically active defects is debated, hence generic states are used) and resulting diffusion of atomic H and molecular H_2 species. CED model (Fig.3) handles forward reaction (k_{F1}) of X-H bond dissociation, which depends on inversion hole density and vertical field, *i.e.*, V_G (and also V_D when applicable) and temperature (T), which includes ambient and SH effect (when applicable). It also depends on mechanical stress coming from raised S/D epi of the FinFET and hence FL (Fig.4), the strain alters the hole bandstructure [7], [8]. All other reaction parameters and diffusivities (Fig.2) are only T activated. The 3-D FinFET is simulated with a backend of 1μm (Fig.4), the latter is necessary for H_2 diffusion. 2D cross-section of the gate stack is also shown depicting IL, High-K, TiN cap, and Tungsten (backend) layers to make it realistic as actual measured device.

IV. NBTI CALIBRATION

Pure NBTI setup is calibrated first. However, the as-measured ΔV_T during NBTI stress is influenced by hole trapping (ΔV_{HT}) and bulk trap generation (ΔV_{OT}) in addition to interface traps (ΔV_{IT}) [10]. Hence, the BTI Analysis Tool (BAT) [10] is used to isolate these and obtain the dominating (primary) component which is ΔV_{IT} due to ΔN_{IT} (Fig.5). The time kinetics of measured ΔV_T at various V_G and T is typically calibrated [10], [11]. Fig.6 shows the modelling of experimental time kinetics of ΔV_T at few V_G and T for fixed FL. Fig.7 shows that of ΔV_T versus V_G at different FL but fixed T, the underlying subcomponents are also shown for the smallest FL. Fig.8 shows the comparison of ΔN_{IT} time kinetics extracted from BAT (symbols) and calibrated pure-NBTI TCAD (lines). Fixed time ΔN_{IT} versus V_G at fixed T (Fig.9 (a)) and versus T at fixed V_G (Fig.9 (b)) are compared for BAT extraction (symbols) and calibrated TCAD (lines). Data in Fig.7 through Fig.9 are shown for different FL. ΔN_{IT} reduces at lower FL due to compressive stress [8], [12].

V. DRAIN BIAS AND SELFHEATING

The above calibrated TCAD deck is used under V_D to evaluate the generation of traps. Non-zero V_D would reduce inversion hole density and vertical field near drain. This in turn would reduce ΔN_{IT} distribution towards the drain, shown in Fig.10 when the SH effect is switched off. However, the SH effect induced T increase in the fin (Fig.11) should be invoked as NBTI is T activated. The thermal resistances of the contacts are calibrated for proper estimation of SH effect induced T rise [5]. SH results in higher ΔN_{IT} magnitude near source compared to pure NBTI (Fig.12), but a smaller ΔN_{IT} distribution towards the drain compared to the no SH effect case of Fig.10, when V_D is applied. The shape of average ΔN_{IT} time kinetics does not change without and with SH as V_D is applied, and stays the same as pure NBTI (Fig.13 (a)). However, the average ΔN_{IT} magnitude monotonically reduces at higher V_D when SH is not invoked, but a turn-around is seen with SH (Fig.13 (b)).

VI. IMPACT OF LOCALIZED ΔN_{IT}

Average ΔN_{IT} is sufficient to use for ΔV_{IT} ($=q.\Delta N_{IT}/C_{OX}$, where C_{OX} is gate capacitance) for pure NBTI as defects are uniform [7]. This is invalid for the case of localized ΔN_{IT} distribution, since average ΔN_{IT} underestimates ΔV_{IT} when compared to that from transfer I-V sweeps before and after NBTI (Fig.14). ΔV_{IT} under different V_G and V_D space for 20nm and 60 nm FL FinFETs are shown respectively in top and bottom panels of Fig.15. These are used to calibrate a compact model (CM) for NBTI, details of which are in [3]. The calibrated CM is further used to decouple NBTI and pure HCD contributions in experiments. Fig.16 shows ΔV_T for measured HCD and modeled subcomponents for 20nm FL device (top panel) where HC is hot carrier degradation (using CM in [3]), IT is from BTI interface trap generation (shown in Fig.15), HT is hole trapping. Fig 16 bottom panel shows the same for the 60nm FL device. Fig 17 shows the time kinetics of measured ΔV_T at different V_G and V_D conditions, for (a) low FL and (b) high FL devices. Note the BTI is being computed from the TCAD shown in Fig 15. Detailed modeling details are mentioned for various V_G and V_D is discussed in [3].

VII. CONCLUSION

TCAD simulation of NBTI under V_D and SH effect is necessary for proper decomposition of experimental HCD data into pure HCD and NBTI components. Although NBTI should reduce under non-zero V_D due to reduction in vertical field near drain, SH effect increases T and arrests the amount of reduction in modern FinFETs. Non-uniform ΔN_{IT} with higher value near the source results in underestimation of ΔV_T when calculated by average ΔN_{IT} based charge impact, and transfer I-V sweeps are necessary. Properly decomposed NBTI and HCD contributions are necessary for compact model calibration and circuit analysis.

Acknowledgment: Narendra Parihar for NBTI measurements, KaranSingh Thakor and Ravi Tiwari for pure NBTI TCAD framework.

VIII. REFERENCE

[1] S. Mahapatra and N. Parihar, "Modeling of NBTI Using BAT Framework: DC-AC Stress-Recovery Kinetics, Material, and Process Dependence," in IEEE Transactions on Device and Materials Reliability, vol. 20, no. 1, pp. 4-23, March 2020, doi: 10.1109/TDMR.2020.2967696.

[2] S. Ramey *et al.*, "Intrinsic transistor reliability improvements from 22nm tri-gate technology," *2013 IEEE International Reliability Physics Symposium (IRPS)*, Anaheim, CA, 2013, pp. 4C.5.1-4C.5.5. doi: 10.1109/IRPS.2013.6532017

[3] U. Sharma, N. Parihar and S. Mahapatra, "Modeling of HCD Kinetics for Full V_G/V_D Span in the Presence of NBTI, Electron Trapping, and Self Heating in RMG SiGe p-FinFETs," in *IEEE Transactions on Electron Devices*, vol. 66, no. 6, pp. 2502-2508, June 2019. doi: 10.1109/TED.2019.2911335

[4] A. Rahman, J. Dacuna, P. Nayak, G. Leatherman and S. Ramey, "Reliability studies of a 10nm high-performance and low-power CMOS technology featuring 3rd generation FinFET and 5th generation HK/MG," *2018 IEEE International Reliability Physics Symposium (IRPS)*, Burlingame, CA, 2018, pp. 6F.4-1-6F.4-6. doi: 10.1109/IRPS.2018.8353648C.

[5] C. Prasad, "A Review of Self-Heating Effects in Advanced CMOS Technologies," in IEEE Transactions on Electron Devices, vol. 66, no. 11, pp. 4546-4555, Nov. 2019, doi: 10.1109/TED.2019.2943744.

[6] D. Son, K. Hong, H. Shim, S. Pae and H. Shin, "New Insight Into Negative Bias Temperature Instability Degradation During Self-Heating in Nanoscale Bulk FinFETs," in IEEE Electron Device Letters, vol. 40, no. 9, pp. 1354-1357, Sept. 2019, doi: 10.1109/LED.2019.2930077.

[7] R. Tiwari et al., "A 3-D TCAD Framework for NBTI—Part I: Implementation Details and FinFET Channel Material Impact," in IEEE Transactions on Electron Devices, vol. 66, no. 5, pp. 2086-2092, May 2019.

[8] R. Tiwari et al., "A 3-D TCAD Framework for NBTI, Part-II: Impact of Mechanical Strain, Quantum Effects, and FinFET Dimension Scaling," in IEEE Transactions on Electron Devices, vol. 66, no. 5, pp. 2093-2099, May 2019

[9] Sentaurus™ Device User Guide, Version N-2017.09, December 2017, Synopsys, CA, USA.

[10] N. Parihar, N. Goel, S. Mukhopadhyay, and S. Mahapatra, "BTI analysis tool—Modeling of NBTI DC, AC stress and recovery time kinetics, nitrogen impact, and EOL estimation," IEEE Trans. Electron Devices, vol. 65, no. 2, pp. 392–403, Feb. 2018, doi: 10.1109/TED.2017.2780083.

[11] N. Parihar, U. Sharma, R. G. Southwick, M. Wang, J. H. Stathis, and S. Mahapatra, "Ultrafast measurements and physical modeling of NBTI stress and recovery in RMG FinFETs under diverse DC–AC experimental conditions," IEEE Trans. Electron Devices, vol. 65, no. 1, pp. 23–30, Jan. 2018, doi: 10.1109/TED.2017.2773122.

[12] N. Parihar, R. Tiwari and S. Mahapatra, "Modeling Channel Length Scaling Impact on NBTI in RMG Si p-FinFETs," 2018 International Conference on Simulation of Semiconductor Processes and Devices (SISPAD), Austin, TX, 2018, pp. 176-180, doi: 10.1109/SISPAD.2018.8551740.

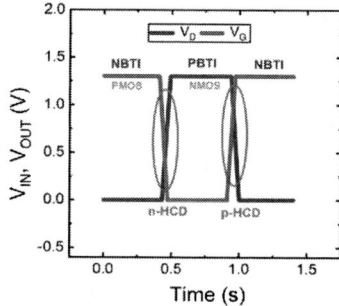

Fig.1. Time dependence of V_G and V_D and degradation mechanism in RO.

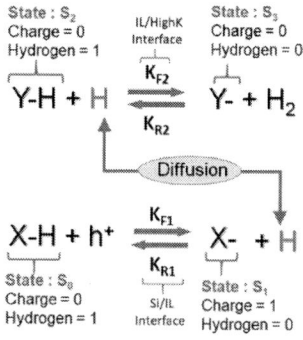

Fig. 2 MSC-HCD TCAD framework for interface trap generation. k_{Fi} is bond dissociation rate and comes from CED model. Details in [7].

$$K_{F1} \sim K_{F10} * \exp(-E_{AKF1}/kT) * \exp(\Gamma_E E_{ox})$$

$$K_{F10} \sim p_H T_H \sigma$$

$$T_H \sim \exp(-\sqrt{m_T \varphi_B})$$

$$\Gamma_E = \Gamma_0 + \alpha/kT$$

$$\Gamma_0 \sim \sqrt{\frac{m_T}{\varphi_B}}$$

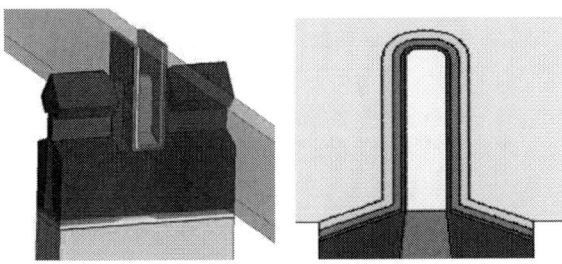

Fig. 3. Schematic of Si-H bond dissociation at the channel/IL interface. Inversion layer holes tunnel into interfacial Si-H bonds aided by the oxide electric field (E_{OX}).

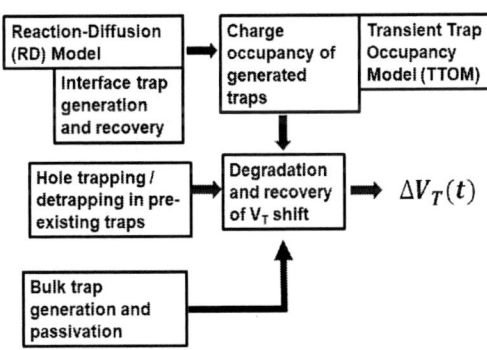

Fig. 4 (a) Isometric view of a 3D p-FinFET structure (FL = 20nm,) having raised epitaxial SiGe S-D and lateral Tungsten backend for H_2 diffusion. (b) 2D cross section of the fin showing IL, High-K, TiN cap and Tungsten layers in the gate stack

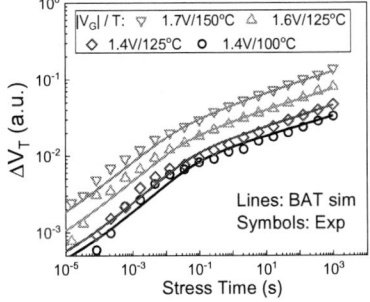

Fig. 5. Schematic of a BAT (BTI Analysis Tool) framework consisting of uncorrelated ΔV_{IT}, ΔV_{HT} and ΔV_{OT} components.

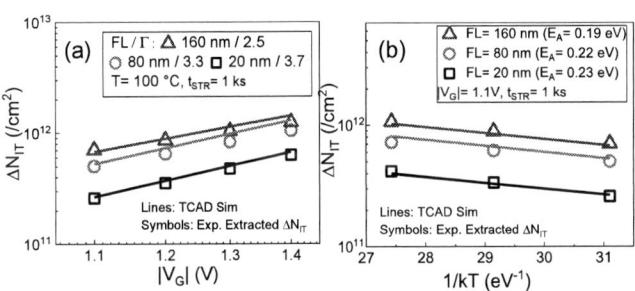

Fig. 6. Time evolution of ultrafast (10μs) measured ΔV_T for DC stress with model estimation at multiple V_G and T for FL 20nm.

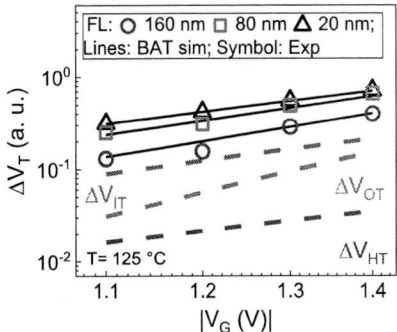

Fig. 7. Modeling of measured fixed time ΔV_T as a function of V_G for a fixed T for different FL. Sub-components of overall ΔV_T for FL 20 nm is shown.

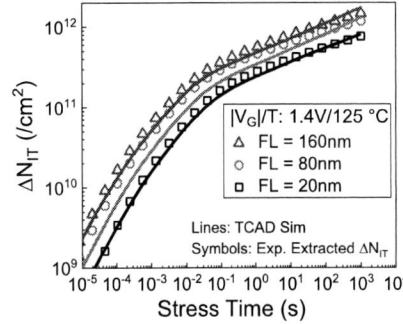

Fig. 8. TCAD modeling of BAT extracted ΔN_{IT} stress time kinetics for different FL for fixed V_G and T conditions.

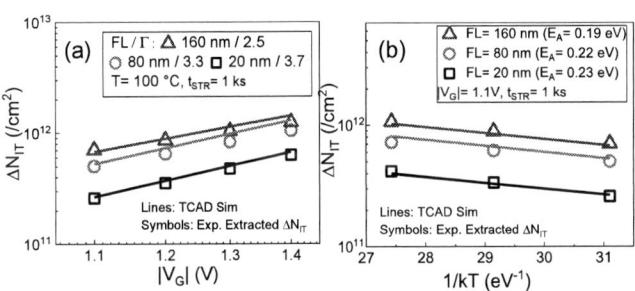

Fig. 9. (a) Fixed time ΔN_{IT} as a function of V_G is shown for a fixed T, (b) plots fixed time ΔN_{IT} as a function of T for a fixed V_G for different FL. Symbol: BAT extracted; Lines: TCAD

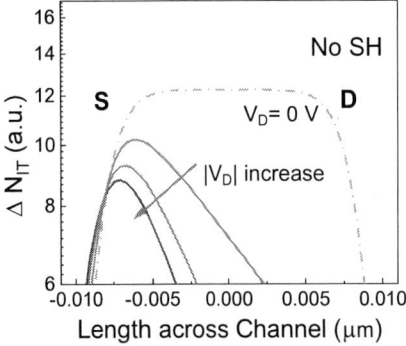

Fig. 10. Spatial distribution of ΔN_{IT} across the channel with varying V_D without SH. It shows the reduction of ΔN_{IT} due to decrease in vertical field (responsible for BTI) as V_D increase. Source (S) and Drain (D) side are marked.

123

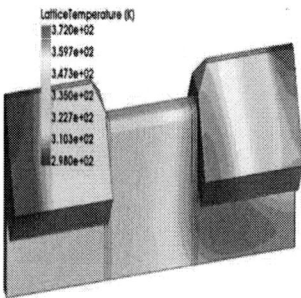

Fig. 11. 3D Lattice Temperature Profile along the channel in presence of V_D. Due to SH localized temperature rise at the drain side.

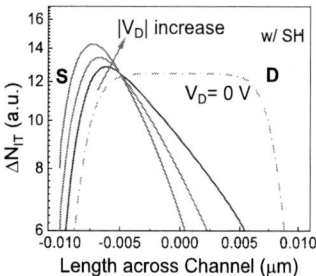

Fig. 12. Spatial distribution of effective ΔN_{IT} across the channel with varying V_D with SH. Source (S) and Drain (D) side are marked.

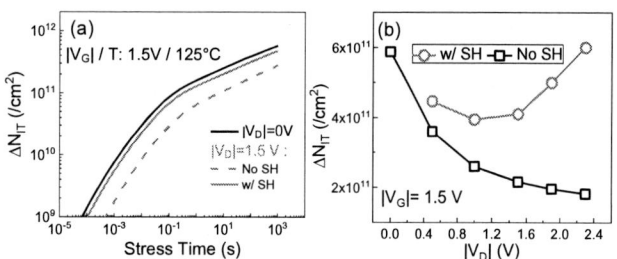

Fig. 13. (a) TCAD extracted ΔN_{IT} time kinetics for different V_D and (b) fixed time ΔN_{IT} as a function of V_D, with and without SH, at fix V_G.

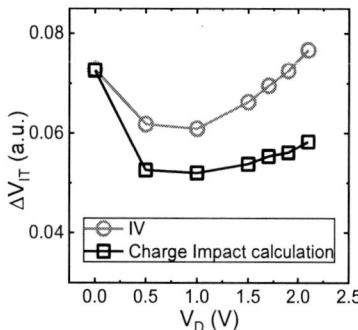

Fig. 14. ΔV_T using avg. N_{IT} and transfer IV under NBTI stress as a function of V_D. Note that charge impact calculations are not valid due to localized trap generation under non-zero V_D. Post stress IV is needed for degradation calculations.

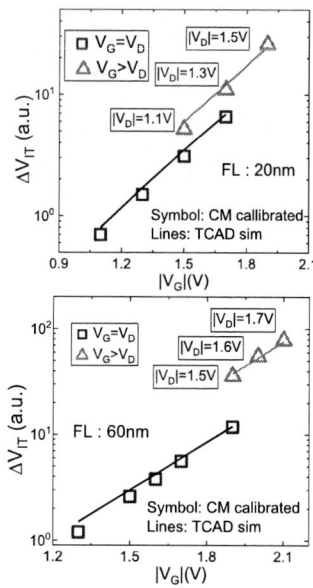

Fig. 15. Fixed time ΔV_{IT} simulated using TCAD under different stress bias (V_G and V_D) for FL of 20nm (top) and 60nm (bottom).

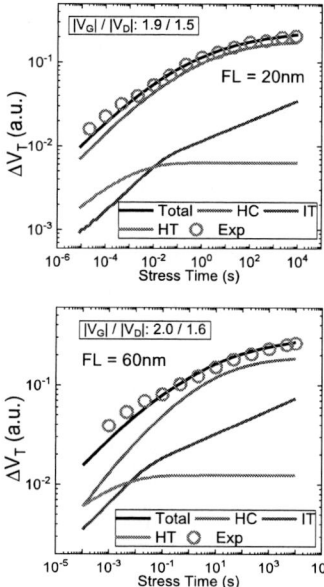

Fig. 16. Overall ΔV_T under $V_G > V_D$ stress for FL of 20nm (top) and 60nm (bottom). ΔV_{IT} kinetics is simulated using TCAD for NBTI with SH. Symbols: Exp Data; Lines: Simulation

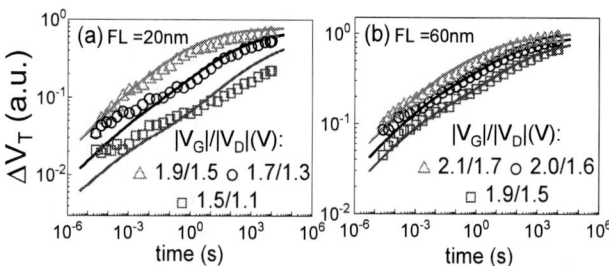

Fig. 17. Time kinetics of measured and modeled ΔV_T at different stress V_G/V_D conditions, for (a) FL= 20 nm and (b) FL= 60 nm devices. Symbols: data, lines: model.

6-5

Model analysis for effects of spatial and energy profiles of plasma process-induced defects in Si substrate on MOS device performance

Takashi Hamano, Keiichiro Urabe, and Koji Eriguchi
Graduate School of Engineering
Kyoto University
Kyoto 615-8540, Japan
Email: hamano.takashi.35c@st.kyoto-u.ac.jp

Abstract—This paper comprehensively discusses impacts of defect profiles in a Si substrate induced by plasma processing on MOS device performance. Both spatial and energy profiles of the defects considering practical plasma parameters were implemented into a conventional device simulation. Unique capacitance–voltage characteristics of MOS capacitors were obtained depending on the energy profiles, which shows good agreement with experimental results. The relationship between the defect profile and device parameter variation was clarified for n- and p-channel MOSFETs. The prediction results suggest the significance of precise control of spatial and energy profiles of defects for future MOS device design and fabrication.

Keywords—plasma process, plasma process-induced defect, defect profile, density-of-state, Si, MOSFET

I. INTRODUCTION

Plasma processing plays an essential role in manufacturing present-day scaled electronic devices [1]. In the fabrication of high-performance devices, suppressing unexpected impurities and defects is a primal concern for both classical and emerging devices. In particular, defects generated during plasma processes—plasma process-induced defects (PD)—have attracted interests in accordance with the device scaling [2]. The PD modifies physical properties of materials, which results in the degradation of device performance and reliability [2]. In the case of Si, the existence of PD changes the effective doping concentration and increases the junction leakage current [3]. The impacts of PD on device parameter variations have been studied in terms of the areal defect density [4]. In the present-day plasma processes, it is expected that PD shows various profiles depending on the process conditions employed for the fabrication of complicated fine structures [5]. However, there have been few reports on the effects of spatial and energy profiles of PD on the device performance. In this study, we performed model prediction for MOS devices with PD in Si substrates by taking into account various conceivable defect profiles. The assumed energy profiles of PD were verified by simulated and experimentally-obtained MOS capacitor characteristics. On the basis of this result, the effects of both spatial and energy profiles of PD on MOSFET performance were analyzed in detail.

II. MODEL ANALYSIS FRAMEWORK

A. Definition of PD profile in Si Substrate

In the course of etch processes, a Si substrate is exposed to plasma and defects are generated on the surface primarily due to ion bombardments. In MOSFETs, defect sites are created in the source/drain (S/D) extension region as shown in Fig. 1. The typical localized structures are displaced-Si atoms and interstitials, which play a role as carrier trapping and detrapping sites at an energy level E in the Si bandgap. Here, considering the group of individual defects, the PD profile $n_{dam}(x,E)$ is defined by,

$$n_{dam}(x,E) = n_{dam}(x) \cdot f_{dam}(E), \qquad (1)$$

where $n_{dam}(x)$ is the depth profile of the defect density and $f_{dam}(E)$ is the normalized energy distribution function of the defect level—the energy profile of PD—in the Si bandgap.

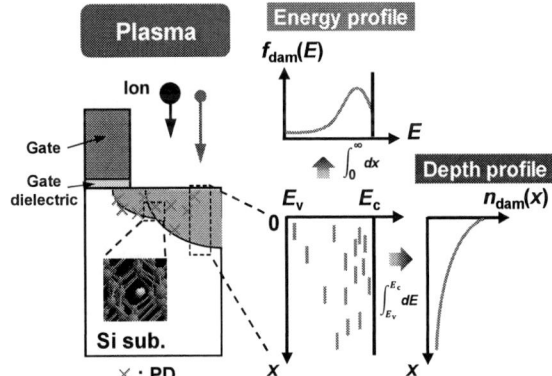

Fig. 1. Schematic illustration of a MOSFET structure with PD in the S/D extension region. PD profile is characterized by both depth $n_{dam}(x)$ and energy $f_{dam}(E)$ profiles.

B. PD profile model

The spatial profiles of PD exhibit an abrupt decrease along the depth depending on the projection range of injected ions. It was reported that the distribution tail of PD is significantly dependent on incident species and the energy [6,7]. First-principle calculations predicted various types of density-of-state (DOS) of local damaged structures, i.e. the energy profile, depending on the species inserted in the Si lattice [8]. In the present model analysis, several PD profiles were assumed to reproduce these features above.

125

The depth profile of PD $n_{dam}(x)$ was assumed to be an exponential distribution expressed by,

$$n_{dam}(x) = n_0 \exp\left(-\frac{x}{\lambda_{dam}}\right), \qquad (2)$$

where n_0 is the peak density and λ_{dam} is the characteristic depth of defects. In this profile, the areal density of PD is $n_0 \cdot \lambda_{dam}$. For the energy profile of PD $f_{dam}(E)$, three profiles: (A)(B) Boltzmann-type and (C) Gaussian-type distributions were assumed as shown in Fig. 2. Profiles (A) and (B) indicate the cases that the major portion of defects is located close to the conduction or valence bands, respectively. Profile (C) corresponds to the case that the DOS has a local maximum value in the bandgap, which is defined by the average energy level $\overline{E}$ and the variance σ_E.

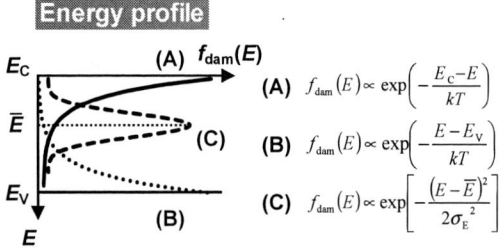

Fig. 2. Energy profiles $f_{dam}(E)$ of PD and their mathematical expressions implemented in the model analysis in this study.

C. Simulation methodology

Device simulations for MOS device performances were carried out solving carrier transport equations in the drift-diffusion approximation with density gradient quantum corrections [9]. The interface traps between Si and a gate dielectric film were also considered, while the trapped and fixed charges in the gate dielectric bulk were neglected.

III. VERIFICATION OF ENERGY PROFILES OF PD USING CAPACITANCE-VOLTAGE CHARACTERISTICS OF MOS CAPACITOR

MOS capacitors with PD on a p-type Si substrate are focused on. The dopant concentration and effective oxide thickness (EOT) are 1×10^{18} cm^{-3} and 7 nm, respectively. The depth profile of PD is fixed ($n_0 = 2 \times 10^{19}$ cm^{-3}, $\lambda_{dam} = 6$ nm). Figures 3(a) and (b) show the predicted modulation frequency f_{mod} dependent C—V curves assuming the profile (A) and (B) with respect to the energy profiles of PD, respectively. f_{mod} dispersions of the C—V curves in the vertical and horizontal directions are observable. This feature is attributed to the change in the f_{mod}-dependent activated defect density and the formation of a shallow n-p junction. Figure 3(c) shows the low-f_{mod} $1/C^2$—V curve assuming the profile (C) ($E_c - \overline{E} = 0.7$ eV and $\sigma_E = 0.2$ eV). A characteristic hump is observable in the depletion region due to the presence of defects located at the deep levels near the mid-gap. Figures 3(d), (e), and (f) show the experimental C—V and $1/C^2$—V curves of damaged devices after Ar, SF$_6$ or HBr/O$_2$ plasma exposure, respectively. The characteristic deviations predicted by the model prediction are seen under the process gas conditions. These results confirm the validity of the energy profiles of PD assumed in the model analysis.

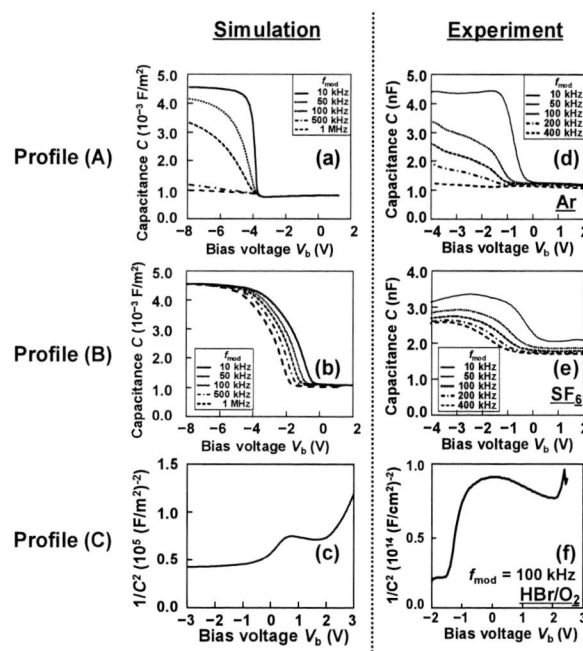

Fig. 3. Simulated C—V curves of MOS capacitors assuming (a) profile (A), (b) profile (B), and (c) profile (C). Experimental C—V curves of damaged devices exposed to (d) Ar, (e) SF$_6$, and (f) HBr/O$_2$ plasmas [10].

IV. PREDICTION OF DAMAGED MOSFET CHARACTERISTICS

A. Effects of spatial profiles of PD

45-nm bulk MOSFETs containing PD are considered to investigate the effects of the spatial profile of PD. EOT is 1.5 nm and the substrate doping is 1×10^{18} cm^{-3}. The peak concentrations of the S/D, S/D extension, and halo regions are 1×10^{20}, 1×10^{19}, and 1×10^{18} cm^{-3}, respectively. Figure 4 shows the I_d—V_g curves at the drain voltage $|V_d| = 0.05$ V in the case of n- and p-channel MOSFETs assuming the energy profile (A) with various λ_{dam} ($n_0 = 1.5 \times 10^{19}$ cm^{-3}). A device without PD is denoted as Ref. The threshold voltage shift ΔV_{th} and the on-current variation were observed mainly due to the modification of the effective doping profile. A sharp contrast in the parameter variations is seen between n- and p-channel devices. These features become remarkable with an increase in λ_{dam}. In the case of the profile (B), in contrast to the profile (A), V_{th} shifts to the positive direction.

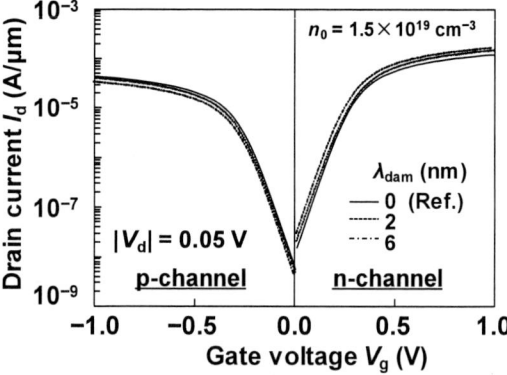

Fig. 4. Simulated I_d-V_g characteristics for n- and p-channel MOSFETs assuming the energy profile (A) with various λ_{dam}.

Figure 5 summarizes the n_0 and λ_{dam} dependences of ΔV_{th} assuming the profile (A) for n-channel MOSFETs. V_{th} is defined at $I_d(V_{th}) = 10^{-6}$ A/μm. It was found that the device characteristics significantly depend on the depth profiles even when the total amount of PD is constant. This result implies that design and control of the depth profiles of PD in response to doping profiles are indispensable to improve the device performance.

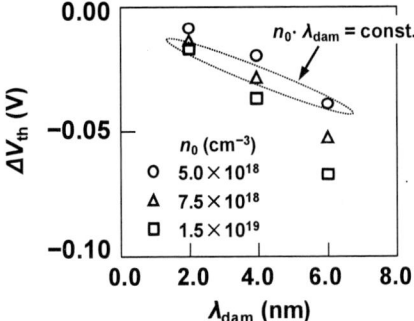

Fig. 5. Simulated ΔV_{th} as a function of λ_{dam} for n-channel MOSFETs assuming the energy profile (A) with various n_0.

B. Effects of energy profiles of PD

32-nm FD-SOI MOSFETs with PD are focused on to investigate the effects of the energy profile of PD. EOT is 1.2 nm, and the thicknesses of the Si body and buried oxide are 7 and 20 nm, respectively. The peak concentration of the S/D region is 5×10^{19} cm^{-3}, while the channel is weakly doped (1×10^{15} cm^{-3}). The depth profile of PD is fixed ($n_0 = 1 \times 10^{19}$ cm^{-3}, $\lambda_{dam} = 5$ nm). Figure 6 shows the I_d—V_g curves at $|V_d| = 0.05$ V for n- and p-channel MOSFETs assuming two energy profiles of PD. In addition to ΔV_{th} and on-current variation, off-leakage current I_{off} increases in the profile (C) in comparison with the profile (A). This feature is attributed to the carrier generation and recombination by PD located near the middle of the Si bandgap.

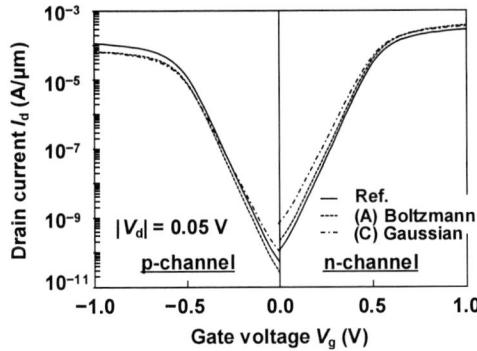

Fig. 6. Simulated I_d-V_g characteristics for n- and p-channel MOSFETs assuming two energy profiles of PD.

Figure 7 shows the $\overline{E}$ dependences of I_{off} assuming the profile (C) ($\sigma_E = 0.2$ eV) for n- and p-channel MOSFETs. As the $\overline{E}$ lowers toward the mid-gap, i.e. the portion of defects at deeper levels increases, the I_{off} value drastically increases for both n- and p-channel devices. This result implies that the precise control of the energy profiles of PD

as well as the reduction of defect density is key to reduce the power consumption of damaged MOSFET devices.

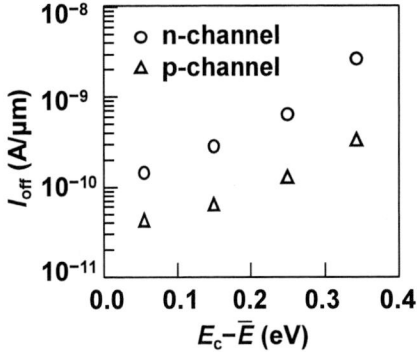

Fig. 7. Simulated I_{off} as a function of E_c-$\overline{E}$ assuming the energy profile (C).

V. CONCLUSION

We clarified by model predictions that the spatial and energy profiles of defects created during plasma processes have significant impacts on MOS device performances. It was found that the defects at deep levels in the Si bandgap degrade the subthreshold characteristics of damaged MOSFETs. Present findings imply that the consideration of the presence of PD after plasma processes is indispensable for designing future high-performance devices.

ACKNOWLEDGMENT

This work was financially supported in part by a Grant-in-Aid for Scientific Research (20J22727) from JSPS.

REFERENCES

[1] H. Abe, M. Yoneda, and N. Fujiwara, "Developments of Plasma Etching Technology for Fabricating Semiconductor Devices," *Jpn. J. Appl. Phys.*, vol. 47, pp. 1435–1455, 2008.

[2] K. Eriguchi, "Defect generation in electronic devices under plasma exposure: Plasma-induced damage," *Jpn. J. Appl. Phys.*, vol. 56, pp. 06HA01, 2017.

[3] Y. Sato, S. Shibata, K. Urabe, and K. Eriguchi, "Evaluation of residual defects created by plasma exposure of Si substrates using vertical and lateral pn junctions," *J. Vac. Sci. Technol.*, vol. B38, pp. 012205, 2019.

[4] K. Eriguchi, Y. Nakakubo, A. Matsuda, Y. Takao, and K. Ono, "Plasma-Induced Defect-Site Generation in Si Substrate and Its Impact on Performance Degradation in Scaled MOSFETs," *IEEE Electron Device Lett.*, vol. 30, pp. 1275–1277, 2009.

[5] V. M. Donnelly and A. Kornblit, "Plasma etching: Yesterday, today, and tomorrow," *J. Vac. Sci. Technol.*, vol. A31, pp. 050825, 2013.

[6] T. Ohchi et al., "Reducing Damage to Si Substrates during Gate Etching Processes," *Jpn. J. Appl. Phys.*, vol. 47, pp. 5324–5326, 2008.

[7] M. Fukasawa et al., "Structural and electrical characterization of HBr/O_2 plasma damage to Si substrate," *J. Vac. Sci. Technol.*, vol. A29, pp. 041301, 2011.

[8] Y. Yoshikawa and K. Eriguchi, "First-principles predictions of electronic structure change in plasma-damaged materials," *Jpn. J. Appl. Phys.*, vol. 57, pp. 06JD04, 2018.

[9] T. Toyabe, "Two and Three Dimensional MOSFETs Simulation with Density Gradient Model", *Proc. SISPAD*, pp. 279, 2004.

[10] T. Hamano, K. Urabe, and K. Eriguchi, "Investigation of spatial and energy profiles of plasma process-induced latent defects in Si substrate using capacitance–voltage characteristics," *J. Phys. D*, vol. 52, pp. 455102, 2019.

7-1

Modeling and Simulation of Si IGBTs

N. Shigyo[1], M. Watanabe[1], K. Kakushima[1], T. Hoshii[1], K. Furukawa[1],
A. Nakajima[2], K. Satoh[3], T. Matsudai[4], T. Saraya[5], T. Takakura[5], K. Itou[5],
M. Fukui[5], S. Suzuki[5], K. Takeuchi[5], I. Muneta[1], H. Wakabayashi[1],
S. Nishizawa[6], K. Tsutsui[1], T. Hiramoto[5], H. Ohashi[1] and H. Iwai[1]
E-mail: n.shigyo@i.softbank.jp
[1]Tokyo Institute of Technology, [2]Nat. Inst. Advanced Industrial Science and Technology,
[3]Mitsubishi Electric Corp., [4]Toshiba Electronic Devices & Storage Corp.,
[5]The University of Tokyo, [6]Kyushu University

Abstract— **Technology CAD (TCAD) has been recognized as a powerful design tool for Si insulated gate bipolar transistors (IGBTs). Here, physical models, such as a mobility model for carrier-carrier scattering, were investigated for a predictive TCAD. Simulated current-voltage characteristics of the trench-gate IGBTs were compared with measurements. The difference between 3D- and 2D-TCAD simulations was observed in a high current region, which was explained by a bias-dependent current flow. A test element group (TEG) for separation of the emitter currents for holes and electrons was also determined as effective for calibration of lifetime model parameters.**

Keywords—IGBT, trench-gate, TCAD simulation, carrier-carrier scattering, three-dimension current flow, injection enhancement effect

I. INTRODUCTION

High-efficiency power electronic systems are necessary for worldwide energy saving. A power device acts as a high voltage power switch in such systems. Therefore, the three main requirements for the power device are a low on-resistance R_{on}, a low switching loss and a sufficient safe operating area (SOA) such as a breakdown voltage.

Basic concepts of Si insulated gate bipolar transistors (IGBTs) were developed in the early 1980s [1]-[3]. Lowering of the on-state voltage V_{CEsat} was achieved by introducing electron injection enhancement (IE) effect for a trench-gate IGBT with non-latch-up structure [4]. Many improvements for IGBTs have been made, which was reviewed in [5]-[7]. Although there has been significant progress with wide bandgap materials such as SiC and GaN [8]-[11], IGBTs will still have an important role for the next years [5]-[7].

Technology CAD (TCAD), especially device simulation, has contributed to improvements in IGBTs [12]. A scaling concept for IGBTs was proposed using TCAD simulations [13], taking the relations between IE effect and scaling scheme into considerations. The first IGBT designed by the scaling scheme was experimentally confirmed [14]-[16]. 3300 V IGBTs were driven by a 5 V gate voltage, which enabled the use flexible and intelligent power control schemes due to its CMOS compatibility [16].

It is well-known that there are tradeoffs among the three stated requirements [5]-[7], [17], [18]. A predictive TCAD is required to analyze tradeoffs and optimize IGBTs. In previous studies [19], good agreements were found between simulations and measurements for trench-gate IGBTs in the *low*-current region, close to V_{CEsat}. In the *high*-current region, only a few studies have been reported that compare the current-voltage characteristics, e.g. [20]. The authors elucidated the difference between 3D- and 2D-simulations for trench-gate IGBTs [21], so that excellent agreement was obtained between 3D-simulations and measurements of the current-voltage characteristics. The importance of 3D-simulation was confirmed in [22], where simulated transient turn-off behavior was also agreed well with measurements.

In this paper, physical models are investigated to achieve predictive TCAD, and a comparison between 3D-simulations and measurements of the current-voltage characteristics is described. A test element group (TEG) [23] for separation of the emitter currents for holes and electrons is described, which provided detailed information on the IGBTs and enabled advances for calibration of model parameters.

II. BASIC OPERATION OF IGBT

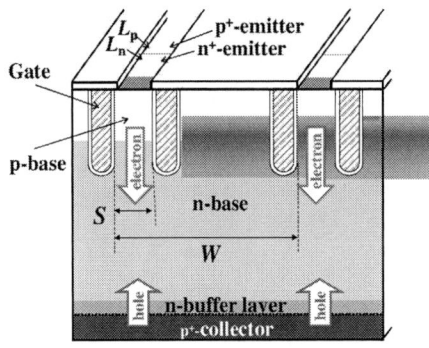

Fig. 1 Schematic diagram of trench-gate IGBT structure.

Fig. 1 shows the basic structure of the trench-gate IGBT [4], where W is the pitch, S is the mesa width between trenches, and L_n and L_p are the lengths of the n+- and p+-emitters, respectively.

The IGBT is a combination of a Si power MOSFET and a bipolar junction transistor (BJT). Electrons are injected into the n-base region by the MOSFET. The injected electrons cause the bottom pn junction to turn on. Holes are then

---

This work was based on results obtained from the project (JPNP10022) commissioned by the New Energy and Industrial Technology Development Organization (NEDO).
N. Shigyo is now with Kioxia Corporation.

© 2020 The Japan Society of Applied Physics

injected into the n-base region from the p$^+$-collector (back emitter), which results in a large amount of electron-hole plasma in the n-base region. Both electrons and holes contribute to the current flow, so that a low R_{on} is a feature of the IGBT.

III. PHYSICAL MODELS FOR IGBT SIMULATION

Selection of the appropriate physical model is essential for predictive TCAD. The IGBT is a combination of a MOSFET and a BJT, so that models for an intrinsic carrier concentration n_i and a bandgap narrowing ΔE_g^{app} were examined, since these models are important for the BJT simulation [24]. A mobility model for carrier-carrier scattering μ_{cc} proposed by Klaassen [25] was also investigated, because the IGBT was operated in the presence of the electron-hole plasma. This μ_{cc} model considers the attractive interaction potential for electron-hole scattering. The Synopsys TCAD software [26] was used for the simulations.

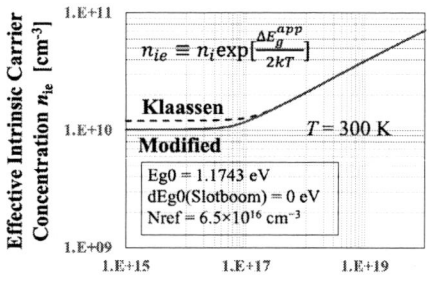

Fig. 2 Models for effective intrinsic carrier concentration n_{ie}.

Fig. 2 shows the effective intrinsic carrier concentration n_{ie} as a function of the impurity concentration. The model for ΔE_g^{app} is coupled with that for n_i [24]. ΔE_g^{app} proposed by Klaassen et al. [27] was based on an n_i of 1.20×10^{10} cm^{-3} at T = 300 K. Here, the model for ΔE_g^{app} was modified to accord with an n_i of 1.00×10^{10} cm^{-3} at T = 300 K proposed by Sproul and Green [28]. The modified parameters are given in the inset of Fig. 2.

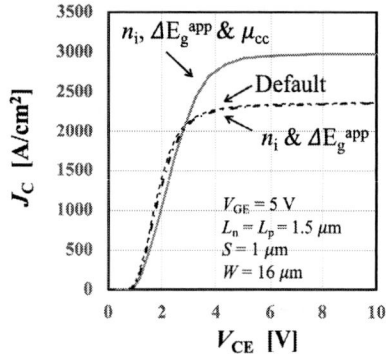

Fig. 3 Influence of physical models (n_i, ΔE_g^{app} and μ_{cc}) on J_C-V_C characteristics.

Fig. 3 shows the influence of the physical model on the J_C-V_{CE} characteristics, where J_C is the collector current density and V_{CE} is the collector-emitter voltage. The change in J_C caused by n_i and ΔE_g^{app} was less than 1%, even under BJT operation. The adoption of μ_{cc} resulted in a decrease in the mobility. However, the collector saturation current J_{Csat}, was increased by 26%. These results are explained as follows.

J_{Csat} consists of the electron and hole current densities, J_e and J_h, which are approximated by the drift current in the n-base region. Thus, J_{Csat} is expressed as

$$J_{Csat} \simeq q(n\mu_e + p\mu_h)\,E \tag{1}$$

where q is the elementary charge, n and p are the electron and hole densities, respectively, μ_e and μ_h are the respective electron and hole mobilities, and E is the electric field in the n-base region. In the high current region, J_e is mainly determined by the electron supply, i.e., the drain saturation current I_{Dsat} of the MOSFET:

$$I_{Dsat} = qn\mu_e E A_C \tag{2}$$

where A_C is the collector area. Thus, the product of nE is expressed as

$$nE = I_{Dsat}/(q\mu_e A_C). \tag{3}$$

The adoption of μ_{cc} resulted in the decrease in μ_e, so that the nE product in the n-base region was increased to satisfy (2), i.e., $I_{Dsat} = J_e A_C$. With consideration of the carrier plasma ($p \simeq n$), (1) can be rewritten using (3);

$$J_{Csat} \simeq (1 + \mu_h/\mu_e)I_{Dsat}/A_C. \tag{4}$$

The μ_{cc} caused an increase in J_h, and therefore J_{Csat}, because the decrease in μ_h was smaller than that in μ_e due to the larger hole effective mass m_h^* compared with m_e^*.

The influence of n_i and ΔE_g^{app} on the J_C-V_{CE} characteristics was negligibly small, because J_e is determined by the drain current I_D of the MOSFET for $J_C > J_{on}$, where J_{on} is the on-state current density. Understanding both the IGBT operation and physical models is thus required for predictive TCAD.

A TEG was developed to monitor the minority carrier lifetime τ_h in the n-base region [29]. The current-voltage characteristics for two diodes consisting of p$^+$-emitter and n-base regions were used, so that τ_h can be measured using the same IGBT wafer.

IV. COMPARISONS WITH MEASUREMENTS

The simulated current-voltage characteristics for the trench-gate IGBTs were compared with measurements.

A. Difference between 3D- and 2D-simulations

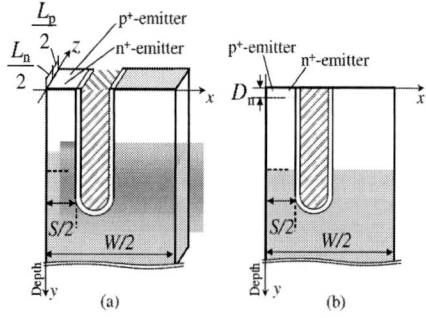

Fig. 4 Half-cell models for (a) 3D- and (b) 2D-device simulations of trench-gate IGBT.

130

Figs. 4(a) and (b) show half-cell models for the 3D- and 2D-simulations, respectively. The most notable difference between the two models is the structure of the n⁺- and p⁺-emitters at the top of the mesa. For the 3D-model, the n⁺- and p⁺-emitters with lengths L_n and L_p are placed alternately, where $L_n + L_p$ is the pitch of the emitter structure. For the 2D-simulation, the p⁺-emitter is embedded in the same plane as the n⁺-emitter.

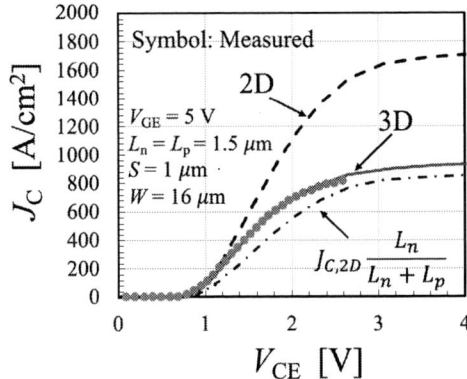

Fig. 5 J_C-V_{CE} characteristics calculated by 3D- and 2D-simulations, and corresponding measurements.

Fig. 5 shows the J_C-V_{CE} characteristics obtained by 3D- and 2D-simulations, and those from the corresponding measurements [14] for a scaling factor $k = 3$ and $L_n = L_p = 1.5$ μm. Here, the resistances of the contact and substrate were taken into account [21]. The 3D-simulation results revealed excellent agreement with the measurements for a wide range of J_C up to 1000 A/cm². However, the 2D-simulation reproduced the measurement only for $J_C < 100$ A/cm². In the high-current region, J_C appeared to be well fitted by the expression $J_{C,2D} \cdot L_n/(L_n + L_p)$, where $J_{C,2D}$ is J_C obtained by the 2D-simulations.

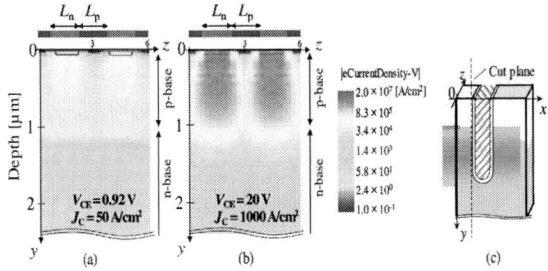

Fig. 6 3D-simulated distribution of electron current density in vertical direction $|J_{ey}|$ on cross-section close to channel surface for $V_{GE} = 5$ V. (a) $J_C = 50$ A/cm², (b) $J_C = J_{Csat} = 1000$ A/cm², (c) position of cross-section.

Fig. 6 shows the distribution of the electron current density component $|J_{ey}|$, for the cross-section close to the channel surface. When J_C was less than J_{on} (= 200 A/cm²), the electric field in the channel was low because the potential difference between the n⁺-emitter and n-base regions was small. Therefore, the injected electrons were distributed uniformly in the channel, as shown in Fig. 6(a). This situation was similar to the 2D-simulation, although the emitter structures were significantly different between the 3D- and

2D-simulations. However, in the high-current region ($J_C = J_{Csat}$), electrons were injected only from the n⁺-emitter, as shown in Fig. 6(b). In this situation, J_C was limited by the electron supply from the MOSFET. Therefore, J_C could be approximated by $J_{C,2D} \cdot L_n/(L_n + L_p)$, as shown in Fig. 5.

Application of the 2D-simulations was limited to only the low-current region for the IGBTs, because of the bias-dependent 3D current flow and carrier distribution.

B. Measurement for electron and hole emitter currents

For the trench-gate IGBT, the on-state voltage V_{CEsat} becomes lower with a reduction of the mesa width S. The holes injected from the p⁺-collector reach the mesa region surrounded by trenches, so that the hole current flow is restricted. The positively charged holes thus enhance electron injection into the n-base region to satisfy charge neutrality, which is referred to as the IE effect [4]. A TEG was fabricated for direct measurement of the IE effect [23].

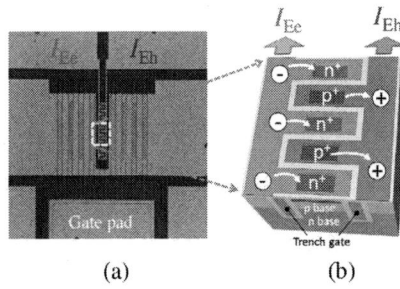

Fig. 7 (a) Plane view of IGBT after fabrication and (b) schematic view of contact arrangement for current-separating emitter.

Fig. 7(a) shows a plan-view image of the TEG, where two separated interdigitated electrodes for the emitter are in contact with each n⁺ and p⁺ region to allow for independent measurement of the emitter currents for electrons I_{Ee} and holes I_{Eh}, as shown in Fig. 7(b).

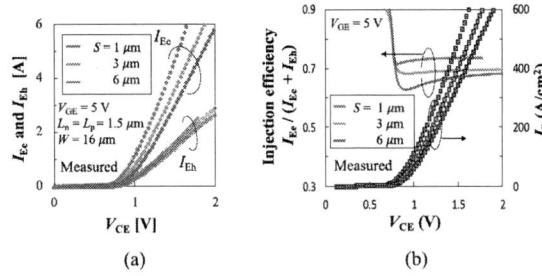

Fig. 8 (a) Measured emitter electron current I_{Ee} and emitter hole current I_{Eh}, and (b) electron injection efficiency $I_{Ee}/(I_{Eh} + I_{Ee})$ (solid lines) and J_E-V_{CE} characteristics (symbols) with various mesa width S.

Figs. 8(a) and (b) show the measured I_{Ee} and I_{Eh} as a function of V_{CE}, and the electron injection efficiency and J_E-V_{CE} characteristics for IGBTs with different mesa widths S, respectively. Fig. 8(a) demonstrates that the increase in I_{Ee} was greater than that in I_{Eh} as S decreased, although both I_{Ee} and I_{Eh} increased. As a result, the electron injection efficiency clearly increased as S narrowed, as shown in Fig. 8(b). This provides direct evidence of the IE effect. In addition, Fig. 8(b) confirms that V_{CEsat} was reduced due to the decrease in S. I_E for low V_{CE} also consisted mostly of I_{Ee}.

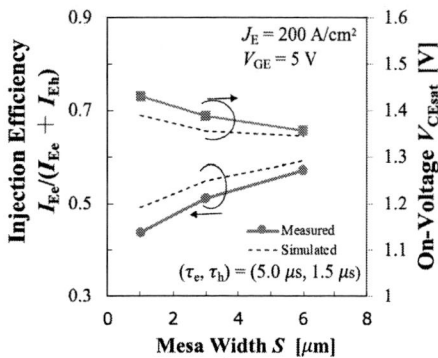

Fig. 9. Electron injection efficiency $I_{Ee}/(I_{Ee} + I_{Eh})$, and on-state voltage V_{CEsat} as function of mesa width S at $J_E = 200$ A/cm².

Fig. 9 shows the dependence of the electron injection efficiency and V_{CEsat} on S at $J_E = 200$ A/cm². The fact that the injection efficiency increases and V_{CEsat} decreases as S is reduced, was consistent with the 3D simulation results.

The individual evaluations of I_{Ee} and I_{Eh} were an additional constraint condition for calibration of the lifetime in the n-base region, where a τ_e of 5.0 μs and a τ_h of 1.5 μs were determined. The information for I_{Ee} and I_{Eh} would thus facilitate TCAD calibration.

V. CONCLUSION

Excellent agreement of the current-voltage characteristics between measurements and TCAD simulations was obtained for trench-gate IGBTs. The adoption of the mobility model for carrier-carrier scattering resulted in an increase in the collector current density J_C, even though the mobility was reduced. The difference between the 3D- and 2D-TCAD simulations in the high-current region could be explained by the bias-dependent 3D-current flow and carrier distribution. The TEG to separate the emitter currents for holes and electrons was determined to be effective for calibration of the lifetime model parameter.

ACKNOWLEDGMENT

The authors would like to thank Prof. I. Omura and Prof. M. Tsukuda of Kyushu Institute of Technology and Prof. W. Saito of Kyushu Univ. for fruitful discussions.

This work was based on results obtained from the project (JPNP10022) commissioned by the New Energy and Industrial Technology Development Organization (NEDO).

REFERENCES

[1] J. D. Plummer, "Monolithic semiconductor switching device," U.S. Patent 4199774, Apr. 22, 1980.

[2] H. W. Becke and C. F. Wheatley, Jr., "Power MOSFET with an anode region," U.S. Patent 4364073, Dec. 14, 1982.

[3] A. Nakagawa et al., "Non-luchup 1200 V 75 A bipolar mode ASO," Tech. Dig. IEDM 1984, pp. 860-861.

[4] M. Kitagawa et al., "A 4500 V injection enhanced insulated gate bipolar transistor (IEGT) operating in a mode similar to a thyristor," Tech. Dig. IEDM 1993, pp. 679–682.

[5] N. Iwamuro and T. Laska, "IGBT history, state-of-the-art, and future prospects," IEEE Transactions on Electron Devices, vol. 64, pp. 741-752, 2017.

[6] N. Iwamuro and T. Laska, "Correction to "IGBT history, state-of-the-art, and future prospects"," IEEE Trans. on Electron Devices, vol. 65, p. 2675-2675, 2018.

[7] T. Laska, "Progress in Si IGBT technology – as on going competition with WBG power devices," Tech. Dig. IEDM 2019, pp. 262-265.

[8] J. W. Palmour, "Silicon carbide power device development for industrial markets," Tech. Dig. IEDM 2014, pp. 1.1.1-1.1.8.

[9] J. B. Casady et al., "New generation 10kV SiC power MOSFET and Diodes for Industrial Applications," Proceedings of PCIM Europe 2015, pp. 1–8.

[10] D. Marcon, Y. N. Saripalli, and S. Decoutere, "200mm GaN-on-Si epitaxy and e-mode device technology," Tech. Dig. IEDM 2015, pp. 16.2.1-16.2.4.

[11] Y. Zhang et al., "Reduction of on-resistance and current crowding in quasi-vertical GaN power diodes," Applied Physics Letters, vol. 111, no. 16, pp. 163-506, 2017.

[12] H. Ohashi and I. Omura, "Role of simulation technology for the progress in power devices and their applications," IEEE Trans. on Electron Devices, vol.60, pp. 528 - 534, 2013.

[13] M. Tanaka and I. Omura, "IGBT scaling principle toward CMOS compatible wafer processes," Solid-State Electronics, vol. 80, pp. 118–123, 2013.

[14] K. Kakushima et al., "Experimental verification of a 3D scaling principle for low Vce(sat) IGBT," Tech. Dig. IEDM 2016, pp. 10.6.1-10.6.4.

[15] T. Saraya et al., "Demonstration of 1200V scaled IGBTs driven by 5V gate voltage with superiorly low switching loss,", Tech. Dig. IEDM 2018, pp. 8.4.1-8.4.4.

[16] T. Saraya et al., "3300 V scaled IGBTs driven by 5 V gate voltage," Proc., ISPSD 2019, pp. 43-46.

[17] K. Satoh et al., "New chip design technology for next generation power module," Proc. PCIM 2008, p.673.

[18] K. Nishi, T. Takahashi and A. Narazaki, "Analysys the complex tradeoff among E_{on}-V_{CEsat}-SCSOA and EMI noise through the single chip evaluation method," Proc., ISPSD 2019, pp. 475-478.

[19] P. Luo et al., "Numerical analysis of 3-dimensional scaling rules on a 1.2-kV trench clustered IGBT," IEEE Tran. Electron Devices, vol. 65, no. 4, pp. 1440-1446, 2018.

[20] H. Feng et al., "A 1200 V-class fin p-body IGBT with ultra-narrow mesas for low conduction loss," Proc., ISPSD 2016, pp. 203-205.

[21] M. Watanabe et al., "Impact of three-dimensional current flow on accurate TCAD simulation for trench-gate IGBTs," Proc., ISPSD 2019, pp. 311-314.

[22] T. Suwa and S. Hayase, "Investigation of TCAD calibration for saturation and tail current of 6.5kV IGBTs," Proc., SISPAD 2019, pp. 101-104

[23] T. Hoshii et al., "Verification of the injection enhancement effect in IGBTs by measuring the electron and hole currents separately," Proc. ESSDERC 2018, pp. 26-29.

[24] N. Shigyo, N. Konishi and H. Satake, "An improved bandgap narrowing model based on corrected intrinsic carrier concentration," Trans. IEICE, vol. E75-C, pp. 156-160, 1992.

[25] D. B. M. Klaassen, "Unified mobility model for device simulation," Solid-St. Electron., vol. 35, pp. 953-959, 1992.

[26] TCAD Sentaurus User Manual, Version M-2016.12, Synopsys, San Jose, CA, USA, 2016.

[27] D. B. M. Klaassen, J. W. Slotboom and H. C. de Graaf, "Unified apparent bandgap narrowing in n and p-type silicon," Solid-St. Electron., vol. 35, pp. 125-129, 1992.

[28] A. B. Sproul and M. A. Green, "Improved value for the silicon intrinsic carrier concentration from 275 to 375 K," J. Appl. Phys., vol. 70, pp. 846-854, 1991.

[29] K. Kakushima et al., "New methodology for evaluating minority carrier lifetime for process assessment," Proc. of Symposia on VLSI Technology and Circuits 2018, pp. 105-106.

Full-Band Monte Carlo simulations of GaAs p-i-n Avalanche PhotoDiodes: What Are the Limits of Nonlocal Impact Ionization Models?

A. Pilotto[1], F. Driussi[1], D. Esseni[1], L. Selmi[2], M. Antonelli[3], F. Arfelli[4,5], G. Biasiol[6], S. Carrato[3], G. Cautero[7,5], D. De Angelis[7], R. H. Menk[7,5,8], C. Nichetti[7,4], T. Steinhartova[4,6], P. Palestri[1]

[1]DPIA, University of Udine, 33100 Udine, Italy. Email: pilotto.alessandro@spes.uniud.it
[2]DIEF, University of Modena and Reggio Emilia, 44100 Modena, Italy.
[3]Department of Engineering and Architecture, University of Trieste, 34128 Trieste, Italy.
[4]Department of Physics, University of Trieste, 34128 Trieste, Italy.
[5]Istituto Nazionale di Fisica Nucleare, INFN Sezione di Trieste, Trieste, 34100, Italy.
[6]IOM CNR, Laboratorio TASC, Area Science Park Basovizza, 34149 Trieste, Italy.
[7]Elettra-Sincrotrone Trieste S.C.p.A, Area Science Park Basovizza, 34149 Trieste, Italy.
[8]Department of Medical Imaging, University of Saskatchewan, Saskatoon, SK S7N 5A2, Canada.

Abstract—We present a Full-Band Monte Carlo (FBMC) investigation of impact ionization in GaAs p-i-n Avalanche Photodiodes (APDs). FBMC simulations have been used to compute the gain and the excess noise factor and a new equation has been derived for the extraction of history-dependent impact ionization coefficients from FBMC simulations. Results from FBMC are then compared with the ones of nonlocal history-dependent impact ionization models. We found that at high reverse bias voltages it is important to take into account the fact that secondary carriers are generated with nonzero kinetic energy. Finally, we propose an improved history-dependent model using an energy-dependent mean free path.

Index Terms—Avalanche Photodiodes, Full-Band Monte Carlo, Impact Ionization

I. INTRODUCTION

Nonlocal History-Dependent (NL-HD) impact ionization models based either on the Dead Space approximation [1] or on effective fields [2]–[4] are often used to estimate the gain (M) and the excess noise factor (F) in avalanche photodiodes (APDs). The strength of NL-HD models lies in their low computational burden, when compared with Monte Carlo simulations, that allows to systematically analyze the trade-offs at the basis of APDs design. The high computational speed, however, comes at the expense of accuracy. In fact, in NL-HD models impact ionization is described at a macroscopic level: the model parameters have to be empirically chosen to fit experimental gain and excess noise factor; furthermore secondary carriers are always generated with zero kinetic energy and can move only in the direction of the applied electric field [1]–[4]. In this paper, we analyze each of these limits of NL-HD models by means of Full-Band Monte Carlo (FBMC) simulations of p-i-n GaAs APDs.

The paper proceeds as follows: the FBMC model, its calibration and the results of FBMC simulations in terms of

This work was supported by Italian MIUR through the PRIN 2015 Project under Grant 2015WMZ5C8.

gain and excess noise factor of thin GaAs p-i-n APDs are presented in Section II. A new equation for the extraction of the history-dependent impact ionization coefficients from FBMC simulations is proposed in Section III and used to analyze the limits of NL-HD models. A new expression for the electron's mean free path to be used in the nonlocal history-dependent model of [4] is also reported in Section III. Finally, conclusions are drawn in Section IV.

II. FBMC MODEL

The band structure of GaAs has been computed by using the Local Empirical Pseudopotentials with the form factors from [5]. Spin Orbit Interaction has been included in the calculations as in [6]. The expressions reported in [5] have been used for the carrier-phonon scattering rates, giving results consistent with other FBMCs (not shown). The deformation potentials for acoustic and nonpolar optical phonons have been adjusted in order to match the experimental results for the drift velocities vs. the applied field [7]–[10] (see Fig. 1).

The impact ionization scattering rates are computed from the band structure embracing the Constant Matrix Element approximation [11] and are consistent with other FBMCs, as shown in Fig. 2.

The impact ionization coefficients corresponding to the scattering rates in Fig. 2 are extracted from FBMC simulations under uniform electric field (infinitely long slabs) with two different methods: namely, either as the reciprocal of the average distance $\langle d \rangle$ between consecutive impact ionization events

$$\alpha = \frac{1}{\langle d \rangle}, \quad (1)$$

or by using the expression (valid for electrons)

$$\alpha = \frac{\int_0^{+\infty} SR_{II,e}(W) F_{occ}(W) dW}{v_e \int_0^{+\infty} F_{occ}(W) dW}, \quad (2)$$

where $SR_{II,e}(W)$ and $F_{occ}(W)$ are, respectively, the electron's impact ionization scattering rate and the occupation function as a function of the energy W, while v_e is the drift velocity. Similar equations are used also for the hole impact ionization coefficient β. The two methods are compared in Fig. 3 and give similar results that are in an overall good agreement with experimental data [15], and with the impact ionization coefficients used in the NL-HD model of [4].

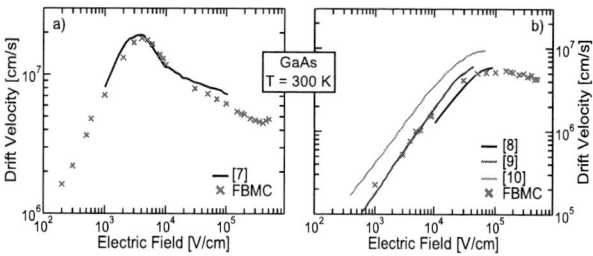

Fig. 1. a) Electron velocity v_e and b) hole velocity v_h versus electric field (E) curves in GaAs at $T = 300$ K. FBMC results (red crosses) are compared with experimental data [7]–[10] (solid lines).

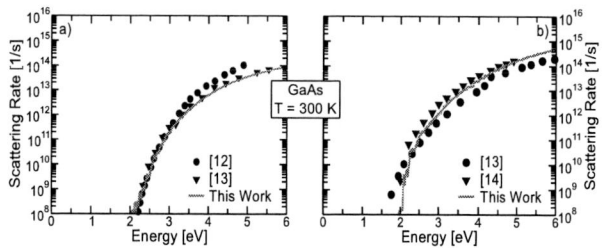

Fig. 2. a) Electron and b) hole impact ionization scattering rates in GaAs as a function of energy. Our results (red lines), obtained by using the Constant Matrix Element approximation [11], are compared with the scattering rates computed by other authors [12]–[14].

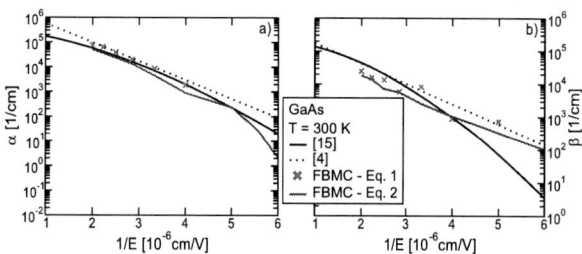

Fig. 3. a) α and b) β versus $1/E$ in GaAs at $T = 300$ K. FBMC (Eq. 1 (red crosses) or Eq. 2 (blue solid)) is compared with experiments [15] (black solid) and with α and β used in [4] (black dotted).

To validate the FBMC transport model, M and F are computed in thin p-i-n diodes by simulating many trials starting from an electron injected at $x = 0$: $M = \langle m \rangle$ and $F = \langle m^2 \rangle / M^2$, with m being the gain obtained in a trial. Figure 4 compares the experimental $M(V_{rev})$ and $F(M)$ curves for GaAs p-i-n diodes of different thicknesses [3] with the results of FBMC. The agreement between experiments and simulations is good with no further parameter adjustments besides the calibration of the phonon deformation potentials carried out to reproduce the $v(E)$ curves of Fig. 1.

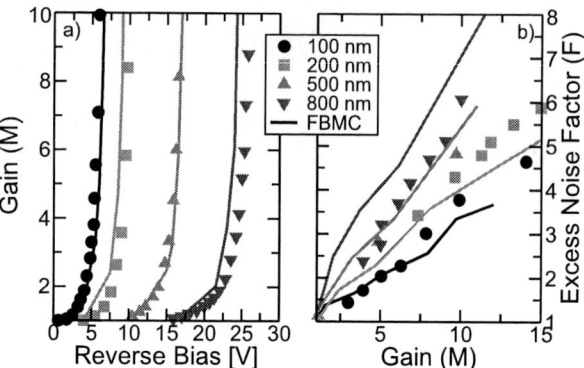

Fig. 4. a) M versus V_{rev} and b) F versus M curves for GaAs p-i-n APDs. FBMC (solid lines) is compared with experiments [3] (symbols). A built-in voltage $V_{bi} = 1.2$ V has been assumed [17].

III. FBMC VS NL-HD MODELS

In NL-HD models α and β are functions of the generation (x) and the ionization (x') points (see Fig. 5) and they have a nonlocal dependence on the complete electric field profile between x and x'. In other words α and β are not given solely by the local electric field, hence it is also necessary an alternative procedure to extract α and β from simulations with non-uniform electric fields, thus going beyond Eqs. 1 and 2 used for Fig. 3.

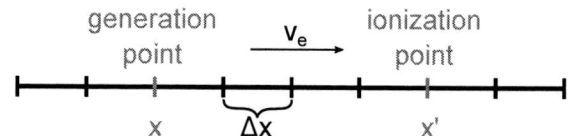

Fig. 5. Discretization of the intrinsic region of a p-i-n diode with a mesh with uniform spacing Δx. x is the point where a carrier is generated, optically or by impact ionization, and x' is the point where it ionizes. Electrons move from left to right with average velocity v_e (while holes move from right to left with average velocity v_h and x and x' are inverted).

To this purpose, we have derived Eq. 3 to extract $\alpha(x|x')$ (and $\beta(x|x')$) from FBMC simulations. In fact, following [2], [4], the electron's impact ionization probability is

$$p(x|x')\Delta x = \frac{\text{No. of el. gen. in } x \text{ that ionize in } x' \pm \Delta x/2}{\text{No. of electrons generated in } x}$$
$$= \alpha(x|x')\exp\left(-\int_x^{x'} \alpha(x|x''')dx'''\right)\Delta x.$$
$$(4)$$

With a similar reasoning, one can also find the probability for an electron to ionize before reaching x', namely:

$$p(x|x'' < x') = \frac{\text{No. of el. gen. in } x \text{ that ionize before } x'}{\text{No. of electrons generated in } x}$$
$$= \int_x^{x'} \alpha(x|x'')\exp\left(-\int_x^{x''} \alpha(x|x''')dx'''\right)dx''$$
$$= 1 - \exp\left(-\int_x^{x'} \alpha(x|x'')dx''\right).$$
$$(5)$$

$$\alpha(x|x') = \frac{\textit{No. of electrons generated in } x \textit{ (by II or a photon) that ionize in } x' \pm \Delta x/2}{\Delta x \cdot (\textit{No. of electrons generated in } x - \textit{No. of electrons generated in } x \textit{ that ionize in } [0, x' - \Delta x/2))} \tag{3}$$

Combining Eqs. 4 and 5, the expression for $\alpha(x|x')$ is found by writing

$$\alpha(x|x') = \frac{p(x|x')}{[1 - p(x|x'' < x')]}, \tag{6}$$

which is equivalent to the proposed Eq. 3. It is worth noting that at the denominator of Eq. 3 we look for all the electrons generated in x that ionize in $[0, x' - \Delta x/2)$, instead of $[x, x' - \Delta x/2)$, to account for those electrons that, due to a scattering event, travel with negative velocity and ionize at points lying before (along the field direction) the location where the carrier was generated.

Sample results are shown in Fig. 6 for an electron generated at $x = 0$ or at $x = 50$ nm. As it can be seen, $\alpha(x|x')$ saturates to a constant value (in the following denoted as α^*) after few tens of nm. The value of α^* differs from the α obtained as a function of the field by using Eqs. 1 and 2. In fact, as explained in [16], α^* is the impact ionization coefficient for an electron which has already traveled over its dead space d_e. Ref. [16] proposes the expression

$$\frac{1}{\alpha} = d_e + \frac{1}{\alpha^*}. \tag{7}$$

Figure 7a compares $\alpha(1/E)$ from Eq. 1 and $\alpha^*(1/E)$ from Eq. 3 when $x' \gg x$ and shows that they can be related by using Eq. 7 with $d_e = W_{th,e}/qE$, where $W_{th,e}$ is the threshold energy for electron's ionization (see Fig. 7b). In Fig. 6b, we notice that at high reverse bias voltages (namely at high gains) $\alpha(x|x') > 0$ also for $x' < x$. In fact, in FBMC secondary carriers are generated with nonzero kinetic energy and in some cases with a velocity going leftward, which is a case neglected by NL-HD models. Such left going electrons can undergo impact ionization events even for $x' < x$.

Figure 8 shows $\beta(x|x')$ for holes extracted from FBMC: also $\beta(x|x')$ saturates to a constant value β^* after few tens of nm. Differently from the case of electrons, the saturation of $\beta(x|x')$ for secondary holes (plot b) takes place after an overshoot that is not present in the NL-HD models. After that overshoot, $\beta(x|x')$ stabilizes to a value independent of x', as long as $x' \ll x$.

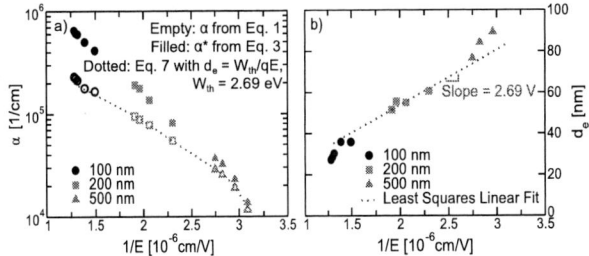

Fig. 7. a) α (empty symbols) and α^* (filled symbols) versus $1/E$ extracted from FBMC simulations of GaAs p-i-n diodes by using Eq. 1 and Eq. 3 (plateau for $x' \gg x$), respectively. b) Electron's dead space d_e versus $1/E$ extracted, point by point, with Eq. 7 from the α and α^* of Fig. 7a with Eq. 7. The blue dotted line in plot b is the least squares linear fit of the d_e values, giving a slope $W_{th,e}/q = 2.69$ V. The same parameter inserted in Eq. 7 gives the blue dotted line in plot a.

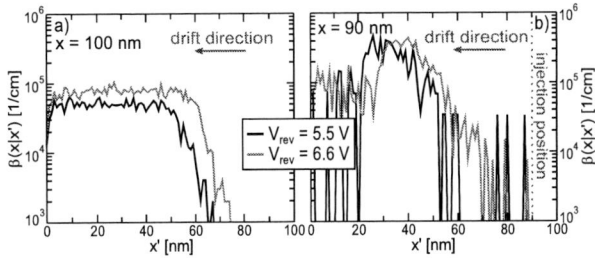

Fig. 8. $\beta(x|x')$ extracted from FBMC simulations by using Eq. 3 in a 100 nm-thick GaAs p-i-n APD with hole injection from the right side at $V_{rev} = 5.5$ V (black) and $V_{rev} = 6.6$ V (red). a) $x = 100$ nm, b) $x = 90$ nm.

Figure 9 compares the $M(V_{rev})$ and $F(V_{rev})$ curves for GaAs p-i-n diodes extracted with FBMC simulations and those obtained with the NL-HD model of [4] by using as input the impact ionization coefficients extracted from the FBMC with Eq. 3. The agreement between the two algorithms for $M(V_{rev})$ is good at low gains and for long diodes. The larger discrepancies at high gains in short devices are due to the fact that the NL-HD model neglects electron impact ionization for $x' < x$ and hole impact ionization for $x' > x$; in fact, such contributions become more relevant as the reverse bias in short devices increases. We also observe that the NL-HD model underestimates $F(V_{rev})$ for short diodes and overestimates $F(V_{rev})$ for long ones.

The NL-HD model we proposed in [4] is based on an energy balance equation with a constant energy relaxation length λ_e. Here we propose the following relation between λ_e and the energy (i.e. the effective field)

$$\lambda_e(x|x') = \frac{A}{B + E_{eff,e}(x|x')}, \tag{8}$$

where the effective field is given by

$$E_{eff,e}(x|x') = \int_x^{x'} \frac{1}{\lambda_e(x|x'')} \frac{dE_C}{dx''} \exp\left(\frac{x'' - x'}{\lambda_e(x|x'')}\right) dx'', \tag{9}$$

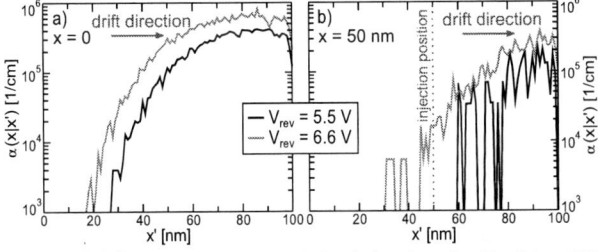

Fig. 6. $\alpha(x|x')$ extracted from FBMC simulations by using Eq. 3 in a 100 nm-thick GaAs p-i-n APD with electron injection from the left side at $V_{rev} = 5.5$ V (black) and $V_{rev} = 6.6$ V (red). a) $x = 0$, b) $x = 50$ nm.

and E_C is the conduction band profile. The parameters $A = 1.4142$ V and $B = 1.32 \times 10^5$ V/cm were chosen so that, for each bias point, the $\alpha(0|x')$ computed with the NL-HD model of [4] and $\lambda(x|x')$ from Eq. 8 could reproduce fairly well the $\alpha(0|x')$ extracted from FBMC simulations. Figure 10 compares $\alpha(0|x')$ for a 100 nm thick GaAs p-i-n diode from FBMC and without hole impact ionization, with the NL-HD impact ionization coefficients corresponding to different approximations. The NL-HD results are shown for the Dead Space model [1] (d_e is taken from Fig. 7b) and for the NL-HD model in [4] using the effective fields from Eq. 9 with $\lambda_e = 18$ nm [4] or with λ_e from Eq. 8. Good agreement is obtained between FBMC results and impact ionization coefficients of the NL-HD model in [4] with energy dependent λ_e (Eq. 8). The Dead Space model and the NL-HD model with constant λ_e are instead found to be less accurate.

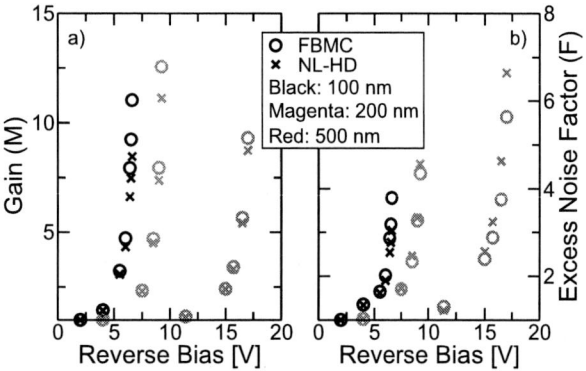

Fig. 9. Comparison between the a) M and b) F versus V_{rev} curves for GaAs p-i-n diodes obtained with the FBMC (circles) or by using the NL-HD model of [4] with $\alpha(x|x')$ and $\beta(x|x')$ extracted from FBMC simulations as inputs (crosses).

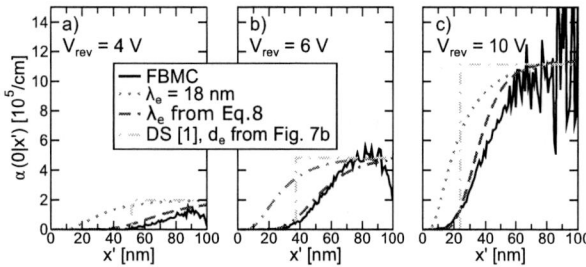

Fig. 10. $\alpha(0|x')$ for a 100 nm-thick GaAs p-i-n diode when hole impact ionization is turned off. FBMC (black solid line) is compared with the NL-HD model for impact ionization in [4] either with $\lambda_e = 18$ nm (red dotted line) or λ_e from Eq. 8 (blue dashed line) and with the Dead Space model (DS) [1], when d_e is taken from Fig. 7b (green dahed-dotted line).

IV. CONCLUSIONS

This work identified some major limits of NL-HD models, that mainly stem from the assumption that secondary carriers are generated with zero kinetic energy and can move only in the direction of the electric field, resulting in impact ionization events only at $x' > x$ for electrons ($x' < x$ for holes). The FBMC simulations point out the role of backscattering

that may lead to impact ionization at positions preceding the generation point along the field direction, in particular at high gains in short devices. Impact ionization coefficients based on an energy dependent λ_e are in better agreement with the FBMC compared to the Dead Space model or the NL-HD model of [4]. As for the $\beta(x|x')$, however, it seems more difficult to derive an expression similar to Eq. 8 for λ_h in GaAs is not straightforward, essentially because $\beta(x|x')$ saturates to a constant value β^* only after the overshoot region illustrated in Fig. 8.

REFERENCES

[1] M. M. Hayat, B. E. A. Saleh and M. C. Teich, "Effect of dead space on gain and noise of double-carrier-multiplication avalanche photodiodes," in IEEE Transactions on Electron Devices, vol. 39, no. 3, pp. 546-552, March 1992, doi: 10.1109/16.123476.

[2] R. J. McIntyre, "A new look at impact ionization-Part I: A theory of gain, noise, breakdown probability, and frequency response," in IEEE Transactions on Electron Devices, vol. 46, no. 8, pp. 1623-1631, Aug. 1999, doi: 10.1109/16.777150.

[3] P. Yuan et al., "A new look at impact ionization-Part II: Gain and noise in short avalanche photodiodes," in IEEE Transactions on Electron Devices, vol. 46, no. 8, pp. 1632-1639, Aug. 1999, doi: 10.1109/16.777151.

[4] C. Nichetti et al., "An Improved Nonlocal History-Dependent Model for Gain and Noise in Avalanche Photodiodes Based on Energy Balance Equation," in IEEE Transactions on Electron Devices, vol. 65, no. 5, pp. 1823-1829, May 2018, doi: 10.1109/TED.2018.2817509.

[5] M. V. Fischetti, "Monte Carlo simulation of transport in technologically significant semiconductors of the diamond and zinc-blende structures. I. Homogeneous transport," in IEEE Transactions on Electron Devices, vol. 38, no. 3, pp. 634-649, March 1991, doi: 10.1109/16.75176.

[6] W. Pötz, P. Vogl, "Theory of optical-phonon deformation potentials in tetrahedral semiconductors", in Phys. Rev. B, vol. 24, no. 4, pp. 2025-2037, Aug. 1981, doi: 10.1103/PhysRevB.24.2025.

[7] P. A. Houston, A. G. R. Evans, "Electron drift velocity in n-GaAs at high electric fields," in Solid-State Electronics, vol. 20, no. 3, pp. 197-204, 1977, doi: 10.1016/0038-1101(77)90184-8.

[8] L. H. Holway, S.R. Steele, M. G. Alderstein, Proceedings of the Seventh Biennial Cornell Electrical Engineering Conference (Cornell University Press, Ithaca, NY, 1979), pp. 199–208.

[9] V. L. Dalal, "Hole velocity in p-GaAs,", in Applied Physics Letters, vol. 16, no. 12, pp. 489-491, 1970, doi: 10.1063/1.1653077.

[10] V. L. Dalal, A. B. Dreeben, A. Triano, "Temperature dependence of hole velocity in p-GaAs," in Journal of Applied Physics, vol. 42, no. 7, pp. 2864-2867, 1971, doi: 10.1063/1.1660641.

[11] N. Sano, A. Yoshii, "Impact-ionization model consistent with the band structure of semiconductors," in Journal of Applied Physics, vol. 77, no. 5, pp. 2020-2025, 1995, doi: 10.1063/1.358839.

[12] H. K. Jung, K. Taniguchi, C. Hamaguchi, "Impact ionization model for full band Monte Carlo simulation in GaAs," in Journal of Applied Physics, vol. 79, no. 5, pp. 2473-2480, 1996, doi: 10.1063/1.361176.

[13] D. Harrison, R. A. Abram, S. Brand, "Impact ionization rate calculations in wide band gap semiconductors," in Journal of Applied Physics, vol. 85, no. 12, pp. 8178-8185, 1999, doi: 10.1063/1.370657.

[14] M. V. Fischetti, N. Sano, S. Laux, K. Natori, "Full-band-structure theory of high-field transport and impact ionization of electrons and holes in Ge, Si, GaAs, InAs, and InGaAs," August 2015, doi: 10.13140/RG.2.1.5004.2088.

[15] G. E. Bulman, V. M. Robbins, K. F. Brennan, K. Hess and G. E. Stillman, "Experimental determination of impact ionization coefficients in (100) GaAs," in IEEE Electron Device Letters, vol. 4, no. 6, pp. 181-185, June 1983, doi: 10.1109/EDL.1983.25697.

[16] J. S. Cheong, M. M. Hayat, X. Zhou and J. P. R. David, "Relating the Experimental Ionization Coefficients in Semiconductors to the Nonlocal Ionization Coefficients," in IEEE Transactions on Electron Devices, vol. 62, no. 6, pp. 1946-1952, June 2015, doi: 10.1109/TED.2015.2422789.

[17] S. A. Plimmer, J. P. R. David, D. S. Ong and K. F. Li, "A simple model for avalanche multiplication including deadspace effects," in IEEE Transactions on Electron Devices, vol. 46, no. 4, pp. 769-775, April 1999, doi: 10.1109/16.753712.

A technique for phase-detection auto focus under near-infrared-ray incidence in a back-side illuminated CMOS image sensor pixel with selectively grown germanium on silicon

Tatsuya Kunikiyo, Hidenori Sato, Takeshi Kamino, Koji Iizuka, Ken'ichiro Sonoda and Tomohiro Yamashita
Renesas Electronics Corporation, 751 Horiguchi, Hitachinaka, Ibaraki, 312-8504, Japan.
Corresponding author: tatsuya.kunikiyo.zn@renesas.com

Abstract—A novel phase-detection auto focus (PDAF) technique for incident 850 nm plane wave is demonstrated using Ge-on-Si layer and deep trench isolation (DTI), which are locally arranged on light receiving surface (LRS) of crystalline silicon (c-Si). No metal light shielding film (LSF) for pupil division is formed. The key concept of the present work for PDAF is to perform the pupil division by the locally arranged Ge-on-Si layer in a pixel according to incident angle. The present pixel is based on a back-side illuminated CMOS image sensor pixel; the pixel pitch is 1.85 μm and the thickness of c-Si is around 3 μm. The simulation, based on three-dimensional finite difference time domain (3D-FDTD) method, shows that the external quantum efficiency (EQE) of the present pixel exhibits above 44.3 % with the maximum of 76.0 % for incident angles of -30° to +30°, owing to the selectively arranged Ge-on-Si layer; it exhibits 3.6 times improvement in the EQE at normal incidence compared to that of current state-of-the-art pixel with half metal-shielded aperture; the EQE is 49.2 % and 13.8 %, respectively. The present technique can enhance the accuracy of AF under low-illuminated condition.

I.   INTRODUCTION

The demand for auto focus in night vision photography requires a technique for phase-detection auto focus (PDAF) to capture near infrared radiation. To achieve a high accuracy of PDAF under low-illuminated condition, it is desired to increase external quantum efficiency (EQE) in a pixel for PDAF fabricated on a cost-effective crystalline silicon (c-Si) substrate.

The PDAF method in digital cameras is a system to which so-called triangulation technique is applied and is a method of obtaining a distance by an angle difference when seeing the same subject from two different points. In a pixel for PDAF, a light shielding film (LSF) for limiting light incident on the micro lens is provided [1], and the amount of light incident on the image pickup element is limited, so that the sensitivity is deteriorated. As a result, there is a possibility that accuracy of detection of the focal position may be lowered.

Moreover, absorption coefficient in c-Si at wavelength of 850 nm is almost one order of magnitude smaller than that at wavelength of 550 nm [2]. While it is possible to obtain a higher EQE by simply having a thicker c-Si absorption layer, the resultant higher power supply voltage for charge transfer may increase power consumption; therefore, it is challenging

to design a high-performance silicon pixel for PDAF under near-infrared-ray incidence.

II.   PUPIL DIVISION DUE TO GE-ON-SI LAYER

The key concept of the present study for PDAF is to realize pupil division by the Ge-on-Si layer [3][4] selectively grown on the light-receiving surface (LRS) of c-Si, as shown in Fig. 1 and Table 1; the key concept utilizes the characteristic that absorption coefficient of germanium is much larger than that of silicon. Due to the ~0.8 eV direct bandgap of Ge, compared to Si, a stronger absorption manifests itself with cut-off wavelength up to 1.6 μm. The Ge film thickness required for a strong interaction with near infrared light is usually more than 200 nm [3]; therefore, the Ge film thickness is set to 200 nm in the present study.

The present pixel is based on a back-side illuminated CMOS image sensor pixel with one photo diode for visible light. The pixel pitch is 1.85 μm and the thickness of c-Si is around 3 μm. In order to suppress crosstalk, a deep trench isolation (DTI) is provided on the pixel boundary. The Ge-on-Si layer is formed by anisotropically etching the c-Si on the LRS to depth of 200 nm and then filling the trench with a crystalline Ge layer so that the Si and Ge layers on the LRS become uniform.

Table 1 summarizes the investigated pixels for PDAF. In the proposed pixel A with a red color filter (CF), we have placed the Ge-on-Si layer selectively so that light with incident angles of +0° to +30° strikes the Ge-on-Si layer. Moreover, we have formed no metal LSF for pupil division, which avoids flare and ghost arising from the LSF; therefore, the present technique is different from the prior art categorized into three types: half-shielding pixel [1], one pixel having two photo diodes [5] and two pixels sharing a micro lens [6].

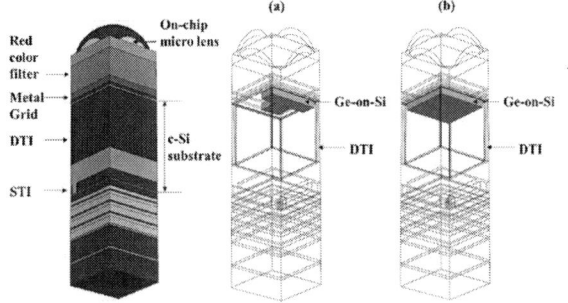

Figure 1. View from below of the present pixel for PDAF. Only the Ge-on-Si layer in (a) pixel A and (b) B (shown in Table 1) are displayed, respectively.

Table 1. Investigated pixels for PDAF.

pixel name		Cross section at metal grid		Cross section at light receiving surface	
A	Ge-on-Si arranged in the shape of T w/o LSF (proposed)	Red Color filter — Metal grid ►		Crystalline silicon DTI	Ge-on-Si unit: μm
B	Ge-on-Si w/ LSF	Light shielding film (LSF)		DTI	Ge-on-Si unit: μm
C	No Ge-on-Si w/ LSF (reference)	Light shielding film (LSF)		Crystalline silicon No DTI	
D	No Ge-on-Si w/o LSF (adjacent to the left side of the pixel A)	Green Color filter Metal grid ►		Crystalline silicon DTI	
E	Same as Pixel A except Ge-on-Si layer	Red Color filter Metal grid ►		Crystalline silicon DTI ►	

The proposed pixel A has neither a metal grid (MG) nor a DTI at the boundary with an adjacent pixel on the left; we omitted them to enhance EQE at an incident angle of +30°. The configuration of the pixels adjacent to the pixel A is the same except the on-chip CF; the DTI and MG of the adjacent pixels are arranged symmetrically with the pixel A in the Bayer format, but the no Ge-on-Si are placed, shown as pixel D in Table 1. The pixel D with the on-chip green CF is used for correcting EQE ratio and contrast of the pixel A, which are relevant to the accuracy of AF.

In the pixel B with half metal-shielded aperture, the Ge-on-Si is grown to cover most region of the LRS. Reference pixel C is a current state-of-the-art pixel for PDAF [1]. The pixel E is same as the pixel A except for the Ge-on-Si layer.

III. RESULTS AND DISCUSSION

We calculated the optical wave propagation in the pixels by rigorously solving the Maxwell's equations, using a three-dimensional finite difference time domain (FDTD) algorithm [7]; we considered both TE- and TM-polarized light and set the periodic boundary condition corresponding to the Bayer format.

Figure 2 displays simulated angular response of power flux density distribution in the pixels (a) A and (b) C, respectively: the distribution under the propagation of TM-polarized light at the wavelength of 850 nm. Fig. 2 (c) illustrates the direction of TM polarization at normal incidence (+0°) for upper figures of Fig. 2 (a) and (b).

In the pixel A, when the incident angle is +10° to +30°, the incident light is absorbed mainly by the Ge-on-Si layer, and the light passing through the Ge-on-Si layer is attenuated and distributed in the Si layer; the pixel A exhibits smaller spread of the distribution for the same incident angles in comparison with the reference pixel C with the half metal-shielded aperture; this is remarkable in the comparison of power flux density distribution in LRS. The light transmitted to the wiring layer of the pixel A is smaller than that of the pixel C for the incident angle of 0° to +20°.

Figure 3 shows that the simulated angular response of the EQE behaves differently depending on the polarization; in the pixel E, reflection occurs on the sidewall of the MG for incident angles of +20° to +30°, so the difference in the EQE due to the polarization is clearly observed: the closer the angle of incidence on the sidewall approaches the Brewster angle, the greater the difference in the EQE due to the

(a) pixel A (proposed) w/o metal LSF

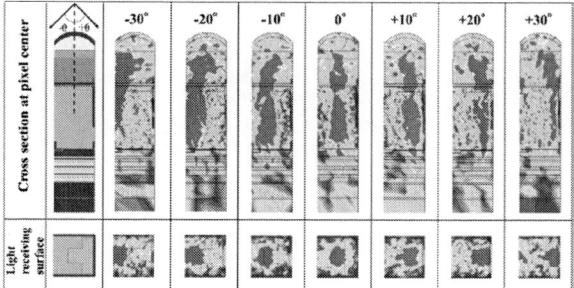

(b) pixel C (reference) w/ half metal-shielded aperture

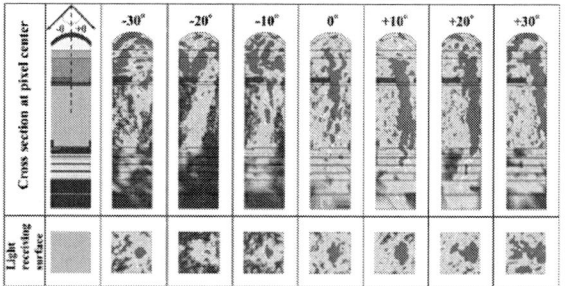

(c)

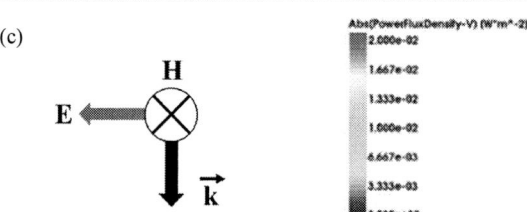

Figure 2. Simulated angular response of power flux density distribution by the propagation of TM-polarized light at wavelength of 850 nm in the cross section at the pixel center (upper figure) and at the light receiving surface (lower figure) of (a) the proposed pixel A, (b) the reference pixel C, respectively; (c) direction of TM polarization at normal incidence (+ 0°) for upper figures

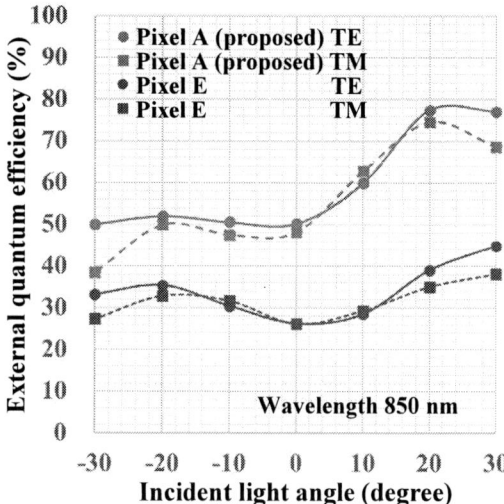

Figure 3. Simulated angular response of external quantum efficiency at wavelength of 850 nm.

polarization. Thus, the average values of the EQE by the propagation of TE- and TM-polarized light for each incident angle are used in the following graphs for simplicity.

Figure 4 displays simulated angular response of the EQE for the pixels shown in Table 1. The EQE of the proposed pixel A exhibits above 44.3 % with the maximum of 76.0 %. The EQE increases from the incident angle of +0° to +30° because the light hits the Ge-on-Si layer and is absorbed.

On the other hand, the reference pixel C exhibits the EQE of 4.9 to 23.0 %. The pixel A exhibits 3.6 times improvement in the EQE at normal incidence (+0°) compared to that of the pixel C; the EQE is 49.2 % and 13.8 %, respectively. The EQE in the pixel B is higher by 5.6 to 22.5 points compared to that in the reference pixel C because of the Ge-on-Si layer.

Even though half of the pixels are shielded by a metal film, the EQE of pixels B and C is larger than 10.5 % and 4.9 %, respectively, at incident angle of -30° to -10°; this is due to light diffraction.

The EQE of the pixel D is lower than that of the pixel E due to the difference in color filters. The curve of the EQE of the pixel E is smallest at the incident angle of 0° because the optical path length in the c-Si layer is the shortest.

The difference in EQE of the pixel E at the incident angle of ±30° is due to the presence or absence of DTI at the pixel boundary; it is suggested that the crosstalk to the adjacent pixels is different. The same is true for the pixel D.

Figure 5 shows the curves of the EQE gain by the selectively arranged Ge-on-Si layer; the EQE gain was calculated by taking the difference between the EQEs of pixels (A and D) or (A and E). The difference in the EQE between pixels A and D shows the EQE gains of 29.6 to 56.5 points for incident angles of +0° to +30°. On the other hand, the difference in the EQE between pixels A and E shows the EQE gains of 23.0 to 39.0 points for incident angles of +0° to +30°.

In addition, the curves of the EQE gain behave in the same manner as the EQE curve of the reference pixel C; they behave similarly to the reference curve in that the curve from the incident angle of -20° to +20° increases monotonically. Therefore, that has numerically confirmed the validity of the pupil division due to the selectively arranged Ge-on-Si layer and the differential operation.

Furthermore, the characteristic curves of EQE gain show EQE values higher than that of the reference pixel C. This suggests that the pupil division works effectively even in a dark environment where the reference pixel C does not operate.

Figure 6 displays simulated EQE as a function of reflectance. In this figure, the horizontal axis of Fig. 4 is changed from the incident angle to the reflectance; the purpose is to clarify the influence of the reflectivity of the Ge-on-Si layer on the EQE.

The optical reflection loss in the pixels A, D and E with no LSF is below 24.0 %; and the EQE of those pixels is more than 16.4 %. The EQE of the pixel A with the Ge-on-Si layer is higher than that of the pixel E despite a slightly higher reflectance on average; therefore, the selectively arranged Ge-on-Si layer is the main factor for the EQE enhancement. The same is true for the comparison of the pixels B and C which have half metal-shielded aperture.

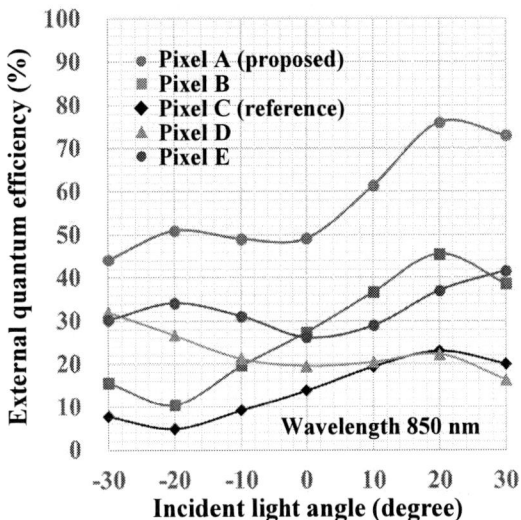

Figure 4. Simulated angular response of external quantum efficiency at wavelength of 850 nm.

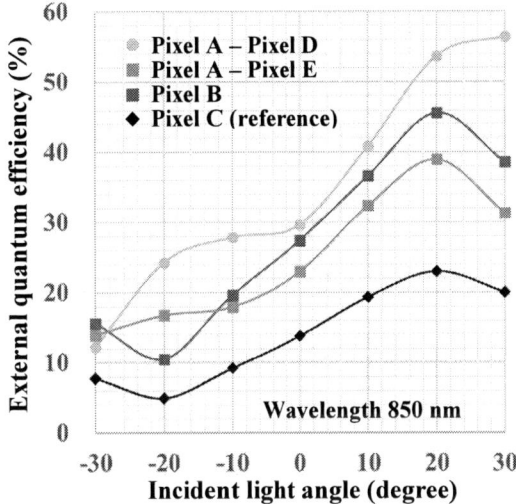

Figure 5. Evaluation of the simulated EQE gain due to the Ge-on-Si layer.

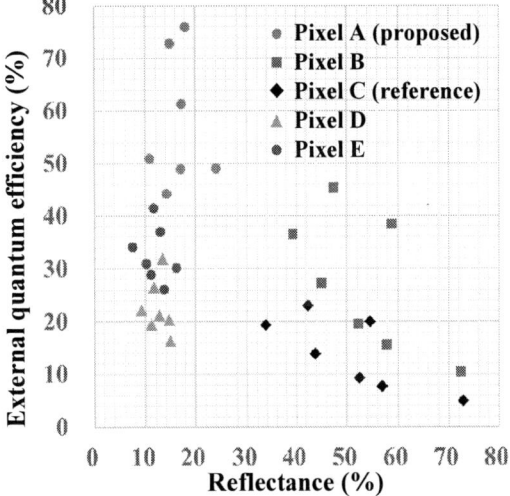

Figure 6. Simulated effective quantum efficiency as a function of reflectance at wavelength of 850 nm.

Figures 7 and 8 show simulated angular response of normalized EQE ratio and the Michaelson contrast, respectively. The Michaelson contrast is defined by the following equation:

$$\text{Michaelson contrast} = \left| \frac{\Psi_{+\theta} - \Psi_{-\theta}}{\Psi_{+\theta} + \Psi_{-\theta}} \right|$$

$\Psi_{+\theta}$: EQE at incident light angle $+\theta$

$\Psi_{-\theta}$: EQE at incident light angle $-\theta$

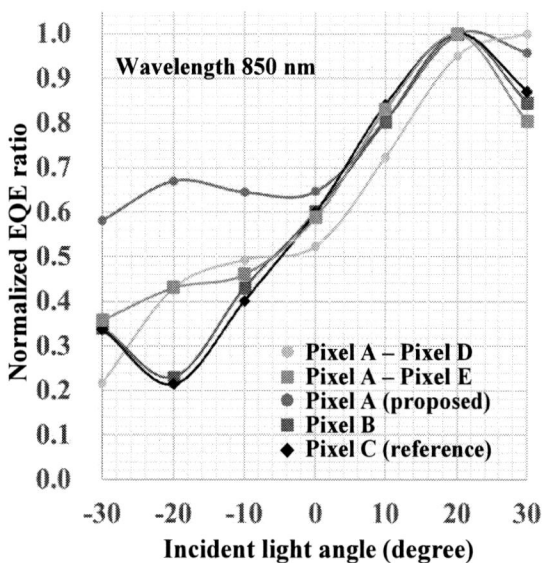

Figure 7. Simulated angular response of normalized external quantum efficiency ratio at wavelength of 850 nm.

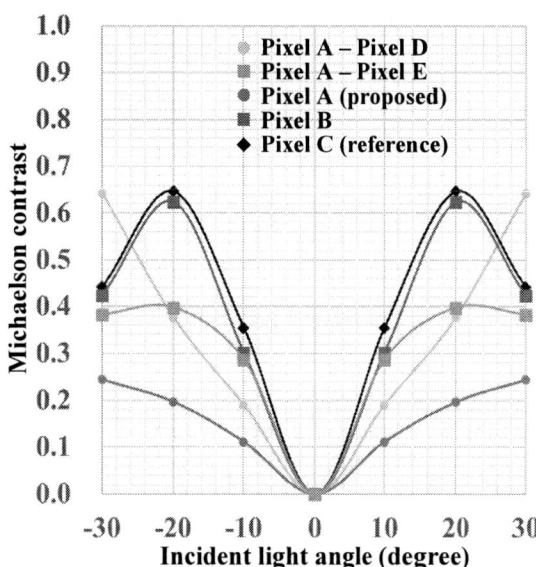

Figure 8. Simulated angular response of Michaelson contrast at wavelength of 850 nm.

The characteristics of the EQE ratio and the contrast of the pixel A deviates from those of the reference pixel C; especially, there is a deviation in the curve from the incident angle of -30° to -10°. This is because there is no light-shielding film that covers half the pixels. To correct the deviation, the EQE ratio and the contrast subjected to the same differential operation as that in Fig. 5 were executed. By taking the difference, the characteristic curve of the EQE gain due to the Ge-on-Si layer locally formed on the LRS can be obtained. The curves based on the difference operation behave in the same manner as the reference curves, showing that the same effect as the pupil division by LSF is obtained by the selectively arranged Ge-on-Si layer and the differential operation.

Conclusions

A novel PDAF technique for incident 850 nm plane wave is demonstrated using the Ge-on-Si layer, which are locally arranged on light receiving surface (LRS) of crystalline silicon (c-Si). No metal light shielding film for pupil division is formed. The key concept of the present study for PDAF is to perform the pupil division by the locally arranged Ge-on-Si layer in a pixel according to incident angle. The present pixel is based on a back-side illuminated CMOS image sensor pixel; the pixel pitch is 1.85 μm and the thickness of c-Si is around 3 μm. The simulation, based on three-dimensional finite difference time domain (3D-FDTD) method, shows that the external quantum efficiency (EQE) of the present pixel exhibits above 44.3 % with the maximum of 76.0 % for incident angles of -30° to +30°, owing to the selectively arranged Ge-on-Si layer; it exhibits 3.6 times improvement in the EQE at normal incidence compared to that of current state-of-the-art pixel with half metal-shielded aperture; the EQE is 49.2 % and 13.8 %, respectively. By taking the difference from the EQE of the similar pixel except for the Ge-on-Si layer, the characteristic curve of the present pixel behaves in the same manner as that of the current state-of-the-art pixel; therefore, that has numerically confirmed the validity of the pupil division due to the selectively arranged Ge-on-Si layer and the differential operation.

REFERENCES

[1] R. Fontaine, "The State-of-the-Art of Mainstream CMOS Image Sensors," Proc. of International Image Sensor Workshop (IISW), 2015.

[2] M. A. Green, "Self-consistent optical parameters of intrinsic silicon at 300 K including temperature coefficients", Solar Energy Materials and Solar Cells, vol. 92, pp. 1305–1310, 2008.

[3] J. Michel, J. Liu, L. C. Kimerling, "High-performance Ge-on-Si photodetectors," Nature Photonics, 4, pp. 527-534, 2010.

[4] N. Na, S.-L. Cheng, H.-D. Liu, M.-J. Yang, C.-Y. Chen, H.-W. Chen, Y.-T. Chou, C.-T. Lin, W.-H. Liu, C.-F. Liang, C.-L. Chen, S.-W. Chu, B.-J. Chen, Y.-F. Lyu, S.-L. Chen, "High-Performance Germanium-on-Silicon Lock-in Pixels for Indirect Time-of-Flight Applications," Tech. Dig. of IEEE International Electron Devices Meeting (IEDM), pp. 751-754, 2018.

[5] M. Kobayashi, M. Johnson, Y. Wada, H. Tsuboi, H. Takada, K. Togo, T. Kishi, H. Takahashi, T. Ichikawa, S. Inoue, "A Low Noise and High Sensitivity Image Sensor with Imaging and Phase-Difference Detection AF in All Pixels," ITE Trans. on Media Technology and Applications (MTA), Vol.4, No. 2, pp. 123-128, 2016.

[6] WIPO Pub. No. WO2016/09864.

[7] Synopsys, 2016.

7-4 Investigation of the relationship between current filament movement and local heat generation in IGBTs by using modified avalanche model of TCAD

Takeshi Suwa

Toshiba Electronic Devices & Storage Corporation, Kawasaki, Japan
Email:takeshi.suwa@glb.toshiba.co.jp

Abstract—For development of high voltage power devices, it is very important to understand local heat generation phenomena of current filaments especially for reliability designs. Current filaments mean high density currents flow only in some parts of active cells and induce large heat generation locally. They appear when excessive current flows for some reasons during device switching. The aim of this paper is to clarify the following by using a modified avalanche model: The local lattice temperature dependence of impact ionization coefficients is a main factor in current filament movements, and the movements significantly suppress local heat generation. In particular, this tendency becomes even stronger when the ambient temperature is low and after the depletion layer reaches the buffer layer on the back surface side of IGBTs.

Keywords—IGBT, current filament, self-heating, avalanche model, Impact Ionization model

I. INTRODUCTION

Since power semiconductor devices such as IGBTs include instability [1], currents may not flow evenly throughout the device under certain conditions. If the currents locally flow and the current density increases, the local lattice temperature of silicon increases mainly due to Joule heat (self-heating), and the risk that the materials of the device melt and break increases. In other words, even under the condition that there is no problem in terms of energy if currents flow uniformly, there is a possibility of thermal breakdown for local and large current density. When developing a device, it is necessary to prevent local heat generation within the guaranteed operating range or to design so that the device will not be destroyed even if it occurs.

The phenomenon in which large local currents flow is called current crowding and creates local hot spots [2][3]. There are two types of current concentration: one in which a large current flows only at structural defects in termination regions or cell regions, and the other in which currents in cell regions gather and flow only in a part of cell regions. The latter, so-called current filaments, is due to device instability and is associated with snapback characteristics. The current filaments may travel within the cell area, hit the termination area, or be trapped in the defect area. The phenomena has been investigated by both actual measurements [4][5] and Technology CAD (TCAD) simulations [6][7].

If it is necessary to take care of short-circuit, overcurrent turn-off and unclamped inductive switching (UIS) phenomena in device developments, the current filament should be given special attention as it is directly related to device destruction. This work focuses on the current filament during overcurrent turn-off phenomenon [8]. The overcurrent turn-off phenomenon is that when IGBT turns off, a current several times the rated current flows for some reasons. Generally, even in that case, if currents flow uniformly through the whole device, the device is not thermally destroyed.

Some current filaments may be formed depending on the device operating conditions, and may freely move three-dimensionally in the cell region and interfere with each other. Since it is difficult to actually observe how they interfere with each other, it is common to investigate them by simulations. If current filaments become pinned or their movements are restricted for some reasons, parasitic NPN structures on the surface region of cells will turn on and the device will be destroyed [8]. Conversely, it is also used to inspect unacceptable defects in the device structure [2].

TCAD simulations of the current filament are performed with a structure in which several tens to several hundreds of two-dimensional cell structures are connected and arranged (so-called multi-cell structure) [9], a structure in which a termination structure is added, or structures in which they are combined in a circuit. Depending on the purpose of the simulation, a defect model may be included in a part of the structure, or a multi-cell structure may be formed in three dimensions. Normally, it is necessary to perform calculation in consideration of self-heating, and set thermal boundary conditions on the front and back sides of the device and boundary conditions on the lateral surface of the device. If the lattice temperature of silicon reaches the melting temperature of silicon even in part, or if the parasitic NPN structure is apparently turned on and feedback is applied [8], it is estimated that the device has been destroyed in TCAD simulation. Therefore, the boundary conditions are very important in current filament simulations and should be determined to reproduce the tendency of operating conditions when the actual device fails.

The trigger for current filamentation in actual devices is subtle differences in snapback characteristics due to subtle structural differences between adjacent cell structures. On the other hand, in TCAD simulation, it is caused by subtle differences in the mesh structures and numerical calculations. Although there are differences in triggers, it is common to generate current filaments in TCAD simulation to investigate the relationship between the behavior of current filaments and the device destruction, and utilize the results for actual device development. However, it is desirable that the gate voltage and lattice temperature dependences of the snapback characteristic in TCAD simulation are almost the same as the actual device.

The movement of the current filament in the device is largely influenced by the lateral electric field caused by the imbalance of the carrier density on the left and right sides of the filament and the difference in the impact ionization coefficient due to the local difference in the lattice temperature [2]. The degree of their influence strongly depends on the circuit conditions and the ambient temperature.

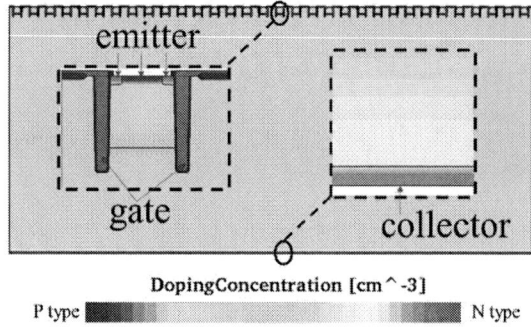

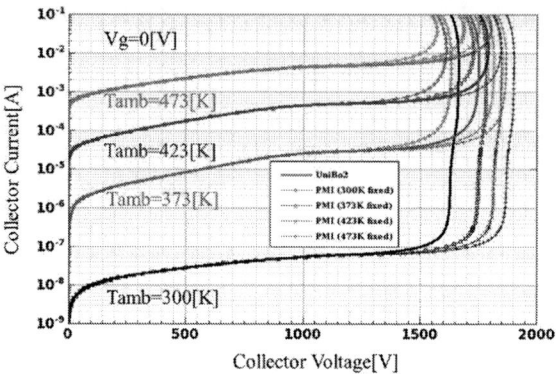

Fig. 1: Schematic view of the simulated structure (IGBT 32cells). The two figures surrounded by dotted lines are enlarged views of the front side surface and the back side surface.

Fig. 2: DC-like breakdown characteristics calculated at each ambient temperature Tamb for gate-emitter voltage Vge=0(V) with UniBo2 model (lines) and modified avalanche model (PMI model, lines with symbol).

II. TARGETS AND SIMULATION APPROCH

In this study, overcurrent turn-off phenomena are simulated by Synopsys TCAD with thermodynamic model and periodic boundary condition for lateral sides. The thermal boundary conditions on the surface and back sides of the device are those that correlate well with the tendency of the breakdown strength of the actual device. The simulated structure has 32 cells arranged side by side as shown in Fig. 1. As the impact ionization model, "New University of Bologna Impact Ionization Model" (UniBo2) [10] is selected, which can reproduce the effect on the electrical characteristics even at high lattice temperatures. The model reads:

$$\alpha(F_{ava}, T) = \frac{F_{ava}}{a(T) + b(T)exp\left[\frac{d(T)}{F_{ava} + c(T)}\right]}$$

where the coefficients a, b, c, and d are polynomials of local lattice temperature T and F_{ava} is the driving force for impact ionization. When current filamentation occurs, the impact ionization coefficients in the current filament region and the impact ionization coefficients in the other regions become significantly different according to this equation. This is because the current filament has high carrier densities and generates a large amount of self-heating.

Furthermore, a modified UniBo2 model is implemented by using Physical Model Interface (PMI) [11] which is an application interface of Sentaurus™ Device. This model is basically UniBo2 model, however the lattice temperature variables T are always forced to be equal to the ambient temperature, and impact ionization coefficients do not depend on the local lattice temperature changes (hereinafter called PMI model). By comparing the overcurrent turn-off results calculated with each model, it is possible to clarify how much the local lattice temperature dependence of the impact ionization coefficients contributes to the current filament movements and local heat generation.

Fig. 2 shows DC-like breakdown characteristics calculated at each ambient temperature Tamb for gate-emitter voltage Vge=0(V) with the above two avalanche models. From the figure, leakage currents are fixed at each Tamb, and the results of the breakdown voltage for each PMI model are

close to the breakdown voltage of the original UniBo2 model according to the fixed lattice temperature. Since the leakage currents increase and the seed currents increase as Tamb rises, the breakdown voltage of each PMI model is slightly smaller than the breakdown voltage of UniBo2 model. From these calculation results, it can be confirmed that the PMI model is implemented as intended. Because a self-heating model is not taken into consideration in this calculation, the results of the UniBo2 model at each Tamb and the results of the PMI model in which the lattice temperature variables T are fixed at the corresponding Tamb are in agreement. Therefore, they are not shown in the figure.

III. RESULTS AND DISCUSSION

Fig. 3 shows simulated overcurrent turn-off waveforms at gate resistance Rg=1(Ω) and Tamb=300(K). From Fig. 3(a), when Vge decreases and reaches about a threshold voltage, the emitter electron current |Ie_electron| becomes almost zero. In this case, the collector-emitter voltage Vce and the collector current Ic are still large, so that dynamic avalanche occurs [12][13]. Fig. 3(b) shows the relationship between the maximum value of Si lattice temperatures at each time (LTmax_Si) and the regions 1 to 4. Each region is divided according to the change of LTmax_Si waveform. The changes are caused by the current filaments generations, movements, and disappearances, which are shown in Fig. 4. From the figures, it can be seen that although current filaments are not generated in the region 2, they are being generated in the region 3. At the beginning of the region 4, thick and low-density current filaments are generated and disappear, and grow into thin and high-density filaments. As the filament moves, the mesh points of high lattice temperature, which are displayed in red, have a distribution that is dragged in the lateral direction. LTmax_Si waveforms in Fig. 3 are jagged as the hottest current filament moves between cells. As can be seen from Figs. 3 and 4, after Vce waveforms enter the plateau, that is, after the depletion layer extends and reaches the buffer layer on the back side, the jaggedness of the waveform, that is, the movement of the current filament becomes blunt.

Fig. 5 shows a comparison of overcurrent turn-off waveforms calculated with UniBo2 and PMI model. LTmax_Si with the PMI model (red line) is less jagged, and the maximum point of LTmax_Si waveform, that is, the

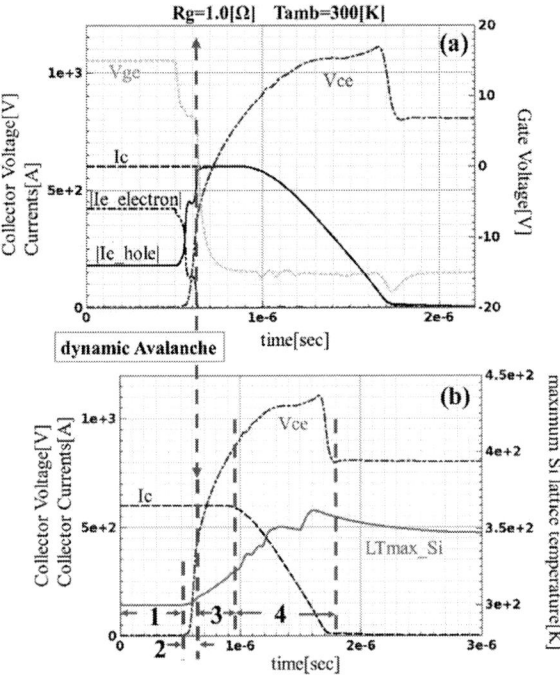

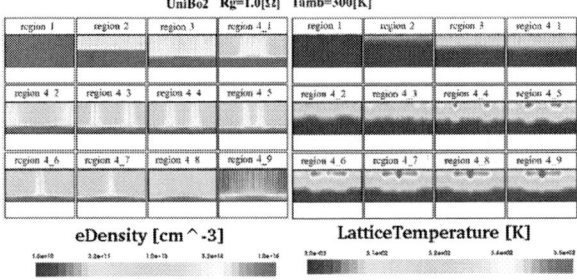

Fig. 3: Overcurrent turn-off simulation waveform calculated with UniBo2 model. The relationship between the waveforms and the cut off timing of the channel current (a) and the relationship between the maximum value of Si lattice temperatures (LTmax_Si) and the regions 1 to 4 (b). Tamb represents ambient temperature.

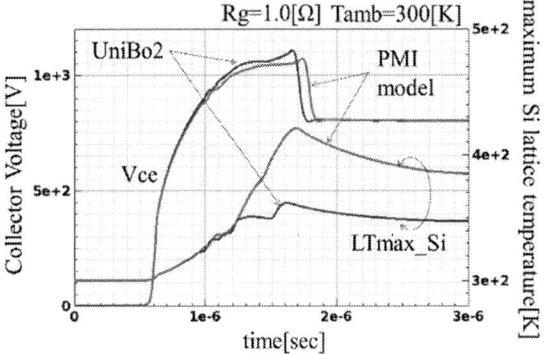

Fig. 4: Distributions of electron density (left) and lattice temperature (right) calculated with UniBo2 model corresponding to regions 1 to 4 in Fig. 3 (b). The figures for Region 4 are shown divided into approximately 9 parts in time (4_1 to 4_9).

maximum lattice temperature in whole turn-off phenomena becomes higher. This means that the local temperature dependency of the impact ionization coefficients greatly contributes to the current filament movements, and the smaller the movements, the larger the local heat generation. In other words, the local increase in the lattice temperature and the phonon scattering leads to the local decrease in the impact ionization coefficients, and as a result, the current filament movements to next cells are greatly promoted. From Fig. 6, it can be seen that the movement of the electron density and the local heat are smaller in PMI model results. As can be seen from Figs. 5 and 6, after Vce waveforms enter the plateau, that is, after the depletion layer extends and reaches the buffer layer on the back side, the difference in LTmax_Si becomes large. Therefore, in the region where the

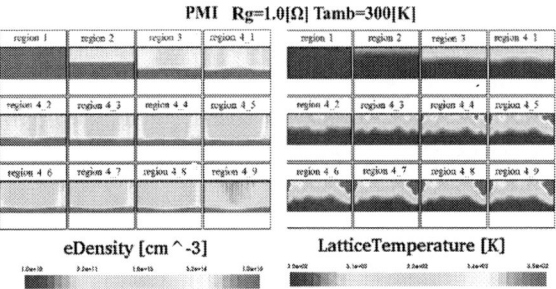

Fig. 5: Comparison of the waveforms calculated with UniBo2 model (blue lines) and an avalanche model created by using PMI (red lines). The latter model is based on UniBo2 model, but is modified to have no local lattice temperature dependence.

Fig. 6: Distributions calculated with PMI model under the same conditions as in Fig. 4.

heat generation is the highest in this phenomenon, the local lattice temperature dependence of the impact ionization coefficients has the greatest influence on the movements of current filaments and the suppression of heat generations.

Figs. 7(a) and 7(b) show Tamb dependences of overcurrent turn-off waveforms for a small and large Rg, respectively. When Rg is small and Tamb is not high, the filaments start moving at an earlier timing and suppress heat generation, however since they travel a long distance, they tend to be caught by structural defects in the actual system. This is one of the reasons that the measurements of overcurrent endurance vary greatly depending on the samples to be measured. In this case, the calculations with PMI model result in large local heat generation because the filament movements are suppressed for a long time. At Tamb=423(K), impact ionization is weak overall, therefore no current filament is generated, and the results are almost the same in both models. From Fig. 8, the hottest filament for a large Rg with PMI model does not move as if it were pinned and the local heat generation is large even for a short time. This is because under this operating condition, since current filaments are formed after Vce waveforms enter the plateau, the movements of current filaments are almost determined by the local lattice temperature dependence of the impact ionization coefficients.

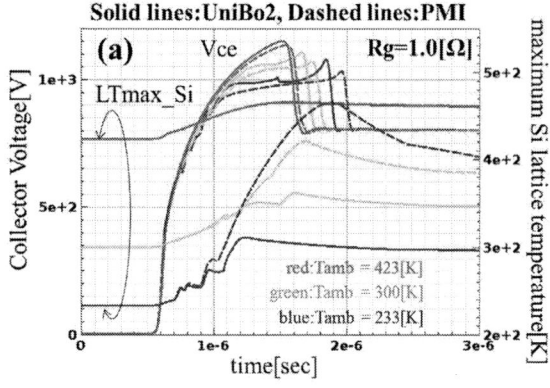

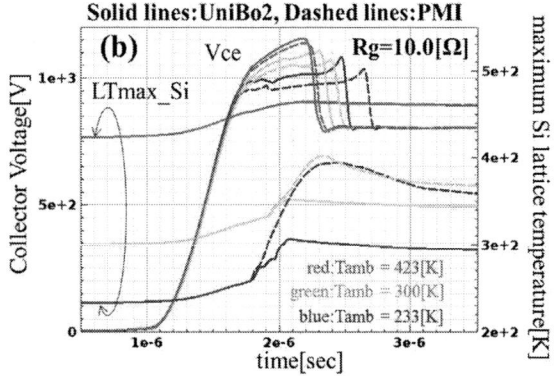

Fig. 7: Ambient temperature dependence of waveforms calculated with each model (Solid lines: UniBo2 model, dashed lines: PMI model) for (a) Rg=1.0(Ω) and (b) Rg=10.0(Ω).

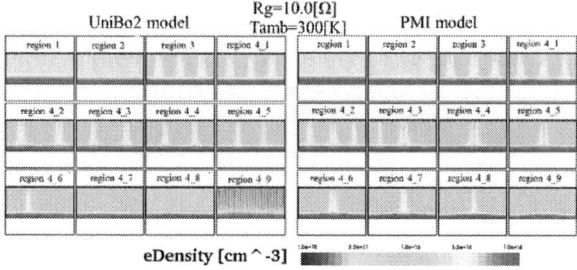

Fig. 8: Comparison of electron density distribution calculated with each model for Rg=10.0(Ω). The acquired timing of each figure is the same filaments conditions as the regions 1 to 4 in Figs. 3 and 4.

ACKNOWLEDGEMENT

I am grateful to K. Nakamura, T. Matsudai and S. Hayase for useful discussion. Company names, product names, and service names may be trademarks of their respective companies.

IV. CONCLUSION

In order to clarify that the local temperature dependence of the impact ionization coefficients is the main factor of current filament movements, the calculation results of the overcurrent turn-off phenomena using UniBo2 model were compared with the calculation results using PMI model. PMI model is based on UniBo2 model, however is a model modified by PMI so that it does not depend on local changes in the lattice temperature. Since current filament movements suppress local heat generation, if the movements are suppressed for some reasons, local heat generation will increase. The current filament movements also depends on the ambient temperature and the gate resistance, and if the ambient temperature is high to some extent, the current filaments are not generated from the beginning. If the ambient temperature is low and the gate resistance is low, current filaments will move a lot. In overcurrent turn-off phenomena, though large heat is generated after the collector voltage enters the plateau, the current filaments hardly move during this period in the calculation using PMI model, and heat generation becomes very large. In other words, the current filaments are almost driven by the difference in impact ionization coefficients due to the difference in lattice temperature between adjacent cells.

REFERENCES

[1] A. Blicher, "Field-Effect and Bipolar Power Transistaer Physics", Academic Press, 1981, pp. 186–190.

[2] T. Tamaki, Y. Yabuuchi, M. Izumi, N. Yasuhara and K. Nakamura, "Numerical study of destruction phenomena for punch-through IGBTs under unclamped inductive switching," Microelectronics Reliability, Vol. 64, September 2016, pp. 469-473.

[3] Z. Chen, K. Nakamura and T. Terashima, "LPT(II)-CSTBT™(III) for High Voltage Application with Ultra Robust Turn-off Capability Utilizing Novel Edge Termination Design," Proc. ISPSD 2012, pp. 25-28.

[4] K. Endo et al., "Direct Photo Emission Motion Observation of Current Filaments in the IGBT under Avalanche Breakdown Condition," Proc. ISPSD 2016, pp. 367-370.

[5] R. Bhojani, J. Kowalsky, J. Lutz, D. Kendig, R. Baburske, H. J. Schulze and F. J. Niedernostheide, "Observation of Current Filaments in IGBTs with Thermoreflectance Microscopy," ISPSD 2018, pp. 164-167.

[6] Y. Mizuno, R. Tagami and K. Nishikawa, "Investigateions of Inhomogeneous Operation of IGBTs under Unclamped Inductive Switching Condition," Proc. ISPSD 2010, pp. 137-140

[7] M. Tanaka and A. Nakagawa, "Simulation studies for short-circuit current crowding of MOSFET-Mode IGBT," Proc. ISPSD 2014, pp. 119-122.

[8] A. Muller-Dauch, F. Pfirsch, M. Pfaffenlehner and D. Silber, "Source Side Thermal Runaway of Trench IGBTs, Dependence on Design Aspects," ISPSD 2006, pp. 1-4.

[9] M. Tanaka and A. Nakagawa, "Growth of short-circuit current filament in MOSFET-Mode IGBTs," Proc. ISPSD 2016, pp. 319-322

[10] Sentaurus™ Device User Guide, Version Q-2019.12, Synopsys, Dec. 2019, pp.431-432.

[11] Sentaurus™ Device User Guide, Version Q-2019.12, Synopsys, Dec. 2019, pp.1049-1324.

[12] T. Ogura, H. Ninomiya, K. Sugiyama and T. Inoue, "Turn-Off Switching Analysis Considering Dynamic Avalanche Effect for Low Turn-Off Loss High-Voltage IGBTs," IEEE Transactions on Electron Devices, Vol. 51, No.4, April 2004, pp. 629-635.

[13] M. Tsukuda, I. Omura, Y. Sakiyama, M. Yamaguchi, K. Matsushita and T. Ogura, "Critical IGBT Design Regarding EMI and Switching Losses," Proc. ISPSD 2008, pp. 185-188.

Verilog-A model for avalanche dynamics and quenching in Single-Photon Avalanche Diodes

Y.Oussaiti*, D.Rideau,
J.R.Manouvrier, V.Quenette,
B.Mamdy, C.Buj, J.Grebot and
H.Wehbe-Alause
STMicroelectronics
Crolles, France
yassine.oussaiti@st.com

A.Lopez, G.Mugny, M.Agnew
and E.Lacombe
STMicroelectronics
Edinburgh, United Kingdom
alexandre.lopez@st.com

M.Pala and P.Dollfus
CNRS, *Université Paris-Saclay
Palaiseau, France
marco.pala@c2n.upsaclay.fr

Abstract— **We present a Verilog-A model accounting for the temporal avalanche buildup and its statistics in Single-Photon Avalanche Diodes (SPADs). This physics-based approach is compared to TCAD mixed-mode analyzing predictions, as well as measurements. The buildup that can be in the order of hundreds picoseconds, affects the statistical pulse width distribution, which is experimentally verified. Furthermore, we address in detail the voltage swing across the device during avalanche and its quenching, studying its impact on power consumption. This model can help a chip designer to optimize circuits for quenching the SPAD photodiode.**

Keywords— Silicon SPAD, avalanche dynamics, quenching circuits, Verilog-A code, TCAD.

I. INTRODUCTION

Single-Photon Avalanche Diodes (SPADs) are reverse-biased p-n junctions used in Geiger mode. In this regime and when a large voltage is applied, the electric field in the depletion layer becomes strong enough that the charge carriers may acquire sufficient energy to trigger an avalanche through impact ionization mechanisms. Hence, when an absorbed photon leads to a photo electron-hole pair generation, a measurable current rises swiftly and keeps flowing rendering the device insensitive to subsequent detections. Therefore, the avalanche must be stopped and the device brought to its quiescent state using ancillary readout electronics: this is the basic principle of quenching. Literature enumerates different types of quenching circuits for which a specific application is best-suited [1]. The passive version is the simplest way which consists of a high-value quench resistor connected in series to the diode (Fig.1). SPADs were developed in the late sixties [2] and related Physics reviewed by Spinelli and Lacaita [3]. Nowadays, these devices are gaining popularity in automotive industry (Light Detection and Ranging systems [4]) and time-of-flight imaging [5].

In this work, we present a Verilog-A model to assess the SPAD buildup and its statistics. For accurate behavior description, we developed a Verilog-A diode model including the underlying physics and calibrated on Technology Computer Aided-Design (TCAD) solver [6]. The buildup rising time statistics impacts the output pulse width, which is confirmed by measurements, as discussed later. Additionally, we investigate the voltage swing on the SPAD sensing node and we confirm that its value can exceed the excess bias as it has already been demonstrated in [7]. Furthermore, this voltage swing directly affects the circuit power consumption.

The paper organization is as follows. After this general introduction, the Verilog-A model incorporating the physics of avalanche is described in Section II. Moreover, the empirical method for the electric field, reproducing TCAD simulations, is outlined. Section III summarizes the main results: the current buildup statistics and its impact on the pulse width on the one hand, and the voltage swing on the sensing node affecting the power consumption, on the other hand. Then, the paper is concluded in Section IV.

II. MODEL DESCRIPTION

As stated previously, a SPAD is a transient behavioral device operating via the avalanche effect and utilizing the fact that a p-n junction can stay stable for a finite time above its breakdown voltage. A photon's arrival initiates a measurable current that increases reaching a peak value, until it is quenched using an ancillary circuit: the initial conditions are then restored. The choice of the quenching resistance R_Q and capacitance C_Q (from a modelling standpoint, we add the intrinsic capacitances of the SPAD, the sensing and quenching transistors, including also parasitic elements from circuit routing) is of a high importance, since it dictates the peak of current/voltage. We consider as a representative device a SPAD with a large collection volume as in [8]. The breakdown voltage at room temperature is of 17.1 volts.

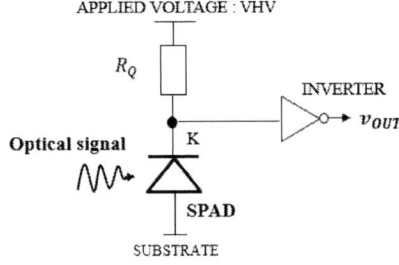

Fig. 1: The cathode terminal of the SPAD diode (sensing node) connected to a quench restor ($R_Q \sim 500\,K\Omega$). During the avalanche, a high generated current gives rise to a detectable swing of the cathode. The inverter allows a square output pulse and reflects the overall time constant of the circuit.

The currents I_e for electrons and I_h for holes follow:

$$\frac{dI_e}{dt} = [M_e . I_e + M_h . I_h - I_e]/\tau_e \quad (1)$$

$$\frac{dI_h}{dt} = [M_e . I_e + M_h . I_h - I_h]/\tau_h \quad (2)$$

Where τ_i is the transit time and M_i the multiplication factor (i represents e for electrons and h for holes). When an electron-hole pair is created, a current $I_0 = \frac{q.v_s}{w}$ (v_s is the carriers saturation velocity, q the elementary charge and w the effective multiplication region width) can initiate the avalanche process. Hence, starting from the assumption that the total current I_T is the sum of both carrier types contributions ($I_T = I_e + I_h$), which is also a succession of ionizations: $I_T = I_0(1+M+M^2+M^3+...)$, it can be written in a more familiar way as: $I_T = \frac{I_0}{1-M}$ where $M = M_e + M_h$.

Theoretically, the coefficient M_i is given by:

$$M_i = \alpha_i.w_i \quad (3)$$

It depends on w_e and w_h [9], as well as on the electric field through the expression:

$$\alpha_i = a_i.exp\left(\frac{-b_i}{F}\right) \quad (4)$$

The ionization coefficients α_i describing the carriers multiplication and increasing with the electric field, express the mean rate of ionization per unit distance and can be also obtained by the inverse of the mean distance a carrier travels before an ionizing collision. a_i and b_i are parameters measured by van Overstraeten and de Man in [10] where different values are tabulated depending on the field ranges.

Regarding previous statements, an accurate estimation of the maximum electric field variation with the diode biasing is compulsory. Even though analytical models exist in certain simple cases (e.g. abrupt junction); we used for this study an empirical method that reproduces well-calibrated TCAD predictions on more complex SPAD architectures. This approximation is based on a polynomial exploiting the low and high field values.

Fig.2 reports the electric field evolution with the applied voltage where we compare the profile obtained by Verilog-A model to TCAD simulation. As can be seen, a change in the slope is observed at ~17V, which corresponds approximately to the breakdown voltage given by TCAD. This is a not surprising effect since it occurs when the device starts to deplete its collection volume with the increasing excess bias.

Once we calculate the ionization coefficients according to equation (4), we deduce the multiplication factor from equation (3).

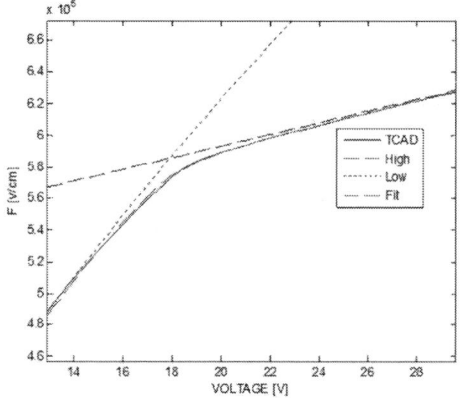

Fig. 2: The empirical model of the electric field (red dashed line) implemented in a Verilog-A code and expressed through a polynomial exploiting the high and low values of the field range. This model is calibrated on TCAD (solid).

Fig.3 illustrates M as a function of applied voltage VHV. When Adjusting the effective multiplication width for electrons at a value $w_e = 0.0736$ µm and for holes at $w_h = 0.092$ µm (fitting parameters), we obtain M=1 for a voltage $V_B = 17.1$ V (at 25°C) which corresponds to the experimental breakdown voltage. The dependence on temperature of the multiplication factor is also attested but in this paper we do not address particularly this issue.

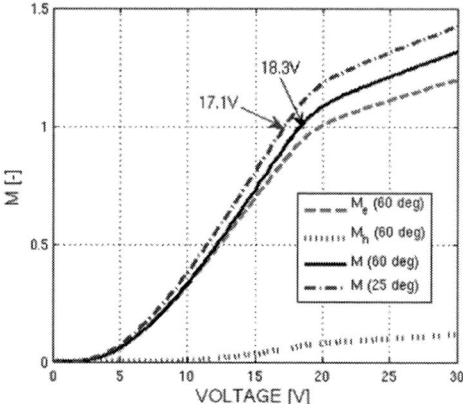

Fig. 3: The mean multiplication factor versus voltage, for two temperatures. M= 1 corresponds to V_B= 17.1 volts at 25°C.

III. SPICE SIMULATION RESULTS

A. Modelling of the avalanche buildup time

Using a Verilog-A code we describe the SPAD behavior. Studying the statistical buildup consists first in determining the conditions in which the device switches from the no-avalanche to the avalanche state (turn-on). In particular, the mechanisms that corroborate the triggering avalanche in SPADs are the photon arrival, the carrier generation and the release of some carriers after they had been captured by a trap in the depletion zone. The latter effect may cause an undesired breakdown event (the so-called afterpulsing phenomena, a major contributor to the Dark-Count Rate (DCR)).

Fig.4 (a) shows the SPAD current and voltage for various applied biasing (different curves refer to a bias increasing from V_B to V_B+5 volts) as a function of time. The buildup takes typically ~ $150ps$ at operational voltage (~20.5V) and can be even larger for lower biasing, which is in line with TCAD simulations (not shown in this paper). Furthermore, as we set a threshold current value (corresponding to the presence of a certain number of carriers in the depletion region) as a criterion for the breakdown event, in some biasing conditions the current does not build up which is also shown on the figure. This behavior occurs especially at low voltages where the electric field is not strong enough to enhance the triggering probability.

B. Description of the quenching operation

After the breakdown event, the bias voltage supply VHV exceeds the breakdown voltage V_B by an amount called the excess bias $V_{EX} = VHV - V_B$. The choice of V_{EX} is directly affecting the device performance. As the electric field increases with the excess bias, other SPAD's Figures-of-Merit -which are drived by the electric field- increases too (time-resolution or jitter, photon-detection probability (PDP), dark-count rate...etc).

Ultimately, the current reaches its maximum value and causes a voltage drop to the asymptotic steady-state mode: the SPAD is rearmed at the end of this regime. In parallel, the choice of R_Q and C_Q values defines the quenching time constant, which is a fundamental metric for the device performance as it limits the total number of detectable photons over one time period.

Fig.4 (c) shows the SPAD transient simulation using a Passive-Quenching Circuit (PQC). The anode current is plotted versus the SPAD cathode voltage. When the quenching circuit components have low values, it induces a non-quench behavior: the current is not completely brought below the triggering threshold and the device starts second avalanches.

C. The voltage swing on the sensing node

As soon as a photon impinges the SPAD surface, its the probability to trigger an avalanche is expressed through the Photon Detection Efficiency (PDE). Upon the detection, the device generates a current pulse and subsequent charges accumulate on capacitive components before they are drained out by the quenching resistance, which results in recharging the SPAD. However, the amount of accumulated charges enables a voltage swing during the avalanche on the SPAD sensing node. This issue is a new topic in the field of modelling and characterization of these devices. Reference [7] addresses in detail this phenomenon and some solutions are proposed to circumvent it, mainly through new design methodologies for preventing carriers overflow by controlling the total number of charges on the total capacitance of the SPAD sensing node. In our simulations and as can be noticed from Fig.4 (b), the voltage swing (that we denote V_{KMAX}) exceeds the excess bias by a factor ~1.4, (remember that the breakdown voltage is of 17.1volts and the excess bias is varying from 1 to 5 volts) which has also been recently reproduced in [7]. This is a drawback to minimizing the power consumption. In fact, an increase of V_{KMAX} results in a growth of the total charge of the SPAD.

Finally, Fig.4 (d) shows the output pulse v_{out} for different applied biasing. The pulse width reflects the circuit recovery time (recharge cycle).

More importantly, the pulse width distribution (Fig. 5) which can be inferred from the statistics of the buildup dynamics itself, is accounted for using a statistical distribution of the ionization rates α_i. The modelling approach reproduces accurately the measurements.

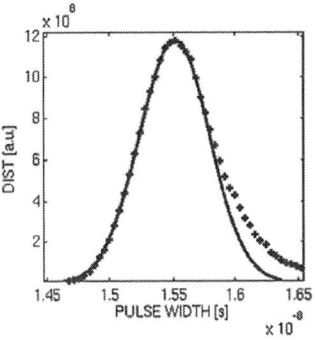

Fig. 5: The pulse width distribution. Solid curve refers to Verilog-A model; symbols to measurements (at VHV= 20.5V and T=60°C).

Furthermore, it is noted that the voltage swing V_{KMAX} affects the charge per pulse C_{PP} (reflecting the power consumption) since:

$$C_{PP} = C_K \cdot V_{KMAX} \quad (5)$$

Where C_K is the total capacitive charge in the cathode node (including parasitic). Generally and for the sake of completeness, we recall that the transient behavior takes into account the parasitic capacitances modelling, as the total charge Q (which is given by the integral of the total current) is directly proportional to parasitic capacitive contribution. Reducing the latter leads to optimizing the power consumption and the quenching time.

To investigate theoretical Verilog-A model accounting for the swing voltage on the SPAD sensing node, we compare SPICE results to the approximation: $V_{KMAX} = VHV - V_B$ (standing for the excess bias). Results are illustrated in Fig.6 and feature the regime where V_{KMAX} exceeds the excess bias. The statement is then valid.

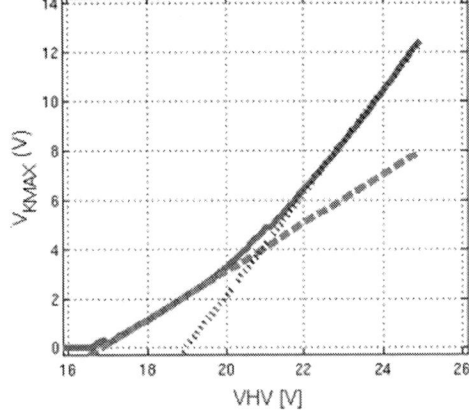

Fig. 6: (a) The swing voltage V_{KMAX} versus *VHV*. Verilog-A model (solid curve), $V_{KMAX} = VHV - V_B$ (dashed line) and the regime where V_{KMAX} exceeds the excess bias (dotted line).

(a)   (b)

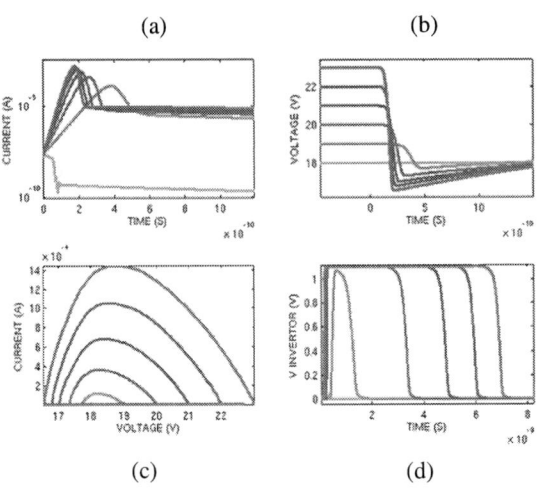

(c)   (d)

Fig. 4: (a) The current buildup evolution with applied voltage. (b) The voltage drop seen from the SPAD cathode exhibiting a swing. (c) SPAD I-V curve. The breakdown voltage is of 17.1 V (d) The output pulse v_{out}.

Fig.7 shows the charge per pulse and the pulse width as a function of VHV voltage: both not only depend on C_K, but also on the dynamics of the avalanche. It can be noticed at first sight that C_{pp} follows the two asymptotic regimes of V_{KMAX} demonstrated in Fig.6. SPICE simulation results reproduce accurately measurements. Therefore, the model can predict the power consumption on complex architectures.

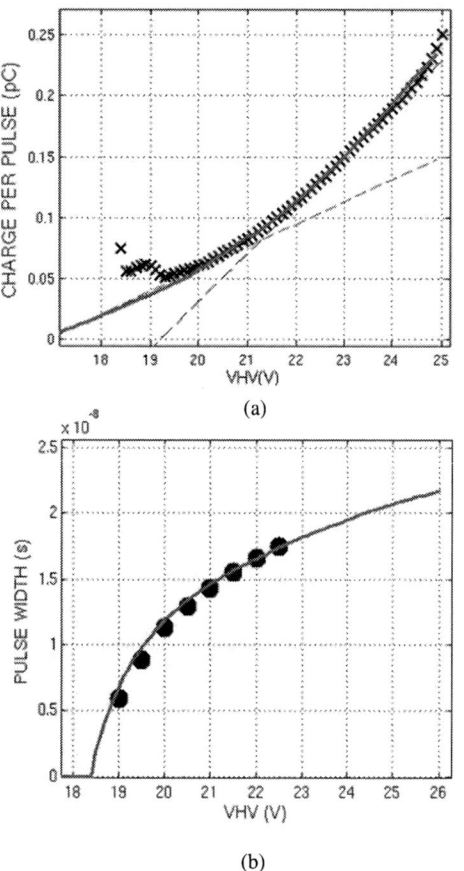

(a)

(b)

Fig. 7: (a) The charge per pulse C_{PP} and (b) the pulse width versus VHV; Verilog-A model (solid line) versus measurements (symbols).

IV. CONCLUSION

We have developed a physics-based Verilog-A model for SPADs accounting for the buildup of the avalanche and its statistics. The model compares favorably with more complex TCAD mixed-mode predictions. Throughout this study, numerical results are confronted with characterization measurements: a good agreement is shown. Moreover, this model allows SPICE simulations of realistic circuits, including quenching and sensing transistors. It provides an accurate way to estimate qualitative outputs such as the power consumption, the voltage swing of the sensing node and the SPAD recovery time, which allow a chip designer to optimize circuits.

REFERENCES

[1] S. Cova, M. Ghioni, A. Lacaita, C. Samori, and F. Zappa, "Avalanche photodiodes and quenching circuits for single photon detection," Applied Optics, vol. 35, no. 12, pp. 1956–1976, 1996.

[2] R. H. Haitz, "Model for the electrical behaviour of a microplasma," J. Appl. Phys., vol. 35, no. 5, pp. 1370–1376, May 1964.

[3] A. Spinelli and A.L. Lacaita, "Physics and numerical simulation of Single Photon Avalanche Diodes," IEEE transactions on electron devices, vol. 44, No. 11, pp. 1931-1943, November 1997

[4] I. Takai, H. Matsubara, M. Soga, M. Ohta, M. Ogawa, T. Yamashita, "Single-photon avalanche diode with enhanced NIR-sensitivity for automotive LIDAR systems", Sensors, vol. 16, no. 4, pp. 459, 2016.

[5] F. Mattioli Della Rocca, T. A. Abbas, N. A. W. Dutton and R. K. Henderson, "A high dynamic range SPAD pixel for time of flight imaging," 2017 IEEE SENSORS, Glasgow, 2017, pp. 1-3.

[6] Synopsys, Inc. Version O-2018.06, June 2018.

[7] Inoue, A., Okino, T.,Koyama, S., Hirose,Y, "Modeling and Analysis of Capacitive Relaxation Quenching in a Single Photon Avalanche Diode (SPAD) Applied to a CMOS Image Sensor", Sensors 2020, 20, 3007.

[8] C. Veerappan and E. Charbon, "A Low Dark Count p-i-n Diode Based SPAD in CMOS Technology," in IEEE Transactions on Electron Devices, vol. 63, no. 1, pp. 65-71, Jan. 2016, doi: 10.1109/TED.2015.2475355.

[9] G.J.Rees and J.P.R David, "Nonlocal impact ionization and avalanche multiplication". Journal of Physics D: Applied Physics, IOP Publishing, 2010, 43 (24), pp.243001.

[10] R. van Overstraeten and H. de Man, "Measurement of the Ionization Rates in Diffused Silicon p-n Junctions," Solid-State Electronics, vol. 13, no. 1, pp. 583–608, 1970.

A Novel Full-Band Monte Carlo Device Simulator with Real-Space Treatment of the Short-Range Coulomb Interactions for Modeling 4H-SiC Power Devices

Chi-Yin Cheng
School of Electrical, Computer and Energy Engineering
Arizona State University
Tempe, AZ 85287-5706, USA
chi-yin.cheng@asu.edu

Dragica Vasileska
School of Electrical, Computer and Energy Engineering
Arizona State University
Tempe, AZ 85287-5706, USA
vasileska@asu.edu

Abstract—In this work, we present a novel full-band Monte Carlo device simulator for modelling 4H-SiC Power electronic devices in which, for the first time, we use real-space molecular dynamics approach for the Coulomb interactions. Proper treatment of the electron-electron interactions is critical for modelling power electronic devices because of the high electron densities. In addition, because of the high applied voltages, the use of a full-band Monte Carlo device simulator is a must. The simulator has been successfully used to explain the steady-state behavior of a 3-D vertical double-diffused MOSFET (VDMOS) fabricated in the 4H-SiC technology.

Keywords—4H-SiC, full-band device simulator, vertical double-diffused MOSFET (VDMOS), Monte Carlo Method, real-space treatment of Coulomb interactions.

I. Introduction

4H-SiC has been widely used in many applications. All of these benefits come from its extremely high critical electric field and good electron mobility [1][2]. 4H-SiC possesses a critical field ten times higher than that of Si, which allows high-voltage blocking layers composed of 4H-SiC to be approximately a tenth of the thickness of a comparable Si device. This reduces the device on-resistance and power losses while maintaining the same high blocking capability [3]. To develop better 4H-SiC devices, good understanding of the transport properties of SiC material is needed and using simulations one can investigate their potential at a low cost. For example, a thorough study of the role of interface traps and self-heating effects on the operation of the SiC power VDMOS device has been performed in [4] and [5]. These authors use technology for computer-aided design (TCAD) tool Silvaco Atlas [6]. Recall that most of the 4H-SiC devices used today are for power electronic applications and are operated under high-voltage conditions. Besides this, the 4H-SiC is an anisotropic material, which features different energy–momentum dispersion profiles along different crystallographic directions. Hence, the drift-diffusion model cannot describe the high-energy behavior of the electron transport precisely. Also, considering the high electron densities in some applications, the impurity scattering issue must be treated properly. In this work, the Coulomb scattering is included via the real-space molecular dynamics (MD) approach. This includes the electron–electron and electron–ion interactions. The molecular dynamics approach is coupled with a full-band ensemble Monte Carlo (FBEMC) simulation framework for the first time. The approach is then used in the simulation of a vertical double-diffused metal-oxide-semiconductor field-effect transistor structure (VDMOS).

II. Theoratical Model

A. Band Structure Calculation

The first and the critical step in the development of a full-band Monte Carlo transport simulator is the calculation of the band structure. In this work, the empirical pseudopotential method is used. Ng [7] applied a genetic algorithm to find the optimal pseudopotential parameter set for the band structure calculation. The parameter set they proposed is used in this work to reproduce the band structure. The lattice constants used in these calculations are $a = 3.032$ Å and $c = 9.928$ Å, and the bond length is $L_z = 1.866$ Å. The irreducible 1st Brillouin zone is discretized into 21000 k-points and all 527 energy levels are calculated.

B. Phonon Scattering Models

Dominant scattering mechanisms in 4H-SiC material system are acoustic phonon, non-polar optical phonon, polar optical phonon and Coulomb scattering. Hjelm [8] presented a complete set of parameters describing the phonon scattering mechanisms in 4H-SiC that are used in this work. The phonon scattering rates are pre-calculated and stored to save run-time execution.

C. Impurity Scattering Model

There are two ways to incorporate Coulomb interaction in particle-based device simulations. The first approach is to include Coulomb scattering in **k**-space. There are several shortcomings to this approach. First, it is difficult to calculate Debye screening length in the case of an inhomogeneous electron gas. Second, it is only valid for binary collisions; the model ignores scattering from multiple impurities at the same time under heavy doping conditions. Third, this scattering approach model assumes that the scattering is instantaneous, and as such, cannot describe the interactions correctly in real space and time domain. Fourth, in power electronic devices electron-electron interaction play very important role and must be accounted for. For the importance of the short-range electron-electron and the electron-plasmon interactions, see the work of Fischetti [9] and Wordelman [10]. To solve most of the above-mentioned difficulties, a real-space molecular dynamics (MD) approach is preferable and is, therefore, adopted in this work. In a real-space Coulomb interaction calculation approach: (1) the non-uniformity in doping is no longer a problem; (2) simultaneous scattering from different impurities is automatically included; (3) the dynamical screening of the Coulomb interaction is also automatically included. Yet, the drawback of the real-space treatment is the calculation time and the scaling ability. The needed time increases dramatically once more particles are considered. To

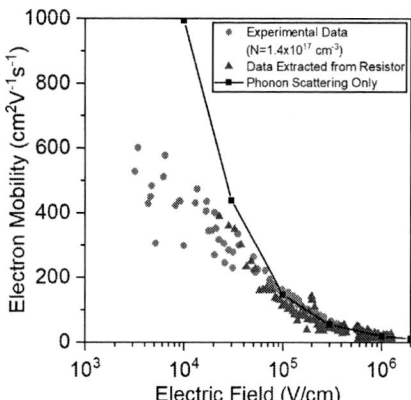

Fig. 1 Comparison of experimental and simulated electron mobility data. Simulation data are extracted from resistor simulations [13]. Also shown here is the phonon-limited mobility (solid black curve). Note that inclusion of the real-space Coulomb interactions accurately describes the field-dependent mobility, in particular at low electric fields, where the importance of this scattering mechanism is most significant. Experimental data are taken from [14].

alleviate the calculation time, the particle-particle-particle-mesh (P^3M) method is used. The concept of the P^3M method is to decompose the interparticle force into two parts: the short-range portion, which is calculated directly from the superimposed Coulomb forces between particles; and the mesh part (called also the Hartree potential), which is calculated through a particle-mesh coupling. Recall that the Coulomb force between two charges is inversely proportional to the square of the distance between them. Hence, for a particle pair, if the distance between them is larger than the threshold, the short-range force calculation can be neglected thus reducing the execution time. To avoid double counting of the mesh force in the short-range region, a reference force is taken from the exact Coulomb force. For a uniform mesh, Hockney and Eastwood [11] proposed that the reference force transitioning smoothly from short range to long range can be described by a sphere with uniformly decreasing density profile $S(r)$. Wordelman [10] modified this approach for the non-uniform mesh case. If the particle pairs are too close, the extremely high and unphysical electric fields must be modified [12]. Otherwise they lead to trapping of the particles (for oppositely charged donors and electrons in the contact region of the device). As adopted in [12], the electric field along a distance shorter than a cut-off range is modified to be linearly decreasing to zero. A cut-off range equal to 1 nm is used in this work.

The details of the methodology for real-space treatment of the Coulomb interaction can be found in [13]. The impact of the Coulomb scattering is illustrated in Fig. 1. The phonon-limited electron mobility along [0001] direction is compared to the data extracted from the example resistor and the experimental data [14]. The doping concentration of the experimental data is $N_d = 1.4 \times 10^{17}\ cm^{-3}$. From the results presented, it is obvious that the inclusion of Coulomb scattering is needed to match the experimental mobility data.

D. Surface Roughness Scattering

Surface roughness scattering at the oxide/SiC interface must be considered, since the channel affects the performance

of the device significantly. The need for incorporating this scattering process in the theoretical model is supported by the fact that in the fabrication process of 4H-SiC MOSFETs, the oxide layer is formed by the oxidation of the SiC substrate. The oxidation process leaves carbon precipitation on site, thus leading to significant interface roughness. In our work surface-roughness scattering is also included in real space. It is assumed that the electrons have 50% chance of specular reflection and 50% chance of diffusive reflection at the oxide/SiC interface (This ratio is exactly true for the Si/SiO₂ interface). The ratios chosen here are just for proof-of-concept; users can calibrate the values depending on the fabrication procedure and experimental data of the oxidation layer. If diffusive reflection occurs, the reflected angle is randomized and the momentum and the energy are changed accordingly.

III. DEVICE STRUCTURE AND SIMULATION DESCRIPTION

A 3-D VDMOS made of 4H-SiC is presented in this work and has the following parameters. The width of the device (into the paper) is only 20 nm to reduce calculation time. The device consists of a thin (40 nm) n^+ substrate layer, on top of which is grown 1.46 μm of n^- drift region. The doping of the n^- drift layer is $5 \times 10^{16}\ cm^{-3}$. The p well is composed of two parts. The p^- region provides low threshold voltage V_{th} and has doping of $1 \times 10^{17}\ cm^{-3}$; on the other hand, the p region acts as a stopper layer to prevent latch-up, and has higher doping of $5 \times 10^{17}\ cm^{-3}$. The channel length is 0.5 μm (if not mentioned elsewhere) and the oxide thickness t_{oxide} is 0.1 μm. In addition, the doping densities of the n^+ source and drain regions are set to $1 \times 10^{18}\ cm^{-3}$ to ensure good ohmic contacts. The workfunction of the gate contact is set to 4.3 eV. The source contact implemented here has two parts (sections). One section is used for grounding the p^- well; the other one is used for the current path on the top of the n^+ source region. Uniform meshing is used in the simulation of the VDMOS to simplify the Monte Carlo calculation. The charges are coupled to the mesh through the nearest-element-center (NEC) method that eliminates the self-force for the case of materials with different dielectric constants. Note that the non-ideal properties, like traps or defects in the oxide, are important for 4H-SiC devices. They are not included in the present version of the theoretical model. The deviations that they lead to are presently being investigated and the results of these investigations will be published elsewhere. Also, impact ionization, responsible for the breakdown in power converters, at present is not included in the model as well. Furthermore, modifications in the theoretical model to properly account for band crossing [15]–[17] are needed and are currently being implemented in the code.

IV. SIMULATION RESULTS

A. Electrical Profiles

The example electron and potential profiles for V_G=40 V and V_D=20 V are shown in Fig. 3. Each dot in Fig. 3(a) represents an electron. Since the short-range interparticle force is of interest, the idea of super particle is not adopted in this work.

B. I_D-V_G Transfer Characteristic

I_D-V_G transfer curve is an important characteristic when describing the operation of MOSFET device. To simulate the I_D-V_G curve, usually a small drain bias (< 0.1 V) is applied.

(a)

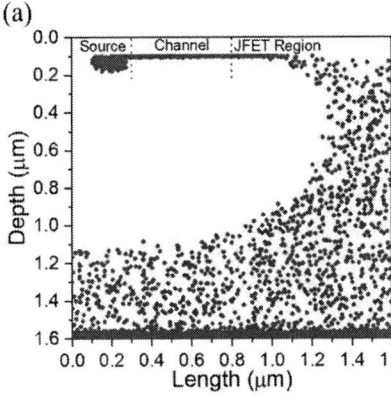

(b)

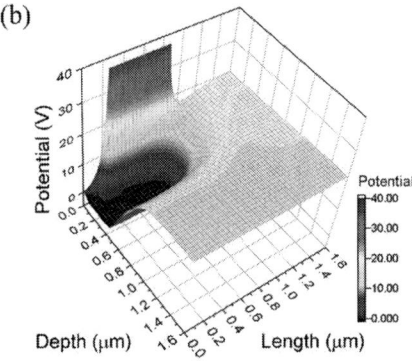

Fig. 3 Electrical profile of example condition. V_G=40 V and V_D=20 V. (a) Electron distribution. Each dot represents an electron. (b) Potential distribution.

Nevertheless, since this voltage level is too small to observe the outcomes in FBEMC device simulations, a voltage V_D=1 V is applied instead. The simulated device parameters are the same as those discussed in the previous section. Fig. 2 shows the I_D-V_G transfer curve of the example VDMOS. Each point is taken as the average of five samples to describe an average device. The threshold voltage V_{th} in this example is about 12 V. It is observed that the drain current I_D gradually gets saturated as $V_G > 20$ V. This is a typical property of a VDMOS, which is attributed to the high resistivity of the thick and low-doped n⁻ drift layer, and the parasitic resistance of the JFET region.

C. I_D-V_D Output Characteristic

Fig. 4 depicts the output I_D-V_D characteristic of the device for three gate voltages. The applied gate voltage is taken to be 20 V, 30 V and 40 V. The linear region of the VDMOS is not as obvious when compared to the regular planar MOSFET. In the regular planar MOSFET, the slope of the I_D-V_D curve in the linear region is proportional to V_G-V_{th}. However, this is not the case here because of the limitation of the high series resistances of the drift region and the JFET. Besides this, due to the voltage drop across the resistances, the actual V_D needed to drive the top channel part is less than the drain voltage V_D applied on the electrode. The downward bending of the curves also indicates the presence of the high-resistance drift region. Furthermore, after carriers pass the channel, they change their direction of motion towards the drain. But, since the gate contact is opposite to the drain contact, the increased V_G results in reduced electron velocity

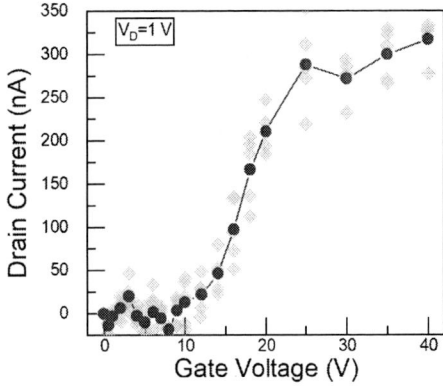

Fig. 2 I_D-V_G transfer curve of the example VDMOS. We use V_D=1 V. Each gray symbol indicates one set of simulation for different distribution of the dopant atoms in the device. (Note that instead of continuum we use atomistic description of the doping.) The blue solid line is the average of the five points of every V_G.

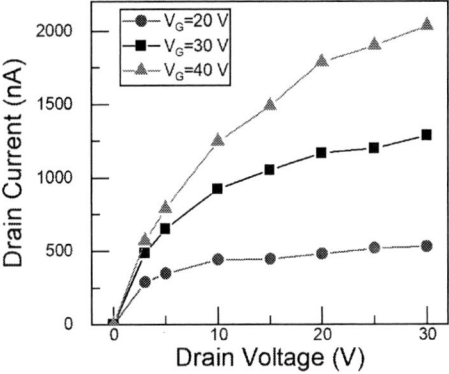

Fig. 4 I_D-V_D (output) characteristics of the example VDMOS. Channel length is 0.5 μm.

when moving toward drain side and, therefore, leads to lower drain current.

D. Channel Length Dependence

Two channel lengths have been investigated in this study. One is the default one (0.5 μm) and the second channel length is taken to be 0.2 μm. The I_D-V_D curves of the device with channel length $L_{channel}$=0.2 μm are depicted in Fig. 5. Comparing these results to the ones presented in Fig. 4, it is observed that the curves shown in Fig. 5 keep increasing with the increase of the drain voltage. Also, all three curves feature the same slope at high drain voltages. This phenomenon implies that the channel resistance is not dominating anymore. In this case, current is limited by the other resistance components.

E. Temperature Dependence

An obvious advantage of using 4H-SiC is the capability of working under high-temperature environment. Note that the intrinsic carrier density increases exponentially with temperature and is inversely exponentially dependent upon the bandgap. SiC, thanks to its wide bandgap, has intrinsic carrier density much lower when compared to Si or GaAs materials. It is so low that even at high temperature, the

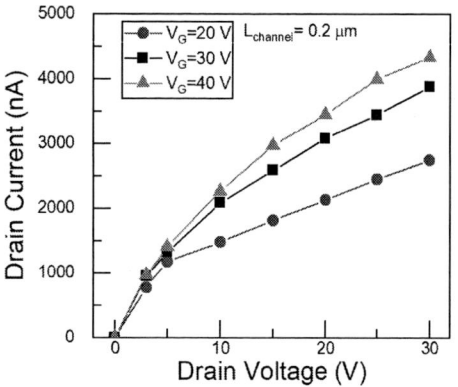

Fig. 5. I_D-V_D output characteristics of example VDMOS. The channel length is 0.2 μm.

concentration of intrinsic carriers is still far lower than of the

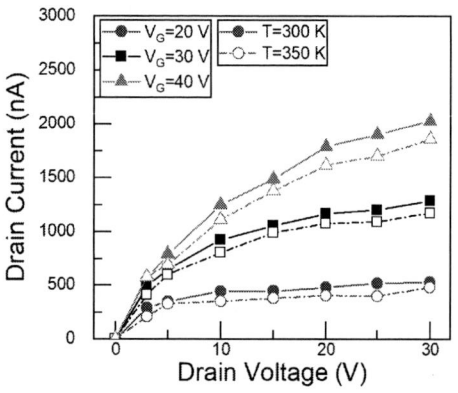

Fig. 6. I_D-V_D output characteristics of the example VDMOS. The channel length is 0.5 μm. Temperature is 300K and 350K.

extrinsic carriers. On the other hand, the concentration of intrinsic carriers in Si or GaAs at high temperature is higher than that of the extrinsic carriers. This results in the malfunction of the device and could damage partially or entirely the system. That is why the temperature dependence is an important characteristic. Fig. 6 shows the temperature dependence. As expected, the currents drop when the temperature is raised to 350 K from nominal 300 K. The degradation of the current is not only attributed to the temperature-related term in the Poisson's equation, but also to the degradation of the mobility resulted from the phonon scattering mechanisms.

V. CONCLUSION

To explore the performances of a 4H-SiC power devices more comprehensively with low cost, a tool that uses the full-band structure and real-space treatment of the electron–electron and electron–ion short-range Coulomb interactions is needed. The band structure of 4H-SiC was obtained using the EPM. The phonon scattering mechanisms included in our theoretical model are due to acoustic, nonpolar optical, and polar optical phonons. A prototypical 3-D VDMOS is simulated in this work. The simulated I_D–V_G and I_D–V_D

curves presented here prove that this device simulator is producing reliable static characteristics. The channel length and temperature dependences are also demonstrated. Since the simulator is particle-based, this provides physical insight not only on the macroscale properties but also the microscale ones. The advantages of this tool lay the foundations for future work on 4H-SiC high power devices. The results of that work will be presented elsewhere.

REFERENCES

[1] H.Matsunami and T.Kimoto, "Step-controlled epitaxial growth of SiC: High quality homoepitaxy," *Mater. Sci. Eng. R*, vol. 20, no. 3, pp. 125–166, 1997, doi: 10.1016/S0927-796X(97)00005-3.

[2] M.Bhatnagar and B. J.Baliga, "Power Devices," vol. 40, no. 3, 1993.

[3] R.Pérez, D.Tournier, A.Pérez-Tomas, P.Godignon, N.Mestres, and J.Millán, "Planar edge termination design and technology considerations for 1.7-kV 4H-SiC PiN diodes," *IEEE Trans. Electron Devices*, vol. 52, no. 10, pp. 2309–2316, 2005, doi: 10.1109/TED.2005.856805.

[4] M. H.Alqaysi, A.Martinez, K.Ahmeda, B.Ubochi, and K.Kalna, "Impact of interface traps/defects and selfheating on the degradation of performance of a 4H-SiC VDMOSFET," *IET Power Electron.*, vol. 12, no. 11, pp. 2731–2740, 2019, doi: 10.1049/iet-pel.2018.5897.

[5] G. D.Licciardo, S.Bellone, and L. DiBenedetto, "Analytical Model of the Forward Operation of 4H-SiC Vertical DMOSFET in the Safe Operating Temperature Range," *IEEE Trans. Power Electron.*, vol. 30, no. 10, pp. 5800–5809, 2015, doi: 10.1109/TPEL.2014.2376778.

[6] Silvaco TCAD. Available: https://www.silvaco.com/.

[7] G.Ng, D.Vasileska, and D. K.Schroder, "Empirical pseudopotential band structure parameters of 4H-SiC using a genetic algorithm fitting routine," *Superlattices Microstruct.*, vol. 49, no. 1, pp. 109–115, 2011, doi: 10.1016/j.spmi.2010.11.009.

[8] M.Hjelm, H. E.Nilsson, A.Martinez, K. F.Brennan, and E.Bellotti, "Monte Carlo study of high-field carrier transport in 4H-SiC including band-to-band tunneling," *J. Appl. Phys.*, vol. 93, no. 2, pp. 1099–1107, 2003, doi: 10.1063/1.1530712.

[9] M.V.Fischetti, "Effect of the electron-plasmon interaction on the electron mobility in silicon," *Phys. Rev. B. Condens. Matter*, vol. 44, no. 11, p. 5527, 1991, doi: 10.1103/PhysRevB.44.5527.

[10] C. J.Wordelman and U.Ravaioli, "Integration of a particle-particle-particle-mesh algorithm with the ensemble Monte Carlo method for the simulation of ultra-small semiconductor devices," *IEEE Trans. Electron Devices*, vol. 47, no. 2, pp. 410–416, 2000, doi: 10.1109/16.822288.

[11] R. W.Hockney and J. W.Eastwood, *Computer simulation using particles*. New York: McGraw-Hill International Book Co., 1981.

[12] W. J.Gross, D.Vasileska, and D. K.Ferry, "Ultrasmall MOSFETs: the importance of the full Coulomb interaction on device characteristics," *IEEE Trans. Electron Devices*, vol. 47, no. 10, pp. 1831–1837, 2000, doi: 10.1109/16.870556.

[13] C.-Y.Cheng and D.Vasileska, "Electron transport analysis of 4H-SiC with full-band Monte Carlo simulation including real-space Coulomb interactions," *J. Appl. Phys.*, vol. 127, no. 15, p. 155702, 2020, doi: 10.1063/1.5144214.

[14] I. A.Khan and J. A.Cooper, "Measurement of High-Field Electron Transport in Silicon Carbide," *IEEE Trans. Electron Devices*, vol. 47, no. 2, pp. 269–273, 2000, doi: 10.1109/16.822266.

[15] J. B.Krieger and G. J.Iafrate, "Time evolution of Bloch electrons in a homogeneous electric field," *Phys. Rev. B*, vol. 33, no. 8, pp. 5494–5500, Apr.1986, doi: 10.1103/PhysRevB.33.5494.

[16] H. E.Nilsson, A.Martinez, and U.Sannemo, "Numerical study of Bloch electron dynamics in wide band-gap semiconductors," *Appl. Surf. Sci.*, vol. 184, no. 1–4, pp. 199–203, 2001, doi: 10.1016/S0169-4332(01)00503-7.

[17] R.Hathwar, M.Saraniti, and S. M.Goodnick, "Modeling of multi-band drift in nanowires using a full band Monte Carlo simulation," *J. Appl. Phys.*, vol. 120, no. 4, 2016, doi: 10.1063/1.4959881.

Tight-binding simulation of optical gain in h-BCN for laser application

Daisuke Maki, Matsuto Ogawa and Satofumi Souma[†]

Department of Electrical and Electronic Engineering, Kobe University, Kobe 657-8501, Japan

[†]email: ssouma@harbor.kobe-u.ac.jp

Abstract—We present a numerical study on the optical gain in semiconductor laser structure with h-BCN as an active layer. By using the tight-binding method along with the drift-diffusion-Poisson equations, we analyze the optical gain spectra for various band gap energies in h-BCN, demonstrating that the largest gain peak of h-BCN is in the mid-infrared region and can be significantly greater than that in the case of conventional semiconductor active layer.

I. INTRODUCTION

In the research field of semiconductor lasers, there are several important key challenges, such as lower power operation and the wider range of wavelength controllability [1]. Especially semiconductor lasers in mid-infrared region is important in various applications such as defense, industry, communication including IoT, medical care, and healthcare. From the material view point, on the other hand, two-dimensional materials have attracted significant attention owing to their extraordinary electronic properties, promising for various applications in optoelectronics. While graphene is a typical 2D material with a planar honeycomb lattice structure with carbon atoms, its zero bandgap nature is a serious limitation in its optoelectronic applications [2]. For the optoelectronic application of a 2D material, materials with a controllable bandgap is desired. One of the possible strategies is to construct a honeycomb lattice structure consisting of C, B and N atoms [3]. Recently, such material (h-BCN) has been successfully synthesized experimentally [4]. With such background, we explore the possibility of using h-BCN as an active layer material for semiconductor lasers, with the special attention to the gain and wavelength controllability by changing the BN composition x in h-$(BN)_x C_{1-x}$.

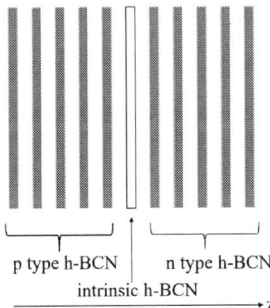

Fig. 1. Schematic illustration of h-BCN multilayer structure assumed in this study.

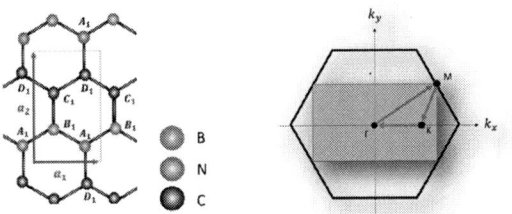

Fig. 2. (Left) Crystal structure of h-BC$_2$N . (Right) The first Brillouin zone of h-BC$_2$N (rectangular region). Hexagonal region in the right panel is the first Brillouin zone in the VCA treatment (see Fig. 4).

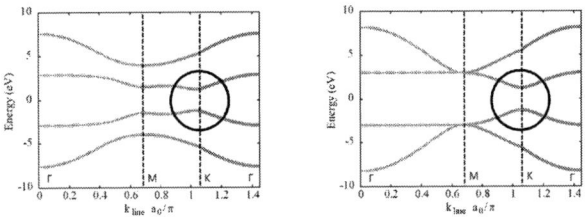

Fig. 3. (Left) Band structure of h-BC$_2$N calculated using conventional TB method. (Right) Band structure of h-BC$_2$N calculated based on VCA-TB method.

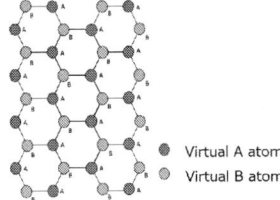

Fig. 4. Crystal structure of h-BCN assumed in VCA.

II. DEVICE STRUCTURE AND CALCULATION METHOD

A. Multilayer h-BCN structure

In this study we assume the h-BCN multilayer structure as shown in Fig. 1, where the central monolayer is intrinsic h-BCN, while the left and the right remaining layers are p- and n-doped h-BCN, respectively. Here assume that different h-BCN layers are quasi independent. That is, although there exist finite overlap of electronic wavefunction between adjacent layers so that the electronic current can flow along the layer direction, the strength of the inter-layer coupling is enough small such that the electronic band structure can be calculated approximately assuming that each layer is independent from others.

153

Such assumption can be justified in the case of turbostratic multilayer structures [5].

B. Tight-binding approximation method for h-BCN

One of the representative crystal structures of h-BCN is h-BC_2N structure shown schematically in Fig. 2 (left). The corresponding energy band structure can be calculated based on the tight-binding (TB) approximation method, where the wave function of an electron is expanded by linear combination of atomic orbitals, taking into account the overlap of atomic orbitals only between neighboring atoms. In Fig. 3 (left) we plotted the energy band structure of h-BC_2N calculated based on the TB method along the Γ, M, K, and Γ points in the first Brillouin zone (rectangular area) shown in Fig. 2 (right), where we employed TB parameters listed in Ref. [6]. In the study of electronic and optical properties, the band structure near the band edges (bottom of the conduction band and the top of the valence band) is essentially important. In the case of h-BCN, the band edges are positioned at the K point as enclosed by circle in Fig. 3 (left).

C. Validity of virtual crystal approximation

Although the above mentioned conventional TB approach is useful to predict precisely the band structures of various specific crystal structures of h-BCN where CC pairs in graphene are replaced by BN pairs in specific ways [7], [8]. On the other hand, for the purpose of understanding how the characteristics of h-BCN-based electronic/optical devices depend on the BN composition x in graphene, the use of virtual crystal approximation (VCA) is useful [9], where the unit cell is kept as the primitive unit cell of graphene (containing two atoms) and the TB parameters (on-site and hopping energies) are replaced by the averaged values between graphene and h-BN with the BN composition ratio x. That is, the on-site energies of boron-like and nitrogen-like virtual atoms [virtual A and B atoms in Fig. 4] are $\varepsilon'_B = x\varepsilon_B + (1-x)\varepsilon_C$ and $\varepsilon'_N = x\varepsilon_N + (1-x)\varepsilon_C$, respectively, and the hopping energy between these boron-like and nitrogen-like virtual atoms is $t'_{BN} = xt_{BN} + (1-x)t_{CC}$, where the on-site energies of pure boron, nitrogen, and carbon are $\varepsilon_B = 2.46$ eV, $\varepsilon_N = -2.55$ eV, and $\varepsilon_B = 0$ eV, respectively, and the B-N and C-C hopping energies are $t_{BN} = -2.16$ eV and $t_{BN} = -2.7$ eV, respectively. Then the electronic properties of each h-$(BN)_xC_{1-x}$ layer can be calculated by solving the eigenvalue problem $H(\boldsymbol{k})|\psi_{l\boldsymbol{k}}\rangle = E_l(\boldsymbol{k})|\psi_{l\boldsymbol{k}}\rangle$ [$l = 1$ (2) corresponds to the valence (conduction) band] with the Hamiltonian

$$H(\boldsymbol{k}) = \begin{pmatrix} \varepsilon'_B & h^*(\boldsymbol{k}) \\ h(\boldsymbol{k}) & \varepsilon'_N \end{pmatrix}, \tag{1}$$

$$h(\boldsymbol{k}) = -t'_{BN}\left(2e^{ik_y a_0/2}\cos(k_x\sqrt{3}a_0/2) + e^{-ik_y a_0}\right). \tag{2}$$

Figure 3 (right) shows the band structure of h-BC_2N (corresponding to the BN composition $x = 0.5$) calculated using VCA-TB method. By comparing the band structures of h-BC_2N calculated using the conventional TB [Fig. 3(left)] and VCA-TB [Fig. 3(right)] near the band edges (enclosed by circles), it can be seen that both have the same band gap at the

K point and the similar band dispersion, indicating the validity of VCA.

D. Difference of quasi Fermi levels for holes and electrons

Next step is to obtain the difference of the quasi Fermi levels between the holes and electrons at the central active layer, calculated by solving the drift-diffusion (DD) equation along with the Poisson's equation:

$$\frac{dc_j(z,t)}{dt} = -\frac{d}{dz}F_j(z,t) + r_j(z,t), \tag{3}$$

$$F_j(z,t) = -D_j\frac{dc_j(z,t)}{dz} + c_j(z,t)q_jM_j\left(-\frac{d\varphi(z,t)}{dz}\right), \tag{4}$$

$$-\frac{d}{dz}\varepsilon(z)\frac{d\varphi(z,t)}{dz} = q_+\left(c_+(z,t) + N_D^+(z)\right) + q_-\left(c_-(z,t) + N_A^-(z)\right). \tag{5}$$

Here $c_j(z,t)$ is the hole/electron density ($j = +/-$), $F_j(z,t)$ is the flux density along the z-direction (layer direction), $r_j(z,t)$ is the reaction (generation+recombination) rate, $\varphi(z)$ is the electrostatic potential, D_j is the diffusion coefficient, $M_j = D_j/k_BT$ is the mobility, $q_\pm = \pm e$, and $N_{D/A}^{+/-}$ are the ionized donor/acceptor densities. We employed the relaxation time approximation for the reaction term with the relation time $\tau = 1$ ps. Other assumed parameters are: the doped charge density ± 0.01 in units of $e(> 0)$ per unit cell per layer for n/p-doped region and the interlayer mobility 10^{-12} m^2/Vs.

In our simulations the hole and electron densities $c_\pm(z,t)$ in the DD equations (3) and (4) are self-consistently calculated not only with the Poisson's equation Eq. (5) but also with the hole and electron densities calculated using the band structure profiles $E_l(\boldsymbol{k}) - e\varphi(z)$ and the local quasi-Fermi levels $E_{Fh/e}(z)$ for holes/electrons, in which $\varphi(z)$ obtained by Poisson's equation and $c_\pm(z,t)$ obtained by DD equations are required, respectively. The above whole self-consistent calculations have been performed based on the Newton's algorithm assuming the steady state.

E. Optical gain in h-BCN active layer

Once the converged quasi Fermi levels $E_{Fh/e}$ for hole/electron are obtained based on the self-consistent procedure mentioned above, the optical gain g in the h-BCN active layer is calculated by the radiation energy W_{rad} per unit area per unit time divided by the incident electromagnetic flux I_{in} as $g = W_{rad}/I_{in}$, where

$$W_{rad} = \frac{2\hbar\omega}{S_{UC}N_k}\sum_{\boldsymbol{k}}\frac{2\pi}{\hbar}|H'_{CV}(\boldsymbol{k})|^2\delta\left(E_C(\boldsymbol{k}) - E_V(\boldsymbol{k}) - \hbar\omega\right) \times [f(E_C(\boldsymbol{k}) - E_{Fh}) - f(E_V(\boldsymbol{k}) - E_{Fe})] \tag{6}$$

and $I_{in} = n\varepsilon_0 cE_0^2/2$. Here $H'_{CV}(\boldsymbol{k}) = \langle\psi_{2\boldsymbol{k}}|H'|\psi_{1\boldsymbol{k}}\rangle$ with H' being the perturbation Hamiltonian due to the coupling between electrons in h-BCN and electromagnetic field (see Ref. [10] for detail).

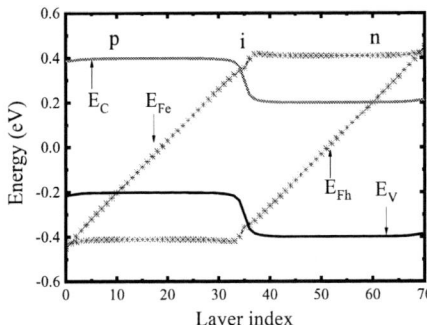

Fig. 5. Band edges (E_V and E_C) and the quasi Fermi levels (E_{Fh} and E_{Fe}) as a function of the layer index.

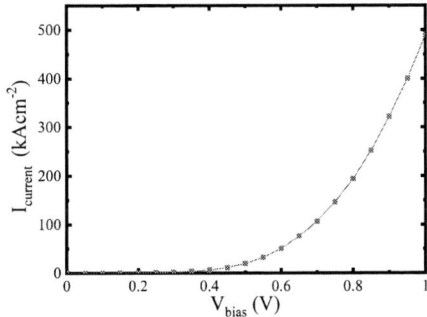

Fig. 6. Forward bias V_{bias} dependence of the current density along the z-direction in Fig. 1 for the BN composition $x = 0.12$.

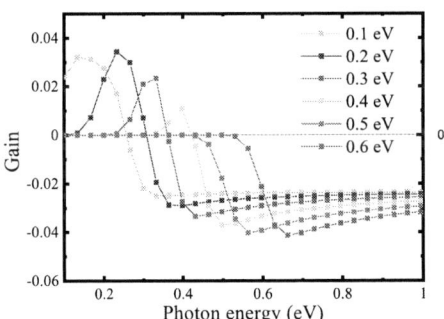

Fig. 7. Photon energy dependence of the optical gain are plotted for various values of the band gap $0.1 \sim 0.6$.

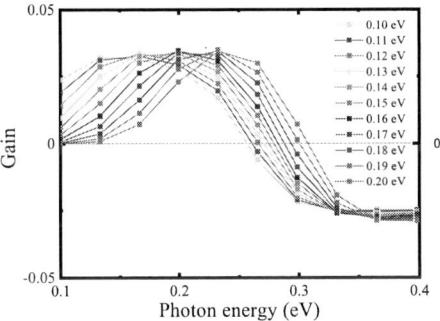

Fig. 8. Photon energy dependence of the optical gain are plotted for various values of the band gap $0.1 \sim 0.2$.

III. RESULTS AND DISCUSSIONS

Figure 5 shows the band edges (E_V and E_C) and the quasi Fermi levels (E_{Fh} and E_{Fe}) as a function of the layer index, where the the band gap was set to 0.6 eV (corresponding to the BN composition $x = 0.12$) and the forward bias voltage $V_{bias} = 1$ V was applied over the whole layers. Calculation were made with a total of 71 layers, including 35 p-type cladding layers, 1 active layer, and 35 n-type cladding layers throughout this study. In this figure it can be seen that the quasi-Fermi level difference between holes and electrons is larger than the band gap at the central active layer. We performed the above mentioned calculations for various values of the band gap, and obtained the corresponding quasi-Fermi level differences at the active layer.

In order to clarify the bias voltage range which can be applied in the assumed layer structure, in Fig. 6 we plotted the forward bias V_{bias} dependence of the current density across the pn junction layers. Here we have confirmed that the typical nonlinear I-V behavior can be maintained at least up to the bias $V_{bias} = 1$ V.

Next we analyze the gain spectra in the proposed structures. In Fig. 7, the photon energy dependence of the optical gain are plotted for various values of the band gap $0.1 \sim 0.6$ corresponding to the BN composition $x = 0.02 \sim 0.12$. As can be seen first from the Fig. 7, finite (positive) gain can be realized for the band gap smaller than 0.5 eV, and for the band gap grater than 0.6 eV all incident light is absorbed. This is basically because the quasi Fermi level difference $E_{Fh} - E_{Fe}$ cannot exceed the band gap for such large band gap.

Figure 8 shows the photon energy dependence of the optical gain are plotted for various values of the band gap $0.1 \sim 0.2$ in 0.01 steps corresponding to the BN composition $x = 0.02 \sim 0.04$. The calculated result shows that the maximum gain is obtained at 0.19 eV.

Figure 9 shows the gain spectra for the above mentioned optimized band gap 0.19 eV for various values of total number of layers (including one active layer at the middle), where we have used the self-consistently obtained values of the Fermi level differences (between holes and electrons) listed in Table 1. In Fig. 9 we can see that the gain spectrum is almost independent of the total number of layers. This is consistent with the fact that the change in the quasi Fermi level differences due to the change in the number of layers is negligibly small. Based on the results obtained so far, it is considered that the gain is determined mainly by the difference of quasi Fermi level with respect to the size of the band gap. It is also important to note that the smaller the band gap, the wider the range of photon energy at which the nonzero gain is possible, although the maximum value of gain becomes smaller when the band gap is larger than 0.19 eV. The resultant values of the 2D gain (dimensionless) can be converted to the 3D gain by using the interlayer distance of graphite (3.85 Å), obtaining the largest gain peak of h-BCN (band gap: 0.19 eV) as being 7.14×10^4 cm^{-1}, which is about 178 times larger than that in the commonly used semiconductor AlGaAs as being

$4.00 \times 10^2 \text{ cm}^{-1}$.

Table. I
RELATIONSHIP BETWEEN THE NUMBER OF LAYERS AND DIFFERENCE OF QUASI FERMI LEVELS.

The number of layers	Difference of quasi Fermi levels (eV)
31	0.29970
41	0.29904
51	0.29896
61	0.29894
71	0.29894

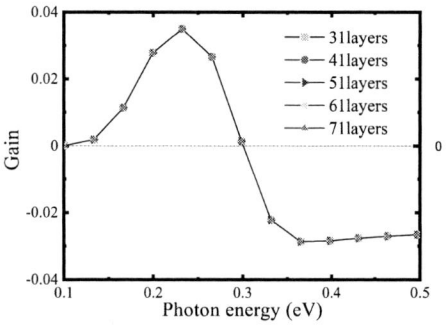

Fig. 9. Photon energy dependence of the optical gain are plotted for various values of layers 31 ~ 71.

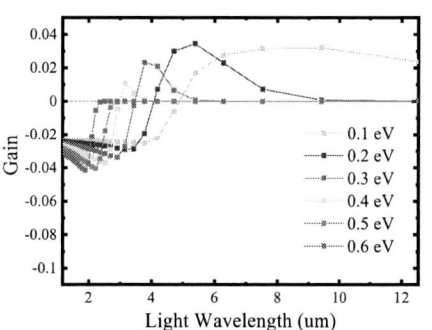

Fig. 10. Wavelength dependence of the optical gain are plotted for various values of the band gap 0.1 ~ 0.6 .

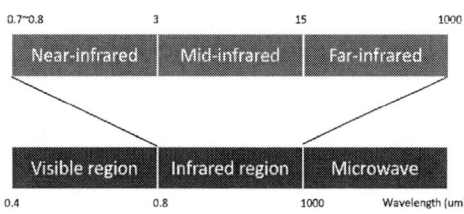

Fig. 11. Wavelength range of light.

Finally in Fig. 10 we plotted the gain spectra for various bandgaps as a function of the light wavelength. From this figure and the schematic diagram in Fig. 11, it is found that the wavelength at which the maximum gain is obtained is in the mid-infrared region, which is important for various applications such as communications including IoT, and health cares.

IV. CONCLUSION

We presented a numerical study on the optical gain in semiconductor laser structure with h-BCN as active an layer. By using the tight-binding formalism, drift-diffusion-Poisson equations, and the Fermi's golden role to calculate the electronic band structure, quasi-Fermi level, and the optical gain, respectively, we analyzed the optical gain spectra for various band gap energies in h-BCN, demonstrating that the largest gain peak of h-BCN can be significantly greater than that in the case of conventional semiconductor active layer. The peak of the gain is in the mid-infrared region, meaning that the proposed material can be applicable for various applications such as communication including IoT and health cares.

ACKNOWLEDGMENT

This work was partially supported by JSPS KAKENHI Grant No. 19H04546.

REFERENCES

[1] Y.-Y. Zhang , Q.-X. Pei , Z.-D. Sha , Y.-W. Zhang, Phys. Lett. A **383**, 2821 (2019).
[2] K. S. Novoselov, V. I. Fal'ko, L. Colombo, P. R. Gellert, M. G. Schwab, and K. Kim, Nature, **490**, 192 (2012).
[3] C.-Z. Ning, Advanced Photonics **1**, 014002 (2019).
[4] S. Beniwal, J. Hooper, D. P. Miller, P. S. Costa, G. Chen, S.-Y. Liu, P. A. Dowben, E. C. H. Sykes, E. Zurek, A. Enders, ACS Nano **11** 2486 (2017).
[5] S. Shallcross, S. Sharma, E. Kandelaki, and O. A. Pankratov, Phys. Rev. B **81** 1 (2010).
[6] P. Giraud, "Study of the Electronic Structure of hexagonal Boron Nitride on Metals Substrates," Master Thesis, Université des Science et Technologies Lille 1 (2012).
[7] G. Fiori, A. Betti, S. Bruzzone, and G. Iannaccone, ACS Nano **6**, 2642 (2012).
[8] Q. Peng, S. De, Phyica E **44** 1662 (2012).
[9] J. Singh, "Electronic and optoelectronic properties of semiconductor structures," Cambridge University Press (2007).
[10] A. Mehdipour, K. Sasaoka, M. Ogawa and S. Souma, Jpn. J. Appl. Phys. **53**, 115103 (2014).

Predictive Compact Modeling of Abnormal LDMOS Characteristics Due to Overlap-Length Modification

Takahiro Iizuka
HiSIM Research Center
Hiroshima University
Higashi-Hiroshima, Japan
iizuka@hiroshima-u.ac.jp

Dondee Navarro
HiSIM Research Center
Hiroshima University
Higashi-Hiroshima, Japan
dondee-navarro@hiroshima-u.ac.jp

Mitiko Miura-Mattausch
HiSIM Research Center
Hiroshima University
Higashi-Hiroshima, Japan
mmm@hiroshima-u.ac.jp

Hidenori Kikuchihara
HiSIM Research Center
Hiroshima University
Higashi-Hiroshima, Japan
kikuchihara532@hiroshima-u.ac.jp

Hans Jürgen Mattausch
HiSIM Research Center
Hiroshima University
Higashi-Hiroshima, Japan
hjm@hiroshima-u.ac.jp

Daniel Nestor Rus
Allegro MicroSystems, LLC,
Buenos Aires, Argentina
DRus@ALLEGROMICRO.com

Abstract—Further compact-model development for LDMOS is reported, enabling concurrent device and circuit optimizations by only varying the ratio between gate-overlap length (L_{over}) and resistive-drift length (L_{drift}). Different from the conventional carrier-dynamics understanding within these two regions, LDMOS shows abnormal characteristics during such a ratio variation. The pinch-off condition occurs under the gate overlap region, and the pinch-off point is found to move along L_{over} with increased drain voltage, even under the accumulation condition. This means that carrier conductivity is no longer controlled by the gate voltage but by the drain voltage. The precise pinch-off condition is determined by the field balancing within gate-overlap and resistive-drift regions. The pinch-off length (ΔL) within L_{over} sustains V_{ds} together with L_{drift}. Thus, the pinch-off region contributes as a part of L_{drift} and improves the device's high-voltage applicability. A new model is developed to describe this balancing phenomenon analytically, where the key physical quantity is ΔL. The developed ΔL model considers the potential distribution along L_{over} together with L_{drift}. At the pinch-off point, the field induced by V_{gs} and that by V_{ds} are assumed to be equal, which derives an analytical description for ΔL. Evaluation results with the developed model are verified with 2D-numerical-device-simulation results.

Keywords—power MOSFETs, LDMOS, compact model, overlap length, conductivity modulation

I. INTRODUCTION

LDMOS (lateral double-diffused MOS) [1] is widely used for power applications. A schematic LDMOS structure is depicted in Fig. 1a. It can be seen that the device consists of an intrinsic MOSFET and a long lightly-doped drift region with the length of L_{drift}, where most of the applied voltage is sustained. Additionally, the gate-overlap length, L_{over}, is relatively long, which controls the carrier flow together with the channel region. Geometrical dependence with L_{over} and L_{drift} varied were studied [2], [3]. This geometrical freedom would lead to an optimization of LDMOS for different voltage-rating, through varying the ratio of the conductive L_{over} and the resistive L_{drift}. For this device technology to be truly usable in actual circuit designing, a good compact model equipped with geometrical scaling capabilities is helpful. Measured I_{ds}-V_{gs} characteristics are shown in Fig. 1b, where L_{drift} is fixed

to 0.4μm and L_{over} is varied from short to extremely long length. Though the measurements show conventional features for low V_{ds} for all lengths L_{over}, the I-V characteristics, especially in case of g_m, show no clear tendency as a function of L_{over}. Current-driving capability is kept high for relatively short L_{over}, which, however, reduces drastically for long L_{over}.

II. ANALYSIS OF OBSERVED PHENOMENON

Simulation experiments using 2D device simulator [4] were undertaken to analyze the measured results. Fig. 2a shows the results of the I_{ds}-V_{gs} characteristics, where L_{drift} was kept to 0.6μm and L_{over} was varied. The simulated set of structures reproduces the experimental features depicted in Fig. 1b. It can be seen that the I_{ds}-V_{gs} characteristics are identical in case of small V_{gs} for all L_{over} lengths. However, a drastic current reduction occurs with increased L_{over} length according to a V_{gs} increase, which can be recognized more clearly in the g_m curves. In summary, the long L_{over} devices show two specific features, as depicted in Fig. 2b. One is a drastic increase of the resistance effect, leading to a reduction of g_m, and the other is a recovery of this reduction, when V_{gs} is still further increased. Fig. 3 shows a parameter extraction result with the existing open-release version of the HiSIM_HV industry-standard compact model [5], which demonstrates that the observed abnormal device features cannot be reproduced. The reason for the first sharp peaks, observed in Fig. 2b, is due to the weak V_{gs}, which cannot induce the accumulation condition yet, so that the applied V_{ds} is mostly consumed at the channel/drain junction as demonstrated in Fig. 4. Once the accumulation condition is formed, the overlap region becomes conductive.

On the left side of Fig. 5, simulated carrier-density distributions of the current characteristics, shown in Fig. 2, are compared at fixed specific V_{gs} values. The corresponding potential distributions along the current-flow contour are depicted on the right side of Fig. 5 together with the injected carrier densities from the channel. Current flows away from the surface for the whole L_{over} region under low V_{gs} condition and thus the whole L_{over} region is under the pinch-off condition. A nearly flat potential distribution is observed for this case and most of the applied potential is consumed at the channel/drift junction. Thus, the pinch-off region contributes as a highly resistive-drift region, together with L_{drift}. However, the current moves towards the surface as V_{gs} increases. With still further increased V_{gs}, carrier injection from the channel increases and the overlap region becomes conductive. The remaining resistive region stays in the pinch-off condition and acts as a part of L_{drift}.

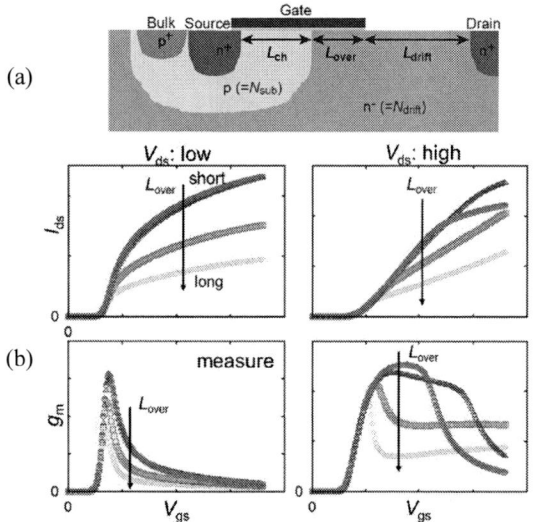

Fig. 1. (a) Schematic of LDMOS structure. (b) Measured I_{ds}-V_{gs} characteristics (upper) and g_{m}-V_{gs} characteristics (lower) for different L_{over} for low (left) and high (right) V_{ds}.

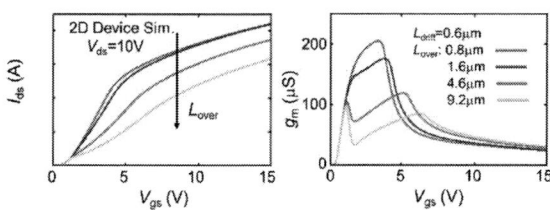

Fig. 2. 2D numerical device simulation results of I_{ds}-V_{gs} characteristics and g_{m}-V_{gs} characteristics for different L_{over} lengths.

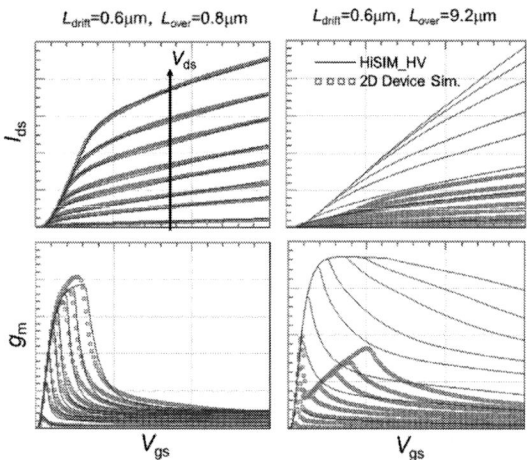

Fig. 3. Parameter extraction results with the industry-standard HiSIM_HV model. A single set of model parameters should predict L_{over} length scaling. For identical LDMOS devices (except for different L_{over} lengths), TCAD-generated data (red dots) of the device with shortest L_{over} were fitted. I_{ds}-V_{gs} and g_{m}-V_{gs} as a function of V_{ds}, are shown on the left to be reproduced well. However, for the longest L_{over} device, accurate representation of the characteristics failed with this extracted model parameter set, i.e., L_{over} scaling is not achieved.

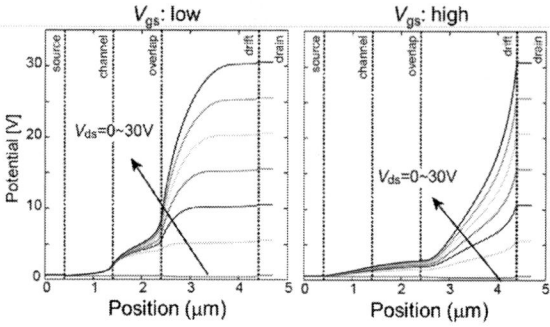

Fig. 4. Comparison of potential distribution along the device small V_{gs} (left) and for large V_{gs} (right).

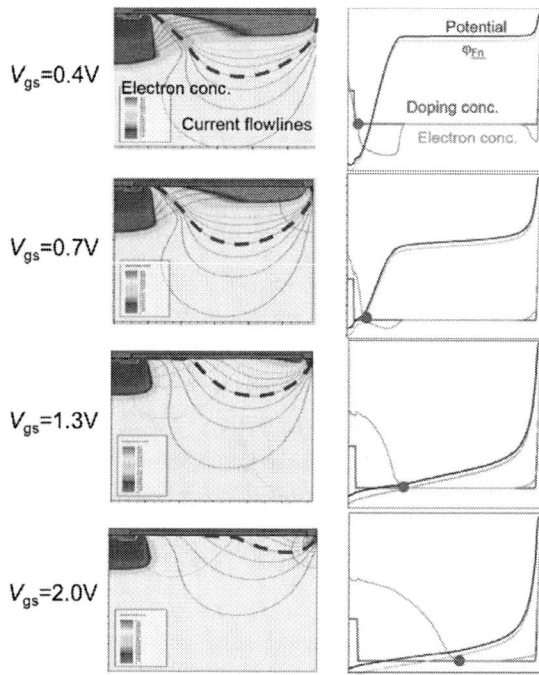

Fig. 5. 2D-device simulation results of (a) the current flow (left side) and (b) the potential as well as the electron-concentration distribution (right side) are compared for different V_{gs} values at V_{ds} =10V along the median current flowline (dashed lines in (a)). The median current flowline falls on the maximum current densities when the current density distribution is peaked and symmetrical around the peak. This is not the case when the distribution is skewed and hence not symmetrical around the peak. The high electron density refers the accumulation condition, which is observed more clearly for large V_{gs} values. The red circles denote an end point of accumulation, where electron concentration has decreased to the doping concentration.

Fig. 5 (left column) shows the maximum current-density path by dashed lines. It can be seen that weak gate control causes a deep current flow. This long way of the current flow increases the resistance effect. The depth of the current flow becomes shallower as V_{gs} increases, referring to a stronger accumulation condition. The main reason for the deep current flow is the pinch-off formation underneath the gate-overlap region in accordance with a diminished V_{gs} control, where no electrons exist (see red circles in the right column of Fig. 5). Since the applied voltage of V_{ds} =10V at the drain contact cannot be sustained only within the short L_{drift} region, the sustaining region extends even into the overlap region for small V_{gs}. As a result, the gate control

158

of the current flow is limited to only within the non-pinch-off region, where injected carriers from the channel prolong the path to the electrode. The details of the observed pinch-off feature are determined by the balancing of the two fields, induced within the conductive gate-overlap region and within the pinch-off region.

III. MODELING OF ΔL AND VERIFICATION

Our task here is to develop a compact model for the pinch-off length ΔL in the lightly-doped resistive region under the gate. The pinch-off occurs at the point where the gate-oxide field becomes equal to the lateral electric field, and thus the lateral electric field starts to dominate beyond this point. Since HiSIM_HV solves the potential distribution along the device iteratively, the key potential value V_{dp}, referring the end of the gate control, is known as demonstrated in Fig. 6 [5]. Here the potential distribution is approximated by a quadratic function within the drift region as

$$\phi = a\left(x + x_0\right)^2 + b \tag{1}$$

$$\begin{cases} a = \dfrac{\Delta V}{L_{\mathrm{total}}^2 - L_0^2} \\[3mm] b = V_{\mathrm{dp}} - \dfrac{\Delta V \cdot x_0^2}{L_{\mathrm{total}}^2 - x_0^2} \end{cases} \tag{2}$$

$$\begin{cases} \Delta V = V_{\mathrm{ds}} - V_{\mathrm{dp}} \\ L_{\mathrm{total}} = L_{\mathrm{over}} + L_{\mathrm{dri}} \\ x_0 = L_{\mathrm{over}} - \Delta L \end{cases} \tag{3}$$

An analytical equation for ΔL is derived as

$$\Delta L = L_{\mathrm{over}} + \frac{\Delta V - \sqrt{\Delta V^2 + E_{\mathrm{ov}}^2 L_{\mathrm{total}}^2}}{E_{\mathrm{ov}}} \tag{4}$$

$$E_{\mathrm{ov}} = \frac{\epsilon_{\mathrm{ox}}}{\epsilon_{\mathrm{si}} T_{\mathrm{ox}}}\left(V_g - \phi_{\mathrm{ov}} - c\right) \tag{5}$$

where E_{ov} is the field strength at the pinch-off point ($x=0$). The surface potential within L_{over} under the gate control ϕ_{ov} is also calculated by solving the Poisson equation iteratively, which determines the electric field within L_{over}. An adjustable parameter c denotes the threshold voltage of the accumulation start. Therefore, the field strengths can be calculated analytically. Thus the current equation within the drift region is written as

$$I_{\mathrm{ddp}} = W_{\mathrm{eff}} X_{\mathrm{ov}} q N_{\mathrm{drift}} \mu_{\mathrm{drift}} \frac{V_{\mathrm{ddp}}}{L_{\mathrm{drift}} + \mathbf{RDRDL1} + \Delta L} \tag{6}$$

where W_{eff} is the device width, X_{ov} is the width of the current flow and μ_{drift} is the carrier mobility within the drift region [5]. As can be seen, the pinch-off length ΔL is treated as an extension of the length of the resistive region [5], and thus ΔL reduces the electric field within the drift region. HiSIM considers the continuity condition between the channel current I_{ds} and I_{ddp} to determine V_{dp} [5], [6].

The pinch-off length ΔL is calculated and compared with 2D device simulations results in Fig. 7. It can be seen from Eq. (6), that a large ΔL for small V_{gs} leads to an increased resistance effect. Agreement of the model calculation results to those of 2D device simulation is quite well. Slight deviation of the V_{gs} dependence might be caused by the quadratic approximation of the potential distribution for deriving the analytical description of ΔL. To compensate this approximation, a model parameter could be introduced to obtain better fittings.

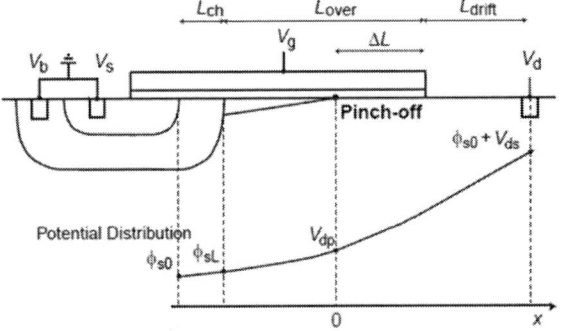

Fig. 6. Potential distribution along the device, as considered in HiSIM_HV. The internal-node potential V_{dp} refers the end of the gate control, namely, the pinch-off point for the presently studied case. Surface accumulation vanishes at the pinch-off point within the gate overlapped region.

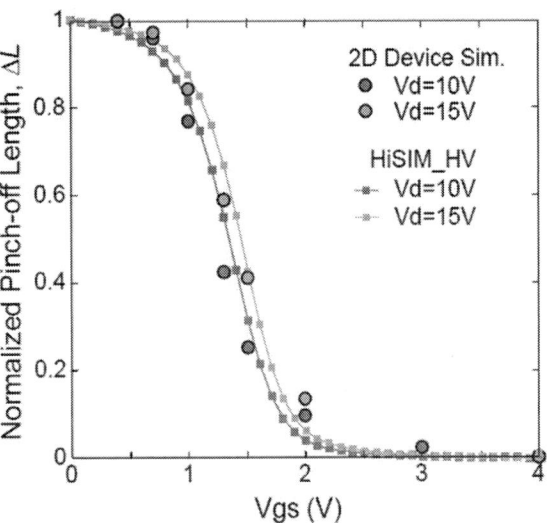

Fig. 7. Comparison of ΔL as extracted with 2D-device simulation and calculated with developed model. The 2D-device simulation results of ΔL are extracted as the points, where the vertical component of electric field turns to negative from positive. The developed model assumes a quasi 1D potential distribution along the device, into which the electron distribution along the depth direction is practically integrated. In spite of the

assumptions introduced for deriving the analytical formulation, agreement is acceptable.

Figs. 8a and b compare the model-calculated I-V characteristics and g_m with those of 2D-device simulation results. The model calculation is done only by introducing the ΔL values extracted in Fig. 7. The first sharp peak in g_m for small V_{gs} refers to the condition that ΔL is nearly equal to L_{over}. According to the V_{gs} increase, ΔL approaches to zero, resulting in a lowering of the resistance effect and the start

of the g_m recovery. After ΔL becomes zero, only the carrier mobility within the drift region determines the feature as demonstrated in Figs. 8c and 8d. In particular, Fig. 8c clearly shows that drain current depletes from that for ΔL=0. However, it reverts to that for ΔL=0 at higher V_{gs}.

REFERENCES

[1] J. D. Plummer and J. D. Meindl, IEEE JSSC vol. SC-11, no. 6, p.809, 1976.
[2] B. Wang et al.,IRPS 2005, p. 654.
[3] Y. Oritsuki et al., Nanotech 2009, Houston, May 3-7, 2009, p. 600.
[4] ATLAS User's Manual, SILVACO, Inc., Santa Clara, CA, 2018.
[5] HiSIM_HV 2.4.0 User's Manual, Hiroshima Univ., Hiroshima, Japan, 2017.
[6] H.J.Mattausch et al., IEEE TED vol.60, no.2, p.653, 2012.

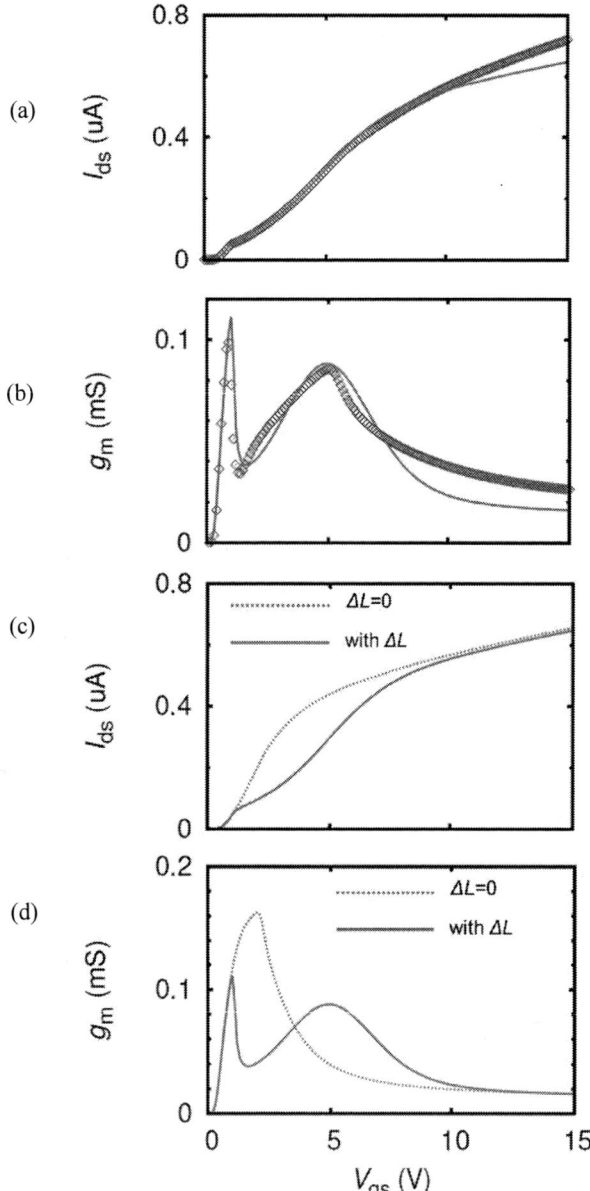

Fig. 8. Comparison of I-V characteristics between the developed model (red line) and 2D-device simulation (blue symbols). a) I_{ds}-V_{gs} and b) g_m-V_{gs}, at V_{ds}=10 V. The observed distinctive modulation in g_m, has been captured by the developed compact model, while in Fig.3 it was completely misrepresented using the industry-starndard HiSIM_HV model, where the developed model for the ΔL modulation is not yet included. c) I_{ds}-V_{gs} and d) g_m-V_{gs}, at V_{ds}=10 V, both with and without ΔL.

8-1 # A TCAD Study on Mechanism and Countermeasure for Program Characteristics Degradation of 3D Semicircular Charge Trap Flash Memory

N. Kariya[1], M. Tsuda[1], T. Kurusu[1], M. Kondo[1], K. Nishitani[1], H. Tokuhira[1], J. Shimokawa[1],
Y. Yokota[1], H. Tanimoto[1], S. Onoue[1], Y. Shimada[1], T. Kato[1], K. Hosotani[1], F. Arai[1],
M. Fujiwara[1], Y. Uchiyama[2], and K. Ohuchi[1,2]

[1]Institute of Memory Technology Research & Development, Kioxia Corporation, Yokkaichi, Mie, Japan
[2]Advanced Memory Development Center, Kioxia Corporation, Yokkaichi, Mie, Japan
Email: nayuta.kariya@kioxia.com

Abstract— A TCAD model to simulate program/erase characteristics of 3D charge trap flash memory cells is constructed and calibrated with the experiment. The mechanism of the program characteristics degradation of the semicircular cells are studied using the TCAD model and it is clarified that the current leakage path due to the fringe parasitic transistor causes the degradation. An initial charge injection technique is proposed to suppress the parasitic leakage current to improve the program characteristics of the semicircular cells.

I. INTRODUCTION

Bit density of 3D flash memories has been grown for years by increasing the number of stacking layers [1]. However, the key process technology of memory hole etching becomes more challenging as the number of stacking layers increases. Shrinking the area per cell is another approach to increase bit density. Especially, bit density can be almost doubled if two strings of cells are formed in each memory hole. Several types of 3D flash memory technologies to shrink the area per cell have been proposed [2], [3]. We have recently developed a new 3D flash memory technology with semicircular shaped cell to boost bit density further [4].

Fig. 1 compares plane view of a conventional circular cell and a new semicircular cell. Fig. 2 illustrates a bird's eye view of an array of semicircular cells. In the circular cell, a control gate (CG) surrounds a cell all around and one NAND string is formed in each memory hole. While in the semicircular cell, the control gate is divided with insulator and two NAND strings are formed in each oval shaped memory hole.

However, experimental results revealed that the program characteristics of the semicircular charge trap cells are degraded compared to the circular cells. In this work, we have clarified the mechanism of the program characteristics degradation and proposed a countermeasure through TCAD simulations.

II. SIMULATOIN MODEL

Fig. 3(a) shows a structure for simulating program/erase characteristics of a semicircular cell. The structure has three CGs on two sides of a memory hole. The cell at the middle layer of front side CG is evaluated. Source and drain electrodes are attached to the lower and upper end of the channel, respectively. Bias conditions for simulating read, program, and erase operations are shown in Table 1. Sufficiently low bias V_{off} is applied to the back side CGs to cut off the current along the back side part of the channel in

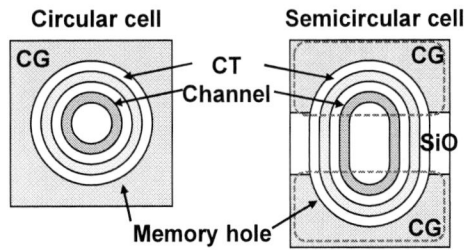

Fig. 1. Plane view of a circular and a semicircular cell.

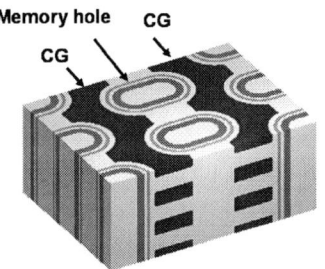

Fig. 2. Bird's eye view of an array of semicircular cells.

read operations. Fig. 3(b) shows enlarged view of a cell. Cell stack films consist of block (BLK), charge trap (CT) SiN, tunnel (TNL), and polycrystalline silicon channel layers.

Fig. 4 illustrates a conduction band diagram and an electron transport model in a program operation. We consider the following four processes [5]: (1) injection from the channel to conduction band of the CT layer according to tunneling effect, (2) drift motion and diffusion in conduction band of the CT layer, (3) capture and emission by traps in the CT layer, and (4) tunneling effect from the CT layer to the CG. The carrier transport model in an erase operation is similar to the program operation, except that injected carrier type is a hole. To obtain accurate current voltage characteristics of the polycrystalline silicon channel, grain boundary mobility and interface/bulk electron traps are considered.

III. SIMULATION AND EXPERIMENTAL RESULTS

First, we calibrated TCAD model parameters associated with CT SiN layer and channel polycrystalline silicon by comparing with the measured electrical characteristics of a conventional circular cell in order to guarantee accuracy of simulations. Fig. 5 shows simulated and experimental program/erase characteristics of a circular cell, and Fig. 6

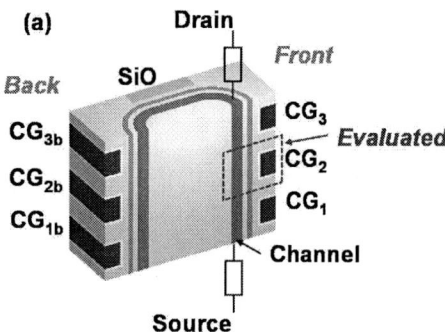

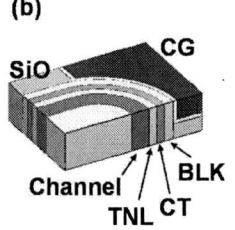

Fig. 3. (a) Simulation structure of a semicircular cell.
(b) Enlarged view of a semicircular cell.

Table 1. Bias conditions for simulating read, program, and erase operations.

	Read	Program	Erase
Source	0	0	0
Drain	V_{DD}	0	0
CG_2	V_t	V_{pgm}	$-V_{era}$
CG_1, CG_3	V_{read}	V_{pass}	$-V_{era}$
CG_{1b}–CG_{3b}	V_{off}	V_{pass}	$-V_{era}$

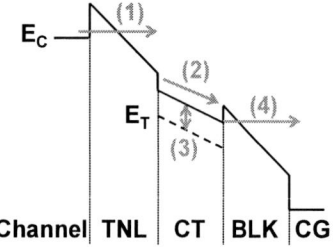

Fig. 4. Conduction band diagram and electron transport model in a program operation.

shows cell current from the initial state to programmed states. The simulated results are in good agreement with the experimental results.

Second, we carried out simulations for a semicircular cell with parameters calibrated by the first step. As shown in Fig. 7, the experimental program/erase characteristics of the semicircular cell are well reproduced again. It is noted that the same model parameters are used in the simulation of both circular and semicircular cells, which indicates that the

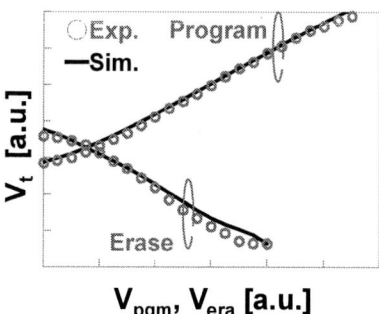

Fig. 5. Experimental and simulated program/erase characteristics of a circular cell.

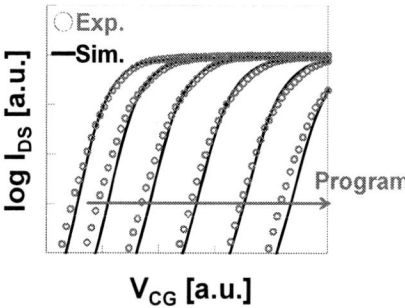

Fig. 6. Experimental and simulated cell current of a circular cell from the initial state to programmed states.

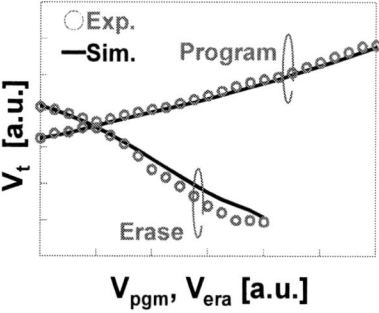

Fig. 7. Experimental and simulated program/erase characteristics of a semicircular cell.

difference in program characteristics results from the difference in the device geometry.

IV. DISCUSSION

A. Mechanism of Program Slope Degradation

Fig. 8(a) compares simulated program characteristics of a circular and a semicircular cell. The program slope, the ratio of the threshold voltage (V_t) change to the program voltage (V_{pgm}) change, of the semicircular cell is extremely lower than that of the circular cell. Fig. 8(b) shows the subthreshold slope (SS) at each programmed state in the circular and semicircular cell. The SS remains almost constant in the circular cell, while the SS increases with V_{pgm} in the semicircular cell. The change in SS indicates that the current path changes as the program state becomes higher.

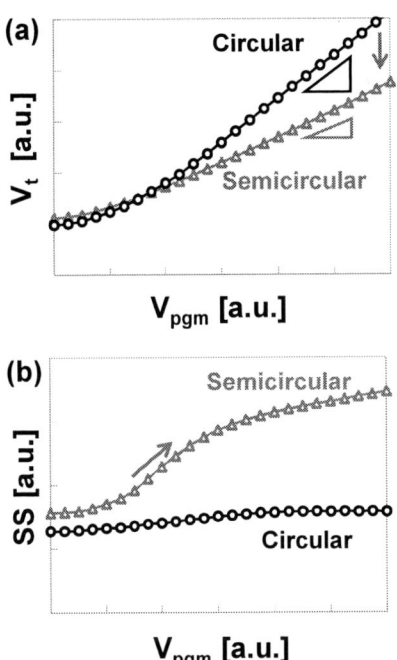

Fig. 8. Simulated program characteristics comparing circular and semicircular cells. (a) V_t and (b) SS.

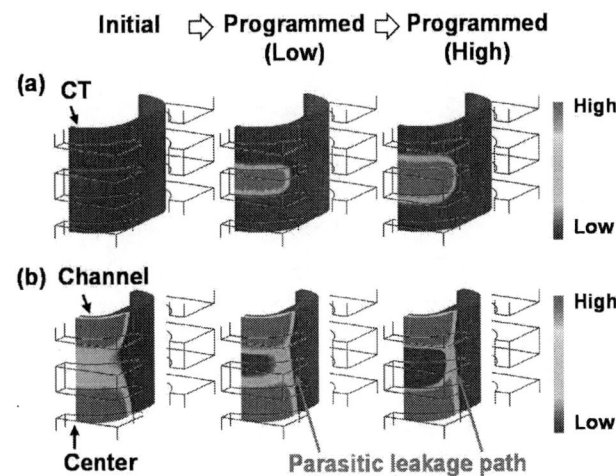

Fig. 9. (a) Simulated trapped electron density in the CT layer and (b) corresponding electron density in the channel at $V_{CG}=V_t$ of the initial state and programmed states of two different program voltages (Low/High).

Fig. 9 shows simulated trapped electron density in the CT layer and electron density in the channel when the CG voltage is equal to V_t. In the initial state with no trapped charge in the CT layer, the current path is formed at the center part of the channel. As the program state becomes higher, electrons are injected mainly at the center part of the CT layer and the current leakage path at the edge part manifests itself.

Based on the above observation, the mechanism of the program slope degradation is explained as follows. A semicircular cell transistor can be expressed as a parallel connection of a main transistor and a parasitic transistor due to fringe field effect as shown in Fig. 10(a). The main transistor has similar characteristics to a circular cell. The parasitic transistor has higher initial Vt and larger SS than the main transistor because of its thick gate dielectric, but it has extremely lower program slope since electrons are not injected around the parasitic transistor. As a result, the cell current and threshold voltage in the high program states are controlled by the parasitic transistor as shown in Fig. 10(b). Thus, the program slope of the semicircular cell is degraded.

B. Program Slope Improvement by Initial Charge Injection

A technique called initial charge injection is proposed to raise the program slope by suppressing the current leakage in the parasitic transistor. Fig. 11(a) depicts the concept. Before normal program operation, a high program voltage is applied to inject electrons to wide part of the CT layer, then a relatively low erase voltage is applied to inject holes so that electrons are extinguished at the center part and localized near the CG edge. The amount of the localized electrons can be controlled by voltage and pulse time of the first program step and the subsequent erase step. The localized electrons switch off the parasitic transistor during subsequent program

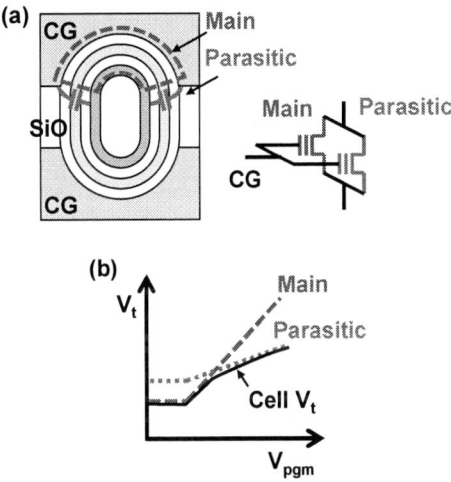

Fig. 10. (a) Cell transistor model in the semicircular cell expressed as a parallel circuit of a main part transistor and a parasitic part transistor. (b) Schematic explanation of program characteristics of a cell transistor, which is controlled by a parasitic transistor in high program states.

operation. Fig. 11(b) shows an example of simulated trapped electron density in the CT layer after the initial charge injection.

Fig. 12(a) shows the simulation results of program characteristics without initial charge injection and with initial charge injection of two different conditions, and Fig. 12(b) shows the simulated SS. In the high injection condition, remarkable improvement of the program slope and weaker dependence of SS on V_{pgm} are predicted. The program slope improvement with the initial charge injection is verified in experiments as shown in the previous report [4].

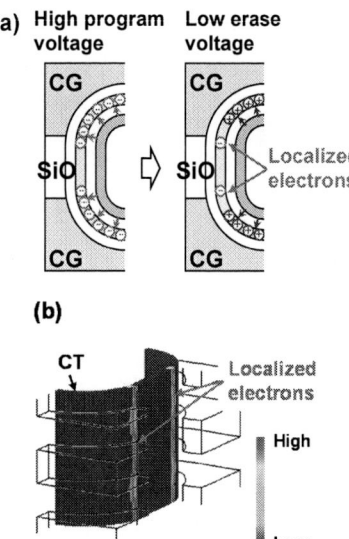

Fig. 11. (a) Procedure of initial charge injection. (b) Trapped electron density in the CT layer after initial charge injection.

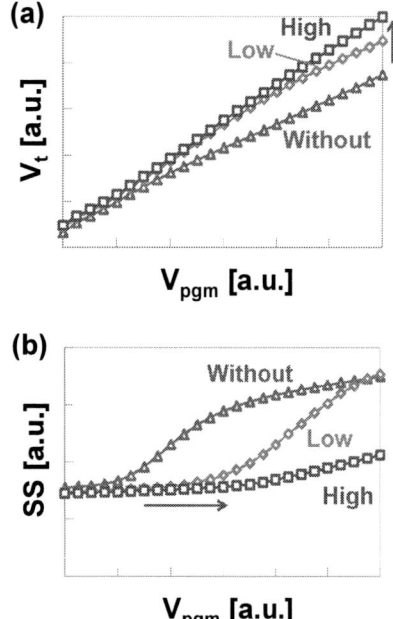

Fig. 12. Simulated program characteristics without initial charge injection and with initial charge injection of two different conditions (Low/High). (a) V_t and (b) SS.

V. SUMMARY

We have constructed a TCAD model to simulate program/erase characteristics of the 3D CT flash memory cells. The model was calibrated to the experiment with high accuracy. We studied the program characteristics of the novel semicircular cells through TCAD simulation and clarified that the current leakage path due to the parasitic transistor causes the program slope degradation. We demonstrated that the initial charge injection technique is effective to suppress the parasitic current leakage to improve the program slope of the semicircular cells.

ACKNOWLEDGMENT

This work was supported by joint development of 3D flash memory with Western Digital Corporation.

REFERENCES

[1] H. Tanaka et al., "Bit Cost Scalable Technology with Punch and Plug Process for Ultra High Density Flash Memory," 2007 IEEE Symposium on VLSI Technology, Kyoto, 2007, pp. 14-15.

[2] H. Lue et al., "A novel double-density, single-gate vertical channel (SGVC) 3D NAND Flash that is tolerant to deep vertical etching CD variation and possesses robust read-disturb immunity," 2015 IEEE International Electron Devices Meeting (IEDM), Washington, DC, 2015, pp. 3.2.1-3.2.4.

[3] H. Lue et al., "A Novel Double-Density Hemi-Cylindrical (HC) Structure to Produce More than Double Memory Density Enhancement for 3D NAND Flash," 2019 IEEE International Electron Devices Meeting (IEDM), San Francisco, CA, USA, 2019, pp. 28.2.1-28.2.4.

[4] M. Fujiwara et al., "3D Semicircular Flash Memory Cell: Novel Split-Gate Technology to Boost Bit Density," 2019 IEEE International Electron Devices Meeting (IEDM), San Francisco, CA, USA, 2019, pp. 28.1.1-28.1.4.

[5] E. Vianello et al., "Experimental and Simulation Analysis of Program/Retention Transients in Silicon Nitride-Based NVM Cells," in IEEE Transactions on Electron Devices, vol. 56, no. 9, pp. 1980-1990, Sept. 2009.

8-2

Impact of Random Phase Distribution in 3D Vertical NAND Architecture of Ferroelectric Transistors on In-Memory Computing

Gihun Choe
School of Electrical and Computer Engineering
Georgia Institute of Technology
Atlanta, GA, USA
gchoe6@gatech.edu

Wonbo Shim
School of Electrical and Computer Engineering
Georgia Institute of Technology
Atlanta, GA, USA
wshim30@gatech.edu

Jae Hur
School of Electrical and Computer Engineering
Georgia Institute of Technology
Atlanta, GA, USA
jhur45@gatech.edu

Asif Islam Khan
School of Electrical and Computer Engineering
Georgia Institute of Technology
Atlanta, GA, USA
asif.khan@ece.gatech.edu

Shimeng Yu
School of Electrical and Computer Engineering
Georgia Institute of Technology
Atlanta, GA, USA
shimeng.yu@ece.gatech.edu

Abstract— **Ferroelectric field-effect transistors (FeFETs) with 3D vertical NAND architecture (3D V-NAND) are investigated for in-memory computing. In polycrystalline ferroelectric Hafnia thin film, there are different phases such as monoclinic (M), and orthorhombic (O) phases. Those are randomly distributed throughout the ferroelectric gate stack. Such positional dispersion of two phases introduces read-out current variation in 3D V-NAND of FeFETs. Herein, we employ TCAD simulations to quantify such variation and optimize bias conditions for improving the accuracy of in-memory computing.**

Keywords— Ferroelectric, mixed phases, variations, nonvolatile memory, in-memory computing

I. INTRODUCTION

Ferroelectric field-effect transistors (FeFETs) using doped HfO_2-based materials have been extensively explored as the next-generation nonvolatile memory and in-memory computing [1]. In previous study, a 4-layer 3D V-NAND with FeFETs has been experimentally demonstrated [2]. 3D V-NAND array has been suggested promising for in-memory computing [3]. Similarly as CMOS devices, scaled FeFETs have intrinsic parameter fluctuations such as line-edge roughness and work function variation [4]. The random distribution of ferroelectric (FE) and dielectric (DE) phases is one of the random variation sources within the polycrystalline ferroelectric thin film, which comes from the physical atomic structure difference when it is deposited on a substrate [5]. FE phase can happen when the atomic structure is orthorhombic (*O*) phase. Other phases (e.g., monoclinic (*M*), and cubic (*C*) phases) exhibit non-FE behaviors [5]. Those FE/DE phases randomly disperse in the ferroelectric gate stack. In addition, the probability of each phase could be varied depending on the deposition method and type of dopant. Therefore, the ferroelectric properties is not uniformly applied for forming the channel. Previous study explored the impact of this variation source on the memory window of planar FeFET for device-to-device variation depending on the probability of each phase, size of grains and dimension parameters [6]. In this paper, we will study how this positional distribution of FE/DE phases affects the read-out current variation of 3D V-NAND architecture of FeFETs. First, we will propose the 3D V-NAND FeFETs structure and simulation method how we generate the FE/DE variation with its result. Second, the bias

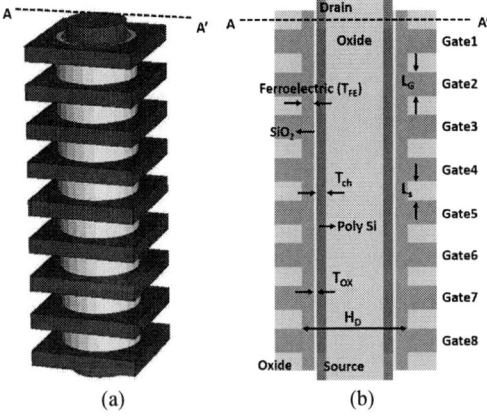

Fig. 1 (a) 3-D view and (b) cross-sectional view of vertical NAND architecture FeFETs.

condition will be discussed to optimize the current distribution and on/off ratio for this architecture. Sequentially, the I_D-V_G characteristic under the adjusted voltage condition will be shown. Last, the read-out current distribution depending on the probability of FE/DE phase distribution will be exhibited under the aforementioned bias scheme.

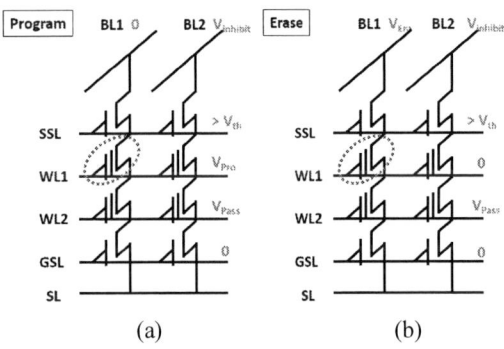

Fig. 2 V-NAND FeFETs (a) gate-program and (b) drain-erase scheme. The circled cell is selected.

165

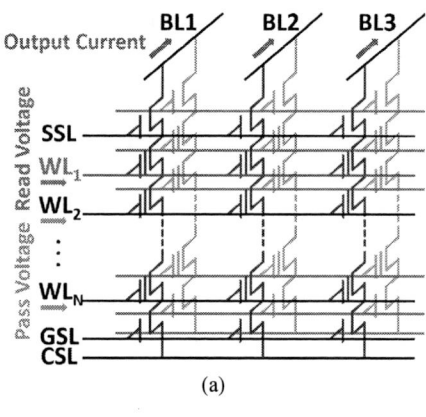

(a)

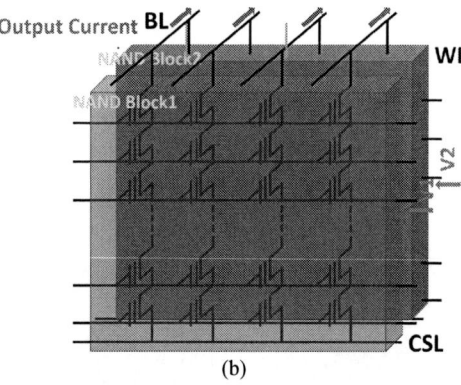

(b)

Fig. 3 (a) Circuit diagram and (b) schematic of V-NAND array consisting of multiple blocks for in-memory computing. V1 is for block 1, and V2 is for block 2.

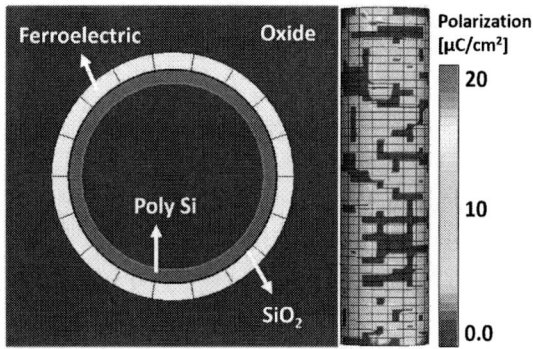

Fig. 4 Generated grains inside the ferroelectric layer for FE/DE phase distribution. Some grains are dielectric, thus no polarization in blue color.

II. SIMULATION METHOD

In this work, 8-layer 3D V-NAND of FeFET is designed using a Sentaurus TCAD to explore the read-out current variation induced by the spatial distribution of FE/DE phase (see **Fig.1**) [7]. The structural and material parameters used in this work are summarized in **Table I**. 1 nm interfacial SiO_2 layer and 8 nm $Hf_{0.5}Zr_{0.5}O_2$ layer are assumed. The source/drain is doped with n-type and the doping concentration is $10^{20}/cm^3$. The doping concentration of channel (p-type) is $10^{10}/cm^3$. Each gate length is 30 nm and

TABLE I. SIMULATION PARAMETERS

Name	Unit
Hole Diameter (H_D)	118 nm
Channel Thickness (T_{ch})	6 nm
Oxide Thickness (T_{OX})	1 nm
FE Thickness (T_{FE})	8 nm
Gate Length (L_G)	30 nm
Space Length (L_S)	20 nm
Grain Size (L_{Gmin})	10 nm
Remnant Polarization (P_R)	15 $\mu C/cm^2$
Saturation Polarization (P_S)	25 $\mu C/cm^2$
Coercive Field (E_C)	1 MV/cm

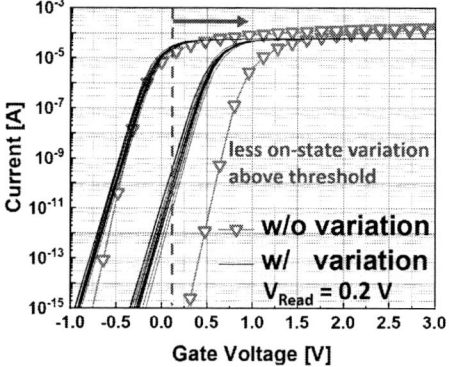

Fig. 5 I_D-V_G of a single cell of V-NAND FeFETs with and without FE/DE phase random variation. The voltage sweep range is from -4 V to 4 V. V_{Read} is designed to be 0.2 V.

space length is 20 nm. As shown in **Fig. 2(a)**, to program the selected cell on V-NAND, the programming voltage (V_{Pro} = ~4 V) will be applied to the word line (WL). The string selection line (SSL), ground selection line (GSL) and the other WLs should be turned on to prevent unselected cells on the same WL from being programmed with an inhibition voltage (~2 V) applied to unselected bit line (BL). To erase the selected cell, the erase voltage (V_{Era} = ~4 V) needs to be delivered from the BL via a drain-erase scheme (see **Fig. 2(b)**) [3]. Similar inhibition voltage (~2 V) needs to be applied to other unselected WLs due to the same reason above. Parallel read-out scheme could be used for in-memory computing in a layer-by-layer manner (**Fig. 3**).

Simulation method for generating random FE/DE phase is as below: 1) Make grains on ferroelectric layer [grain size = 10 nm×($2\pi \cdot R_{FE}/N_g$), N_G is the number of grains per L_{grain}, R_{FE} is $T_{FE}+T_{ch}+R_{BOX}$]. 2) Generate each grain's parameter for FE/DE phases as many as the total number of grains. 3) Apply random grain phases to ferroelectric parameter file. 4) Designate a phase of each grain using generated parameter file on the Sentaurus TCAD [7]. **Fig. 4** shows the polarization of generated grains and its positional FE/DE fluctuation. The blue color on the ferroelectric layer means those region does not have polarization. In other words, those grains are under dielectric phase.

Simulations are also comprised of doping-dependent model, thin-layer model, and high-field saturation model for mobility. Shockley-Read-Hall model and Auger model are used for recombination. For ferroelectric polarization, the Preisach model is employed [7].

166

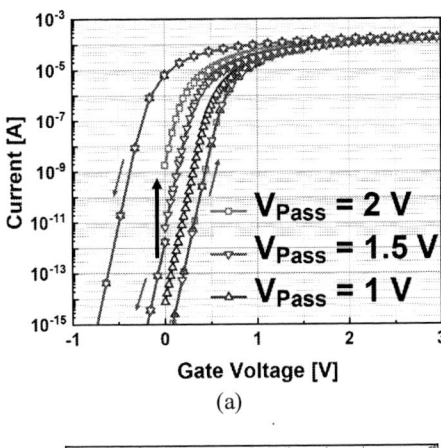

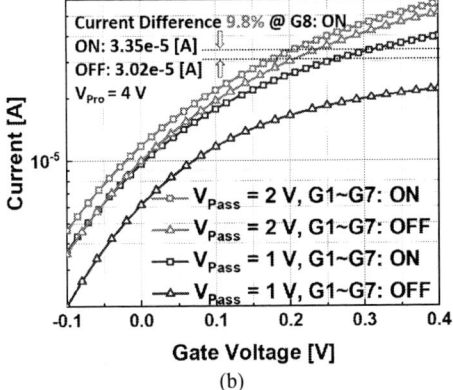

Fig. 6 (a) I_D-V_G of a V-NAND string: After applying V_{Pass}, off-state is disturbed with increasing current. (b) Read-out current of G8 $vs.$ the data pattern of G1-G7.

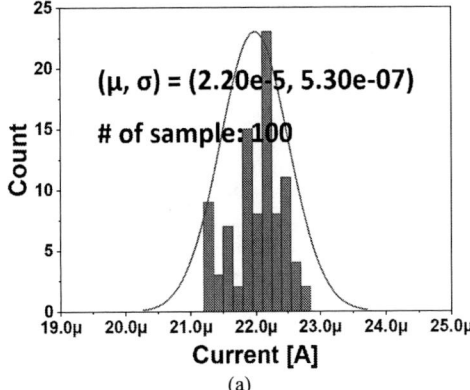

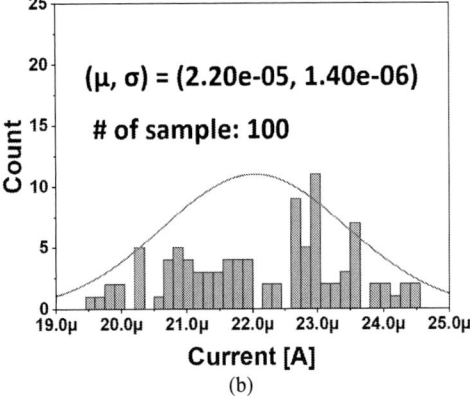

Fig. 8 Distribution of V-NAND read-out currents with (a) 10% and (b) 30% of DE phase probability. Here, μ denotes the mean value and σ denotes the standard deviation.

III. BIAS OPTIMIZATION FOR READ-OUT CURRENT

Fig. 5 illustrated the I_D-V_G of a single cell of 3D V-NAND FeFETs with/without FE/DE phase distribution at V_{read} = 0.2 V. The gate voltage was swept from -4 V to 4 V and returned to -4 V to investigate the on-state current and off-state current. Red line I_D-V_G does not have FE/DE variation but has same

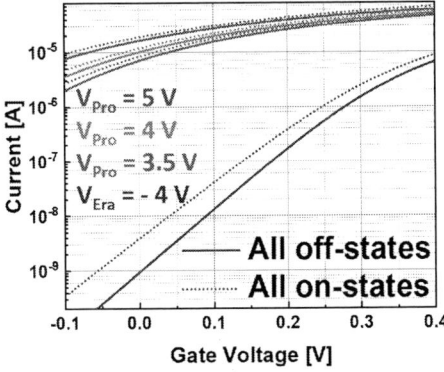

Fig. 7 Read-out current difference of G8 $vs.$ gate voltage depending on V_{Pro}. With higher programming voltage, on/off ratio could be improved. Dot is G1 ~ G7 (all on-states) and straight line is G1 ~ G7 (all off-states).

structural dimension and polarization properties. As comparing with the result without phase variation, the memory window with phase variation is much smaller. This is because only 50% of total area has ferroelectric phase, whereby the leakage current could pass via the channel area under the other 50% of dielectric phase on the ferroelectric layer. Even though the FE/DE phase distribution, which is the result of the spatial phase variation, causes the read-out current distribution, the impact of this variation on on-state current is alleviated above the threshold where the read voltage (V_{Read}) should be biased at (see **Fig. 5**). As shown in **Fig 6**, there is a trade-off between on/off ratio of a single cell (G8) and its read-out current's dependence on the data pattern of other cells along the same vertical string. A higher the pass voltage V_{Pass} on unselected cells (G1 to G7) could minimize the current difference between the worst-case (all cells are off-states) vs. the best-case (all cells are on-states) since the other cells could much fully flow the current. However, a higher V_{Pass} may disturb the cell, whereby the on/off ratio is reduced. On the other hand, a lower the V_{Pass} on the unselected cells undesirably increases the read current variation for both on-state and off-state. Hence, by optimizing V_{Pass}, the current difference becomes less than 10% at V_{Read} = 0.2 V with V_{Pass} = 2 V. In addition, the other solution is using a higher programming voltage (V_{Pro} = 4 V). **Fig. 7** shows that the higher V_{Pro} helps achieve larger on/off ratio and reduced current difference between the worst-case and the best-case.

IV. IMPACT OF VARIATION ON READ-OUT CURRENT

The probability of DE phase can be varied from 0% ~ 70% depending on the types and concentrations of the cation- and anion-based dopants used for the HZO deposition [5]. **Fig. 8** shows the read-out current distribution of the on-state G8 cells. The DE phase probabilities are 10% and 30%, respectively. The number of sample is hundred for each phase probability case. As shown in **Fig. 8**, when the probability of DE phase decreases, the read-out current variation decreases. Although both 10% and 30% of cases have the same mean value, the 10% DE probability of V-NAND shows 62% smaller standard deviation in read-out current than the 30% DE probability. This means that lowering the DE phase probability, in other words increasing the orthorhombic phase, is critical to ensure the read-out accuracy for in-memory computing.

V. CONCLUSION

3D V-NAND FeFETs for in-memory computing have been simulated. We optimized the bias conditions for mitigating the read-out current's data pattern dependency while maintaining more than at least 10 times of on/off ratio. The read-out current variation caused by the positional fluctuation of FE/DE phases is quantified, suggesting future material engineering is required to improve the purity of the FE phases.

ACKNOWLEDGMENT

This work is supported by ASCENT, one of the SRC/DARPA JUMP centers.

REFERENCES

[1] S. Mueller *et al.*, "From MFM Capacitors Toward Ferroelectric Transistors: Endurance and Disturb Characteristics of HfO_2-Based FeFET Devices," *IEEE Transactions on Electron Devices*, vol. 60, no. 12, pp. 4199–4205, Dec. 2013.

[2] K. Florent *et al.*, "First demonstration of vertically stacked ferroelectric Al doped HfO2 devices for NAND applications," in *2017 Symposium on VLSI Technology*, Kyoto, Japan, Jun. 2017, pp. T158–T159.

[3] P. Wang *et al.*, "Drain-Erase Scheme in Ferroelectric Field Effect Transistor—Part II: 3-D-NAND Architecture for In-Memory Computing," *IEEE Trans. Electron Devices*, vol. 67, no. 3, pp. 962–967, Mar. 2020.

[4] K. Ni, A. Gupta, O. Prakash, S. Thomann, X. S. Hu, and H. Amrouch, "Impact of Extrinsic Variation Sources on the Device-to-Device Variation in Ferroelectric FET," in *2020 IEEE International Reliability Physics Symposium (IRPS)*, Dallas, TX, USA, Apr. 2020, pp. 1–5.

[5] L. Xu, T. Nishimura, S. Shibayama, T. Yajima, S. Migita, and A. Toriumi, "Kinetic pathway of the ferroelectric phase formation in doped HfO_2 films," *Journal of Applied Physics*, vol. 122, no. 12, p. 124104, Sep. 2017.

[6] Y.-S. Liu and P. Su, "Variability Analysis for Ferroelectric FET Nonvolatile Memories Considering Random Ferroelectric-Dielectric Phase Distribution," *IEEE Electron Device Lett.*, vol. 41, no. 3, pp. 369–372, Mar. 2020.

[7] Sentaurus Device User Guide Version: N-2017.09, Synopsys, 2017.

8-3 TCAD Modeling and Optimization of 28nm HKMG ESF3 Flash Memory

Alban Zaka
GLOBALFOUNDRIES Dresden
Dresden, Germany
alban.zaka@globalfoundries.com

Tom Herrmann
GLOBALFOUNDRIES Dresden
Dresden, Germany
tom.herrmann@globalfoundries.com

Ralf Richter
GLOBALFOUNDRIES Dresden
Dresden, Germany
ralf.richter@globalfoundries.com

Stefan Duenkel
GLOBALFOUNDRIES Dresden
Dresden, Germany
stefan.duenkel@globalfoundries.com

Ruchil Jain
GLOBALFOUNDRIES Dresden
Dresden, Germany
ruchil.jain@globalfoundries.com

Abstract—**The paper presents a TCAD modeling approach of the 28nm HKMG ESF3 Flash Cell. The methodology encompasses both DC and transient simulations with focus on hot carrier injection modeling. The ensuing Floating Gate Spacer optimization presents the trade-off between the various figures of merit and highlights the need for a comprehensive DC/transient simulation approach during Flash cell optimization.**

Keywords—ESF3, TCAD, Hot Carrier Modeling

I. INTRODUCTION

Split-gate cells have been object of numerous studies in the last 30 years [1-10]. In the recent years, the third generation of SuperFlash® Cell has been successfully demonstrated on planar 28nm Gate-First HKMG [1, 10]. State of the art TCAD has been employed to model and optimize the 28nm HKMG ESF3 device. Section II introduces the flash cell under study, whereas Sections III, IV present the DC and hot-carrier modeling of the cell, respectively. Finally Section V discusses the optimization metrics of the Floating Gate Spacer (FGSpc) within the Cell.

II. CELL DESCRIPTION

The split-gate ESF3 cell is integrated together with HV-MOS devices on the 28nm HKMG gate-first Super Low Power platform [1]. The ESF3 fabrication process includes the formation of the Floating Gate (FG), Control Gate (CG) and of the Erase Gate (EG) as highly-doped Polysilicon regions, followed by the Wordline (WL) formation which is composed of the medium-thick oxide and the HKMG module of the baseline technology (Fig.1). The Silicon doping below the FG, WL as well as around BL and SL can be individually tuned.

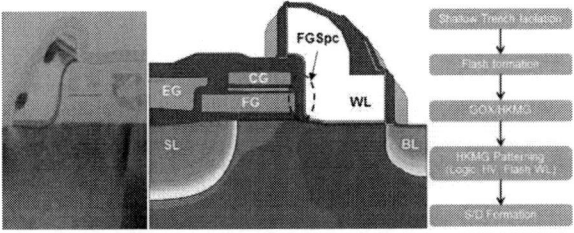

Fig.1: TEM and TCAD structure of the investigated 28nm HKMG ESF3 cell as well as the major fabrication steps, whereby the Wordline (WL) contains the HKMG stack, while Floating Gate (FG), Control Gate (CG) and Erase Gate (EG) are built with Polysilicon. The encircled are "FGSpc" is the object of the optimization section.

Table 1 summarizes the biasing schemes employed to program and erase the cell as well as to retrieve the threshold voltage of the cell (VtCG) and the BL current after program and erase operations (IR0 and IR1, respectively). While the erase operation is included in the present modeling, it is not object of a detailed analysis.

	Program	Erase	V_TCG (P/E)	V_TWL	IR (0:P, 1:E)
time	10us	10ms	-	after erase	-
SL	4.5V	0	0	0	0
BL	1uA	0	0.1V	0.1V	0.8V
EG	4.5V	11V	0	0	0
CG	10.5	0	ramp	1.8	1.8V
WL	0.8	0	1.8	ramp	1.8V

Table1: Nominal bias conditions used in this study. During programming, BL voltage is set such as to have 1uA flowing in the cell at the program start and then kept constant. The Programmed and Erased states are characterized by their threshold voltage (VtCG-Prog, VtCG-Erase) and the current which can flow in the channel after the operation (IR0, IR1).

III. CELL DC MODELING

Process and device simulations have been put in place using Sentaurus TCAD tools. The first step of the modeling concerns the correct description of the capacitances and doping in the cell. To achieve this, 2D Process simulations including the relevant fabrication steps and many TEMs taken after different process steps allow achieving the structural matching of the cell. The calibration of the dopant profiles in Silicon is facilitated by having several implant splits affecting the different Silicon areas of the device, i.e. below the FG, below the WL, etc.

Additionally, there are two aspects to consider. Firstly and more importantly, since the cell is a 3D device, an additional capacitance between FG and CG has to be added in the electrical simulations to match the CG-FG coupling ratio (separately measured, defined as C_{FG-CG}/C_{tot}), which is a key parameter for the cell operation. Secondly, one needs to adjust the charge in the FG electrode such as to have the V_TCG-Erase match the HW readout for a given initial structure. This value corresponds to a deep erase state (V_TCG-Erase ~ -8V) of the FG and it is important to note that this is a unique value, after fixing the additional FG-CG capacitance.

TCAD DC device simulations are performed using a drift-diffusion transport approach with doping, field and surface mobility degradation components. The FG terminal remains floating throughout the modeling, i.e. its potential is defined from the capacitive coupling to the other cell terminals and from the amount of charges flowing in/out of the FG.

The overall goodness of the electrostatic modeling of the cell has been established by comparing the simulation results to Silicon in terms of VtWL and IR1 (Fig.2) as well as to cell IV curves (Fig.3) for a range of doping split experiments.

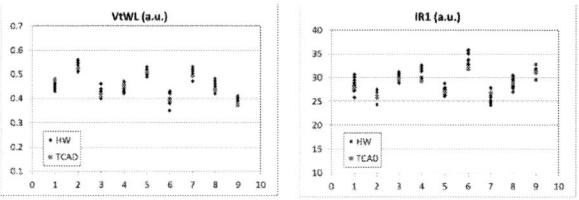

Fig.2: HW vs. TCAD comparison of VtWL and IR1 (c.f. definition in Table.1) for different Silicon doping splits affecting the cell electrostatics. Good agreement has been achieved around typical cell operation range.

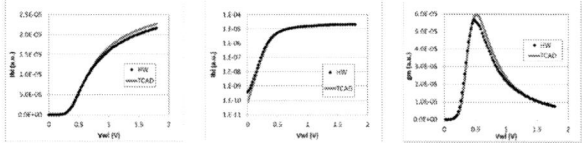

Fig.3: Typical example of HW vs TCAD comparison in terms of I_{BL} vs. V_{WL} and g_m vs. V_{WL} of the cell at V_{BL}=0.1V and V_{CG}=1.8V at erased state. Reasonable agreement is achieved below and above the threshold voltage.

The agreement achieved in these comparisons illustrate that the 2D TCAD methodology is able to capture reasonably well the 3D cell behavior and is ready to be used in transient simulations.

IV. HOT CARRIER INJECTION MODELING IN 28NM HKMG ESF3 CELLS

The analysis of the carrier transport and in particular of the hot carrier population and the ensuing injection current into the FG is performed using the Spherical Harmonics Expansion (SHE) method proposed in [11]. The accuracy of this method compared to Monte Carlo reference has been discussed in [12] and in the references therein. It is worth noting that the SHE method used in this study does not include Carrier-Carrier Scattering (CCS) phenomena.

At program start, the cell is in the erased state with the considerations of Section III. The initial programming current of I_{BL}=1uA is set by adjusting the V_{BL} (typically in a range of 0.1 – 0.5V), whereas the other terminals follow Table 1. Under these bias conditions, the portion of the channel under the WL is in saturation since V_{WL}~ V_{BL}+V_{tWL}. Since FG is in deep erase state and the adjacent terminals are high, the VFG is also pulled high. Hence, the channel under the FG is in strong inversion since it satisfies $V_{FG} >> V_{SL} + V_{tFG}$, where Vt_{FG} is the threshold voltage of the channel area under the FG.

As a consequence, at the conjunction of these two regions both a high lateral and vertical field are present (Fig.2).

Fig.5 show the simulated Electron Distribution Functions (EDF) obtained with SHE along the channel of the cell at the beginning (t=0) and at the end of the programming (t=10us). The former are the object of the following discussion.

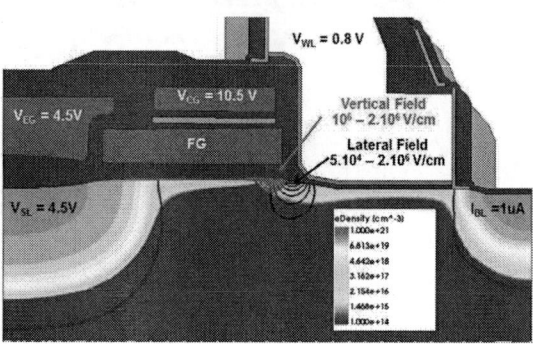

Fig.4: Electron density in the channel of the cell and Electric Fields at the Program start condition. The peak lateral field (along carrier transport) is located below the FG Spacer; the peak vertical field is located at the edge of the FG.

Analyzing the electron flow from BL to SL one can distinguish several regions of interest. As the "cold" electrons (Pos1) of the BL enter into and travel in the channel, they acquire the kinetic energy under the work of the Lateral Field. Note that nothing really happens till ~2/3 of the WL (Pos2), but then a rapid shift of the electron population towards the higher kinetic energies is observed (Pos3, 4, 5). This closely follows the pattern of the Lateral Field (Fig.4), albeit with ~10-20nm spatial delay, whereby the most heated electron population is located a few nm inside the FG (Pos5).

The "heated" distributions are described by a "plateau" and a Maxwellian slope (~1/kT), whereby i) the plateau is a result of the rearrangement in energy of the electrons after being subject to isotropic elastic and inelastic phenomena (such as acoustic and optical phonon scattering and impact ionization), ii) the steep slope is a reflection of the scattering phenomena which involve energy absorption (such as optical phonons – included, or CCS – not included here). The virtual intersection of these two parts is also called the knee point. It is interesting to notice that the knee point roughly corresponds to the amount of the available kinetic energy in the system under consideration ~q.(V_{SL}-V_{BL}).

This means that there is a non-negligible amount of electrons which were subject to only a few inelastic processes up to Pos5. As the electrons move further towards the SL (Pos6 and 7), the "hot" plateau starts cooling off, even though hot electrons are still present in the SL. At the same time the cold electrons population of the SL is clearly visible at the lower energies.

Fig.6 shows the distributions at Pos5 for different V_{SL}/V_{CG} biases at program start. The overall shape remains the same, but the knee point is shifted according to the available energy with a clear dependence to V_{SL}. It is interesting to notice that the V_{CG} also affects the maximum available carrier energy, by modulating the strong inversion under the FG, which modulates the distance needed to drop the lateral potential. However the vertical field shows a second-order effect compared to the lateral field in the hot carrier generation.

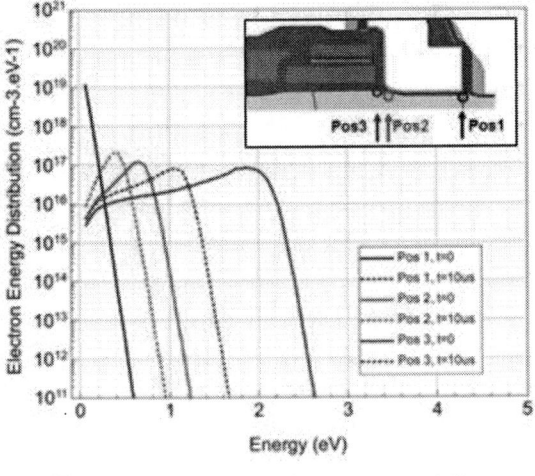

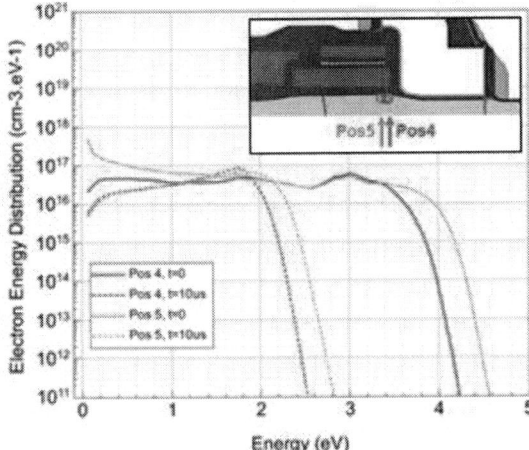

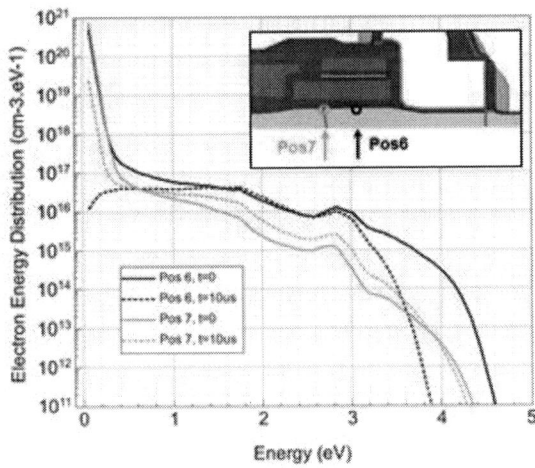

Fig.5: Electron Distribution Function (EDF) at the Silicon / Oxide interface plotted at the beginning (plain) and at the end (dashed) of the programming sequence (Table 1), respectively, *a)* in the channel region from BL to FGSpc, *b)* below the FG corner, *c)* below the FG and at the SL junction.

Assuming an isotropic carrier energy distribution, the current density tunneling into the FG is then calculated following [11]. Electrons having a kinetic energy higher than ~3eV will have a higher probability of being injected into the FG due to thermionic emission process. This

illustrates the importance of a correct description of the energy distributions.

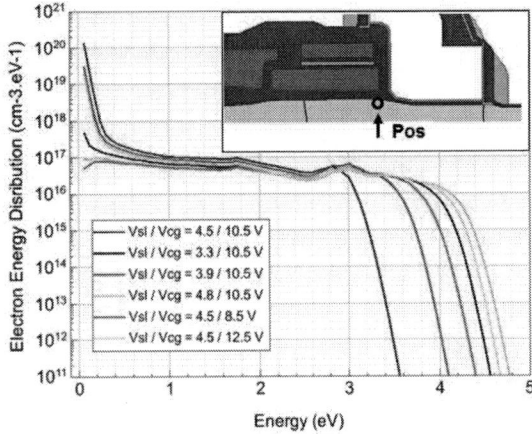

Fig.6: EDF plotted below the FG corner for varying Vsl/Vcg bias conditions at program start.

Fig.7 shows the calculated injected current density together with the electric fields along the channel position beginning (t=0) and at the end of the programming (t=10us). Initially the hot carrier injection into the FG occurs at the FG edge with a high injection length of ~30nm, and then falls exponentially towards the SL. As more electrons are injected into the FG, its potential drops which continually brings the Silicon channel region below the FG in saturation. This simultaneously causes a decrease i) of the Iprog (current flowing in the channel), ii) of the lateral field, iii) of the vertical field. All these effects limit the hot carrier injection. As the fields are modified, the EDF in Fig.5 (dashed) are also modified and at the end of the programming at 10us, the hottest distribution is found at around Pos.6. Eventually, the injection location is also shifted towards the SL, while still conserving the overall shape.

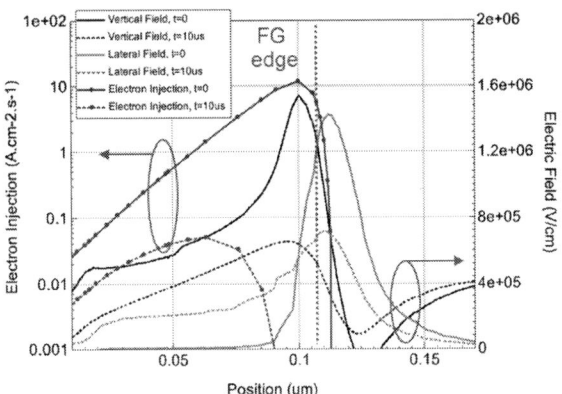

Fig.7: Lateral and vertical electric fields and electron injection into the FG along the Silicon / Oxide interface.

Finally Fig.8 shows the HW vs. TCAD comparison of the V_{tCG} after 10us of programming for a wide range of SL and CG biases. In particular, the SL bias sensitivity is well captured. This validates the injection and the overall cell simulation methodology.

171

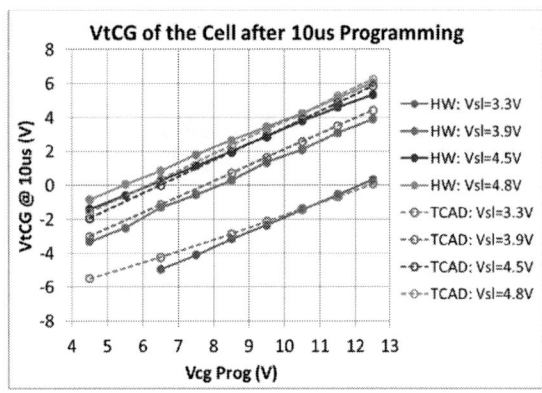

Fig.8: Vt$_{CG}$ as a function of the V$_{CGprog}$ at different V$_{SLprog}$ for HW (solid) and TCAD (dashed).

V. CELL OPTIMIZATION

The impact of the FGSpc thickness on the Cell performance is analyzed hereafter. BV aspect, as discussed in [9] is outside of our scope. Fig.9 shows that from a purely injection perspective, the injection efficiency (I$_{FG}$/I$_{BL}$) is rather constant or slightly reducing when increasing the FG Spacer thickness, at a given V$_{FG}$. This is in agreement with the MC findings of [7].

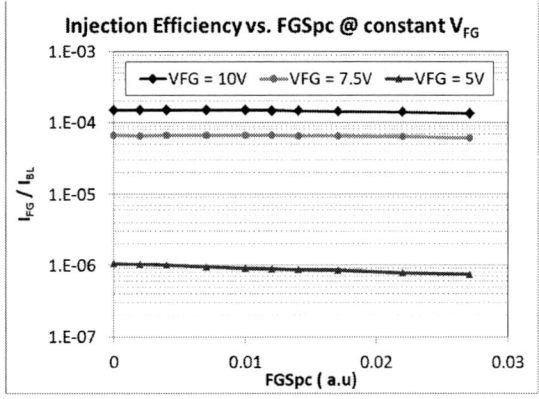

Fig.9: Injection efficiency for various FGSpc Thk. and FG biases.

However, applied to our investigated cell and bias conditions (Fig.1, Table1), a thicker FGSpacer results in an increased CG-FG coupling ratio (Fig.10), since the capacitance to the WL is reduced. This in turn caused the FG potential to be higher during the programming, which in turn increases the injection efficiency (Fig.9). Eventually a higher number of electrons are injected into the FG, thus resulting in a higher V$_{tCG}$. On the other hand, a thicker FGSpc degrades the read current of the cell in the ereased state (Fig.11), due to an increased resistance between channel areas under WL and FG. Inversely, having a too thin FGSpc, despite reliability concerns, also increases the SL-BL leakage after programming. This leads to a trade-off value for the FGSpc.

The discussed application highlights the need to consider comprehensive simulation methodologies which capture the interactions between FEoL processing, DC device operation and transient hot carrier injection modeling.

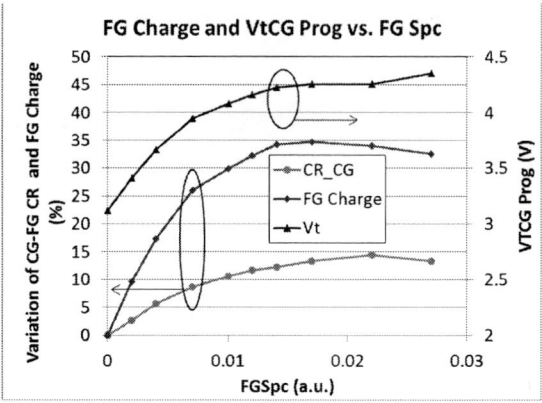

Fig.10: Variation of FG-CG coupling ratio,FG charge (left) as well Vt$_{CG}$ (right) vs. FGSpc Thk.

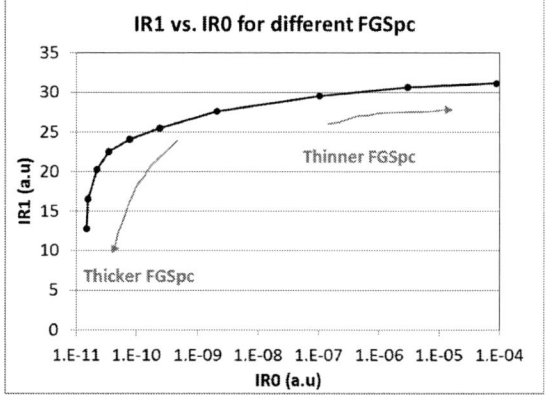

Fig.11: Cell read current after Erase (IR1) and Program (IR0) for different FGSpc thicknesses.

VI. CONCLUSION

28nm HKMG ESF3 Flash Cell operation has been modeled with TCAD in DC and transient operation with focus on hot carrier injection mechanism. The developed methodology has been employed to illustrate on a specific case the various metrics which are interwoven and which should be considered during cell architecture optimization.

ACKNOWLEDGMENT

The authors would like to thank SST team for providing the presented measurements as well as for the very insightful discussions.

REFERENCES

[1] R. Richter et. al., IEDM, 2018, pp. 18.5.1-18.5.4.

[2] Y. K. Lee et al., VLSI, 2017, pp. T202-T203.

[3] L.Q. Luo et. al., IMW, 2017, pp. 1-4.

[4] N. Do et al., IMW, 2018, pp. 1-3.

[5] Y. Tkachev, et al., IMW, pp. 1-4, 2017.

[6] W. Stefanutti, et al., IEEE TED, vol. 53, no. 1, pp. 89-96, Jan. 2006.

[7] P. Palestri, et al., IEEE TED, vol. 53, no. 3, pp. 488-493, March 2006.

[8] J. Van Houdt et al., IEEE TED, vol. 39, no. 5, pp. 1150-1156, May 1992.

[9] M. Slotboom et al., ESSDERC, 2003, pp. 159-162.

[10] S.Jourba et al., IMW, 2020.

[11] S. Jin, et al., SISPAD, 2009, pp. 1-4.

[12] A.Zaka et al., 2014, in Grasser T. (ed.), ISBN 978-3-319-08994-2, pg.151-196

Coupling the Multi Phase-Field Method with an Electro-Thermal Solver to Simulate Phase Change Mechanisms in Ge-rich GST based PCM

Raphaël Bayle
STMicroelectronics
Ecole Polytechnique, CNRS, IP Paris
CEA, LETI, Univ. Grenoble Alpes

Olga Cueto
CEA, LETI, Univ. Grenoble Alpes
Grenoble, France
olga.cueto@cea.fr

Serge Blonkowski
CEA, LETI, Univ. Grenoble Alpes
Grenoble, France

Thomas Philippe
Lab. Physique de la Matière Condensée
Ecole Polytechnique, CNRS, IP Paris
Palaiseau, France

Hervé Henry
Lab. Physique de la Matière Condensée
Ecole Polytechnique, CNRS, IP Paris
Palaiseau, France

Mathis Plapp
Lab. Physique de la Matière Condensée
Ecole Polytechnique, CNRS, IP Paris
Palaiseau, France

Abstract—**The ternary alloy GeSbTe is widely used as material for phase-change memories. Thanks to an optimized Ge-rich GeSbTe alloy, the crystallizion temperature of the alloy is increased and the stability requirements of high working temperature required for automotive applications are fullfilled, but the crystallization of the Ge-rich alloy proceeds with a composition change and a phase separation. We have developed a multi-phase-field model for the crystallization of the Ge-rich GeSbTe alloy and we have coupled it to an electro-thermal solver. This model is able to capture both the emergence of a two-phase polycristalline structure starting from an initially amorphous material, and the melting and recrystallization during the device operations.**

Index Terms—**Phase Change Memories, simulation, crystallization with segregation, multi phase-field method**

I. INTRODUCTION

This work is part of the development of the co-integration of 28nm Fully Depleted Silicon on Insulator (FD-SOI) advanced CMOS technology with an electrical non volatile Phase Change Memory (PCM) based on a chalcogenide ternary material for automotive micro-controller applications [1]. Thanks to an optimized Ge-rich GeSbTe alloy, a good trade-off between set speed and data retention performances is obtained [2]. In particular, the crystallization temperature of the Ge-rich GeSbTe (Ge-rich GST) alloy can be raised up to 200°C in comparison with that of the stoichiometric alloy $Ge_2Sb_2Te_5$ (GST). This helps to fulfill the stability requirements of high working temperature required for automotive applications. The Ge-rich alloy does not crystallize congruently: the crystallization proceeds with a composition change and a phase separation. The additional Ge is rejected into the amorphous phase upon crystallization of GST, which leads to the nucleation and growth of an additional crystalline phase that is almost pure Ge. Since Ge and GST phase have strongly different physical properties, it is of critical importance to understand the crystallization process of Ge-rich GST. The

Multi Phase-Field Method (MPFM) is an extension of the Phase-Field Method (PFM) widely used for alloys with non congruent crystallization. Phase Field Models are continuum models based on non equilibrium thermodynamics used in materials science to simulate the evolution of microstructures in a wide variety of processes. In a previous work, we had implemented the PFM for the congruent GST alloy [3]. Here, we present a new implementation of the MPFM for the Ge-rich GST alloy and its coupling with a finite element electro-thermal solver.

II. MPFM WITH ORIENTATION FIELDS

We assume that our alloy can be described in a quasibinary approximation: instead of considering that the concentration of Sb, Te and Ge are independent, we assume that the three concentrations evolve on a predetermined segment of the ternary diagram Sb-Te-Ge. The chosen segment links stoichiometric GST to pure Ge and we use as a reference the pseudobinary phase diagram along a line reaching from Sb_2Te_3 to pure Ge given in [8] which is close to the segment we consider. The parameter c in the model accounts for the position on this segment: $c = 0$ corresponds to stoichiometric phase (GST) and $c = 1$ to quasi-pure Ge phase (Ge). Three phases are considered: the crystalline GST, the crystalline Ge, and the disordered phase (amorphous/melted). The model relies on three phase fields p_i, $i = 1, 2, 3$, that correspond to the local volume fractions of the three phases and thus satisfy $p_i \in [0, 1] \, \forall i$ and $\sum_i p_i = 1$.

In addition, our model takes into account the orientations of the crystalline grains through two fields θ_1 and θ_2 encoding the orientation in phase Ge and GST. This adds complexity to the model and can lead to numerical singularities. Despite increased complexity, taking into account the grain orientation allows to track grain boundaries and to follow their evolution. This helps to understand the role of grain boundaries in the complex electrical behaviour of GST alloys (e.g. drift of the Set state).

Following the insight that the appropriate thermodynamical potential for systems with exchanges of matter is the grand-potential rather than the free-energy [5] [6], a grand-potential functional is used:

$$\Omega = \int_V \omega \, dV, \tag{1}$$

defined as the volume integral of a grand-potential density.

$$\omega(\mathbf{p}, \vec{\nabla}\mathbf{p}, \mu, T) = K\omega_{\text{grad}}(\vec{\nabla}\mathbf{p}) + H\omega_p(\mathbf{p}) + \omega_{\text{th}}(\mathbf{p}, \mu, T) \tag{2}$$

where μ is the chemical potential defined by $\mu = \partial f/\partial c$, T is the temperature and $\mathbf{p}$ is the vector (p_1, p_2, p_3).

$$\omega_{\text{grad}} = \frac{1}{2}\sum_i \left(\vec{\nabla p_i}\right)^2, \tag{3}$$

ω_{grad} sets a free-energy cost for gradients in p, forcing interfaces to have a finite width, and K has dimensions of energy per unit length.

$$\omega_p = \omega_{\text{TW}} = \sum_i p_i^2(1-p_i)^2. \tag{4}$$

ω_p depends only on the phase fields and provides an "energy landscape" in $\mathbf{p}$ and H is a constant with dimensions of energy per unit volume.

ω_{th} is the thermodynamic part of the grand-potential

$$\omega_{th}(\mu, T) = \sum_i g_i(\mathbf{p})\omega_i(\mu, T), \tag{5}$$

where the ω_i are given by (6) and the $g_i(p_i, p_j, p_k)$ are interpolation functions taken as proposed by [5].

The grand-potential ω is defined as the Legendre transform of the free-energy f, that is,

$$\omega_i(\mu, T) = f_i(c, T) - \mu c_i(\mu, T) \tag{6}$$

where $c_i(\mu, T)$ is the inverse of $\mu_i(c, T) = \partial f_i/\partial c$. The invertibility of the μ_i functions is required here; this condition is equivalent to the convexity of free energy functions at any temperature.

The p_i, i=1,2 are assumed to evolve towards the minimum of Ω,

$$\frac{\partial p_i}{\partial t} = -\Gamma \left.\frac{\delta\Omega}{\delta p_i}\right|_{p_1+p_2+p_3=1}, \quad \text{i=1,2} \tag{7}$$

where Γ is a relaxation rate constant.

The interface width W and the relaxation time τ are defined by

$$W = \sqrt{\frac{K}{H}}, \tag{8}$$

$$\tau = \frac{1}{\Gamma H} \tag{9}$$

An implementation of the orientation model was chosen taking benefit of the model proposed in [4] for a single cristalline phase case. A coupling term is added to (7):

$$\frac{\partial p_i}{\partial t} = -\Gamma \left.\frac{\delta\Omega}{\delta p_i}\right|_{p_1+p_2+p_3=1} - \Gamma C_\theta q'(p_i)(\vec{\nabla}\theta_i)^2, \quad \text{i=1,2} \tag{10}$$

where C_θ is a constant associated to the energy of the grain boundaries. This coupling term is chosen in order to restore the model proposed in [4] when one of the 2 cristallines phases is missing. The function q is the coupling function introduced in [4]:

$$q(p_i) = \frac{7p_i^3 - 6p_i^4}{(1-p_i)^3}. \tag{11}$$

Finally, the non-conserved phase-fields evolve as:

$$\begin{aligned}
\tau \frac{\partial p_i}{dt} &= W^2\nabla^2 p_i + \frac{2}{3}F_{\text{TW}}(p_i) \\
&\quad - \frac{1}{H}\sum_j \omega_i(\mu)\left.\frac{\partial g_j}{\partial p_i}\right|_{\sum_i p_i = 1} \\
&\quad - \frac{C_\theta}{H}q'(p_i)(\vec{\nabla}\theta_i)^2, \quad \text{i=1,2}
\end{aligned} \tag{12}$$

with $F_{\text{TW}}(p_i) = -2p_i(1-p_i) + \sum_{j\neq i} p_j(1-p_j)(1-2p_j)$

The orientation fields evolve as:

$$\tau_{\theta_i}\partial_t\theta_i = \frac{1}{q(p_i)}\nabla.(q(p_i)\vec{\nabla}\theta_i), \quad \text{i=1,2} \tag{13}$$

where τ_{θ_i} are relaxation times for the orientation fields.

The concentration c is a conserved field, and thus obeys the continuity equation

$$\frac{\partial c}{\partial t} + \nabla \cdot \vec{J} = 0, \tag{14}$$

where the flux $\vec{J}$ is driven by the gradient of the diffusion potential

$$\vec{J} = -M(\mu, p)\vec{\nabla}\mu \tag{15}$$

with $M(\mu, p)$ the mobility of Ge. In order to ensure the conservation of matter with the best numerical accuracy, the chemical diffusion (14) is solved using c as a variable rather than μ. Due to the convexity of the free-energies, μ can be easily calculated from c by inversion of $c = \sum_i g_i(p)c_i(\mu, T)$.

III. THERMODYNAMIC AND MATERIAL MODEL

The PFM requires the choice of a free energy for each phase as a function of the concentration and temperature. Our energy model has been obtained by assuming a regular solution model and by merging two lens-shape phase diagrams to get an eutectic phase diagram with reasonable value for the eutectic composition and temperature (Fig. 1).

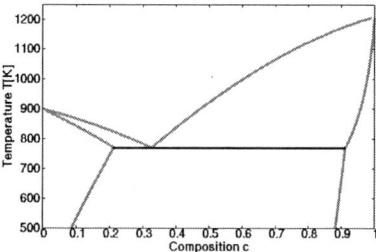

Fig. 1: Phase diagram obtained from the free energy functions

Among the parameters to be fixed, the constants H and K are related to the interface width W by (8) and to the

174

interface free energy σ by $\sigma = \sqrt{2KH}/3$ [5]. The parameter interface width is chosen to be $W = 5\ 10^{-10}m$, which is a physically realistic value for solid-liquid interfaces. The value $\sigma = 0.4J/m^2$ is extracted for the surface energy near 700K from data about surface tension for amorphous-crystalline interfaces in germanium [10].

The relaxation time τ is evaluated from kinetic data for GST225 [7] and the relaxation times for orientation fields τ_{θ_i} are deduced from τ.

$M(\mu, p)$, the mobility of Ge, is used to evaluate the diffusion flux of the equation (15). It is evaluated from $D(\mu, p)$, the diffusivity of Ge, using the relation $M(\mu, p) = D(\mu, p)\partial c/\partial\mu$. Diffusivity of Ge has been chosen for both the crystalline phases and the amorphous phases according to the one used in GST225 [9]. Near 700K, Ge diffusivity is found to be close to $10^{-9}m^2/s$ in the amorphous and close to $10^{-12}m^2/s$ in the crystalline phases.

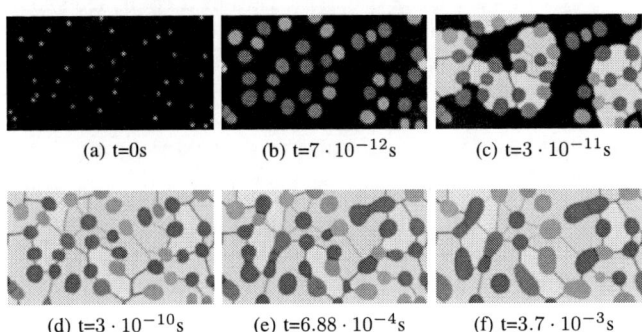

(a) t=0s   (b) t=$7 \cdot 10^{-12}$s   (c) t=$3 \cdot 10^{-11}$s

(d) t=$3 \cdot 10^{-10}$s   (e) t=$6.88 \cdot 10^{-4}$s   (f) t=$3.7 \cdot 10^{-3}$s

Fig. 2: Temporal evolution of the crystalline state in 2D simulation of the crystallization of an amorphous layer of Ge-rich GST. Amorphous areas are represented in black. The different shades of pink/red stand for crystalline Ge phase and green/blue for GST phase.

IV. CRYSTALLIZATION OF AN AMORPHOUS LAYER

In the production of PCM devices, the PCM material is deposited amorphous, but crystallizes during subsequent heat treatments. In order to model the initial state of the material before operations of the device, we must therefore simulate the crystallization of an amorphous layer. Accordingly, the MPFM model is used to simulate an isothermal crystallization at 673K during 3.7ms. The simulation domain is 150 nm wide and 100 nm high. A nucleation scenario inferred from experimental observations is used [11]. During the annealing of an amorphous layer of Ge-rich GST, it is observed that the Ge crystallizes first, followed by nucleation of the GST phase, for temperature above 673K. Thus in the simulation, grains of the Ge phase are introduced initially randomly in the amorphous matrix (Fig. 2a). Their subsequent growth drains Ge from the matrix. When the latter is sufficiently impoverished in Ge, the GST phase can nucleate. The GST nuclei are randomly introduced as soon as a critical concentration of 0.35 is reached locally. After many nucleation events and some growth of the nuclei, grain boundaries are formed between the GST grains (Fig. 2c). Even when the crystallization has completed (Fig. 2d), the microstructures are still evolving (Fig. 2e and Fig. 2f). This reorganization occurs mainly by two ways. The first one is the Ostwald ripening: small grains dissolve while bigger ones grow. Since Ge grains pin the grain boundaries between GST grains, as soon as a Ge grain disappears, the GST grain boundary network is modified. The second way of reorganization is due to the coalescence of Ge grains and to the motion of grain boundaries which evolve so that their local curvature is reduced. Indeed, when two Ge grains coalesce nothing can hinder the grain boundary motion. Those two mechanisms can be seen in Fig. 2e and Fig. 2f.

V. COUPLING WITH AN ELECTRO-THERMAL SOLVER

To simulate Joule heating during device operations, an electro-thermal solver is used; temperature fields obtained from the electro-thermal simulations are then used by the

MPFM model. The electro-thermal solver relies on the coupled system of partial differential equations formed by the charge conservation and the heat transfer equations :

$$\nabla \cdot (-\sigma\nabla V) = 0 \tag{16}$$

$$\rho C_p \frac{\partial T}{\partial t} + \nabla \cdot (-k_{th}\nabla T) = \sigma(\nabla V)^2 \tag{17}$$

where σ, ρ, C_p and k_{th} stand for the materials electrical conductivity, density, heat capacity and thermal conductivity. Electrical and thermal conductivity of the Ge rich GST are modeled as function of temperature:

$$\sigma_{\text{pcm}} = \frac{\sigma_0}{2}\left(\tanh(B_eT + C_e) + D_e\right) \tag{18}$$

$$k_{th,pcm} = \frac{k_{th}^0}{2}\left(\tanh(B_{th}T + C_{th}) + D_{th}\right) \tag{19}$$

A set of parameters for electrical and thermal conductivity for Ge-rich GST was fixed (see Table I) in order to obtain current-voltage curves close to experimental results.

TABLE I: Parameters for Ge-rich GST electrical and thermal conductivity

σ_0	B_e	C_e	D_e
$2.8 \cdot 10^4 S/m$	$0.0022K^{-1}$	-1.8	1
k_{th}^0	B_{th}	C_{th}	D_{th}
$2.566W/K/m$	$0.0051K^{-1}$	-48.359	1.418

The system of electro-thermal equations is solved using the Finite Element method with COMSOL Multiphysics®. The MPFM equations are discretized using the Finite Difference method on a cartesian grid with an Euler explicit time discretization. The coupling between the electro-thermal model and the MPFM model has been implemented first with a simplified approach; the electro-thermal equations are solved without taking into account phase changes. At each time t^n, the electro-thermal equations are computed with a time step of

$10^{-9}s$. Then one nanosecond of the MPFM is computed using the value of temperature field at t^n. This is actually CPU-time expensive because of the explicit time scheme which implies a stability condition on the time step $(dt < 7.8\,10^{-13}s)$.

VI. SIMULATION RESULTS AND DISCUSSION

The device simulated is a state-of-the-art PCM device [1] comprising a wall storage element (Fig. 3a) to which a trapezoidal pulse of $6\mu s$ with a plateau of $100ns$ is applied. With a current of the order of the melting current, Joule heating is high enough, during the plateau of the trapezoidal pulse, for the melting of the active zone of the PCM material as illustrated by the temperature field (Fig. 3b and 3c). For this

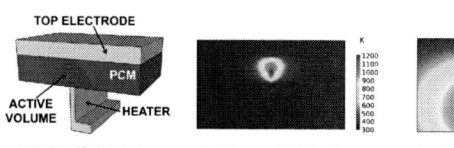

(a) Wall Structure   (b) Tmax/Wall Structure   (c) Tmax/Active PCM

Fig. 3: Joule heating simulation of the Wall Structure

simulation, the PCM layer is initially in a polycristalline state as obtained from a isothermal recrystallization (fig. 4a). As illustrated by Fig. 4c, the active domain of PCM is partially melted during the pulse. When the current is ramped down, the temperature decreases, and crystallization takes place with the coupled growth of the two crystalline phases: the Ge phase and the GST phase. The exchange of Ge atoms between the two phases is a determining factor of the crystallization kinetics. From $1.6\mu s$ the Ge phase is not growing any more (the grains of the Ge phase have the same shape in Fig. 4e and Fig. 4f). When the GST phase is the only one growing, the remaining melted domain gets richer in germanium and this leads to the nucleation of the Ge phase in the central domain (Fig. 4f). Among the main differences between the microstructures at the initial state (Fig. 4a) and at the end of the recrystallization (Fig. 4f), we see clearly a redistribution of germanium. Our simulation confirms qualitatively the impoverishment of germanium of the active domain and the accumulation of germanium on the periphery of this active domain that was already highlighted by experimental results [12].

VII. CONCLUSION AND PERSPECTIVES

We have implemented a Multi-Phase Field Model to simulate the non congruent phase change of a Ge-rich GST alloy. This model takes into account the possibility of crystallization into two different phases, the stoichiometric GST225 and an almost pure Ge phase, and resolves the orientation of the crystal grains. We have used a simplified thermodynamic description in terms of a pseudobinary phase diagram, and we have adjusted the parameters of this diagram as well as the kinetic parameters of the model to experimental data available in the literature. This work brings a novel implementation of the Multi-Phase Field Model and its coupling to an electrothermal solver. As of now, the method is clearly only qualitative, and

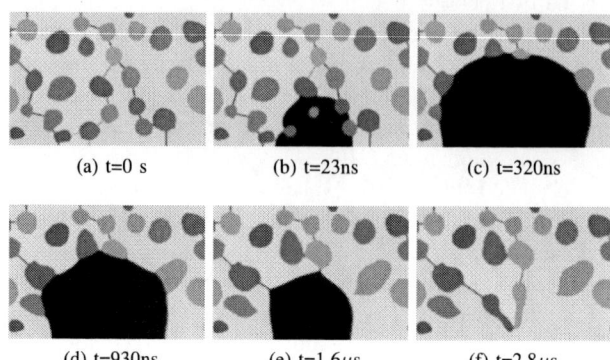

(a) t=0 s   (b) t=23ns   (c) t=320ns

(d) t=930ns   (e) t=1.6μs   (f) t=2.8μs

Fig. 4: Temporal evolution of the microstructures during the application of a trapezoidal current pulse to the Wall Structure with the PCM initially in polycrystalline state. The color code is the same as in Fig.2

numerical optimization is still needed to allow for intensive simulations, but our work already paves the way to a better understanding of the complex phenomena (including germanium segregation and grain boundary evolution) that take place during operations in GST-based PCM. This work was funded by the Association Nationale Recherche Technologie (ANRT), France, and STMicroelectronics through the CIFRE contract number 2016/1237.

REFERENCES

[1] F. Arnaud et al, "Truly Innovative 28nm FDSOI Technology for Automotive Micro-Controller Applications embedding 16MB Phase Change Memory," IEDM Tech. Dig. Dec. 2018
[2] G. Navarro et al, "Trade-off Between SET and Data Retention Performances Thanks to Innovative Materials for Phase-Change Memory," IEDM Tech. Dig. Dec. 2013, pp. 21.5.1-21.5.4
[3] O. Cueto, V. Sousa, S.Blonkowski, G.Navarro, "Coupling the Phase-Field Method with an Electrothermal Solver to Simulate Phase Change Mechanisms in PCRAM Cells," proceedings of IEEE conf. SISPAD 2015
[4] H. Henry, J. Mellenthin, and M. Plapp, "Orientation-field model for polycrystalline solidification with a singular coupling between order and orientation," Phys. Rev. B **86**, 054117 (2012)
[5] R. Folch and M. Plapp, "Quantitative phase-field modeling of two-phase growth," Phys. Rev. E **72**, 011602 (2005)
[6] M. Plapp, "Unified derivation of phase-field models for alloy solidification from a grand-potential functional," Phys. Rev. E **84**, 031601 (2011)
[7] J. Orava, A. L. Greer, B. Gholipour, D. W. Hewak, C. E. and Smith, "Characterization of supercooled liquid Ge2Sb2Te5 and its crystallization by ultrafast-heating calorimetry," Nat. Mater. 2012, 11, 279–283 DOI: 10.1038/nmat3275
[8] S. Bordas, M.T Clavaguer-Mora, B. Legendre and C. Hancheng, "Phase diagram of the ternary system Ge-Sb-Te: II. The subternary Ge-GeTe-Sb2Te3-Sb," Thermochim. Acta, 107 (1986), pp. 239-265
[9] G. Novielli, A. Ghetti, E. Varesi, A. Mauri, R. Sacco, "Atomic migration in phase change materials," IEDM Tech. Dig. Dec. 2013
[10] C. Reina, L. Sandoval, J. Marian, "Mesoscale computational study of the nanocrystallization of amorphous Ge via a self-consistent atomistic phase-field model, ", Acta Mater. 77, 2017, 335-351
[11] M. Agati, M. Vallet, S.Joulié, D. Benoit and A. Claverie, "Chemical phase segregation during the crystallization of Ge-rich GeSbTe alloys,", Journal of Materials Chemistry C, 2019, 7,8720
[12] P. Zuliani et al, "Overcoming Temperature Limitations in Phase Change Memories With Optimized GeSbTe," IEEE Trans. Electronics Devices v60 n12, pp.4020, 2013

Efficient partitioning of surface Green's function: toward *ab initio* contact resistance study.

Guido Gandus[1,2], Youseung Lee[2], Daniele Passerone[1] and Mathieu Luisier[2]

[1]nanotech@surfaces (EMPA); [2]Integrated Systems Laboratory (ETH Zurich)

Abstract—In this work, we propose an efficient computational scheme for first-principle quantum transport simulations to evaluate the open-boundary conditions. Its partitioning differentiates from conventional methods in that the contact self-energy matrices are constructed on smaller building blocks, principal layers (PL), while conventionally it was restricted to have the same lateral dimensions of the adjoining atoms in a channel region. Here, we obtain the properties of bulk electrodes through non-equilibrium Green's function (NEGF) approach with significant improvements in the computational efficiency without sacrificing the accuracy of results. To exemplify the merits of the proposed method we investigate the carrier density dependency of contact resistances in silicon nanowire devices connected to bulk metallic contacts.

I. INTRODUCTION

In recent years, quasi one-dimensional (1D) materials such as nanowires and nanotubes have been at the forefront of nanotechnology research. Their unique electronic properties make them promising candidates as next-generation logic switches. However their performance, is often limited by their high contact resistance [1]. First-principles quantum transport simulations are ideally suited to design contacts with low resistances based on atomistic models as close as possible to reality. Such *ab initio* investigations require heavy computational burden as compared to, e.g., tight-binding or k·p models. One of the major bottlenecks arises from calculating the boundary conditions. Usually, the electrodes employed in 1D devices are bulk like, i.e. they consist of a repeating unit cell. The standard method in transport

calculations is to build a supercell (SC)-sized contact which is constraint to have at least the same lateral dimensions as the channel (see Fig. 1). Because the computational cost to evaluate the boundary self-energies is in general proportional to the cube of the system size, large-scale electron-transport characterization are often impractical.

In this work, we propose an efficient algorithm to address this issue, partitioning the contact self-energy into smaller building blocks called principal layers (PL). This method allows to fully retain the electronic information of the contacts, while significantly improving the computational efficiency, on the basis of Bloch's theorem. As an application, we demonstrate the strength of the proposed approach for large-scale quantum transport simulations of silicon nanowires with metallic electrodes and report the resulting contact resistances as a function of the nanowire diameters and carrier densities.

II. ALGORITHM

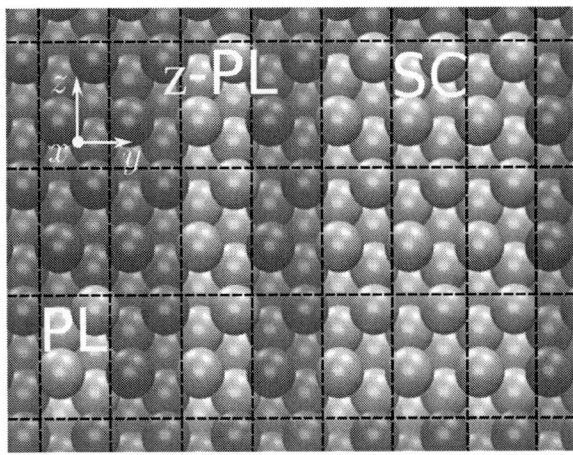

	$n_x \times n_y \times n_z$	$k_x \times k_y \times k_z$
PL	$1 \times 1 \times 1$	$3 \times n_y \times n_z$
z-PL	$1 \times 1 \times n_z$	$3 \times n_y \times 1$
SC	$1 \times n_y \times n_z$	$3 \times 1 \times 1$

Fig. 2: Schematic representation of a PL, a z-PL and a SC on Au(111) surface. The table summarizes the number of PL repetitions in real space (left column) and the corresponding k-space mesh (right column) used in the electronic structure calculation to ensure a correct inclusion of the minimum image convention.

Our self-energy partitioning algorithm is summarized in

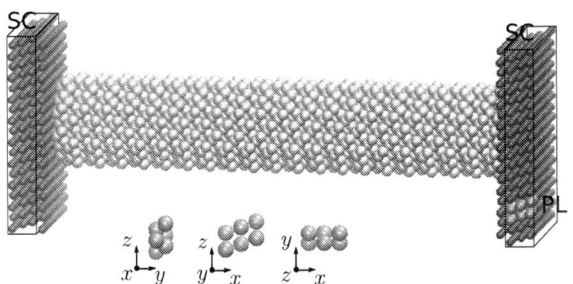

Fig. 1: Schematic view of a circular Silicon nanowire (SiNW) structure with bulk-like metallic contacts surrounding it. Yellow spheres represent Si and H atoms (9 and 5 orbitals per atom), the brown ones Au (9 orbitals per atom). The black rectangles mark the SC contacts, while the pink spheres highlight one PL composing the SC contacts. The latter is constructed by repeating a single PL in the transverse directions perpendicular to the nanowire axis.

Fig. 2. We consider a SC contact as a stack of principal layers (PLs) that are repeated $n_y \times n_z$ times along the y- and z-directions transverse to the transport x-direction. We assume that: i) the PL has only nearest neighbor couplings ($n_x = 1$) in the x-direction, with n_y and n_z being odd numbers; ii) the SC is periodic in the y- and z-directions. First, the electronic structure of the periodic PL is computed with a local basis set of orbitals. The Brillouin zone is sampled with a $3 \times n_y \times n_z$ k-space grid. This ensures that projecting the SC into the PL leads to the corresponding Bloch-space Hamiltonian. We then carry out a partial Bloch sum over the 3 wavevectors in the x-direction,

$$H_0(k_t) = \sum_{k_x} H(k) e^{i(k_x, k_t) \cdot (0,0,0)}, \tag{1}$$

$$H_1(k_t) = \sum_{k_x} H(k) e^{i(k_x, k_t) \cdot (1,0,0)}, \tag{2}$$

where k_t includes the transverse components (k_y and k_z) of k. The $H_i(k)$ are the k-dependent Hamiltonian matrices. The overlap matrices $S_0(k_t)$ and $S_1(k_t)$ are obtained in the same way. Eqs. (1) and (2) transform a set of fully-delocalized Bloch wavefunctions into a set of hybrid wave-functions which remain extended (Bloch-like) along the transverse directions, but are localized in the transport direction. We note that the aforementioned transformation enables the usage of an iterative scheme [2] to compute the surface Green's function $G(E, k_t)$. Hereafter, we drop the energy dependence E for ease of readability. By performing a Bloch summation over the transverse wavevectors, the distance-dependent Green's functions $G_{n,m}$ can be finally obtained as

$$G_{n,m} = \sum_{k_t} G(k_t) e^{i(k_t) \cdot (n,m)}, \tag{3}$$

where n, m refer to the position of the PL within the SC. The surface Green's function within one PL is simply given by $\sum_{k_t} G(k_t)$. As last step, we construct the SC Green's function G as the circulant matrix

$$G = \begin{bmatrix} G_0^z & G_1^z & \cdots & G_{-1}^z \\ G_{-1}^z & G_0^z & \cdots & G_1^z \\ \vdots & & \ddots & \vdots \\ G_1^z & G_{-1}^z & \cdots & G_0^z \end{bmatrix}, \tag{4}$$

with matrix elements that are themselves circulant

$$G_i^z = \begin{bmatrix} G_{i,0} & \cdots & G_{i,-1} \\ \vdots & \ddots & \vdots \\ G_{i,1} & \cdots & G_{i,0} \end{bmatrix}. \tag{5}$$

The indices of the first row in G and G_i^z are: $[0, 1, ..., (n_y - 1)/2, -(n_y - 1)/2, ..., -1]$ and $[0, 1, ..., (n_z - 1)/2, -(n_z - 1)/2, ..., -1]$, respectively. The boundary self-energies can then be derived from the matrix products [3]

$$\Sigma(z) = (zS_1^\dagger - H_1^\dagger) G(z)(zS_1 - H_1), \tag{6}$$

where $z = i\eta + E$ with η a small positive number

while H_1 and S_1 are the Hamiltonian and overlap matrices, respectively, connecting two consecutive SC in the x-direction. They are evaluated in the same way as G from the elements of Eq. (2).

Here we highlight the computational efficiency of the proposed algorithm. Indeed, the iterative scheme to compute the surface Green's function requires to calculate the inverse of a matrix with the same dimensions as G. The computational cost is then $O((nN)^3)$, where $n = n_y \times n_z$ and N is the number of basis functions in one PL. Thanks to the proposed partitioning, the computational complexity of the recursive algorithm is reduced to $O(nN^3)$ as one has to invert n distance-dependent Green's functions $G_{n,m}$, each of dimension N. This scheme is therefore particularly advantageous to simulate 1D devices with large channels. We also note that it is the odd condition on n_y and n_z that ensures a correct inclusion of the minimum-image convention. Careful attention must be payed when constructing the SC contacts which must be able to be partitioned into an odd number of PLs. As an alternative, one can consider a stack of PLs as the smallest building block for which the distant-dependent Green's functions are evaluated. This is the case when either n_y or n_z is an even number.

III. RESULTS

As benchmark example, we consider a Silicon nanowire (SiNW) attached to bulk Au(111) surfaces [4]. The SC are modeled by a three-layer-thick Au(111) and are composed of 7×5 PLs, each having 6 atoms with 9 orbitals per atom. The scattering region includes the nanowire extended to each terminal with four Au(111) slabs (see Fig. 1). The H and S matrices are computed from density functional theory (DFT) with GPAW [5] using single zeta polarized (SZP) basis functions and the Perdew, Burke, Ernzerhof (PBE) exchange-correlation functional. To validate the accuracy of our approach, we compare the conductance spectra obtained with the entire SC, a stack of PLs with $n_y \times n_z = 1 \times 5$ (z-PL), and a single PL. Fig. 3 shows that the proposed algorithm successfully reproduces the SC conductance spectra within 3×10^{-2} absolute errors.

To investigate the influence of the discrepancy in electronic structure results on the conductance spectra, we compare the H_0 and H_1 matrices obtained by the SC and PL approaches. For the former, the matrices are directly obtained by Eqs. (1) and (2) since the wavefunctions are already real in the transverse directions, whereas for the latter, they are constructed following the same steps outlined for G Eqs. (1) to (6). Figs. 4(a-b) report the absolute errors for each entry of the matrices. As expected, the large errors appear in the entries connecting neighboring PLs and are comparable with the average absolute error in conductance (see black line in the bottom panel of Fig. 3). We conclude that it is already at the DFT level that the majority of the errors are accumulated. When constructing H_0 with Eqs. (4) and (6) we assume that the electronic potential in the PLs is incommensurate with the SC structure, i.e., the diagonal entries H_0^z and $H_{i,0}$ are

178

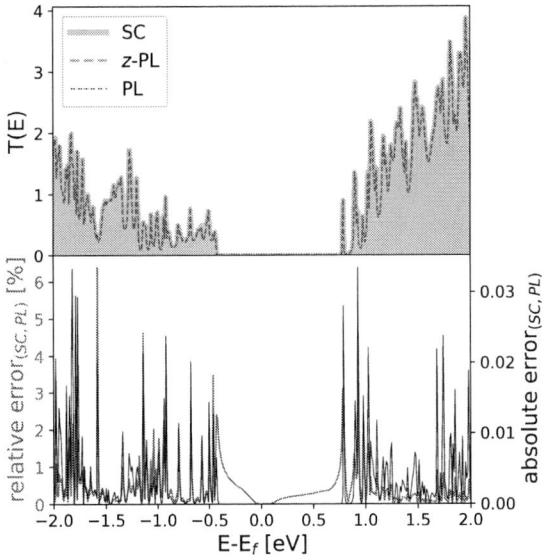

Fig. 3: (Top panel) Conductance spectra of a SiNW obtained with the entire SC, a z-PL, and a single PL. The nanowire has a length of 10 nm and a diameter of 2 nm. (Bottom panel) numerical difference between the SC and PL transmissions.

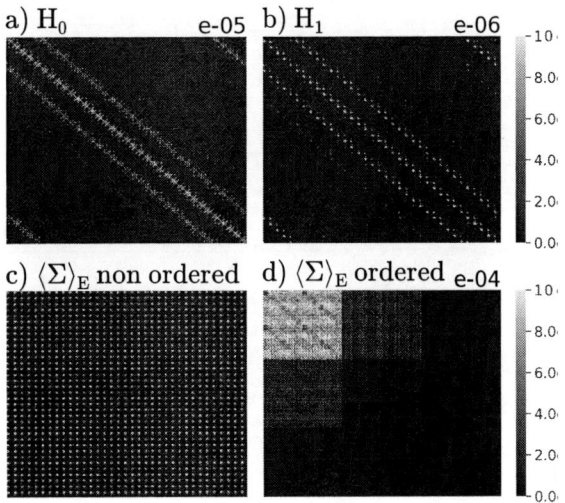

Fig. 4: Absolute error in (a) $\boldsymbol{H}_0$ and (b) $\boldsymbol{H}_1$ between the Hamiltonian matrices obtained by an electronic structure calculation of a single PL with respect to the reference SC. (c) Absolute error for each entry in $\Sigma(z)$ averaged over an energy range from -4eV to 4eV with steps of 0.5eV. (d) Same as (c), but with matrix elements arranged in such a way that the SC atoms are ordered along the transport direction.

equivalent. However, the Hamiltonian of the SC does not reflect this symmetry.

To see how the discrepancies introduced by the DFT calculation propagate in the recursive algorithm for computing the surface Green's function, we evaluate the absolute error between the boundary self-energies obtained with the Hamiltonian and overlap matrices corresponding to the PL and SC structures. To achieve this without loss of generality for energy points, we compute the average error for each entry in $\Sigma(z)$ using 16 energy points in the range from -4eV to 4eV with a regular grid spacing of 0.5eV. The result is illustrated in Fig. 4(c) where it can be observed that the absolute error overall increases by one order of magnitude when compared to Figs. 4(a-b). This is attributed to the average number of iterations required to converge $\boldsymbol{G}$, which is found to be 12. The error adds up linearly at each iteration step. To gain further insights into the distribution of the error, we reorder the matrix elements from Fig. 4(c) so that the atoms in the SC structure are aligned along the transport direction [1]. The resulting matrix is plotted in Fig. 4(d). It can be seen that the majority of the errors appear in the entries in the upper-left corner which couples the electrodes to the adjoining atoms in the scattering region closer to the interface.

The computational efficiency of our approach is summarized in Fig. 5 where we report the SC/PL Speed Up in generating the self-energies terms for several SC sizes. The result shows that the CPU time could be accelerated

---

[1] The reordering is applied consistently in the conductance calculations where the adjoining atoms in the scattering region are ordered for increasing x-coordinate so as to maintain a block tridiagonal structure of the scattering Hamiltonian [6].

by a factor 280 for the largest SC, reducing the time from 554.43 to 1.98 sec.

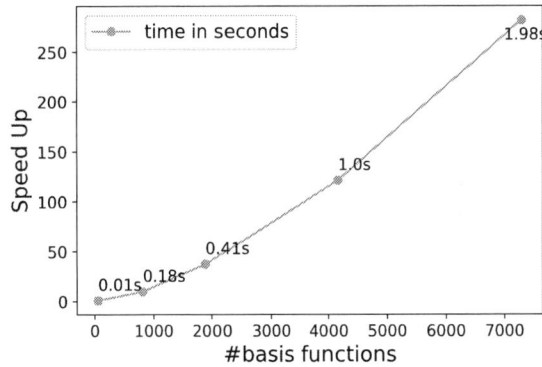

Fig. 5: Scalability tests to generate the self-energy matrix of a Au-SiNW-Au structure as a function of the number of basis functions. The data points refer to SCs with $n_y \times n_z = 1 \times 1$, 5×3, 7×5, 11×7 and 15×9. The speed up factor is obtained as [SC time / PL time]. In addition, the CPU time required by the proposed algorithm is reported on top of the data points.

To verify that increasing the SC size does not deteriorate the accuracy, we evaluate the density of states (DOS) for a selection of SC sizes. The DOS is given by the formula:

$$DOS(E) = Tr\left[\boldsymbol{G}(E)\boldsymbol{S}_0\right]. \tag{7}$$

From Eq. (7) we can see that this quantity includes both diagonal and off diagonal errors into one number. The energy-resolved DOS is shown in the top panel of Fig. 6(a-c) for varying $n_y \times n_z$. In all considered cases, the DOS

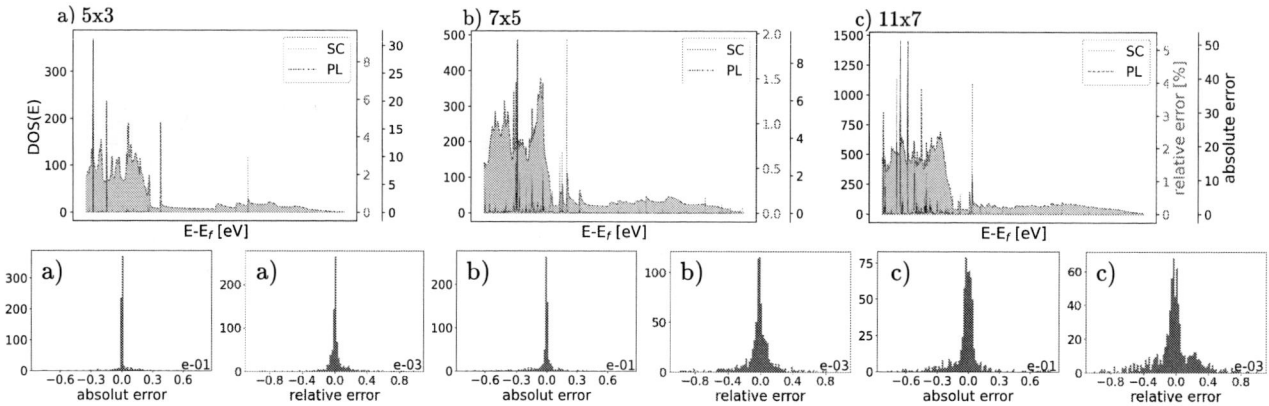

Fig. 6: (Top panel) Energy-resolved DOS of the device contacts in Fig. 1 (left axis) and absolute and relative errors between the proposed method and the reference SC (right axis). (Bottom panel) Histogram of the error for the considered contact models.

obtained by a single PL agrees almost perfectly with the SC reference, thus confirming the strength and validity of our scheme. The absolute and relative errors (with the error-axis on the right side of top panel in Fig. 6(a-c)) are overlaid on the DOS curves. Both errors are more pronounced at the position of the peaks in the DOS curves. From the distribution of the error summarized in the histograms in the bottom panel of Figs. 6(a-c) we observe that the variance increases with the SC size. However, the maximum error does not show a SC-size dependence as can be seen in Table I.

Maximum error		
$n_y \times n_z$	absolute error	relative error
5x3	31.13	0.092
7x5	9.22	0.019
11x7	42.21	0.04

TABLE I: Table summarizing the maximum absolute and relative errors found for the DOS curves in Fig. 6. All maximum errors are found correspond to peaks in the DOS curves.

As an application of the proposed method, we investigate the electrical resistance R between SiNWs and metal contacts as a function of the carrier density $\langle n_{2d} \rangle$ for various diameters. We consider three SiNWs models with 2, 3 and 4 nm diameters. Fig. 7 shows that R i) exhibits a clear $\langle n_{2d} \rangle^{-\alpha}$ dependence (with $\alpha = 1.068$) and ii) it is weakly affected by the diameter. This confirms that R is dominated by the quasi-Fermi level drop near the contact regions [7].

IV. CONCLUSIONS

We developed an algorithm to efficiently compute the self-energy matrices of large quasi-1D channels without sacrificing accuracy. By partitioning the contacts into PLs, we are able to significantly reduce the simulation time and to evaluate the electrical resistance of SiNWs with metallic contacts. This algorithm is key to simulate more realistic nanostructures with bulk-like contacts.

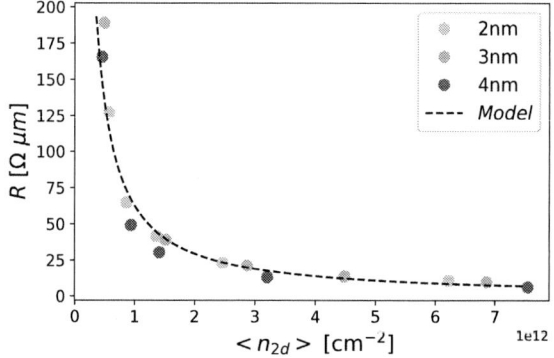

Fig. 7: Electrical resistance of the SiNW model in Fig. 1 as a function of the average carrier density for various nanowire diameters.

REFERENCES

[1] L. Bourdet et al., *J. Appl. Phys.* 119, 084503 (2016).

[2] M. L. Sancho et al., *J. Phys. F* 14, 1205 (1984).

[3] K.S. Thygesen et al., *Phys. Review B* 73, 035309 (2006).

[4] S. E. Mohney et al., *Solid-State Electron.* 49, 227-232 (2005).

[5] J.Enkovaara et al., *J. Phys.* 22, 253202 (2010).

[6] A. R. Dmitry et al., *Theory of Quantum Transport at Nanoscale: An Introduction.*

[7] D. Rideau et al., *SISPAD*, 101-104 (2014).

Quantum Transport in Si:P δ-Layer Wires

Juan P. Mendez
Cognitive & Emerging Computing
Sandia National Laboratories
Albuquerque, USA
jpmende@sandia.gov

Denis Mamaluy
Cognitive & Emerging Computing
Sandia National Laboratories
Albuquerque, USA
dnmamal@sandia.gov

Xujiao Gao
Electrical Models & Simulation
Sandia National Laboratories
Albuquerque, USA
xngao@sandia.gov

Evan M. Anderson
Multiscale Fab. Sci. & Tech. Dev.
Sandia National Laboratories
Albuquerque, USA
emander@sandia.gov

DeAnna M. Campbell
Biological & Chemical Sensors
Sandia National Laboratories
Albuquerque, USA
dmlope@sandia.gov

Jeffrey A. Ivie
Multiscale Fab. Sci. & Tech. Dev.
Sandia National Laboratories
Albuquerque, USA
jaivie@sandia.gov

Tzu-Ming Lu
Quantum Phenomena
Sandia National Laboratories
Albuquerque, USA
tlu@sandia.gov

Scott W. Schmucker
Multiscale Fab. Sci. & Tech. Dev.
Sandia National Laboratories
Albuquerque, USA
swschmu@sandia.gov

Shashank Misra
Multiscale Fab. Sci. & Tech. Dev.
Sandia National Laboratories
Albuquerque, USA
smisra@sandia.gov

Abstract—We employ a fully charge self-consistent quantum transport formalism, together with a heuristic elastic scattering model, to study the local density of state (LDOS) and the conductive properties of Si:P δ-layer wires at the cryogenic temperature of 4 K. The simulations allow us to explain the origin of shallow conducting sub-bands, recently observed in high resolution angle-resolved photoemission spectroscopy experiments. Our LDOS analysis shows the free electrons are spatially separated in layers with different average kinetic energies, which, along with elastic scattering, must be accounted for to reproduce the sheet resistance values obtained over a wide range of the δ-layer donor densities.

Index Terms—quantum transport, Si:P δ-layer systems, contact block reduction, NEGF, elastic scattering

I. INTRODUCTION

The electronic structure and conductive properties of Si:P δ-layer systems have been a subject of experimental [1]–[5] and computational [6], [7] works due to the high potential for beyond-Moore computational applications. However, many discrepancies persist among these studies, and many questions still remain open, such as how many conductive modes exist, or the influence of the δ-layer thickness and doping density on these conductive modes. Previous studies were based on traditional closed-system quantum approaches, either tight-binding [6] or DFT [7]. However, these approaches rely on

This work is funded under Laboratory Directed Research and Development Grand Challenge (LDRD GC) program, Project No. 213017, at Sandia National Laboratories. Sandia National Laboratories is a multimission laboratory managed and operated by National Technology and Engineering Solutions of Sandia, LLC., a wholly owned subsidiary of Honeywell International, Inc., for the U.S. Department of Energy's National Nuclear Security Administration under contract DE-NA-0003525. This paper describes objective technical results and analysis. Any subjective views or opinions that might be expressed in the paper do not necessarily represent the views of the U.S. Department of Energy or the United States Government.

two essential approximations: 1) the closed-system approximation that attempts to represent the density of states of a highly conductive system as $DOS(E) = \sum_\alpha \delta(E - E_\alpha)$ and thus neglecting the quantum wire (continuum) states; 2) the (semi)classical approximations for the extraction of systems' conductive properties assuming that the current is proportional to the electron density $j \sim n$ instead of the quantum-mechanical flux $j \sim \Psi\nabla\Psi^* - \Psi^*\nabla\Psi$ (that is zero for any closed system wave-function Ψ). In this work, we have employed a fully charge self-consistent open-system quantum transport (QT) formalism [8]–[10], which is free of the above mentioned approximations, together with an elastic scattering model, to study the conduction sub-band structure and the corresponding conductive properties of Si:P δ-layer systems.

II. METHODOLOGY AND MODEL

A. Quantum transport method

To conduct this study, we have employed a quantum-mechanical transport framework for open systems [8]–[10] that relies on a fully charge self-consistent solution of the Poisson-open system Schrödinger equation in the single-band (Γ-valley) effective mass approximation. Fig. 1 shows a detailed flow chart of the method implemented in the QT simulator. For a numerically efficient implementation of the Non-Equilibrium Green's Function (NEGF) formalism we utilized the Contact Block Reduction (CBR) method [8], [9]; for the charge self-consistent solution of the non-linear Poisson equation we employed a combination of the open-system predictor-corrector approach and Anderson mixing scheme [10], [11]. The standard values of electron effective masses $m_l = 0.98 \times m_e$, $m_t = 0.19 \times m_e$ and the dielectric constant

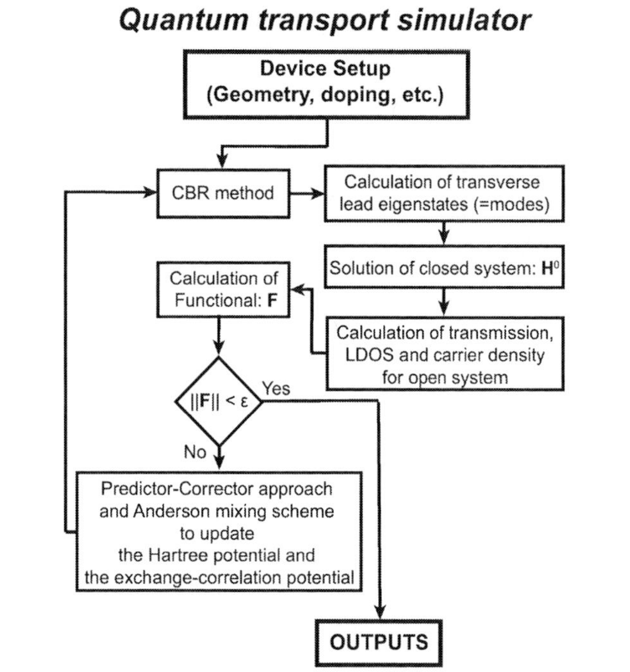

Quantum transport simulator

Device Setup
(Geometry, doping, etc.)

↓

CBR method

→ Calculation of transverse lead eigenstates (=modes)

↓

Solution of closed system: $\mathbf{H}^0$

↓

Calculation of transmission, LDOS and carrier density for open system

Calculation of Functional: $\mathbf{F}$

↓

$\|\mathbf{F}\| < \epsilon$ — Yes →

No ↓

Predictor-Corrector approach and Anderson mixing scheme to update the Hartree potential and the exchange-correlation potential

OUTPUTS

Fig. 1. Flow chart of the self-consistent QT method. ϵ is the Poisson residuum tolerance.

$\epsilon_{Si} = 11.7$ of Silicon were employed in our calculations, as well as the cryogenic temperature of 4 K.

B. Elastic scattering model

We can distinguish two types of elastic scattering that can occur in a device: 1) geometry scattering, due to the ohmic contacts, doping profile, device geometry, etc.; and 2) defect scattering, due to defects, vacancies, impurities, etc. [12], [13]. The former one is already taken into account by the charge self-consistent QT framework. However, the second one needs explicitly to be included in this framework. Thus, we introduce a heuristic elastic defect scattering model for meso- and macroscopic scale, which treats the defects as abstract scatterers. We consider that scatterers are spatially equally distributed along the conductor channel and can be defined by a linear defect density ν. Fig. 2 shows the schematic elastic scattering model, formed by M modes with N_m scatterers, where $m = 1, ..., M$. In general, each transmission mode can possess a different linear defect density $\nu_m = N_m/L$, where L is the length of the channel device. If the electronic transmission for mode m across the conductor without scatterers is giving by $T_{mm}(E)$ at energy E, then the effective electronic transmission for mode m including N_m scatterers can be computed as [14]

$$\frac{1 - T_{mm}^{\text{eff}}(E)}{T_{mm}^{\text{eff}}(E)} = \frac{1 - T_{mm}(E)}{T_{mm}(E)} + \sum_{i=1}^{N} \frac{1 - t_{mm}^{\prime(i)}(E)}{t_{mm}^{\prime(i)}(E)}, \quad (1)$$

where $t_{mm}^{\prime(i)}(E)$ is the defect transmission probability due to scatterer i in mode m. The first term in (1) accounts for geometry scattering, whereas the second term encompasses

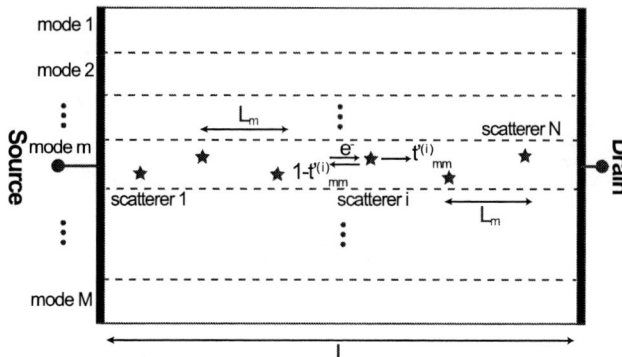

Fig. 2. Schematic elastic scattering model.

for defect scattering. In absence of defects, i.e. $t_{mm}^{\prime(i)}(E) = 1$ with $i = 1, ..., N$, the effective transmission function $T_{mm}^{eff}(E)$ reduces to $T_{mm}(E)$, as expected. If we assume that the defect transmission probability is the same for all scatterers, i.e. $t_{mm}^{\prime(i)}(E) = t_{mm}^{\prime}(E)$, the effective transmission probability can be rewritten as

$$T_{mm}^{\text{eff}}(E) = \frac{1}{1 + \frac{1 - T_{mm}(E)}{T_{mm}(E)} + L\nu_m \frac{1 - t_{mm}^{\prime}(E)}{t_{mm}^{\prime}(E)}}. \quad (2)$$

The term $\nu_m(1 - t_{mm}^{\prime})/t_{mm}^{\prime}$ is of the order of $1/L_m$ [14], where L_m is the mean free path. Therefore, it can be approximated as α/L_m, where α is an adjustable parameter that is proportional to the linear defect density in the system. More complex elastic scattering models can be employed in this framework, e.g., models in which the defect transmission probability is energy-dependent.

The total current density J from source to drain can be then calculated from the Landauer formula as

$$J = \frac{2e}{h} \int \sum_{m=1}^{M} T_{mm}^{eff}(E)\big(f_S(E) - f_D(E)\big)dE, \quad (3)$$

where e is the electron charge and h is the Planck's constant. The equilibrium distribution functions for source and drain leads are $f_S(E)$ and $f_D(E)$, respectively.

C. Computational model

The geometry of the Si: P δ-layer wire is shown in Fig. 3, which is composed of a Si body, a very high P-doped layer, and a Si cap. The conductor channel is in contact with two semi-infinite leads, the source and drain, respectively. For simplicity, we only focus on symmetric configurations, where the widths, W, and acceptor doping concentrations, N_A, of the body and cap are chosen to be the same. Asymmetric doping would result in an asymmetric electron distribution around the δ-layer plane. Similarly, we chose the body and cap widths as large as possible to avoid additional border effects on the electron confinement around the δ-layer ($W = 10$ nm). The wire length is set to $L = 50$ nm. An acceptor doping density in the Si body/cap of 1.0×10^{17} cm^{-3} was used throughout this work.

182

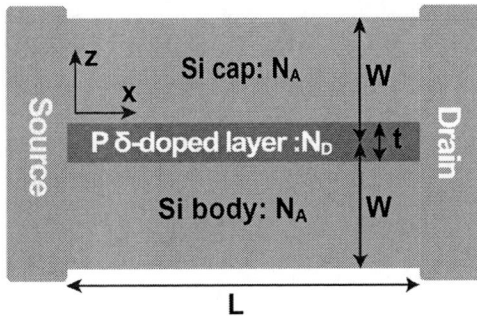

Fig. 3. Schematic model for Si: P δ-layer wires.

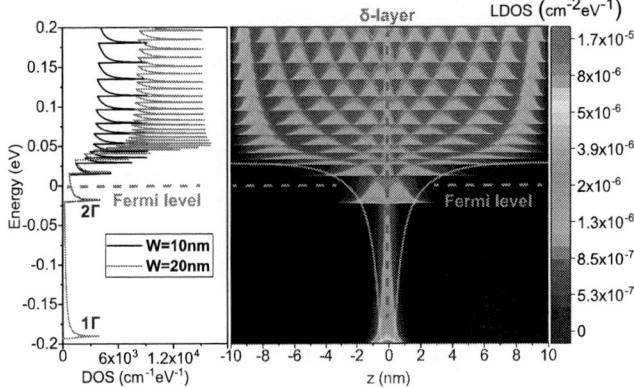

Fig. 4. Left panel: DOS for Si:P δ-layer wires with different widths, $W = 20$ nm and 40 nm. Right panel: LDOS(z,E) for a 20 nm-width Si:P δ-layer wire. For both panels, $t = 0.2$ nm, $N_D = 1.2 \times 10^{14}$ cm^{-2} and $N_A = 1.0 \times 10^{17}$ cm^{-3}.

III. RESULTS AND DISCUSSION

A. Electronic structure

The DOS of the system at equilibrium conditions is included in the left panel of Fig. 4. The DOS results reveal the presence of conducting sub-bands (below the Fermi level) corresponding to 1Γ and 2Γ valleys, respectively, recently observed in high-resolution angle-resolved photoemission spectroscopy experiments (ARPES) [4], [5]. To get a better insight of the electronic structure in Si: P δ-layer systems, we take a look at the LDOS(E,z) in the YZ plane, included in the right panel of Fig. 4 for a 20 nm-width wire. The LDOS presents a peculiar quantization of the modes in space and energy, which we have referred to as "Quantum Menorah". The occupied modes below the Fermi level are independent of the encapsulation depth of δ-layer from the surfaces of the Si cap/body, as shown in the left panel of Fig. 4 for $W = 20$ and 40 nm. The LDOS also reveals that the free electrons are strongly confined spatially around the δ-layer, however, they are distributed in layers with different kinetic energies.

Our simulations also indicate that the number of sub-bands and the corresponding structure is strongly influenced by the thickness and doping density of the δ-layer, as shown in Fig. 5 (upper and middle panel). The effect of the Si body/cap doping on the conduction sub-bands turns out to be secondary, as shown in the lower panel of Fig. 5. For a fixed δ-layer thickness, the increment of the sheet doping increases the number of conducting modes, as well as the splitting energy between them. In contrast, for a fixed sheet doping, the increment of the δ-layer thickness increases the number of modes, but decreases the energy splitting between them, which is in agreement with the ARPES observation in [5]. This is the result of the electronic confinement around the δ-layer: smaller layer thickness and higher doping create a stronger potential well (see right panel of Fig. 4) that leads to higher electronic confinement around the δ-layer and, therefore, an increased energy difference between sub-bands.

B. Conductive properties

With our QT framework and the elastic scattering model for meso- and macroscopic scale, we can compute the sheet conductance of the system. The results are included in Fig. 6 for a wide range of donor doping concentration, from $4.0 \times$

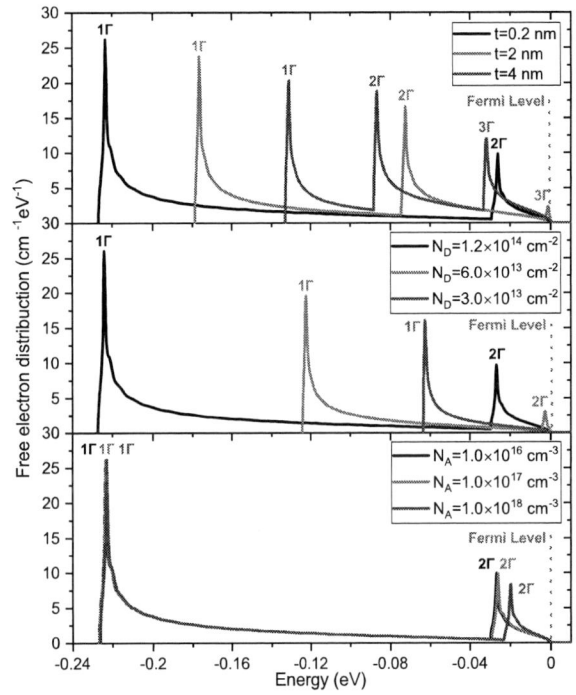

Fig. 5. Upper panel: influence of δ-layer thickness t on the free electron distribution for a fixed sheet doping density of $N_D = 1.2 \times 10^{14}$ cm^{-2} ($N_A = 1.0 \times 10^{17}$ cm^{-3}). Middle panel: influence of sheet doping density N_D on the free electron distribution for a fixed 0.2 nm δ-layer thickness ($N_A = 1.0 \times 10^{17}$ cm^{-3}). Lower panel: influence of the Si cap/body doping N_A on the free electron distribution for a fixed 0.2 nm δ-layer thickness and a donor doping of $N_D = 1.2 \times 10^{14}$ cm^{-2}.

10^{11} cm^{-2} to 5.0×10^{14} cm^{-2}, and δ-layer thicknesses, from 2 nm to 5 nm. To account for the linear increase of the mean free path L_m with the sheet doping density reported in [15], we approximated the term $\nu_m (1 - t'_{mm})/(t'_{mm})$ in (2) as $\alpha/L_m(N_D)$, where the values of $L_m(N_D)$ were taken from [15] and α was set to 1.0. Moreover, our simulations reproduce the experimental sheet resistance data (see Fig. 7) obtained by several groups [15]–[18]. From our conductive

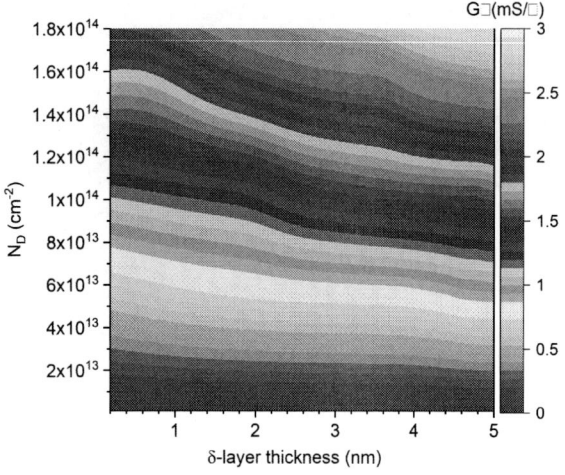

Fig. 6. Sheet conductance for Si:P δ-layer wires in function of the δ-layer thickness and doping density (for the linear impurity density $\alpha = 1.0$)

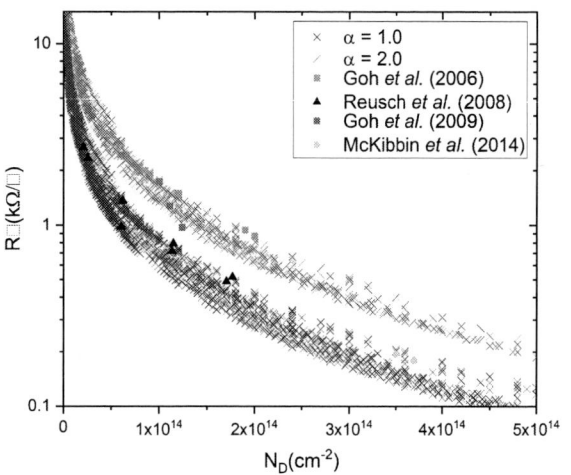

Fig. 7. Comparison of the sheet resistance for different δ-layer thicknesses, from 2 nm to 8 nm, and doping densities against experimental data [15]–[18].

results, we can distinguish two regimes, corresponding to high P doping densities (above 5.0×10^{13} cm^{-2}) and low doping density (below 1.0×10^{13} cm^{-2}), which present dissimilar behavior. The conductance of the system raises with the δ-layer thickness for a fixed sheet doping density. However, our simulations indicate that the slope of the conductance vs thickness becomes steeper as we move to higher doping densities. In the limits, the conductance is strongly dependent on the δ-layer thickness for high doping densities, whereas it remains approximately invariant for low doping densities.

IV. CONCLUSION

Our simulations allow us to explain the origin of shallow conducting sub-bands, recently observed in ARPES experiments. The LDOS analysis reveals a peculiar structure that we have termed as "quantum menorah", as well as the free electrons are spatially separated by layers with different average kinetic energies. The number of conductive sub-bands is

mainly determined by the thickness and sheet doping density of the δ-layer. The effect of the acceptor doping density in the Si body/cap is secondary and negligible. Furthermore, by applying an elastic scattering model in our QT framework, we reproduce the sheet resistance values measured by various experimental groups. Finally, we report that the conductance of the system increases with the increment of the δ-layer thickness for high sheet doping densities.

REFERENCES

[1] D. Ward, S. Schmucker, E. Anderson, E. Bussmann, L. Tracy, T.-M. Lu, L. Maurer, A. Baczewski, D. Campbell, M. Marshall, and S. Misra, "Atomic precision advanced manufacturing for digital electronics," *Electronic Device Failure Analysis*, vol. 22, no. 1, pp. 4–10, 2020.

[2] Y. He, S.K.Gorman, and D. Keith et al., "A two-qubit gate between phosphorus donor electrons in silicon," *Nature*, vol. 571, pp. 371–375, 2019.

[3] J. Wyrick, X. Wang, R. V. Kashid, P. Namboodiri, S. W. Schmucker, J. A. Hagmann, K. Liu, M. D. Stewart Jr., C. A. Richter, G. W. Bryant, and R. M. Silver, "Atom-by-atom fabrication of single and few dopant quantum devices," *Adv. Funct. Mater.*, vol. 29, no. 52, p. 1903475, 2019.

[4] F. Mazzola, C.-Y. Chen, R. Rahman, X.-G. Zhu, C. M. Polley, T. Balasubramanian, P. D. King, P. Hofmann, J. A. Miwa, and J. W. Wells, "The sub-band structure of atomically sharp dopant profiles in silicon," *npj Quantum Mater.*, vol. 5, no. 34, 2020.

[5] A. J. Holt, S. K. Mahatha, R.-M. Stan, F. S. Strand, T. Nyborg, D. Curcio, A. K. Schenk, S. P. Cooil, M. Bianchi, J. W. Wells, P. Hofmann, and J. A. Miwa, "Observation and origin of the Δ manifold in Si:P δ layers," *Phys. Rev. B*, vol. 101, p. 121402, 2020.

[6] S. Lee, H. Ryu, H. Campbell, L. C. L. Hollenberg, M. Y. Simmons, and G. Klimeck, "Electronic structure of realistically extended atomistically resolved disordered Si:P δ-doped layers," *Phys. Rev. B*, vol. 84, p. 205309, 2011.

[7] D. J. Carter, O. Warschkow, N. A. Marks, and D. R. McKenzie, "Electronic structure models of phosphorus δ-doped silicon," *Phys. Rev. B*, vol. 79, p. 033204, 2009.

[8] D. Mamaluy, M. Sabathil, and P. Vogl, "Efficient method for the calculation of ballistic quantum transport," *J. Appl. Phys.*, vol. 93, no. 8, pp. 4628–4633, 2003.

[9] D. Mamaluy, D. Vasileska, M. Sabathil, T. Zibold, and P. Vogl, "Contact block reduction method for ballistic transport and carrier densities of open nanostructures," *Phys. Rev. B*, vol. 71, p. 245321, 2005.

[10] X. Gao, D. Mamaluy, E. Nielsen, R. W. Young, A. Shirkhorshidian, M. P. Lilly, N. C. Bishop, M. S. Carroll, and R. P. Muller, "Efficient self-consistent quantum transport simulator for quantum devices," *J. Appl. Phys.*, vol. 115, no. 13, p. 133707, 2014.

[11] H. R. Khan, D. Mamaluy, and D. Vasileska, "Quantum transport simulation of experimentally fabricated nano-finfet," *IEEE T. Electron Dev.*, vol. 54, no. 4, pp. 784–796, 2007.

[12] J. Mendez, F. Arca, J. Ramos, M. Ortiz, and M. Ariza, "Charge carrier transport across grain boundaries in graphene," *Acta Mater.*, vol. 154, pp. 199 – 206, 2018.

[13] F. Arca, J. P. Mendez, M. Ortiz, and M. P. Ariza, "Charge-carrier transmission across twins in graphene," *J. Phys. Condens. Matter*, vol. 32, no. 42, p. 425003, 2020.

[14] S. Datta, *Electronic transport in mesoscopic systems*. Cambridge university press, 1997.

[15] K. E. J. Goh, L. Oberbeck, M. Y. Simmons, A. R. Hamilton, and M. J. Butcher, "Influence of doping density on electronic transport in degenerate Si:P δ-doped layers," *Phys. Rev. B*, vol. 73, p. 035401, 2006.

[16] K. E. J. Goh and M. Y. Simmons, "Impact of si growth rate on coherent electron transport in Si:P delta-doped devices," *Appl. Phys. Lett.*, vol. 95, no. 14, p. 142104, 2009.

[17] S. R. McKibbin, C. M. Polley, G. Scappucci, J. G. Keizer, and M. Y. Simmons, "Low resistivity, super-saturation phosphorus-in-silicon monolayer doping," *Appl. Phys. Lett.*, vol. 104, no. 12, p. 123502, 2014.

[18] T. C. G. Reusch, K. E. J. Goh, W. Pok, W.-C. N. Lo, S. R. McKibbin, and M. Y. Simmons, "Morphology and electrical conduction of Si:P δ-doped layers on vicinal si(001)," *J. Appl. Phys.*, vol. 104, no. 6, p. 066104, 2008.

Analytical Formulae for the Surface Green's Functions of Graphene and 1T' MoS$_2$ Nanoribbons

Hans Kosina and Heribert Seiler
Institute for Microelectronics
TU Wien, Vienna, Austria
{kosina,seiler}@iue.tuwien.ac.at

Viktor Sverdlov
Christian Doppler Laboratory for Nonvolatile Magnetoresistive
Memory and Logic at the Institute for Microelectronics
TU Wien, Vienna, Austria
sverdlov@iue.tuwien.ac.at

Abstract—Surface Green's functions describe the coupling of the device region with the attached leads. A lead represents a semi-infinite region with uniform properties such as cross section and electrostatic potential. The scattering states in the leads can be determined in different ways. In this work we exploit the uniformity of the system and formulate the problem in reciprocal space where the Green's function takes on a simple form. A Fourier transformation yields the elements of the Green's function in real space. We present the principal steps of this calculation and discuss results for nanoribbons. The 2D materials considered are graphene and MoS$_2$ in the 1T' phase, their electronic structure is represented by k·p Hamiltonians.

Index Terms—surface Green's functions, ballistic transport, nanoribbon, NEGF method, k*p method

I. INTRODUCTION

In the non-equilibrium Green's function formalism, open system boundary conditions are introduced by means of the retarded contact self energies. These quantities are calculated from the retarded surface Green's functions of the semi-infinite leads which are attached to the spatially finite device. There exist both iterative [1] and direct methods [2] [3] for the numerical calculation of the surface Green's function (GF). The direct method is generally more efficient than the iterative one [3].

In this work we present a k-space method for the direct evaluation of the surface GF. If the size of the matrix elements of the Hamiltonian is small, which can be the case, for instance, in 1D systems described by a k·p Hamiltonian, explicit expressions for the surface GF can be derived.

Discretization of the Hamiltonian of the semi-infinite lead yields an infinite system of difference equations.

$$(E\,\underline{I} - \underline{H}_{00})\,\underline{g}_{00} - \underline{H}_{01}\,\underline{g}_{10} = \underline{I}$$
$$-\underline{H}_{10}\,\underline{g}_{n-1,0} + (E\,\underline{I} - \underline{H}_{00})\,\underline{g}_{n0} - \underline{H}_{01}\,\underline{g}_{n+1,0} = \underline{0}$$
$$n \geq 1 \qquad (1)$$

Here, E is the energy. The underlined symbols represent matrices of order N. It holds $\underline{H}_{10} = \underline{H}_{01}^{\dagger}$. The element $\underline{g}_{00}$ associated with the left surface of the lead is sought.

This work has been partly funded by the Austrian Science Fund (FWF), project P33151.

II. THEORY

In the first step we consider the infinite lead which is described by a system of difference equations of the form

$$-\underline{H}_{10}\,\underline{y}_{n-1} + (E\,\underline{I} - \underline{H}_{00})\,\underline{y}_n - \underline{H}_{01}\,\underline{y}_{n+1} = \underline{r}_n. \qquad (2)$$

The index n ranges from $-\infty$ to ∞. The right hand side is defined as $\underline{r}_n = \delta_{n,0}\,\underline{I}$ and represents an impulse applied at point $n = 0$. The Fourier transforms of the discrete functions $\underline{y}_n$ and $\underline{r}_n$ are given by the respective Fourier series.

$$\underline{Y}(k) = \sum_{n=-\infty}^{\infty} \underline{y}_n\,\mathrm{e}^{inka} \qquad (3)$$

$$\underline{R}(k) = \sum_{n=-\infty}^{\infty} \underline{r}_n\,\mathrm{e}^{inka} = \underline{I} \qquad (4)$$

Conversely, the unknowns $\underline{y}_n$ are obtained as the Fourier coefficients of $\underline{Y}(k)$.

$$\underline{y}_n = \frac{1}{k_0} \int_0^{k_0} \underline{Y}(k)\,\mathrm{e}^{-ikna}\,\mathrm{d}k \qquad (5)$$

Here $na = x_n$ is the coordinate of the n-th grid point and a the grid spacing. The period is $k_0 = 2\pi/a$. The Fourier transform of the equation system (2) gives a single equation in reciprocal space,

$$\underline{A}(k, E)\,\underline{Y}(k, E) = \underline{I}, \qquad (6)$$

where

$$\underline{A}(k, E) = \underline{A}_{-1}\,\mathrm{e}^{-ika} + \underline{A}_0(E) + \underline{A}_1\,\mathrm{e}^{ika} \qquad (7)$$
$$\underline{A}_{-1} = -\underline{H}_{01}, \ \underline{A}_0(E) = E\,\underline{I} - \underline{H}_{00}, \ \underline{A}_1 = -\underline{H}_{10}$$

Equation (6) has the formal solution

$$\underline{Y} = \underline{A}^{-1}. \qquad (8)$$

Defining the characteristic polynomial $D(k, E) = \mathrm{Det}\,(\underline{A})$, and designating the adjugate of $\underline{A}$ as $\underline{B} = \mathrm{adj}(\underline{A})$, the inverse of $\underline{A}$ can be written as

$$\underline{Y}(k, E) = \frac{1}{D(k, E)}\,\underline{B}(k, E) \qquad (9)$$

185

It holds $\underline{A}\,\underline{B} = D\underline{I}$. Note that $\underline{A}$, $\underline{B}$, and D are complex trigonometric polynomials: $\underline{A}$ of degree one according to (7), D of degree N, and $\underline{B}$ of degree $N-1$:

$$\underline{B} = \sum_{l=1-N}^{N-1} \underline{B}_l\, e^{ila} \qquad (10)$$

In analytical calculations the adjugate matrix can be conveniently determined using the Cayley-Hamilton theorem. From the Fourier expansion of the reciprocal of the determinant

$$\frac{1}{D(k,E)} = \sum_{n=-\infty}^{\infty} c_n\, e^{inka}, \quad c_n = \frac{1}{k_0}\int_0^{k_0} \frac{e^{-inka}}{D(k,E)}\, dk \quad (11)$$

we obtain the $\underline{y}_n$ by a convolution of the form

$$\underline{y}_n = \sum_{l=1-N}^{N-1} c_{n-l}\, \underline{B}_l \qquad (12)$$

Introducing the complex variable $u = \exp(ika)$ the integral in (11) turns into a contour integral over the unit circle in the complex plane. For given E, the characteristic equation $D(k,E) = 0$ has $2N$ solutions $k_1, k_2, \ldots k_{2N}$ which always occur in pairs $\pm k_i$. Thus the poles in the integral in (11) also occur in pairs u_i and u_i^{-1}. The contour integral is evaluated by virtue of the residue theorem. The system (2) has two solutions, known as the retarded and the advanced GF, respectively. For a propagating mode in the lead the wave number k_i is real and the pole u_i is located on the unit circle. To select the retarded GF one adds a small damping term $i\eta$ to the Hamiltonian and keeps only the convergent terms. The damping term moves the poles u_i associated with a positive group velocity into the unit circle, such that their resides contribute to the integral, whereas the complementary poles u_i^{-1} move outwards and have no effect.

A. Impulse Response in the Semi-infinite Lead

The coefficients (12) represent the impulse response in the infinite domain. To describe a semi-infinite lead one has to truncate the domain by imposing the boundary condition (BC) $\underline{g}_{-1,0} = \underline{0}$. This BC can be matched by adding to the particular solution $\underline{y}_n$ a homogeneous solution. The latter can be created by applying an impulse outside the domain of interest, that is at some negative m. Any impulse within the negative domain will emit a right traveling wave into the positive domain. This wave satisfies the homogeneous difference equations for all $n \geq 0$. The homogeneous solution is specified up to a free, constant matrix which is determined from the BC $\underline{g}_{-1,0} = \underline{0}$. This procedure yields the GF for the left surface.

$$\underline{g}_{00} = \underline{y}_0 - \underline{y}_m\, \underline{y}_{m-1}^{-1}\, \underline{y}_{-1} \qquad (13)$$

The GF for the right surface can also be calculated from the impulse response $\underline{y}_n$. In this case the system has to be truncated at $n = 1$ by imposing the BC $\underline{g}_{10} = \underline{0}$.

III. GRAPHENE NANORIBBON

We apply the method to the Dirac-like Hamiltonian of graphene:

$$\underline{H}(k_x, k_y) = \begin{pmatrix} U_0 & \hbar v_F(k_x - ik_y) \\ \hbar v_F(k_x + ik_y) & U_0 \end{pmatrix}$$

In a nanoribbon the transverse momentum is quantized. Thus k_y can be treated as a constant parameter that represents a particular subband. Substituting $k_x = -i\partial_x$ and approximating the derivative by a central difference quotient yields a system of difference equations of the form (1) with the matrices

$$\underline{H}_{00} = ck_y \begin{pmatrix} 0 & -i \\ i & 0 \end{pmatrix}, \quad \underline{H}_{01} = \frac{c}{2ia}\begin{pmatrix} 0 & 1 \\ 1 & 0 \end{pmatrix} \qquad (14)$$

Here we have set the on-site potential to $U_0 = 0$ and introduced the parameter $c = \hbar v_F$. The characteristic equation is quadratic and has two solutions. According to the residue theorem only that solution (k_0) associated with a positive group velocity has to be considered for the retarded GF. The result from our method is

$$\underline{g}_{00} = \frac{4a^2}{c^2}\frac{u_0^2}{u_0^2 - 1}\left(E\underline{I} + \underline{H}_{00}\right) \qquad (15)$$

In [4] the very same expression has been obtained in a different way. Because of the simple structure of $\underline{H}_{01}$, the iteration series of the Sancho-Rubio method can be summed up analytically.

IV. MoS$_2$ NANORIBBON

Recently it has been discovered that MoS$_2$ in the 1T' phase is a topological insulator [5]. The inverted band structure is well described by a parabolic dispersion relation. The effective Hamiltonian is of the form [5]

$$\underline{H}(k_x, k_y) = \begin{pmatrix} \delta - c_1 k_x^2 - c_2 k_y^2 & c_6 k_y - \alpha E_z + ic_5 k_x \\ c_6 k_y - \alpha E_z - ic_5 k_x & -\delta + c_3 k_x^2 + c_4 k_y^2 \end{pmatrix}$$

with the coefficients

$$c_1 = \frac{\hbar^2}{2m_x^p}, \quad c_2 = \frac{\hbar^2}{2m_y^p}, \quad c_3 = \frac{\hbar^2}{2m_x^d}, \quad c_4 = \frac{\hbar^2}{2m_y^d}$$

$$c_5 = \hbar v_1, \quad c_6 = \hbar v_2$$

Here the x-axis represents the longitudinal direction, the y-axis the transverse direction, and the z-axis the direction perpendicular to the nanoribbon.

A. Bulk Dispersion Relation

The characteristic polynomial is of second degree in E.

$$p(k_x^2, k_y, E) = -\big[c_5^2 k_x^2 + (c_6 k_y - \alpha E_z)^2 + \big(c_1 k_x^2 + c_2 k_y^2 - \delta + E\big)\big(c_3 k_x^2 + c_4 k_y^2 - \delta - E\big)\big] \quad (16)$$

Its roots $E_{1,2}(k_x, k_y)$ represent the two branches of the bulk dispersion relation. In the following we use the parameter values reported in [5]. Wavenumbers are scaled with k_0,

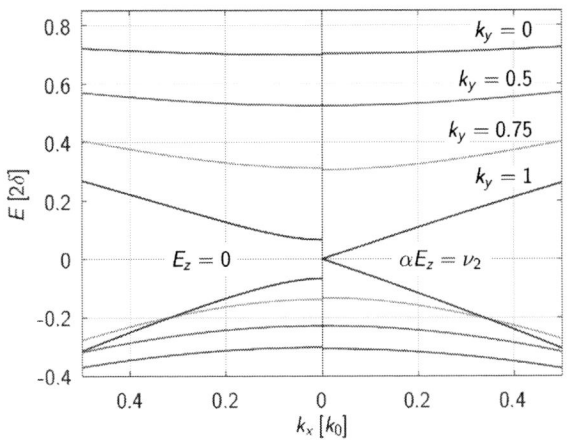

Fig. 1. Cuts of the bulk dispersion of 1T' MoS$_2$ at different values k_y. The bandgap at $E_z = 0$ (left panel) gets closed at $\alpha E_z = v_2$ (right panel).

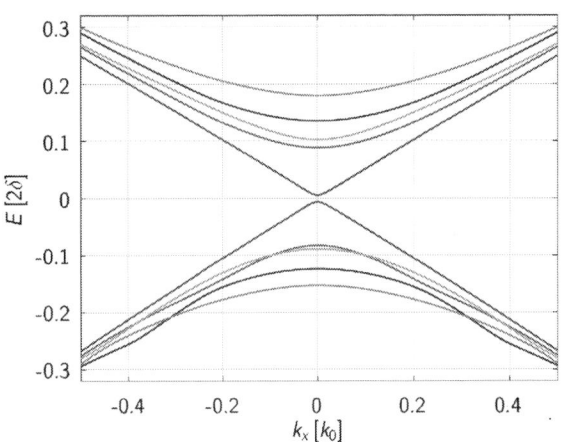

Fig. 2. Subband structure of the nanoribbon at $E_z = 0$. The subbands with a nearly linear dispersion correspond to the topologically protected edge states.

whereas energies are scaled with 2δ and offset by ΔE as suggested in [6].

$$k_0 = \sqrt{\frac{4\delta}{\hbar^2} \frac{m_y^d m_y^p}{m_y^d + m_y^p}}, \qquad \Delta E = \frac{1}{2}\frac{m_y^d - m_y^p}{m_y^d + m_y^p} \qquad (17)$$

In Fig. 1 cuts of the bulk dispersion at different values of k_y are shown. The bandgap at $(k_x, k_y) = (0, 1)$ which is finite at $E_z = 0$ (left panel) closes at $\alpha E_z = v_2$ (right panel) and reopens when E_z is further increased [6].

B. The Subband Structure

The characteristic polynomial (16) is of fourth degree in k_y. The roots are designated as k_y^j with $j = 1, 2, 3, 4$. We choose a mode space approach and express the transverse modes as linear combinations of the four basis functions $\exp(ik_y^j y)$. The two spinor components have to satisfy homogeneous Dirichlet BCs at the two edges of the nanoribbon. These BCs yield a system of four homogeneous equations with a coefficient matrix $\underline{M}$ [6]. The Schrödinger equation and the characteristic equation for the BCs,

$$\mathrm{Det}\,(E\underline{I} - \underline{H}) = 0,$$
$$\mathrm{Det}\,(\underline{M}) = 0,$$

are solved simultaneously using the Newton method. In this way the discrete eigen-energies E and the related transverse momenta k_y^j for a given k_x can be computed.

We assume a nanoribbon width of $d = 40/k_0 = 28.86$ nm. Five electron and hole subbands at $E_z = 0$ are depicted in Fig. 2. The dispersions of the lowest electron and topmost hole subbands are linear. The energies of these subbands lie in the bulk bandgap (Fig. 1, left panel). These subbands correspond to the topologically protected edge states as the related wave functions are localized at the edges.

Fig. 3 shows the subband structure at an electric field of $\alpha E_z = v_2$. While the bandgap of the bulk material would

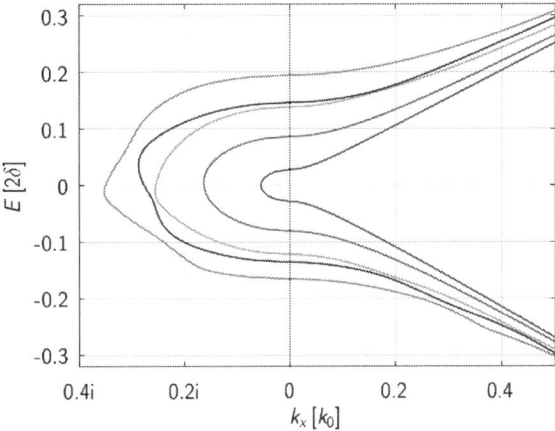

Fig. 3. Complex subband structure of the nanoribbon at $\alpha E_z = v_2$.

be closed at this field strength, it remains finite in a narrow nanoribbon. Transistor action can be achieved by applying a vertical field over a finite region of the nanoribbon, which represents the channel region. In the off-state, electrons in the leads ($E_z = 0$) would be blocked by the finite bandgap in the channel. In this situation tunneling through the gap has to be considered, which require knowledge of the complex band structure (Fig. 3, left panel).

C. Contact Self Energy

After the substitution $k_x = -i\partial_x$ and applying the central difference approximation the Hamiltonian of the nanoribbon becomes a block-tridiagonal matrix with the following 4x4 blocks.

$$\underline{H}_{00} = \begin{pmatrix} \underline{T} & \underline{S} \\ \underline{S}^\dagger & \underline{T} \end{pmatrix}, \qquad \underline{H}_{01} = \begin{pmatrix} \underline{R} & \underline{0} \\ \underline{S} & \underline{R} \end{pmatrix} \qquad (18)$$

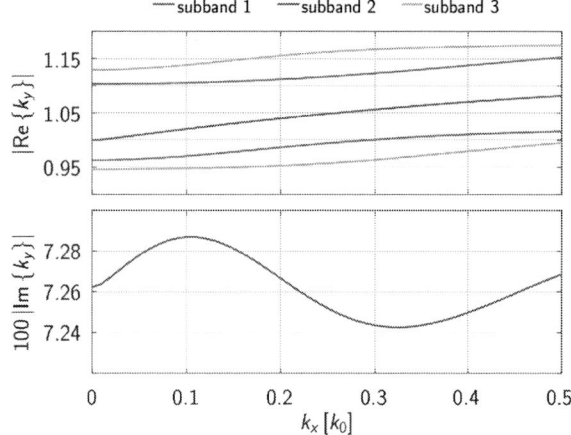

Fig. 4. Absolute values of $\mathrm{Re}\{k_y\}$ and $\mathrm{Im}\{k_y\}$ as functions of the longitudinal wave number k_x. Only the lowest subband develops an imaginary part which causes the formation of edge states. The higher subbands have real wave numbers which describe simple standing waves in the transverse direction.

Here,

$$\underline{R} = \frac{1}{4a^2}\begin{pmatrix} c_1 & 0 \\ 0 & -c_3 \end{pmatrix}, \qquad \underline{S} = \frac{c_5}{2a}\begin{pmatrix} 0 & 1 \\ -1 & 0 \end{pmatrix} \qquad (19)$$

and $\underline{T} = \underline{H}(0, k_y) - 2\underline{R}$.

To describe the discrete system one has to replace k_x with $\tilde{k}_x$ defined as

$$\tilde{k}_x = k_x\,\mathrm{sinc}(k_x a) \qquad (20)$$

We introduce the complex variable $u = e^{2ik_x a}$ and express $\tilde{k}_x^2$ as $\tilde{k}_x^2 = (2 - u - u^{-1})/(4a^2)$. The characteristic polynomial of the discrete system is related to (16) by the following transformation.

$$f(u, k_y, E) = u^2 p((2 - u - u^{-1})/(4a^2), k_y, E) \qquad (21)$$

The characteristic equation $f(u) = 0$ has to be solved for a given energy E and a wavenumber k_y characteristic of a given subband. The transverse wavenumbers k_y depend weakly on the longitudinal wavenumber k_x as shown in Fig. 4. To enable one-dimensional transport calculations in mode space one has to neglect this dependence and assume a constant k_y. For energies close to the subband minimum one can use the approximation $k_y(k_x) \approx k_y(0)$.

For an energy E_0 inside a given subband (outside the bandgap) the equation $p(k_x^2, ky, E_0) = 0$, or equivalently the equation $E(k_x, k_y) = E_0$, has two real solutions $\pm k_x^{(1)}$. Due to the monotonicity of the dispersion relation, see Fig. 2, there can exist only two real solutions, and thus the remaining two solutions $\pm k_x^{(2)}$ must be be complex.

The roots $u_{1,3} = \exp(\pm 2ik_x^{(1)}a)$ are located on the unit circle. The root associated with positive group velocity we designate as u_1. One of the two complex solutions $\pm k_x^{(2)}$ gives a root u_2 inside the unit circle, whereas the complementary root $u_4 = u_2^{-1}$ is located outside.

Having identified the roots relevant to the retarded GF we can now apply the residue theorem to determine the coefficients (11).

$$c_n = \frac{u_1^{1+|n|}}{f'(u_1)} + \frac{u_2^{1+|n|}}{f'(u_2)} \qquad (22)$$

Following the procedure outlined in Section II we arrive at the surface GF. We summarize the result for contact self energy which is defined as $\underline{\Sigma} = \underline{H}_{01}\,\underline{g}_{00}\,\underline{H}_{10}$. First, one has to set up four matrices:

$$\underline{U} = \underline{H}_{01}\left[(E - \mathrm{Tr}(\underline{T}))\,\underline{I} + \underline{H}_{00}\right]\underline{H}_{10}$$
$$\underline{\sigma}_1 = \mu_1\,\underline{U} + D(\mu_2\,\underline{H}_{01} + \underline{H}_{10})$$
$$\underline{\sigma}_2 = \underline{U} + \mu_1 D(\underline{H}_{01} + \underline{H}_{10})$$
$$\underline{\sigma}_3 = \mu_1\,\underline{U} + D(\underline{H}_{01} + \mu_2\,\underline{H}_{10})$$

where $\mu_1 = c_1/c_0$, $\mu_2 = c_2/c_0$, and $D = -\mathrm{Det}(\underline{R})$. Then the contact self energy is obtained as

$$\underline{\Sigma} = c_0\left[\underline{\sigma}_2 - \underline{\sigma}_3\,\underline{\sigma}_2^{-1}\,\underline{\sigma}_1\right] \qquad (23)$$

The results of this derivation have been verified numerically. The computed surface GF satisfies the quadratic matrix equation

$$\left(E\underline{I} - \underline{H}_{00} - \underline{H}_{01}\,\underline{g}_{00}\,\underline{H}_{10}\right)\underline{g}_{00} = \underline{I}, \qquad (24)$$

whereas the computed matrices $\underline{y}_n$ satisfy the difference equations (2).

REFERENCES

[1] M. P. L. Sancho, J. M. L. Sancho, J. M. L. Sancho, and J. Rubio, "Highly convergent schemes for the calculation of bulk and surface green functions," *Journal of Physics F: Metal Physics*, vol. 15, no. 4, pp. 851–858, apr 1985.

[2] R. C. Bowen, W. R. Frensley, G. Klimeck, and R. K. Lake, "Transmission resonances and zeros in multiband models," *Phys. Rev. B*, vol. 52, pp. 2754–2765, Jul 1995.

[3] M. Luisier, A. Schenk, W. Fichtner, and G. Klimeck, "Atomistic simulation of nanowires in the $sp^3d^5s^*$ tight-binding formalism: From boundary conditions to strain calculations," *Phys. Rev. B*, vol. 74, p. 205323, Nov 2006.

[4] S.-K. Chin, K.-T. Lam, D. Seah, and G. Liang, "Quantum transport simulations of graphene nanoribbon devices using Dirac equation calibrated with tight-binding π-bond model," *Nanoscale Research Letters*, vol. 7, no. 1, p. 114, 2012.

[5] X. Qian, J. Liu, L. Fu, and J. Li, "Quantum spin Hall effect in two-dimensional transition metal dichalcogenides," *Science*, vol. 346, no. 6215, pp. 1344–1347, 2014.

[6] V. Sverdlov, A.-M. B. El-Sayed, H. Kosina, and S. Selberherr, "Conductance in a nanoribbon of topologically insulating MoS_2 in the 1T' phase," *IEEE Transactions on Electron Devices*, 2020 (submitted).

Numerical Solution of the Constrained Wigner Equation

Robert Kosik, Johann Cervenka, and Hans Kosina

Institute for Microelectronics, TU Wien, Vienna, Austria

http://www.iue.tuwien.ac.at

{kosik,cervenka,kosina}@iue.tuwien.ac.at

Abstract—**Quantum electron transport in modern semiconductor devices can be described by a Wigner equation which is formally similar to the classical Liouville equation. The stationary Wigner equation has a singularity at zero momentum (k=0). In order to get a non-singular solution it is necessary to impose a constraint for the solution at k=0 which gives the constrained Wigner equation. We introduce a Petrov-Galerkin method for the solution of the corresponding constrained sigma equation. The constraint in the Wigner equation is interpreted as an extra test function and is naturally incorporated in the method.**

Index Terms—**Wigner function, sigma function, constrained equation, quantum transport, device simulation, resonant tunneling diode, Galerkin method**

I. Introduction

The Wigner function method [1] is based on a mathematical formulation of quantum mechanics which is close to a classical phase space description [2], [3]. This allows for flexible mixed quantum-classical models and makes it an attractive approach for many applications where only parts of the system need to be modeled fully quantum mechanically [4].

Quantum electron transport in modern semiconductor devices can be described by a Wigner equation which is formally similar to the classical Liouville equation (also called Vlasov equation). The existence and uniqueness of solutions to the stationary Wigner equation with classical inflow boundary conditions is a long-standing open problem even in a single spatial dimension [5], [6].

Solution methods for the Wigner equation can be divided into deterministic and stochastic methods. Much of the practical work using the Wigner equation is based on Monte Carlo methods [7]. In the classical case, Monte Carlo methods are more versatile than deterministic methods. In particular, they can solve the Vlasov equation with inflow boundary conditions.

In previous work [8] we have argued that the constrained Wigner equation is a viable option for a deterministic solution of the stationary Wigner equation which is well-behaved at $k=0$. Below we introduce a deterministic numerical method for the solution of the constrained Wigner equation.

This work was partly funded by FWF Austrian Science Fund, project number P33151 "Numerical Constraints for the Wigner and the Sigma Equation".

II. Sigma and Wigner Function

Stationary quantum transport is described by the Liouville-von Neumann equation for the density matrix $\rho(x,y)$

$$-\frac{\hbar^2}{2m}\left(\frac{\partial^2}{\partial x^2} - \frac{\partial^2}{\partial y^2}\right)\rho + \left(V(x) - V(y)\right)\rho = 0 \qquad (1)$$

where $V(x)$ is the potential energy.

From the density matrix $\rho(x,y)$ the Wigner function $f(r,k)$ is defined as the result of two consecutive transformations:

1) Introduce new coordinates for the quantum density

$$r = \frac{x+y}{2}, \qquad s = x - y.$$

Using these coordinates the density matrix transforms into the sigma function

$$\sigma(r,s) = \rho(r + \frac{s}{2}, r - \frac{s}{2}). \qquad (2)$$

2) The Wigner function $f(r,k)$ is derived from the sigma function $\sigma(r,s)$ via a Fourier transform in coordinate s

$$f(r,k) = \frac{1}{2\pi}\int \sigma(r,s)\, e^{-iks}\, ds. \qquad (3)$$

The sigma function has the symmetry property

$$\sigma(r,s) = \overline{\sigma(r,-s)} = a(r,s) + ib(r,s). \qquad (4)$$

The real part $a(r,s)$ is an even function, the imaginary part $b(r,s)$ is an odd function of variable s.

III. Sigma and Wigner Equation

Using coordinates (r,s) the stationary von Neumann equation transforms into the stationary sigma equation

$$\frac{\hbar^2}{m}\frac{\partial^2 \sigma}{\partial rs} = U(r,s)\sigma(r,s) \qquad (5)$$

where the potential term $U(r,s)$ is defined by

$$U(r,s) = V\left(r + \frac{s}{2}\right) - V\left(r - \frac{s}{2}\right). \qquad (6)$$

Applying the Fourier transform (3) to the sigma equation (5) gives the stationary Wigner equation

$$\frac{\hbar k}{m}\frac{\partial f(r,k)}{\partial r} = \int f(r, k-k')V_w(r,k')dk'. \qquad (7)$$

189

Here the Wigner potential $V_w(r,k)$ is defined as the Fourier transform of $U(r,s)$ divided by $i\hbar$

$$V_w(r,k) = \frac{1}{i\hbar}\frac{1}{2\pi}\int U(r,s)\,e^{-iks}\,ds. \tag{8}$$

Classical inflow boundary conditions are imposed on the stationary Wigner equation at the spatial boundaries $r_{\min}$, $r_{\max}$.

IV. CONSTRAINED WIGNER AND SIGMA EQUATION

In a single spatial dimension the Wigner equation (7) can be rewritten for $k \neq 0$ as

$$\frac{\partial f(r,k)}{\partial r} = \frac{1}{k}\frac{m}{\hbar}\int f(r,k-k')V_w(r,k')dk'. \tag{9}$$

The form (9) emphasizes that the equation becomes singular at $k=0$. In order to avoid the singularity one has to impose two equations for $k=0$

1) Putting $k=0$ in (7) gives the following integrability constraint [5], [6]

$$\int f(r,k')V_w(r,-k')dk' = 0. \tag{10}$$

2) In the limit $k \to 0$ equation (9) reduces to the "transport" equation at $k=0$

$$\frac{\partial f(r,0)}{\partial r} = \frac{m}{\hbar}\int f_k(r,-k')V_w(r,k')dk'. \tag{11}$$

Together these equations define an overdetermined system which we call the constrained Wigner equation.

By inverse Fourier transform a corresponding constrained sigma equation can be derived [8] which consists in equation (5) plus double homogeneous boundary conditions

$$\sigma(r,s_{\max}) = \sigma(r,-s_{\max}) = 0. \tag{12}$$

With double homogeneous boundary conditions the sigma equation is overdetermined as well.

V. SPECTRAL METHOD

In previous work [8] we solved the constrained sigma equation directly in a least squares sense using finite differences. Here we solve the constrained sigma equation by a spectral approach with sine and cosine ansatz functions. This ansatz amounts to a solution in the Fourier domain, i.e., a solution of the constrained Wigner equation. Improving on [8] we also incorporate a complex absorbing potential

$$\tilde{U}(r,s) = U(r,s) + iW(s) \tag{13}$$

in order to diminish unphysical reflections of outgoing waves as advocated in [9].

For the numerical solution we first discretize the sigma equation (5) in coordinate r using finite differences

$$\frac{\hbar^2}{m}\frac{\sigma_s(r_{i+1},s) - \sigma_s(r_i,s)}{\triangle r} =$$
$$\frac{\tilde{U}(r_i,s)\sigma(r_i,s) + \tilde{U}(r_{i+1},s)\sigma(r_{i+1},s)}{2}. \tag{14}$$

The semi-discrete equation (14) is then discretized in coordinate s by employing a Petrov-Galerkin approach where the set of ansatz functions is different from the set of test functions. The ansatz functions for a, b from (4) fulfill the homogeneous boundary conditions (12)

$$a(r_i,s) = \sum_{n=1}^{N_k} a_n(r_i)\cos\left(\frac{2n-1}{2}\triangle ks\right) \tag{15}$$

$$b(r_i,s) = \sum_{n=1}^{N_k} b_n(r_i)\sin\left(n\triangle ks\right). \tag{16}$$

Our test functions are the union of two families

$$T_n^a = \sin\left(\frac{2n-1}{2}\triangle ks\right)\quad (n=1\dots N_k) \tag{17}$$

$$T_n^b = \cos(n\triangle ks)\qquad (n=0\dots N_k) \tag{18}$$

which were chosen because they are the derivatives $\frac{\partial}{\partial s}$ of the ansatz functions. With this we have $2N_k$ ansatz functions and $2N_k+1$ test functions. The test function $T_0^b = 1$ is extra and the corresponding equations represent the integrability constraint. Inflow boundary conditions and an equation enforcing exact conservation of mass are imposed as constraints.

VI. IMPLEMENTATION

The overdetermined system is solved in a least squares sense with constraints exactly fulfilled. Mathematically this represents a quadratic programming problem with linear equality constraints. For the solution we introduce the error vector e

$$e = Ax - b \tag{19}$$

where matrix A and vector b represent the system matrix and right hand side from the Petrov-Galerkin equations. (In our case the right hand side b is zero.) The vector x represents the unknown coefficients of the ansatz functions.

In order to find the constrained least squares solution we minimize $\|e\|^2$ using the method of Lagrange multipliers. The equations (19) are interpreted as the definition of e and are imposed as constraints (introducing one Lagrange multiplier for each equation). We also add the constraints for the boundary conditions and for the exact discrete conservation of mass. This method gives a linear system in variables x and e plus the Lagrange multipliers λ. The resulting system has an increased number of unknowns but avoids building the normal equations. This is a trade-off between numerical accuracy (condition number) and system size.

To deal with high memory requirements a specialized solver for the solution of the banded linear system is deployed.

VII. NUMERICAL EVALUATION

The spectral method is evaluated by simulating a GaAs/AlGaAs resonant tunneling diode (RTD) and comparison with the quantum transmitting boundary method (QTBM) which we take as the reference model.

The RTD model used in the simulation is depicted in Fig. 1, which consists of a 4.5 nm-wide GaAs quantum well and 2.8 nm-wide AlGaAs barrier layers. Doped GaAs electrode layers of 30 nm are included on each side of the device. The potential drops linear between barriers. Outside the

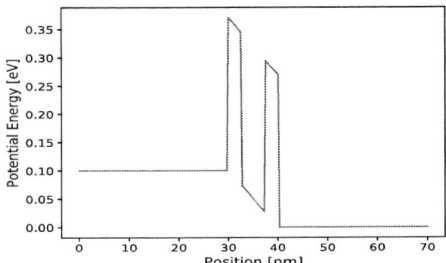

Fig. 1: Structure of the resonant tunneling diode with two AlGaAs barrier layers as used in the simulation. In the plotted example a bias of -0.1 V is assumed.

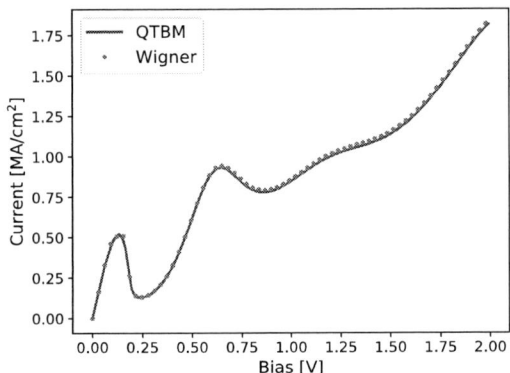

Fig. 2: Simulation of a resonant tunneling diode and comparison with QTBM. A weak second resonance near 0.7 V is shown.

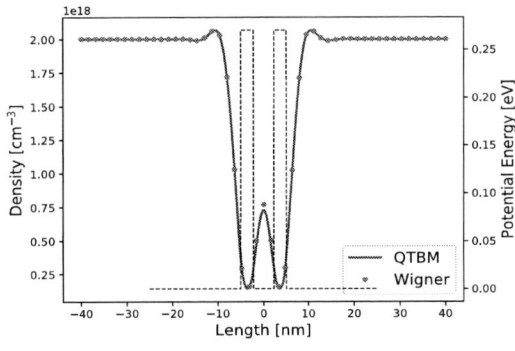

(a) Bias = 0.0 V

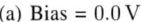

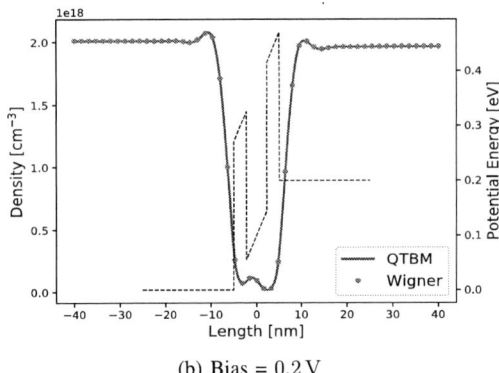

(b) Bias = 0.2 V

Fig. 3: Particle density in the active region of the RTD for different biases. The dashed black lines indicate potential energy.

electrodes, reservoirs characterized by the thermal equilibrium distributions of carriers are assumed. The conduction band discontinuity is taken to be 0.27 eV.

We demonstrate the numerical capability of the method in Fig. 2 by simulating up to a high bias which is beyond the usual operating conditions of the device. The QTBM and the Wigner function model fit almost perfectly up to a bias of 2 V. In Fig. 3 we plot the particle density in the active region of the device for varying biases. Again, the fit between QTBM and Wigner is reasonably good.

In order to achieve high accuracy, a coherence length of at least 60 nm must be assumed and a fine mesh with $N_r \sim 800$ points has to be used for the simulation. Consequently, numerical costs are high as the Wigner equation is dense in variable k and the incorporation of constraints complicates the solution.

VIII. CONCLUSION

Using a large coherence length and a sufficiently fine mesh the constrained Wigner equation reproduces results from the QTBM reliably and with good accuracy.

REFERENCES

[1] D. K. Ferry and M. Nedjalkov, *The Wigner Function in Science and Technology*. IOP Publishing, 2018.

[2] M. A. D. Gosson, *The Wigner Transform*. World Scientific, 2017.

[3] C. Zachos, D. Fairlie, and T. Curtright, *Quantum Mechanics in Phase Space*. World Scientific, 2005.

[4] J. Weinbub and D. K. Ferry, "Recent advances in Wigner function approaches," *Applied Physics Reviews*, vol. 5, no. 4, p. 041 104, 2018.

[5] A. Arnold, H. Lange, and P. F. Zweifel, "A discrete-velocity, stationary Wigner equation," *Journal of Mathematical Physics*, vol. 41, no. 11, pp. 7167–7180, 2000.

[6] R. Li, T. Lu, and Z. Sun, "Parity-decomposition and moment analysis for stationary Wigner equation with inflow boundary conditions," *Frontiers of Mathematics in China*, vol. 12, no. 4, pp. 907–919, 2017.

[7] D. Querlioz, P. Dollfus, and M. Mouis, Eds., *The Wigner Monte Carlo Method for Nanoelectronic Devices*. Wiley, 2013.

[8] R. Kosik, J. Cervenka, M. Thesberg, and H. Kosina, "A revised Wigner function approach for stationary quantum transport," in *Large-Scale Scientific Computing*, Springer International Publishing, 2020, pp. 403–410.

[9] L. Schulz and D. Schulz, "Complex absorbing potential formalism accounting for open boundary conditions within the Wigner transport equation," *IEEE Transactions on Nanotechnology*, vol. 18, pp. 830–838, 2019.

9-5

Calibrated Si Mobility and Incomplete Ionization Models with Field Dependent Ionization Energy for Cryogenic Simulations

Hiu Yung Wong
Electrical Engineering
San Jose State University
San Jose, USA
hiuyung.wong@sjsu.edu

Abstract— **Cryogenic silicon CMOS operating between 77K and 4.2K is becoming more popular in high-speed server applications and the periphery of quantum computers. In the cryogenic regime, dopant incomplete ionization and field enhanced ionization become dominating physical phenomena. Therefore, it is important to use accurate and well-calibrated mobility and incomplete ionization models in cryogenic TCAD simulations. In this paper, we present a Philips Unified Mobility Model (PhuMob) and Altermatt's incomplete ionization model calibrated between 300K and 20K for boron and arsenic dopants in silicon across 5 orders of magnitude in doping concentration. A novel method is proposed to include field-dependent ionization energy in Altermatt's model, which results in good convergence even in 3D TCAD simulations at 4K.**

Keywords—Cryogenic, Field Dependent Ionization, Incomplete Ionization, TCAD

I. INTRODUCTION

Cryogenic electronics operating between 77K (the boiling point of liquid-nitrogen) and 4.2K (the boiling point of liquid-helium) are attracting considerable attention due to the emergence of quantum computers [1][2], the quest for higher performance servers for big data processing using existing technologies [3], and the revival of deep space exploration [4][5]. For example, cryogenic CMOS devices are indispensable for the control and readout circuits in quantum computers. To provide low latency, CMOS circuits need to be in close proximity to the qubits and thus need to operate at cryogenic temperatures [1][2]. Cryo-CMOS also delivers improved performance in terms of on-state/leakage current ratio, subthreshold swing, and transconductance and is expected to improve the high-performance servers substantially [6][7]. In space applications, temperatures can drop to 44K on the surface of Pluto, or 3K in interstellar space [5].

TCAD has been the powerhouse technique for semiconductor device development in the last few decades [8]. However, TCAD is not ready to be fully deployed for cryogenic CMOS development. Firstly, cryogenic physics models and parameters have not been fully understood (e.g. subthreshold swing, [9]), developed, and calibrated for effective TCAD simulation. Secondly, numeric convergence is challenging in low-temperature TCAD simulation due to the vanishingly small intrinsic carrier concentration [10]. Even with simple models, convergence is still very difficult below 15K [11].

In this paper, we self-consistently calibrated some of the most important cryogenic CMOS models, namely the incomplete ionization model for Boron and Arsenic and low-field carrier (both hole and electron) mobility models from 300K to 20K based on experimental data [12]. A new method of simulating incomplete ionization is also proposed which achieves excellent convergence at 4K even for 3D simulations that include field-dependent ionization.

II. MODEL SELECTION

At cryogenic temperatures, dopants are not fully ionized due to their finite ionization energies (E_A). However, when the doping concentration is high, impurity bands are formed (Mott transition [13][14]) (Fig. 1). Moreover, E_A is a function of doping concentration and given by the following equation in [13] and [14],

$$E_A = \frac{E_{A0}}{1 + \left(\dfrac{N_{dop}}{N_{ref}}\right)^c} \tag{1}$$

where E_{A0} is the ionization energy of isolated dopants in the absence of field enhancement, N_{dop} is the dopant concentration, and N_{ref} and c are fitting parameters. N_{ref} can be regarded as the doping concentration above which E_A is significantly reduced.

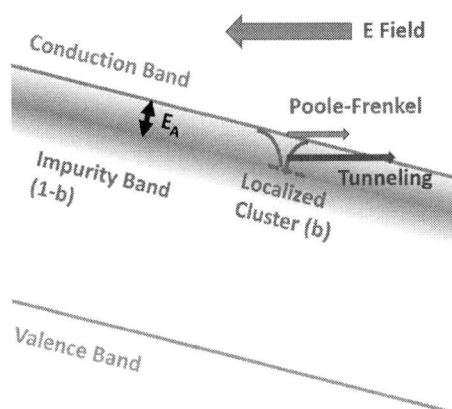

Figure 1: Illustration of n-type dopant ionization in Silicon. Impurity band just merging with the conduction band is illustrated. While all donor electrons in the impurity band (fraction *1-b*) of total) can contribute to the electrical conduction, the fraction *b* of the donor electrons can only contribute to electrical conduction through ionization. The effective ionization energy E_A depends on doping concentration and can be lower when Poole-Frenkel or tunneling effects are strong.

193

According to [13][14], there are two sources of free carriers contributing to the conduction current. One is localized states (including localized dopant clusters) whose carriers are thermally ionized into the conduction band. Their ionization rate depends on E_A, temperature, and electric field. Another is the delocalized carriers (e.g. those residing in the impurity bands). The fraction of the delocalized carriers is given by the following equation ([13] and [14]),

$$b = \frac{1}{1 + \left(\frac{N_{dop}}{N_b}\right)^d} \qquad (2)$$

where N_b and d are fitting parameters. N_b can be regarded as the doping concentration above which clusters merge to form the impurity band at the expense of isolated carriers or clusters.

In [13][14], an equation was derived for TCAD simulations and calibrated at 300K to account for these effects (dubbed Altermatt's model). We implemented this model using the Physical Model Interface (PMI) in Sentaurus Device [15] and calibrated to experimental data from the literature [12] down to cryogenic temperatures. Reference [12] is relatively old. The experimental data accuracy may not be as high as desired (in particular, the compensation doping concentration). However, it has the most extensive data in terms of doping concentration and temperature ranges available in the literature.

To model the carrier mobility at cryogenic temperature, we need to take into account that dopants are not fully ionized and thus the impurity scattering is reduced. Moreover, screening of ionized impurities by carriers should also be included for accurate simulations. Philips Unified Mobility Model (PhuMob) [16] is chosen because it includes carrier screening and temperature effects.

Under high electric field, the effective dopant ionization energy is reduced due to tunneling and Poole Frenkel ionization [17]. This will greatly affect the dopant ionization

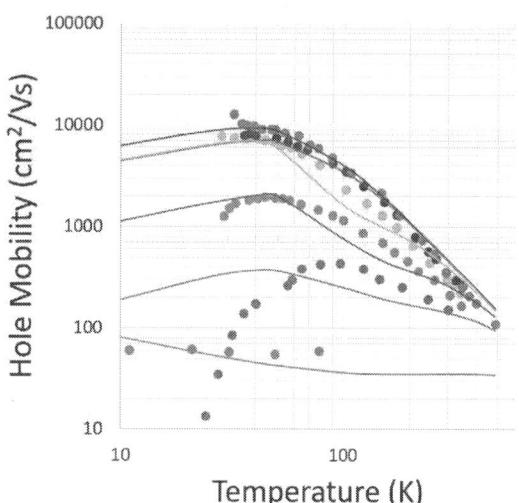

Figure 3: Hole mobility as a function of temperature and boron concentration. Markers: experimental data in [12]. Lines: TCAD simulation. Net impurity concentrations from *bottom to top* are 1.5×10^{19}cm^{-3}, 10^{18}cm^{-3}, 2×10^{17}cm^{-3}, 2.4×10^{16}cm^{-3}, 7×10^{14}cm^{-3} and 3.1×10^{14}cm^{-3}, respectively.

behavior and should be considered in TCAD simulations at cryogenic temperatures.

III. CALIBRATION OF IONIZATION AND MOBILITY MODELS

Due to the interplay between the aforementioned physics, it is important to calibrate carrier concentrations and mobilities self-consistently as carrier mobility depends strongly on ionized dopant concentration at low temperatures. Boron and arsenic doped silicon data at various doping concentrations and temperatures from [12] are used for calibrations. Fig. 2 to Fig. 5 show the calibrated free carrier concentrations and their mobilities using PhuMob and Altermatt's models. It should be noted that there may be some

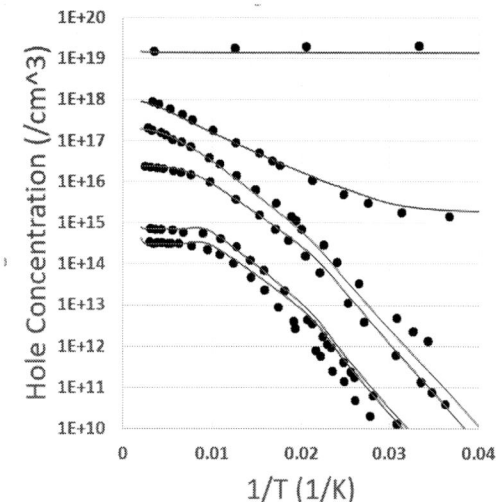

Figure 2: Free hole concentration as a function of temperature and boron concentration. Markers: experimental data in [12]. Lines: TCAD simulation. Net impurity concentrations from top to bottom are 1.5×10^{19}cm^{-3}, 10^{18}cm^{-3}, 2×10^{17}cm^{-3}, 2.4×10^{16}cm^{-3}, 7×10^{14}cm^{-3} and 3.1×10^{14}cm^{-3}, respectively.

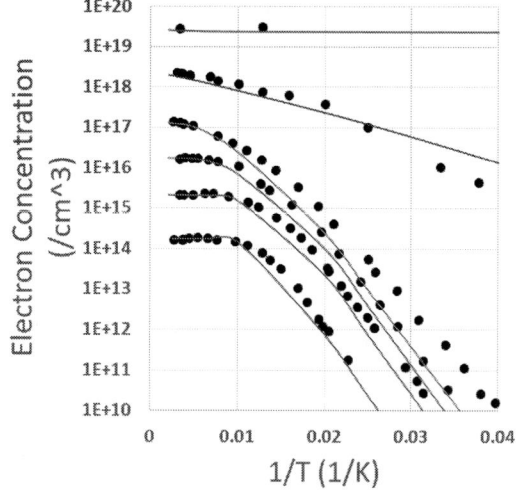

Figure 4: Free electron concentration as a function of temperature and arsenic concentration. Markers: experimental data in [12]. Lines: TCAD simulation. Net impurity concentrations from top to bottom are 2.7×10^{19}cm^{-3}, 2.2×10^{18}cm^{-3}, 1.3×10^{17}cm^{-3}, 1.75×10^{16}cm^{-3}, 2.1×10^{15}cm^{-3} and 1.75×10^{14}cm^{-3}, respectively.

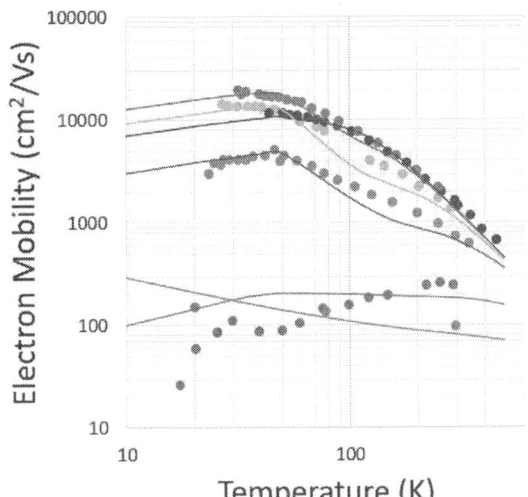

Figure 5: Electron mobility as a function of temperature and boron concentration. Markers: experimental data in [12]. Lines: TCAD simulation. Net impurity concentrations from *bottom to top* are $2.7\times10^{19} \text{cm}^{-3}$, $2.2\times10^{18} \text{cm}^{-3}$, $1.3\times10^{17} \text{cm}^{-3}$, $1.75\times10^{16} \text{cm}^{-3}$, $2.1\times10^{15} \text{cm}^{-3}$ and $1.75\times10^{14} \text{cm}^{-3}$, respectively

measurement errors in reference [12] especially due to the uncertainty of compensation doping levels. Reference [12] also mentions that a few of their low-temperature mobility data points are anomalous. Therefore, it is difficult to obtain a perfect fit to their data. However, with a single set of parameters, we are able to obtain good overall fittings between 300K and 20K for doping concentrations from 10^{14}cm^{-3} to 10^{19}cm^{-3}.

The fitting is obtained by using the default parameters with the following modifications. In Altermatt's model [14], N_b is $3.6\times10^{18} \text{cm}^{-3}$ instead of $6\times10^{18} \text{cm}^{-3}$ for boron (Equation 2). N_{ref} is 10^{18}cm^{-3} instead of $4\times10^{18} \text{cm}^{-3}$ for Arsenic (Equation 1). These modifications are important to fit the intermediate doping level curves ($10^{17} \sim 10^{18} \text{cm}^{-3}$).

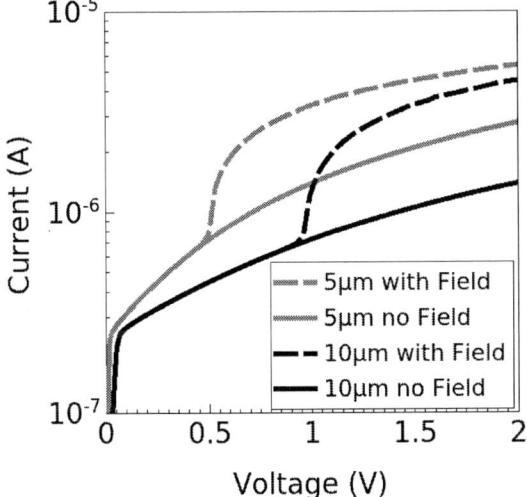

Figure 6: IV curves of 3D resistors of various lengths with and without field dependent model turned on.

In the PhuMob model, the default parameters are kept unchanged as much as possible. The major modification is changing the μ_{min} of Boron to $14 \text{cm}^2/\text{V}\cdot\text{s}$ instead of $44.9 \text{cm}^2/\text{V}\cdot\text{s}$ [15][16]. It is also important to adjust the compensation dopant concentration to obtain good fitting of the mobility curves at low temperatures in the low dopant concentration samples. This is because at low temperatures, dopants are mostly not ionized and the compensation dopants contribute strongly to the scattering effect. Compensation doping concentrations are chosen based on [12] with reasonable adjustments. It should also be noted that each curve in Fig. 2 to 5 corresponds to a certain *net impurity* concentration, which is the difference between the dopant concentration and the compensation dopant concentration. The compensation dopant concentrations used in the fittings are given in Table I.

TABLE I.   COMPENSATION DOPING IN FIG. 2 TO FIG. 5

p-type (cm^{-3})[a]		n-type (cm^{-3})[a]	
Experiment	TCAD	Experiment	TCAD
4.1×10^{14}	3.1×10^{14}	1.0×10^{14}	8.0×10^{14}
2.2×10^{14}	5.2×10^{14}	5.25×10^{14}	3.0×10^{14}
2.3×10^{15}	5.0×10^{15}	1.48×10^{15}	5.0×10^{14}
4.9×10^{15}	4.0×10^{15}	2.2×10^{15}	3×10^{15}
Unknown	5.0×10^{16}	Unknown	5×10^{17}
Unknown	2.9×10^{15}	Unknown	5×10^{17}

[a.] Each row corresponds to each curve in Figs. 2 to 4 with increasing net impurity concentration from top to bottom

IV. SIMULATION WITH FIELD DEPENDENT IONIZATION

To include field-dependent ionization in Altermatt's model, one may modify equation (1) by including field-dependent terms. However, the incomplete ionization PMI used converges very poorly when field-dependent effects are included because the PMI is not designed for this purpose (e.g. lack of Jacobian for electric field). To improve convergence, we propose to model the delocalized dopant states as conventional dopants not subject to incomplete ionization (i.e. all of these dopants are ionized and they are a fraction (1-b) of the total dopants). The localized dopant states are modeled using traps with the same final E_A given by Altermatt's model that are a fraction b of the total dopants (Fig. 1). The partition between these two types of dopants is calculated using Altermatt's model (i.e. Equation (2)) and the calibrated Altermatt parameters are used. The Trap Energy Shift PMI [15] is then employed to include the field-dependent effect on E_A. This is dubbed "*b-method*".

This model is applied to 3D resistors of various lengths (L) doped with 10^{14}cm^{-3} arsenic at 4K sandwiched by two regions with fully ionized dopants of about 10nm length. The short fully ionized regions are contacted with Ohmic contact boundaries. Altermatt's model is included using the proposed method. The calibrated PhuMob model and ionization energy lowering due to the Poole-Frenkel effect are also incorporated. The field dependency is set up such that significant E_A reduction occurs when the electric field is about $0.1 \text{V}/\mu\text{m}$ [17]. Fig. 6 shows the IV curves of L = 5 μm and 10 μm. It can be seen that when the electric field is less than the critical field (0.5V for 5 µm and 1V for 10 µm), without and without field ionization the model give the same results. Once the critical electric field is reached, field ionization increases the current substantially.

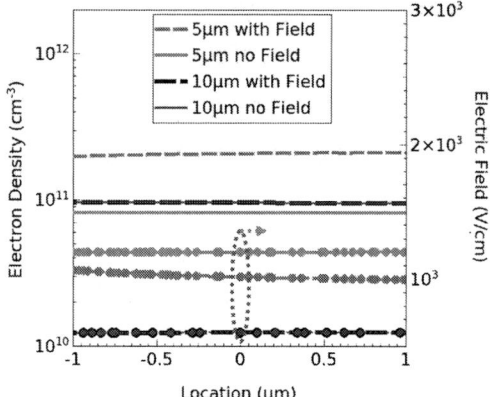

Figure 7: Electron density and electric field profiles along the resistors in Fig. 6 in the transport direction when V = 0.6V. The midpoint of the resistor is at 0µm. Curves with markers are electric field. The two 10 µm curves overlap with each other.

Fig. 7 shows the electron density and electric field plots along the transport direction of the resistors at V = 0.6V. Since the electric field is still below the critical field in the L = 10 µm case, with and without field dependent ionization give the same result. For L = 5 µm, with field dependence has higher current density. Note that it also has lower field because it is more conductive.

It is well-known that convergence in TCAD simulations is problematic at cryogenic temperatures. It typically becomes more problematic when including the impact of field dependent ionization. With the proposed method the convergence is excellent. Although not shown, good convergence is also achieved for transistor simulations at 4K.

V. CONCLUSIONS

An incomplete ionization model (Altermatt's model) and mobility model (PhuMob) are calibrated to experimental data from 300K to 20K for a wide range of doping concentrations (10^{14}cm^{-3} to 10^{19}cm^{-3}) for boron and arsenic doped silicon. Resulting simulations capture important features such as the plateau in free carrier curves at medium doping levels.

A new method is also proposed to perform effective TCAD simulations at cryogenic temperature in which field-dependent ionization is included in the Altermatt's model. It shows excellent convergence in a 3D resistor simulation at 4K.

ACKNOWLEDGMENT

The author thanks Dr. Gibson for his suggestion of *b-method* to improve convergence.

REFERENCES

[1] E. Charbon, F. Sebastiano, A. Vladimirescu, H. Homulle, S. Visser, L. Song, and R M. Incandela, "Cryo-CMOS for Quantum Computing," 2016 IEEE International Electron Devices Meeting (IEDM), San Francisco, CA, 2016, pp. 13.5.1-13.5.4, doi: 10.1109/IEDM.2016.7838410.

[2] M. J. Gong, U. Alakusu, S. Bonen, M. S. Dadash, L. Lucci, H. Jia, et al., "Design Considerations for Spin Readout Amplifiers in Monolithically Integrated Semiconductor Quantum Processors," 2019 IEEE Radio Frequency Integrated Circuits Symposium (RFIC), Boston, MA, USA, 2019, pp. 111-114, doi: 10.1109/RFIC.2019.8701847.

[3] G. Lee, D. Min, I. Byun and J. Kim, "Cryogenic computer architecture modeling with memory-side case studies," 2019 ACM/IEEE 46th Annual International Symposium on Computer Architecture (ISCA), Phoenix, AZ, USA, 2019, pp. 774-787.

[4] D. Clery, "U.S. lawmaker orders NASA to plan for trip to Alpha Centauri by 100th anniversary of moon landing," Science, May 23, 2016 [Online]. DOI: 10.1126/science.aag0558

[5] R. L. Patterson, A. Hammoud, J. E. Dickman, S. Gerber, M. Elbuluk and E. Overton, "Electronics for Deep Space Cryogenic Applications," Proceedings of the 5th European Workshop on Low Temperature Electronics, Grenoble, France, 2002, pp. 207-210, doi: 10.1109/WOLTE.2002.1022482.

[6] A. Beckers, F. Jazaeri and C. Enz, "Cryogenic MOS Transistor Model," in IEEE Transactions on Electron Devices, vol. 65, no. 9, pp. 3617-3625, Sept. 2018, doi: 10.1109/TED.2018.2854701.

[7] H. Homulle, F. Sebastiano and E. Charbon, "Deep-Cryogenic Voltage References in 40-nm CMOS," IEEE Solid-State Circuits Letters, vol. 1, no. 5, pp. 110-113, May 2018, doi: 10.1109/LSSC.2018.2875821.

[8] R. Minixhofer, "TCAD as an Integral Part of the Semiconductor Manufacturing Environment," 2006 International Conference on Simulation of Semiconductor Processes and Devices, Monterey, CA, 2006, pp. 9-16, doi: 10.1109/SISPAD.2006.282827.

[9] A. Beckers, F. Jazaeri and C. Enz, "Revised Theoretical Limit of Subthreshold Swing in Field-effect Transistors," arXiv:1811.09146, 2019.

[10] L. Luo, "Physics, Compact Modeling and TCAD of SiGe HBT for Wide Temperature Range Operation," Ph.D. dissertation, ECE, Auburn University, Auburn, AL, 2011.

[11] F. A. Mohiyaddin, F. G. Curtis, M. N. Ericson and T. S. Humble, "Simulation of Silicon Nanodevices at Cryogenic Temperatures for Quantum Computing," Proceedings of the 2017 COMSOL Conference in Boston.

[12] F. J. Morin and J. P. Maita, "Electrical properties of silicon containing arsenic and boron," Phys. Rev. 96, 28, 1954.

[13] P. P. Altermatt, A. Schenk and G. Heiser, "A Simulation model for the density of states and for incomplete ionization in crystalline Silicon. I. stablishing the model in Si:P," Journal of Applied Physics, vol. 100, 113714 (2006).

[14] P. P. Altermatt, A. Schenk, B. Schmithüsen and G. Heiser, "A simulation model for the density of states and for incomplete ionization in crystalline silicon. II. Investigation of Si:As and Si:B and usage in device simulation," Journal of Applied Physics, vol. 100, 113715 (2006).

[15] Sentaurus™ Device User Guide Version Q-2019.12, December 2019.

[16] D. B. M. Klaassen, "A unified mobility model for device simulation— I. Model equations and concentration dependence," Solid-State Electronics, Vol. 35, No. 7, pp.953-959, 1992.

[17] D. P .Forty, "Impurity ionization in MOSFETs at very low temperatures," Cryogenics, Volume 30, Issue 12, December 1990, Pages 1056-1063.

Monte Carlo Simulation of a Three-Terminal RRAM with Applications to Neuromorphic Computing

Akhilesh Balasingam
Department of Electrical Engineering
Stanford University
Stanford, California
avb03@stanford.edu

Akash Levy
Department of Electrical Engineering
Stanford University
Stanford, California
akashl@stanford.edu

Haitong Li
Department of Electrical Engineering
Stanford University
Stanford, California
haitongl@stanford.edu

Priyanka Raina
Department of Electrical Engineering
Stanford University
Stanford, California
praina@stanford.edu

Abstract—We developed a Monte Carlo simulator to compute the state-dependent I-V characteristics of three-terminal (3T) RRAM devices. State switching in these devices is modeled using a combination of vacancy migration and trap-assisted-tunneling mechanisms. We describe key elements of the simulator, compute hysteresis curves under typical voltage cycling conditions, and demonstrate agreement with experimental results. We then study the response of 2T- and 3T-RRAMs under pulsed operation and show that 3T-RRAM conductance values have both greater dynamic range than 2T-RRAMs and the potential to deliver superior inference accuracy in neuromorphic applications.

Keywords—RRAM, Three-Terminal, Non-Volatile Memory, Monte Carlo Simulator, Neuromorphic

I. INTRODUCTION

Two-terminal Resistive Random-Access Memory (2T-RRAM) devices have been researched extensively for high-density memory and neuromorphic computing applications. Recently, a three-terminal variant (3T-RRAM) of this device family has been investigated experimentally, which, by separating the terminals used for *read* and *write* operations, seeks to enable efficient resistive state update and flexible neuromorphic architectures [1, 2]. State switching in these devices has been theorized to occur via a non-filamentary mechanism, involving a combination of (1) field-driven oxygen vacancy migration and (2) electron transport based on trap-assisted-tunneling [1]. We develop a Monte Carlo simulator for 3T-RRAMs which captures these two mechanisms and makes predictions that show broad agreement with published experimental results. Using this simulator, we also study the response of 3T-RRAMs under pulsed operation and investigate their potential for use as synapses in neuromorphic applications.

The device structure we model is motivated by the experimental structure from [1], which is excerpted in Fig. 1a. The simulated structure is an idealization, consisting of a top electrode (TE), a bottom electrode (BE) and a gate electrode (GE), all of which are in contact with the switching layer, as shown in Fig. 1b. The dots in Fig. 1b depict oxygen vacancies, whose spatial distribution is modulated during *write* by a gate voltage V_G applied on GE, with TE and BE grounded. During *read*, GE floats, BE is grounded, and a

voltage V_D applied on TE draws a current I_D, which depends on the effective distance between the defect population and the channel region between TE and BE. While the discussion below focuses on the geometry shown in Fig. 1b, alternative multi-terminal configurations and biasing arrangements for *read* and *write* can also be studied using this simulator.

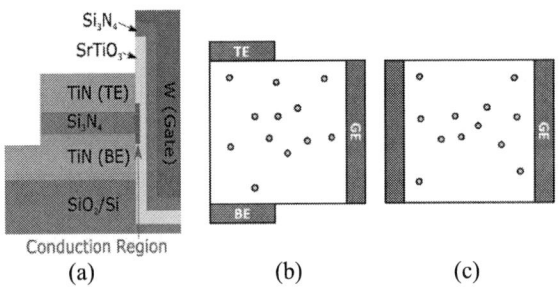

Fig. 1 (a) Schematic of 3T-RRAM structure, excerpted directly from [1], (b) 2D schematic of the 3D simulation domain, in which vacancy dynamics within the switching layer (e.g., TiO₂, SrTiO₃) are captured, and (c) the 2T-RRAM structure for comparison.

In Fig. 1c, we show, for comparison, a schematic of a 2T-RRAM device where the switching layer is sandwiched between two electrodes. In the 2T device, current is maximized when the centroid of the defect population is approximately half-way between the two electrodes. In the 3T configuration, conductance between TE and BE is maximized when the defect population is pushed closer to the TE-BE channel region. Thus, we can expect the 3T-RRAM to exhibit a greater dynamic range of conductance values than the 2T-RRAM. This is confirmed by the simulations presented in this paper.

II. MONTE CARLO SIMULATOR

Our model for this device extends an earlier two-terminal, non-filamentary vacancy modulation-based RRAM model by adding support for multiple terminals [3]. In this approach, the simulations are initialized with a population of N vacancies distributed at random within the switching layer. While the vacancies have the freedom to move within the simulation domain during device operation, their total number is assumed to be conserved. For the results reported below, the simulation domain is a rectangular prism with dimensions $10nm \times 10nm \times 12nm$, containing $N = 30$ vacancies, corresponding to a concentration of $n_v = 2.5 \times 10^{19} cm^{-3}$. The current flowing within the device is calculated using a fully-connected non-linear resistor network constructed by linking each vacancy m to each of the electrodes, and to every other vacancy n within the oxide, as illustrated in Fig. 2a. The tunneling current I_{mn} between vacancy m and vacancy n is approximated, as in [3], by an empirical function which is strongly nonlinear in the distance d_{mn} and the potential difference $V_{mn} = (V_m - V_n)$ between them:

$$I_{mn} = g_0 V_{mn}[2 - e^{c_1(d_{mn}-c_2)}]e^{\frac{c_3 d_{mn}}{V_{mn}}} \quad (Eqn. 1)$$

where $g_0 = 10nA.V^{-1}$, $c_1 \approx 0.069\ nm^{-1}$, $c_2. = 1\ nm$, $c_3 = 1\ V\ nm^{-1}$ are parameters of the model [3]. The potential V_m at the location of each vacancy m is then computed by iteratively solving a system of current continuity equations. The electric field $\vec{E}_m$, obtained from the local gradient of the potential distribution $\{V_m\}$, is used to calculate the vacancy migration rates:

$$r_m = ve^{-\Phi_b/k_B T} \sinh\left(\frac{\gamma q E_m}{k_B T}\right) \quad (Eqn. 2)$$

where k_B is the Boltzmann constant, $|q| = 1.6 \times 10^{-19} C$ is the charge on an electron, T is the absolute temperature, and $v = 10^{13} s^{-1}$ is the migration attempt frequency [4]. The energy barrier for defect movements $\Phi_b = 0.8eV$ and the local field enhancement factor $\gamma = 12$ are empirical parameters of the model, which can depend both on the properties of the materials system under study and on process conditions [5]. The vacancy m that moves is selected with probability proportional to the rates r_m, and is moved a distance corresponding to a typical lattice spacing in transition metal oxides, $\delta = 0.2nm$, in the direction of the local electric field $\vec{E}_m$. The elapsed time for the event is then computed using the sum of all the defect migration rates:

$$\Delta t_i = -(\textstyle\sum_m r_m)^{-1}\ln(s_i) \quad (Eqn. 3)$$

where s_i is a random number distributed uniformly on the unit interval [4]. The key elements of the simulator are shown in Fig. 2b.

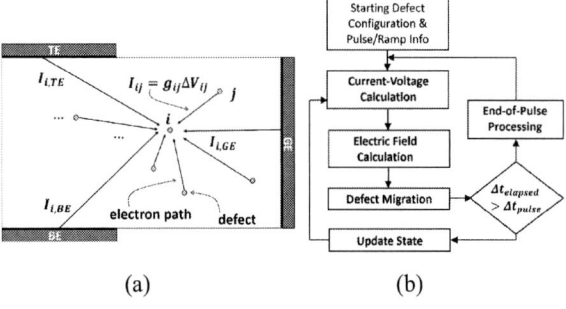

(a)   (b)

Fig. 2 (a) Continuity condition imposed at one of the vacancies forming the non-linear resistor network used to model current flow via trap-assisted tunneling (TAT), and (b) flowchart showing key functional units of the simulator.

During programming, TE and BE are grounded, and a voltage stimulus is applied on GE as a series of pulses, each of duration $\Delta t = 10\mu s$. The pulse height V_G is cycled twice between $-5V$ and $+5V$ in $|\Delta V_G| = 50mV$ increments per pulse, as depicted in the triangular profile in the upper graph

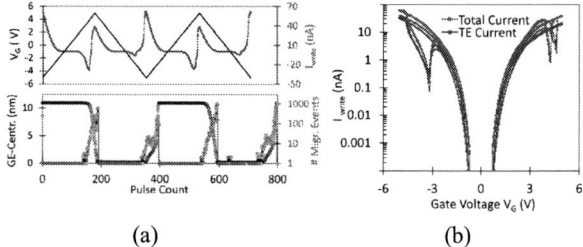

(a)   (b)

Fig. 3 (a) The upper graph shows the heights of the voltage pulses and the total 'write' current I_write drawn through the GE, and the lower graph shows the distance between the centroid of the defect population and GE and the number of defect migration events occurring during each pulse, (b) 3T-RRAM hysteresis loop, showing I_G from (a) as a function of V_G, which splits approximately evenly between the TE and BE. The TE current I_TE is also shown; the BE current I_BE= I_G − I_TE is not shown to preserve clarity of figure.

of Fig. 3a, which also shows the corresponding write current I_{write}. As V_G is cycled, the defects within the switching layer migrate in response to the local electric field, which is oriented primarily in a direction perpendicular to GE. The distance between GE and the centroid of the defect distribution, and the number of defect migration events during each pulse are shown in the lower graph of Fig. 3a. The same *write* current I_{write} is shown as a function of V_G in Fig. 3b, which shows the standard hysteresis behavior.

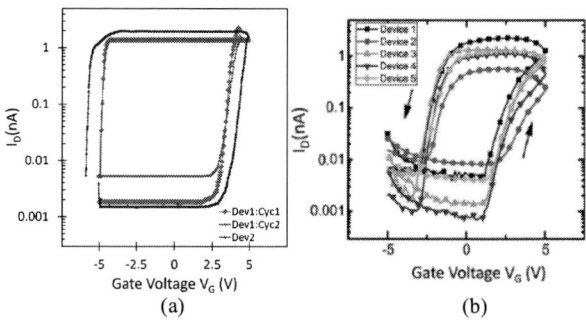

(a)   (b)

Fig. 4. The curves labeled Dev1:Cyc1 and Dev1:Cyc2 show the 'read' current corresponding to the two 'write' cycles shown in Fig. 3. The curve Dev2 corresponds to a second device instance with V_G swept with a different pulsing schedule: $\Delta t = 10\mu s$, and $|\Delta V_G| = 10mV$. The 'read' is performed nondestructively with GE floating, BE grounded and TE biased to $V_D = 2V$, and (b) Graph directly excerpted from experimental paper [1] shows box-like hysteresis loops comparable to our simulation results. The upper and lower bounds of I_D in the simulation results shown in Fig. 6 are within the band of variability seen experimentally.

As the number of defects within reach of the electric field between the TE and BE rises or falls, the drain current increases or decreases, respectively. So, at the conclusion of each gate pulse, we apply a voltage pulse on TE, grounding BE and floating GE, to *read* the drain current I_D as a function of V_G, which is shown in Fig. 4a. The box-like hysteresis loop in this graph agrees qualitatively with the experimental results shown in Fig. 4b (excerpted from [1]), in shape and magnitude.

III. NEUROMORPHIC BENCHMARKING

With this validation and demonstration of model capabilities, we explore the suitability of 3T-RRAMs for neuromorphic applications by computing their drain conductance as a function of the number of fixed-height pulses M applied on the gate. Fig. 5a and 5b show characteristics for the case of $M = 100$ for 2T- and 3T-RRAMs. It is evident that the 3T-RRAM exhibits better linearity and greater on/off ratio. Unlike the 2T-RRAM, in which the conductance peaks when the centroid of the defect population is roughly equidistant from the two electrodes, the conductance of the 3T-RRAM steadily increases as the defect population is pushed closer to the BE and TE.

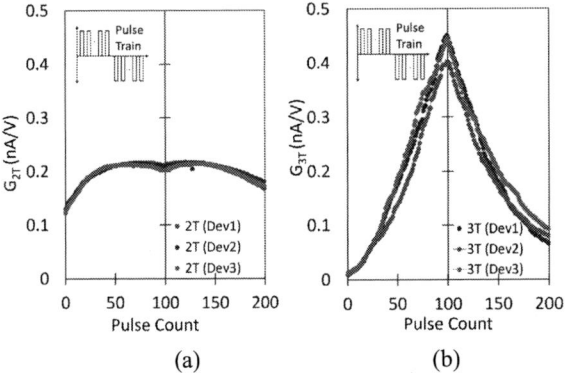

(a)   (b)

Fig. 5 Comparison of the response of 2T-RRAM and 3T-RRAM devices when a sequence of gate pulses of constant height and duration are applied during the 'write' phase. The first M=100 pulses have a height of $V_G = 3V$ (potentiation) and the next M=100 pulses have a height of $V_G = -3V$ (depression). (a) 2T-RRAM conductance with read voltage of $V_G = 2V$, for three devices (b) 3T-RRAM conductance with a read voltage of $V_D = 2V$, between TE and BE, for three devices.

Next, we model the mapping of an offline-trained fully-connected neural network (NN) to two crossbar circuits, with synapses based on 2T- and 3T-RRAMs, respectively, and simulate inference accuracy in MATLAB. The NN consists of 400 input neurons, 25 neurons in the hidden layer, and 10 output neurons, and it is trained in software on the MNIST data [6]. The weights of the NN are then mapped to discrete RRAM conductance levels attainable by applying a fixed number of gate pulses. With increasing M, it is clear that the 3T-RRAM will yield a greater number of discrete levels than the 2T-RRAM, leading to a mapping with higher fidelity. Fig. 6b shows the inference accuracy achieved when NN weights are mapped to 2T- and 3T-RRAM conductance levels, as well as the best accuracy of 98.86% attained in software (SW). This analysis shows that as M increases, the inference accuracy of the 3T-based circuit approaches levels achieved in SW more quickly than the 2T-based circuit.

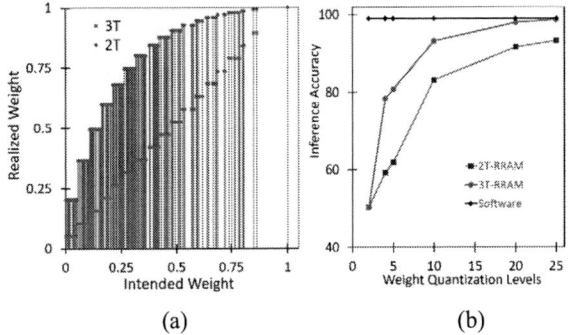

(a)   (b)

Fig. 6 Comparison of the neuromorphic performance of crossbar circuits configured with 2T- and 3T-RRAM synapses with characteristics shown in Fig. 5. (a) Weights realized with the conductances of 2T- and 3T-RRAM vs. scaled weights determined in SW (intended weights), with 20-level quantization, and (b) inference accuracy when NN weights are represented by 2T- and 3T-RRAMs at different quantization levels.

IV. CONCLUSION

We have developed a new Monte Carlo simulator for multi-terminal, non-filamentary RRAM devices. The model is simple, capturing only the dynamics within the switching layer of the device, with only a few fitting parameters for calibration. Yet it produces results that are in qualitative agreement with behavior observed experimentally in a 3T-RRAM device. Using the simulator, we demonstrated that 3T-RRAMs show promise for use as synapses in neuromorphic crossbar circuits, warranting further experimental study of multi-terminal RRAMs. Future work includes model enhancements such as the addition of vacancy generation and annihilation, support for additional material layers and interfaces and an exploration of the performance of multi-terminal RRAMs when used in NN training.

ACKNOWLEDGMENTS

Akash Levy is supported by the National Science Foundation Graduate Research Fellowship under Grant No. 1650114. We thank Weier Wan from Stanford University for insightful discussions.

REFERENCES

[1] Herrmann, E., et al. (2018). Gate Controlled Three-Terminal Metal Oxide Memristor. *IEEE Electron Device Letters,* 39(4), 500–503. doi: 10.1109/led.2018.2806188

[2] Rush, A. J., Jones, A., Herrmann, E., & Jha, R. (2019). Gated-ReRAM Based Strategies for On-Chip Supervised Learning. *2019 IEEE National Aerospace and Electronics Conference (NAECON).* doi:10.1109/naecon46414.2019.9057906

[3] Subhechha, S., et al. (2017). Kinetic defect distribution approach for modeling the transient, endurance and retention of a-VMCO RRAM. *2017 IEEE International Reliability Physics Symposium (IRPS).* doi: 10.1109/irps.2017.7936322

[4] Voter A.F. (2007) Introduction to the Kinetic Monte Carlo Method. In: Sickafus K.E., Kotomin E.A., Uberuaga B.P. (eds) Radiation Effects in Solids. NATO Science Series, vol 235. Springer, Dordrecht

[5] Guan, X., Yu, S., & Wong, H. P. (2012). A SPICE Compact Model of Metal Oxide Resistive Switching Memory With Variations. *IEEE Electron Device Letters,* 33(10), 1405-1407. doi:10.1109/led.2012.2210856

[6] Le Cun, Y., et al. (2013). THE MNIST DATABASE. Retrieved from http://yann.lecun.com/exdb/mnist/

200

10-2

Fully Analog ReRAM Neuromorphic Circuit Optimization using DTCO Simulation Framework

Anh Nguyen
Electrical Engineering
San Jose State University
San Jose, USA
anh.d.nguyen@sjsu.edu

Hoi Nguyen
Electrical Engineering
San Jose State University
San Jose, USA
hoi.nguyen@sjsu.edu

Sruthi Venimadhavan
Electrical Engineering
San Jose State University
San Jose, USA
sruthi.venimadhavan@sjsu.edu

Ayyaswamy Venkattraman
Mechanical Engineering
University of California
Merced, USA
vayyaswamy@ucmerced.edu

David Parent
Electrical Engineering
San Jose State University
San Jose, USA
david.parent@sjsu.edu

Hiu Yung Wong*
Electrical Engineering
San Jose State University
San Jose, USA
hiuyung.wong@sjsu.edu

Abstract— Neuromorphic inference circuits using emerging devices (e.g. ReRAM) are very promising for ultra-low power edge computing such as in Internet-of-Thing. While ReRAM synapse is used as an analog device for matrix-vector-multiplications, the neuron activation unit (e.g. ReLU) is generally digital. To further minimize its power and area consumption, fully analog neuromorphic circuits are needed. This requires Design-Technology Co-Optimization (DTCO). In this paper, we use our Software+DTCO framework for fully analog neuromorphic inference circuit optimization using ReRAM as an example. The interaction between software machine learning, ReRAM, current comparator, and ReLU are studied. It is found that the neuromorphic circuit is very robust to the variation of ReLU, which confirms the importance of DTCO simulation.

Keywords—ReLU, ReRAM, DTCO, Neuromorphic, Machine Learning, Circuit Simulation, Verilog-A

I. INTRODUCTION

Machine learning (ML) using von Neumann architecture is facing the bottleneck issue of inefficient power consumption and large latency [1]. The issue is due to the physical separation of memory and computing units, causing most clock cycles to be used for intensive data exchange between the units rather than processing the information. Recently, various architectures have been proposed to overcome the bottleneck [2][3]. Among them, neuromorphic circuits which are analog circuits inspired by human brain architecture have been demonstrated to be a potential solution to the bottleneck. The major difference is to break the physical separation between memory and computing units by combining them at the same location to obviate data movement. The neuromorphic circuit can be regarded as an analog neural network in a memory crossbar array. The key component in most neuromorphic circuits is the analog memory device implemented using emerging memories. In recent years, emerging memories such as Resistive Random-Access Memory (ReRAM) [4][5], Phase Change Memory (PCM) [6], Ferro-Electric FET (FeFET) [7] have gained significant attention.

The neuromorphic circuit performance, however, depends strongly on various parameters such as input voltage range, loading impedance, temperature and sneak current in the cross-bar array interconnection [8] due to the non-linearity of both the memory device and the peripheral circuits. Since ML algorithms have built-in fault tolerance, the requirement of the

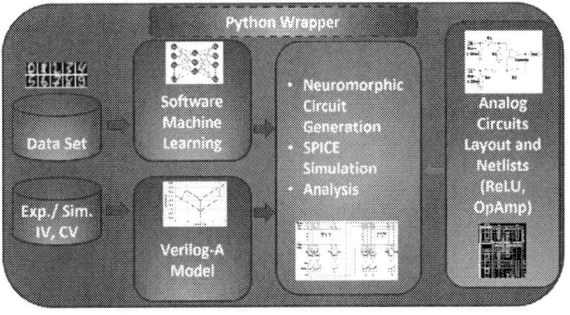

Fig. 1: Software+DTCO framework used. Verilog-A module is used when emerging devices do not have SPICE compact model.

circuit precision can be less stringent. So, it is important to have a Design-Technology-Co-Optimization (DTCO) framework to understand the interaction between the emerging memory, the circuit, and the ML algorithm to obtain the best trade-off. Moreover, the circuit performance is expected to be also dependent on the ML algorithm being used. Therefore, to achieve the ultimate optimization for various applications, it is necessary to co-optimize the ML algorithm (software) and circuits and devices (hardware).

In [3], a MATLAB framework has been built to study the temperature, loading resistance, and input voltage range effect independently. It only studies the precision of circuit behavior instead of the accuracy of the final ML outputs. In [9], a system-level simulator is built using behavior models. Despite its lower speed, SPICE simulation provides more insight into circuit design and avoids the use of behavior models. In [10], SPICE is used but only for studying the behavior of loading resistance and has no interaction with ML algorithm design.

In this paper, we improve and use our Software+DTCO framework [11] to study the effect of various ReLU circuits on a ReRAM inference circuit performance (Fig. 1). The circuit is a neural network (NN) for hand-written digit recognition using ReRAM. Its accuracy in predicting hand-written digits is studied as a function of ReLU circuit designs.

*Corresponding Author: hiuyung.wong@sjsu.edu

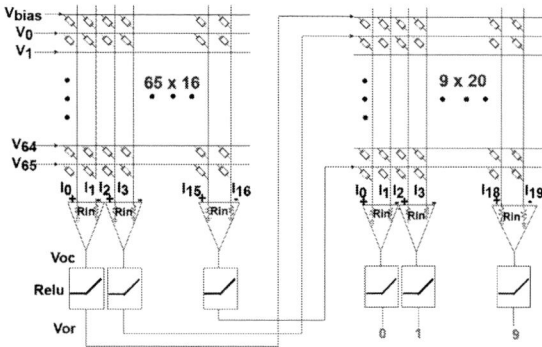

Fig. 2: Schematic showing the generated circuit for hand-written character recognition (only 1-layer case is shown for clarity.)

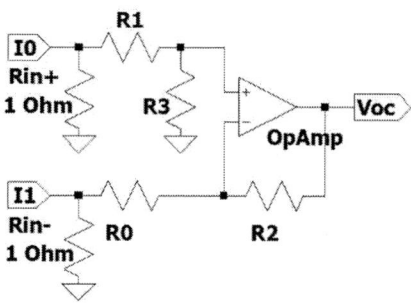

Fig. 4. Current subtractor and current-to-voltage convertor used in this study.

II. SOFTWARE+DTCO FRAMEWORK

Fig. 1 shows the Software+DTCO simulation framework for the neuromorphic inference circuit. The framework is written in python. Users can design a neuron network, NN, (by choosing the number of hidden layers and internal nodes) to train a machine in software for their purpose (e.g. hand-written character recognition). A neuromorphic SPICE circuit will be generated based on the NN designed in software (Fig. 2). Users can pick the emerging memory to be tested (e.g. ReRAM) and the weights in NN found in software training will be loaded to the SPICE compact model or Verilog-A model instances of the emerging memory. Users can also choose the type of sub-circuits (e.g. various ReLU circuit designs) to be included in the simulation. The sub-circuits may be post-layout netlists in which layout-dependent parasitic resistances and capacitances are included. Users can also choose various levels of accuracy for the components (e.g. whether an OpAmp is described by an ideal model or built from transistors of certain technologies). The framework will then invoke a commercial SPICE simulator to simulate the circuit to extract the desired metrics (e.g. inference accuracy) by performing a statistically meaningful number of simulations. The simulations can be performed in parallel. It is important to note that for certain NN design that gives the best accuracy in software (e.g. larger number of layers) might perform worse in neuromorphic circuits due to error propagation [11]. Therefore, Software+DTCO simulation is

important for the ultimate optimization to not just select an NN that gives the best performance in software but also find a neuromorphic circuit that has the best performance for a given emerging memory and peripheral circuit technologies.

As an example, in this paper, hand-written digit recognition using NN is used [12]. The ReRAM model is obtained, modified, and calibrated based on [13] (Fig. 3). The ReRAM device and circuit are co-optimized so that the ReRAM filament gap has the required resistance range ($2.5k\Omega$ and $300k\Omega$) for the neuromorphic circuit (Fig. 3). Users can then choose different circuits and layouts to realize various functions in the neuromorphic circuit. Here, the effects of the current comparator and ReLU designs on the accuracy of the inference circuit are studied. 180 images are tested for inference accuracy. In the neuromorphic circuit simulation, it is found that 3-layer NN gives the highest accuracy [11]. So only 3-layer neuromorphic circuits are simulated unless specified.

III. CURRENT SUBTRACTOR AND CURRENT-VOLTAGE CONVERTOR

Since ReRAM can only have positive conductance, to encode the negative weight from the NN, duplicated ReRAM strings are used followed by the current subtractor (Fig. 2). The current subtractor also converts the current to a voltage signal (Fig. 4). The input resistors to the subtractor, R_{in+} and

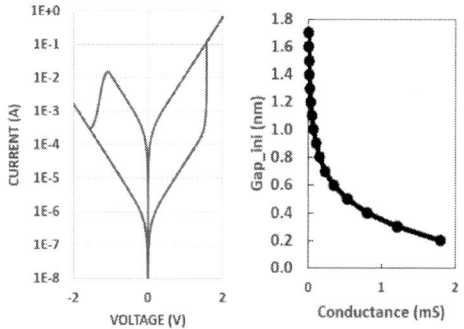

Fig. 3: Simulated ReRAM IV (left) and filament gap size as a function of ReRAM conductance (Right).

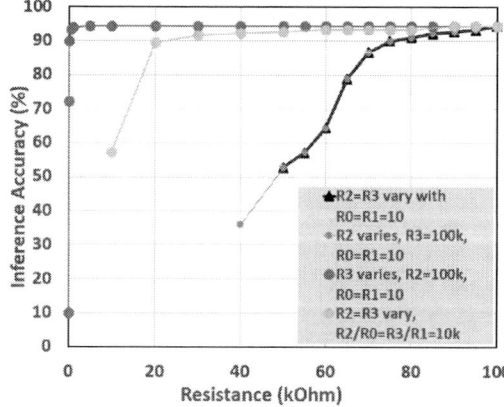

Fig. 5: Accuracy of the neuromorphic circuit (3-layer) as a function of resistor variation in the subtractor and converter.

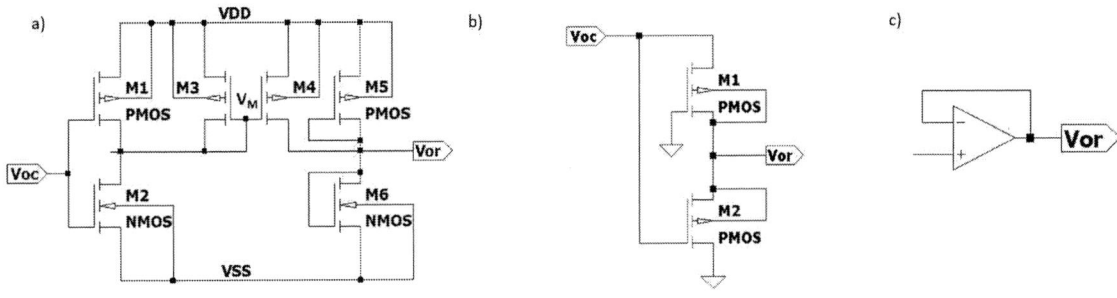

Fig. 6: Analog ReLU circuits used. a) 6T-ReLU based on [14]. b) 2T-ReLU. c) Unity-gain buffer is added to a) and b) to form 6T-ReLU-Buf and 2T-ReLU-Buf, respectively.

R_{in-}, are chosen to be 1 Ω so that it is low enough compared to the ReRAM resistance.

For an ideal OpAmp, it can be derived that $V_0 = A(I_0 R_{in+} - I_1 R_{in-})$. With $R_0 = R_1 = 10Ω$, $R_2 = R_3 = 100kΩ$, it can be showed that $A = R_2/R_0 = 10^4$. The effect of resistor variations in the subtractor and convertor is then studied. Fig. 5 shows the results. When R_2 and R_3 are reduced together, the prediction accuracy of the circuit is higher than 90% until the resistance is dropped by 25% to 75kΩ. Further study shows that this is limited by R_2. If only R_3 is varied, the accuracy is still high even R_3 reaches 100Ω. This is because of the large value of A being used. If A is kept constant but the absolute values of R_0, R_1, R_2, and R_3 are changed proportionally, it is found that the accuracy decreases significantly when $R_{2,3}=20kΩ$ and $R_{0,1}=2Ω$. This is because $R_{0,1}$ is now close to $R_{in+,in-}$, and the equation derived is no longer valid and the subtractor cannot function as it is supposed to be. This example shows that while the impact of resistance variation can be predicted qualitatively by circuit analysis, the inference prediction accuracy can only be obtained by using DTCO simulation.

IV. ANALOG ReLU DESIGN

ReLU function is critical in NN and it retains the value if the input is positive and gives zero otherwise (Fig. 7). An analog ReLU circuit with 6 transistors (6T-ReLU) is adopted from [14] using 45nm design rules. The circuit is shown in Fig. 6a. The input inverter (formed by M1 and M2) inverts the input voltage V_{oc} into V_m, the gate voltage of M4, to turn on/off M4. When V_{oc} is negative, V_m is positive and M4 is in the cut-off region. The output voltage V_{or} is tuned to 0V by adjusting M5 and M6 of the voltage divider.

When V_{oc} is positive, V_m is low and M4 is in the saturation region. The output voltage V_{or} is derived from the small-signal model as in [14],

$$\Delta V_{or} = g_{m1}g_{m2}R_1R_2\Delta V_{oc} \quad (1)$$

with,

$$g_{m1}=(g_{m,M1}+g_{m,M2}) \quad (2)$$
$$R_1=(r_{o,M1}||r_{o,M2}||1/g_{m,M3}||r_{o,M3}) \quad (3)$$
$$g_{m2}=g_{m,M4} \quad (4)$$
$$R_2=(r_{o,M4}||1/g_{m,M5}||r_{o,M5}||1/g_{m,M6}||r_{o,M6}) \quad (5)$$

where g_m's and r_o's are the transconductances and output impedances of the corresponding transistor, respectively. By

sizing the transistors so that $g_{m1}g_{m2}R_1R_2 = 1$, the circuit in Fig. 6a will then function as a ReLU circuit.

This circuit was designed for CMOS synapses which have large input impedance (due to small gate current in CMOS). Therefore, its output impedance is too large for ReRAM synapses because of the low ReRAM input impedance (Note that the output of the ReLU is used as the input of the next NN layer, Fig. 2). Fig. 7 shows that when the loading to the ReLU circuit is in the order of the maximum ReRAM conductance of the technology used, it does not behave as a ReLU function. However, the accuracy still reaches 73% in the 1-layer case but degrades quickly in the 3-layer case due to error accumulation (Fig. 8). Therefore, a unity gain buffer (Fig. 6c) is added and the accuracy can reach 94% in the 3-layer case.

To simplify the circuit, based on [15] and [16], a new 2T-RELU is proposed (Fig. 6b). Assume the threshold voltages ($V_{th,M1}$, $V_{th,M2}$) of M1 and M2 are zero, when V_{oc} is negative, M2 is turned on because $V_{GS,M2} = V_{oc} - V_{or}$ and V_{or} can be discharged to 0V, while M1 is turned off as $V_{GS,M1} = 0V - V_{oc} > 0V$. However, the threshold voltages of M1 and M2 are not set to zero but with finite negative values. Therefore, for small negative (or positive) values of V_{oc}, V_{or} is determined by the voltage divider formed by the two off-state transistors, M1 and M2 (Fig. 7). When V_{oc} is positive, M2 is turned off because $V_{GS,M2} > V_{th,M2}$. But M1 is turned on as $V_{GS,M1} = - V_{oc} <$

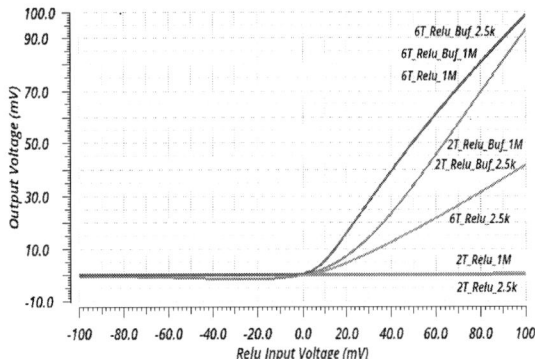

Fig. 7. Performance of the ReLU circuits in Fig. 6 with 1MΩ and 2.5kΩ loadings.

$V_{th,M2}$. Therefore, the behavior of 2T-ReLU can be summarized as

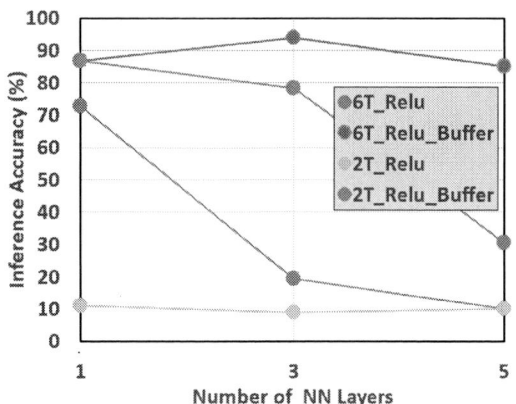

Fig. 8. Inference accuracy as a function of ReLU designs and number of NN layers.

$$V_{or} \approx 0V \text{ if } V_{oc} \leq 0V \qquad (2)$$
$$V_{or} \approx V_{oc} \text{ if } V_{oc} > 0V \qquad (3)$$

which is essentially a ReLU circuit with non-idealities.

However, the analysis above is only true if there is no loading. When there is loading with small input resistance (such as that of the next layer ReRAM), V_{or} will be pulled down substantially. As can be seen in Fig. 7, the 2T-ReLU does not behave as a ReLU even with loading impedance is as large as $1M\Omega$ and gives very bad inference accuracy (Fig. 8). Therefore, an output buffer needs to be added (Fig 6c). Although with a buffer, it still has non-idealities, the neuromorphic circuit performs well with an accuracy of 87% in the 1-layer case. Therefore, by using the framework, we found a simpler analog ReLU that can achieve 87% accuracy for this task which is impossible without this framework.

V. CONCLUSIONS

A software+DTCO simulation framework is constructed in which users can perform co-optimization of synapses (traditional CMOS or emerging devices) and neuron and software ML. A fully analog ReRAM neuromorphic inference circuit is optimized using the framework. It is found that the inference accuracy is very robust. Moreover, the circuit can tolerate up to 25% variation of current comparator resistance value and is robust against impedance mismatching between ReLU neuron and ReRAM synapse. A simplified 2T-ReLU design is proposed and verified by using this framework and can achieve 87% accuracy despite the non-idealities.

ACKNOWLEDGMENT

This project was supported by San Jose State University College of Engineering Small Group Project Team Fund (2019).

REFERENCES

[1] E. García-Martín, C. F. Rodrigues, G. Riley, and H. Grahna, "Estimation of energy consumption in machine learning," Volume 134, December 2019, Pages 75-88.doi: 10.1016/j.jpdc.2019.07.007

[2] H. Tsai, S. Ambrogio, P. Narayanan, R. M. Shelby and G. W. Burr, "Recent progress in analog memory-based accelerators for deep learning", 2018 J. Phys. D: Appl. Phys. 51 283001.

[3] M. Hu, J. P. Strachan, Z. Li, E. M. Grafals, N. Davila, C. Graves, S. Lam, N. Ge, J. Yang, and R. S. Williams, "Dot-product engine for neuromorphic computing: Programming 1T1M crossbar to accelerate matrix-vector multiplication," 2016 53rd ACM/EDAC/IEEE Design Automation Conference (DAC), Austin, TX, 2016, pp. 1-6. doi: 10.1145/2897937.2898010

[4] S. Yu, B. Gao, Z. Fang, H. Yu, J. Kang, and H.-S. P. Wong, "A low energy oxide-based electronic synaptic device for neuromorphic visual systems with tolerance to device variation," Adv. Mater. 2013, 25, 1774–1779. https://doi.org/10.1002/adma.201203680.

[5] H. -S P. Wong, H.-Y. Lee, S. Yu, Y.-S. Chen, Y. Wu, P.-S. Chen, B. Lee, F. T. Chen, and M.-J. Tsai, "Metal–Oxide RRAM," Proceedings of the IEEE, vol.100, no.6, pp.1951,1970, June 2012, doi: 10.1109/JPROC.2012.2190369

[6] S. Kim, M. Ishii, S. Lewis, T. Perri, M. BrightSky, W. Kim, R. Jordan, G. W. Burr, N. Sosa, A. Ray, J.-P. Han, C. Miller, K. Hosokawa, and C. Lam, "NVM neuromorphic core with 64k-cell (256-by-256) phase change memory synaptic array with on-chip neuron circuits for continuous in-situ learning," 2015 IEEE International Electron Devices Meeting (IEDM), Washington, DC, 2015, pp. 17.1.1-17.1.4. doi: 10.1109/IEDM.2015.7409716

[7] X. Sun, P. Wang, K. Ni, S. Datta, and S. Yu, "Exploiting Hybrid Precision for Training and Inference: A 2T-1FeFET Based Analog Synaptic Weight Cell," 2018 IEEE International Electron Devices Meeting (IEDM), San Francisco, CA, 2018, pp. 3.1.1-3.1.4. doi: 10.1109/IEDM.2018.8614611

[8] M. Hu, H. Li, Q. Wu and G. S. Rose, "Hardware realization of BSB recall function using memristor crossbar arrays," DAC Design Automation Conference 2012, San Francisco, CA, 2012, pp. 498-503.

[9] P. Chen, X. Peng and S. Yu, "NeuroSim+: An integrated device-to-algorithm framework for benchmarking synaptic devices and array architectures," 2017 IEEE International Electron Devices Meeting (IEDM), San Francisco, CA, 2017, pp. 6.1.1-6.1.4, doi: 10.1109/IEDM.2017.8268337.

[10] P. Gu, B. Li, T. Tang, S. Yu, Y. Cao, Y. Wang, H. Yang, "Technological exploration of RRAM crossbar array for matrix-vector multiplication," The 20th Asia and South Pacific Design Automation Conference, Chiba, 2015, pp. 106-111. doi: 10.1109/ASPDAC.2015.7058989

[11] H. Cao, T. Lam, H. Nguyen, A. Venkattraman, D. Parent, and H. Y. Wong, " Study of ReRAM Neuromorphic Circuit Inference Accuracy Robustness using DTCO Simulation Framework," Accepted by IEEE Workshop on Microelectronics and Electron Devices, 2020.

[12] Dua, D., and Graff, C. (2019). UCI Machine Learning Repository [http://archive.ics.uci.edu/ml]. Irvine, CA: University of California

[13] Jiang, Z., et al., (2014). Stanford University Resistive-Switching Random Access Memory (RRAM) Verilog-A Model. nanoHUB.

[14] J. Zhu, Y. Huang, Z. Yang, X. Tang and T. T. Ye, "Analog Implementation of Reconfigurable Convolutional Neural Network Kernels," 2019 IEEE Asia Pacific Conference on Circuits and Systems (APCCAS), Bangkok, Thailand, 2019, pp. 265-268, doi: 10.1109/APCCAS47518.2019.8953177.

[15] M. Yilmaz, B. A. Tunkar, S. Park, K. Elrayes, M. A. E. Mahmoud, E. Abdel-Rahman, and M. Yavuz, "High-efficiency passive full wave rectification for electromagnetic harvesters ," Journal of Applied Physics 116, 134902 (2014); https://doi.org/10.1063/1.4896668.

[16] Priyanka P., et al., CMOS Implementations of Rectified Linear Activation Function in VLSI Design and Test. VDAT 2018. Communications in Computer and Information Science, vol 892.

Effect of Shape Deformation by Edge Roughness in Spin-Orbit Torque Magnetoresistive Random-Access Memory

Jihun Byun, Doo Hyung Kang, and Mincheol Shin*
School of Electrical Engineering
Korea Advanced Institute of Science and Technology
Daejeon 34141, Republic of Korea
*mshin@kaist.ac.kr

Abstract— We present a micromagnetic simulation study of shape deformation and edge roughness effect in the spin orbit torque-magnetic random access memory (SOT-MRAM). The two different write schemes, magnetic field induced SOT write scheme and SOT-spin transfer torque (STT) hybrid write scheme, were studied in the presence of the stray field from the reference layer. We found that for conventional magnetic field induced SOT, shape deformation can cause non-deterministic switching even at a relatively high gilbert damping constant of 0.08. Higher Gilbert damping constant (α) of 0.09 is needed to ensure deterministic switching under the shape deformation effect. The SOT-STT hybrid write scheme showed deterministic switching even at lower damping constant with relatively low device variations due to the constant $-z$ directed torque of the STT. However, with higher damping constant of $\alpha = 0.1$, device variation with the SOT-STT hybrid write scheme increases while the SOT-magnetic field write scheme successfully compensates the most of the variation caused by the edge deformation.

Keywords—Magnetic Random Access Memory, Spin Orbit Torque, Spin Transfer Torque, Damping constant, Micromagnetic simulation

I. INTRODUCTION

Spin-orbit torque magnetoresistive random-access memory (SOT-MRAM) is a promising novel non-volatile memory device due to sub-ns operation speed and higher reliability compare to spin-transfer torque magnetoresistive random-access memory (STT-MRAM). Writing operation involves the current flowing through the heavy metal beneath the magnetic tunnel junction (MTJ). This current induces in-plane torque driven by spin Hall effect (SHE) and/or Rashba effect. Unlike STT, of which the direction is perpendicular, SOT has the in-plane directed torque therefore it only lays down the initially perpendicular magnetization (z-direction) towards the spin-polarized direction in the x-y plane. Additional torque is required to complete the switching. For this purpose, one of the most commonly used methods is to apply the in-plane magnetic field. The in-plane field would tilt the magnetization towards the targeted hemisphere and relax towards the targeted state upon turning off the SOT current. Although it required an external magnetic field, newly proposed methods suggest that the external field-free SOT-MRAM is possible by imposing a magnetic hard mask [1] or using antiferromagnet/ferromagnet bilayer [2] or ferromagnet/heavy metal/ferromagnet trilayer [3]. This write scheme using the in-plane magnetic field will be referred as the SOT-magnetic field write scheme in the rest of the paper for simplicity. Another well-known write scheme is the SOT-STT hybrid write scheme, which uses STT current to induce

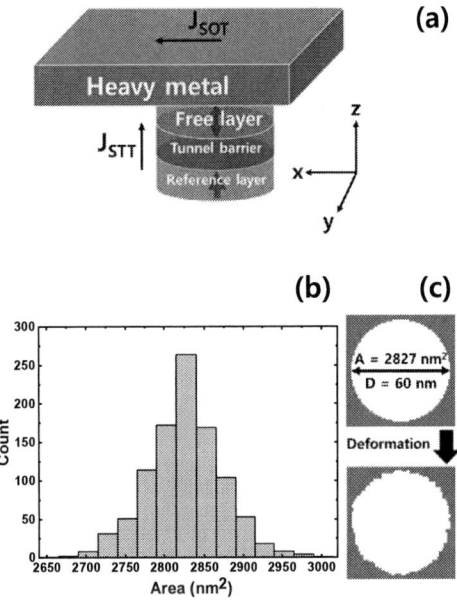

Fig. 1. (a) Typical structure of SOT-MRAM and (b) the distribution of the area of the generated MTJ samples and (c) deformation process

perpendicular torque instead of the magnetic field [4]. It requires additional STT current but it tends to have a simpler MRAM structure as there is no need for additional process to provide the effective in-plane magnetic field. Both write schemes demonstrated sub-ns switching with lower writing power compared with the MRAM using the STT write scheme alone [1, 4]. Despite the outstanding performance of the SOT-MRAM, it lacks the study of the expected variation in the switching characteristics as the SOT-MRAM has not been yet commercially mass-produced.

The magnetization reversal process can be affected by boundary roughness [5] and the shape deformation of MTJ has an influence on switching time in STT-MRAM due to the in-plane demagnetizing tensor induced from deformed MTJ [6]. The MTJs have suffered from critical dimension variations during the manufacturing process [7] and fabricated MTJs indicate that they are not in perfect circular shape [8]. The shape deformation of MTJ is expected to affect the magnetization switching behavior of the SOT-MRAM but this is not yet elucidated to the best our knowledge. Earlier studies analyzed shape deformation of SOT-MRAM modeled the

TABLE I. PARAMETERS USED IN THE SIMULATION

Symbol	Parameters	Value
A	Damping constant	$0.02 \sim 0.1$
θ_{SH}	Spin hall angle	0.3
H	Spin torque efficiency	0.7
$Ku1_{free}$	Anisotropy constant of the free layer	9.8×10^5 J/m^3
$Ku1_{reference}$	Anisotropy constant of the reference layer	1.5×10^6 J/m^3
A_{ex}	Exchange constant	20×10^{-12} A/m
M_{sat}	Saturation magnetization	1.2×10^6 A/m
t_F	Free layer thickness	1 nm
t_R	Reference layer thickness	1 nm
t_T	Tunnel barrier thickness	1 nm
D	MTJ Diameter	60 nm
	Mesh size	$1 \times 1 \times 1$ nm^3
T_{SOT}	SOT pulse width	1 ns
T_{STT}	STT pulse width	5 ns

shape deformation by varying the area of the MTJ rather than considering altered edge and shape of MTJ [9].

In this work, we present a micromagnetic simulation study of shape deformation due to edge roughness effect in the SOT-MRAM. Two different writing schemes of SOT-MRAM were studied; the SOT-magnetic field write scheme and the SOT-STT hybrid write scheme.

II. METHODOLOGY

To consider the shape deformation, micromagnetic simulations were performed by using object oriented micromagnetic framework (OOMMF) [10]. Simulated SOT-MRAM has a structure shown in Fig. 1. Ideal MTJ's are of the cylindrical shape and consist of three layers of free layer, tunnel barrier, and reference layer, each of which has a thickness of 1 nm and diameter of 60 nm. We assumed that the heavy metal on the top of the free layer provides SOT current with the spin polarization ($\vec{\sigma}$) along the –y-direction. Perpendicularly (z-direction) magnetized MRAM was considered with the initial free layer and the reference layer magnetization in +z-direction. We treated 1000 samples each with x-y plane random edge roughness on ideal circular-shaped MTJ to have ±5% of area difference within 3σ distributions. Field-like torque induced by STT and SOT was not considered in this simulation.

The SOT-magnetic field write scheme needs SOT current pulse under external magnetic field in x-direction (H_x) and the SOT-STT hybrid write scheme uses both SOT current (J_{SOT}) and STT current (J_{STT}) in the absence of magnetic field. Parameters used in the simulations are shown in Table. 1. H_x of 30 mT was applied for the SOT-magnetic field write scheme case. As can be seen in Fig. 2., the SOT critical current

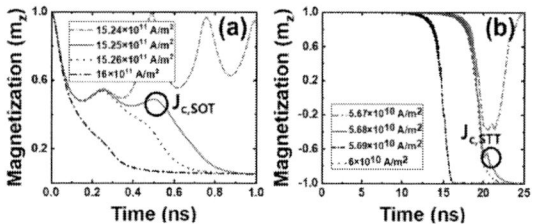

Fig. 2. Critical current of the ideal circular device with the α = 0.02. (a) SOT critical current ($J_{c,SOT}$) and (b) STT critical current ($J_{c,STT}$)

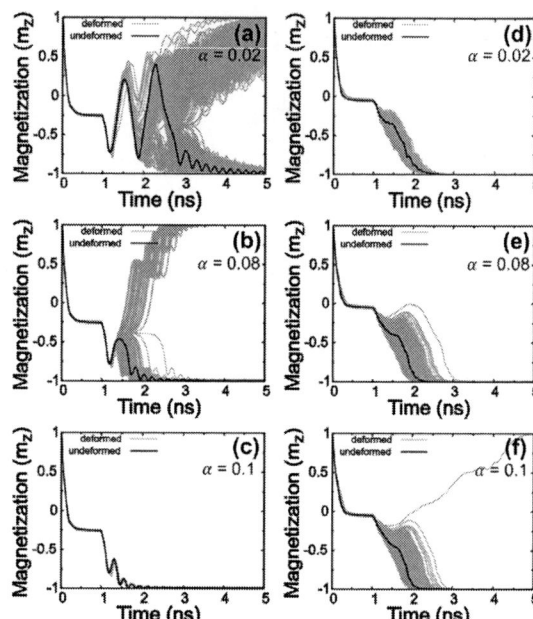

Fig. 3. Time evolution of perpendicular component of the free layer magnetization m_z of SOT-magnetic field write scheme with $J_{SOT} = 16 \times 10^{11}$ A/m^2, $H_x = 30$ mT, $t_{SOT} = 1$ ns with damping constant (a) α = 0.02, (b) α = 0.08, and (c) α = 0.1. SOT-STT hybrid write scheme with $J_{SOT} = 16 \times 10^{11}$ A/m^2, $J_{STT} = 6 \times 10^{10}$, $t_{SOT} = 1$ ns, $t_{STT} = 5$ ns with damping constant (d) α = 0.02, (e) α = 0.08, and (f) α = 0.1. Bold and gray lines represent the circular MTJ without deformation and the deformed samples with random edge roughness, respectively.

($J_{c,\,SOT}$) was 15.25×10^{11} A/m^2 and STT critical current ($J_{c,\,STT}$) was 5.68×10^{10} A/m^2. Therefore, the operation currents were chosen to have a higher value than the critical current to ensure switching, which were $J_{SOT} = 16 \times 10^{11}$ A/m^2 and $J_{STT} = 6 \times 10^{10}$ A/m^2. Square current pulse was assumed.

III. RESULTS AND DISCUSSION

Fig. 3 represents the z-direction component of the magnetization (m_z) under the SOT-magnetic field write scheme when the Gilbert damping constants (α) were 0.02, 0.08, and 0.1. As shown in Fig. 3. (a), only 42.3% of samples are switched for the SOT-magnetic field write scheme with low α of 0.02 and the ideal circular MTJ shows a highly oscillatory magnetization trajectory. At α as high as 0.08, the sample without the deformation shows stable damping with precession which occurs only in the lower hemisphere. However, α is not high enough that the switching probability of 100% was still not achievable (see Fig. 3 (b)).

The relatively low α induces unstable magnetization trajectory with the high precession amplitude which crosses to the upper hemisphere ($m_z > 0$) and damps toward the stable initial state. The earlier work explains such unstable switching as the hysteric behavior of the transition to the intermediate position, where the behavior intensifies with the lower α [11]. Such an unstable magnetization trajectory makes the switching sensitive to various field terms, that demagnetization field alternation due to ±5% deformation in the area within 3σ is enough to cause more than half of the samples to switch back to the initial state.

Another possible influence from the shape deformation is the edge roughness affecting the nucleation and the

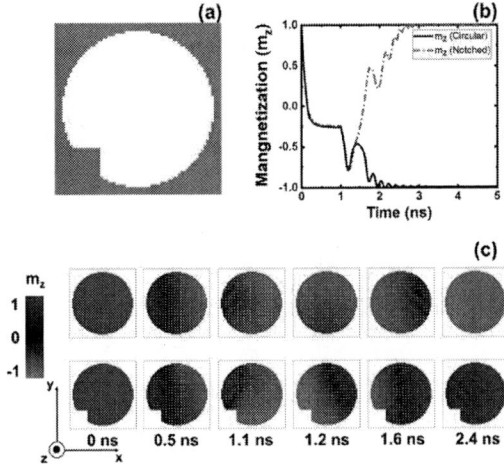

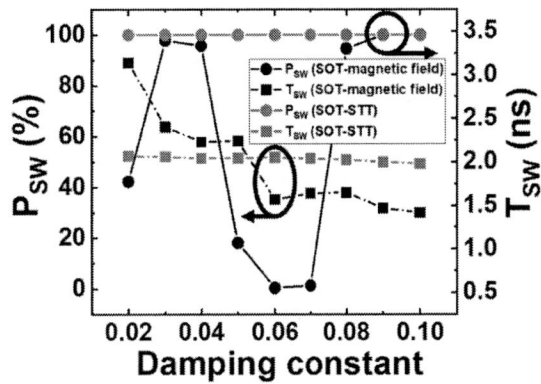

Fig. 4. (a) sample with 10 nm × 10 nm edge notch (b) Time evolution of perpendicular component of the free layer magnetization m_z of SOT-magnetic field write scheme with $\alpha = 0.08$, compared with the ideal circular MTJ. (c) magnetization snapshots of the free layer. Blue inditcates the parallel state, red indicates the antiparallel state and black indicates the in-plane magnetization. Top line is the snapshots of the circular MTJ and the Bottom line is the snapshots of the edge notched MTJ.

propagation of the domain. Fig. 4. (b) depicts the magnetization trajectory of the edge notched MTJ and the ideal circular MTJ using the SOT-magnetic field write scheme with with $\alpha = 0.08$. From Fig. 4. (c), it can be seen that the switching behavior of the MTJs involves asymmetric nucleation of the edge and the propagation. Such behavior is known to be seen in the presence of the Dzyaloshinskii-Moriya interaction (DMI) causing the inward or outward magnetization canting of the edge [12]. We estimate that the asymmetric edge nucleation and propagation were derived by the stray field. This is because the reference layer can induce a similar effect as the edge canting of the DMI due to the inward or outward in-plane stray magnetic field at the edge [13, 14].

With the edge notch, the propagation characteristics can be altered from the ideal case as the propagation of the domain is disturbed by the notch therefore fails to completely propagate to the opposite end. The magnetic domain nucleation and propagation can be largely affected by the edge roughness [5] and we estimate that they are the origin of the switching characteristics variations and switch failure derived from the edge roughness.

As shown in Fig. 3. (c), 100% switching is possible for a higher α of 0.1. Also, using α of 0.1 results in lower device variations of maximum switching time difference of 3.57% with the average switching time of 1.42 ns. Those for the $\alpha = 0.02$ are 87.74% and 3.14 ns, respectively. This is because the higher α induces less oscillatory magnetization so that the precession only occurs in the lower hemisphere ($m_z < 0$) and does not cross the equator ($m_z = 0$). Higher α induces higher $-z$ directed effective magnetic field in the $-z$-direction magnetized domain region due to the perpendicular anisotropy. Since the domain is larger in size after the SOT pulse is turned off, higher α will more favor this domain to propagate towards the $+z$-direction magnetized domain region side. However, the widely used MTJ structure of the CoFeB/MgO layer has α of $0.01 \sim 0.03$ [15]. Within these values, one cannot achieve deterministic switching

Fig. 5. Switching probability (P_{SW}) and switching time (T_{SW}) with damping constant α variation. Red lines indicate the SOT-STT hybrid write scheme and black lines indicate the SOT-magnetic field write scheme.

independent of shape deformation caused by process variations.

Fig. 3. (d) shows the results of the SOT-STT hybrid write scheme at $\alpha = 0.02$. The SOT-STT hybrid write scheme enables the switching of all samples even at relatively low α. This is because the perpendicular ($-\mathbf{m}\times(\mathbf{m}\times\hat{\mathbf{z}})$) torque of the STT suppresses the precession towards the upper hemisphere and induces less oscillatory behavior of magnetization. Less switching time variation of 27.3% and lower switching time of average 2.06 ns were achieved. The SOT-magnetic field write scheme with the same α have an average switching time of 3.14 ns. Also, the switching speed of the SOT-STT hybrid write scheme is 7.63 times faster than the STT write scheme in the absence of the SOT current, which has 15.81 ns of switching time as can be seen in Fig. 2.(b).

The switching trajectory variation in the SOT-STT hybrid write scheme induced by the deformation might be due to the altered in-plane demagnetization tensor, which affects the switching time [6] and the edge roughness, which can slow down the domain propagation [5].

Using the SOT-STT hybrid write scheme, with increasing the α, the device variation increases with almost constant switching time. Also, switching probability degrades to 99.9% when $\alpha = 0.1$. We analyzed that this is because the intermediate state of the SOT-STT hybrid write scheme has close proximity to the x-y plane. The m_z of the SOT-STT hybrid write scheme at the 1 ns, which is the end of the intermediate state, is only -0.05. Due to this magnetization lying near the equator, after the J_{SOT} is turned off, both parallel ($m_z = 1$) and the antiparallel ($m_z = -1$) states of the SOT-STT hybrid write scheme are highly stable. The magnetization switching is also affected by the out-of-plane component of the stray field generated from the reference layer. Therefore, after t = 1 ns, STT in the absence of the SOT has to compete against the damping towards the initial parallel state and the stray field. This hindrance of the switching enlarges with increasing the α, resulting in higher variation. They lead to the degradation of switching characteristics and switching back behavior in $\alpha = 0.1$.

However, for the SOT-magnetic field write scheme, when t = 1 ns, the intermediate state is located at $m_z = -0.26$. In this state, the antiparallel state is clearly more stable and the

207

magnetization can damps towards the targeted state with faster switching time when using higher α.

These tendencies can be seen in Fig. 5 which shows switching probability (P_{SW}) and switching time (T_{SW}) varying with the damping constant α. The T_{SW} of the SOT-magnetic field write scheme decrease with the increasing α, so that its switching probability reaches to 100% at α = 0.09, while the SOT-STT hybrid write scheme switching time is almost independent on α. The oscillatory behavior in P_{SW} of the SOT-magnetic field at lower α agrees with the earlier macrospin work [16].

One might argue that decreasing the in-plane field and increasing J_{STT} can result in better performance of the SOT-STT hybrid write scheme at high α. Note that the $J_{STT} = 6 \times 10^{10}$ A/m^2 used in the simulation shows the switching time of 15.81 ns which satisfies the target write speed for STT-MRAM used in an L4 cache application of sub 20 ns [17]. While the $H_x = 30$ mT is also commonly used value for the in-plane magnetic field in the SOT-MRAM [1, 18]. Therefore, the tendency shown in Fig. 5 shows the results using a common magnitude of H_x and J_{STT} rather than the exceptional case.

IV. CONSCLUSION

The performance effect of shape deformation on SOT-MRAM with two different write schemes has been studied by using micromagnetic simulation. It is found that the edge roughness and deformation of MTJ in SOT-MRAM can result in severe variation in switching behavior such as write failure or switching time degradation. The source of the variation might be the alteration of demagnetization tensor and the disturbance in the domain propagation by the rough edge.

The simulation results show that the SOT-magnetic field write scheme may not be practical up to high α of 0.09 because of the low switching probability. The SOT-STT hybrid write scheme shows 100% switching even with α of 0.02, having faster switching and lower device variation compared to the SOT-magnetic field write scheme. However, with the stray field from the reference layer, using high α the SOT-magnetic field write scheme shows better performance with less device variation and enhanced switching speed whereas it results in the degradation of the SOT-STT hybrid write scheme with higher device variation with write failure occurs. A reason for the failure is that the SOT-STT hybrid write scheme has its intermediate state located near the x-y plane. Higher α induces damping towards both parallel and anti-parallel state and the out-of-plane component of the stray field results as the disturbance of the intended switching.

Our results suggest that in the presence of the stray field, the SOT-magnetic field write scheme shows better performance for MTJ with high α whereas the SOT-STT hybrid write scheme shows better performance with low α material.

ACKNOWLEDGMENT

This work was supported by Samsung Research Funding & Incubation Center of Samsung Electronics under Project Number SRFC-IT1901-11

REFERENCES

[1] K. Garello et al., "Magnufacturable 300mm platform solution for Field-Free Switching SOT-MRAM," in 2019 Symposium on VLSI Circuits, Kyoto, Japn, 2019

[2] Y. Oh et al., "Field-free switching of perpendicular magnetization through spin-orbit torque in antiferromagnet/ferromagnet/oxide structures," Nat. Nanotech, vol. 11, 878-884, 2016

[3] W. Chen, L. Quain, and G. Xiao, "Deterministic Current Induced Magnetic Switching Without External Field using Giant Spin Hall Effect of β-W," Nat. Sci. Rep., vol. 8, 8144, 2018

[4] S. Pathak, C. Youm, and J. Hong, "Impact of Spin-Orbit Torque on Spin-Transfer Torque Switching in Magnetic Tunnel Junctions," Nat. Sci. Rep., 10, 2799, 2020

[5] D. Meyners, H. Bruckl, and G. Reiss, "Influence of boundary roughness on the magnetization reversal in submicron sized magnetic tunnel junctions," J. Appl. Phys., vol. 93, 2676, 2003.

[6] C. Engel, S. Goolaup, H. Teoh, and W. Lew, "Effect of Geometrical Modulation on pMTJ Magnetization Reversal," IEEE Trans. Magn., vol. 53, no.12, 2017.

[7] K. Lee et al., "1Gbit High Density Embedded STT-MRAM in 28nm FDSOI Technology," in 2019 IEEE International Electron Devices Meeting (IEDM), San Francisco, CA, USA, 2019.

[8] M. Pak et al., "LCDU optimization of STT-MRAM 50nm pitch MTJ pillars for process window improvement," in Proc. Extreme Ultraviolet (EUV) Lithography X, San Jose, CA, USA, 2019.

[9] B. Zeinali, J. Madsen, P. Raghaan, and F. Moradi., "Ultra-Fast SOT-MRAM Cell with STT Current for Deterministic Switching," in 2017 IEEE International Conference on Computer Design (ICCD), Boston, MA, USA, 2017.

[10] R. McMichael and M. Donahue, "OOMMF User's Guide, Version 1.0," National Institute of Standards and Technology, Gaithersburg, MD, USA, Interagency Report NISTIR 6376, Sept. 1999.

[11] B. chen, S. Lim, and M. Tran, "Magnetization Switching of a Thin Ferromagnetic Layer by Spin-Orbit Torques," IEEE Magn. Lett., vol. 7, 2016

[12] N. Mikuszeit, et al., "Spin-orbit torque driven chiral magnetization reversal in ultrathin nanostructures," Phys. Rev. B., Vol.2, 14424, 2015

[13] Y. Wang, et al., "Impact of the stray field on the switching properties of perpendicular MTJ for scaled MRAM," in 2012 IEEE International Electron Devices Meeting (IEDM), San Francisco, CA, USA, 2012.

[14] H. Jiancheng, et al., "Effect of the stray field profile on the switching characteristics of the free layer in a perpendicular magnetic tunnel junction," J. Appl. Phys., vol. 117, 17B721, 2015

[15] S. Ikeda, et al., "A Perpendicular-anisotropy CoFeB-MgO magnetic tunnel junction," Nat. Mat., 9, 721-724, 2010

[16] K. Lee, S. Lee, B. Min, and K. Lee, "Threshold current for switching of a perpendicular magnetic layer by spin Hall effct," Appl. Phys. Lett., vol. 102, 112410, 2013

[17] J.G. Alzate, et al., "2 MB Array-Level Demonstration of STT-MRAM Process and Performance Towards L4 Cache Applications," in 2019 IEEE International Electron Devices Meeting (IEDM), San Francisco, CA, USA, 2019.

[18] N. Sato, F. Xue, R. M. White, C. Bi, and S. X. Wang, "Two-terminal spin–orbit torque magnetoresistive random access memory", Nat. Electron., vol. 1, 508-511, 2018

Computation of Torques in Magnetic Tunnel Junctions through Spin and Charge Transport Modeling

Simone Fiorentini
Christian Doppler Laboratory
for NovoMemLog at the
Institute for Microelectronics
TU Wien
Vienna, Austria
fiorentini@iue.tuwien.ac.at

Johannes Ender
Christian Doppler Laboratory
for NovoMemLog at the
Institute for Microelectronics
TU Wien
Vienna, Austria
ender@iue.tuwien.ac.at

Mohamed Mohamedou
Christian Doppler Laboratory
for NovoMemLog at the
Institute for Microelectronics
TU Wien
Vienna, Austria
mohamedou@iue.tuwien.ac.at

Roberto Orio
Institute for Microelectronics
TU Wien
Vienna, Austria
orio@iue.tuwien.ac.at

Siegfried Selberherr
Institute for Microelectronics
TU Wien
Vienna, Austria
selberherr@iue.tuwien.ac.at

Wolfgang Goes
Silvaco Europe Ltd
Cambridge, United Kingdom
wolfgang.goes@silvaco.com

Viktor Sverdlov
Christian Doppler Laboratory
for NovoMemLog at the
Institute for Microelectronics
TU Wien
Vienna, Austria
sverdlov@iue.tuwien.ac.at

Abstract—Spin-transfer torque based devices are among the most promising candidates for emerging nonvolatile memory. Reliable simulation tools can help understand and improve the design of such devices. In this paper, we extend the drift-diffusion approach for coupled spin and charge transport, commonly applied to determine the torque in metallic valves, to the case of magnetic tunnel junctions, which constitute the cell of modern spin-transfer torque memories. We demonstrate that, by introducing a magnetization dependent conductivity and properly choosing the spin diffusion coefficient in the tunnel barrier, the expected behavior of both, the electric current and the spin accumulation, is properly reproduced. The spin torque values' dependence on the system parameters is investigated. As a unique set of equations is used for the entire memory cell, this constitutes the basis of an efficient finite element based approach to rigorously describe the magnetization dynamics in emerging spin-transfer torque memories.

Keywords—*Spin and charge drift-diffusion, spin-transfer torque, magnetic tunnel junctions, STT-MRAM*

I. INTRODUCTION

Nonvolatility is an emerging solution to the high stand-by power leakages due to the downscaling of the dimensions of traditional semiconductor components. Spin-transfer torque magnetoresistive random access memory (STT-MRAM) is a viable nonvolatile component, thanks to its simple structure and compatibility with CMOS technology. It possesses both high speed and excellent endurance, as well as low fabrication costs, and is thus promising for applications ranging from IoT and automotive uses to embedded DRAM, and last level caches [1]- [7].

The binary information in modern magnetic memories is stored as the relative orientation of the magnetization in the free and reference layers of a magnetic tunnel junction (MTJ) schematized in Fig.1. The magnetization in the free layer can be switched by a current through the structure. The electrons

get spin polarized by the reference layer and, when entering the free layer, transfer their polarization to the magnetization, providing the torque.

Accurate simulations of the magnetization dynamics in these structures provide a way of improving the design of future devices. Our aim is to demonstrate a finite element method (FEM) simulation tool for efficiently predicting the behavior of spintronic devices. The focus of this work is to extend the drift-diffusion approach for the computation of the torques to the case of a magnetic tunnel junction.

II. MAGNETIZATION DYNAMICS

Accurate simulation tools are a powerful support in the design of STT-MRAM devices. The description of the magnetization $\mathbf{m}$, subject to the spin-transfer torque, is given by the Landau-Lifshitz-Gilbert equation

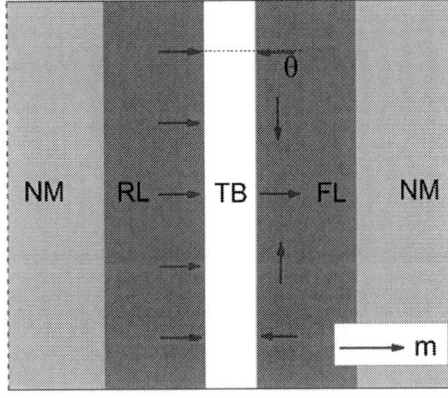

Figure 1. MTJ structure with magnetization in the free layer going from parallel in the center to antiparallel on the sides. The structure is composed of reference layer (RL), tunnel barrier (TB), free layer (FL) and two non-magnetic contacts (NM).

$$\frac{\partial \mathbf{m}}{\partial t} = -\gamma \mu_0 \mathbf{m} \times \mathbf{H}_{\text{eff}} + \alpha \mathbf{m} \times \frac{\partial \mathbf{m}}{\partial t} + \frac{1}{M_S} \mathbf{T_S}, \qquad (1)$$

where $\mathbf{m} = \mathbf{M}/M_S$ is the position-dependent normalized magnetization, M_S is the saturation magnetization, α is the Gilbert damping constant, γ is the gyromagnetic ratio, μ_0 is the vacuum permeability, and $\mathbf{H}_{\text{eff}}$ is the effective magnetic field, containing various contributions such as the external field, the exchange interaction, the anisotropy field, and the demagnetizing field.

Modeling of STT switching can be performed by assuming Slonczewski-like torque expressions [8]. This, however, allows to approximately simulate the magnetization dynamics of the free layer only. A more complete description of the process can be obtained by computing the non-equilibrium spin accumulation $\mathbf{S}$ across the whole structure. In this case, the torque takes the form [9], [10]

$$\mathbf{T_S} = -\frac{D_e}{\lambda_J^2} \mathbf{m} \times \mathbf{S} - \frac{D_e}{\lambda_\varphi^2} \mathbf{m} \times (\mathbf{m} \times \mathbf{S}). \qquad (2)$$

The torque $\mathbf{T_S}$ is created by the spin accumulation acting on the magnetization. λ_J is the exchange length, λ_φ is the spin dephasing length, and D_e is the electron diffusion coefficient in the ferromagnetic layers. The spin accumulation is generated, when an electric current passes through the structure, thanks to the polarizing effect of the background magnetization. In order to compute $\mathbf{S}$, coupled spin and charge transport must be resolved. This requires taking into account that the cells in STT-MRAM devices consist of *magnetic tunnel junctions* (MTJ), a sandwich of two ferromagnets separated by a *tunnel barrier*. The tunnel barrier is essential to achieve a high tunneling magnetoresistance ratio (TMR), related to the large difference in the conductances G_P and G_{AP} in the parallel/anti-parallel MTJ configuration.

$$\text{TMR} = \frac{G_P - G_{AP}}{G_{AP}} \qquad (3)$$

III. CURRENT DENSITY COMPUTATION

The resistance of the tunnel barrier defines the current through the MTJ, as it is much larger than the resistances of the conductive layers. To compute the current through the structure, we model the tunnel barrier as a poor conductor whose low conductivity depends on the ***relative orientation of the magnetization*** in the ferromagnetic layers as

$$\sigma(\theta) = \sigma_0 \left(1 + \left(\frac{\text{TMR}}{2+\text{TMR}}\right)\cos(\theta)\right), \qquad (4)$$

where σ_0 is the average between the conductivities in the P and AP state, and θ is the local angle between magnetic vectors in the free and reference layers. To obtain the current, we solve

$$\nabla \cdot (\sigma \nabla V) = 0, \qquad (5a)$$

$$\mathbf{J_C} = \sigma \nabla V, \qquad (5b)$$

where σ is the conductivity, V is the electrical potential, and $\mathbf{J_C}$ is the current density. The equation is solved with the magnetization configuration schematized in Fig.1. The potential is fixed by Dirichlet conditions on the left and right boundaries, which are defined by the applied voltage. The conductivity is constant in the ferromagnetic layers and in the

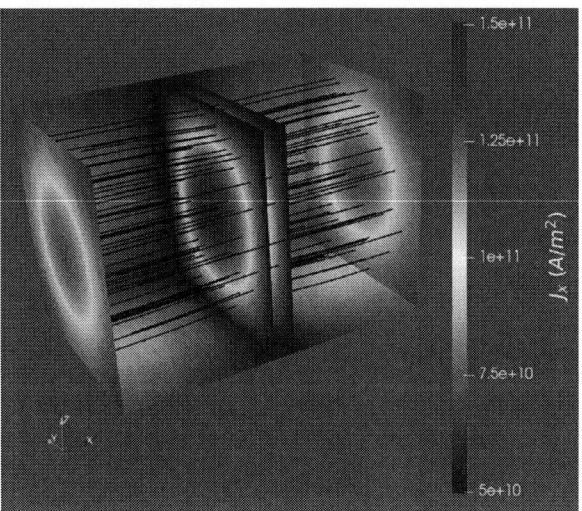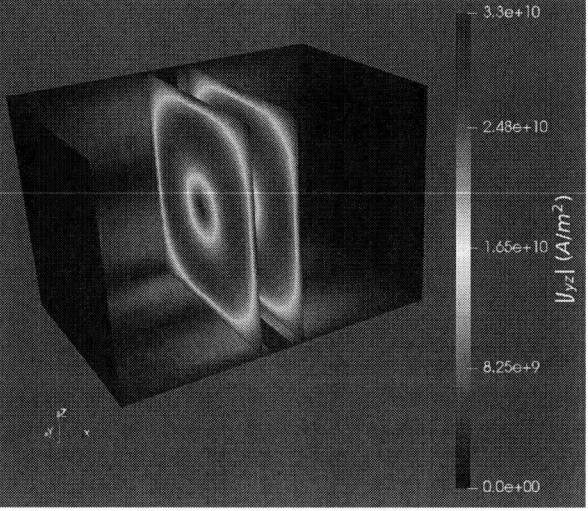

Figure 2. Current density distribution through a square MTJ with a non-uniform magnetization. The left panel shows the x-component (perpendicular) of the current density, while the right panel shows the modulus of the y- and z- (in-plane) components. The x-component flow is higher for aligned magnetizations because of the lower resistance. Due to conservation of the current flow, it is redistributed in the yz-plane in the metal contacts (right panel).

210

TABLE I: DRIFT-DIFFUSION PARAMETERS OF THE MAGNETIC REGIONS

Parameter	Value
Current spin polarization, β_σ	0.9
Diffusion spin polarization, β_D	0.8
Electron diffusion coefficient, D_e	2×10^{-3} m²/s
Spin-flip length, λ_{sf}	10 nm
Exchange length, λ_J	2 nm
Spin dephasing length, λ_φ	5 nm

non-magnetic leads, while it is described by (4) in the tunneling layer. The solution for the current in this scenario is computed via the finite element method and is reported in Fig.2 for a TMR of 200%. This TMR value is comparable to those reached in typical present devices [11]. The computed current is highly non-uniform and is redistributed to accommodate the varying conductivity in the middle layer. In particular, the x-component of the current is highest in the center of the structure, where the magnetization vectors are parallel, and lowest on the sides, where the vectors are antiparallel. The in-plane components try to make the current converge towards the center. This shows that, for non-uniform relative magnetization, characteristic to switching, the conductivity and current in an MTJ depend strongly on the position. By the described method, we can obtain the current at every time step, and use it to compute the spin accumulation.

IV. SPIN ACCUMULATION COMPUTATION

Obtaining a good approximation of the current density $\mathbf{J_C}$ is not sufficient to model spin transport in MTJs. Once the current $\mathbf{J_C}$ is known, the spin accumulation and the spin current density $\mathbf{J_S}$ are found as [10], [12]

$$\mathbf{J_S} = \frac{\mu_B}{e}\beta_\sigma\left(\mathbf{J_C}+\beta_D D_e\frac{e}{\mu_B}[(\nabla S)m]\right)\otimes m - D_e\nabla S, \quad (6a)$$

$$-\nabla\mathbf{J_S} - D_e\left(\frac{\mathbf{S}}{\lambda_{sf}^2}+\frac{\mathbf{S}\times m}{\lambda_J^2}+\frac{m\times(\mathbf{S}\times m)}{\lambda_\varphi^2}\right) = 0, \quad (6b)$$

where μ_B is the Bohr magneton, e is the electron charge, β_σ and β_D are polarization parameters, λ_{sf} is the spin-flip length, and $\otimes$ stands for the tensor product.

One must also ensure that the spin accumulation is preserved across the barrier, as in the case of ideal tunneling with no spin-flips. Equations (6a) and (6b) in the middle layer reduce to

$$D_S\nabla^2\mathbf{S} - D_S\frac{\mathbf{S}}{\lambda_{sf}^2} = 0, \quad (7)$$

where we indicate with D_S the diffusion coefficient in the middle layer. The first step to prevent $\mathbf{S}$ from decaying in the middle layer is to assume the spin flip length is infinite in the tunnel layer. However, this is not sufficient. The *spin diffusion coefficient* in the barrier region, which is a parameter one is free to choose, must also be taken considerably *larger* compared to the electron diffusion coefficient in the ferromagnetic layers, as shown in Fig.3. If the two coefficients have the same value, the spin accumulation decays linearly through the tunnel barrier, while the choice of a large spin diffusion coefficient reduces the slope in the middle layer to the point that the spin accumulation is practically preserved. The parameters of the magnetic regions used in the simulations are reported in Table 1.

With $\mathbf{S}$ known, we can compute the torques acting on the magnetization. Fig.4 shows the dependence between the torque acting on the free layer and the choice of the spin diffusion coefficient. Provided that the value of the latter is

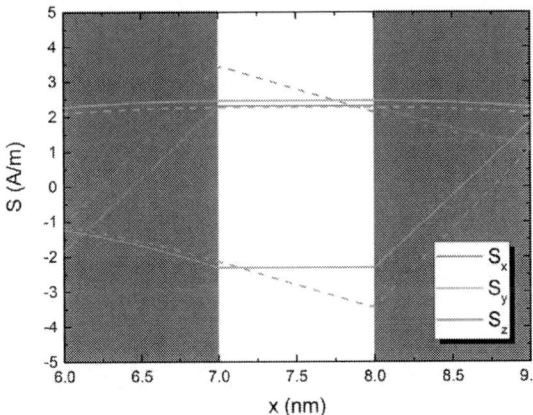

Figure 3. Spin accumulation across the middle layer, with magnetization along z in the FL and along x in the RL. The dashed lines use the same value for D_S in TB and D_e in the FL/RL, while solid lines use a very high value of D_S.

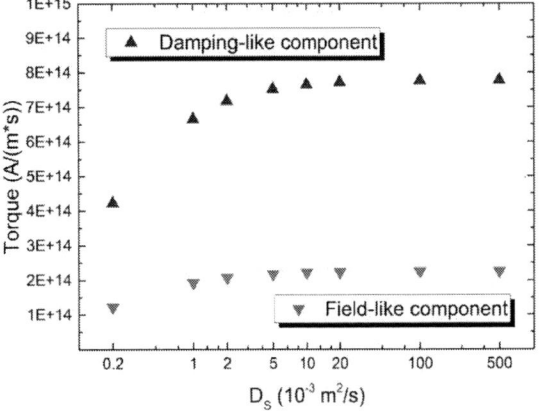

Figure 4. Magnitude of the torque in the FL as a function of the spin diffusion coefficient in the TB. At high values of the coefficient, the torque does not depend on it.

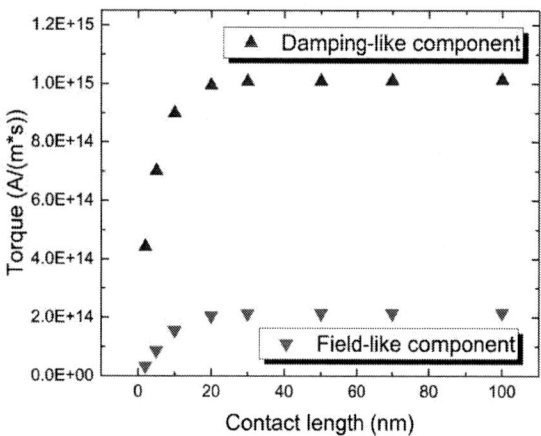

Figure 5. Magnitude of the torque in the FL as a function of the non-magnetic contact length. For a length of 30 nm or more, the torque becomes independent of it.

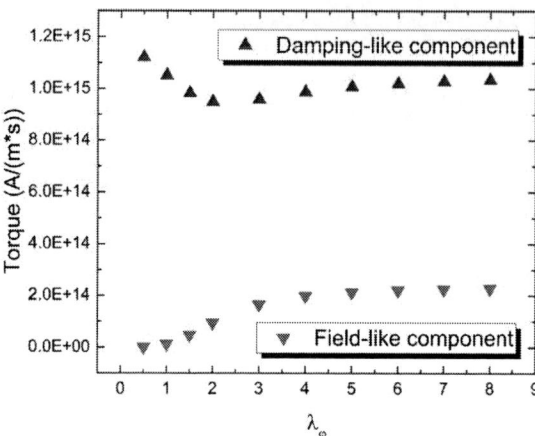

Figure 6. Magnitude of the torque in the FL as a function of the spin dephasing length. For values lower than 3 nm, the contribution of λ_φ to the total torque becomes substantial.

chosen large enough, the torque does not depend on it. In Fig.5 we report the dependence of the torque on the length of the non-magnetic contacts. The results show that a contact length of at least 30 nm is required to let the spin accumulation relax to zero and obtain a torque value independent of this parameter. Finally, we investigated the influence of the usually neglected [9] spin dephasing length λ_φ on the computation of the torque. Fig. 6 shows, that for values of λ_φ less than 3 nm, its contribution to the torque is substantial and cannot be neglected for accurately describing the magnetization dynamics.

V. CONCLUSION

In this work we presented a method of applying the spin drift-diffusion approach for computing the torques acting on the magnetization in an MTJ structure. The current is successfully described by modeling the tunnel barrier as a poor conductor, with an electrical conductivity locally depending on the relative magnetization orientation in the ferromagnetic layers. The spin accumulation is preserved through the middle layer, as is the case in an ideal barrier, by taking an infinite spin-flip length and a diffusion coefficient much larger than the one in the ferromagnetic layers. We showed that the contacts must be chosen long enough to not influence the torque values, and that the spin dephasing length contribution becomes significant only, if the corresponding length is below 3 nm. The generalized spin and charge drift-diffusion approach can be successfully applied to determine the torques acting in an STT-MRAM cell.

ACKNOWLEDGMENT

This work was supported by the *Austrian Federal Ministry for Digital and Economic Affairs* and the *National Foundation for Research, Technology and Development*.

REFERENCES

[1] S. Aggarwal, H. Almasi, M. DeHerrera, B. Hughes, S. Ikegawa *et al.*, "Demonstration of a Reliable 1 Gb Standalone Spin-Transfer Torque MRAM for Industrial Applications," *Proceedings of the IEDM*, pp. 2.1.1–2.1.4, 2019.

[2] K. Lee, J. H. Bak, Y. J. Kim, C. K. Kim, A. Antonyan *et al.*, "1Gbit High Density Embedded STT-MRAM in 28nm FDSOI Technology," *Proceedings of the IEDM*, pp. 2.2.1–2.2.4, 2019.

[3] V. B. Naik, K. Lee, K. Yamane, R. Chao, J. Kwon *et al.*, "Manufacturable 22nm FD-SOI Embedded MRAM Technology for Industrial-Grade MCU and IOT Applications," *Proceedings of the IEDM*, pp. 2.3.1–2.3.4, 2019.

[4] G. Hu, J. J. Nowak, M. G. Gottwald, S. L. Brown, B. Doris *et al.*, "Spin-Transfer Torque MRAM with Reliable 2 ns Writing for Last Level Cache Applications," *Proceedings of the IEDM*, pp. 2.6.1–2.6.4, 2019.

[5] W. J. Gallagher, E. Chien, T. W. Chiang, J. C. Huang, M. C. Shih *et al.*, "22nm STT-MRAM for Reflow and Automotive Uses with High Yield, Reliability, and Magnetic Immunity and with Performance and Shielding Options," *Proceedings of the IEDM*, pp. 2.7.1–2.7.4, 2019.

[6] S. Sakhare, M. Perumkunnil, T. H. Bao, S. Rao, W. Kim *et al.*, "Enablement of STT-MRAM as Last Level Cache for the High Performance Computing Domain at the 5nm Node," *Proceedings of the IEDM*, pp. 18.3.1-18.3.4, 2018.

[7] J. G. Alzate, U. Arslan, P. Bai, J. Brockman, Y. J. Chen *et al.*, "2 MB Array-Level Demonstration of STT-MRAM Process and Performance Towards L4 Cache Applications," *Proceedings of the IEDM*, pp. 2.4.1–2.4.4, 2019.

[8] J. C. Slonczewski, "Currents, Torques, and Polarization Factors in Magnetic Tunnel Junctions," *Physical Review B*, vol. 71, p. 024411, 2004.

[9] C. Abert *et al.*, "A Three-Dimensional Spin-Diffusion Model for Micromagnetics," *Scientific Reports*, vol. 5, p. 14855, 2015.

[10] S. Lepadatu, "Unified Treatment of Spin Torques using a Coupled Magnetisation Dynamics and Three-Dimensional Spin Current Solver," *Scientific Reports*, vol. 7, p. 12937, 2017.

[11] W. Skowronski, M. Czapkiewicz, S. Zietek, J. Checinski, M. Frankowski *et al.*, "Understanding Stability Diagram of Perpendicular Magnetic Tunnel Junctions," *Scientific Reports*, vol. 7, p. 10172, 2017.

[12] S.Fiorentini, R. L. de Orio, W. Goes, J. Ender and V. Sverdlov, "Comprehensive Comparison of Switching Models for Perpendicular Spin-Transfer Torque MRAM Cells," *Proceedings of the SISPAD*, pp. 57–60, 2019.

Efficient Demagnetizing Field Calculation for Disconnected Complex Geometries in STT-MRAM Cells

Johannes Ender
Christian Doppler Laboratory for NovoMemLog at the Institute for Microelectronics TU Wien
Vienna, Austria
ender@iue.tuwien.ac.at

Mohamed Mohamedou
Christian Doppler Laboratory for NovoMemLog at the Institute for Microelectronics TU Wien
Vienna, Austria
mohamedou@iue.tuwien.ac.at

Simone Fiorentini
Christian Doppler Laboratory for NovoMemLog at the Institute for Microelectronics TU Wien
Vienna, Austria
fiorentini@iue.tuwien.ac.at

Roberto Orio
Institute for Microelectronics TU Wien
Vienna, Austria
orio@iue.tuwien.ac.at

Siegfried Selberherr
Institute for Microelectronics TU Wien
Vienna, Austria
selberherr@iue.tuwien.ac.at

Wolfgang Goes
Silvaco Europe Ltd
Cambridge, United Kingdom
wolfgang.goes@silvaco.com

Viktor Sverdlov
Christian Doppler Laboratory for NovoMemLog at the Institute for Microelectronics TU Wien
Vienna, Austria
sverdlov@iue.tuwien.ac.at

Abstract—**Micromagnetic simulations of MRAM cells are a computationally demanding task. Different methods exist to handle the computational complexity of the demagnetizing field, the most expensive magnetic field contribution. In this work we show how the demagnetizing field can efficiently be calculated in complex memory structures and how this procedure can be further used to simulate spin-transfer torque switching in magnetic tunnel junctions.**

Keywords—*Micromagnetics, LLG, spin-transfer torque, MRAM, demagnetizing field*

I. INTRODUCTION

Spin-transfer torque magnetic random access memory (STT-MRAM) is an emerging non-volatile memory compatible with CMOS technology rapidly conquering the market. STT-MRAM possesses an endurance higher than flash memory, and it is suitable for both stand-alone and embedded applications. STT-MRAM is fast and its access time can be tuned in a broad range. This positions STT-MRAM as a universal memory capable to replace both static random access memory and flash memory [1]-[7].

Magnetic tunnel junctions (MTJ) are the basic means to store the information of a single bit and lie at the heart of every MRAM cell. As information has to be read and written many times and the magnetization in the MTJ undergoes frequent changes, the dynamic behavior of this process is of great interest. Here, micromagnetic simulations provide deeper insight and many simulation tools have been developed for this purpose in recent years [8].

Our aim is to develop a simulation tool based on the finite element method (FEM) in order to efficiently perform magnetization dynamics simulations of spintronic devices. The focus in this work lies on the efficient calculation of the demagnetizing field, for which special means are required to prevent the excessive use of memory and computation time.

II. THE COMPUTATIONAL CHALLENGES

Calculating magnetization dynamics in the magnetic layers of an STT memory cell involves the solution of the extended Landau-Lifshitz-Gilbert (LLG) equation

$$\frac{\partial \mathbf{m}}{\partial t} = -\gamma\mu_0 \mathbf{m} \times \mathbf{H}_{\text{eff}} + \alpha\mathbf{m} \times \frac{\partial \mathbf{m}}{\partial t} \qquad (1)$$

$$+ \frac{1}{M_s}\mathbf{T}_S \, ,$$

where $\mathbf{m}=\mathbf{M}/M_S$ is the position-dependent magnetization normalized by the saturation magnetization M_S, γ is the gyromagnetic ratio, μ_0 is the vacuum permeability, α is the Gilbert damping constant, $\mathbf{H}_{\text{eff}}$ is the effective magnetic field, and $\mathbf{T}_S$ is the spin-torque exerted by a spin-polarized current flowing through the memory cell.

In a FEM discretization, the computationally most demanding contribution of the effective field is the demagnetizing field $\mathbf{H}_d$, which describes the long-range dipole interaction of the magnetic moments. To compute the demagnetizing field by FEM, it is convenient to introduce a scalar magnetic potential u defining

$$\mathbf{H}_d = -\nabla u \qquad (2)$$

with u being the solution of the following Poisson equation:

$$\nabla^2 u = \nabla \cdot \mathbf{M} \qquad (3)$$

As the potential $u(\mathbf{x})$ slowly decays to zero as $|\mathbf{x}| \to \infty$, a large computational domain surrounding the magnetic material is required in a FEM simulation to accurately describe this behavior. Various solutions to this so-called open boundary problem have been suggested [8]-[11]. The

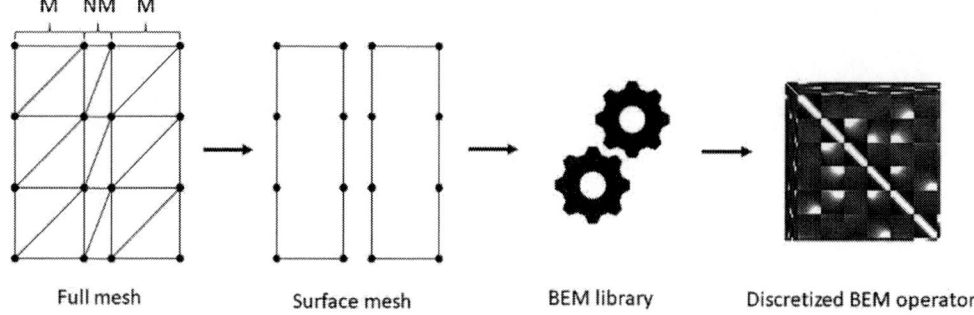

Figure 1: Depiction of the process for discretizing a boundary integral operator, which considers the interaction between disconnected magnetic parts.

truncation of the external computational domain surrounding the magnetic material at a certain distance is usually called truncation approach. As computational efficiency is desired, the external domain cannot be made too large in order to keep the number of degrees of freedom small. An external domain of around five times larger than the magnetic domain has shown to be an acceptable trade-off between computational efficiency and accuracy [12].

To achieve high accuracy and reduce the computational costs, a hybrid approach, where the finite element method is coupled to the boundary element method (hybrid FEM-BEM), is employed to restrict computation to the magnetic domain [13]. Here, the potential is split into two parts u_1 and u_2, calculated by solving a Poisson and a Laplace equation with respect to the corresponding boundary conditions. The potentials u_1 and u_2 are related by the integral over the boundary $\partial\Omega$ of the magnetic domain [13].

$$u_2 = \int_{\partial\Omega} u_1 \frac{\partial}{\partial \mathbf{n}} \frac{1}{|\mathbf{x}-\mathbf{x}'|} d^2\mathbf{x} \qquad (4)$$

This relation in its discretized form reduces to the following matrix-vector multiplication:

$$u_2 = \mathbf{B} \cdot u_1 \qquad (5)$$

Using (5), u_2 is evaluated at the boundary and set as Dirichlet boundary condition for the Laplace equation. Despite the benefit of the hybrid FEM-BEM method to drastically reduce the size of the computational domain, it comes at the cost of having to deal with dense matrices, arising from the discretization of the boundary operator $\mathbf{B}$ in (5). To reduce the high memory demands, matrix compression algorithms are applied [14]. So-called hierarchical matrix compression algorithms approximate the original matrix by hierarchically decomposing it into low-rank submatrices, which are then represented as a factorization of two matrices. These compression algorithms

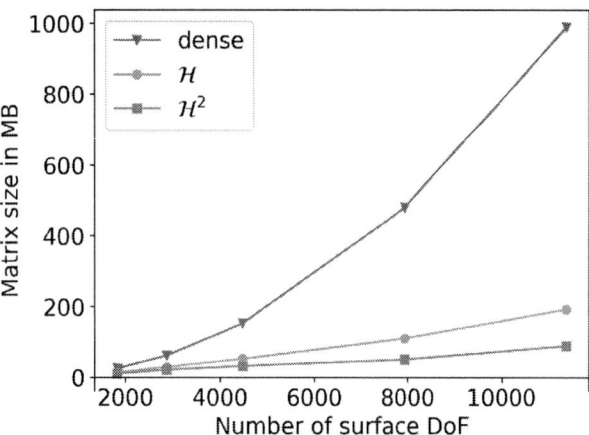

Figure 2: Comparison of the memory consumption of the uncompressed boundary operator matrix and when applying H and H^2 matrix compression algorithms, respectively.

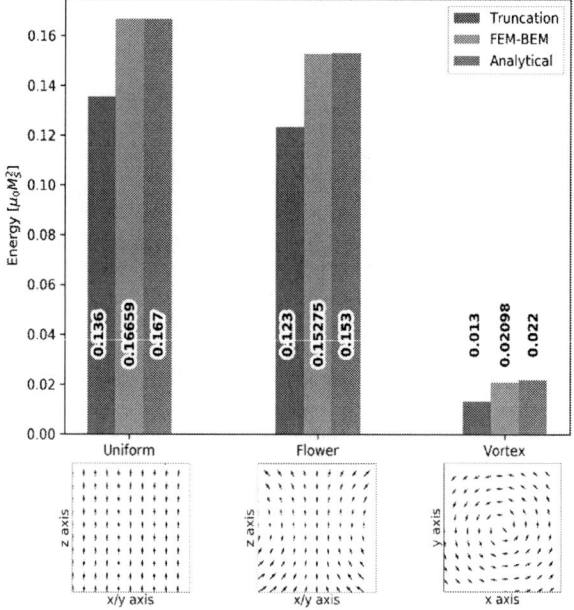

Figure 3: Comparison of the demagnetizing energy of a unit cube with different magnetization configurations calculated with the truncation approach and the hybrid FEM-BEM approach.

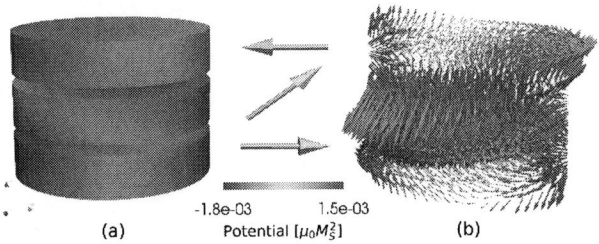

(a)   -1.8e-03   1.5e-03   (b)
Potential $[\mu_0 M_S^2]$

Figure 4: Magnetic potential (a) and demagnetizing field (b) calculated for a three-layer structure. The arrows indicate the magnetization orientation in the respective layers. The color-coding in both figures indicates the magnetic potential value.

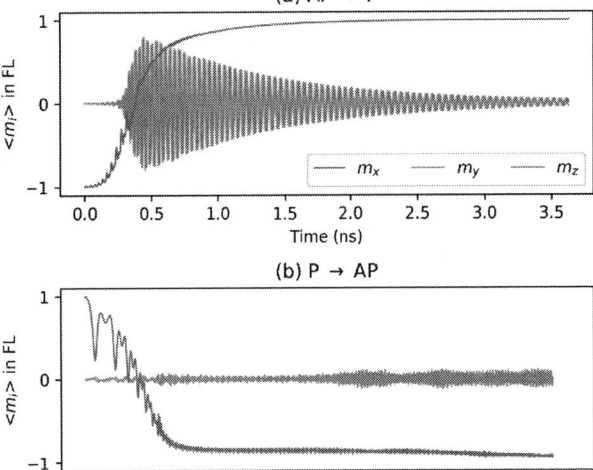

Figure 5: Average of the magnetization components during the switching process of a perpendicularly magnetized STT-MRAM cell from the (a) anti-parallel (AP) to parallel (P) configuration and from the (b) parallel to the anti-parallel configuration. The x direction is the direction of perpendicular magnetic anisotropy.

can significantly reduce the memory requirements of the simulation, as can be seen in Figure 2.

We implemented the demagnetizing field calculations into a micromagnetic simulation environment written in C++, which uses finite elements to solve (1) in STT-MRAM devices. In order to enhance the performance and accuracy, the open-source third party dependencies of the tool are the FEM library MFEM [15] and H2Lib [16]. The latter library contains matrix compression algorithms and basic BEM functionality.

Simulating magnetization dynamics in an STT-MRAM cell requires the calculation of the demagnetizing field of more complex magnetic structures. These consist of multiple disjoint magnetic layers whose stray fields act on each other. The boundary operator must be set up properly to deal with magnetic regions only for the hybrid FEM-BEM approach. In order to do this, we implemented the strategy depicted in Figure 1, where a simplified two-dimensional scenario is shown. As it is usually found in magnetic tunnel junctions, the three-layered structure consists of two magnetic (M) layers sandwiching a non-magnetic (NM) layer. In a first step the surface mesh of the magnetic regions is extracted from the full volume mesh. Treating this as a single mesh - even if disconnected – it is subsequently fed into the BEM library that performs the discretization of the boundary integral operator (5). This densely populated matrix incorporates the interaction between the magnetic layers and is then used in the calculation of the values of u_2. The procedure of generating the boundary operator **B** is computationally demanding, but must be performed only once, as it solely depends on the geometry of the structure to be calculated.

III. RESULTS

Using a set of micromagnetic standard problems [17], we verified the correctness of our results for the demagnetizing field calculation and can now confirm that the hybrid FEM-BEM approach is superior to the truncation approach with respect to accuracy, as can be seen in Figure 3. Here, the demagnetizing energy of a unit cube with several initial magnetizations, computed with the two above-mentioned approaches, was compared to the analytical value. Cross-sections of the different magnetization scenarios can be seen on the bottom of Figure 3. A uniform magnetization as well as a so-called flower state and a vortex-like magnetization are compared. While the relative error for the truncation approach ranges from around 19% in the uniform and flower scenario to up to 39% in the vortex scenario, the hybrid FEM-BEM approach on the other hand achieves a relative error of below 1% for the first two scenarios and remains under 5% for the vortex scenario.

Figure 4 shows the result obtained for computing the magnetic potential (Figure 4a) and the demagnetizing field (Figure 4b) for a magnetic structure consisting of three disjoint magnetic layers with the magnetization in the respective layers indicated by the arrows in the middle. Without interaction, the potential would be varying linearly along the direction of the magnetization in the corresponding layer. When applying the strategy depicted in Figure 1 this is not the case and the magnetic potential is shifted due to the different orientations of the magnetization in the three layers and their mutual interaction.

Applying the developed technique for calculating the demagnetizing field, we successfully simulated the switching of a perpendicularly magnetized STT-MRAM cell.

TABLE I: SIMULATION PARAMETERS

Parameter	Value
Gilbert damping, α	0.02
Gyromagnetic ratio, γ	$1.76 \cdot 10^{11}$ rad s^{-1} T^{-1}
Vacuum permeability, μ_0	$4\pi \cdot 10^{-7}$ H m^{-1}
Saturation magnetization, M_S	$8 \cdot 10^5$ A m^{-1}
Exchange constant, A	$1.3 \cdot 10^{-11}$ J m^{-1}
Anisotropy constant, K	$2 \cdot 10^5$ J m^{-3}

The magnetization in the reference layer is fixed in the positive x direction and the contributions to the effective magnetic field $\mathbf{H}_{\text{eff}}$ include the exchange field, the anisotropy field and the demagnetizing field. By applying a current through the MTJ a switching from an anti-parallel (AP) to a parallel (P) configuration was simulated as well as switching from P to AP. Figure 5 shows the average of the magnetization components in the free layer (FL). The parameters used for the simulations can be seen in Table I.

IV. CONCLUSION

In this work we presented how, by the use of open-source libraries, efficient micromagnetic simulations can be carried out. Through a drastic reduction of the computational domain and compression of large, dense matrices, the overall performance of the demagnetizing field calculation can be optimized. Results were compared with reference problems for uniform and non-uniform magnetization and a strategy was demonstrated that enables the computation of the demagnetizing field in complex geometries with disconnected magnetic regions. In addition, results of a simulation showing the switching between the two states of an STT-MRAM cell were presented.

ACKNOWLEDGMENT

This work was supported by the *Austrian Federal Ministry for Digital and Economic Affairs* and the *National Foundation for Research, Technology and Development*.

REFERENCES

[1] D. Apalkov, B. Dieny, J. M. Slaughter, "Magneto-resistive Random Access Memory," *Proceedings of the IEEE*, *104*(10), pp. 1796-1830, 2016.

[2] S. Aggarwal, H. Almasi, M. DeHerrera, B. Hughes, S. Ikegawa *et al.*, "Demonstration of a Reliable 1 Gb Standalone Spin-Transfer Torque MRAM for Industrial Applications," *Proceedings of the IEDM*, pp. 2.1.1-2.1.4, 2019.

[3] K. Lee, J. H. Bak, Y. J. Kim, C. K. Kim, A. Antonyan *et al.*, "1Gbit High Density Embedded STT-MRAM in 28nm FDSOI Technology," *Proceedings of the IEDM*, pp. 2.2.1-2.2.4, 2019.

[4] V. B. Naik, K. Lee, K. Yamane, R. Chao, J. Kwon *et al.*, "Manufacturable 22nm FD-SOI Embedded MRAM Technology for Industrial-Grade MCU and IOT

Applications," *Proceedings of the IEDM*, pp. 2.3.1-2.3.4, 2019.

[5] J. G. Alzate, U. Arslan, P. Bai, J. Brockman, Y. J. Chen *et al.*, "2 MB Array-Level Demonstration of STT-MRAM Process and Performance Towards L4 Cache Applications," *Proceedings of the IEDM*, pp. 2.4.1-2.4.4, 2019.

[6] G. Hu, J. J. Nowak, M. G. Gottwald, S. L. Brown, B. Doris *et al.*, "Spin-Transfer Torque MRAM with Reliable 2 ns Writing for Last Level Cache Applications," *Proceedings of the IEDM*, pp. 2.6.1-2.6.4, 2019.

[7] W. J. Gallagher, E. Chien, T. W. Chiang, J. C. Huang, M. C. Shih *et al.*, "22nm STT-MRAM for Reflow and Automotive Uses with High Yield, Reliability, and Magnetic Immunity and with Performance and Shielding Options," *Proceedings of the IEDM*, pp. 2.7.1-2.7.4, 2019.

[8] J. Leliaert, J. Mulkers, "Tomorrow's Micromagnetic Simulations," *Journal of Applied Physics*, *125*(18), p. 180901, 2019.

[9] J. Imhoff, G. Meunier, X. Brunotte, and J. Sabonnadiere, "An Original Solution for Unbounded Electromagnetic 2d- and 3d-Problems Throughout the Finite Element Method," *IEEE Transactions on Magnetics*, *26*(5), pp. 1659–1661, 1990.

[10] X. Brunotte, G. Meunier, and J.-F. Imhoff, "Finite Element Modeling of Unbounded Problems using Transformations: a Rigorous, Powerful and Easy Solution," *IEEE Transactions on Magnetics*, *28*(2), pp. 1663–1666, 1992.

[11] F. Henrotte, B. Meys, H. Hedia, P. Dular, and W. Legros, "Finite Element Modelling with Transformation Techniques," *IEEE transactions on magnetics*, *35*(3), pp. 1434-1437, 1999.

[12] Q. Chen, A. Konrad, "A Review of Finite Element Open Boundary Techniques for Static and Quasi-Static Electromagnetic Field Problems," *IEEE Transactions on Magnetics*, *33*(1), pp. 663-676, 1997.

[13] D. R. Fredkin, T. R. Koehler, "Hybrid Method for Computing Demagnetizing Fields," *IEEE Transactions on Magnetics*, *26*(2), pp. 415-417, 1990.

[14] N. Popović, D. Praetorius, "Applications of H-Matrix Techniques in Micromagnetics," *Computing*, *74*(3), pp. 177-204, 2005.

[15] T. Kolev, V. Dobrev, "MFEM: Modular Finite Element Methods Library," http://mfem.org, 2010.

[16] N. Albrecht, C. Börst, D. Boysen, S. Christophersen, S. Börm, "H2Lib," http://www.h2lib.org, 2016.

[17] C. Abert, L. Exl, G. Selke, A. Drews, T. Schrefl, "Numerical Methods for the Stray-Field Calculation: A Comparison of Recently Developed Algorithms," *Journal of Magnetism and Magnetic Materials*, *326*, pp. 176-185, 2013.

Properties of Conductive Oxygen Vacancies and Compact Modeling of IV Characteristics in HfO$_2$ Resistive Random-Access-Memories

Junsung Park, Min-Jae Kim, Jae-Hyung Jang, and Sung-Min Hong

School of EECS, Gwangju Institute of Science and Technology, Gwangju, Republic of Korea

email: smhong@gist.ac.kr

Abstract—The HfO$_2$-based resistive random-access-memory (RRAM) is studied. In the first part, two parameters of oxygen vacancies are extracted. The migration barrier of the oxygen vacancy (or the extended Frenkel pair) is calculated. The resistivity of a filament is also calculated. In the second part, an existing compact model for the RRAM is implemented and its results are compared with the experimental data.

I. INTRODUCTION

The resistive random-access-memory (RRAM) has been studied intensively. It has been regarded as a promising candidate for future non-volatile memory and neuromorphic applications. They show fast operation (nanosecond operation), scalability (scaling down to tens of nanometer), and multi-bit topology. Structural simplicity is another advantage of the RRAM devices.

It has been reported that various materials can be used as switching layers. Among them, especially, the HfO$_2$-based RRAMs are widely adopted, because it can be successfully integrated into the modern CMOS technology. From the viewpoint of power, stability, and operation speed, the HfO$_2$-based RRAM exhibits superior performance.

It is known that the conductive filament plays an important role in the HfO$_2$-based RRAMs. Shape of a filament, which is a conduction path made of oxygen vacancies, determines the electrical properties of the RRAM. Since the states are characterized by their resistance values, it is very important to understand the change of filament shape under a given operation condition. The modeling approach can be very helpful in this aspect.

In order to model the HfO$_2$-based RRAMs, various approaches with varying complexities (from the first-principles calculation to compact models) have been reported. For example, previous reports on the FEM/KMC approaches [1], [2] are available. Parameters are quite important in those device simulations. A systematic procedure to extract the related parameters is highly desirable.

In this work, properties of oxygen vacancies are calculated with the density-functional theory code (Vienna Ab initio Simulation Package, VASP [3], [4]). Moreover, a measured IV curve of a HfO$_2$-based RRAM is calculated by employing an existing compact model. The organization of this extended abstract is as follows: In Section II, the migration barrier of oxygen vacancies are calculated. In Section III, the resistivity

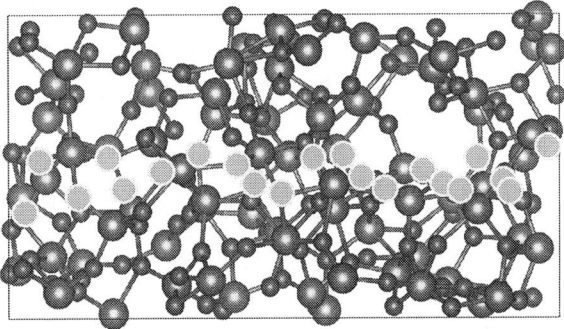

Fig. 1. One example of the atomistic structure of an amorphous HfO$_2$ system. Yellow dots represent the possible defect locations, considered in Fig. 2. In the actual calculation, several amorphous structures are generated.

of a HfO$_2$ supercell is calculated. In Section IV, an existing compact model is implemented and its results are compared with the experimental IV curve. The conclusion is drawn in Section V.

II. MIGRATION BARRIER OF OXYGEN VACANCIES

The switching operation of the HfO$_2$-based RRAM is explained by the migration of oxygen vacancies. The migration barrier of oxygen vacancies, the difference between the local maximum and minimum formation energies, is an important material parameter to affect the vacancy migration.

In Fig. 1, an atomistic structure of an amorphous HfO$_2$ system is shown. The structure is obtained by the DFT-MD optimization. In the cell optimization, the PBEsol exchange-correlation functional is used. In addition to an amorphous structure, a tetragonal supercell structure is also considered. Each structure (amorphous or tetragonal) has 108 atoms.

After the atomic structure is prepared, a defect is introduced. Two defect types – an oxygen vacancy and the extended Frenkel pair, which is a pair of an oxygen vacancy and an oxygen interstitial – are considered. In the case of the tetragonal supercell, the defect is generated at each symmetric position inside a unit cell. However, for an amorphous structure, it is difficult to find symmetric positions. Therefore, we have manually selected possible defect locations as shown in Fig. 1.

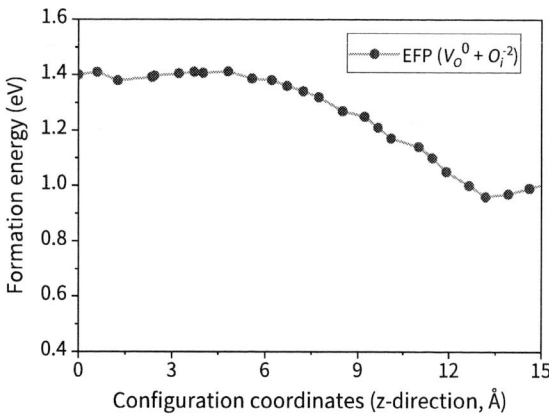

Fig. 2. Formation energy of the extended Frenkel pair for the structure shown in Fig. 1.

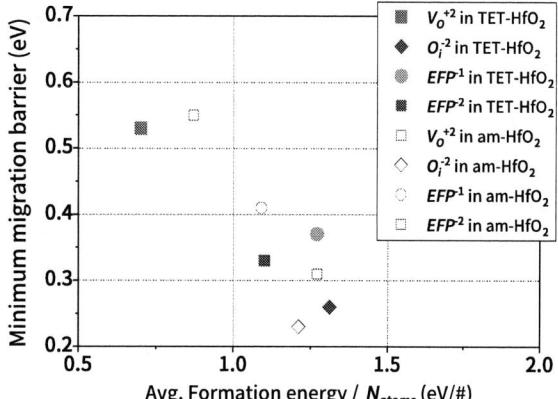

Fig. 3. Migration barriers of various defects in HfO_2. Both the crystalline and amorphous structures are considered.

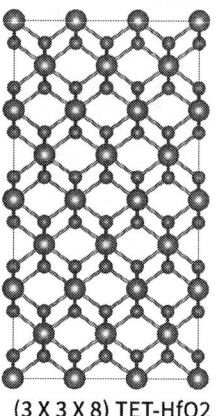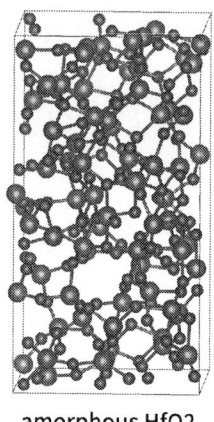

(3 X 3 X 8) TET-HfO2   amorphous HfO2

Fig. 4. HfO_2 supercells for the conductivity calculation. (Left) A crystalline structure and (Right) an amorphous structure.

III. ELECTRICAL PROPERTY OF OXYGEN VACANCY FILAMENT

The migration barrier calculated in Section II is related with how the filament shape changes. On the other, when the filament shape is fixed at a given time instance, the conduction property is of interest.

The resistivity of a HfO_2 supercell is calculated. First, a structure is prepared. In Fig. 4, the atomistic structures are shown. Both the crystalline structure (the tetragonal structure with $3 \times 3 \times 8$ unit cells) and the amorphous one are considered. 296 atoms are included in each supercell. The procedure to generate a structure is similar with that in Section II.

Starting from these structures, oxygen vacancies are introduced to generate an oxygen vacancy filament. The resistivity along the vertical direction is calculated with the Mott formula [6]. The conductivity tensor is obtained by using the linear response function of the VASP package. The k-point sampling is done with a $4 \times 4 \times 20$ grid. In Fig. 5, the energy spectrum of the calculated conductivity is drawn for HfO_2 nanorods. In the case of a defect-free crystalline system, it is an insulator because no contribution in the forbidden zone. On the other hand, when defects are introduced, the defect states in the forbidden zone significantly contribute to the conductivity.

In Fig. 6, the resistivity is drawn as a function of the defect density. Each symbol represents a distinct sample. Amorphous samples have relatively higher resistivity that the crystalline ones. When the number of oxygen vacancies increases from 2 to 8, the resistivity of the crystalline HfO_2 increases almost six orders-of-magnitude. Similar trends are found for different samples. Since the atomistic simulation model is much limited in its size (0.8 nm $\times$ 0.8 nm $\times$ 1.5 nm), the defect density is unrealistically high. A direct comparison with the experimental results is not possible. Instead, in the right subfigure of Fig. 6, the experimental results in previous reports [1], [2] are drawn with their original defect density without scaling. A good qualitative agreement is obtained.

At every defect position, the formation energy is calculated. When the formation energy is calculated, a hybrid exchange-correlation functional (optimized HSE) is used. The formation energy of the extended Frenkel pair [5] is drawn as a function of the position in Fig. 2. In this example, the extracted migration barrier is about 0.45 eV.

For several amorphous structures (5-10 samples), the same calculation has been performed in order to improve the fidelity. As stated above, the crystalline HfO_2 (the tetragonal structure) has been also considered for comparison. In Fig. 3, the results are summarized. The minimum value of the extracted migration barrier is reported. Filled symbols are results for the crystalline structure, while empty ones are results for amorphous structures. They exhibit good qualitative agreement. The extended Frenkel pair has a much lower migration barrier than the oxygen vacancy. An oxygen ion interstitial is lowering the migration barrier of the corresponding oxygen vacancy.

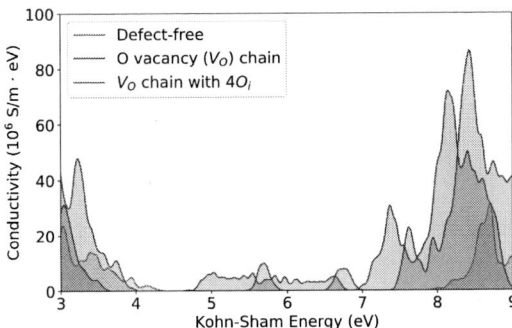

Fig. 5. Energy spectrum of the calculated conductivity of HfO_2 nanorods. The x-axis is Kohn-Sham energy.

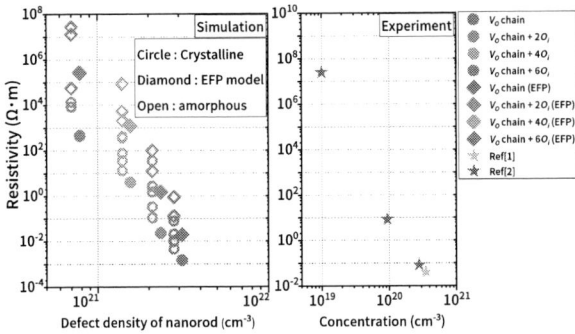

Fig. 6. (Left) Calculated resistivities of various HfO_2 structures. When the number of oxygen vacancies increases, the resistivity decreases exponentially. (Right) Experimental results in [1], [2].

IV. COMPACT MODELING OF IV CHARACTERISTICS

A HfO_2-based RRAM is fabricated. A 90-nm-thick SiO_2 layer is made on top of the p^+ silicon substrate. After the Ti/Pt (10/100 nm) bottom electrode is formed, the switching layer (HfO_2 15 nm) is deposited. The Ti/Pt (30/100 nm) layer acts as the top electrode. The top electrode has a circular shape and its diameter is 300 μm. In Fig. 7, the IV characteristics of a HfO_2-based RRAM measured at room temperature is shown. The compliance current is 10 mA. The SET and RESET voltages are 2.5 V and -1.1 V, respectively.

By adopting an existing compact model reported in [7], switching behaviors are calculated. In this model, the filament shape is simply characterized by a gap between the top electrode and the filament end point. The model equations are not repeated here and the interested readers are referred to [7]. The compliance current is implemented with an ideal MOSFET in the saturation mode. As shown in Fig. 8, it is shown that the compact model with adjusted parameters can reproduce the SET and RESET voltages considerably well.

Based on the calibrated compact model, the switching behaviors are predicted at various temperatures (down to 250 K), as shown in Fig 9. For lower temperatures, the absolute values of the SET and RESET voltages increase. A similar trend can be found in a previous report [8]. Sensitivities of

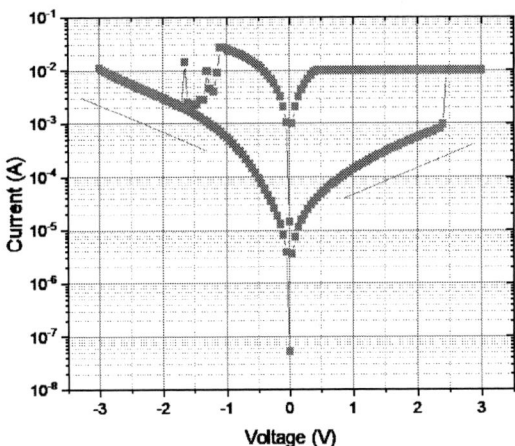

Fig. 7. Measured IV switching characteristics of a HfO_2-based RRAM. Thickness of the HfO_2 layer is 15 nm. The SET and RESET voltages are 2.5 V and -1.1 V, respectively.

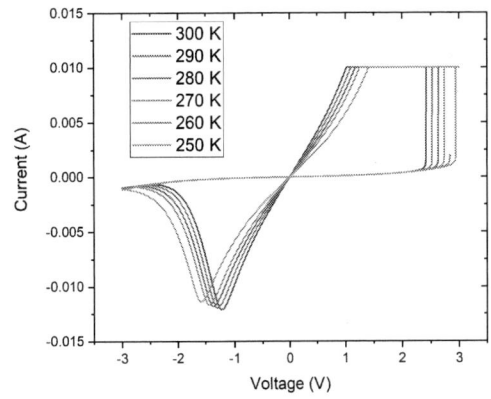

Fig. 8. Simulated switching characteristics of a HfO_2-based RRAM at various temperatures.

the SET/RESET voltages with respect to the temperature are 0.5 V / 50 K and -0.36 V / 50 K, respectively.

V. CONCLUSION

In summary, our recent research efforts on the HfO_2-based RRAM are presented. The material modeling (in Sections II and III) and the compact modeling (in Section IV) presented in this work will be combined with the FEM/KMC approaches to construct a fully functional modeling environment for the RRAM.

ACKNOWLEDGEMENT

This work was supported by the National Research Foundation of Korea (NRF) grant funded by the Korea government (NRF-2019R1A2C1086656) and Samsung Electronics.

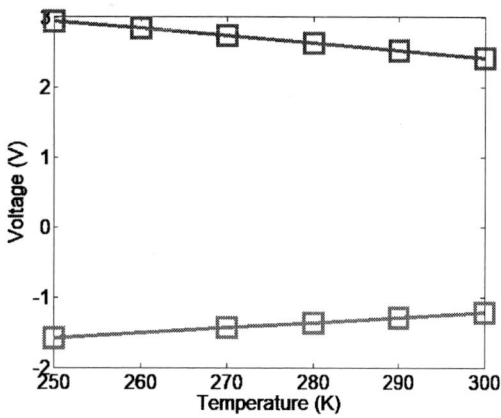

Fig. 9. SET (Blue) and RESET (Red) voltages of the RRAM in Figs. 7 and 8 as functions of the temperature.

REFERENCES

[1] S. Ambrogio, S. Balatti, A. Cubeta, A. Calderoni, N. Ramaswamy, and D. Ielmini, "Statistical fluctuations in HfOx resistive-switching memory: Part I - Set/Reset variability," *IEEE Transactions on Electron Devices*, vol. 61, no. 8, pp. 2912–2919, 2014.

[2] A. Padovani, L. Larcher, O. Pirrotta, L. Vandelli, and G. Bersuker, "Microscopic modeling of HfOx RRAM operations: From forming to switching," *IEEE Transactions on Electron Devices*, vol. 62, no. 6, pp. 1998–2006, 2015.

[3] G. Kresse and J. Hafner, "Ab initio molecular dynamics for liquid metals," *Phys. Rev. B*, vol. 47, pp. 558–561, Jan 1993. [Online]. Available: https://link.aps.org/doi/10.1103/PhysRevB.47.558

[4] ——, "Ab initio molecular-dynamics simulation of the liquid-metal–amorphous-semiconductor transition in germanium," *Phys. Rev. B*, vol. 49, pp. 14 251–14 269, May 1994. [Online]. Available: https://link.aps.org/doi/10.1103/PhysRevB.49.14251

[5] B. Traoré, P. Blaise, and B. Sklénard, "Reduction of monoclinic HfO_2: A cascading migration of oxygen and its interplay with a high electric field," *The Journal of Physical Chemistry C*, vol. 120, no. 43, pp. 25 023–25 029, 2016.

[6] S. J. Blundell and K. M. Blundell, *Concepts in Thermal Physics*. OUP Oxford, 2009.

[7] P. Chen and S. Yu, "Compact modeling of RRAM devices and its applications in 1T1R and 1S1R array design," *IEEE Transactions on Electron Devices*, vol. 62, no. 12, pp. 4022–4028, 2015.

[8] R. Fang, W. Chen, L. Gao, W. Yu, and S. Yu, "Low-temperature characteristics of HfOx-based resistive random access memory," *IEEE Electron Device Letters*, vol. 36, no. 6, pp. 567–569, 2015.

MOS-like approach for compact modeling of High-Electron-Mobility Transistor

Adrien Vaysset
CEA, LETI, Univ. Grenoble Alpes,
38000 Grenoble, France
adrien.vaysset@cea.fr

Sébastien Martinie
CEA, LETI, Univ. Grenoble Alpes,
38000 Grenoble, France
sebastien.martinie@cea.fr

François Triozon
CEA, LETI, Univ. Grenoble Alpes,
38000 Grenoble, France
francois.triozon@cea.fr

Olivier Rozeau
CEA, LETI, Univ. Grenoble Alpes,
38000 Grenoble, France
olivier.rozeau@cea.fr

Marie-Anne Jaud
CEA, LETI, Univ. Grenoble Alpes,
38000 Grenoble, France
marie-anne.jaud@cea.fr

René Escoffier
CEA, LETI, Univ. Grenoble Alpes,
38000 Grenoble, France
rene.escoffier@cea.fr

Thierry Poiroux
CEA, LETI, Univ. Grenoble Alpes,
38000 Grenoble, France
thierry.poiroux@cea.fr

Abstract—**High-Electron-Mobility Transistor (HEMT) with Al-GaN/GaN gate stack is a promising candidate for high-speed and high-power applications. Recent HEMT compact modeling works have proposed threshold-based [1] and surface-potential-based models [2]. In the latter approach, inversion charge is calculated from the quantum expression of a 2-dimensional electron gas (2DEG). Here, we investigate the possibility to model HEMTs with a MOSFET-like approach whereby quantum confinement is included as an effective bandgap widening in the surface potential equation. We evidence that such a MOSFET-like approach leads to a more accurate description over the whole polarization range, especially in the moderate inversion regime. This analytical model is validated by Poisson-Schrödinger numerical simulations. Furthermore, to address a specific feature of HEMT devices, a field plate model is also presented.**

Index Terms—**Power device, GaN, HEMT, compact model, SPICE**

I. SURFACE POTENTIAL CALCULATION: QUANTUM CONFINEMENT CORRECTION

In the surface-potential-based ASM-HEMT model [3], the HEMT channel is modeled as a 2DEG. Its quantum expression is directly included in the surface potential equation, assuming charge sheet approximation with 2D degeneracy and Fermi-Dirac statistics [4]. The complexity of this model contrasts with the usual approach of MOS compact models that build upon classical physics and include quantum effects as a correction.

Here, we propose an approach similar to MOS models to describe HEMT devices. First, the Poisson equation is solved classically, using Boltzmann statistics. Then, quantum confinement is added as an effective bandgap widening through

The research leading to these results has received funding from the collaborative research program between CEA-Leti and SILVACO.

triangular-well approximation [5]. The resulting surface potential equation is similar to the one used in MOS compact models. Therefore, current, charge and other features can be readily implemented with proven methods.

The device studied here is shown in Fig. 1. Firstly, to derive the surface potential Ψ_s, we consider only the AlGaN/GaN stack.

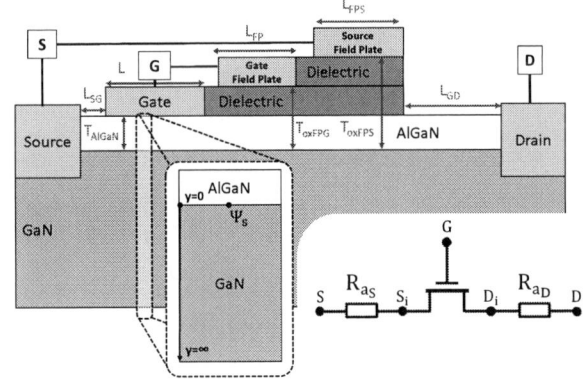

Fig. 1. HEMT structure definition for n-type configuration including Gate and Source field plate description. Inset shows the equivalent circuit in-cluding non-linear resistance (Si and Di are internal nodes).

Considering only the inversion charge and no GaN doping, the Poisson equation is derived in the classical case.

$$\frac{d^2x}{dy^2} = \frac{1}{\Phi_T\,\epsilon_{GaN}\,q\,n_i\exp\left(x - x_n\right)} \tag{1}$$

where Φ_T is the thermal voltage, $x = \Psi/\Phi_T$ is the normalized potential, y is the vertical coordinate, ϵ_{GaN} is the GaN permittivity, q is the electron charge, n_i is the intrinsic concentration and $x_n = V_D/\Phi_T$ is the normalized quasi-Fermi level.

Then, using the boundary conditions

$$\frac{dx}{dy}(y=0) = \frac{C_{AlGaN}}{\epsilon_{GaN}(x_g - x)} \tag{2}$$

$$\frac{dx}{dy}(y \to \infty) = 0 \,, \tag{3}$$

the implicit surface potential equation is obtained:

$$(x_g - x_s)^2 = G^2 \exp\left(x_s - x_n\right). \tag{4}$$

Then, following the same approach as [5], an energy shift is introduced to account for quantum effects,

$$(x_g - x_s)^2 = G^2 \exp\left(x_s - x_n - K_Q\left(x_g - x_s\right)^{2/3}\right) \tag{5}$$

$G^2 = 2qn_i\epsilon_{GaN}/C_{AlGaN}^2\Phi_T$, K_Q is the quantum parameter and C_{AlGaN} is the AlGaN capacitance. The shift $-K_Q\left(x_g - x_s\right)^{2/3}$ corresponds to the difference between the bottom of the "classical" conduction band and the first occupied sub-band, assuming a triangular energy well. The limitations of this approximation have been discussed extensively [6]. According to the consensus, this is sufficient to capture the global behavior of the quantum effect, particularly in moderate inversion. For strong inversion, accuracy decreases due to non-consideration of Fermi-Dirac statistics. However, fitting can still be improved by slightly tuning the related charge parameters.

The numerical solution of (4) and (5) is plotted in Fig. 2(a). The exact solution of (4) is expressed via the Lambert $\mathcal{W}_0$ function.

$$x_{s,0} = x_g - 2\,\mathcal{W}_0\left(\frac{G}{2}\exp\left(\frac{x_g}{2}\right)\right) \tag{6}$$

An analytical approximation of the $\mathcal{W}_0$ function exists.

$$\mathcal{W}_0(x) \sim \ln(1+x)\left(1 - \frac{\ln(1+\ln(1+x))}{2+\ln(1+x)}\right) \tag{7}$$

It can be used as an initial guess in the implicit surface potential equation (4). Then, an error correction based on a 2nd-order Taylor expansion is applied twice.

$$x_{s,1} = x_{s,0} + e_2(x_{s,0}) \tag{8}$$

$$x_{s,2} = x_{s,1} + e_2(x_{s,1}) \tag{9}$$

where the function $e_2(x)$ is defined as

$$e_2(x) = \frac{a(x)}{b(x) + \frac{a(x)\,c(x)}{b(x)}}(x_g - x). \tag{10}$$

and the functions $a(x)$, $b(x)$ and $c(x)$ are defined as

$$a(x) = \ln(x_g - x) - \ln(G) - \frac{x}{2} + \frac{K_Q}{2}(x_g - x)^{2/3} \tag{11}$$

$$b(x) = 1 + \frac{1}{2}(x_g - x) + \frac{K_Q}{3}(x_g - x)^{2/3} \tag{12}$$

$$c(x) = \frac{1}{2} + \frac{K_Q}{18}(x_g - x)^{2/3} \tag{13}$$

The error is brought below the numerical noise, as shown in Fig. 2(b,c).

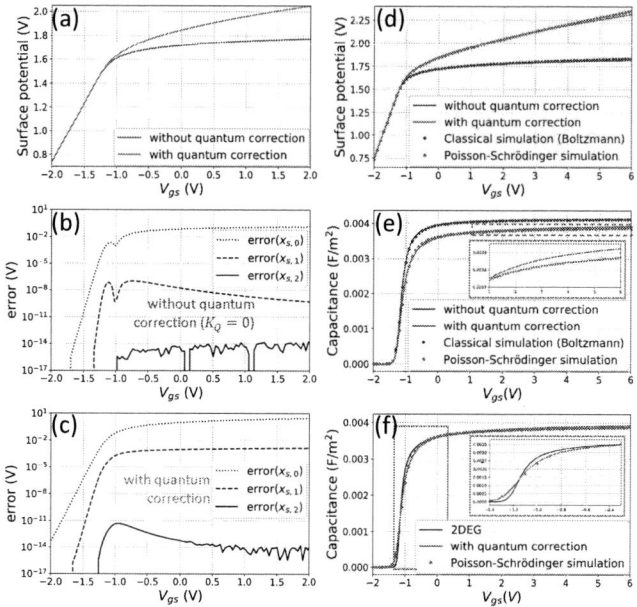

Fig. 2. (a) Surface potential as a function of gate voltage in the case of (4) (without quantum correction) and (5) (with quantum correction); (b,c) Absolute error of the initial guess $x_{(s,0)}$ and the explicit solutions $x_{(s,1)}$ and $x_{(s,2)}$ of (8) and (8). Surface potential (d) and associated capacitance (e) for model (solid lines) and simulations (dots). (f) Comparison with the 2DEG model with inset in the moderate inversion regime.

To validate our model, we have developed a dedicated Poisson-Schrödinger numerical solver for AlGaN/GaN stack. Figs. 2(d,e) illustrate the comparison between the Poisson-Schrödinger solver and the model. It also includes numerical simulations with Boltzmann statistics. The full model from (5) and the classical model from (4) are in good agreement with the Poisson-Schrödinger solver and the Boltzmann simulations, respectively.

For a complete comparison, in Fig. 2(f), the capacitance is also plotted for an analytical 2DEG model. This model is obtained with charge sheet approximation and Fermi-Dirac statistics in a 2-dimensional electron gas, considering only the first energy level. It obeys the following equation:

$$x_g - x_s = \alpha \ln\left(1 + \exp\left(x_s' - K_Q(x_g - x_s)^{2/3}\right)\right) \tag{14}$$

where $\alpha = \frac{qd}{\epsilon_{AlGaN}}\frac{m_e kT}{\pi\hbar^2\Phi_T}$ (where d is the AlGaN layer thickness) and $x_s' = x_s + \frac{E_c - E_g/2}{q\Phi_T}$ (where E_c is the bottom of the energy well).

In moderate inversion, our MOS-like model is in good agreement with the Poisson-Schrödinger solution, while the 2DEG model exhibits a more abrupt transition (inset of Fig. 2f). Since the energy well is shallow in moderate inversion, it is similar to an energy continuum. Therefore, it is better described by the MOS-like model based on classical physics. In strong inversion, the 2DEG model is more accurate than the MOS-like model. This is due to greater confinement. However, this discrepancy can be compensated by slightly tuning the effective AlGaN layer thickness, as already mentioned.

Finally, one of the main advantages of the proposed model is its simplicity in terms of implementation. Moreover, it builds upon previous MOS models, which means that implementation of current, charge and other features can be done in a similar way.

II. CURRENT AND CHARGE CALCULATION INCLUDING FIELD PLATES

Implementation of the current is similar to the Leti-NSP model [7]. The main features have been adapted from this model, such as mobility degradation (including Coulomb scattering, phonon and surface roughness), saturation velocity, channel length modulation and self-heating effect. One of the main differences is the description of source/drain access resistance and the addition of field plates. Access resistance was modeled similarly to previous work [8], thus adding an internal node and an extra resistance with nonlinear current dependence [9], as illustrated in the inset of Fig. 1.

$$R_{a_{S,D}} = R_{L_{S,D}} + \frac{R_{NL_{S,D}}}{\left(1 - \left(\frac{I_{ds}}{\text{ISAT}_{S,D}}\right)^{\beta_{S,D}}\right)^{1/\beta_{S,D}}} \quad (15)$$

where $R_{L_{S,D}}$, $R_{NL_{S,D}}$, $\text{ISAT}_{S,D}$ and $\beta_{S,D}$ are the nonlinear access parameters and I_{ds} is the drain current. Fig. 3 illustrates the model versus experimental data from [9].

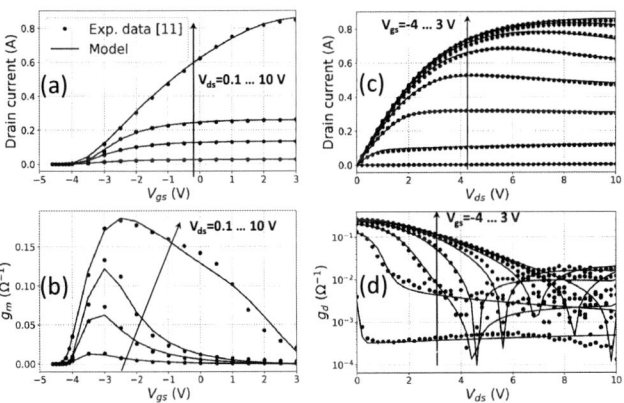

Fig. 3. Comparison between experimental data from reference [9] and model for I_d vs V_{gs} (a), $g_m = dI_d/dV_{gs}$ vs V_{gs} (b), I_d vs V_{ds} (c) and $g_d = dI_d/dV_{ds}$ vs V_{ds} (d). Dots show experimental data. Solid lines represent the model.

Field plate modeling is challenging. Former work [10] uses duplication of internal node to capture each field plate effect. Here, we propose to add only a parasitic field plate charge without adding any internal node, in the same way a parasitic charge is induced by an overlap capacitance [11]. In practice, the charge calculation reuses the surface potential procedure with dedicated parameters for field plate workfunction, interface state and geometry. In HEMT, the field plate dielectric thickness is larger than the field plate length. Therefore, the dependence of field plate charge with longitudinal field is

neglected. Equations (16) and (17) express gate and source field plate charges.

$$Q_{FP_G} = C_{ox,FP_G} (x_{g,FP_G} - x_{s,FP_G}) \quad (16)$$

where C_{ox,FP_G} is the gate field plate dielectric capacitance, $x_{g,FP_G} = \frac{V_{GD} - V_{off,FP_G}}{\Phi_T}$ and V_{off,FP_G} is the cut-off voltage of the gate field-plate.

$$Q_{FP_S} = C_{ox,FP_S} (x_{g,FP_S} - x_{s,FP_S}) \quad (17)$$

where C_{ox,FP_S} is the source field plate dielectric capacitance, $x_{g,FP_S} = \frac{V_{SD} - V_{off,FP_S}}{\Phi_T}$ and V_{off,FP_S} is the cut-off voltage of the source field-plate.

Fig. 4 illustrates model versus experimental data from reference [10]. Note that drain and depletion capacitances are added in order to capture C_{oss} (or C_{sd}) capacitance.

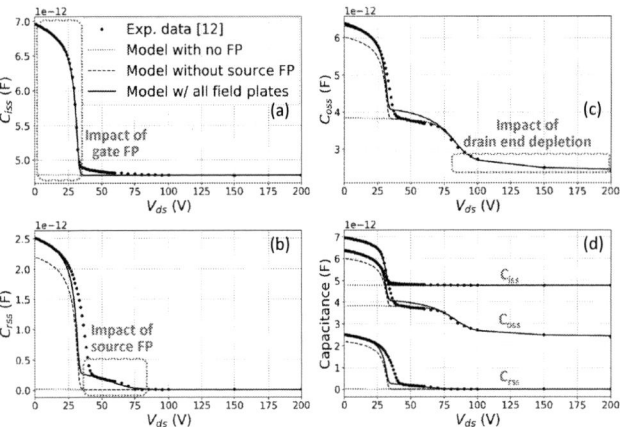

Fig. 4. Comparison between experimental data from reference [10] and model for C_{iss} (a), C_{rss} (b), C_{oss} (c) and these three capacitances (d). Dots are for experimental data; solid line is for the model including all the field plates (FP); dashed line is the model without source FP; dotted line is the model without any FP.

III. CONCLUSION

We have developed a surface-potential-based compact model for HEMT devices. Similarly to MOS models, quantum confinement is described as a simple energy shift, corresponding to the difference between the bottom of the "classical" conduction band and the first occupied sub-band. Above this first energy level, the conduction band is treated as a continuum, using Boltzmann statistics. Thankfully, a good initial guess exists for the entire polarization range. Applying a Taylor-based error correction, the accuracy can go beyond numerical precision. Our model matches very well Poisson-Schrödinger simulations in the moderate inversion regime. In strong inversion, the slight difference can be compensated with fitting parameters. One main advantage of the model, beside its simplicity, is its close resemblance to other surface-potential-based MOS models such as PSP. This means that charge, current and other features can be readily implemented with proven methods.

Finally, we have adressed a specific feature of HEMT devices: the field plates. In order to minimize the number of fitting parameters and avoid additional internal nodes, field plate charge is added similarly to a parasitic charge induced by an overlap resistance. Capacitance and currents are fitted on experimental data, demonstrating model capability in a real case.

ACKNOWLEDGMENT

The authors thank Bogdan Tudor and Eric Guichard from SILVACO for fruitful discussions and collaboration.

REFERENCES

[1] MVSG model. [Online]. Available: https://uwaterloo.ca/waterloo-emerging-integrated-systems-group/mvsg-model

[2] ASM-HEMT model. [Online]. Available: http://iitk.ac.in/asm/

[3] S. Khandelwal, Y. S. Chauhan, T. A. Fjeldly, S. Ghosh, A. Pampori, D. Mahajan, R. Dangi, and S. A. Ahsan, "Asm gan: Industry standard model for gan rf and power devices—part 1: Dc, cv, and rf model," *IEEE Transactions on Electron Devices*, vol. 66, no. 1, pp. 80–86, 2019.

[4] S. Kola, J. M. Golio, and G. N. Maracas, "An analytical expression for fermi level versus sheet carrier concentration for hemt modeling," *IEEE Electron Device Letters*, vol. 9, no. 3, pp. 136–138, 1988.

[5] G. Gildenblat, T. L. Chen, and P. Bendix, "Closed-form approximation for the perturbation of MOSFET surface potential by quantum-mechanical effects," *Electronics Letters*, vol. 36, no. 12, pp. 1072–1073, 2000.

[6] R. van Langevelde, A. Scholten, and D. Klaassen. MOS Model 11. [Online]. Available: https://www.nxp.com/wcm_documents/models/mos-models/model-11/nltn2004_00085.pdf

[7] O. Rozeau, S. Martinie, T. Poiroux, F. Triozon, S. Barraud, J. Lacord, Y. M. Niquet, C. Tabone, R. Coquand, E. Augendre, M. Vinet, O. Faynot, and J. . Barbé, "NSP: Physical compact model for stacked-planar and vertical gate-all-around mosfets," in *2016 IEEE International Electron Devices Meeting (IEDM)*, 2016, pp. 7.5.1–7.5.4.

[8] D. R. Greenberg and J. A. del Alamo, "Velocity saturation in the extrinsic device: a fundamental limit in HFET's," *IEEE Transactions on Electron Devices*, vol. 41, no. 8, pp. 1334–1339, 1994.

[9] S. Ghosh, S. A. Ahsan, Y. S. Chauhan, and S. Khandelwal, "Modeling of source/drain access resistances and their temperature dependence in GaN HEMTs," in *2016 IEEE International Conference on Electron Devices and Solid-State Circuits (EDSSC)*, 2016, pp. 247–250.

[10] S. Aamir Ahsan, S. Ghosh, K. Sharma, A. Dasgupta, S. Khandelwal, and Y. S. Chauhan, "Capacitance modeling in dual field-plate power GaN HEMT for accurate switching behavior," *IEEE Transactions on Electron Devices*, vol. 63, no. 2, pp. 565–572, 2016.

[11] PSP model. [Online]. Available: http://www.cea.fr/cea-tech/leti/pspsupport

Compact modeling of gate leakage phenomenon in GaN HEMTs

1st Kexin Li
Holonyak Micro & Nanotechnology Laboratory
University of Illinois at Urbana-Champaign
Urbana, IL 61801, USA
kexinli4@illinois.edu

2nd Eiji Yagyu
Advanced Technology R&D Center
Mitsubishi Electric Corporation
Hyogo 661-8661, Japan
Yagyu.Eiji@cb.MitsubishiElectric.co.jp

3rd Hisashi Saito
Advanced Technology R&D Center
Mitsubishi Electric Corporation
Hyogo 661-8661, Japan
Saito.Hisashi@bc.MitsubishiElectric.co.jp

4th Koon Hoo Teo
Mitsubishi Electric Research Labs
Cambridge, MA 02139, USA
teo@merl.com

5th Shaloo Rakheja
Holonyak Micro & Nanotechnology Laboratory
University of Illinois at Urbana-Champaign
Urbana, IL 61801, USA
rakheja@illinois.edu

Abstract—**This paper implements a physically derived compact model of current conduction and gate leakage in AlGaN/GaN high-electron mobility transistors (HEMTs). The drain-source current conduction through the device is described using the surface potential based virtual-source model applicable for scaled gate length devices. The gate leakage model includes contributions from thermal emission (TE), trap-assisted tunneling (TAT), Poole Frenkel (PF) emission, and Fowler-Nordheim (FN) tunneling. The full I-V model is applied to fabricated AlGaN/GaN HEMTs with SiN passivation and excellent agreement of the model against measured data is demonstrated over a broad bias and temperature range from 298 K to 573 K.**

Index Terms—**Compact model, GaN HEMTs, Gate leakage, High temperature operation**

I. INTRODUCTION

AlGaN/GaN-based high electron mobility transistors (HEMTs) are an excellent candidate to realize high-power and high-frequency electronic devices that could also operate in extreme environment. In particular, AlGaN/GaN HEMTs can support a high concentration of the two-dimensional electron gas (2DEG) at the interface due to polarization engineering, thus obviating the need of additional impurity doping [1]. With a high 2DEG mobility (> 1250 cm^2/Vs), high maximum carrier drift velocity ($\sim 2.5 \times 10^7$ cm/s), and high breakdown field strength (330 MV/cm), AlGaN/GaN HEMTs surpass their III-V counterparts in terms of achieving the desired Johnson figure of merit for microwave and millimeter wave applications [2]. However, despite these advantages, the excessive gate leakage current is a limiting factor in these devices, especially when the devices are subject to high-field and high-temperature operating conditions [3]. Incorporating gate leakage into compact device models is therefore important to accurately assess the technology-device-circuit interaction and further use this assessment to fine-tune the device technology to achieve lower leakage.

When the device is forward biased, the dominant leakage current is due to thermal emission (TE), which increases exponentially with the bias across the gate Schottky contact. In the case of GaN HEMTs in which localized trap centers are often found in the AlGaN layer, the trap-assisted tunneling (TAT) current also becomes important to consider in the forward biased regime of the gate Schottky contact. On the other hand, the leakage through the gate contact is due to the Poole-Frenkel (PF) emission and the Fowler-Nordheim (FN) tunneling when the contact is reverse biased.

While prior works have focused on the physics of gate leakage in AlGaN/GaN HEMTs, there are limited reports of modeling gate leakage within a compact device model that would facilitate accurate circuit simulations. In [3], gate leakage is computed without including the effect of the drain bias and, therefore, is not applicable over a broad bias range. In [4], gate leakage in GaN HEMTs was studied for temperature less than 320 K. However, during nominal device operation in RF circuits, the channel temperature of GaN HEMTs could exceed 450K. Reference [5] uses a simplified Schottky-diode theory to represent the gate-source and gate-drain leakage currents. In [6], the gate leakage current due to the existence of surface traps is reported to be the primary mechanism and is modeled physically by solving the Poisson's and current density equations self-consistently. However, the validity of these models over broad electrical bias and temperature range in short-channel HEMTs has not been established. In this work, a physically motivated model for gate leakage in sub-500 nm gate-length GaN HEMTs is developed. The gate leakage current and the drain-source current are computed self-consistently using the surface-potential-based virtual-source model developed by the co-authors Li and Rakheja. Moreover, the compact model comprehends the nonlinear effects pertinent to GaN technology including nonlinear access region resistances and self-heating under high current density. The model validity is demonstrated by applying it to measured device data over broad bias and temperature range between 298 K to 573 K.

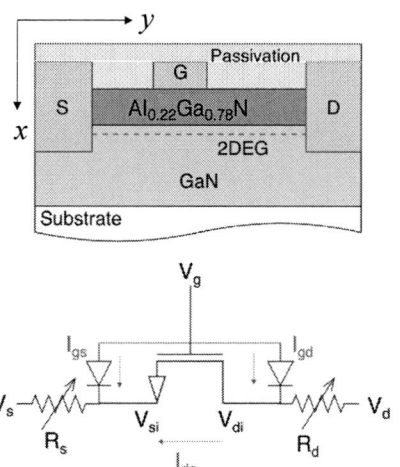

Fig. 1. Illustration of the cross-section of a prototypical AlGaN-GaN HEMT. Below is the equivalent circuit diagram of the GaN HEMT. The gate-source leakage is I_{gs}, gate-drain leakage is I_{gd}, and nonlinear source/drain resistances are represented as R_s and R_d. The voltages at the intrinsic source and drain nodes are represented as V_{si} and V_{di}, respectively. The gate length of the fabricated devuce is 400 nm, and its width is 50 μm. The device is built on SiC substrate.

II. MODEL DESCRIPTION

A prototypical AlGaN/GaN HEMT device structure and its equivalent circuit model are shown in Fig. 1. The Schottky diodes between the gate metal and the source/drain are shown in the equivalent circuit model. This circuit also indicates the non-linear source and drain access-region resistances, R_s and R_d, respectively. The intrinsic voltage at the source and drain nodes is given as V_{si} and V_{di}, respectively. Since the voltage drop across the access regions depends on the drain-source current I_{ds}, the intrinsic voltage drops from the gate-to-source, V_{gsi}, and the gate-to-drain, V_{dsi}, are first obtained from the core drain-source I-V model. These voltage drops are then used as inputs to compute the gate leakage self-consistently with I_{ds}. The core drain-source I-V model is based on the virtual-source (VS) approach to current conduction, where in the charges and the velocity of charges at VS are obtained in terms of the Landauer's channel transmission coefficient. The channel transmission is formulated differently for drift-diffusive and quasi-ballistic devices as discussed in [7], [8]. Further, we consider that only the first quantized energy sub-band in the channel is occupied. We have verified the validity of this assumption by conducting coupled Schrödinger-Poisson simulations for the devices examined here. Additional details regarding the core model formulation can be found in [8].

Figure 2 shows the various components of gate leakage in both reverse and forward bias modes of the gate Schottky contact. The TE current component at the gate-source and the gate-drain Schottky contacts in the forward biased mode is given as

$$I_{g(s,d)}^{TE} = I_{TE0}T^2 \exp \frac{-\phi_{TE}}{\phi_t} \left[\exp \frac{V_{g(si,di)}}{\eta_{TE}\phi_t} - 1 \right], \quad (1)$$

where I_{TE0} is the product of the Richardson's constant and the contact area A_{con}, T is the operating temperature, ϕ_{TE} is the Schottky barrier height, η_{TE} is the non-ideality parameter

of the junction, and $\phi_t = k_B T/q$ (k_B is Boltzmann constant) is the thermal voltage. The TAT current is modeled similarly as (1) but with different values of the non-ideality parameter (η_{TAT}), Schottky barrier height (ϕ_{TAT}), and the Richardson's constant (I_{TAT}).

Due to the spatial inhomogeneities at the metal-semiconductor interface, which lead to a spatial distribution of the Schottky barriers within the interface plane [9], the Schottky barrier height (ϕ_B) is expressed as

$$\phi_B = \overline{\phi_B} - \frac{\sigma_s^2}{2k_B T/q}, \quad (2)$$

where $\overline{\phi_B}$ is the fundamental or mean Schottky barrier, while σ_s is the standard deviation. Assuming that $\overline{\phi_B}$ is temperature independent and σ_s is either temperature-independent or varies linearly with temperature, (2) shows that ϕ_B increases with an increase in temperature. Experimental data analyzed in this work and others [10] confirm this temperature dependency of ϕ_B. Accordingly, we model $\phi_{TE/TAT}$ as

$$\phi_{TE/TAT} = \phi_{TE/TAT,300} + \gamma_{TE/TAT}(T - 300), \quad (3)$$

where $\phi_{TE/TAT,300}$ is the Schottky barrier height at 300 K, and $\gamma_{TE/TAT}$ captures the linear temperature dependency of the Schottky barrier height.

Under the statistical model of the Schottky contacts, the non-ideality parameter results from the deformation of the spatial barrier distribution when a bias is applied. We restrict ourselves to the case that the non-ideality of a given process is bias independent. This assumption is justified based on the experimental data of AlGaN/GaN HEMTs, as well as other Schottky contacts such as PtSi/Si measured in previous works. For such contacts, the temperature-dependent non-ideality is modeled as

$$\eta = \frac{1}{1 - \rho_2 + \rho_3/(2k_B T/q)}, \quad (4)$$

where ρ_2 (ρ_3) quantifies the bias-dependence of $\overline{\phi_B}$ and σ_s^2. When $-\rho_2 + \rho_3/(2k_B T/q) \ll 1$ and $\rho_2 \ll \rho_3/(2k_B T/q) \ll 1$, we can simplify the non-ideality as

$$\eta = 1 - \frac{\rho_3}{2k_B T/q}. \quad (5)$$

For the purpose of compact device modeling, we represent (5) for TE and TAT processes as

$$\eta_{TE/TAT} = \eta_{0(TE/TAT)} + \kappa_{\eta(TE/TAT)} \left[\frac{1}{T_0} - \frac{1}{T} \right], \quad (6)$$

where $\eta_{0(TE/TAT)}$ is the non-ideality for TE and TAT processes at $T_0 = 300$ K, while $\kappa_{\eta(TE/TAT)}$ is the temperature coefficient of the non-ideality.

The FN tunneling which dominates gate leakage in reverse - biased condition for a given electric field $\mathcal{E}$ across the insulator (AlGaN layer) is given as [11]

$$I_{g(s,d)}^{FN} = A_{con}J_{FN0}(T)\mathcal{E}^2 \exp \left(-\frac{B(T)}{\mathcal{E}} \right),$$
$$\mathcal{E} = \frac{V_{gsi} - \psi_s}{t_{ins}}, \quad (7)$$

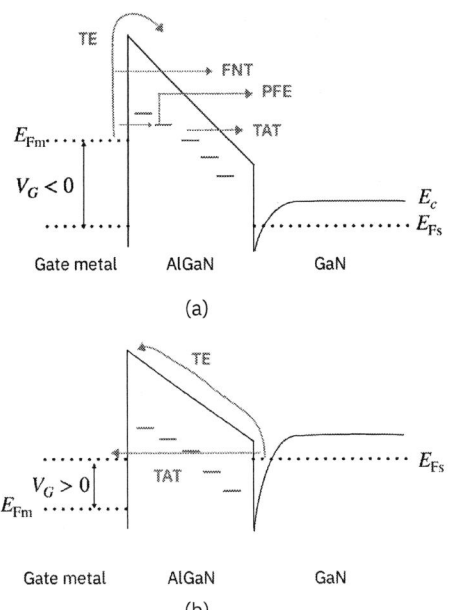

(a)

(b)

Fig. 2. Illustration of gate leakage mechanisms when the gate-source Schottky contact is (a) reverse biased ($V_G < 0$) and (b) forward biased ($V_G > 0$). TE: thermionic emission, PFE: Poole-Frenkel emission; TAT: trap-assisted tunneling; FNT: Fowler Nordheim tunneling tunneling. $E_{\rm fm}$ and $E_{\rm fs}$ denote the Fermi level in the metal and GaN channel, respectively. E_c denotes the bottom of the conduction band.

where the pre-factor $J_{\rm FN0}(T)$ and the coefficient $B(T)$ are determined from examining the measured gate leakage data. To evaluate the electric field, we need the information on the surface potential, ψ_s, at VS, which is readily provided by the core drain-source I-V model.

From an analysis of experimental data, we find that the temperature dependence of $J_{\rm FN0}(T)$ can be modeled as

$$J_{\rm FN0} = J_{\rm FN00} + \gamma_{\rm FN}(T - T_0)^2. \tag{8}$$

where $I_{\rm FN00}$ is evaluated at $T_0 = 300$ K, while $\gamma_{\rm FN}$ is the second-order temperature sensitivity coefficient of $J_{\rm FN0}$. The exponent $B(T) \propto \sqrt{m_{\rm eff,t}\phi_1}$ depends on the tunneling mass of electrons through the AlGaN layer ($m_{\rm eff,t}$) and the barrier height for tunneling (ϕ_1). Previous works have shown that the barrier height for tunneling in AlGaN/GaN HEMTs reduces with an increase in temperature, implying that $B(T)$ reduces with temperature. Similar temperature-dependent behavior is also observed in our samples (see Sec. III). Hence, we model $B(T)$ as

$$B(T) = B_0 - \gamma_B(T - T_0), \tag{9}$$

where B_0 is the value of $B(T)$ at $T = T_0 = 300$ K, and γ_B measures its temperature sensitivity.

The PF emission gate leakage is due to trap-assisted hopping transport, where the traps located closer to the bottom of the conduction band are involved [12]. In its simplified form, the PF emission gate leakage current is given as

$$I_{\rm PF} = A_{\rm con}C\mathcal{E} \exp\left[\frac{-q(\phi_{\rm PF} - \sqrt{\beta_{\rm PF}\mathcal{E}})}{k_B T}\right], \tag{10}$$

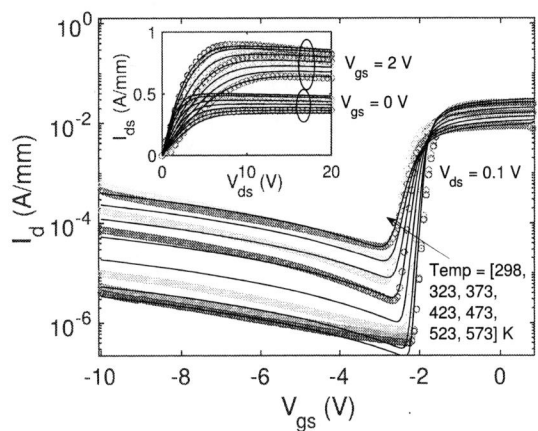

Fig. 3. Validation of the core static I-V model against measured data. Main plot shows transfer curves, while output curves are shown in the inset. Symbols are measured data, while model fits are shown in solid lines.

where C is a constant of proportionality, $\phi_{\rm PF}$ is the emission barrier height with reference to the trap level of the electron, and $\beta_{\rm PF} = q/\pi\epsilon_{\rm ins}$.

III. RESULTS

The cross-section of the fabricated device is shown in Fig. 1. The device consists of a 20-nm thick $Al_{0.22}Ga_{0.78}N$ on top of 900-nm unintentionally doped GaN channel followed by 300-nm GaN buffer. We first match the measured I_d versus V_{ds} and V_{gs} using our core static I-V model. Once the core model parameters are determined, we calibrate the gate leakage model against measurements. Figure 3 shows the results of output and transfer curves of the HEMT obtained from measurements and from model-generated results. We can see that the model does an excellent job to capture the full I-V plane of the device over a broad range of temperature from 298 K to 573 K.

Figure 4 shows a plot of $\ln(J_g^{\rm meas}/\mathcal{E}^2)$ versus $1/\mathcal{E}$ in the reverse-biased mode. Here, $J_g^{\rm meas}$ is the measured gate leakage current density. From the slope of the plot and the y-axis intercept at different temperatures, we can extract $B(T)$ and $J_{\rm FN0}(T)$, respectively (see inset of Fig. 4).

To confirm that the PF emission is insignificant in our devices, in Fig. 5, we show $\ln(J_g^{\rm meas}/\mathcal{E})$ versus $\sqrt{\mathcal{E}}$. The slope of the straight line in this plot is $q\sqrt{\beta_{\rm PF}}/k_B T$. However, the extracted value of $\beta_{\rm PF}$ leads to a significantly higher value of $\epsilon_{\rm ins}$, which for the devices under examination is physically unfeasible. As we discuss later, the gate leakage current from measurements can indeed be fully explained by neglecting contribution from PF emission.

Figure 6 shows the measured gate leakage data versus gate bias at 100 mV drain-to-source bias. We see the model-generated results match the measurements quite well over a broad range of bias and temperature. The model parameters used to obtain the model fits are listed in Table I.

IV. SUMMARY

In this work, a physically motivated compact model is proposed for gate leakage current in GaN HEMTs. The gate

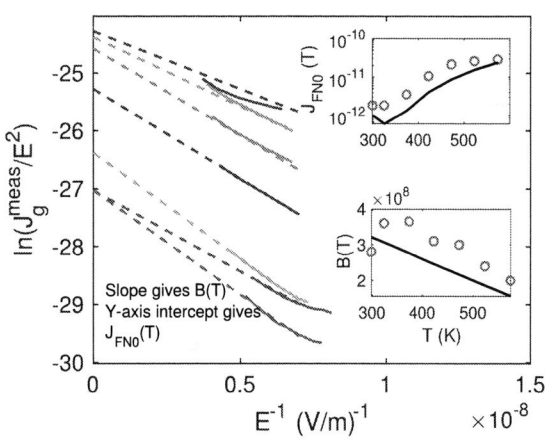

Fig. 4. Main plot shows data on $\ln(J_g^{\mathrm{mes}}/\mathcal{E}^2)$ versus $\mathcal{E}^{-1}$ from measurements. Assuming solid line fit to the experimental data allows us to extract temperature-dependent FN tunneling parameters as shown in inset. Solid lines show model fits while symbols are extracted from measurements.

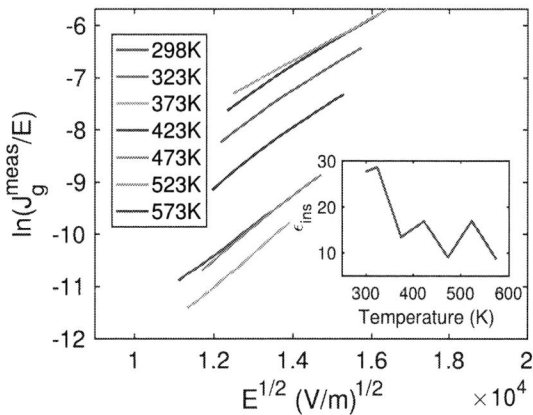

Fig. 5. Assuming PF emission model, the permittivity of the insulator layer is extracted as a function of temperature. As shown in inset, only at very high temperatures the permittivity approaches physically meaningful values, indicating that for our samples PF emission is not dominant.

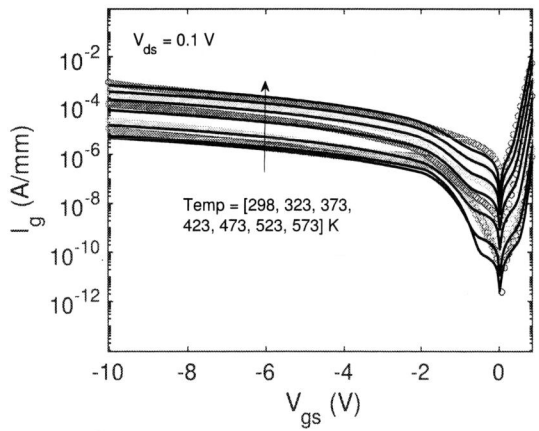

Fig. 6. Validation of the gate leakage model against measured data for broad temperature range. Symbols are measured data points, while solid line corresponds to model fit.

TABLE I
MODEL PARAMETERS USED FOR VALIDATING GATE-LEAKAGE MEASUREMENTS.

Parameter	Value
Key core static I-V model parameters	
Device width, W	50 μm
Gate length, L_G	400 nm
Source access region length, $L_{\mathrm{acc,s}}$	0.6 μm
Source access region length, $L_{\mathrm{acc,d}}$	3.3 μm
Effective mobility, μ_{eff}	1275 cm^2/Vs
Thermal resistance, R_{th}	120 K/W
Contact resistance, R_c	3×10^{-4} Ω-m
Non-ideality parameter for the channel, n_0	1.4
TE model parameters	
Schottky barrier height for TE at 300 K, $\phi_{\mathrm{TE,300\,K}}$	0.67 V
Temperature sensitivity of TE Schottky barrier height, γ_{TE}	0.0007 V/K
Non-ideality parameter for TE at 300 K, $\eta_{0,\mathrm{TE}}$	15
Temperature sensitivity of TE non-ideality, $\kappa_{\eta,\mathrm{TE}}$	100 K
TAT model parameters	
Schottky barrier height for TAT at 300 K, $\phi_{\mathrm{TAT,300\,K}}$	1 V
Temperature sensitivity of TE Schottky barrier height, γ_{TE}	0.00005 V/K
Non-ideality parameter for TAT at 300 K, $\eta_{0,\mathrm{TAT}}$	1.5
Temperature sensitivity of TAT non-ideality, $\kappa_{\eta,\mathrm{TAT}}$	100 K
FN tunneling model parameters	
Pre-factor in FN tunneling current at 300 K, J_{FN00}	6×10^{-13} A/V^2
Temperature sensitivity of FN tunneling pre-factor, γ_{FN}	3×10^{-16} A/V^{2}K^2
FN tunneling exponent factor at 300 K, B_0	2.3×10^8 V/m
Temperature sensivity of FN tunneling exponent factor, γ_B	3.4×10^5 V/mK

leakage model is coupled to our previously developed core static I-V model, a surface-potential based model that can be adapted to handle transport physics of both drift-diffusive with velocity saturation and quasi-ballistic devices. Our gate leakage model is applied to a 400-nm Al$_{0.22}$Ga$_{0.78}$N/GaN HEMTs from 298 K to 573 K and excellent match of the model against measured data is demonstrated. Our future work will include gate leakage modeling in shorter gate length devices as well as in different epitaxial heterostructures.

ACKNOWLEDGMENT

The authors acknowledge the support of the Mitsubishi Electric Research Labs (Cambridge) and Mitsubishi Electric Corporation (Japan) for providing the experimental data and sponsoring this research.

REFERENCES

[1] Wu, S. et al. (2019). IEEE Elec. Dev. Lett., 40(6), 846-849.
[2] Ranjan, K. et al. (2014). Appl. Phys. Exp., 7(4), 044102.
[3] del Alamo, J. A., & Joh, J. (2009). Microelec. Rel., 49(9-11), 1200-1206.
[4] Turuvekere, S. et al. (2013). IEEE Trans Elec. Dev., 60(10), 3157-3165.
[5] Saadaoui, S. et al. (2019). Jour. of Phys. and Chem. of Solids, 132, 157-161.
[6] Radhakrishna, U. et al. (2014). physica status solidi (c), 11(34), 848-852.
[7] Goswami, A. et al. (2014). IEEE Trans. Elec. Dev., 61(4), 1014-1021.
[8] Rakheja, S. et al. In 2014 IEEE IEDM (pp. 35-1).
[9] Li, K., & Rakheja, S. (2018). Jour. of Appl. Phys., 123(18), 184501.
[10] Werner, J. H. et al., (1991). Jour. of Appl. Phys., 69(3), 1522-1533.
[11] Werner, J. H. et al. (1992). MRS Online Proc., 260.
[12] Turuvekere, S. et al. (2014). IEEE Trans. Elec. Dev., 61(12), 4291-4294.
[13] Debnath, A. et al. (2020). IEEE Trans. Elec. Dev., 67(3), 834-840.

11-3

Effect of Atomic Interface on Tunnel Barrier in Ferroelectric HfO₂ Tunnel Junctions

Junbeom Seo and Mincheol Shin[*]

School of Electrical Engineering

Korea Advanced Institute of Science and Technology

Daejeon 34141, South Korea

[*]E-mail: mshin@kaist.ac.kr

Abstract— We have demonstrated the dependence of the atomic terminations on ferroelectric tunnel junctions (FTJs) based on ferroelectric HfO₂ using density functional theory calculation. The atomistic structures of HfO₂ FTJs with various interfaces are constructed and their device performances are calculated. We have found that the potential barrier is significantly tailored by atomic species of the terminating atom of HfO₂. In particular, the atomistic effect contributes to the electric field across the tunnel barrier, which leads to asymmetric behavior. We demonstrate that the ON/OFF current ratio of FTJs can be improved by adjusting the atomic terminations, albeit without the external asymmetric structure such as dissimilar metal electrodes and additional composite layers.

Keywords—Ferroelectric tunnel junction, Schottky barrier, Ferroelectric HfO₂

I. INTRODUCTION

The recent discovery of the ferroelectric phase (space group: $Pca2_1$) of HfO₂ has provided a breakthrough for emerging memory and neuromorphic devices based on ferroelectric materials [1, 2]. Compared with perovskite oxide ferroelectric materials such as BaTiO₃ and PbTiO₃, excellent ferroelectric properties of HfO₂ can be retained with the thickness of few nanometers [2, 3]. Since HfO₂ is used in the commercial electronic devices, it is compatible with the Si-based CMOS process.

As one of the ferroelectric-based emerging memory devices, ferroelectric tunnel junctions (FTJs) have attracted much attention due to the high scalability and speed [4-7]. Since the concept of FTJs was proposed by L. Esaki in 1971 [4], the device based on BaTiO₃ has been demonstrated in the experiment [6]. A typical FTJ is composed of the ferroelectric thin film sandwiched between two metal electrodes as shown in Figure 1(a). The resistance states of FTJs can be achieved by varying the tunnel barrier with the polarization of the ferroelectric (See Figure 1(b)). This phenomenon is called the tunneling electro-resistance effect (TER).

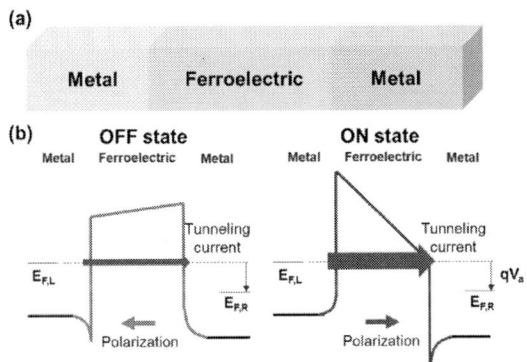

Figure 1. (a) Schematic structure of FTJ. (b) Band diagram of the FTJ under OFF and ON states.

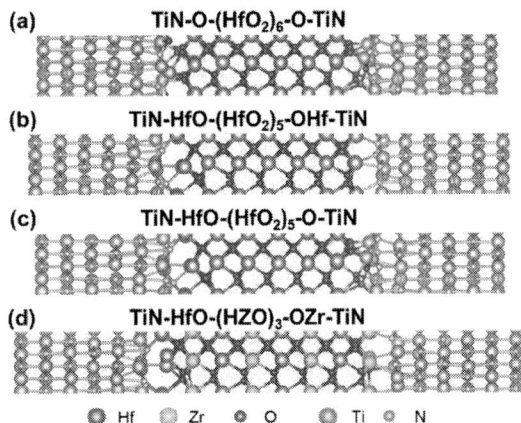

Figure 2. Atomistic structures of HfO₂-based FTJs. (a) O-terminated HfO₂, (b) Hf-terminated HfO₂, (c) Hf- and O-terminated HfO₂, and (d) Hf- and Zr-terminated HZO.

Recent studies have demonstrated FTJs with a HfO₂ film owing to excellent ferroelectricity in thin ferroelectric HfO₂ film [3, 7]. To utilize FTJs in practical applications, a high ON/OFF ratio of FTJs is required. For this purpose, asymmetric structures such as different metal electrodes, semiconductors, and additional composite layers have been considered [8, 9]. However, HfO₂-based FTJs still have shown a poor ON/OFF current ratio compared with perovskite oxide-based FTJs [7]. To further enhance the ON/OFF current ratio of HfO₂-based FTJs, the interface engineering between metal and ferroelectric layers can play a key role in modifying the barrier properties by adjusting the interface configuration [10]. However, to the best of our knowledge, there are few works which studied the effect of the interface on the tunnel barrier of HfO₂-based FTJs.

In this work, we investigate the effect of the terminated layers on the tunneling barrier of FTJs with a ferroelectric HfO₂ film using density functional theory (DFT) calculation. We focus on the influence of the atomic termination on the device performances. We construct the atomistic structures of FTJs based on HfO₂ and Zr-doped HfO₂ (HZO) with various atomic terminations. We reveal that the tunnel barrier is varied with the type of the atomic terminations of HfO₂, even leading to the asymmetric behavior of the tunnel barrier. Also, we report that without asymmetric metal electrodes and a composite layer, the ON/OFF current ratio of FTJs can be improved by adjusting the atomic termination at the interface.

II. SIMULATION APPROACH

All DFT calculations were performed by the SIESTA package [10]. We used the generalized gradient approximation (GGA) for exchange-correlation functional.

229

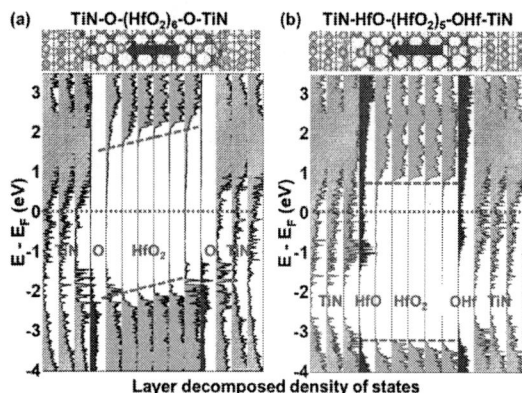

Figure 3. Layer-decomposed density of states of (a) OO and (b) HH structures. Blue thick arrows represent the polarization direction of HfO_2.

The Brillouin zone was sampled with the Monkhorst-Pack scheme with 5×5×1 *k*-points for the atomistic structures. Since the polarization of HfO_2 and HZO is oriented along the [001] direction, (001) surface of HfO_2 and HZO slab structures is in contact with the TiN electrodes. The atomistic structures are relaxed until the forces on each atom are less than 0.02 eV/Å. To explore the influence of the atomic termination of HfO_2 and HZO on the device performance, we constructed four types of the atomistic structures, namely, TiN-HfO-$(HfO_2)_5$-OHf-TiN, TiN-O-$(HfO_2)_6$-O-TiN, TiN-HfO-$(HfO_2)_5$-O-TiN, and TiN-HfO-(HZO)-OZr-TiN denoted by HH, OO, HO, and HZ, respectively, as shown in Figure 2. After the structural relaxation, we calculated the current-voltage characteristics using non-equilibrium Green's function implemented in the SIESTA package.

III. RESULTS AND DISCUSSION

Depending on the atomic termination of the HfO_2, there are noticeable differences in the tunnel barrier. Figure 3 represents the layer-decomposed density of states (L-DOS) of the OO and HH structures under the same polarization state ($P_\leftarrow$). $P_{\leftarrow(\rightarrow)}$ represents the polarization pointing toward the left

(right) direction. From this figure, the OO structure has a higher and steeper tunnel barrier compared with that of the HH structure. These discrepancies in the tunnel barriers between OO and HH structures are due to the dissimilar properties of interfacial charges.

A native polar surface of HfO_2 plays a key role in determining the interfacial charge properties, resulting in the modulation of the tunnel barrier. Figures 4(a) and (b) show the layer-resolved charges at the left and right interfaces between TiN and HfO_2 of the OO and HH structures. The atomic charges are calculated by Bader charge analysis [11]. As the electrons are transferred between interfacial atoms of TiN and HfO_2, non-zero charges are induced at the interfacial atoms. In the OO (HH) structure, the interfacial layers of HfO_2 are negatively (positively) charged, while the layers of TiN become positive (negative) (See Figures 4(a) and (b)). Note that such charge transfer generates the interface dipoles between the terminated atoms of HfO_2 and TiN [12], so that the electric field induced by the interface dipole affects the heights of the tunnel barrier. As shown in Figure 3, this explains that the O-terminated structure with negative charged interfacial layers shows a higher barrier height than that of the Hf-terminated structure with positively charged interfacial layers.

To explain the dependence of the electric field across the tunnel barrier on the terminating atoms of HfO_2, atomistic effects should be considered. The spontaneous polarization is induced by the movement of O atoms from the center of HfO_2, leading to the mismatch of the charges between the left and right terminated layers of HfO_2 and TiN. The polarization bound charges are incompletely screened by the metal electrodes inducing the depolarization field (E_{dep}). For simplicity, we assumed that E_{dep} depends on the metal properties, irrespectively of the terminated atoms of HfO_2. Also, the displacement of O atoms provides a discrepancy in the atomic charges between two terminated layers of HfO_2 (See Figures 4(a) and (b)). As we impose the additional intrinsic electric field (E_{int}) which is induced by this atomic

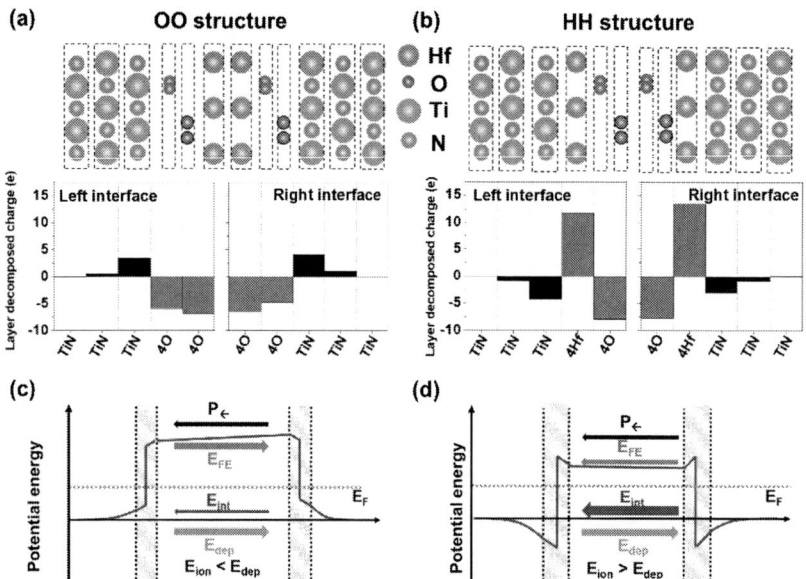

Figure 4. (a) and (b) Layer resolved atomic charges of the OO and HH structures, respectively. (c) and (d) schematic potential energy diagrams. The charge transfer occurs between the terminated layers of HfO_2 and TiN, which generate the interface dipoles. The polarization of the ferroelectric is imperfectly screened by TiN electrodes.

230

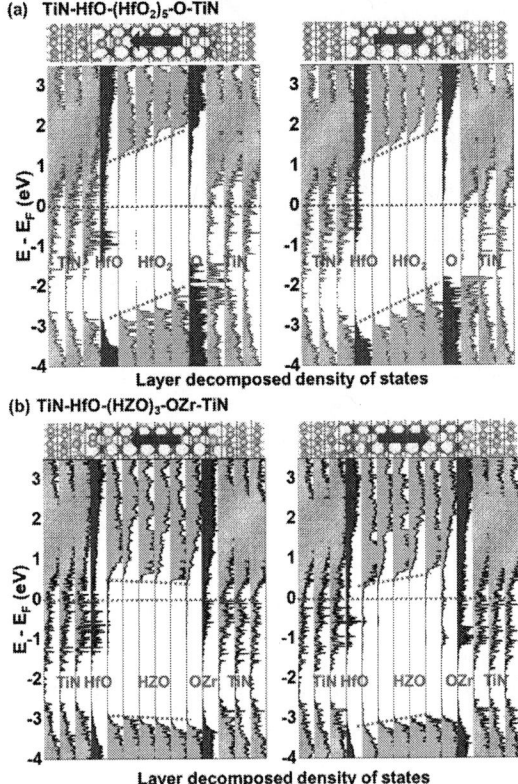

Figure 5. Layer-decomposed density of states of (a) HO and (b) HZ structures under P$_\leftarrow$ and P$_\rightarrow$ states.

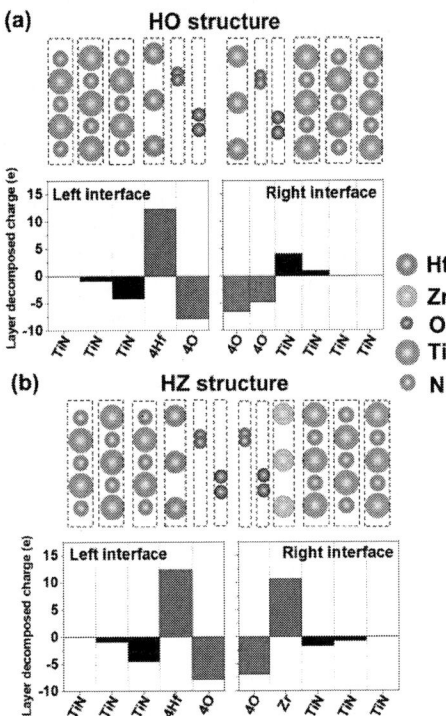

Figure 6. Layer-resolved atomic charges at the interfaces of (a) OO and (b) HZ structures, respectively.

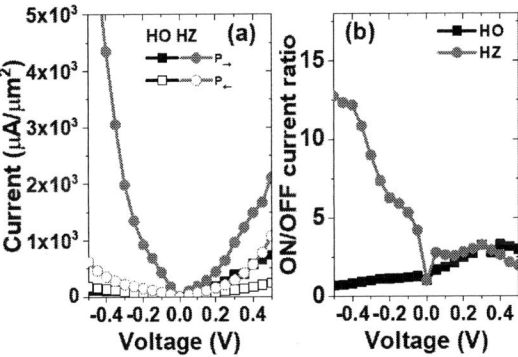

Figure 7. (a) Current-voltage characteristics and (b) ON/OFF current ratio of HO and HZ structures.

charge difference, the total electric field (E$_{FE}$) across the tunnel barrier is given as

$$E_{FE} = E_{dep} + E_{int} + E_{bi} + E_a \quad (1)$$

where E$_{bi}$ is the built-in field induced by the difference in the work function between metal and ferroelectrics and E$_a$ is the external field. E$_{bi}$ is zero because we considered the symmetric metal electrodes. If the atomistic effect is not considered, E$_{int}$ is ignored and hence E$_{FE}$ is dominated by E$_{dep}$. The large mismatch of the charges between two interfacial layers of HfO$_2$ leads to the large E$_{int}$ of which the direction is opposite to E$_{dep}$. From Figures 4(a) and (b), it can be seen that the charge mismatch of the HH structure is larger than that of the OO structure. It means that the HH structure has a stronger E$_{int}$ than that of the OO structure. E$_{FE}$ of the HH structure becomes weak and opposite to the direction of E$_{dep}$ because E$_{int}$ is large enough to overcompensate E$_{dep}$, but in the OO structure, E$_{FE}$ is the same direction to E$_{dep}$ due to weak E$_{int}$.

Based on the atomic charge properties of the terminated layer of HfO$_2$, an asymmetry between the left and right barrier heights can be achieved, which is analogous to the effects of the disparate metal electrodes. Figure 5 shows the L-DOS of the HO and HZ structures. As shown in Figure 5(a), the HO forms asymmetric terminated layers of which the tunnel barrier is not modulated by varying the polarization. However, as seen in Figure 5(b), although the interfacial layers of the HZ structure are asymmetric, the electric field across the ferroelectric layer is switched by the polarization reversal.

Figure 6 represents the layer resolved atomic charges of the HO and HZ structures. As shown in Figure 6(a), the HO

structure has a large difference in the interfacial charges between Hf and O layers, leading to large E$_{int}$ which is not switched by the polarization reversal. Non-switchable E$_{int}$ of the HO structure makes the tunnel barrier independent of the polarization states. However, as shown in Figure 6(b), in the case of the HZ structure, atomic charges of both terminated layers are positive, but their difference is small enough so that E$_{int}$ can be switched by the polarization reversal.

Figure 7 represents the current-voltage characteristics and ON/OFF current ratio of the HO and HZ structures. Although HO and HZ structures form the asymmetric tunnel barrier, the HO structure leads to a low ON/OFF current ratio of about 3, but the ON/OFF current ratio of the HZ structure reaches to 12. This is because the tunnel barrier of the HO structure is not altered by the polarization reversal. In contrast, the left and right barrier heights of the HZ structure are asymmetrically varied with the polarization. This means that the difference in tunneling current between ON and OFF state can be achieved, resulting in the large ON/OFF current ratio. This result shows that without the symmetric metal electrodes or additional composite layers, the ON/OFF current ratio can be improved.

IV. CONCLUSION

We investigated the effect of the interface configuration of HfO_2 on the tunnel barrier using DFT simulation. We have found that the interface charge properties are highly dependent on the terminated layer of HfO_2. Therefore, by adjusting the type of the terminated layer, one can produce the large ON/OFF ratio, albeit without either asymmetric metal electrodes or an additional composite barrier. Our findings may provide the guideline for the device fabrication to achieve a high ON/OFF ratio in FTJ devices.

ACKNOWLEDGMENT

This work was supported by Samsung Research Funding and Incubation Center of Samsung Electronics under Project Number SRFC-TA1703-10.

REFERENCES

[1] T. S. Böscke, J. Müller, D. Bräuhaus, U. Schröder, and U. Böttger, "Ferroelectricity in hafnium oxide thin films," Appl. Phys. Lett. vol. 99, 102903, 2011.

[2] S. J. Kim, J. Mohan, S. R. Summerfelt, and J. Kim, Jom, "Ferroelectric $Hf_{0.5}Zr_{0.5}O_2$ thin films-A review of recent advances," vol. 71, pp. 246-255, 2019.

[3] S. S. Cheema, et al., "Enhanced ferroelectricity in ultrathin films grown directly on silicon," Nature, vol. 580, no. 7804, pp. 478–482, 2020

[4] L. Esaki, R. B. Laibowitz, and P. J. Stiles, "Polar switch," IBM Tech. Discl. Bull, 1971

[5] E. Y. Tsymbal and H. Kohlstedt, "Tunneling across a ferroelectric," Science, vol. 313, no. 181, 2006.

[6] A. Gruverman, D. Wu, H. Lu, Y. Wang, H. W. jang, C. M. Folkman, M. Ye. Zhuravlev, D. Felker, M. Rzchowski, C. B. Eom, and E. Y. Tsymbal, "Tunneling Electroresistance Effect in Ferroelectric Tunnel Junctions at the Nanoscale," Nanoletters, vol. 9, no. 9, pp. 3539-3543, 2009.

[7] Z. Wen and D. Wu, "Ferroelectric tunnel junctions: Modulations on the Potential Barrier," Adv. Mater., 1904123, 2019

[8] J. Yoon, S. Hong, Y. W. Song, J. Ahn, and S. Ahn, "Understanding tunneling electroresistance effect through potential profile in $Pt/Hf_{0.5}Zr_{0.5}O_2/TiN$ ferroelectric tunnel junction memory," Appl. Phys. Lett., 115, 153502, 2019

[9] B. Max, M. Hoffmann, S. Slesazeck, and T. Mikolajick, "Ferroelectric tunnel junctions based on ferroelectric-dielectric $Hf_{0.5}Zr_{0.5}.O_2/Al_2O_3$capacitor stacks," European Solid-State Device Research Conference, pp. 142–145, Sep. 2018

[10] J. M. Soler, E. Artacho, J. D. Gale, A. García, J. Junquera, P. Ordejón, and D. Sánchez-Portal, D, "The SIESTA method for ab initio order-N materials simulation," Journal of Physics: Condensed Matter, vol. 14, no. 11, 2745, 2002.

[11] G. Henkelman, A. Arnaldsson, and H. Jónsson, "A fast and robust algorithm for Bader decomposition of charge density," Computational Materials Science, 36(3)," pp. 354-360, 2006.

[12] R. T. Tung, "Recent advances in Schottky barrier concepts," Materials Science and Engineering: R: Reports, vol. 35, pp. 1-138, 2001

11-4 Surge Current Capability in lateral AlGaN/GaN Hybrid Anode Diodes with p-GaN/Schottky Anode

Gökhan Atmaca, Marie-Anne Jaud*, Julien Buckley, Jérôme
Biscarrat, Romain Gwoziecki, Marc Plissonnier and Thierry
Poiroux
CEA, LETI, MINATEC Campus,
Univ. Grenoble Alpes, Grenoble, France
*marie-anne.jaud@cea.fr

Arnaud Yvon and Emmanuel Collard
STMicroelectronics
10 rue Thalès de Milet, 37071 Tours, France

Abstract—In lateral power diodes, the conductivity modulation mechanism can pave the way to the demonstration of surge current capability. In a Hybrid Anode Diode concept with a p-GaN layer, an anode contact on p-GaN layer can be a source of hole injection that increases the electron density at AlGaN/GaN interface. The role of p-GaN layer on the surge current capability and its demonstration are investigated through TCAD simulations that explain the role of hole barrier tunneling at anode metal/p-GaN interface. These simulations show that surge current can occur in case of Ohmic p-GaN contact as the injected holes can lead to create additional electron density in the channel as well as a hole current to support the total diode current.

Keywords—hybrid anode diodes, hole injection, surge current, tcad

I. INTRODUCTION

Gallium Nitride (GaN) became a very attractive material for high-power and high frequency device applications as a wide band gap semiconductor [1,2]. Due to its superior properties such as high critical electric field, high electron velocity, large band gap, high electron mobility and density etc., GaN based transistors and diodes are emerging as promising candidates especially for new-generation power switching. An AlGaN/GaN High Electron Mobility Transistor (HEMT) is inherently normally on device, having a negative threshold voltage (v_{th}) due to presence of the two-dimensional electron gas (2DEG) at heterostructure interface. For many power device applications, it is desirable to use normally off devices for safety reasons and system reliability. In an AlGaN/GaN heterostructure topped by a p-GaN layer, p-GaN layer raises the conduction band edge at AlGaN/GaN interface leading to a depletion of the 2DEG channel, which induces a positive V_{th} shift that is necessary for the normally-off operation [1,2]. Moreover, above a certain gate voltage in HEMTs, the p-GaN layer can allow the hole injection from the gate metal to p-GaN/AlGaN interface. Due to electrostatic equilibrium, this results in additional electron in the channel and consequently increasing of drain current. This is called conductivity modulation [1,3]. In AlGaN/GaN Schottky Barrier Diodes (SBDs) based on a HEMT structure described in Fig. 1a), a p-GaN layer can be used to obtain surge current due to the hole injection phenomena. In a study reported by Hsueh et al. [4], a p-GaN layer with a Schottky contact was inserted into SBDs in order to deplete the 2DEG and increase breakdown voltage in reverse operation. In case of Schottky contact, the depletion width in the p-GaN below anode is large and limits hole barrier tunneling at metal/p-GaN interface [5].

In this paper, investigated Hybrid Anode Diode (HAD) device consists of a Mg doped p-GaN layer on an AlGaN/GaN heterostructure on a Si substrate. The p-GaN layer is located at the anode contact side between nitride and anode. A second

a) AlGaN/GaN Schottky Barrier Diode (SBD)

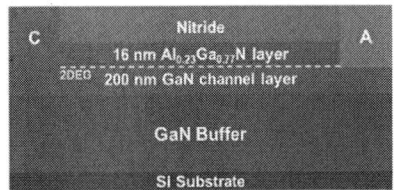

b) AlGaN/GaN Hybrid Anode Diode (HAD) with p-GaN

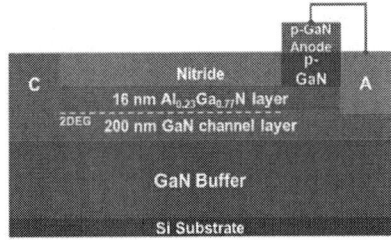

Fig. 1: Device structures used in TCAD simulations for a) a standard AlGaN/GaN Schottky Barrier Diode (SBD) without p-GaN layer and b) an Hybrid Anode Diode (HAD) with a p-GaN layer.

anode metal (p-GaN anode) is added onto p-GaN layer as shown in Fig. 1b). Total diode current is sum of anode and p-GaN anode currents. In order to demonstrate the surge current capability in HADs, we report TCAD simulations that include a tunneling model that explains the role of hole injection on the presence of additional electrons in the channel [6]. In order to achieve surge current capability in SBDs, the improvement of hole barrier tunneling and use of the hole injection can be significant.

In TCAD simulations, the electrical characteristics are simulated solving the Poisson and drift-diffusion equations. Polarization charges are calculated using a piezoelectric strain model with Vurgaftman values [7]. Hole barrier tunneling is considered at p-GaN anode metal/p-GaN interface and is simulated through Non-Local Tunneling Model [6]. In this model, the hole tunneling mass is also set to 0.3 m_0 [5].

II. THE EFFECT OF MG CONCENTRATION

Fig. 2 describes the conductivity modulation mechanism in a metal/p-GaN/AlGaN/GaN heterostructure. When applied voltage to p-GaN anode metal (V_A) is higher than a certain value, the hole injection begins from the p-GaN anode metal to p-GaN layer and it increases the accumulated hole density at p-GaN/AlGaN interface, where the 2-dimensional hole gas (2DHG) is formed, as V_A increases. Due to preserving electrostatic equilibrium, additional electrons are accumulated in 2DEG channel through the conductivity modulation mechanism.

Fig. 3 presents drain current as a function of anode voltage and V_{th} as a function of Mg concentration in p-GaN layer. Due to p-GaN layer, 2DEG carriers in the channel are depleted, HAD with p-GaN anode has a lower total diode current than SBD. For low Mg concentrations, increasing of Mg in p-GaN layer depletes more importantly the 2DEG, inducing a V_{th} increase. At high Mg concentrations, p-GaN layer still depletes 2DEG electrons but also allows holes to accumulate at p-GaN/AlGaN interface. V_{th} tends to decrease as Mg concentration increases above $5x10^{17}$ cm^{-3} as can be seen in Fig. 3 b). Due to electrostatic equilibrium illustrated by Fig. 2 [8], the 2DHG thus formed at p-GaN/AlGaN interface induced additional electrons in the channel responsible for V_{th} decreasing. Since low V_{th} or turn-on voltage is desirable for SBDs due to minimize power loss during operation [9, 10], use of high Mg concentration values is necessary in this device concept according to Fig. 3 b). Note that in these simulations, the hole injection from the p-GaN anode metal is limited because TiN metal induces a large depletion width. As can be seen in Fig. 2, if tunneling holes at anode metal/p-GaN interface increase, electron density in the channel can increase via conductivity modulation. To change this interface or increase the hole injection from the p-GaN anode metal, the effect of metal work-function on I-V curve can be examined.

III. THE EFFECT OF METAL WORK-FUNCTION

Fig 4 presents effects of p-GaN anode metal work-function considering Mg concentration is kept at $1x10^{19}$ cm^{-3} to ensure lower V_{th}. Fig. 4 a) shows total diode current for various p-GaN anode metal work-functions. Increasing work-function induces an increasing of current by the conductivity modulation. In case of Ideal Ohmic contact, after V_A= +9.0 V increase of total diode current is amplified. In cases of TiN, Ni and Pt metals, in range of V_A, total diode currents are lower than the reference device. As illustrated by Fig. 4 b), this is due to reduction of the depletion width that allows higher hole injection. One thing is interesting that in case of Pt metal, an increase of total diode current is obtained at very high V_A owing to the hole injection since relatively narrow depletion width allows the hole barrier tunneling at metal/p-GaN interface at very high V_A. On the other hand, Fig 4. a)

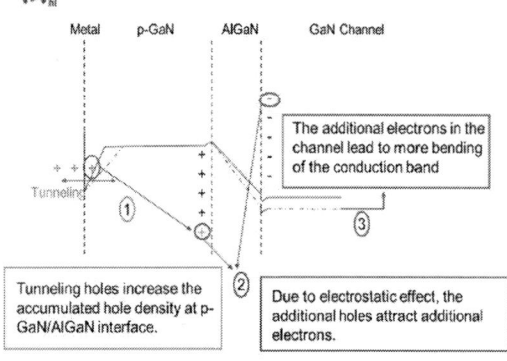

Fig. 2: The conductivity modulation mechanism in a Metal/p-GaN/AlGaN/GaN heterostructure. Hole injection begins when applied voltage to metal/p-GaN is higher than hole injection point (V_{hi}). Injected holes are accumulated at p-GaN/AlGaN interface. Due to electrostatic equilibrium, these holes cause additional electrons in the channel.

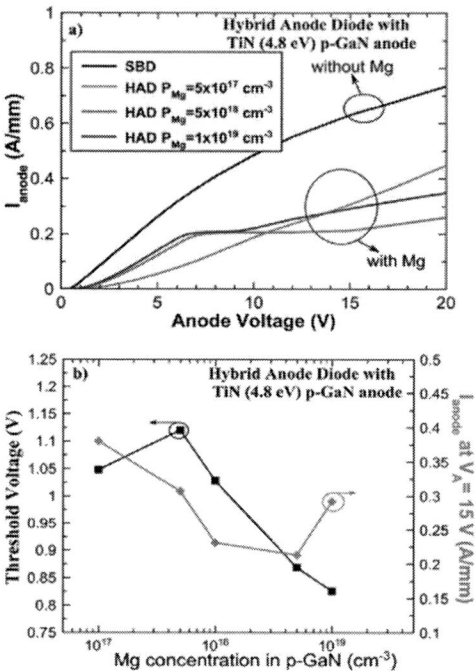

Fig. 3: a) The forward current characteristics and b) the Vth change and maximum anode current at V_A=+15.0 V for different Mg concentrations in p-GaN layer in case of HAD with Schottky p-GaN anode. The Vth is defined as V_A at I_A= 1 mA/mm.

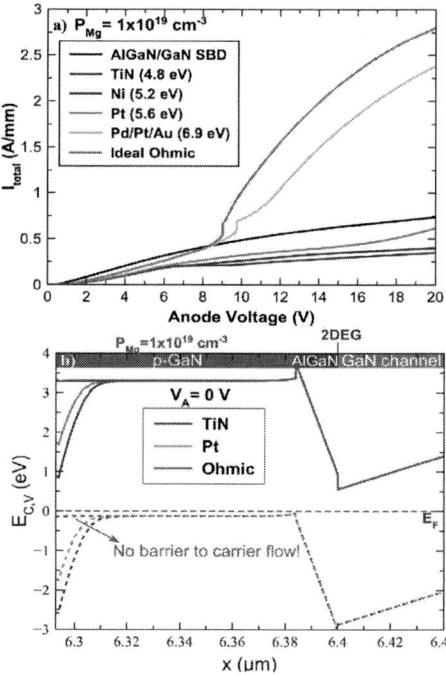

Fig. 4: a) The forward current characteristics for metal work-functions of anode metal on p-GaN and b) the conduction and valence band energy diagrams for several metal work-functions at V_A=0V.

234

highlights that surge current is possible in case of very high p-GaN anode metal workfunction (close to ohmic contact) owing to no barrier to carrier flow as can be seen in Fig. 4 b).

Fig. 5 a) summarizes total diode currents for investigated devices. If the anode and p-GaN anode metals are TiN, total diode current is lower than standard AlGaN/GaN SBD due to depletion of 2DEG electrons by p-GaN layer. If the anode metal is TiN and p-GaN anode metal is ideal Ohmic, this time total diode current is higher than standard AlGaN/GaN SBD. Moreover, a surge current is obtained due to injected holes that create additional electrons in the channel via the conductivity modulation mechanism described in Fig. 2. Fig.5 b) shows current components, I_{anode} (electron) and I_{pGaN} (hole) for HADs with Ohmic and TiN p-GaN anode. In case of TiN p-GaN anode, total diode current only consists of electron current. In case of Ohmic p-GaN anode, I_{anode} increases as V_A increases. After V_A= +9.0 V, a dramatic increase of I_{anode} is obtained due to the conductivity modulation mechanism. At the same voltage, a hole current (I_{pGaN}) starts to contribute to total diode current since the hole barrier tunneling is dramatically increased. On the other hand, main contribution to total diode current is still I_{anode} at high V_A and it increases as V_A increases.

When electron and hole density cartographies at V_A=15.0 V for each device are compared in Fig. 6 and 7, the conductivity modulation leads to a higher electron density in the channel in case Ohmic p-GaN anode than TiN p-GaN anode case. As V_A increases, the injected hole density from Ohmic anode metal increases and leads to more electrons in the channel due to conductivity modulation effect. On the other hand, surge current capability appears due to an increase in mostly electron current of I_{anode} via conductivity modulation effect and a hole current of I_{pGaN} contributes to the surge current.

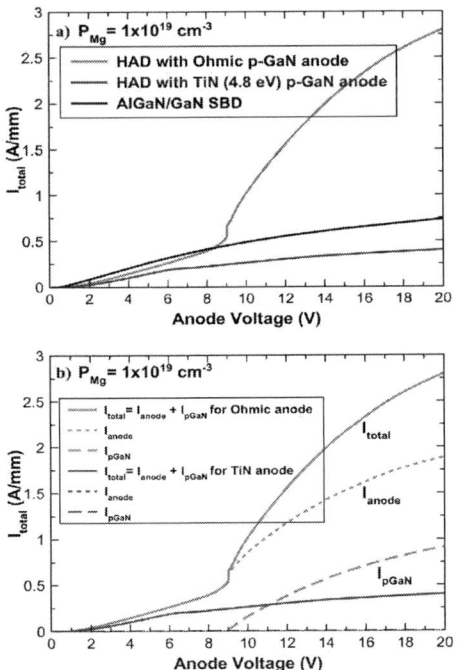

Fig. 5: a) The forward current characteristics for three different type devices and b) a demonstration of current components of HAD with Ohmic and TiN p-GaN anode device.

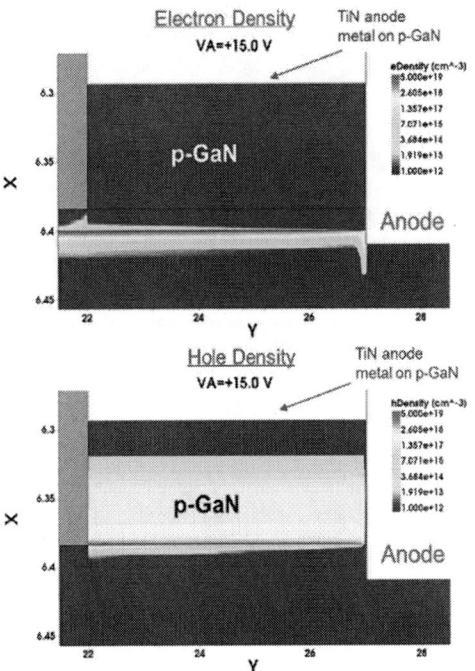

Fig. 6: Electron density and hole density distributions at V_A=15.0 V for HAD with TiN p-GaN anode.

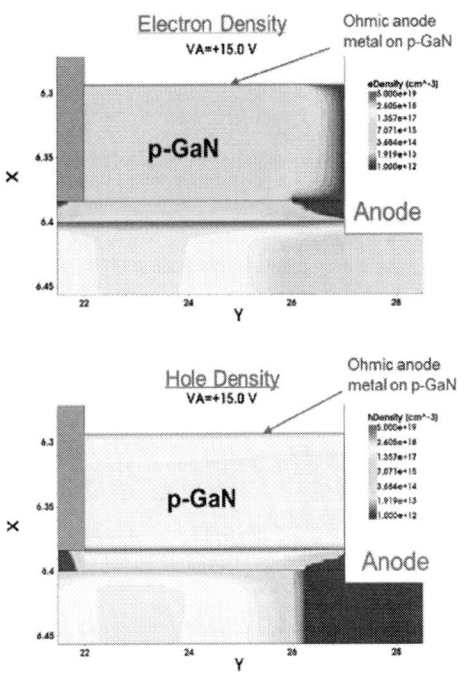

Fig. 7: Electron density and hole density distributions at V_A=15.0 V for HAD with Ohmic p-GaN anode.

IV. CONCLUSION

In this paper, the forward characteristics of AlGaN/GaN SBDs and HADs with Ohmic and Schottky p-GaN contacts were studied via TCAD simulations in order to demonstrate the surge current capability in lateral power diodes. It was shown that use of high Mg concentrations in p-GaN layer for HADs provides lower turn-on voltage owing to the conductivity modulation which was described for HEMTs. The surge current can occur in case of Ohmic p-GaN contact as the injected holes from anode can lead to create additional electron density in the channel and even holes can create a hole current to support the total diode current. Recently, several studies [11,12] show possible Ohmic contact formation on p-GaN. The findings obtained in our work can pave the way to the use of conductivity modulation to obtain surge current capability in lateral power diodes.

ACKNOWLEDGMENT

This work was funded by the French national program "programme d'Investissements d'Avenir IRTNanoelec" ANR-10-AIRT-05.

REFERENCES

[1] Y. Uemeto, M. Hikita, H. Ueno, H. Matsuo, H. Ishida, M. Yanagihara, T. Ueda, T. Tanaka, and D. Ueda, "Gate Injection Transistor (GIT)—A Normally-Off AlGaN/GaN Power Transistor Using Conductivity Modulation" IEEE Trans. Electron. Dev. 54, 2007, pp. 3393-3399.

[2] N. E. Posthuma, S. You, H. Liang, N. Ronchi, X. Kang, D. Wellekens, Y. N. Saripalli, S. Decoutere, "Impact of Mg out-diffusion and activation on the p-GaN gate HEMT device performance", Proc. Int. Symp. Power Semicond. Devices ICs, Jun. 2016, pp. 95-98.

[3] Y. Liu, S. Han, S. Yang, and K. Sheng, "Surge Current Capability of GaN E-HEMTs in Reverse Conduction Mode", Proc. Int. Symp. Power Semicond. Devices ICs, May 2019, pp. 439-442.

[4] K. -P. Hsueh, Y.-S. Chang, B.-H. Li, H.-C. Wang, H.-C. Chiu, C.-W. Hu, R. Xuan, "Effect of the AlGaN/GaN Schottky Barrier Diodes Combined with a Dual Anode Metal and a p-GaN Layer on Reverse Breakdown and Turn-on Voltage", Mater. Sci. Semicond. Process. 2019, 90, pp. 107–111.

[5] L. Efthymiou, G. Longobardi, G. Camuso, T. Chien, M. Chen, F. Udrea, "On the physical operation and optimization of the p-GaN gate in normally-off GaN HEMT devices", Appl. Phys. Lett. 110, 2017, 123502.

[6] Synopsys, TCAD tools, SDevice User Guide N-2017.09, p. 722.

[7] I. Vurgaftmann, JR. Meyer, "Band parameters for nitrogen-containing semiconductors", J. Appl. Phys. 94, 2003, 3675.

[8] B.Bakeroot, S. Stoffels, N. Posthuma, D. Wellekens, S. Decoutere, "Trading Off between Threshold Voltage and Subthreshold Slope in AlGaN/GaN HEMTs with a p-GaN Gate", Proc. Int. Symp. Power Semicond. Devices ICs, May 2019, pp. 419-422.

[9] M.D. Zhu, B. Song, M. Qi, Z.Y. Hu, K. Nomoto, X.D. Yan, Y. Cao, W. Johnson, E. Kohn, D. Jena et al. "1.9-kV AlGaN/GaN Lateral Schottky Barrier Diodes on Silicon", IEEE Electron. Device Lett. 36, 2015, pp.375–377.

[10] J.G. Lee, B.R. Park, C.H. Cho, K.S. Seo, H.Y. Cha, Low Turn-on Voltage AlGaN/GaN-on-Si Rectifier with Gated Ohmic Anode. IEEE Electron. Device Lett. 34, 2013, pp. 214–216.

[11] L. Xiao-Jing, D-G. Zhao, D-Sh. Jiang, P. Chen, J-J. Zhu, Z-S. Liu, L-C. Le, J. Yang and X-G. He, "Low contact resistivity between Ni/Au and p-GaN through thin heavily Mg-doped p-GaN and p-InGaN compound contact layer", Chin. Phys. B 24, 2015, 116803.

[12] Y. Liu, "Recent research on ohmic contacts on GaN-based materials", In IOP Conference Series: Materials Science and Engineering,Vol. 738, No. 1, p. 012007. .

Dynamic Simulation of Write '1' Operation in the Bi-stable 1-Transistor SRAM Cell

Tapas Dutta*, Fikru Adamu-Lema*, Asen Asenov*, Yuniarto Widjaja[†], and Valerii Nebesnyi[‡]

*Semiwise Ltd., Rankine Building, Glasgow G12 8LT, Scotland, UK

[†]Zeno Semiconductor Inc., Cupertino, CA 95014 USA

[‡]MCPG

e-mail: tapas.dutta@glasgow.ac.uk

Abstract—For the first time, physical insights into the writing process in the bi-stable 1-transistor SRAM cells are provided using dynamic (time dependent) TCAD simulations. The simulations are based on 28 nm planar CMOS technology, and the setup is carefully calibrated against available experimental data. Based on the simulations, we were able to identify clearly the mechanisms involved in the write '1' operation. The dependence of the writing process on drain and gate bias conditions was also investigated.

Index Terms—1T-SRAM; Bi-stability; Dynamic Simulation; TCAD; Impact-ionization

I. INTRODUCTION

6T-SRAM is a key component in microprocessors and system on chip (SoC) applications as embedded memory occupies a significant fraction of the chip area. The SRAM real estate is even larger in Artificial Intelligence (AI) chips. However, the 6T-SRAM cell scaling has become increasingly challenging due to the increasing stability and leakage current problems. A novel bi-stable, one transistor (1T) SRAM structure has been previously disclosed [1], which has the potential to dramatically reduce the SRAM real estate and to revolutionize the in-memory AI computing.

In this paper for the first time we present dynamic (time dependent) simulations of the write operation of the 1T SRAM cell based on 28nm planar CMOS technology.

II. METHODOLOGY

The process simulation for the 1T SRAM cell was carried using Sentaurus Process [2]. The corresponding simulation domain is illustrated in Fig. 1. In order to create the SRAM cell in a standard 28nm CMOS process flow of the *n*-channel MOSFET a heavily doped *n*-type 'boost' region is created by ion implantation beneath the source/drain, having peak concentration located deep in the substrate. This creates a floating *p*-type charge storage region isolated by the shallow trench isolations (STIs) on the sides of the transistor and by a *p*-*n* junction from below. The Drain and Gate terminals of the MOSFET are used for reading and writing into the cell and are connected to the Bit Line and Word Line, respectively, while the vertical *npn* transistor formed by the Drain/Source (Emitter), *p*-well (Base) and Boost (Collector) regions plays the key role in holding the data.

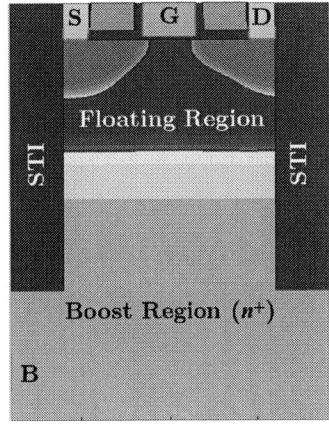

Fig. 1. 1T SRAM simulation domain.

The simulated current-voltage characteristics of the MOSFET at low and high drain bias in steady state conditions are calibrated to fit the experimentally measured data, as illustrated in Fig. 2. The important physical models used in the simulation are: Masetti model for doping dependence and Lombardi and Caughey-Thomas models for field dependence of mobility, doping dependent Shockley-Read-Hall (SRH) recombination model, and the Schenk model for band-to-band tunneling.

Further, impact ionization plays a critical role in the operation of the of 1T SRAM cell. After a number of test run simulations using available models, we found that the Okuto model [3], is the most suitable to compute the ionization coefficient, reproducing the three experimentally observed dips [1], [4] in the base (the floating *p*-well region) current versus base-emitter voltage (*p*-well potential relative to the source) *i.e.* the I_p-V_p characteristics of the vertical *npn* bipolar transistor.

In Fig. 3, we show the impact of different boost voltages on the I_p-V_p characteristics at zero Gate and Drain bias. Here the dips in the I_p-V_p characteristic represent zero net *p*-well current due to reversal of the base current at specific *p*-well potentials (V_p). This mechanism has been studied earlier in conventional bipolar junction transistors and explained in context of the 1T SRAM cell in [1]. In essence, the electrons emitted from the Source arrive at the *p*-well/Boost junction and

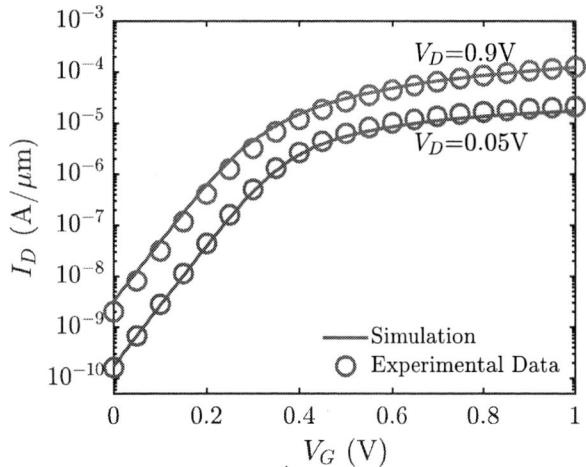

Fig. 2. Calibration of the MOSFET transfer characteristics.

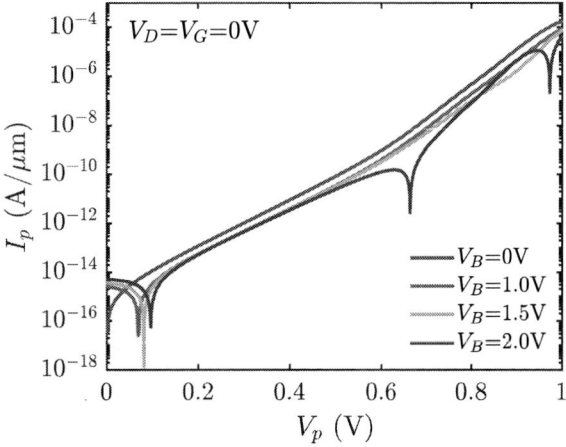

Fig. 3. Base current Vs base-emitter voltage (I_p-V_p) characteristics of the vertical bipolar transistor at different boost voltages (V_B) with V_G=V_D=0V. Here the subscript 'p' refers to the floating p-well region.

the Drain regions and if the electric field at these junctions are sufficiently high, the electrons will cause impact ionization. The resulting excess holes flow to the base terminal (the p-well region). This reduces the net p-well current, and hence depending on the bias conditions, it can be negative or positive. This gives rise to two stable states: state '0' at low V_p and state '1' at high V_p, with an intermediate metastable state, and this is the origin of the memory effect. Depending on the applied Drain and Gate voltages (V_D and V_G respectively), the cell will be latched to one of the stable states. At low boost voltage, the electric field at the p-well/boost junction is not strong enough for avalanche multiplication and hence we do not observe this bi-stable phenomenon.

III. RESULTS AND DISCUSSION

The Write '1' operation of the 1T SRAM cell is associated with the charging of the floating p-well. The dynamic (time dependent) simulation of this process involves fine balance between carrier generation and carrier recombination. In this

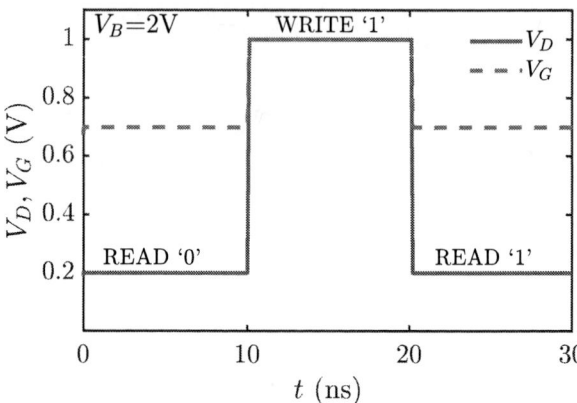

Fig. 4. V_G and V_D waveform for switching sequence comprised of READ '0'→ WRITE '1' → READ '1'. The boost voltage is constant throughout at V_B=2V.

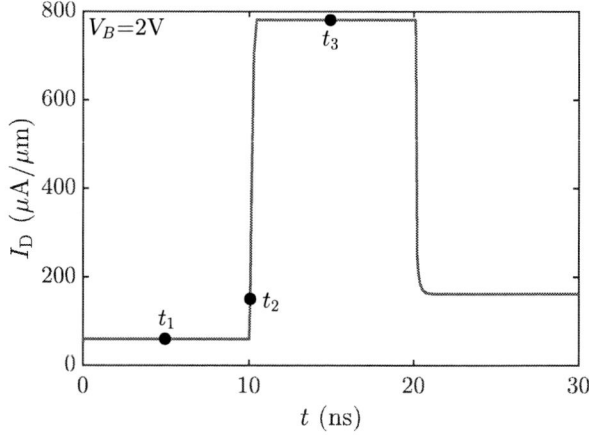

Fig. 5. Switching profile of the 1T-SRAM cell consisting of the READ '0', WRITE '1', and READ '1' operations.

section, we present simulation results and discussion on the write '1' operation in the 1T-SRAM.

The Drain and Gate terminal input waveform corresponding to read '0', write '1', and read '1' operations is shown in Fig. 4. The nominal Drain and Gate voltages used for writing '1' are 1V each while for reading, the voltages used are 0.2V and 0.7V, respectively. All the ramp times for the transients were 100 ps. The impact ionization rate is calibrated by tuning the 'b' parameter of the Okuto model in order to achieve desired operation.

Fig. 5 shows the cell switching process in terms of the Drain terminal current. The drain current is assessed for reading the cell state. We can see that the read '1' current is higher than the read '0' current. The stability of state '1' over time is achieved without refresh operation, indicating the bi-stability of the 1T-SRAM cell.

In order to understand the switching dynamics, in Fig. 6 we present the two-dimensional distributions of the electron current density, hole current density, potential, and impact ionization at three instants of time during the dynamic simulation:

before the start of the writing process (t_1= 4.9ns), during the application of the write signals (t_2=10.05ns) and after completion of the write '1' process (t_3=14.9ns). These three times are marked in Fig. 5. The MOSFET current increases as the Drain and Gate bias increase. In addition, as a result of impact ionization at the p-well/Boost junction and drain end of the channel there are excess holes in the floating p-well which is continuously charged during the switching period resulting in the increase of the drain current. Subsequently, a positive feedback mechanism kicks off between increased impact ionization and increased p-well/Source junction forward bias which results in sharp switching of the terminal currents.

To examine the effect of impact ionization on the writing process, we performed simulations with varying values of the parameter 'b' in the Okuto impact ionization model. The impact ionization coefficient in this model is given by [3]:

$$\alpha\left(\xi\left(x\right)\right) = a\xi\left(x\right)^n \exp\left[-\left(b/\xi\left(x\right)\right)^m\right] \quad (1)$$

where $\alpha\left(\xi\right)$ is the position dependent ionization coefficient, ξ is electric field, n and m are fitting parameters, a and b are linear functions of the lattice temperature. It can be readily seen from (1) that the ionization coefficient decreases with increasing values of b. Accordingly, we can observe from Fig. 7 that a very low b value leads to high impact ionization and the cell cannot sustain '0' state. On the other hand, if a very high value of b is used, the amount of impact ionization is not sufficient to achieve the required level of majority carriers in the p-well and the associated positive feedback action is not strong enough to execute stable writing. The dashed waveform corresponds to the b value used in this work.

Next, we have investigated the impact of recombination on the cell switching by changing the carrier lifetimes. Full switching simulations were performed with different electron and hole recombination rates. As can be seen from the results illustrated in Fig. 8, the switching process is greatly affected by the electron and hole lifetimes. We have considered a range of electron and hole lifetimes that are different from the default values for silicon for the purpose of investigating the bi-stability and write operation mechanism. Firstly, increased electron lifetime with hole lifetime set close to the default value leads to an undesirable increase in the currents right at the start of the read '0' period, implying that the '0' state is not stable. In other cases where the lifetimes of both the carriers are reduced, the initial currents remain low, and read '0' is executed properly. However, the write '1' and read '1' currents are significantly lower due the increased recombination rates leading to write failure for extremely short lifetimes. This demonstrates the importance of carrier lifetimes in ensuring bi-stability of the 1-transistor memory cell.

Finally, we simulated the write '1' process with different drain and gate voltages used for writing ($V_{D,W}$, $V_{G,W}$ respectively). As shown in Fig. 9, the write '1' current decreases with reduction in $V_{D,W}$ and $V_{G,W}$. Regardless, successful write '1' operation can be performed even at $V_{D,W} = V_{G,W} = 0.6$V. This shows that the 1T-SRAM cell is viable at scaled supply

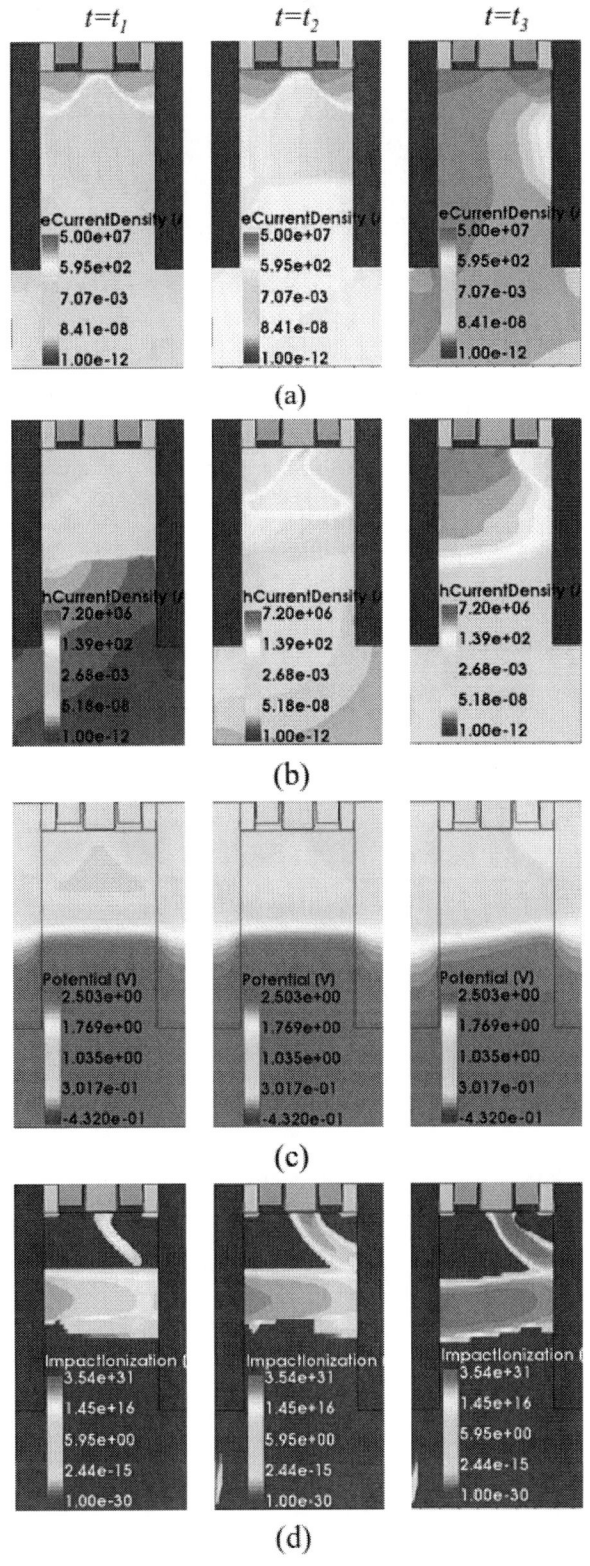

Fig. 6. Temporal snapshots of 2D distributions of (a) electron current density (b) hole current density (c) Electrostatic potential and (d) Impact ionization during the switching. V_B=2V. The three time instants (t_1, t_2, t_3) are marked in Fig. 5.

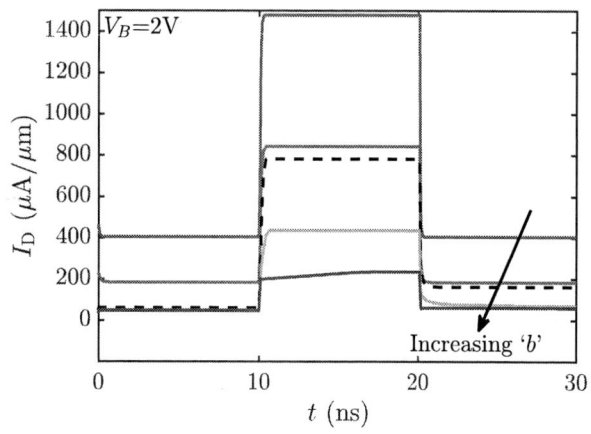

Fig. 7. Impact of the Okuto model parameter 'b' on the switching characteristics.

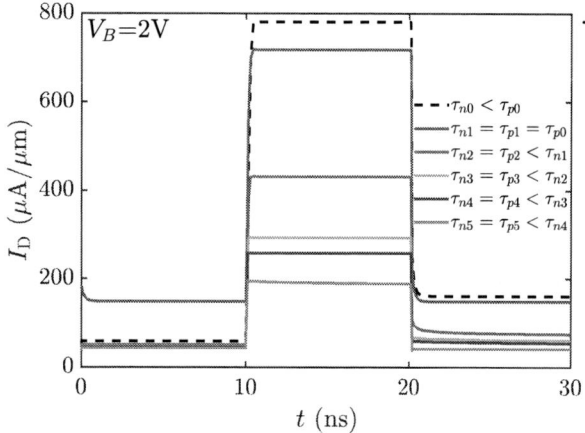

Fig. 8. Time dependent simulations for different carrier lifetimes at V_B=2V. The dashed waveform corresponds to the carrier lifetimes used in this work.

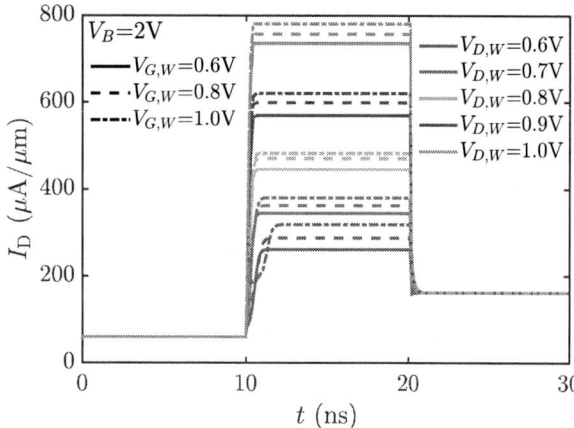

Fig. 9. Switching characteristics at V_B=2V for different values of drain and gate voltages used for writing.

voltages. The read '1' current is not affected by the amplitude of the write pulse.

IV. CONCLUSION

We have simulated and investigated the time dependent switching characteristics of the bi-stable 1T SRAM cell and have clarified the switching mechanism and the impact of generation and recombination phenomena in the switching process. We also demonstrated that writing '1' can be executed successfully at supply voltages as low as 0.6V.

REFERENCES

[1] J. Han, B. Louie, N. Berger, V. Abramzon, S. K. Lai, Z. Or-Bach, P. Lee, R. Chang, W. Lee, Y. Nishi, and Y. Widjaja, "A novel bi-stable 1-transistor SRAM for high density embedded applications," in *2015 IEEE International Electron Devices Meeting (IEDM)*, 2015, pp. 26.7.1–26.7.4.
[2] "Sentaurus process and device simulation tools," *Synopsys Inc., Mountain view, CA, USA*, 2019.
[3] Y. Okuto and C. Crowell, "Threshold energy effect on avalanche breakdown voltage in semiconductor junctions," *Solid-State Electronics*, vol. 18, no. 2, pp. 161–168, 1975.
[4] P.-F. Lu and T.-C. Chen, "Collector-base junction avalanche effects in advanced double-poly self-aligned bipolar transistors," *IEEE Transactions on Electron Devices*, vol. 36, no. 6, pp. 1182–1188, 1989.

11-6

Simulation of gated GaAs-AlGaAs resonant tunneling diodes for tunable terahertz communication applications

V. P. Georgiev, *SMIEEE*, A. Sengupta, *SMIEEE*, P. Maciazek, O. Badami, C. Medina-Bailon,
T. Dutta, F. Adamu-Lema, A. Asenov, *FIEEE*

Device Modelling Group
James Watt School of Engineering
University of Glasgow, Glasgow G12 8QQ, United Kingdom
E-mail: vihar.georgiev@glasgow.ac.uk

ABSTRACT

In this work, we report simulations on a GaAs-AlGaAs gated nanowire resonant tunneling diode (RTD) for tunable terahertz communication applications. All calculations are performed with the self-consistent Non-Equilibrium Green's Function (NEGF) quantum transport formalism implemented in our in-house Nano-Electronic Simulation Software (NESS). Our simulations successfully capture the detailed picture of the quantum mechanical effects such as quantum confinement and resonant tunneling of electrons through barriers in such structures. Moreover, we report for the first time the correlation between the gate-bias voltage and the position of the resonant peak (V_R) in the current – voltage characteristics. Such V_R, which is associated with tunneling effects in RTD, could lead to tunable terahertz generation and detection for communication applications.

Keywords: nanowire, resonant tunneling diodes, quantum device simulator, physically terahertz applications

INTRODUCTION

An emerging area related to 5G technology development and application is terahertz communications [1,2]. The resonant tunneling diode (RTD) [3,4,5] is a very promising device for wireless and optical terahertz generation and detection. The current planar RTD frequencies of operation and performance are determined by the layer design and growth of the RTD structures. However, there is a possibility to tune further the RTD behaviour by patterning and introducing gate control. Due to the simple design and the quantum phenomena associated with their operation, RTDs have been a very popular test-bed for the study of quantum mechanical effects in semiconductor devices. Theoretically, these devices have been a rather successful model for testing and benchmarking device simulation software and suites.

As for any exploratory study on electron devices and applications, computer simulation tools are the ones to pave the way for experimental device fabrication. The Device Modelling group at the University of Glasgow is actively developing a device simulation framework known as Nano-Electronic Simulation Software (NESS) [6] [7], which simulates charge transport in ultra-scaled nanoelectronic devices with various integrated modules and solvers, such as

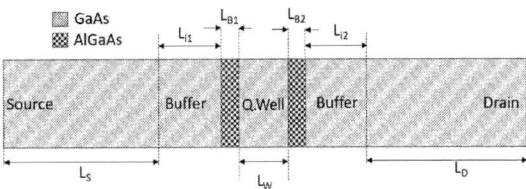

Figure 1. Schematic representation of the simulated RTD device where $L_s=L_d=19$nm, $L_{i1}=L_{i2}=3$nm, $L_{b1}=L_{b2}=3$nm and $L_w=5$nm. The barrier regions are considered to be made of $Al_{0.3}Ga_{0.7}As$. The source and drain regions are n-type doped with concentration of 2×10^{18} cm^{-3}, while in the rest of the device the doping is n-type with concentration of 1×10^{15} cm^{-3}.

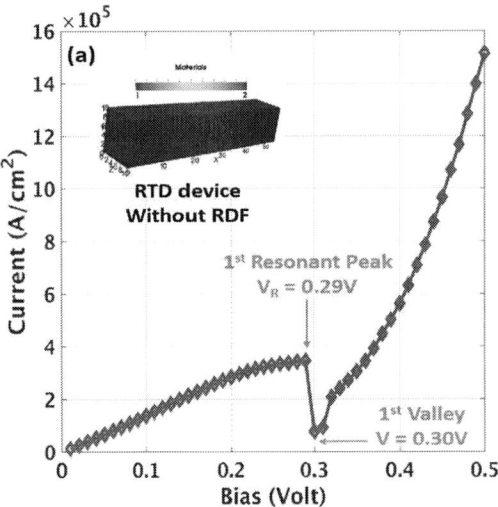

Figure 2. Current-voltage (I_D-V_D) curve for the simulated 'smooth' device and the positions on the first resonant peak – V_R. 'Smooth' device is the device without any source of statistical variability, such as Random Dopants Fluctuation (RDF). The inner plot shows the device structure implemented in NESS where material 1 is GaAs and material 2 is $Al_{0.3}Ga_{0.7}As$. The dimensions of the device in nm is also shown on the same plot. The I_D-V_D curve shows the 1st resonant peak at 0.29V drain bias and after this there is the so called negative differential resistance region (NDR) where the current drops with increasing of the bias. The first valley is 0.30V drain bias.

non-equilibrium Green's function (NEGF), drift-diffusion, and Kubo-Greenwood methods [7]. In this paper, we report a simulation study on tunable gated RTD device-based GaAs materials generated by creating two AlGaAs barriers in a nanowire structure. For the first time, we demonstrate that by applying gate voltage we can tune the RTD resonant peak position.

241

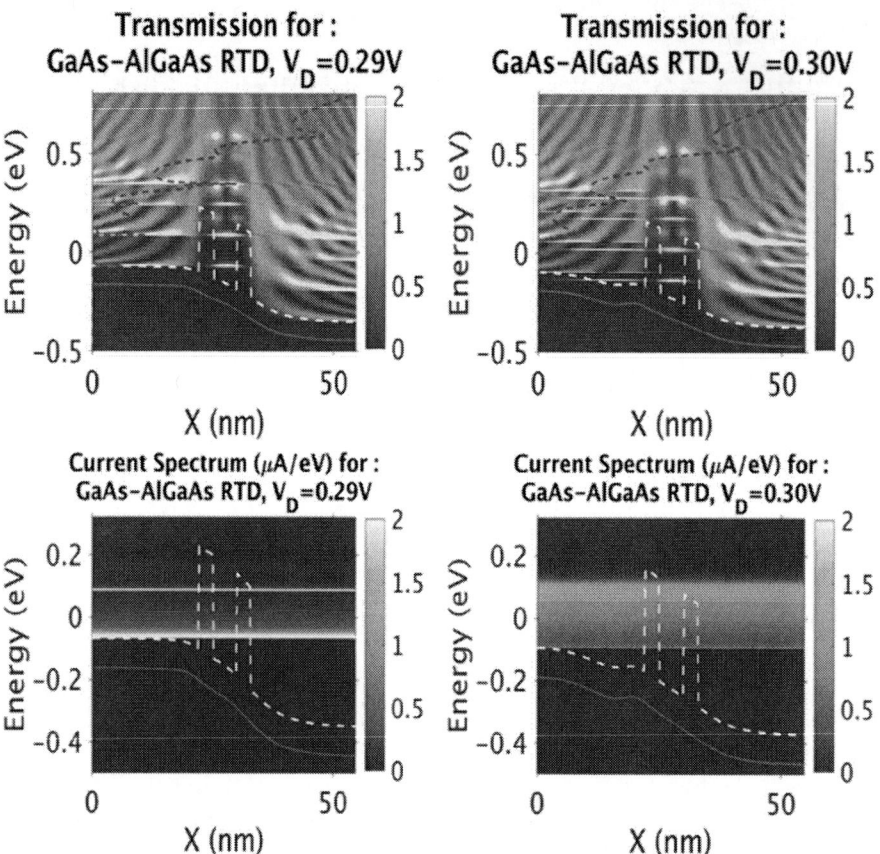

Figure 3. Top row: Local Density of States (LDOS), band diagram (white line), average potential (red solid line), transmission curves (red dashed lines (top). Bottom row: current spectra of the device at the two values of the drain voltage (V_D): $V_D = 0.29V$ and $0.30V$.

DEVICE DESCRIPTION AND SIMULATION METHODOLOGY

The schematic view of the device considered herein is shown in Fig. 1. GaAs nanowire RTD has a square cross-section of 10nm and a total length of 55nm. The dimensions with reference to Fig. 1(a) are $L_s=L_d=19nm$, $L_{i1}=L_{i2}=3nm$, $L_{b1}=L_{b2}=3nm$, and $L_w=5nm$. The barrier regions are considered to be made of $Al_{0.3}Ga_{0.7}As$. The source and drain regions are considered n-type doped with concentration of 2×10^{18} cm^{-3}, whereas n-type concentration of 1×10^{15} cm^{-3} is taken in the rest of the device.

The simulation results presented herein are obtained using the NEGF solver implemented in the NESS framework [7]. For carrier transport, the recursive Green's function algorithm is employed considering effective mass approximation in the mode-space. The quantum transport is carried out self-consistently with a numerical Poisson solver. The total carrier density determines the new potentials which are then fed back into the NEGF solver for a new current and charge distribution within the active region of the device. The self-consistent loop is repeated until the specific current convergence criterion is reached. For more details about the NESS software and capabilities please see [7].

RESULTS AND DISCUSSION

The output characteristic of the nanowire RTD provided in Fig. 2 shows the first resonant tunneling peak having a current value of 3.8×10^5 A/cm^2 at a bias voltage of 0.29V, followed by a negative differential resistance region where the current drops to 7.7×10^4 A/cm^2 and thereafter increases. The results for the nanowire RTD device are qualitatively consistent with theoretical and experimental reports of bulk-devices made of similar materials, with similar thickness and doping profiles [3,4,5]. The LDOS plots shown in Fig. 3 for the resonance condition show the alignment of the quantum well energy levels with the source states at -0.068, 0.088 and 0.346eV for the resonance voltage (V_R) of 0.29V. When the LDOS of the source and quantum dot are aligned energetically, the transmission curves (the red dashed lines in Fig. 3) reveal sharp transmission peaks. This is an indication of tunneling of electrons trough the barriers from the source to the quantum dot.

This is in agreement with the current spectra, which is shown in the bottom part of Fig. 3, and it has two very well pronounced channels at the same energy where the transmission peaks are located. From Fig. 3, it is evident that the current is tunneling through the barrier. The LDOS alignment between the states from the source and the quantum dot is more disrupted as the bias changes to 0.30V. The current spectra in Fig. 3 also shows a more smeared nature at 0.30V as compared to that at the resonance peak. Furthermore, the first two peaks of the transmission spectra are of much lower magnitude. This is also clearly visible in the 1D transmission plots in Fig. 3. This is due to the fact that the LDOS in the source and the quantum dot region are not aligned anymore. As a result, there is around one-order of magnitude decrease of the current from 0.29V to 0.30V.

242

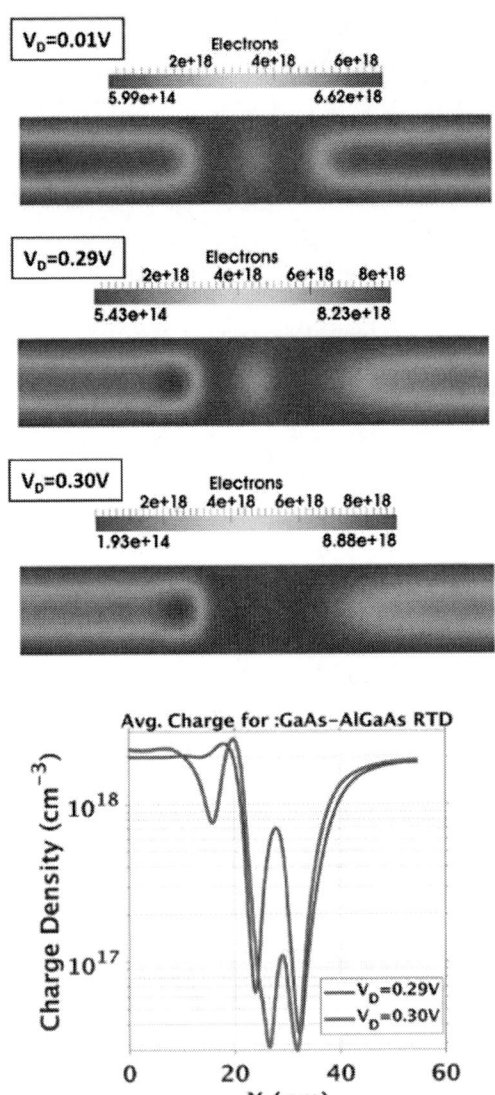

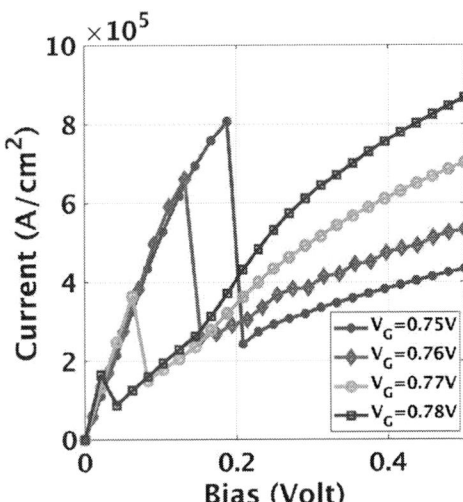

Figure 5. I-V characteristics of the proposed gated RTD with an ideal work-function matched gate material showing the modulation of the resonance peak position with applied gate voltage.

Figure 4. Top: charge density plots of the device at bias of 0.01V (top), 0.29V (middle), 0.30V (bottom); Bottom: the average 1-D charge along the transport directions of the device for V_D = 0.29V and 0.30V.

Top part of Fig. 4 shows the 2D plots of the charge distribution in the XY cut-plane in the middle of the device at different bias conditions. In particular, it shows how the charge distribution in the device changes as a function of the applied bias and the quantum well gets charged up to the resonance voltage of 0.29V followed by a discharge at 0.30V. This is consistent with the data presented in Fig. 3 where it is clear that, at 0.29V, there are very pronounced transmission peaks at around -0.1 eV and 0.1eV. At 0.30V drain bias, due to change of the alignment between the LDOS in the contacts and the quantum well, the electrons cannot tunnel through the barrier and this leads to a significant drop in the current.

This is also manifested in the 1D profile of the charge density presented in Fig. 4. The charge inside of the well decreases with around one order of magnitude when the drain bias increases from 0.29V to 0.3V and, more interestingly, there is a significant charge drop just before the first barrier. The drop corresponds to the constructive interference coming from the

wavefunctions that are reflected from the barrier because they cannot tunnel through. With the results simulated in NESS showing qualitative consistency in predicting output characteristics and capturing quantum phenomena in the RTD structure, we study the impact of the ideal work-function matched gate on the resonance voltage. For this purpose, a 2nm thick SiO_2 gate oxide was considered wrapped around the nanowire under an ideal surround gate of 17nm in length (covering the buffer regions, the barriers and the quantum well).

From the simulation results presented in Fig. 5, it can be concluded that for a gate work function of 4.07eV, the RTD behaviour may be only tuned on for voltages above 0.75V. Subject to a slight variation of V_G in the ON condition, there can be a significant shift in V_R. As V_G is changed from 0.75 to 0.78V, the value of V_R is drastically shifted from 0.185 V to 0.02V. Such wide variation of the resonance peak with small changes in the gate voltage and the ability to control the RTD behaviour with gate bias advocates the potential use of such devices for tunable terahertz communication applications. The position of the resonant peak is determined by the alignment of the LDOS in the source and the quantum dot. The latter can be controlled by the gate voltage and therefore the RTD behaviour can be tuned.

Fig. 6 reveals simulations of 125 RTD devices, each of them with a unique distribution of random dopants fluctuation (RDF), introduced in the device. From this data, it is clear that there is a direct correlation between the number and the position of RDFs in the device, the position of the V_R, and the current-voltage characteristics (I_D-V_G). The key figures of merit that can be extracted from the I_D-V_G simulations are presented in Fig. 7. Fig 7. (a) reveals the correlation between V_R and the number of dopants. It can be concluded that devices with the same number of dopants have a huge variation of V_R. For example, devices with 10 dopants have variation of V_R from 0.05V all the way up to 0.61V. This can be explained by the specific position of each dopant in the 3D volume of the device.

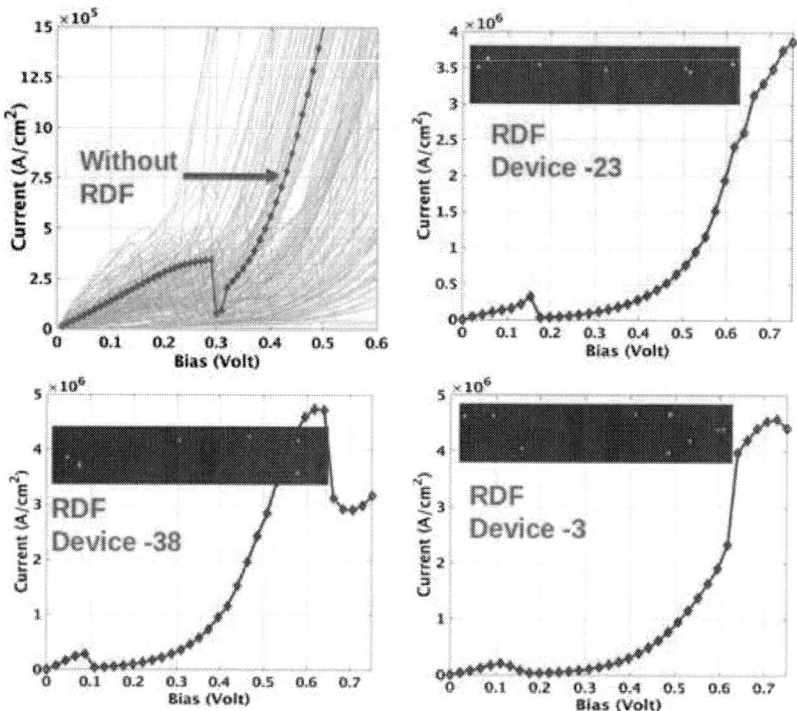

Figure 6. Current-voltage (I_D-V_D) plots of the devices with the random dopant distribution (RDD). The device without any RDD (smooth device) is highlighted in orange for comparison purposes. I_D-V_D characteristics of RDF devices labelled as (b) device 23 (c) device 38 and (d) device 3 with inset device screenshots, where the exact location of the RDD in the structure can be seen as the grey spheres.

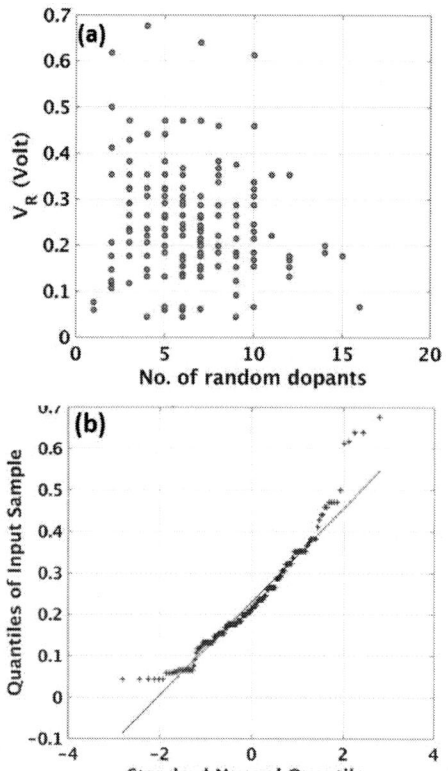

Figure 7. (a) Distribution of the first resonant peak voltage (V_R) for the 125 devices as a function of the number of the random dopants. (b) Quantile-quantile (Q-Q) plot for the V_R data.

Fig. 7. (b) is a QQ plot that shows the distribution of V_R for all 125 RTD devices. It shows that there is a significant deviation from the normal distribution (red line) in the tail regions. This plot shows that it is not possible to use analytical models to describe the behaviour of automictically different RTDs.

CONCLUSION

Our work shows that quantum mechanical effects have a significant impact on the RTD devices. There is a direct link between the position of the resonance peak and the number and the position of the random dopants in the device. Indeed, these unique relationship opens a possibility for these devices to be used for tunable terahertz communication applications.

REFERENCES

[1] M. Feiginov, "Frequency Limitations of Resonant-Tunnelling Diodes in Sub-THz and THz Oscillators and Detectors," J. Infrared, Millimeter, Terahertz Waves, pp. 365–394, 2019.

[2] K. Kasagi, S. Suzuki, and M. Asada, "Large-scale array of resonant-tunneling-diode terahertz oscillators for high output power at 1 THz," J. Appl. Phys., vol. 125, no. 15, 2019.

[3] B. Biegel and J. Plummer, "Comparison of self-consistency iteration options for the Wigner function method of quantum device simulation," Phys. Rev. B - Condens. Matter Mater. Phys., vol. 54, no. 11, pp. 8070–8082, 1996.

[4] B. A. Biegel and J. D. Plummer, "Applied bias slewing in transient Wigner function simulation of resonant tunneling diodes," IEEE Trans. Electron Devices, vol. 44, no. 5, pp. 733–737, 1997.

[5] K. L. Jensen and F. A. Buot, "Numerical simulation of intrinsic bistability and high-frequency current oscillations in resonant tunneling structures," Phys. Rev. Lett., vol. 66, no. 8, pp. 1078–1081, 1991.

[6] S. Berrada, et al., "Quantum Transport Investigation of Threshold Voltage Variability in Sub-10 nm JunctionlessSi Nanowire FETs," 2018 International Conference on Simulation of Semiconductor Processes and Devices (SISPAD): Sept. 2018.

[7] S. Berrada et al., Nano-electronic Simulation Software (NESS): a flexible nano-device simulation platform. Journal of Computational Electronics, 19, pp. 1031—1046, 2020.

11-7   **Theoretical Study of Double-Heterojunction AlGaN/GaN/InGaN/δ-doped HEMTs for Improved Transconductance Linearity**

Tsung-Hsing Yu

Inforsight Computing
E-mail: thyu@inforsight-computing.com

Abstract—The aim of this study is to propose a novel double-heterojunction high electron mobility transistor (DH-HEMT) structure, $Al_{0.3}Ga_{0.7}N/GaN/In_{0.15}Ga_{0.85}N/δ$-doped, to improve transconductance linearity. A theoretically based quasi-two-dimensional model is well calibrated with experiments and is used to project the transistor performance. It is found that a thin $In_{0.15}Ga_{0.85}N$ back barrier and δ-doped layer significantly enhance carrier confinement and increase carrier concentration in the channel. It is the combination effect of enhanced carrier confinement and increased carrier concentration that leads to a larger voltage swing. A wider linear range of transconductance can be achieved on account of the larger voltage swing. Moreover, this novel structure not only improves the transconductance linearity but also increases its maximum transconductance and the corresponding drain current, which is beneficial to high power and high frequency applications.

Keywords—transconductance; g_m; linearity; HEMT; double-heterojunction; InGaN; δ-doped; polarization

I. INTRODUCTION

Due to the superior material properties such as high breakdown voltage, high saturation velocity, and high thermal conductivity, III-Nitride devices are expected to offer better high frequency, high power, and high temperature performance compared to conventional Si and GaAs devices. However, GaN high electron mobility transistors (HEMTs) have been found that their transconductance g_m, current gain, and power gain drop off at high current levels [1]. To solve this issue, several methods have been presented and reported in the literature [1-4]. These are (1) distributed three-dimensional (3D) channel electrons induced by a highly scaled graded channel [1], (2) transverse electric field reduction through the use of a composition channel [2], (3) enhanced electron confinement with InGaN back barriers [3-4]. Though much progress has been made by these researchers, to the author's knowledge this is the first theoretical study of a novel structure $Al_{0.3}Ga_{0.7}N/GaN/In_{0.15}Ga_{0.85}N/δ$-doped double-heterojunction HEMT (DH-HEMT) to improve transconductance linearity.

Since polarization fields can induce a larger sheet charge and alter the band bending at the GaN/InGaN heterointerface, the carrier confinement in the DH-HEMT is significantly enhanced. Furthermore, an inserted δ-doped layer at the backside of the InGaN layer plays an important role in supplying carriers to the channel. Both enhanced carrier confinement and carrier increment allow a better modulation of the channel electrons by the gate voltage, which effectively improve transconductance linearity. Generally, the approaches to improve linearity usually generate undesirable features such as g_m and power gain reduction at high current levels [2]. However, by the combined action of the δ-doped layer and InGaN back barrier, the proposed DH-HEMT not only increases g_m and its corresponding power gain but also raises driving current.

II. TRANSISTOR DESIGN

The proposed DH-HEMT structure with a gate length of

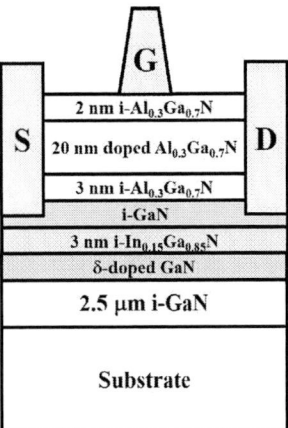

Fig.1 Schematic cross section of the proposed double-heterojunction HEMT.

0.3μm is schematically illustrated in fig.1, which is used in the following simulation. In fig.1, the channel region is composed of an undoped GaN channel, a 3-nm $In_{0.15}Ga_{0.85}N$ back barrier layer, and a 1-nm δ-doped layer with doping concentration of $5x10^{19}$ cm^{-3}.

Instead of being used as the major channel, the $In_{0.15}Ga_{0.85}N$ layer is inserted below the GaN channel. The 3-nm thickness and 15% Indium composition in the InGaN layer are designed to achieve two goals. (1) The conduction band offset and strain-induced polarization charge are 391meV and $1.39x10^{13}$ cm^{-2}, respectively. The strain-induced polarization field in the InGaN layer raises the conduction band profile of the GaN buffer, thus significantly enhancing the carrier confinement. (2) The thickness of the $In_{0.15}Ga_{0.85}N$ layer is designed as thin as 3-nm to reduce the risk of indium cluster formation [5]. In addition, Si δ-doping in a GaN layer is able to reach a sheet charge density as high as $1x10^{12}$ cm^{-2} to $2x10^{13}$ cm^{-2} [6]. Therefore, a δ-doped layer is designed to insert at the backside of the InGaN layer and is treated as a carrier supplier.

III. MODEL CALIBRATION

It is clear that the polarization fields play a critical role in AlGaN/GaN HEMTs and are the primary source of the sheet charge density in the devices. Therefore, in order to model the sheet charge density correctly in the heterostructure under an applied gate voltage, the Poisson and Schrodinger equations are solved self-consistently at the heterojunction. The 3D free carriers and the partial neutralization of donors are directly taken into account in place of the usual assumption of fully ionized donors in the doped layer. Moreover, the polarization effects can be incorporated into the model by the boundary condition at the heterointerface. That is the electric displacement must be continuous at the heterointerface [7]. The field-dependent mobility in the GaN HEMT is obtained from an analytical formula extracted from Monte Carlo simulation [8]. The calculation of drain-current voltage characteristics is made using a quasi-two-

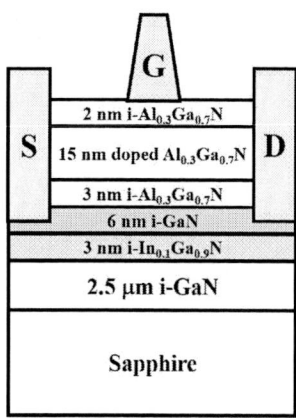

Fig.2 Schematic cross section of a GaN HEMT with a thin InGaN back barrier in ref [4] is used for model calibration.

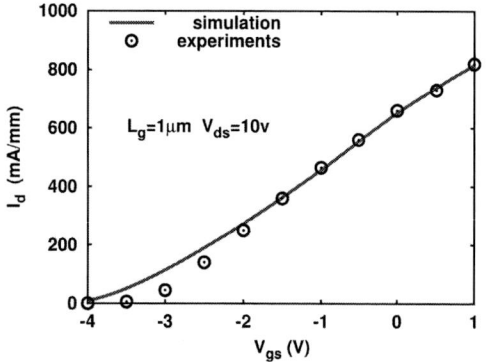

Fig.3 Comparison of I_d versus V_{gs} characteristics between simulation and measurements (symbols) for the $Al_{0.3}Ga_{0.7}N/GaN/In_{0.1}Ga_{0.9}N$ HEMT. The experimental data are from ref. [4].

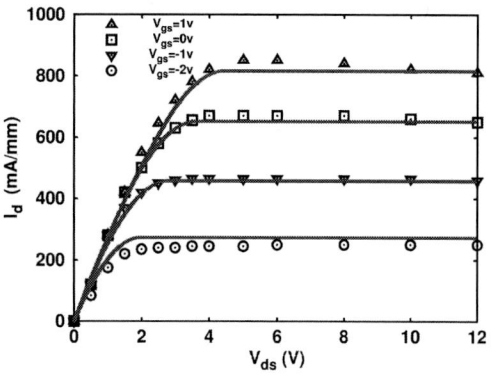

Fig.4 Calculated current-voltage characteristics (solid lines) compared to experimental data (symbols) at various gate biases for the $Al_{0.3}Ga_{0.7}N/GaN/In_{0.1}Ga_{0.9}N$ HEMT structure in ref.[4].

dimensional model [7] based on the self-consistent charge control and field-dependent mobility models.

To calibrate this model, a GaN HEMT with a thin $In_{0.1}Ga_{0.9}N$ back barrier structure [4] as schematically shown in fig.2 is used to simulate and compare to experiments. Excellent agreement between theory and experiments for both I_d-V_{gs} and I_d-V_{ds} characteristics is obtained over the full range of applied gate biases as can be seen from figs.3-4. The field-dependent mobility parameters and drain and source contact resistance are used to adjust to fit the measurements. The relevant polarization and material parameters are obtained from ref. [9].

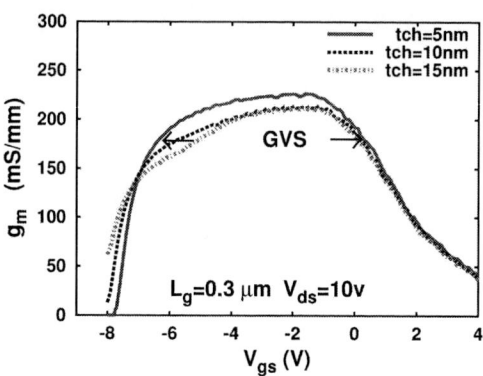

Fig.5 Calculated transconductance for $Al_{0.3}Ga_{0.7}N/GaN/In_{0.15}Ga_{0.85}N/\delta$-doped HEMTs under three different GaN channel thickness: tch=5nm,10nm, and 15nm.

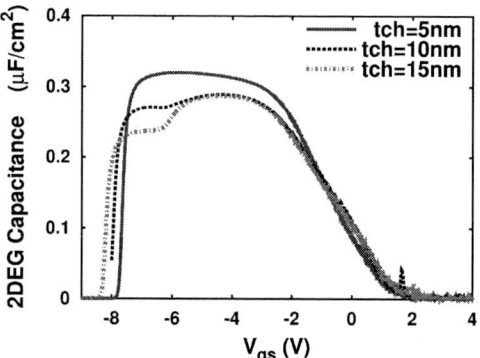

Fig.6 Calculated 2DEG capacitance under the gate in $Al_{0.3}Ga_{0.7}N/GaN/In_{0.15}Ga_{0.85}N/\delta$-doped HEMTs. As tch increases to 10nm and15nm, the minor channel formed in the InGaN layer leads to double-hump behavior.

IV. GaN Channel Thickness Optimization

Gate voltage swing (GVS) defined as the range of the gate voltage where the transconductance drops 80% of its peak value [6] is adopted to quantitatively measure the transconductance linearity. The channel thickness, tch, is defined as the distance between AlGaN and InGaN heterojunctions. Fig.5 plots the transconductance variations as the channel thickness, tch, increases from 5nm to 10nm and15nm. In fig.5, their corresponding GVS values decrease slightly from 6.6v to 5.8v. To further investigate the carrier dynamic behavior with respect to the gate voltage sweep in the channel, the calculated two-dimensional electron gas (2DEG) capacitance per unit area is visible in fig.6. The 2DEG capacitance is calculated by differentiating the sheet charge density under the gate with respect to the applied gate voltage. As the gate voltage increases, the capacitance increases quickly because of the band bending and an increase in the confined electrons induced by the gate voltage. Once the 2DEG is formed, the effective position of the 2DEG from the heterointerface is no longer noticeably influenced. Hence, the capacitance remains almost constant around the maximum capacitance.

However, the 2DEG capacitance in fig.6 clearly illustrates that there are two humps in the HEMTs with tch=10nm and tch=15nm. This is because two separated channel are formed in the GaN channel and InGaN layer, respectively. The major part of the channel is formed by the GaN channel between AlGaN and InGaN heterojunctions. The minor part of the channel is generated from the InGaN layer. This double-hump characteristics result from two separated

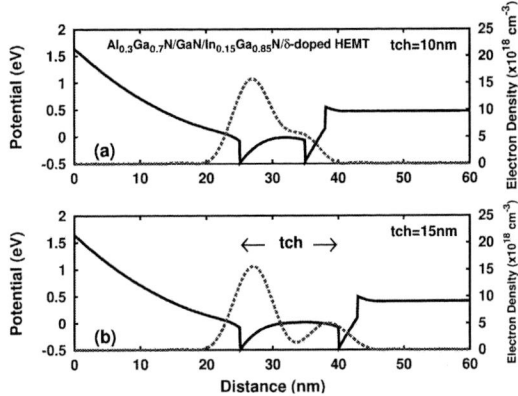

Fig.7 Conduction band profiles and electron density at a gate bias of zero voltage for GaN channel thickness (a) tch=10nm and (b) tch=15nm. The Fermi-level is set to zero.

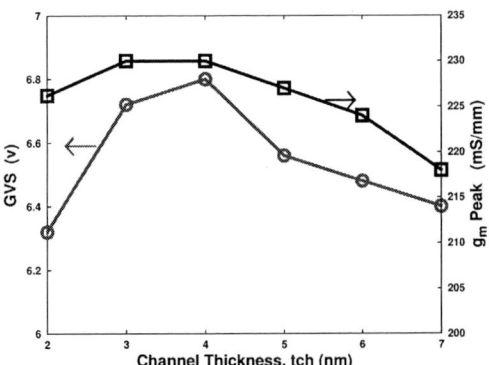

Fig.8 Channel thickness, tch varies from 2nm to 7nm and their corresponding GVS (circle symbols) and g_m peak (square symbols). With tch=4nm, a DH-HEMT has the highest GVS and g_m peak.

channels and lead to a secondary transconductance peak that degrades the device's linearity [4][6]. Fig.7 gives the example of two separated channels from the self-consistent results of DH-HEMTs. The conduction band profile and confined carrier concentration are illustrated in fig.7(a) channel thickness tch=10nm and fig.7(b) tch=15nm, respectively. It is noted that the Fermi-level is set to a reference level and set equal to zero in fig.7. Also the applied gate bias is 0v. Fig.8 displays the channel thickness optimization results. It is observed that with tch=4nm, a DH-HEMT has the highest GVS and its corresponding g_m peak. As a result, a 4-nm thick GaN channel is employed in the following HEMT discussions.

V. COMPARISON BETWEEN PROPOSED AND CONVENTIONAL HEMTs

A comparison of the dc transfer characteristics for the proposed HEMT and a conventional one is shown in fig.9. The drain-to-source voltage is biased at 10V. The maximum transconductance of 230 mS/mm in the proposed device with a Vgs=-1.8v is larger than that of 203 mS/mm in the conventional HEMT with a Vgs=-2.4v. The maximum g_m is influenced by two factors: capacitance due to 2DEG and low-field mobility. The 2DEG capacitance demonstrates how much charge can be supplied to the 2D channel with respect to a variation of the gate voltage. The excess supply charge directly contributes to an increase of the drain current. In addition to the modulation of charge in the 2D channel, the electrons in the channel are not all traveling at the saturated or peak electron velocity except those at the

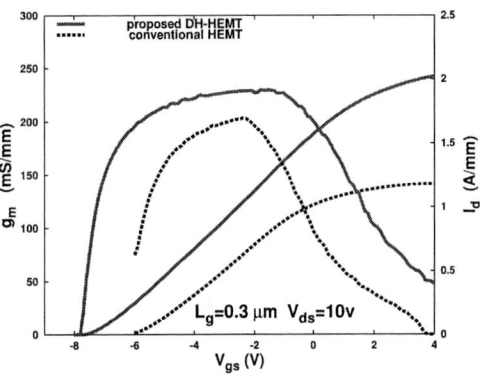

Fig.9 DC transfer characteristics comparison between the proposed DH-HEMT and a conventional HEMT. The proposed HEMT significantly improves linearity compared to the conventional one.

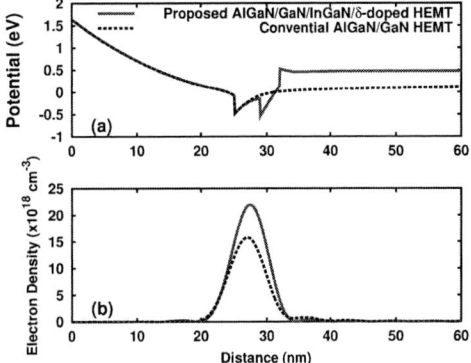

Fig.10 Two major features cause the linearity improvement in the proposed HEMT structure. (a) Raise the conduction band profile in the GaN buffer layer. (b) Increase electron density in the channel. The applied gate bias is 0v and the Fermi-level is set to zero.

drain edges. Hence, g_m is also proportional to the low-field mobility. To simplify the calculation, the same magnitude of low-field mobility is assumed in fig.9 simulation. Therefore, the g_m in fig.9 indicates that the modulation efficiency of 2D channel charge is significantly improved in the proposed device. The maximum drain current of 2023 mA/mm at a gate bias of 4.0v and a drain bias of 10v in the proposed HEMT is 72% larger than that of the conventional device with the same bias.

As can be seen from fig.9, the proposed device's transconductance with respect to that in the conventional HEMT is quite flat and remains close to its peak value at high current levels. Their GVS values are 6.8v and 3.8v, respectively. The substantial linearity improvement results from two reasons: One, the carrier confinement is enhanced by the raised conduction band induced by the InGaN layer as evident from the fig.10(a). Two, the δ-doped layer significantly increases 2D carrier density in the channel as illustrated by fig.10(b).

VI. INGAN AND Δ-DOPED EFFECTS

Fig.11 reveals the calculated sheet carrier concentration as a function of gate voltage under three epilayer conditions: the proposed DH-HEMT, without δ-doped, and without InGaN layer structures, respectively. As the gate voltage increases, the free electrons in the 2D channel tend to go into the doped layer as 3D free electrons or to neutralize the donors. As a result, the sheet carrier concentration in the 2D channel becomes saturated, which limits the maximum

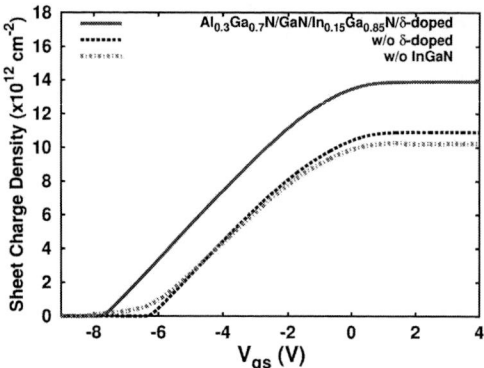

Fig.11 Calculated sheet charge density in 2D channel for a DH-HEMT under three different layer conditions: the proposed DH-HEMT, without δ-doped, and without $In_{0.15}Ga_{0.85}N$ layer structures.

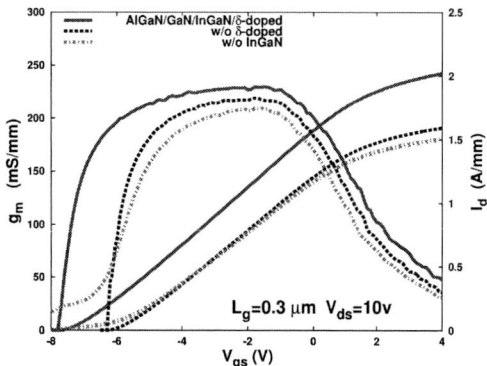

Fig.12 DC transfer characteristics for a DH-HEMT under three different layer conditions: the proposed DH-HEMT, without δ-doped, and without $In_{0.15}Ga_{0.85}N$ layer structures.

voltage swing in the HEMTs. The δ-doped layer in the proposed HEMT demonstrates a 29% of the sheet charge density increase compared to the HEMT without this layer at a gate bias of 0v in fig.11. Similarly, poor carrier confinement cannot sustain larger charge density in the channel as shown in the device without InGaN case. Therefore, the δ-doped layer, together with a thin InGaN layer, obtains a wide linear range of transconductance as can be seen from fig.12. Also, this wider linear range of transconductance results from the wider voltage swing in fig.11. Consequently, The effects of the carrier increment and enhanced carrier confinement act on the channel simultaneously to enlarge a wider voltage swing corresponding to the g_m linearity improvement.

In addition, the influence of AlGaN top barrier thickness on g_m is given in fig.13. Two top barrier thickness of 25nm and 18nm are examined in this plot. As the top barrier thickness decreases, the modulation of the 2D channel charge by the gate voltage is increased. Therefore, the g_m in 18nm thickness is higher than that in 25nm case. Hence, the AlGaN thickness can be optimized to obtain the best trade-off between power gain and linearity.

VII. CONCLUSIONS

This article presents a new DH-HEMT structure to improve transconductance linearity. The simulation model is well calibrated with the experiments indicating that the

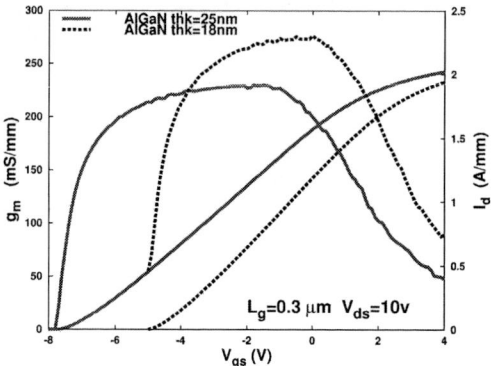

Fig.13 Top barrier AlGaN thickness impacts on DC transfer characteristics. Thinner top barrier thickness increases g_m but its corresponding GVS becomes small.

model can be successfully employed to project DH-HEMT performance. It is found that to maintain transconductance linearity at high current levels, the conducting channel needs to sustain sufficient larger 2D charge. This can be achieved by using the InGaN and δ-doped layers together to enhance carrier confinement and increase carrier density in the channel. These two factors result in a larger sheet charge in the channel. To deplete this larger sheet charge at the heterointerface, the threshold voltage in the HEMT device becomes more negative, corresponding to a larger voltage swing. Moreover, compared to the conventional HEMT, the transconductance in the proposed device decreases slowly as the gate voltage switches to a forward bias. This also indicates that the proposed DH-HEMT can be employed in a larger voltage swing. Consequently, the proposed DH-HEMT exhibits the potential for high power and high frequency applications.

REFERENCES

[1] S. Bajaj, Z. Yang, F. Akyol, P. S. Park, Y. Zhang, A. L. Price, S. Krishnamoorthy, D. J. Meyer, and S. Rajan, "Graded AlGaN Channel Transistors for Improved Current and Power Gain Linearity," IEEE Trans. Electron Devices, vol. 64, no. 8, pp. 3114-3119, Aug. 2017.

[2] J. Liu, Y. Zhou, R. Chu, Y. Cai, K. J. Chen, and K. M. Lau, "Highly Linear $Al_{0.3}Ga_{0.7}N$-$Al_{0.05}Ga_{0.95}N$-GaN Composite-Channel HEMTs," IEEE Electron Device Lett., vol. 26, no. 3, pp. 145-147, Mar. 2005.

[3] T. Palacios, A. Chakraborty, S. Heikman, S. Keller, S. P DenBaars, and U. K. Mishra, "AlGaN/GaN High Electron Mobility Transistors with InGaN Back-Barriers," IEEE Electron Device Lett., vol. 27, no. 1, pp. 13-15, Jan. 2006.

[4] J. Liu, Y. Zhou, J. Zhu, Y. Cai, K. M. Lau, and K. J. Chen, "DC and RF Characteristics of AlGaN/GaN/InGaN/GaN Double-Heterojunction HEMTs," IEEE Trans. Electron Devices, vol. 54, no. 1, pp. 2-9, Jan. 2007.

[5] J. Liu, Y. Zhou, J. Zhu, K. M. Lau, and K. J. Chen, "AlGaN/GaN/InGaN/GaN DH-HEMTs with an InGaN Notch for Enhanced Carrier Confinement," IEEE Electron Device Lett., vol. 27, no. 1, pp. 10-12, Jan. 2006.

[6] C. Tang, K. H. Teo, and J. Shi, "Simulation of GaN HEMT with Wide-Linear-Range Transconductance," in Proc. Electron Device and Solid-State Circuit Conference, 2017

[7] T.-H. Yu and K. F. Brennan, "Theoretical Study of a GaN-AlGaN High Electron Mobility Transistor Including a Nonlinear Polarization Model," IEEE Trans. Electron Devices, vol. 50, no. 2, pp. 315-323, Feb. 2003.

[8] T.-H. Yu and K. F. Brennan, "Monte Carlo Calculation of Two-dimensional Electron Dynamics in GaN-AlGaN Heterostructures," J. Appl. Phys., vol. 91, pp. 3730-3736, Mar. 2002.

[9] C. Wood and D. Jena, Polarization Effects in Semiconductors, Springer, 2008

11-8 # Nanoscale FET: How To Make Atomistic Simulation Versatile, Predictive, and Fast at 5nm Node and Below

Philippe Blaise
Silvaco Inc.
Montbonnot, 38330 France
philippe.blaise@silvaco.com

Udita Kapoor
Silvaco Inc.
Santa Clara, CA 95054, USA

Mark Townsend
Silvaco Inc.
Santa Clara, CA 95054, USA

Eric Guichard
Silvaco Inc.
Santa Clara, CA 95054, USA

James Charles
School of Electrical and Computer
Engineering
Purdue University
West Lafayette, IN 47907, USA

Daniel Lemus
School of Electrical and Computer
Engineering
Purdue University
West Lafayette, IN 47907, USA

Tillmann Kubis
School of Electrical and Computer
Engineering
Purdue University
West Lafayette, IN 47907, USA

Abstract—**Ultra-scaled FET technology requires simulations at the atomic scale. We present the Victory Atomistic tool inherited from Nemo5. Thanks to a combination of non-equilibrium Green's functions and state-of-the-art band structure calculations, versatile, predictive, and fast simulations become accessible within the self-consistent Born approximation, optimized by a generalized low-rank projection.**

Keywords—*Atomistic, Simulation, CMOS, FET, NEGF, Scattering, Tight-Binding*

I. INTRODUCTION

Field Effect Transistor (FET) miniaturization is the workhorse of the microelectronic industry. The FET physical dimensions continue to shrink to five nm node and below, characterized by new types of architectures with nanosheet (NS) and nanowire (NW) shapes [1]. The present choice of material is made of Si, Ge, or SiGe alloy due to their high carrier concentrations. In compliment to III-V technology envisaged for a while, new 2D materials are also investigated, for example the TMDs monolayers. Such nanomaterials and nano-architectures require atomistic simulations for at least two crucial reasons: 1) bulk parameters like the effective masses and forbidden bandgap are no longer pertinent quantities, and 2) the wave nature of charge carriers becomes predominant for predicting I(V) characteristics. Non-equilibrium Green's functions (NEGF) allow us to perform quantum transport simulation of such a device made of two leads and an active region, where carriers move according to a custom band structure [2]. In practice, one needs to build the device's Hamiltonian with a suited set of wave functions, in order to solve the NEGF and Poisson-Schrödinger coupled equations. This comes at the price of a high computational burden, which becomes almost untractable when scattering self-energies of inelastic carrier interactions are included [3]. Thankfully, by generalizing a low-rank approximation (LRA) with a mode space scheme [4], we show how the calculation load can drastically be reduced while preserving accuracy.

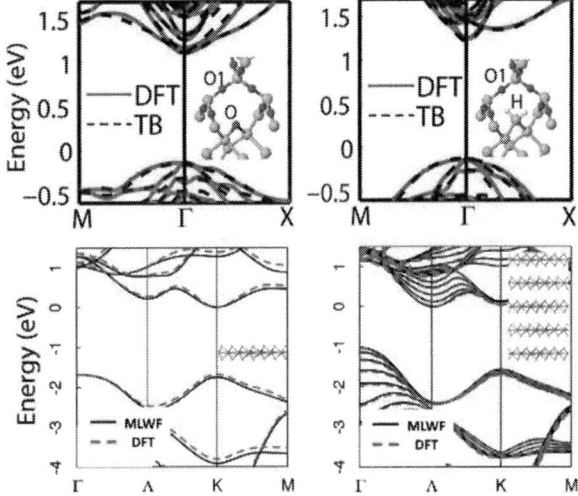

Fig. 1. Top: Tight Binding band structure of Si reproducing the DFT results of the influence of the Si/SiO_2 interface – with a bridging oxygen at the interface (left), with H passivation of Si dangling bonds (right). Bottom: MLWF band structure of MoS_2 TMD reproducing DFT results – for one monolayer, direct gap (left), for five monolayers, indirect band gap (right)

II. ATOMISTIC SIMULATION USAGE

A. Materials and Electronic Bandstructure

One of the most efficient ways to describe the band structure of a material by taking into account its atomic arrangement is the Slater-Koster (SK) tight-binding (TB) model [5]. The set of SK-TB parameters is usually based on accurate experimental measurements and can be extended to state-of-the-art density functional theory (DFT) data, including the most accurate results of hybrid functionals. Passivation chemistry, species alloying, specific mechanical stress can then be included [6]. Eventually, matrix elements of new materials like TMDs 2D layers, topological insulators,

249

and conductive oxides are derived thanks to maximally localized Wannier functions (MLWF) obtained in hybrid-DFT [7]. As an illustration, band structure results with efficient parametrizations of silicon in contact with SiO_2 and of MoS_2 are shown in Fig. 1.

B. Quantum Transport

We simulate a Si NW of different square sections from 2x2 nm^2 to 5x5 nm^2. The contacts are made of highly-doped Si with an intrinsic channel surrounded by SiO_2, see Fig. 2. NEGF calculations are performed using a recursive Green's function [2] and an $sp^3d^5s^*$ Si TB custom basis [6]. I(V) simulations go from ballistic transport up to inelastic scattering by including the self-energies of acoustic and optical phonons, evaluated in the self-consistent sc-Born approximation. This requires computing the G^R (retarded) and $G^<$ (lesser) Green's functions [8]. Note that Victory Atomistic allows treating explicit electrostatic boundary conditions, Ohmic or Schottky contacts, and systematic passivation of the Si dangling bonds.

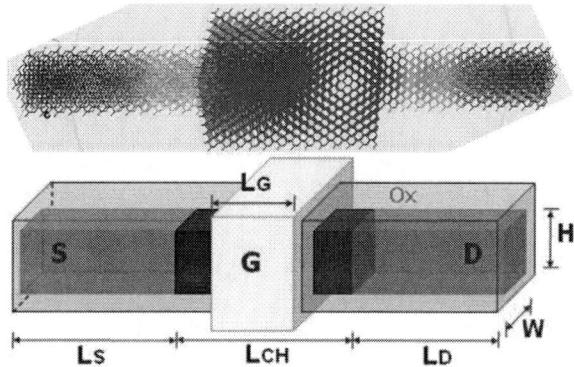

Fig. 2. Si nanowire FET structure of square section. Top: atomistic structure of a 2x2 nm2 nanowire with source and drain of six nm length, and seven nm for the grid. Bottom: drawing of the NW geometry and various characteristic lengths.

C. Optimization Strategy

Roughly speaking, the order N of NEGF matrices comes as the product of the number of atoms times the size of the TB basis. In order to reduce the matrix operations of $O(N^3)$ complexity, the scattering mechanisms are often neglected, relegating such simulations to a pure ballistic regime. For keeping the sc-Born approximation, the system's Hamiltonian is reduced by filtering of degrees of freedom that are unlikely to contribute to device operation. This way, our generalized LRA reduces the rank N of the system by a factor of ten or more [9]. We performed a benchmark by qualifying the method against full calculations, see Fig. 3. The obtained speed-up of more than two orders of magnitude is detailed in Table 1.

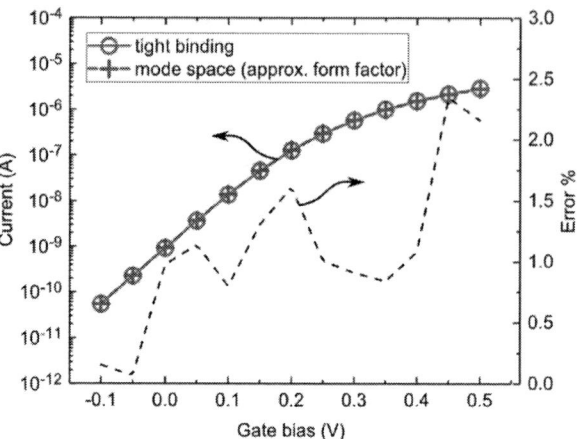

Fig. 3. Benchmark of low rank approximation in mode space including sc-Born inelastic scattering for a nanowire of 3x3 nm^2 and 20 nm length. The LRA (in red) matches the I(V) curve of full band calculation (in blue), with relative error (in black).

TABLE I.   SIMULATION TIMES (S) FOR ONE SC-BORN ITERATION FOR VARIOUS SECTIONS, IN FULLBAND AND LRA, WITH SPEEDUP.

Method	Size nm^2			
	2x2	3x3	4x4	5x5
Full Band	150	990	4200	17200
LRA	8	9	30	80
Speedup	19	110	140	215

III. SILICON NW FET

Thanks to the LRA approximation, we can study the quantum transport of a silicon nanowire FET transistor, with an affordable computer.

We start with a 2x2 nm^2 structure, see Fig. 2, with a gate-all-around (GAA) structure, for which we draw the I(V) characteristics of the ballistic mode, as shown in Fig. 4.

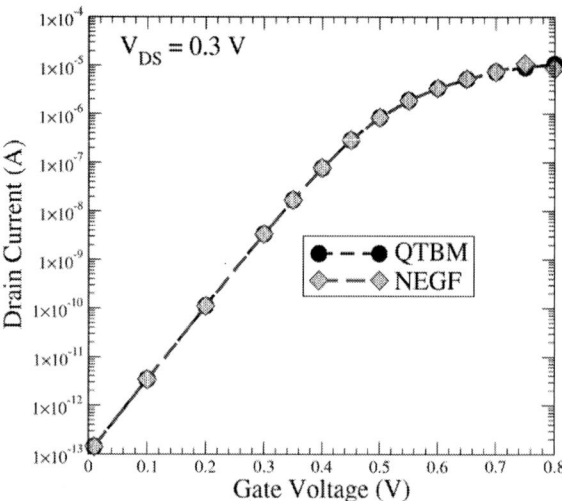

Fig. 4. Id(Vg) characteristics of a GAA Si-nanowire of 2x2 nm^2 and 20 nm long, in balistic mode (QTBM, black circle) and NEGF (blue diamond) at fixed drain voltage Vds.

We use our NEGF with the LRA, and a full quantum transmission boundary method (QTBM) for comparison. Both methods show an excellent agreement for an applied gate voltage Vg of less than 0.7 V. The slight difference observed for larger Vg is mainly due to numerical convergence which is more difficult to achieve with a high ON ballistic current. (Note that our NEGF+LRA is already two times more efficient in terms of simulation time than the QTBM).

Fig. 5 shows the simulation of a 2x2 nm2 Si NW with a double-grid (DG) in ballistic mode with various drain voltages applied, up to saturation. As expected the maximum drain current obtained is lower for a double grid than for a GAA nanowire, due to a less efficient electrostatic control of the gate.

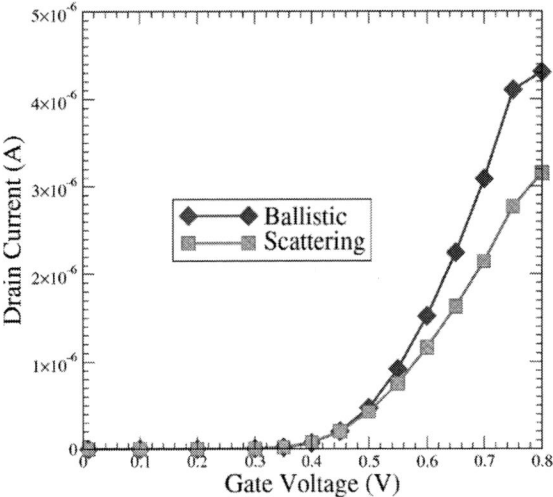

Fig. 6. Id(Vg) characteristics of a DG Si nanowire of 2x2 nm² and 20 nm long, in NEGF + LRA a fixed Vds of 0.3 V. With no scattering (blue dimaond), and including acoustic and optical phonons (red square).

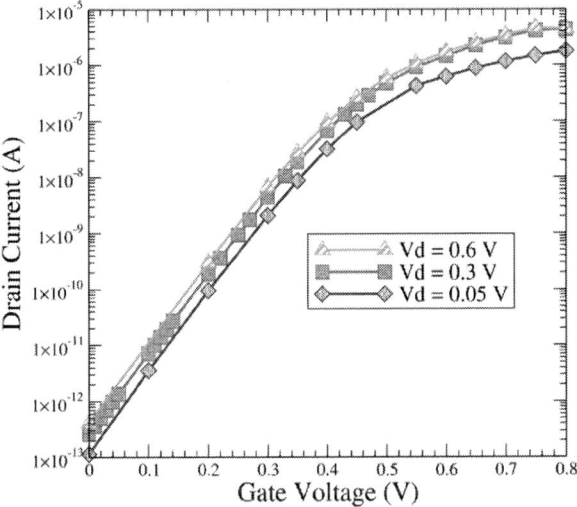

Fig. 5. Id(Vg) characteristics of a DG Si nanowire of 2x2 nm² and 20 nm long, in NEGF + LRA for various drain voltages (blue, red, yellow). The simulation is ballistic without including scattering effects.

We performed the simulation of a DG Si NW of 2x2 nm² and 20 nm long, by including two scattering effects: the acoustic phonons and optical phonons within the Sc-Born approximation. We limit the Sc-Born number of iterations to ten, which is sufficient to achieve the necessary self-consistency with the outer loop of the Poisson solver. Results are shown Fig. 6 with a net decrease of ON current by 25% approx..

For such a size of a silicon nanowire, the simulation has been performed with a relatively modest number of thirty CPU cores, Intel, (Skylake architecture). One voltage point of the I(V) curve is obtained in 120 minutes, increasing the time of the initial ballistic simulation by a factor no more than 20.

We now turn to a larger structure of 5x5 nm² and 20 nm long, with DG electrostatic control. As before, we include the acoustic and optical phonons in the Sc-Born approximation, using NEGF+LRA. For performance comparison, one voltage point of the I(V) curve is now obtained in five hours using 240 CPU cores Intel (Skylake), which is consistent with the results of Table I for one Sc-Born iteration.

The energy profile of the conduction and valence bands along the transport direction is visualized Fig. 7.

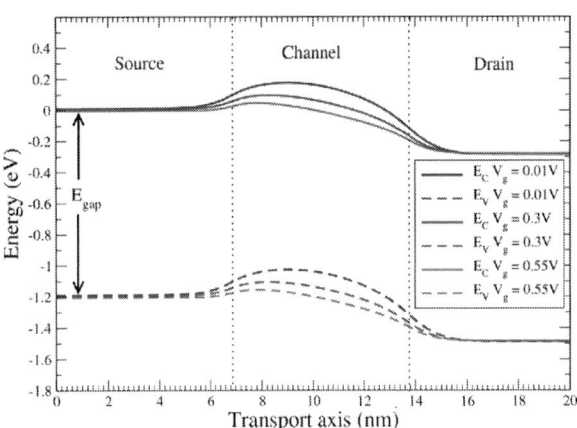

Fig. 7. 5x5 nm² Si nanowire, double grid, NEGF+LRA with acoustic and optical phonons. The effective 1-D energy band profiles (conduction Ec and valence Ev) along the transport direction are visualized for different gate voltages and fixed drain voltage of 0.3V. The source is taken as a reference (0 eV).

The band profile of Fig. 7 is directly related to the electrostatic profile of the device. The Poisson potential of the NW-FET is shown in Fig. 8 in three dimensions for illustrative purposes.

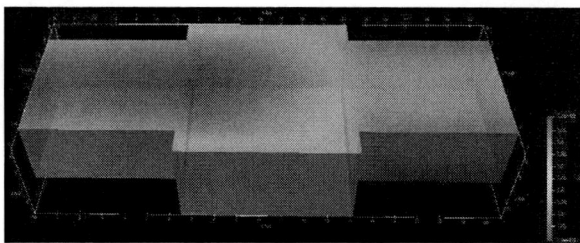

Fig. 8. 5x5 nm² Si nanowire, double grid, NEGF+LRA with acoustic and optical phonons. The 3D Poisson potential is visualized with a cut-plane in the middle of the device, where current flows. Voltage in V, dimensions in nm.

Eventually, knowing all our simulations are performed using an atomistic description of the silicon material, this allows us to visualize the electron charge repartition onto each atom. The sp³d⁵s* Si TB basis components are summed and visualized using VictoryVisual tool, Fig. 9. At Vd = 0.3V, and Vg = 0.55V, this shows a non-intrinsic charge distribution inside the channel, right below the grid (in red) by transparency.

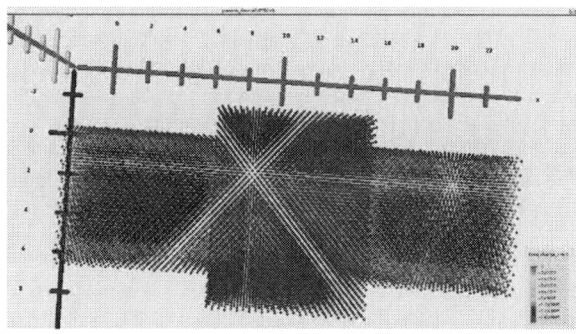

Fig. 9. 5x5 nm² Si nanowire, double grid, NEGF+LRA with acoustic and optical phonons. The charge distibution is visualized onto each atomic site, thanks to the VictoryVisual tool. Drain bias is fixed at 0.3 V, and gate voltage at 0.55 V.

IV. CONCLUSION

Thanks to our new developments, the simulation of a nanoscale FET combining atomic accuracy and inelastic interactions of carriers becomes a daily routine. Moreover, relevant quantities in quantum transport can be safely extracted and visualized. This obviously opens up new perspectives to engineers for simulation of innovative devices in atomistic TCAD. In doing so, the electrostatic of the FET device is revealed at the atomic scale, paving the way to fast and accurate extraction of transistor parameters..

ACKNOWLEDGMENT

Calculations were performed using the Purdue RCAC facilities, Brown's machine.

REFERENCES

[1] A. Razavieh, P. Zeitzoff and E. J. Nowak, "Challenges and Limitations of CMOS Scaling for FinFET and Beyond Architectures," in *IEEE Transactions on Nanotechnology*, vol. 18, pp. 999-1004, 2019

[2] R. Lake, G. Klimeck, R. C. Bowen and D. Jovanovic, "Single and multiband modeling of quantum electron transport through layered semiconductor devices", *J. Appl. Phys*, vol. 81, pp. 7845-7869, 1997.

[3]] S. Steiger, M. Povolotskyi, H.-H. Park, T. Kubis and G. Klimeck, "Nemo5: a parallel multiscale nanoelectronics modeling tool", *IEEE Trans Nano,* vol. 10, p. 1464, 2011.

[4] G. Mil'nikov, N. Mori and Y. Kamakura, "Equivalent transport models in atomistic quantum wires*", Phys. Rev. B* 85, 035317, 2012.

[5] D.A. Papaconstantopoulos, "Handbook of the Band Structure of Elemental Solids", Springer, Boston, MA, 2015.

[6] Y. Tan, M. Povolotskyi, T. Kubis, T.B. Boykin, and G. Klimeck, "Transferable tight-binding model for strained group IV and III-V materials and heterostructures", *Phys. Rev. B* 94, 045311, 2016.

[7] K. Wang, D. Valencia, J. Charles, Y. He, M. Povolotskyi, G. Klimeck, J. Maassen, M. Lundstrom, T. Kubis, "NEMO5: Predicting MoS2 heterojunctions," *2016 International Conference on Simulation of Semiconductor Processes and Devices (SISPAD)*, pp. 221-224, 2016.

[8] L. P. Kadanoff and G. Baym, "Quantum Statistical Mechanics", Adison-Wesley, New York, 1989

[9] D.A. Lemus, J. Charles and T. Kubis, "Mode-space-compatible inelastic scattering in atomistic nonequilibrium Green's function implementations", *arXiv*, 2003.09536, 2020

TCAD-Assisted MultiPhysics Modeling & Simulation for Accelerating Silicon Quantum Dot Qubit Design

F. A. Mohiyaddin, G. Simion, N. I. Dumoulin Stuyck[*], R. Li, A. Elsayed[#], M. Shehata[#], S. Kubicek, C. Godfrin,
B. T. Chan, J. Jussot, F. Ciubotaru, S. Brebels, F. M. Bufler, G. Eneman, P. Weckx, P. Matagne, A. Spessot,
B. Govoreanu & I. P. Radu

imec, Kapeldreef 75, Leuven, B-3001, Belgium
[*] also with Department of Materials Engineering (MTM), KU Leuven, Kasteelpark Arenberg 44, Leuven, B-3001, Belgium
[#] also with Department of Physics and Astronomy, KU Leuven, Celestijnenlaan 200d, Leuven, B-3001, Belgium
fahd.ayyalil.mohiyaddin@imec.be

Abstract — **We summarize the design parameters and modeling techniques for silicon quantum dot qubit devices. A general overview on the operation of the devices - including various methods of qubit readout, control, and interaction - is provided with relevant parameters. With these blocks forming the backbone of silicon quantum computation, the paper provides a guideline to aid and accelerate the design and optimization of silicon qubit devices.**

Keywords— Silicon Quantum Computation, Device Design, Qubit Readout, Control & Interaction, Multiphysics Modeling.

I. INTRODUCTION

Over the last decade, Quantum Computers (QCs) have undergone a rapid transformation from lab-based devices satisfying academic curiosity to much larger-scale fab devices integrated with sophisticated control electronics [1-4]. Widespread interest in the development of quantum processors is evident, with several qubit platforms and novel operation methods being investigated [5]. While remarkable progress has been achieved towards the development of QC, the requirements of scalable fault-tolerant quantum computers [6] pose huge breakthroughs in the engineering of materials, process-integration, cryogenic control-electronics, and software. Device design methodologies are crucial for scaling up existing qubit demonstrations to larger-scale quantum computers, where the impact of several design parameters on the qubits need to be systematically investigated for better device operation, reproducibility, and yield.

Of several implementations, qubits based on silicon Quantum Dots (QDs) offer the advantages of compactness, long coherence times, elevated operation temperatures, and fabrication compatibility with techniques used in the semiconductor industry [7]. Silicon QD devices consist of "multi-gated transistors", where the gates are biased to confine a single electron in a quantum well potential, formed in the proximity of a material interface, typically Si/SiO$_2$ or Si/SiGe. The electron spin orientation with respect to an external static magnetic field B$_0$ constitutes the qubit. To be functional, spin-based QCs require blocks that demonstrate high-fidelity spin readout, control, and interaction over different length scales [8]. In this work, we present an overview on the operation and design of silicon QD qubits, including the various building blocks, design parameters and modeling techniques.

II. DESIGN OF SILICON QUBIT DEVICES

A. Requirement for a MultiPhysics Modeling Strategy

The operation of silicon qubit devices covers various aspects, and cannot be captured with a single design tool, requiring a multiphysics simulation approach [9]. For instance, the electron confinement is dependent on the potential landscape in the device (Section II.B), whereas spin readout is based on capacitances and tunnel interactions (Section II.C). Spin control is based on magnetic resonance with electromagnetic fields arising from transmission line antennas and magnets, requiring RF and magnet design respectively (Section II.D). The default interaction between electron spins is the short-range exchange interaction, necessitating sophisticated band-structure calculations (Section II.E). Medium- and long-range coupling employs dipolar interaction and microwave resonators respectively, requiring detailed understanding of the coupling mediators (Section II.E). Furthermore, additional effects such as noise (Section II.F), thermal strain and superconductivity (Section II.G) must be taken into consideration for device design. Fig. 1 shows a summary of all the modeling parameters for designing spin qubits, and the physical relations to build the models. We will now sequentially discuss devices that can host QD spin qubits, emphasizing key parameters and design considerations.

B. Electrostatic Confinement & Electron Energy Levels

As mentioned before, silicon qubits are defined in a quantum well potential at a material interface (Fig. 2a) [7]. The electron confinement and electrostatic potential can be estimated by solving the Schrödinger-Poisson (SP) equations self-consistently at low temperature with TCAD simulators. We note that appropriate choice of boundary conditions, meshes, truncation of extremely

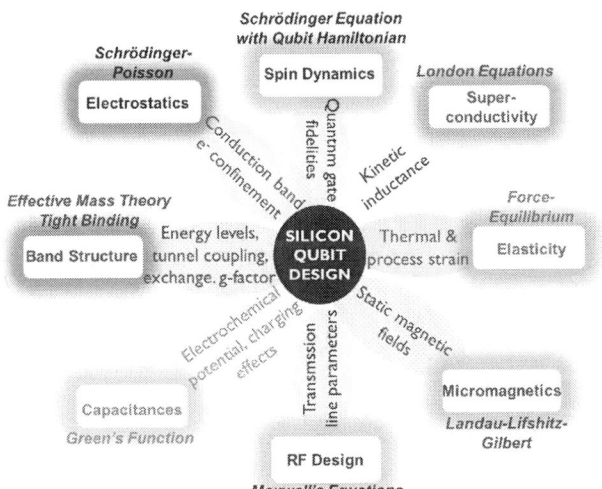

Fig. 1. Summary of all modeling parameters and techniques for designing silicon qubits devices, indicating the requirement of a multiphysics simulation approach.

small electron densities, and ramping down solutions from higher temperatures, can improve convergence significantly at low (~ 10 K) temperatures [10]. The spatial extent of the dots in the single-electron regime typically spans ~ 5 nm in the vertical direction perpendicular to the interface, and few tens of nm in the lateral direction [9]. This confinement leads to energy separations $\delta_0 \sim 3 - 5$ meV between the ground (s-orbital) and excited (p-orbital) states (Fig. 2c), and is obtainable from SP solvers.

The degeneracy of the silicon conduction band minima leads to additional states (valley states) that negatively interfere with qubit operation (Fig. 2b). These states and the resultant splitting (valley splitting) between them are not captured fully in most SP solvers that rely on conventional effective mass theory. To capture such effects, the TCAD potentials are fed into detailed band-structure solvers that consider atomistic details at the interface. These solvers include (i) atomistic tight-binding methods [11] or (ii) modified versions of effective mass theory, where the TCAD wave-functions are combined with the silicon lattice Bloch functions (obtained from Density Functional Theory) to estimate valley splitting [12]. The valley splitting can be shown to be dependent on the interface step potential, vertical electric field in the dot, atomistic imperfections such as roughness and step edges, and typically varies between 50 – 200 μeV and 300 – 700 μeV in Si/SiGe and Si/SiO₂ devices respectively [13,14].

In the presence of a static magnetic field B_0, the spin (qubit) states of the electron in the dot are split in energy by the Zeeman Splitting $E_{ZS} = g_e \mu_B B_0$, where $g_e \sim 2$ is the electron g-factor and μ_B is the Bohr magneton (Fig. 2d). All the relevant energy separations are summarized in the table of Fig. 2, with values chosen to ensure that the qubit states are well separated from the excited orbital & valley states, for an isolated two-level qubit system.

C. Electron Spin Readout

The charge state of the electron in the dot acts as a handle for reading out the electron spin qubit. The process of spin readout starts with spin-dependent tunneling of the dot electron to an adjacent reservoir or to an ancilla dot, and the subsequent detection of the presence or absence of the dot electron with a charge sensor.

The first step of spin-dependent tunneling to a *reservoir* requires E_{ZS} to be much larger than the thermal broadening $\delta(T)$ of the reservoir's Fermi level, with a suitable choice of $B_0 \sim 1$ T [15]. This provides a good contrast between a spin ↑ electron tunneling to the reservoir, when compared to a spin ↓ electron remaining in the dot (Fig. 3b). The reservoir can also be replaced with an *ancilla dot*, containing an electron with spin initialized as ↓ [16]. The Pauli spin-exclusion principle then facilitates spin-dependent tunneling of the dot electron to the ancilla dot (Fig. 3b). The relevant design parameter here is the energy spacing ($\Delta_{ST} \sim \delta_{VS}$) in the ancilla dot, which needs to be sufficient to prevent thermal excitations.

A schematic of a charge sensor for charge detection in the dot is shown in Fig. 3a. The sensor is biased in a regime, where the current passing through the sensor is strongly sensitive to its electrostatic vicinity [15]. In this regime, the resulting change (Δ_μ) in the electrochemical potential (and thereby the sensor current) with the presence or absence of the QD electron (Fig. 3c) allows single-charge detection, and hence spin readout. A critical requirement here is a large capacitance between the QD and sensor, such that Δ_μ is not screened out by thermal broadening $\delta(T)$ of the Fermi level in the sensor [9]. For good charge detection fidelities, a sufficient on current ($I_{DS} \sim 1$ nA) is also required across the sensor, and is achievable with large tunnel rates ($\Gamma_S \sim 10$ GHz) and source drain bias ($V_{DS} \sim 1$ meV). Furthermore, the tunnel rate ($\Gamma_{dot} \sim 10$ kHz) of the QD electron to the reservoir or ancilla dot needs to be in the right range for good fidelities in reasonably short timescales.

A resonator (lumped or transmission line), directly connected to the gate above the readout dot [17], can also be used for charge detection (Fig. 3d). By appropriately biasing the dots, the resonator capacitance (and hence its resonant frequency) can be made dependent on the electron occupation in the readout dot, which further depends on the spin. The electron spin state can then be read out via the resonator reflectance. The change (Δv_0) in the resonator

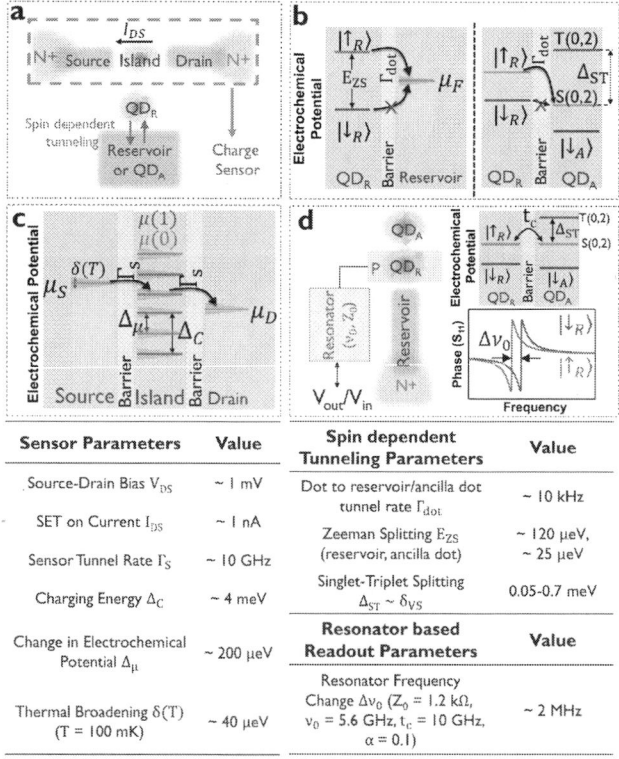

Sensor Parameters	Value
Source-Drain Bias V_{DS}	~ 1 mV
SET on Current I_{DS}	~ 1 nA
Sensor Tunnel Rate Γ_S	~ 10 GHz
Charging Energy Δ_C	~ 4 meV
Change in Electrochemical Potential Δ_μ	~ 200 μeV
Thermal Broadening $\delta(T)$ (T = 100 mK)	~ 40 μeV

Spin dependent Tunneling Parameters	Value
Dot to reservoir/ancilla dot tunnel rate Γ_{dot}	~ 10 kHz
Zeeman Splitting E_{ZS} (reservoir, ancilla dot)	~ 120 μeV, ~ 25 μeV
Singlet-Triplet Splitting $\Delta_{ST} \sim \delta_{VS}$	0.05-0.7 meV
Resonator based Readout Parameters	**Value**
Resonator Frequency Change Δv_0 (Z_0 = 1.2 kΩ, v_0 = 5.6 GHz, t_c = 10 GHz, α = 0.1)	~ 2 MHz

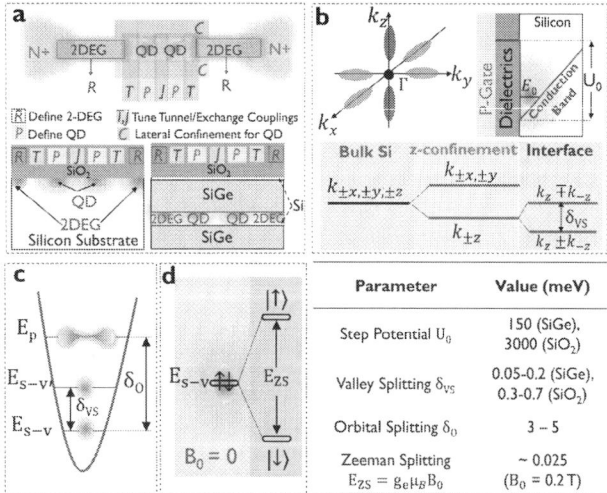

Fig. 2 (a) Schematic of accumulation-mode qubit devices based on Si/SiO₂ and Si/SiGe interfaces. (b) Multivalley nature of silicon conduction band and breaking of valley-degeneracies when an electron is confined at the interface. (c) Device potential sketch with different orbital energy levels. (d) Qubit energy levels in a static magnetic field B_0. (Table) Summary of relevant energy separations in the QD.

Fig. 3. (a) Schematic indicating the position of the charge sensor, quantum dots and reservoir for spin readout. (b-c) Alignment of electrochemical potentials for (b) spin-dependent tunneling, and (c) charge sensing. μ(N) is the electrochemical potential in the island, N is electron number in the dot (d) Spin readout with a resonator via reflectometry. (Tables) Summary of design parameters for readout.

254

frequency can be shown to depend on the difference (α) of the lever arms of the resonator to the dots, the resonator characteristic impedance Z_0, frequency v_0, and the interdot tunnel coupling t_c [17]. For sufficient readout contrast, these parameters are chosen such that Δv_0 is comparable to or larger than the resonator line width v_0/Q, where Q is the resonator quality factor.

All readout methods involve capacitances, which can be extracted with standard capacitance simulation software. These capacitances can be fed into Monte-Carlo circuit simulators to model current transport across charge sensors. For the tunnel rates, a WKB approximation provides the required barrier heights (few 100 μeV) and barrier dimensions (few tens of nm) [9]. Estimation of interdot tunnel couplings involve modeling energy levels in a double QD system, obtainable with standard SP solvers. For dispersive readout, it is crucial to model the resonator frequencies and impedances, for which the Maxwell's equations are solved.

D. Electron Spin Control

For spin control, the electron is subject to an RF magnetic field, whose direction is perpendicular to that of the static field B_0. Under the resonance condition, where the frequency of the RF field is equivalent to E_{ZS}, the electron spin state rotates between $\uparrow$ and $\downarrow$ states with a Rabi frequency f_R proportional to the RF field amplitude B_{ac}. The effective RF field can be generated either by a transmission line (resulting in Electron Spin Resonance), or by displacing the electron electrically in a magnetic field gradient b_c (resulting in Electric Dipole Spin Resonance). The former method involves a microwave transmission line with a node, such that the RF magnetic fields experienced by the electron is maximized, while ensuring minimal RF electric fields (Fig. 4a). The transmission line also needs to be impedance matched, for maximal B_{ac} at the qubit location. Assuming an input power $P_{ac} \sim 1\mu$W, a field amplitude $B_{ac} \sim 80\ \mu$T is achievable at the qubit position, yielding Rabi frequencies $f_R \sim 1$ MHz [18].

The second method involves placing magnets near the electron qubits, and electrically shifting the electron in the resultant magnetic field gradient (Fig. 4b) [19]. The magnetic fields and their gradients are obtainable by solving the Landau-Lifshitz-Gilbert equations. The field gradients will be present in different directions, some (b_c) facilitating qubit control, while the others (b_d) result in spin dephasing. The design needs to ensure a large value of b_c (typically $\sim$ 0.35 mT/nm), while keeping b_d minimal. The shift in the electron position with gate voltages can be modeled with SP solvers, and is $\sim$ 1 nm for voltage amplitudes $V_{ac} \sim$ 200 mV in single QDs, yielding $f_R \sim 5$ MHz [9].

To achieve similar or larger Rabi frequencies with lower drive voltages, the electron can be further delocalized in a double QD (Fig. 4c) [20]. Quantum mechanically, this strongly couples the electron spin with its position via the magnetic field gradient, in addition to creating a charge dipole that is strongly sensitive to gate voltages. This picture can be simplified with a transformation (Schrieffer-Wolff (SW)) to yield a Rabi frequency f_R dependent on field gradient b_c, interdot tunnel coupling t_c, and difference α in lever arms. With an appropriate choice of parameters, $f_R \sim 25$ MHz is achievable at much smaller voltage amplitudes $V_{ac} \sim 30\ \mu$V.

Note that the electron spin resonance frequency and coherence depend on the stability of g-factor g_e, which is related to the spin-orbit interaction (SOI). Tight-binding and extensions of effective mass theory have been extensively used to estimate SOI in qubit devices [21]. It can be shown that SOI is strongly sensitive to atomistic interface details, when B_0 is aligned parallel to the interface, leading to non-uniform (by $\sim$ 1%) qubit frequencies. This is overcome by aligning B_0 perpendicular to the interface.

Finally, we highlight that the ability to readout and control the qubit will also provide the capability for deterministic qubit initialization.

E. Coupling Electron Spins

The interaction between electron spins in a double QD is the exchange coupling J_e, which is a short-range mechanism depending on the overlap between the orbital wave-functions in the dots (Fig. 5a) [22]. J_e is estimated by first calculating the device potential with SP solvers, feeding the potential into band-structure calculators to estimate orbital wave functions, and combining them with computational chemistry techniques. For two qubit operations, it is essential to tune J_e by either (i) varying the tunnel barrier between the two dots, or by (ii) shifting the relative energies (i.e. by a detuning) between the dots. Both methods yield several orders of magnitude tunability (1 MHz - 1 GHz) of J_e, when the tunnel barriers and relative energies are tuned by several hundred μeV [9].

For medium range coupling, the spin-spin interaction can be mediated by the charge dipoles. Each electron is then delocalized in a double QD in the presence of a magnetic field gradient (Fig. 5b). The electron spin is then coupled to its position that constitutes a charge dipole (Fig. 4c) [20]. Two such dipoles are then coupled via the dipolar interaction, yielding an indirect spin-spin interaction [23]. The strength D_e of the interaction can be estimated with a SW transformation, and depends on the field B_0, field gradient b_c, interdot tunnel coupling V_t, interdot distance L, and dipole separation r [23]. An optimal choice of parameters yields $D_e \sim 3$ MHz between qubits that are separated by as far as 500 nm. Note that the interaction can also be turned off by localizing the electron qubits in a single QD.

As charge dipoles are very sensitive to electric fields and voltages, they can be coupled to a resonator mode via the resonator vacuum voltage V_{vac} (Fig. 5c) [24]. This leads to an indirect

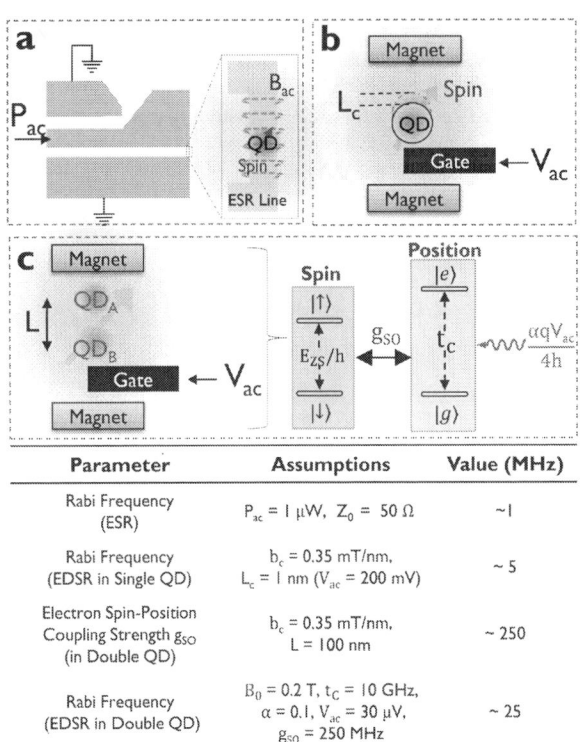

Parameter	Assumptions	Value (MHz)
Rabi Frequency (ESR)	$P_{ac} = 1\ \mu$W, $Z_0 = 50\ \Omega$	~1
Rabi Frequency (EDSR in Single QD)	$b_c = 0.35$ mT/nm, $L_c = 1$ nm ($V_{ac} = 200$ mV)	~5
Electron Spin-Position Coupling Strength g_{SO} (in Double QD)	$b_c = 0.35$ mT/nm, $L = 100$ nm	~250
Rabi Frequency (EDSR in Double QD)	$B_0 = 0.2$ T, $t_C = 10$ GHz, $\alpha = 0.1$, $V_{ac} = 30\ \mu$V, $g_{SO} = 250$ MHz	~25

Fig. 4. Spin control methods based on (a) Electron Spin Resonance (ESR) with a transmission line antenna, (b-c) Electric Dipole Spin Resonance (EDSR) in a (b) single QD and (c) double QD. h is the Planck constant and q is elementary charge. (Table) Summary of all design parameters and Rabi frequencies for the control methods. Note : Energy splittings and coupling strengths are in units of frequency.

255

coupling between the spin and resonator. Two spins coupled to the resonator can then be coupled with each other. The effective spin-spin interaction strength P_e via the resonator can be estimated with SW transformations and depends on several parameters summarized in the table of Fig. 5 [24]. An appropriate choice of parameters, obtained by using high-impedance resonators, yields $P_e \sim 0.3$ MHz between spins that are separated by few mm. The resonator can also be used to readout the spin via the reflectometry techniques described in Section II.C. The difference here is that an ancilla spin is not necessary for readout, as the spin states are directly coupled to the resonator. The design of structures involving dipolar- or resonator-based coupling requires modeling magnetic field gradients, electrostatic potentials, tunnel couplings and resonator parameters in addition to transformations, such as SW.

F. Electromagnetic Noise

Noise arises from several sources including Johnson noise, imperfections in the magnets, charge defects, thermal fluctuations, and magnetic (e.g. Si-29) impurities [14]. Popular noise minimization techniques involve attenuation of DC lines, reducing magnet vibrations, driving qubits at smaller RF powers, and using isotopically purified Si-28 substrates [14, 25]. However, a detailed investigation of all noise sources, especially electric field noise E_n, is impractical, due to lack of experimental evidence on their origin. Hence, empirical values of E_n (~ 100 V/m) and spin-coherence times ($\sim 1 - 100$ μs), along with realistic parameters summarized in all the tables, are often fed into simplified qubit Hamiltonians, where the Schrodinger and Master Equations are solved to estimate the qubit dynamics and fidelity of quantum gate operations.

The sensitivity of qubits to noise can be reduced with proper device design. For example, aligning B_0 perpendicular to the interface renders the g-factor g_e insensitive to vertical electric fields, and thereby a component of noise [21]. Optimal magnet design to ensure an extremely small and robust dephasing gradient b_d can also enhance coherence times [24]. This is critical when the qubits are delocalized in double QDs for fast control or long-distance interaction, where they are made sensitive to the electric field.

Under such operation regimes, the design parameters are also chosen to minimize spin dephasing effects via the position and resonator. Operating the qubits in certain biasing conditions may also render the qubit in-sensitive to noise, as noise affects the qubit frequency via different pathways, that could cancel out [20]. In addition to device design, utilizing sophisticated pulsing protocols can further mitigate the impact of noise on the qubits [26].

G. Other Low Temperature Considerations

Thermal strain has been shown to modify the conduction band by few meV in the QDs and their vicinity [27], and can be modeled with TCAD process and device simulators. The requirement of small barrier heights ($\sim$ few 100 μeV) between the QDs/reservoirs for qubit operations necessitates extremely small values of strain gradients in the barriers. This is achievable with appropriate choice of gate materials (e.g. Polysilicon) and device geometries [9].

Another cryogenic effect is superconductivity, which plays a significant role in the design of transmission lines and resonators for qubit readout, control, and coupling. The key parameter is the kinetic inductance L_k of materials and can be included in RF simulations by assuming an analytical derived value of the sheet inductance [28]. A large value of L_k (few tens of nH) is generally chosen for good readout contrast and strong qubit coupling via high-impedance resonators. On the contrary, L_k needs to be minimized for the qubit to be controlled with maximal B_{ac}.

III. SUMMARY

Based on a multiphysics modeling strategy, we have provided a general guideline on the interplay between various design parameters for silicon qubit structures. These guidelines can be used for designing optimal silicon qubit devices with spin readout, control, and coupling. We foresee that the modeling techniques constitute a design framework that will play a pivotal role in scaling up existing qubit prototypes to larger-scale quantum processors.

REFERENCES

[1] F. Arute et al., *Nature*, 574, 505-510, 2019.
[2] R. Pillarisetty et al., *Proceedings of IEDM*, 2018.
[3] B. Govoreanu et al., *Proceedings of SNW*, 2019.
[4] D. J. Dean et al., Quantum Information Science, *Factual Document for the Office of Basic Energy Sciences at the Department of Energy*, 2017.
[5] T. S. Humble et al., *IEEE Design & Test*, 36, 3, 2019.
[6] A. Fowler et al., *Physical Review A*, 86, 032324, 2012.
[7] L. M. K. Vandersypen et al., *npj Quantum Information*, 3, 34, 2017.
[8] D. Divincenzo, *Fortschritte der Physik:Progress of Physics*, 48, 2000.
[9] F. A. Mohiyaddin et al., *Proceedings of IEDM*, 2019.
[10] F. A. Mohiyaddin et al., *Proceedings of COMSOL Conference*, 2017.
[11] R. Rahman et al., *Physical Review B*, 83, 239904, 2011.
[12] J. K. Gamble et al., *Applied Physics Letters*, 109, 253101, 2016.
[13] C. H. Yang et al., *Nature Communications*, 4, 2069, 2013.
[14] F. A. Zwanenburg et al., *Reviews of Modern Physics*, 85, 961, 2013.
[15] M. Veldhorst et al., *Nature Nanotechnology*, 9, 981-985, 2014.
[16] N. W. Hendrickx et al., *Nature Communications*, 11, 3478, 2020.
[17] A. C. Betz et al., *Nano Letters*, 15, 7, 4622-4627, 2015.
[18] J. P. Dehollain et al., *Nanotechnology*, 24, 1, 2013.
[19] J. Yoneda et al., *Nature Nanotechnology*, 13, 102-106, 2018.
[20] M. Benito et al., *Physical Review B*, 100, 125430, 2019.
[21] R. Ferdous et al., *Physical Review B*, 97, 241401, 2018.
[22] Q. Li et al., *Physical Review B*, 81, 085313, 2010.
[23] J. Cayao et al., *Physical Review B*, 101, 195438, 2020.
[24] F. A. Mohiyaddin et al., *Proc. of IITC*, 2019.
[25] J. P. G. Van Dijk et al., *Physical Review Applied*, 12, 044054, 2019.
[26] S. Machnes et al., *Physical Review A*, 84, 022305, 2011.
[27] T. Thorbeck et al., *AIP Advances*, 5, 087107, 2015.
[28] E. F. C. Driessen et al., *Physical Review Letters*, 109, 107003, 2012.

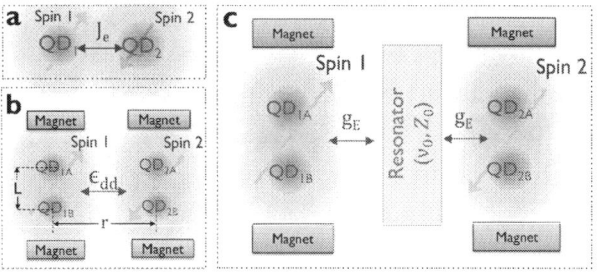

Parameter	Assumptions	Value (MHz)
Spin-Spin Coupling Strength J_e via Exchange Interaction	Qubit pitch r ~ 80 nm	~ 1 – 1000
Electron Spin - Position Coupling Strength g_{SO} in a double QD	b_c ~ 0.35 mT/nm, Dot pitch L = 100 nm	~ 250
Dipolar Coupling Strength ϵ_{dd} between Electron Positions	L = 100 nm, r ~ 500 nm	~ 600
Spin-Spin Coupling Strength D_e via Dipolar Interaction	B_0 = 0.2 T, t_c = 10 GHz, g_{SO} = 250 MHz, ϵ_{dd} = 600 MHz	~ 3
Coupling Strength g_E between Electron Position and Resonator	α = 0.1, v_0 = 5.65 GHz, Z_0 ~ 1.2 kΩ, (V_{vac} ~ 10 μV)	~ 60
Spin-Spin Coupling Strength P_e via Resonator-Mediated Interaction	B_0 = 0.2 T, t_C = 10 GHz, g_{SO} = 250 MHz, g_E = 60 MHz	~ 0.3

Fig. 5. Coupling of spin qubits via (a) short-range exchange interaction, (b) medium-range dipolar interaction, and (c) resonators facilitating long-range interaction. (Table) Summary of design parameters and coupling strengths for the three methods. Note : All coupling strengths are expressed in units of frequency.

12-2

Physics-augmented Neural Compact Model for Emerging Device Technologies

Yohan Kim, Sanghoon Myung, Jisu Ryu, Changwook Jeong, and Dae Sin Kim
Data & Information Technology Center
Samsung Electronics
Hwaseong-si, Gyeonggi-do 18448, Korea
johnn.kim@samsung.com

Abstract— **This paper proposes a novel compact modeling framework based on artificial neural networks and physics informed machine learning techniques. This physics-augmented neural compact model shows highly accurate fitting abilities and physically consistent inferences even at the unseen data. It is also scalable and technology independent, and consequently, is suitable for electrical modeling of new emerging devices. In addition, this neural compact model is able to cover both digital and analog circuit analysis due to the weight decay regularization as well as high order derivative losses. Finally, it is applied to promising DRAM and Logic technologies to be evaluated in terms of its scalability and fitting accuracy. The CMC's (Compact Model Coalition) standard model API (Application Programming Interface) supports the custom model implementation for SPICE. Therefore, this framework enables the circuit simulators to assess technology-independent PPA (Power, Performance, Area) and early-stage DTCO (Design Technology Co-optimization) for new emerging devices.**

Keywords—Compact Model, PDK, SPICE Circuit Simulation, Machine Learning, Artificial Neural Networks, Physical theory augmentation, Emerging device modeling, PPA and DTCO

I. INTRODUCTION

Over the past decade, CMOS technology is aggressively scaled down and the physical process limit is reached. As a result, various emerging devices with rapid changes in geometry, architecture, and material need to be evaluated through PPA assessments and DTCO activities in a timely manner. However, existing CMC's standard compact models are inappropriate for these emerging devices because developing new models is very time-consuming and strongly depends on each technology expertise. Therefore, we propose an artificial intelligence assisted compact model framework which is highly accurate, extremely general, and physics-embedded. It is going to change the paradigm of the expertise intensive model research and early-stage DTCO environments.

II. LIMITS OF THE EXISTING COMPACT MODELS

The existing compact models are at extreme ends of parametric (physics based) and non-parametric (empirical lookup table based) models. The analytical equations, represented by standard compact models, are physics-based and computationally efficient, however it takes very long time and requires special expertise to develop a reliable model for new technologies. In addition, the parametric analytical models support increasing number of empirical fitting parameters [1] even though they are built on solid physical theories because of various secondary effects and non-monotonic characteristics from modern complex process flows as shown in Fig. 1.

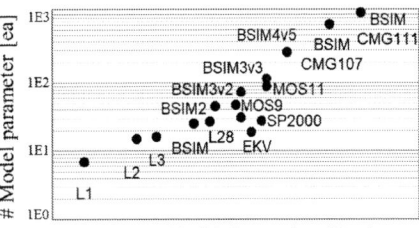

Fig. 1. The number of standard compact model's fitting parameters. It is rapidly increasing because of modern device's secondary effects and non-monotonic characteristics

III. PHYSICS-AUGMENTED NEURAL COMPACT MODEL

Artificial intelligence is providing effective alternatives to various Electronic Design Automation (EDA) technologies, and outperformed the existing competitors. Therefore, combining these data science techniques (artificial neural networks) and scientific theories (device physics) are able to have the best of both worlds [2-4] for SPICE application as shown in Fig. 2.

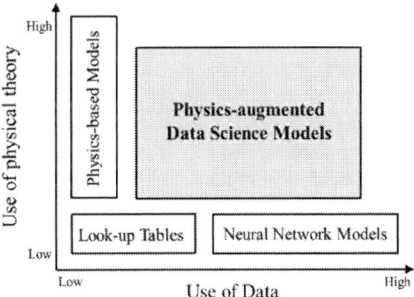

Fig. 2. Physics-augmented neural network model for circuit simulations

This modeling framework includes two main steps, 1) data preprocessing according to various physical origins and machine learning training with each physical constraint loss, 2) model validation (accuracy test), model compression (model scaling for computational efficiency) and charge-conservative model implementations as shown in Fig. 3. First of all, electrical measurement data of various circuit elements could be componentized using the scientific domain knowledges and physical origins as shown in Fig. 4. To briefly demonstrate the component-wise modeling flow, we select a modern MOSFET device as an example as shown in Fig. 3(b). Drain current (drift-diffusion transport), gate current (tunneling), and body current (gate-induced drain/source leakage) are main target components according to design instance (W/L/T), bias (V_{gs}/V_{ds}/V_{bs}) and technology (T_{ox}/WF) parameters.

257

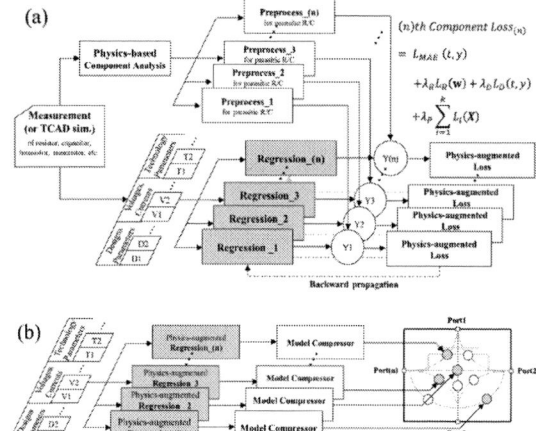

(a)

(b)

Fig. 3. Physis-augmented neural compact modeling framework. (a) Component-wise data prepocessing and physics-augemented machine learning training, (b) Component-wise model validation and compression for accuracy and computational efficiency in SPICE applications

Elements (e.g.)		Components examples
Capacitor	Passive	Area cap + Fringing cap
	MOS	Intrinsic charge cap + Extrinsic cap
Transistor	Bipolar	Hole current + Electron current
	Unipolar	Transport + body + tunneling current

Fig. 4. An example of circuit element's electrical and physical components

Parameter	Description
X	Number of input features in the neural networks
Y	Number of physical components in the equvalent circuit
L	Number of hidden layer in the neural networks
Nl	Number of neurons of lth layer
λ_p	Physics augmentation loss coefficient
λ_a	Derivative loss coefficient for analog models
λ_d	Weight decay parameter for digital models
wji.l	Weight from ith neuron of the (l-1)th layer to the jth neuron of the lth layer
bi	Bias of the ith neuron of the lth layer

Fig. 5. Model parameters to demonstrate the component-wise neural network modeling flow in this work

These component data pass through each hypothesis space to find optimal model parameters, and in this demonstration, fully connected neural networks are used to capture each target function as shown in Fig. 5. The trained models are implemented as an equivalent-circuit form to meet the charge conservation raw, and the SPICE result shows a good agreement between the model inferences and corresponding targets as shown in Fig. 6.

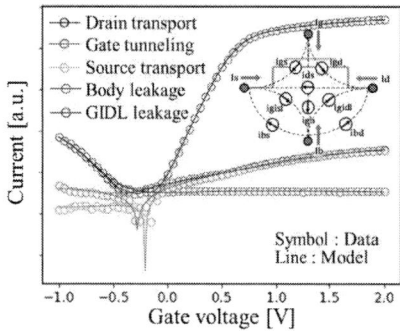

Fig. 6. Model inference results which meet the charge conservation raw

A. Physics-augmented model

In this paper, physics-augmented neural network model is introduced to provide an interpretable and physically consistent circuit simulations. For example, MOSFET's I-V transfer characteristics has a distinct temperature dependency (i.e. trend reversion by an zero temperature coefficient point) because dominant transport mechanisms are changed according to the bias and electric field conditions as shown in Fig. 7.

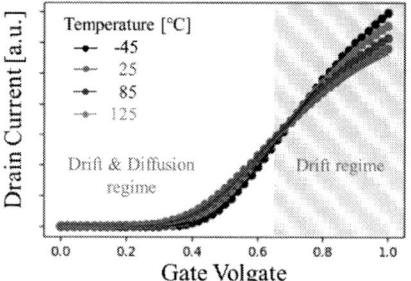

Fig. 7. MOSFET's I-V transfer curve which has an reversed temperature dependency according to the bias conditions

However, measurements with various temperatures are time-consuming and the data is costly. Therefore, most model parameter extraction procedures are based on few temperature sweeps. Fig. 8 shows the ordinary neural network modeling results which are trained using these costly data at -25 and 125 degree Celsius only, and it has physically wrong inferences around the unseen data. These 'black-box' neural network models are prone to small data set and unintended noises which is most common at SPICE modeling environments.

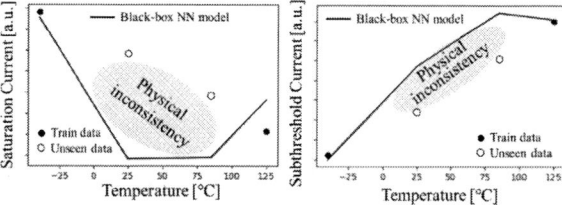

Fig. 8. Neural network model training results (left: saturation currents, right: subthreshold currents) based on cool and hot temperatures only. These black-box models are vulnerable to the unseen data, and it has physically reversed temperature coefficients around 25~85 degrees Celsius

Therefore, we need to make the most of the physical relationships between sample data (transport currents) and input features (bias voltages, temperatures) as follows.

Temperature (T) is closely related to the carrier mobility (μ) and Fermi potential (Φ_F) in the saturation and subthreshold regions respectively as below,

$$log(\mu) = \tau_c \cdot log(T)^{-\frac{3}{2}} \quad (if\ V \gg V_{th}) \quad (1)$$

$$\phi_F = \frac{kT}{q} \cdot ln\left(\frac{N_a}{n_i}\right) \quad (if\ V \ll V_{th}) \quad (2)$$

τ_c is an impurity scattering related constant, N_a is substrate doping concentration, n_i is intrinsic carrier concentration and k is Boltzmann constant. The carrier mobility (μ) in the saturation region is closely related to drift

258

currents and Fermi potential (Φ_F) in subthreshold region also has a physics-based relationship with diffusion currents as shown in Fig. 9-10.

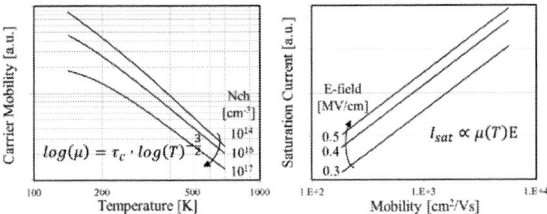

Fig. 9. The temperature effects on the carrier mobility (left) and saturation currents (right)

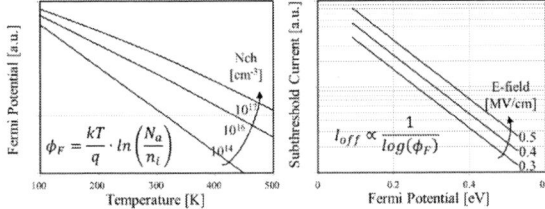

Fig. 10. The temperature effects on the Fermi potentials (left) and subthreshold currents (right)

Therefore, the carrier mobility and Fermi potential at two different temperatures ($T_1 < T_2$) on the same device size (W, L) are related to each other in the following manner,

$$\mu[T_2, W, L, V_{dH}] - \mu[T_1, W, L, V_{dH}] \le 0 \quad (V_{dH} \gg V_{th}) \quad (3)$$

$$\phi_F[T_2, W, L, V_{dL}] - \phi_F[T_1, W, L, V_{dL}] \le 0 \left(V_{dL} \ll V_{th} \right) (4)$$

It also means we are able to compute the differences in mobility and Fermi potential of a neural network model, based on any pair of consecutive temperatures, T_i and T_{i+1} ($T_i < T_{i+1}$) in a device size (W, L) as below,

$$\Delta\mu = \mu[T_{i+1}, W, L, V_{dH}] - \mu[T_i, W, L, V_{dH}] \quad (5)$$

$$\Delta\phi_F = -\phi_F[T_{i+1}, W, L, V_{dL}] + \phi_F[T_i, W, L, V_{dL}] \quad (6)$$

A positive value of these differences are considered as a physical consistency based on the theory equation (1-2) on the temperature and device size. Therefore, we can also construct physics-augmented loss functions to guide the neural networks toward a negative mean of all consecutive temperature pairs (T_i, T_{i+1}) using the Rectified Linear Unit activation function (ReLU) [5].

Physics augmented loss of drift components

$$= \frac{\lambda_P}{n_w n_l (n_t - 1)} \sum_{w=1}^{n_w} \sum_{l=1}^{n_l} \sum_{i=1}^{n_t-1} ReLU(\Delta\mu[T_i, W, L, V_{dH}]) \quad (7)$$

Physics augmented loss of diffusion components

$$= \frac{\lambda_P}{n_w n_l (n_t - 1)} \sum_{w=1}^{n_w} \sum_{l=1}^{n_l} \sum_{i=1}^{n_t-1} ReLU(\Delta\phi_F[T_i, W, L, V_{dL}]) \quad (8)$$

A neural network model is trained using same limited data (-25 and 125℃ only), however in this case, the device physics-informed losses are also used to search the optimum solution. Fig. 11 shows the physically correct inference results at both drift and diffusion regions, and the fitting

accuracies are greatly improved by introducing these scientific theory-guided learning algorithms.

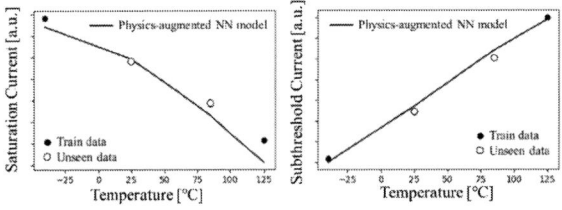

Fig. 11. Neural network model training results (left: saturation currents, right: subthreshold currents) based on cool and hot temperatures. In this case, physics augmented losses are applied, and it has physically correct inferences in the unseen data regions

B. Circuit-friendly model

Semiconductor data which pass through the neural networks are inherently vulnerable to inevitable measurement noises and electrical fluctuations. In addition, robust compact models require to satisfy the high order derivatives and symmetry characteristics for various digital and analog circuit applications. A transistor's on-off ratio is one of the most important performance factors in digital circuit analysis, and the accurate leakage current modeling is critical to PPA assessments. Therefore, the weight decay techniques are essential to these digital models as shown in Fig. 12, and the regularization strengths need to be properly adjusted according to the data qualities as below,

$$Regularization\ loss = \frac{\lambda_R}{2} \sum_{i=0}^{m} W_i^2 \quad (9)$$

λ_R is regularization coefficient, and W_i stands for weight parameters of a regression model.

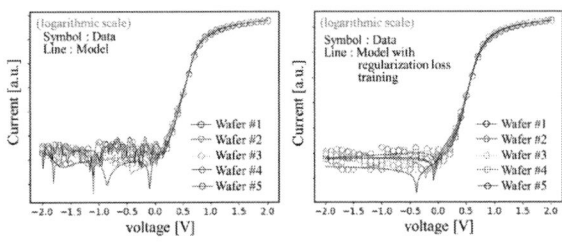

Fig. 12. Semiconductor measurement data which include random noises and electrical fluctuations. Practical neural models require to use effective regularization losses (right) for avoiding the overfitting to noises (left)

The relationships between terminal current and conductance or terminal charge and capacitance are also important features in device modeling. An accurate inference of transconductances from current model or capacitances from charge model are all dependent on derivative considerations. These derivative characteristics are critical, not only for analog circuit analysis but also for iterative solving algorithm of SPICE simulators. Therefore, high order derivatives need to be introduced in the loss calculation as shown in Fig. 13.

$$Deriv\ loss = \frac{\lambda_D}{2} \sum_{n=1}^{\delta} \sum_{i=0}^{m} \left(\frac{d^{(n)} t_i}{dt^{(n)}} - \frac{d^{(n)} y_i}{dy^{(n)}} \right)^2 \quad (10)$$

λ_D is derivative coefficient. t_i and t_i are true and estimation value, respectively.

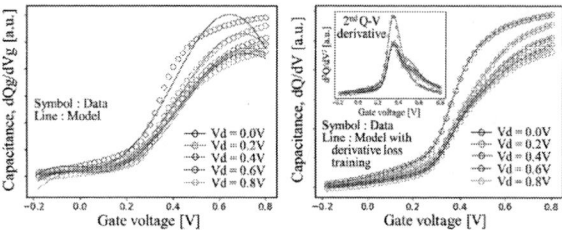

Fig. 13. An ordinaly neural charge model's inaccurate inference results of gate capacitance (left). Higher order derivtive loss is able to dramatically impoves the accuracy of derivative characteristics.

IV. MODEL VALIDATIONS

A. Scalablility in DRAM applicatoins

This modeling framework is applied to a DRAM technology. Fig. 14 shows a highly scalable and accurate fitting ability of neural compact model, which is contrast to a standard compact model (scalable, inaccurate) and empirical binning approach (non-scalable, accurate).

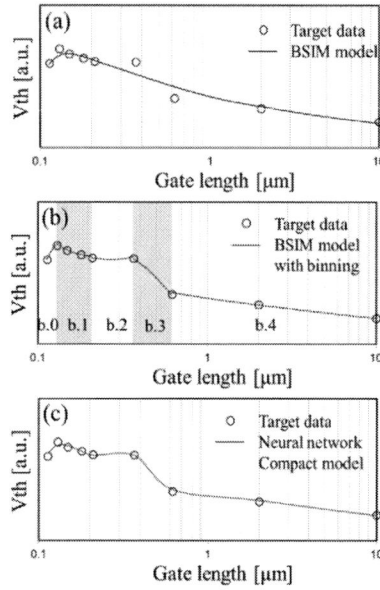

Fig. 14. Model scalability comparison between standard compact model, empirical binnging and neural compact model.

B. Technology-independence in Logic applications

It is also applied to various state-of-the-art emerging devices [6-10] as shown in Fig. 15. These new technologies are not supported by existing standard compact models. Fig. 16 shows a highly general fitting ability to cross-sectional size dependencies such as multiple humps and peaks in C-V and steep switching characteristics in I-V curves.

	Emerging Device (a)	Emerging Device (b)	Emerging Device (c)	Emerging Device (d)
Technology	NanoWire	Nano Wire	NCFET	Tunneling FET
Structure	Gate-All-Around	Gate-All-Around	Tri-gate	Vertical
Ch. Material	Si	InGaAs	Si	Si
EOT	1nm	1nm	-	0.6nm
Dielectrics	SiO2	SiO2	HfO/TaN	SiO2
Device Features	1-D cross sectional QME in Si	1-D cross sectional QME in III-V	Low SS < 60mV/Dec	Low On/Off ratio <1e-9

Fig. 15. Various state-of-the-art emerging devices for early stage PPA and DTCO benchmarks in this work

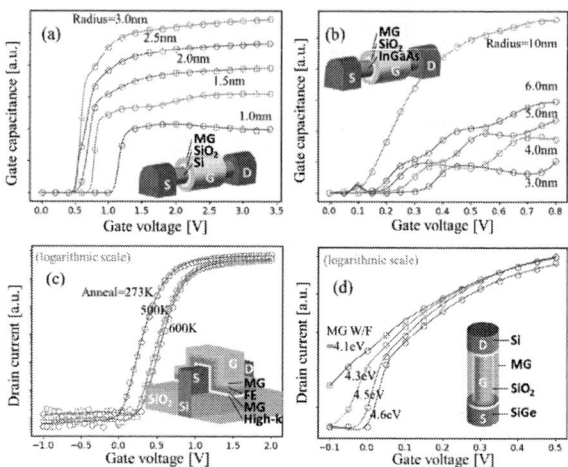

Fig. 16. Neural SPICE model results of various emerging devices. (a) Ultra narrow silicon nanowire, (b) Ultra narrow InGaAs nanowire, (c) Tri-gate negative capacitance transistor and (d) Vertical tunneling transistor

V. CONCLUSION

A series of neural network modeling algorithms and physics informed machine learning techniques are proposed. Semiconductor data passes through each neural networks according to its physical origin, and parameters are automatically updated in backpropagation with each physics-based constraint loss. This physics-augmented neural compact model has the best of two worlds, neural network model's fitting accuracy and theory-based model's physical consistency. It is also applied to recent DRAM and promising Logic technologies, and evaluated in terms of its scalability and technology independency. This new approach is expected to be an effective alternative for early stage PPA and DTCO activities for new emerging devices.

REFERENCES

[1] Yohan Kim et al., "The efficient DTCO Compact Modeling Solutions to Improve MHC and Reduce TAT", SISPAD, September 2018.

[2] Anuj Karpatne et al.,"Theory-Guided Data Science: A New Paradigm for Scientific Discovery from Data", IEEE Transs on Knowledge and Data Engineering, vol. 29, pp. 2318-2331, June 2017.

[3] Anuj Karpatne et al., "Physics-guided Neural Networks (PGNN): An Application in Lake Temperature Modeling", arXiv preprint arXiv:1710.11431, 2017.

[4] Xiaowei Jia et al., "Physics-Guided Machine Learning for Scientific Discovery: An Application in Simulating Lake Temperature Profiles", arXiv preprint arXiv:2001.11086, 2020.

[5] Xavier Glorot et al., "Deep sparse rectifier nueral networks", International Conference on Artificial Intelligence and Statistics, vol. 15, pp. 315-323, 2011.

[6] Avirup Dasgupta et al., "Unified Compact Model for Nanowire Transistors Including Quantum Effects and Quasi-Ballistic Transport" IEEE Trans on electron devices, vol. 64, no. 4. pp. 1837-1845, April 2017.A

[7] Avirup Dasgupta et al., "Compact Modeling of Cross-Sectional Scaling in Gate-All-Aroud FETs: 3-D to 1-D Transition", IEEE Trans on electron devices, vol. 65, no. 3, pp.1094-1100, March 2018.

[8] Kai-Shin Li et al., "Sub-60mV-Swing Negative-Capacitance FinFET without Hysteresis", IEDM, December 2015.

[9] S. Khandelwal et al., "Circuit Performance Analysis of Negative Capacitance FinFETs", IEEE Symposium on VLSI Technology, June 2016.

[10] Eunah Ko et al., "Vertical Tunneling FET: Design Optimization With Triple Metal-Gate Layers", IEEE Trans on Electron Devices, vol. 63, December 2016.

A Modeling Study on Performance of a CNOT Gate Devices based on Electrode-driven Si DQD Structures

Hoon Ryu[§][*] and Ji-Hoon Kang[§]

[§]Korea Institute of Science and Technology Information (KISTI), Daejeon 34141, Republic of Korea
[*]Electronic mail: elec1020@kisti.re.kr

Abstract— **Behaviors of quantum bits (qubits) encoded to electron spins in silicon double quantum dot (Si DQD) systems are examined with a multi-scale modeling approach that combines electronic structure simulations and Thoas-Fermi calculations. Covering the full-stack functionality of Si DQD devices from electrode-driven charge controls to logic operations, we investigate the sensitivity of exchange interaction between two initialized qubits and its effect on the fidelity of controlled-NOT gate operations to understand the experimental reported feature. This preliminary work not only presents a theoretical clue for understanding the major control factors for the gate fidelity, but opens the possibility for further exploration of the engineering details of qubit logic gate devices that is hard to be uncovered with experiments due to the time and the expense.**

I. INTRODUCTION

Silicon(Si) has obtained attention for universal quantum gate designs due to its strong potential of easy integration with classical control hardwares using industry-standard fabrication processes. Recently, experimentalists reported a successful implementation of the controlled-NOT (CNOT) gate based on electrode-driven silicon (Si) double quantum dot (DQD) structures [1] with the theoretical foundation presented by Russ *et al.* [2]. Detailed design guidelines for Si DQD qubit gates are not still clear due to the lack of modelling studies that escalate device simulations to the level of logic gate operations. In this preliminary work, we study the effects of physical design factors of Si DQD structures on CNOT operations with a full-stack modeling approach that covers device simulations as well as corresponding time responses of spin-qubits.

II. METHODS

Figure 1(a) shows the simulated DQD structure, which is 2D as the dimension along [001] direction is assumed to be very long as the experimentally reported structure is [1]. At T=1.5K, the bias-dependent electrostatic of Si DQD is simulated with Thomas-Fermi (TF) model [3] assuming the Fermi-energy (E_F) of source-drain electrodes is 0. The QD region near the middle Si layer that confines electrons is simulated with an single-band effective mass model [4] to correct TF-driven charges. A DC magnetic field is applied along [010] direction (B_Y) with a maximal magnitude of 300mT and a $\sim 40\mu$T/nm gradient along [100] direction, to mimic the distribution generated by a cobalt miromagnet [5]. Frequency and

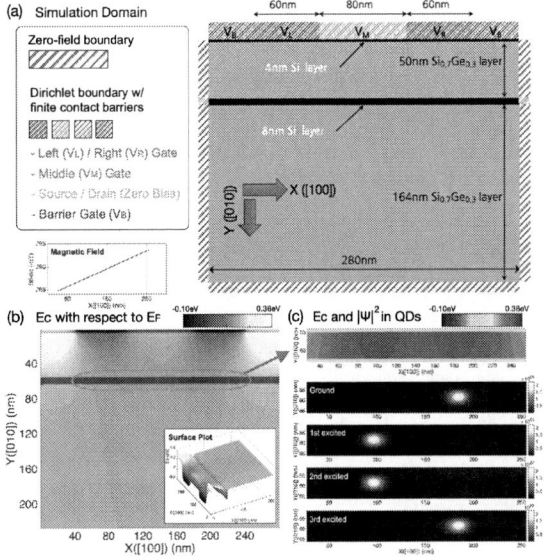

Fig. 1. **Si DQD device and electronic structures.** (a) A 2D simulation domain describing a Si double quantum dot (DQD) structure that is assumed be very long along [001]direction. Quantum confinement along [100] (X) direction is formed by controlling biases applied on top electrodes (V_L,V_M,V_R). Confinement along [010] (Y) direction is created by the conduction band offset between Si and $Si_{0.7}Ge_{0.3}$ layers. (b) Local band structure (E_C, conduction band diagram) of the DQD system when $V_L = V_R = 542$mV and $V_M = 400$mV, and (c) Electron density ($|\Psi|^2$) at the 4 lowest conduction band states. Due to the magnetic field (B_Y) gradient along X direction, Zeeman-splitting in right QD is slightly larger than the one in left QD.

magnitude of an AC magnetic field that is applied along [100] direction (B_X) are used as control variables. Time responses of spin-qubits are simulated with the Heisenberg Hamiltonian of two neighboring spins in a varying magnetic field [2].

III. RESULTS AND DISCUSSION

Wavefunctions in empty DQD: Figure 2(b) shows the conduction band diagram of the Si DQD structure that are simulated with $V_L = V_R = 542$mV and $V_M = 400$mV. Due to the conduction band offset between Si and $Si_{0.7}Ge_{0.3}$ layers, vertical confinement (along [010] direction) is created near the middle Si layer and lateral confinement (along [100]) is driven by biases of the three electrodes. Though QDs are empty in this bias-condition, it is still possible to explore the confinement-driven energy states.

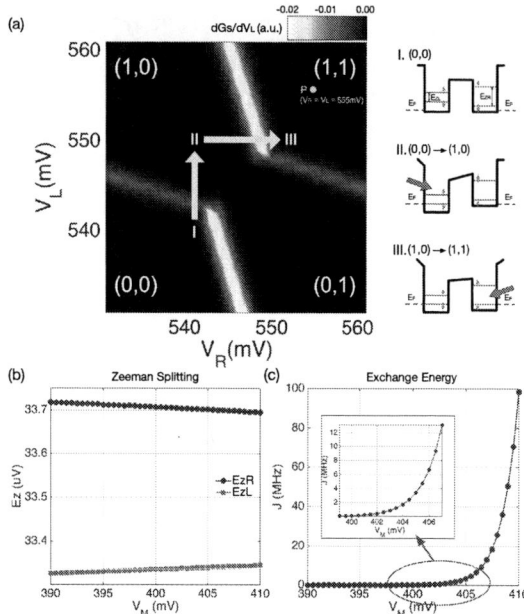

Fig. 2. **Charge and spin-qubit controllability.** (a) A charge stability diagram shown as a function of (V_L, V_R) biases (V_M is set to 400mV) The first step needed for a CNOT operation is to initialize the double quantum dot (DQD) system so both the left and right QD downspin ($|\downarrow\rangle$) state are occupied by a single electron ((1,1) region in the stability diagram). The conceptual band diagrams given in the right side explain the variation of QD potential along the path of charge transfer marked with yellow arrows in the stability diagram. (b) Zeeman-splitting of the right (EzR) and left (EzL) QD and (c) Exchange constant (J) of the two downspin states shown as a function of V_M ($V_L = V_R = 555$mV). ΔV_M of 20mV does not drive remarkable changes of EzR and EzL, but J is very sensitive to V_M.

Electron densities ($|\Psi|^2$) of the 4 lowest conduction band states are given in Figure 1(c). Here, the 1st (ground) and 4th (3rd excited) state are down/upspin ($|\downarrow\rangle/|\uparrow\rangle$) state of the right QD, where the remaining two states are those of the left QD. Zeeman-splitting in right and left QD are calculated as 33.7μeV and 33.3μeV, respectively, and show a 0.4μeV difference due to B_Y gradient along [100] direction.

Control of charge/energy-splitting/exchange constant: A full charge stability diagram is shown in Figure 2(a) and the left arrows indicate the path of charge transfer $(0.0)\rightarrow(1,0)\rightarrow(1,1)$, where the two numbers in parentheses mean the number of electrons filled in left/right QD. Corresponding variation of DQD conduction band diagram is conceptually illustrated in the right subfigures. The starting step of qubit operations is to initialize the DQD system such that the $|\downarrow\rangle$ state of each QD is filled with a single electron ($|\downarrow\rangle_L \otimes |\downarrow\rangle_R = |\downarrow\downarrow\rangle$). When V_M is set to 400mV, our results show that $V_R = V_L > 500$meV are required to perform this initialization process. Once the initialization is done, the energy-level of four 2-qubit states ($|\downarrow\downarrow\rangle/|\downarrow\uparrow\rangle/|\uparrow\downarrow\rangle/|\uparrow\uparrow\rangle$) is determined with Zeeman-splitting in each QD (EzR/EzL) and exchange constant

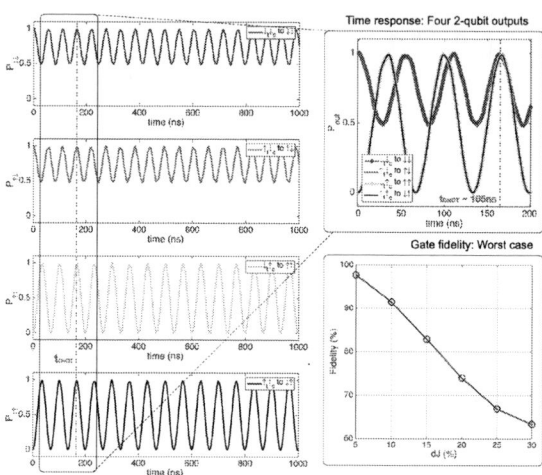

Fig. 3. **CNOT gate operation and fidelity of the worse case.** The left 4 subfigures show time responses of our DQD system against the 4 2-qubit inputs at $V_L = V_R = 555$mV and $V_M = 407$mV. Frequency of the left qubit oscillation depends on the right qubit so, oscillations in the first and last 2 subfigures have slightly different frequencies. At $t_{CNOT} = 165$(ns), however, the oscillations are synchronized and a CNOT operation is conducted. The zoomed-in time responses are given in the right-top subfigure. Fidelity of a CNOT gate operation strongly depends on the charge noise whose effect is considered with unexpected deviation of the exchange constant (dJ).

(J) between $|\downarrow\rangle_L$ and $|\downarrow\rangle_R$ state wavefunction. As shown in Figure 2(b), controlling V_M in a 20mV range (directly related to the barrier height between two QDs) does not drive remarkable changes of EzL and EzR. However, the sensitivity of J to ΔV_M is extremely large and ΔV_M of 10mV is large enough to increase J by a factor of 100. J is the reason why the spin-states in two QDs can be entangled. However, its strong sensitivity to control signals becomes the very issue that must be overcome to procure the stability of gate operations.

Logic operation of CNOT gate: The real-time CNOT gate operation is simulated at $V_L = V_R = 555$mV and $V_M = 407$mV, and time responses of the four 2-qubit inputs are shown in left subfigures of Figure 3, where the top and bottom two subfigures show an oscillation of the left qubit given that the right qubit is $|\downarrow\rangle$ and $|\uparrow\rangle$, respectively. Under an AC magnetic field (B_X) of $\sim$1mT amplitude, the oscillations have slightly different frequencies but are synchronized at $\sim$165ns, conducting a CNOT operation. The right-bottom subfigure in Figure 3 shows the gate fidelity as a function of unexpected deviation of J (dJ), where we observe 20% of dJ already reduces the fidelity under 80%.

IV. CONCLUSION

A preliminary modeling study on the controlled-NOT logic gate device based on the silicon double quantum dot structure is presented. We find that the exchange interaction between a ground-state electron in each quantum dot, which turns out to be extremely sensitive to the bias applied to the barrier gate electrode, would become the major factor that can explain why the experimentally reported fidelity [1] is quite lower than other competitors such as 2-qubit logic gate devices based on superconductors [6] and ion traps [7].

ACKNOWLEDGMENTS

This work has been supported by the Korea Institute of Science and Technology Information (KISTI) institutional R&D program (K-20-L02-C09) and the Institute for Information & communication Technology Promotion (IITP) grant (2019-0-0003) funded by Korea government(MSIP). The NURION high performance computing resource has been extensively used for all the simulations.

REFERENCES

[1] D. M. Zajac, A. J. Sigillito, M. Russ, F. Borjans, J. M. Taylor, G. Burkard, and J. R. Petta, "Resonantly driven CNOT gate for electron spins," *Science* 359, pp. 439-442 (2018).

[2] M. Russ, D. M. Zajac, A. J. Sigillito, F. Borjans, J. M. Taylor, J. R. Petta, and G. Burkard, "High-fidelity quantum gates in Si/SiGe double quantum dots," *Physical Review B* 97, p. 085421 (2018).

[3] V. Grimalsky, L. M. Gaggero-Sager, and S. Koshevaya, "Electron spectrum of δ-doped quantum wells by the Thomas-Fermi method at finite temperatures," *Physica B: Condensed Matter* 406, pp. 2218-2223 (2011).

[4] J. Wang, A. Rahman, A Ghosh, G. Klimeck, and M. Lundstrom, "On the Validity of the Parabolic Effective-Mass Approximation for the I-V Calculation of Silicon Nanowire Transistors," *IEEE Transactions on Electron Devices* 52, pp. 1589-1595, (2005).

[5] R. Neumann and L. R. Schreiber, "Simulation of micro-magnet stray-field dynamics for spin qubit manipulation," *Journal of Applied Physics* 117, p. 193903 (2015).

[6] S. Li, A. D. Castellano, S. Wang, Y. Wu, M. Gong, Z. Yan, H. Rong, C. Zha, C. Guo, L. Sun, C. Peng, X. Zho, and J.-W. Pan, "Realisation of high-fidelity nonadiabatic CZ gates with superconducting qubits," *npj Quantum Information* 5, p. 84 (2019).

[7] T. Choi, S. Debnath, T. A. Manning, C. Figgatt, Z.-X. Gong, L.-M. Duan, and C. Monroe, "Optimal quantum control of multimode couplings between trapped ion qubits for scalable entanglement," *Physical Review Letters* 112, p. 190502 (2014).

12-4 Simulation and Evaluation of Plasmonic Circuits

Mitsuo Fukuda
Electrical & Electronic information Engineering
Toyohashi University of Technology
Toyohashi, Aichi, Japan
fukuda.mitsuo.af@tut.jp

Yasuhiko Ishikawa
Electrical & Electronic information Engineering
Toyohashi University of Technology
Toyohashi, Aichi, Japan
ishikawa@ee.tut.ac.jp

Abstract— This paper presents and discusses the modeling and evaluation of plasmonic circuits including plasmonic devices such as a multimode interferometer, mode converter, multiplexer, logic circuits, and signal transmission networks.

Keywords— plasmonic waveguide, plasmonic mode convertor, plasmonic multiplexer, plasmonic integrated circuit, silicon integrated circuit

I. INTRODUCTION

High-performance information communication systems are now indispensable for daily life in various fields, and silicon integrated circuits (ICs) are key devices in such systems. The information processing speed of silicon integrated circuits, however, gradually saturate. To enhance silicon IC operating speeds, optical interconnects have begun to be introduced [1]–[3]. Plasmonic circuits are also promising candidates for this IC enhancement, although the development of these circuits is still in its primitive stage.

In the nanoscale range where silicon photonic waveguides are scarcely applied, plasmonic circuits are strong tools to construct high-speed circuits merged with silicon ICs [4]. Although plasmonic circuits generally exhibit heavy signal-transmission loss via ohmic loss, we have clarified that the circuits formed with SiO_2 stripes on a metal film within the range of less than a few hundred square micrometers exhibit lower transmission loss per single bit when compared with electric circuits even at a bit error rate of 10^{-30} [5]. In addition, we indicated the feasibility of a full adder operating at light speed without large loss or heat generation by applying the configuration developed for a cascadable plasmonic full adder to silicon waveguides [6]. For the plasmonic full adder, the basic operation of plasmonic circuits were numerically and experimentally demonstrated, and the output signals were connected to plasmonic detector-integrated metal-oxide semiconductor field effect transistors. These circuits were fabricated by combining only single- and multi-mode waveguides, the latter acting as multimode interferometers (MMIs). The circuits operated at light speed via interference of transmitted plasmonic signals. Through the plasmonic circuits, various kinds of plasmonic signals were transmitted in nano/micrometer-scale area using the techniques currently used in commercial optical fiber communication systems [4].

The design and modeling techniques for the waveguides have been relatively well established until now, but more precise design and simulation techniques are needed to fabricate the above-mentioned complicated plasmonic circuits operating via interference. Without precise simulations adjusting the waveguide shapes and the propagating modes of signals, the working nano/micrometer-scale plasmonic circuits cannot be practically fabricated. This paper presents and discusses the simulation procedures of the plasmonic circuits. The simulations provided guidelines for design and fabrication of nano/microscale plasmonic circuits, and these

simulation results were experimentally confirmed by scanning near-field optical microscopy (SNOM).

II. MODELING OF PLASMONIC CIRCUITS

We used the 3D finite-difference time-domain method using electromagnetic wave analysis software for the circuit simulations after calculating waveguide parameters such as propagation constants (i.e. complex refractive indices). This was done by solving the characteristic equation of a waveguide for a multilayer structure (e.g., SiO_2 film/metal film/Si substrate) and then for a mesa structure (e.g., SiO_2 mesa structure/metal film/Si substrate). For the calculation, the mesh spacing was changed from a few nanometers to a few tens of nanometers corresponding to the structural complexity to be simulated. The time spacing was set at a range of a few tens of attosecond. Perfectly matched layer (PML) absorbing boundary conditions were applied to minimize the influence of light/plasmon reflection. The large-scale integrated circuits were divided into a few sections and then precisely simulated with small mesh spacing. The circuit designs and simulations were performed taking into consideration the actual fabricated waveguide shapes and the propagating modes of plasmonic signals. The simulated circuits were actually fabricated and then evaluated using SNOM.

III. PLASMONIC WAVEGUIDES AND COMPONENTS

Surface plasmons (SPs) are quanta of an electromagnetic wave coupled with the collective oscillation of electrons and propagate along the interface between a metal and a dielectric layer (or air) at light speed. To construct plasmonic circuits, waveguides comprising dielectric stripes deposited on a metal film (i.e., dielectric-loaded surface plasmon (DLSP) waveguides) were used in this study because the structure is technically simple and easy to translate to waveguide components. Fig.1 shows an example of a SiO_2-loded waveguide. This acts as a plasmonic single-mode waveguide in 1300 and 1550 nm wavelength bands.

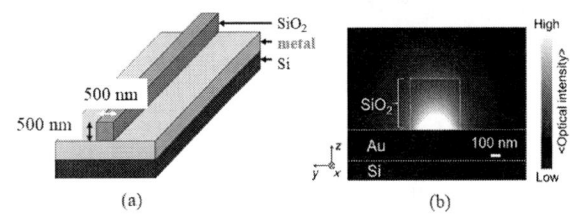

(a)   (b)

Figure 1. (a) Schematic of plasmonic single-mode waveguide in 1300 and 1550 nm wavelength bands, (b) simulated cross sectional view of optical power distribution.

This work was supported in part by JSPS KAKENHI Grant number JP18K04282.

Plasmonic circuits are planar lightwave circuits, and thus bending and crossing waveguides [7] are necessary to increase circuit flexibility. An example of simulated and experimental results for a bending single-mode waveguide are shown in Fig. 2. For a bending waveguide, an optical power distribution simulation was performed along the upper surface of the mesa-shaped SiO_2 waveguide, and the optical power along the surface was monitored using SNOM. The simulated and experimental results coincided, and simulation was used to identify an optimal structure for a bending waveguide using various conditions, curvatures and degrees of bending. Here, the waveguide samples were fabricated using focused ion beam milling etching, thus modifying the rectangular cross-sectional shapes for the etched mesa-structure. In such cases, the simulation was carried out after modifying the shape of the cross-section to match the fabricated sample. This modification was important to more exactly simulate some complex circuits, although the results of simulation and experiment were not significantly different when only a simple waveguide was examined.

(a) SEM image

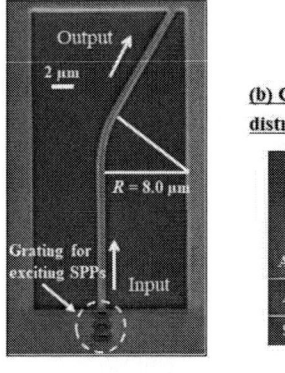

(b) Cross section of optical field distribution (FDTD simulation)

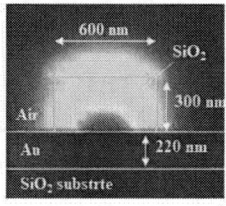

(c) FDTD simulation   **(d) Near-filed optical image**

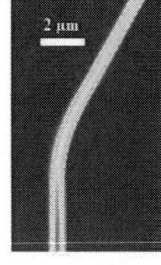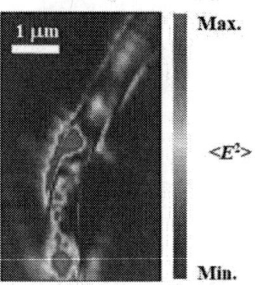

Figure 2. (a) Fabricated bending waveguide, (b) the cross section of plasmonic optical field distribution calculated with FDTD method, (c) simulated plasmonic power distribution along with the bending waveguide, and (d) plasmonic power distribution monitored by SNOM. The wavelength of light incident to the grating is set at 1300 nm.

These waveguide simulations were extended to functional components such as multiplexers, demultiplexers, and mode-converters. These components were constructed by combining single- and multimode waveguides. An example of a plasmonic multiplexer is indicated in Fig. 3. The 1310 and 1550 nm-wavelength plasmonic signals were inputted to the MMI from the different input ports and outputted to the same port through the MMI. In the FDTD simulation and SNOM image in Fig. 3, the plasmonic signal emission from the MMI

part is recognized, and some tapered structures were introduced into the MMI to suppress the signal leak [8].

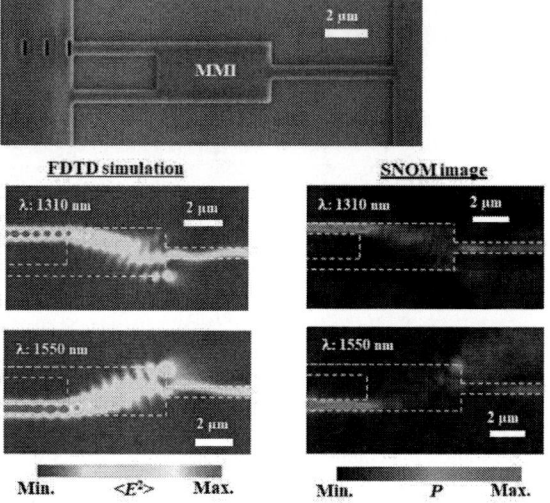

Figure 3. Fabricated multiplexer for 1310- and 1550-nm-wavelength band. The plasmonic signals are converted from highly coherent light beams at the grating set just before the input port.

SEM image

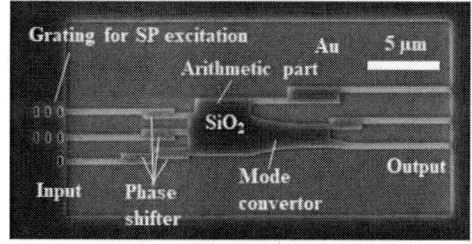

FDTD simulation

SNOM image

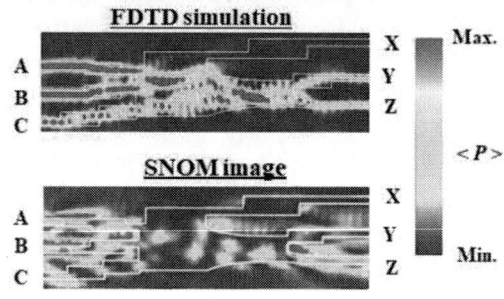

Fig. 4. Fabricated plasmonic logic circuit (half adder). The surface plasmon signals are converted from a highly coherent light beam with a 1310 nm wavelength in the air.

The waveguide and multiplexer discussed above are composed of relatively simple structures. For such components, the performances of the fabricated components well coincide with each other without confirming the distribution of the plasmonic signal phase and propagating modes in single and multimode (MMI) waveguides. In complicated plasmonic circuits, however, the distribution of signal phase and propagating mode in the waveguide has to be considered in the design, if interference of signals is employed in the components such as MMI. Fig. 4 shows the simulation

266

and experimental results for a logic circuit, wherein a SiO_2 film was deposited and patterned on a Au film. The input signals A and B were computed (i.e., interfered in MMI) and outputted to Z through a mode converter. The output Y was an emission port for unnecessary plasmon signals after computing and the key port for maintaining correct interference without noise [9]. The half adder operated well as a Boolean logic circuit. Here, the interference patterns in the MMI (phase shifters, arithmetic part, and mode converter) had to be precisely controlled by optimizing the MMI shapes to suppress the introduction of unnecessary higher-order- mode signals to the next MMI. Without this interference mode control, working circuits with complicated networks could not be fabricated. These procedures are very important for developing nano/microscale circuits and components because experimental examination is difficult for such small components and circuits.

In addition, the propagation modes should be also paid attention at the output port of MMIs. Just after interference in MMI, some modes are introduced to the single-mode waveguide. After transmitting to some extent in the single-mode waveguide, only signals with single-mode remain. If the two MMIs are connected with a single-mode waveguide, signals with higher-order modes outputted from the first MMI are introduced to the next MMI, and the interference in the second MMI is not correctly carried out. This mixing of higher-order modes was suppressed by narrowing the waveguide or lengthening the single-mode waveguide between the two MMIs. By taking these phase and mode control into the consideration, the design and fabrication of working plasmonic cascadable full adder [6] and complicated wavelength-division-multiplexing circuits with MOSFETs [5] was accomplished.

IV. SUMMARY

For simulation and modeling of complicated plasmonic circuits and networks, modeling with reference to the actual fabricated shapes of the individual components and waveguides is indispensable for fabricating working optical networks. Controlling the transmitted optical mode by modifying the shape of the circuit and network is also a key design and fabrication factor to suppress unnecessary interference within the components. Using these techniques, functional and complicated plasmonic logic circuits and networks were obtained.

ACKNOWLEDGMENT

The authors would like to thank Dr. M. Ota, Ms. A. Sumimura, R. Watanabe, and Mr. T. Hirano for their supports and discussions on the design and fabrication of plasmonic waveguides while enrolled at Toyohashi University of Technology.

REFERENCES

[1] M. Haurylau, G. Chen, H. Chen, J. Zhang, N. A. Nelson, D. H. Albonesi, E. G. Friedman, P. M. Fauchet, "On-chip optical interconnect roadmap: challenges and critical direction," IEEE J Selected Topics Quantum Electron, vol.12, pp.1699–705, 2006.

[2] K. Ohashi, K. Nishi, T. Shimizu, M. Nakada, J. Fujikata, J. Ushida, S. Torii, K. Nose, M. Mizuno, H. Yukawa, M. Kinoshita, N. Suzuki, A. Gomyo, T. Ishii, D. Okamoto, K. Furue, T. Ueno, T. Tsuchizawa, T. Watanabe, K. Yamada, S. Itabashi, J. Akedo, "On-chip optical interconnect," Proc. IEEE, vol.97, pp.1186–98, 2009.

[3] D.A.B. Miller, "Optical interconnects to electronic chips," Appl. Opt., vol. 49, pp. F59-F70, 2010.

[4] M. Fukuda, S. Okahisa, Y. Tonooka, M. Ota, T. Aihara, Y. Ishikawa, "Feasibility of Plasmonic Circuits in Nanophotonics," IEEE Access, vol.8, pp.142495-142506, 2020.

[5] M. Fukuda, Y. Tonooka, T. Inoue, M. Ota, "Feasibility of plasmonic circuits for on-chip interconnects," Solid-State Electron. vol. 156, pp. 33-40, 2019.

[6] M. Fukuda, R. Watanabe, Y. Tonooka, M. Ota, "Feasibility of Cascadable Plasmonic Full Adder," IEEE Photon. J., vol. 11, No. 4, Aug. 4801612, 2019.

[7] M. Ota, M. Fukuhara, A. Sumimura, M. Ito, T. Aihara, Y. Ishii, M. Fukuda, "Dielectric-loaded surface plasmon polariton crossing waveguides using multimode interference," Opt. Lett., vol.40, No.10, pp.2269-2272, 2015.

[8] A. Sumimura, M. Ota, K. Nakayama, M. Ito, Y. Ishii, M. Fukuda, "Low return loss plasmonic multiplexer with tapered structure," IEEE Photon. Technol. Lett., vol.28, No.21, pp.2419-2422, 2016.

[9] M. Ota, A. Sumimura, M. Fukuhara, Y. Ishii, M. Fukuda, "Plasmonic-multimode-interference-based logic circuit with simple phase adjustment," Scientific Reports, vol. 6, 24546, 2016.

Numerical study of surface chemical reactions in 2D-FET based pH sensors

A. Toral-Lopez
Depto. de Electrónica
Univ. Granada
Granada, Spain
atoral@ugr.es

E.G. Marin
Depto. de Electrónica
Univ. Granada
Granada, Spain
egmarin@ugr.es

J. Cuesta
Depto. de Electrónica
Univ. Granada
Granada, Spain
f52cuelj@correo.ugr.es

F.G. Ruiz
Depto. de Electrónica
Univ. Granada
Granada, Spain
franruiz@ugr.es

F. Pasadas
Dept. d'Enginyeria Electrònica
Universitat Autònoma de Barcelona
Barcelona, Spain
Francisco.Pasadas@uab.cat

A. Medina-Rull
Depto. de Electrónica
Univ. Granada
Granada, Spain
amedinarull@ugr.es

A. Godoy
Depto. de Electrónica
Univ. Granada
Granada, Spain
agodoy@ugr.es

Abstract—This work numerically evaluates the impact of surface chemical reactions on the performance of 2D-FET based pH sensors. More precisely, we focus on the adsorption of chlorine ions and the expulsion of protons at the sensing interface of FET sensors. This analysis is performed through numerical simulations encompassing the modelling of both the semiconductor device and the liquid solution to be analysed. In the semiconductor region the 2D Poisson - 1D Continuity equations are self-consistently solved, while in the electrolyte region we deal with the modified Poisson - Boltzmann system [1]. The simulator also includes the interactions taking place at the electrolyte-sensing layer interface through: i) the non-constant profile of water permittivity, and ii) the steric effects in the surface ions concentration by means of the Potentials of Mean Force (PMFs) [2], [3]. This comprehensive description of the electrolyte-device interface provides a suitable framework to unveil the relevance of multiple chemical reactions, such as the adsorption of chlorine ions, on the behaviour of 2D-FET based pH sensors.

Index Terms—pH, ISFET, Site-Binding, modified Site-Binding, Potential of Mean Force

1. Introduction

The Ion-Sensitive FET (ISFET) was introduced back in the 70s by P.Bergveld et al. [4] as a device to detect neural activity. Since then, it has also been widely used as a chemical sensor. Very recently, 2D-materials have shown up as promising candidates to implement this kind of sensors, due to their atomic thickness [5], deeply impacting on their design and performance. Compared with the plethora of experimental results provided by numerous research groups [6]–[9], the activities carried out to develop analytical and numerical approaches able to describe these devices have been quite limited [10]–[12]. In the case of pH sensing, previous studies only consider the chemical reactions between the sensing interface and protons, neglecting the impact of additional reactions in the adsorption of hydrogen ions. In order to assess them, we have integrated the generalised approach presented in [13], [14] in our numerical simulator allowing us to evaluate the impact of the adsorption of chlorine ions on the pH sensor performance.

2. Simulation of 2D-ISFETs

The semiconductor device and electrolyte models are combined to achieve a comprehensive description of the whole device [12]. To do so, we first employ different models to estimate the charge in each region and then the 2D Poisson equation is solved to estimate the electrostatic potential.

A. Semiconductor device modelling

In the semiconductor region we consider a diffusive transport [15]. Under equilibrium conditions, the electron (hole) density profile, n_L (p_L), is calculated using the 2D density of states. This information is then used to solve the non-equilibrium situation. In that case, n_L (p_L) is calculated using the 1D Drift-Diffusion transport equation.

B. Electrolyte and sensing interface modelling

Regarding the electrolyte region, the local concentration of the i-th ion is related to the potential by the equation [1]:

$$c_i = \frac{c_{i,0} \exp\left(-V_{\text{PMF},i} - qz_i \frac{V-V_{\text{Ref}}}{k_B T}\right)}{1 - 2\frac{c_{i,0}}{c_{\max}}\left(1 - \cosh\left(|z_i|\frac{V-V_{\text{Ref}}}{k_B T}\right)\right)} \quad (1)$$

where $c_{i,0}$ is the bulk concentration, z_i the valence, $c_{max,i}$ the maximum allowed concentration and $V_{PMF,i}$ is the PMF profile. This potential is combined with a non-constant water permittivity profile at the device - electrolyte interface. The parameters required for these models are extracted from [2], [3]. As for the reactions taking place at the sensing interface, we integrate the approach reported in [13], [14]. There, each reaction is defined by a simple relation between the active element concentration f_i and the product concentration, f_j:

$$A + B \rightleftharpoons AB, \ [AB] = \frac{[B]}{K}[A] \Rightarrow f_j = \frac{[B_j]}{K_{ij}}f_i \quad (2)$$

Once the concentrations of the different components are obtained, the net charge in the surface

with an active site density N_s can be calculated as:

$$Q_{\text{net}} = qN_{\text{s}}\sum_i z_i f_i \quad (3)$$

The reactions considered in the Site-Binding (SB) model and its modified version (mSB) are depicted in Figure 1. In both models the behaviour of the sensing interface is captured by the reaction constants K_a and K_b. In the case of the mSB model, there is a third reaction constant K_c concerning the adsorption of chlorine ions.

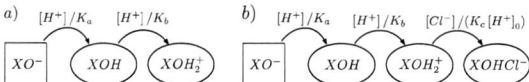

Fig. 1: Reactions considered in the SB model (a) and its modified version, the mSB model (b). The material of the sensing layer is indicated as X and K_a, K_b and K_c are the reaction constants. $[H^+]$ is the hydrogen ion concentration near the interface, $[H^+]_0 = 10^{-pH}$ is the bulk hydrogen concentration and $[Cl^-]$ is the chlorine concentration near the interface.

3. Results

In all the simulations, the structure considered, depicted in Figure 2, is a MOSFET whose channel is formed by a 100nm long monolayer of MoS_2 sandwiched between a 20nm-thick SiO_2 layer, which acts as a substrate, and a 5nm-thick Al_2O_3 layer acting as the sensing layer. The electrolyte considered is based on KCl along with NaOH and HCl, whose concentration change according to the pH. We assume pH values ranging from 4 up to 10, along with four different KCl concentrations, [KCl]: 10mM, 50mM, 100mM, 500mM. Concerning the reactions constants used for the SB models, they are summarised in Table I.

TABLE I: Al_2O_3 parameters. The reaction constants pK_i are $pK_i = -\log_{10}(K_i)$

Material	ϵ	pK_a	pK_b	pK_c	N_s (cm^{-2})
Al_2O_3	$14\epsilon_0$	6	10	2.48	$8 \cdot 10^{14}$

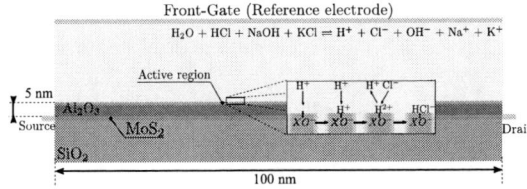

Fig. 2: Structure of the device considered in the simulations. The channel is defined by a mono-layer MoS_2 semiconductor sandwiched between a SiO_2 layer, acting as substrate, and an Al_2O_3 layer interface with the electrolyte.

Those parameters give rise to the $N_{it} - V_{it}$ profiles depicted in Figure 3 when using the SB model (a) or the mSB model (b).

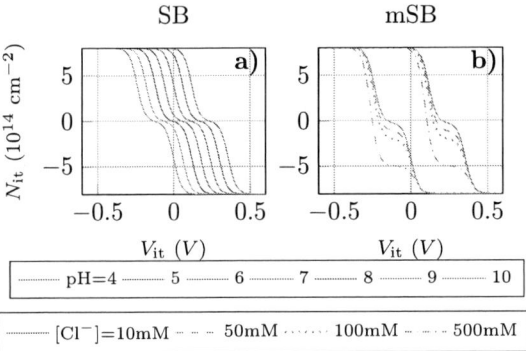

Fig. 3: $N_{it} - V_{it}$ when using the SB (a) or the mSB (b) models. In (b) the data for pH values between 5 and 9, both included, were omitted as they show the same shift than in (a) when the pH is reduced.

The results for both SB models show a shift towards positive interface potentials as the pH is reduced. These results could be expected as reduction of pH give rise to an increase in $[H^+]$, and higher potential is required to obtain a null net charge at the interface. In (b) the mSB considers the presence of $[Cl^-]$ and its effect becomes evident. As the bulk Cl^- concentration increases, the light plateau of the profiles moves down towards negative charge values, and it might even disappear for higher values of $[Cl^-]$. º

This preliminary analysis shows clear differences in the $N_{it} - V_{it}$ profile achieved with each model. So that, it would be interesting to shed light on the impact of these differences on the device response.

A. Site-Binding

Figure 4 depicts the sensor transfer characteristic making use of the basic SB model, showing a decrement of the threshold voltage V_{th} as the pH is reduced: the lower the pH, the higher the concentration of positive H^+ ions, and thus the lower the potential required to switch on the device. To evaluate the influence of KCl, we calculate the change in the current ΔI with respect to the minimum concentration [KCl]=10mM, Figure 4 (bottom). The maximum values of ΔI are quite low compared with I. The results show an interesting feature with pH as ΔI changes from negative to positive at pH=8. This can be associated to the characteristics of the active layer, Figure 3, as it shows a neutral charge for $V_{FG} = 0V$ at that pH value.

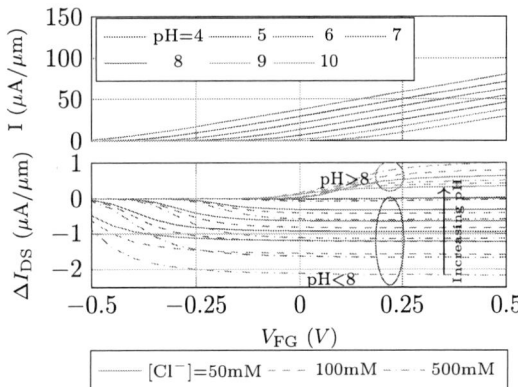

Fig. 4: Sensor I-V_{FG} curves for [KCl]=10mM using the SB model (top) and increment of current when [KCl] is increased (bottom).

B. Modified Site-Binding

The results obtained for that scenario are depicted in Figure 5. The magnitude of ΔI is almost seven times larger than the SB model, and the trend with V_{FG} is monotonic, i.e. there is not a change in the ΔI sign showing that the current always decreases as [KCl] increases, independently of the pH considered. This is consistent with the additional reaction considered in the model: the chlorine ions generate a negative charge in the active layer that reduces the amount of the electrons in the channel. Moreover, the higher the concentration of these ions, the more acute is this effect.

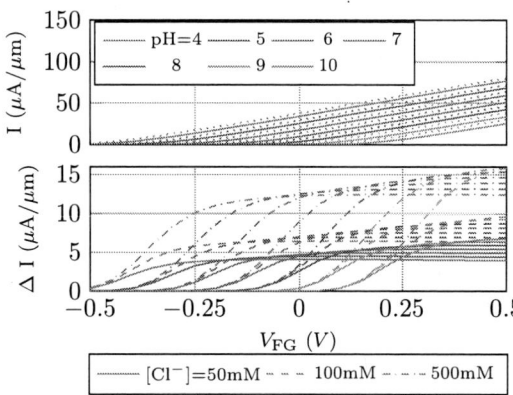

Fig. 5: Sensor I-V_{FG} curves for [KCl]=10mM using the mSB model (top) and ΔI when [KCl] is increased (bottom). Dotted lines show the previous results with the SB model.

4. CONCLUSIONS

In this work we have evaluated the transfer response of a MoS$_2$ FET based pH sensor making use of two models for the active layer. The first model only considers the adsorption of hydrogen ions, while the second one includes an additional reaction concerning the adsorption of chlorine ions.

Relevant differences have been revealed comparing both models, highlighting the importance of side reactions on the modelling of chemical sensors. Once the side reactions are included through the mSB model, the impact of the changes of [KCl] is assessed: as the concentration of chlorine ions increases, the output current diminishes notably. These changes make possible

to obtain the same response by either increasing the pH or the KCl concentration. Therefore, it becomes critical to take this effect into account when analysing pH sensors, due to the "crosstalk" between hydrogen and chlorine ions adsorption.

ACKNOWLEDGMENT

The authors thank the financial support from the Spanish Government under projects TEC2017-89955-P (MINECO / AEI / FEDER, UE) and EQC2018-004963-P (MINECO / AEI / FEDER, UE), and the European Commission under H2020 project WASP (contract no. 825213). A. Toral-López acknowledges the FPU program (FPU16/04043). E. G. Marín acknowledges Juan de la Cierva Incorporación IJCI-2017-32297 (MINECO/AEI).

REFERENCES

[1] F. Pittino, P. Palestri, P. Scarbolo, D. Esseni, L. Selmi, *Solid-State Electronics* **98**, 63 (2014).

[2] N. Schwierz, D. Horinek, R. R. Netz, *Langmuir* **26**, 7370 (2010).

[3] N. Schwierz, D. Horinek, R. R. Netz, *Langmuir* **29**, 2602 (2013).

[4] P. Bergveld, *IEEE Transactions on Biomedical Engineering* **BME-17**, 70 (1970).

[5] D. Sarkar, *et al.*, *ACS Nano* **8**, 3992 (2014).

[6] P. K. Ang, W. Chen, A. T. S. Wee, K. P. Loh, *Journal of the American Chemical Society* **130**, 14392 (2008).

[7] M. H. Lee, *et al.*, *Nanoscale* **7**, 7540 (2015).

[8] H. Wang, P. Zhao, X. Zeng, C. D. Young, W. Hu, *Nanotechnology* **30**, 375203 (2019).

[9] W. Liao, W. Wei, Y. Tong, W. K. Chim, C. Zhu, *ACS Applied Materials & Interfaces* **10**, 7248 (2018).

[10] R. Narang, M. Saxena, M. Gupta, *IEEE Transactions on Electron Devices* **64**, 1742 (2017).

[11] A. Shadman, E. Rahman, Q. D. Khosru, *Sensing and Bio-Sensing Research* **11**, 45 (2016).

[12] A. Toral-Lopez, *et al.*, *Nanoscale Advances* **1**, 1077 (2019).

[13] L. J. Mele, P. Palestri, L. Selmi, *Proceedings of SISPAD* (2019).

[14] L. J. Mele, P. Palestri, L. Selmi, *IEEE Transactions on Electron Devices* **67**, 1149 (2020).

[15] A. Toral-Lopez, *et al.*, *Nanomaterials* **9**, 1027 (2019).

12-6

A Combined First Principles and Kinetic Monte Carlo study of Polyoxometalate based Molecular Memory Devices

P. Lapham, O. Badami, C. Medina-Bailon, F. Adamu-Lema, T. Dutta, D. Nagy, V. Georgiev
and A. Asenov
Device Modelling Group
James Watt School of Engineering
University of Glasgow, Glasgow G12 8QQ, United Kingdom
E-mail: vihar.georgiev@glasgow.ac.uk

Abstract--In this paper, we combine Density Functional Theory with Kinetic Monte Carlo methodology to study the fundamental transport properties of a type of polyoxometalate (POM) and its behaviour in a potential flash memory device. DFT simulations on POM molecular junctions helps us demonstrate the link between underlying electronic structure of the molecule and its transport properties. Furthermore, we show how various electrode-molecule contact configurations determine the electron transport through the POM. Also, our work reveals that the orientation of the molecule to the electrodes plays a key role in the transport properties of the junction. With Kinetic Monte Carlo we extend this investigation by simulating the retention time of a POM-based flash memory device. Our results show that a POM based flash memory could potentially show multi-bit storage and retain charge for up to 10 years.

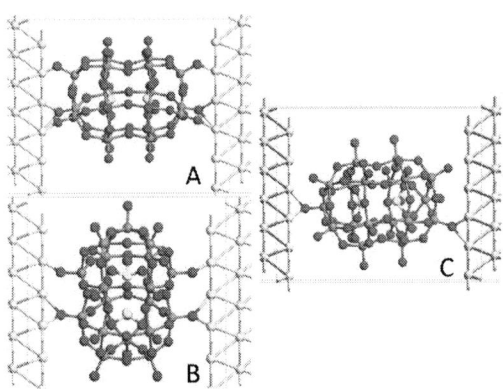

Figure 1. Au-POM-Au molecular junctions with different orientations (A&C vs B) and different number of contacts. The POM molecule is [W18O54(SO3)2]-4 where S=Yellow, W=Blue and O=Red.

INTRODUCTION

Molecular electronics is an exciting interdisciplinary field that is attempting to overcome the inherent limit of miniaturization of conventional CMOS electronic devices [1]. One of the main challenges in fabricating commercial molecular electronic devices is a lack of fundamental understanding of quantum transport at the molecular level [2]. Using first principle methods and higher-level device simulation techniques the transport properties of molecules and realistic molecular devices can be understood further [3]. One such potential molecular device is based on Wells-Dawson Polyoxometalates (POMs). Due to their extremely small size and the fact they are highly redox active, POMs are being explored for potential use as the charge storage component in flash memory [4]. In addition to their size and redox properties, the fact that they are oxygen rich suggest they will have good compatibility with the CMOS fabrication process [5]. They are also thermally stable allowing for the device fabrication and operation [6].

Much of the previous work has focused on high level device simulation of POM based devices. The aim of this work is to fundamentally understand the transport properties of a POM molecule and link its underlying electronic structure to the current flow (electron transport). The most extensively studied single molecule device, both theoretically and experimentally, is the molecular junction where a single molecule is placed between two metal contacts. For that reason, gold-POM-gold (Au-POM-Au) devices were studied using Density Functional Theory (DFT).

in combination with non-equilibrium Green's function (NEGF) to explore the electron transport properties of a specific POM molecule, $[W_{18}O_{54}(SO_3)_2]^{-4}$

Finally, to assess the use of POM molecules as the charge storage component in a flash cell, the retention time of a realistic POM-based flash cell is investigated using Kinetic Monte Carlo (KMC) approach. Some important input parameters (such as energy positions of molecular levels) are taken from DFT, thus this work combines both simulation methodologies.

SIMULATION METHODOLOGY

In this work the simulations on the POM molecular junctions are carried out using DFT in combination with NEGF approach as implemented by QuantumATK-2018.06 software [7]. To accurately describe the electronic properties of the molecule, SGGA.BP86 functional was employed as it has been shown to successfully describe POMs in previous publications [8]. The SG15 pseudopotential and a medium basis set - comparable to DZP (Double Zeta Polarised)- were also used and satisfactorily reproduced the geometry and electronic properties of $[W_{18}O_{54}(SO_3)_2]^{-4}$ [4],[7],[9]. Due to computational cost, SZP (Single Zeta Polarised) basis-set was used to adequately describe the gold electrodes.

The geometry of the molecule was optimized in vacuum and the energy levels calculated using DFT. The devices were then built by placing the geometrically optimized POM cluster between two Au electrodes which are used as source and drain contact. In this way we were able to calculate the current flow in the POM molecule using NEGF computational method. Finally, a drain bias was applied incrementally between -1.5V

273

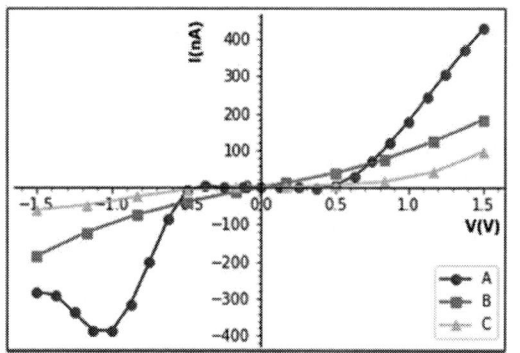

Figure 2. Simulated drain current – drain Bias curves computed for all configurations: A,B and C using DFT and NEGF methodology.

to +1.5V for all devices and the transmission spectra and current at each bias point were calculated.

The retention time was studied using Kinetic Monte Carlo (KMC) module of the in-house simulator – NESS[10],[11]. The molecule's energy levels and geometry obtained from the DFT calculations were used as input parameters for NESS. The KMC module calculates the changes in threshold voltage (V_T) due to the emission of electrons as a function of time. The KMC simulations were repeated 75 times to capture the statistical variability of the device.

RESULTS AND DISCUSSION

One of the biggest challenges in the field of molecular electronics is the lack of detail at the atomic level. The current approach for theoretically studying molecular junctions is to simulate several reasonable junction geometries to gain an average picture of a realistic device where there are intrinsic geometric changes in the environment [12]. The geometry and bonding of the molecule to the electrode/surface largely dictates the transport properties of molecular devices [13].

Fig. 1 shows three different of many possible configurations for the molecular junctions aimed at probing the effect of molecule orientation and contact with the gold electrodes. The working models show the bulk electrodes act as a "reservoir of electrons" whilst the transport properties are dominated by the molecule and its connection to the electrodes.

The drain current - drain voltage (I_D-V_D) characteristics of all configurations have been calculated between the bias of - 1.5V and +1.5V and are illustrated in Fig. 2. Using the Landauer-Buttiker formalism combined with NEGF and DFT approaches, the theoretical transmission spectra for the three different configurations presented in Fig.1 can be calculated. These are shown in Fig.3 at zero drain bias.

The transmission spectrum highlights energies at which electrons can pass through the system via scattering and describes the probability of this transmission T(E). The computed spectra are also linked with the underlying symmetry of the molecular orbitals of the POM cluster. The position of each transmission peak corresponds to the conduction channel and hence molecular orbital. For instance, the first peak above the Fermi level (E_f=0.0 eV) is directly linked to the Lowest

Transmission Spectrum

Figure 3. Transmission spectra for the three molecular junction configurations computed using the combination of DFT and NEGF

Unoccupied Molecular Orbital (LUMO). The peak below the E_f is due to the Highest Occupied Molecular Orbital (HOMO).

The coupling of the molecule to the electrodes in the molecular junction results in a shift in the position of the energy levels compared to the isolated molecule. Depending on the strength of the coupling, the discrete energy levels are broadened into band-like states. This broadening affects the transmission coefficient which correspond to a discrete energy level(or levels) of the molecule [14].

Fig. 3 shows how the intensity of the peaks at similar energies are significantly different for each configuration. This highlights the impact that contact geometry and the strength/number of contacts between the molecule and electrodes has on the transport. The bonding between molecule and the electrode determines how the molecular energies are distorted and hence influences the transmission coefficients of the POM junctions.

Another important feature is the lack of peaks between -1eV and 1eV which corresponds to the HOMO-LUMO gap of the molecule. This gap influences the electron transport properties of the device with larger gaps resulting in insulator-like properties. In addition, the symmetry of the HOMO-LUMO levels-in the device- have a very similar profile as those of the isolated molecule in vacuum [4].

The I_D-V_D curves presented in Fig. 2 are consistent with the computed transmission spectra shown in Fig.3. Clearly, configuration A has higher drain current as compared to B and C. This is because, in configuration A, the POM cluster has six oxygen atoms in direct contact to the gold electrodes, whereas configurations B and C have four and two, respectively (please see Fig. 1). Larger contact area between the molecule and the electrodes leads to more conduction channels and hence higher probability for charge transport between the source and the drain.

The contact geometry of A seems to favorably align the HOMO and LUMO peaks within the bias window resulting in considerably higher current than configurations B & C. In addition, the "horizontal" geometry of A & C suggests better

274

alignment of the transmission peaks as their intensity (T(E)) are higher than that of B. This is visualized in the electron density difference calculations in Fig. 4. The electron density difference was calculated and visualized for the three configurations by subtracting the values of the electron density at drain bias 1.5V and at 0.0V. The differences are visualized with the same isovalues.

It is evident from Fig.4 the following assertions. Firstly, the difference in electron density is the most pronounced at the drain side (right hand side is the drain electrode) hence where there is the maximum change of the charge. This can be directly linked to the most probable conduction pathways. Secondly, structure A shows the biggest clouds of the electron density difference hence the movement of electrons from source to drain is bigger than that for B and C which is consistent with the I_D-V_D curves in Fig.2.

Thirdly, Fig.4b highlights the link between the molecule's discrete energy levels and the transport. The blue surface represents where the density has moved from and is centered around the SO_3^{2-} moieties, which resembles the structure of the HOMO. This is a strong indication that the movement of electrons from the source to the drain is predominately dominated by the HOMO level which has the same symmetry and shape as in the isolated POM molecule.

When a bias is applied across a molecular junction, the transmission peaks are shifted to higher energy with increased drain bias. Fig. 5 gives insight to how the transmission spectra changes under drain bias for configuration A. The blue line represents the drain bias window. The spectrum with no blue line represents zero bias transmission. From top to bottom the drain bias goes from -2.0V, -1.5V, -1.0V -0.5V 0.0V, 0.5V, 1.0V, 1.5V, 2.0V. The figure demonstrates how the intensity of the transmission peaks decrease within the bias window. As a result, the change in the transmission spectra with increased applied bias explains why less current flows than expected from the zero-bias transmission spectra.

Fig. 5 also provides an insight to why the current decreases between the bias of -1V to -1.5V. The intensity of the peaks at a bias of -1V is higher than that of the peaks in -1.5V. The current has been calculated over a larger energy window and so the current goes down. This is consistent with the Landauer formalism for calculating current [15].

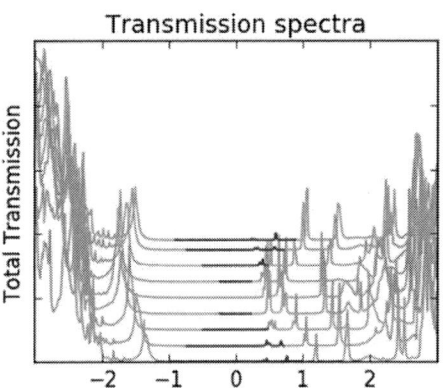

Figure 5. Series of transmission plots for Configuration A under different applied bias from -2.0V to 2.0V. From top to bottom, the blue line illustrates the applied bias window at -2.0V, -1.5V, -1.0V, -0.5V, 0V, +0.5V, +1.0V, +1.5V, +2.0V.

$$I = \frac{2e}{h} \int T(\varepsilon)[f_L(\varepsilon) - f_R(\varepsilon)]\, d\varepsilon \qquad (1)$$

where h is Planck's constant, $T(\varepsilon)$ is the transmission function, $f_{L/R}$ is the Fermi function for left and right contacts respectively, and ε is the energy window due to the applied bias. This demonstrates how the transmission spectra at different bias can be a useful tool for explaining the I_D-V_D characteristics of molecular junctions.

Fig. 6 shows a schematic for a conceptual flash storage device. To investigate the potential application of POM based molecular memories, the DFT simulations were linked to the device simulations by extracting the LUMO level positioning (in terms of energy position) and molecular geometry as input parameters for our in-house device simulator -NESS. The energy position of the LUMO level with respect to the conduction band of SiO_2 is vital for calculating the retention time of molecular flash memories which is described in Fig.6 The POM layer consists of a 3 x 3 array of $[W_{18}O_{54}(SO_3)_2]^{-4}$ molecules which stores the charge in the working device.

Fig. 7 shows the link between the V_T and emission time of POM based flash memory. The KMC simulation models the change in V_T over time due to the emission of electrons from the POM storage centre. Fig. 7 shows the results of 75 KMC simulations where we consider a different probability and order of discharging each POM in the gate oxide.

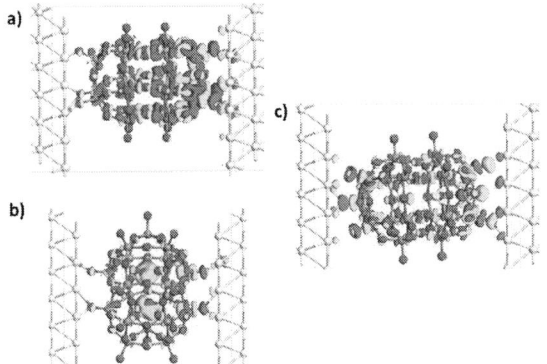

Figure 4. Electron density difference for the A, B and C configurations. Illustrating the change in electron density from 0.0V drain bias to 1.5V. (Isovalues= 0.0004 e/bohr³)

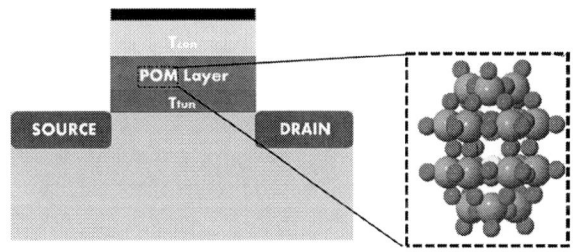

Figure 6. Conceptual device architecture for molecule-based charge trap flash memory cell. The POM cluster ($[W_{18}O_{54}(SO_3)_2]^{-4}$) is also illustrated where S=Yellow, W=Blue and O=Red.

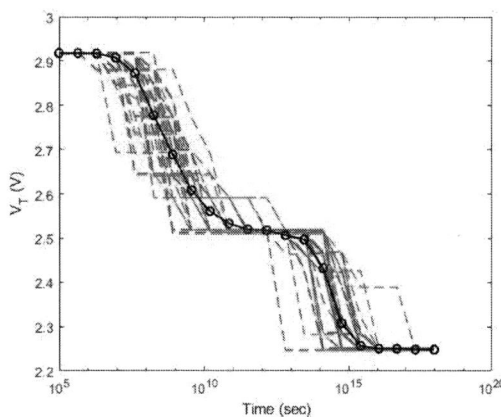

Figure 7. Threshold Voltage (V_T) as a function of time using Kinetic Monte Carlo in order to estimate retention time. The red is for each of the 75 devices. The black is the mean (average) change in V_T vs time.

The retention time is the time it takes for the device to emit 90% of the stored electrons in the POM layer. These simulations suggest that the device will retain charge for approximately 10^8 seconds until emissions occur and 10^{15} seconds to empty the storage layer. This simulated retention time meets the industry standard ten-year criteria for commercial flash memory. However, it must be noted that there is considerable statistical variation between individual KMC simulations with regards to retention time. Further study is required to optimize the device. Nonetheless these results show promise for suitable flash memory retention time.

In addition to the retention time, the plateau like feature in Fig.7 illustrates where all the POMs have lost an electron and each POM is in a 1x oxidised state. This highlights the stable redox capabilities of POM and shows potential for multi-bit storage.

CONCLUSIONS

This work has combined first-principle modelling techniques with mesoscopic device simulation (KMC approach) to give insight into the transport properties of a POM molecule and their potential application as a floating gate in modern flash memory devices.

By using DFT and NEGF approaches the link between molecular junction geometry, contact strength and the transport properties of a POM molecule have been explored. It is evident that stronger contact with electrodes increases the available conductance channels and thus increases the current under an applied bias. It was also highlighted how different molecule geometry affects the alignment of the energy levels and thus the transmission modes leading to different current characteristics.

The calculations showed that the "horizontal" A configuration (see Fig. 1) of POM is most favorable for electron transport and shows the lowest resistance and the highest current. The theoretically calculated current is in the nA region which highlights the relatively low current flow through POM molecule in comparisons to solid state semiconductors or conductors. However, this relative higher resistance is advantageous for holding electron and thus consistent with their potential application in memory devices.

In addition to the calculations from first principles, we have combined it with mesoscopic device calculations using our in-

house simulator NESS. Kinetic Monte Carlo simulations were carried out to estimate the retention time of POM based Molecular Flash memory. From an average of 75 simulations, our results demonstrate that the flash cell can hold data for up to 10 years as well as showing promising multi-bit storage.

REFERENCES

[1] R. K. Cavin, P. Lugli, and V. V. Zhirnov, "Science and engineering beyond moore's law," *Proc. IEEE*, vol. 100, no. SPL CONTENT, pp. 1720–1749, 2012, doi: 10.1109/JPROC.2012.2190155.

[2] M. Thoss and F. Evers, "Perspective: Theory of quantum transport in molecular junctions," *J. Chem. Phys.*, vol. 148, no. 3, 2018, doi: 10.1063/1.5003306.

[3] J. Maassen, M. Harb, V. Michaud-Rioux, Y. Zhu, and H. Guo, "Quantum transport modeling from first principles," *Proc. IEEE*, vol. 101, no. 2, pp. 518–530, 2013, doi: 10.1109/JPROC.2012.2197810.

[4] L. Vilà-Nadal *et al.*, "Towards polyoxometalate-cluster-based nano-electronics," *Chem. - A Eur. J.*, vol. 19, no. 49, pp. 16502–16511, 2013, doi: 10.1002/chem.201301631.

[5] V. P. Georgiev, S. M. Amoroso, L. Vila-Nadal, C. Busche, L. Cronin, and A. Asenov, "FDSOI molecular flash cell with reduced variability for low power flash applications," *Eur. Solid-State Device Res. Conf.*, pp. 353–356, 2014, doi: 10.1109/ESSDERC.2014.6948833.

[6] C. Busche *et al.*, "Design and fabrication of memory devices based on nanoscale polyoxometalate clusters," *Nature*, vol. 515, no. 7528, pp. 545–549, 2014, doi: 10.1038/nature13951.

[7] S. Smidstrup *et al.*, "QuantumATK: An integrated platform of electronic and atomic-scale modelling tools," *J. Phys. Condens. Matter*, vol. 32, no. 1, 2020, doi: 10.1088/1361-648X/ab4007.

[8] X. López, J. J. Carbó, C. Bo, and J. M. Poblet, "Structure, properties and reactivity of polyoxometalates: A theoretical perspective," *Chem. Soc. Rev.*, vol. 41, no. 22, pp. 7537–7571, 2012, doi: 10.1039/c2cs35168d.

[9] M. Schlipf and F. Gygi, "Optimization algorithm for the generation of ONCV pseudopotentials," *Comput. Phys. Commun.*, vol. 196, pp. 36–44, 2015, doi: 10.1016/j.cpc.2015.05.011.

[10] S. Berrada *et al.*, "Nano-electronic Simulation Software (NESS): a flexible nano-device simulation platform," *J. Comput. Electron.*, vol. 19, no. 3, pp. 1031–1046, 2020, doi: 10.1007/s10825-020-01519-0.

[11] O. Badami *et al.*, "Multiscale modeling of charge trapping in molecule based flash memories," *Int. Conf. Simul. Semicond. Process. Devices, SISPAD*, vol. 2019-Septe, pp. 6–9, 2019, doi: 10.1109/SISPAD.2019.8870518.

[12] H. Basch, R. Cohen, and M. A. Ratner, "Interface geometry and molecular junction conductance: Geometric fluctuation and stochastic switching," *Nano Lett.*, vol. 5, no. 9, pp. 1668–1675, 2005, doi: 10.1021/nl050702s.

[13] Y. Xue and M. A. Ratner, "Theoretical principles of single-molecule electronics: A chemical and mesoscopic view," *Int. J. Quantum Chem.*, vol. 102, no. 5 SPEC. ISS., pp. 911–924, 2005, doi: 10.1002/qua.20484.

[14] D. Xiang, X. Wang, C. Jia, T. Lee, and X. Guo, "Molecular-Scale Electronics: From Concept to Function," *Chem. Rev.*, vol. 116, no. 7, pp. 4318–4440, 2016, doi: 10.1021/acs.chemrev.5b00680.

[15] S. Datta, *Quantum Transport: Atom to Transistor*. New York: Cambridge Univ. Press, 2005.

12-7

Modeling Assisted Room Temperature Operation of Atomic Precision Advanced Manufacturing Devices

Xujiao Gao
Electrical Models & Simulation
Sandia National Laboratories
Albuquerque, USA
xngao@sandia.gov

Lisa A. Tracy
Quantum Phenomena
Sandia National Laboratories
Albuquerque, USA
latracy@sandia.gov

Evan M. Anderson
Multiscale Fab Sci & Tech Dev
Sandia National Laboratories
Albuquerque, USA
emander@sandia.gov

DeAnna M. Campbell
Biological & Chemical Sensors
Sandia National Laboratories
Albuquerque, USA
dmlope@sandia.gov

Jeffrey A. Ivie
Multiscale Fab Sci & Tech Dev
Sandia National Laboratories
Albuquerque, USA
jaivie@sandia.gov

Tzu-Ming Lu
Quantum Phenomena
Sandia National Laboratories
Albuquerque, USA
tlu@sandia.gov

Denis Mamaluy
Cognitive & Emerging Computing
Sandia National Laboratories
Albuquerque, USA
dnmamal@sandia.gov

Scott W. Schmucker
Multiscale Fab Sci & Tech Dev
Sandia National Laboratories
Albuquerque, USA
swschmu@sandia.gov

Shashank Misra
Multiscale Fab Sci & Tech Dev
Sandia National Laboratories
Albuquerque, USA
smisra@sandia.gov

Abstract—One big challenge of the emerging atomic precision advanced manufacturing (APAM) technology for microelectronics application is to realize APAM devices that operate at room temperature (RT). We demonstrate that semiclassical technology computer aided design (TCAD) device simulation tool can be employed to understand current leakage and improve APAM device design for RT operation. To establish the applicability of semiclassical simulation, we first show that a semiclassical impurity scattering model with the Fermi-Dirac statistics can explain the very low mobility in APAM devices quite well; we also show semiclassical TCAD reproduces measured sheet resistances when proper mobility values are used. We then apply semiclassical TCAD to simulate current leakage in realistic APAM wires. With insights from modeling, we were able to improve device design, fabricate Hall bars, and demonstrate RT operation for the very first time.

Keywords—atomic precision advanced manufacturing (APAM), room temperature operation, current leakage, semiclassical TCAD, ionized impurity scattering

I. INTRODUCTION

Atomic Precision Advanced Manufacturing (APAM) is a process of area-selective chemical incorporation of dopants at an atomic scale using the scanning tunneling microscope (STM) technique [1]. Expanding what APAM can do unlocks the door to revolutionary opportunities in microelectronics from the very physical limit of atoms [2]. Current APAM devices can only work at cryogenic temperatures which is sufficient for quantum computing applications [3]. However, to make any potential impact to microelectronics, APAM devices must work at room temperature (RT). To address this challenge, an interdisciplinary experimental and modeling

team at Sandia National Laboratories work closely together to enable APAM devices towards RT operation.

A typical APAM wire designed to operate at cryogenic temperatures is schematically shown in Fig. 1, where a phosphorous (P)-doped delta-layer using STM is capped by a silicon epitaxy layer to activate the P dopants and is contacted to heavily doped implant regions. The leakage paths through the substrate and cap layers are freeze-out at 4 Kelvin, but they become important at RT. How much leakage the parallel paths may have is a critical question to address in order to design APAM device for RT operation. TCAD device modeling can be very useful to help answer this question.

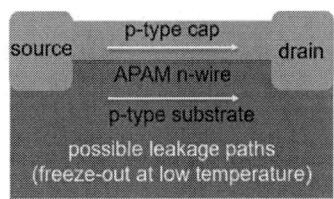

Fig. 1. Schematic of APAM wire device with leakage paths denoted by yellow lines. Leakage paths through cap layer and substrate are freeze-out at 4 Kelvin but become important at RT.

II. APPLICABILITY OF SEMICLASSICAL SIMULATION

To assess the applicability of semiclassical TCAD simulation to APAM devices, we first modeled electron mobility in a highly doped semiconductor using semiclassical scattering theory; then we simulated APAM wire sheet resistance using semiclassical TCAD simulation and compared with experimental data.

Electron mobility extracted from Hall measurements of APAM n-type wires are known to have low values on the order of 50 to 100 cm^2/(V.s) or less [5-8]. Hwang and Das Sarma [9] used a somewhat sophisticated two-dimensional (2D) treatment to model the electron mobility and achieved qualitative agreement with measured data. Nonetheless, we will show that by using the simple, semiclassical, 3D ionized impurity scattering theory, we can obtain electron mobility values that are not only qualitatively consistent with measured data, but also can reproduce the data with a single parameter.

This work is funded under Laboratory Directed Research and Development Grand Challenge (LDRD GC) program, Project No. 213017, at Sandia National Laboratories. Sandia National Laboratories is a multimission laboratory managed and operated by National Technology and Engineering Solutions of Sandia, LLC., a wholly owned subsidiary of Honeywell International, Inc., for the U.S. Department of Energy's National Nuclear Security Administration under contract DE-NA-0003525.

This paper describes objective technical results and analysis. Any subjective views or opinions that might be expressed in the paper do not necessarily represent the views of the U.S. Department of Energy or the United States Government.

277

We started with the famous mobility Drude model [10], i.e., $\mu = q\langle\langle\tau_m\rangle\rangle/m_c^*$, where $\langle\langle\tau_m\rangle\rangle$ denotes a specially defined averaged momentum relaxation time, m_c^* is the conductivity effective mass [11]. The momentum relaxation time τ_m is

$$\frac{1}{\tau_m} = \sum_{\vec{k}'} S(\vec{k}, \vec{k}') \left(1 - \frac{\vec{k}' \cdot \vec{k}}{k^2}\right)\left(1 - f(\vec{k}')\right), \quad (1)$$

where $f(\vec{k}')$ is the probability that the state at $\vec{k}'$ is occupied. The transition rate from an initial wave vector $\vec{k}$ to a final $\vec{k}'$ can be computed using the Fermi's Golden Rule as

$$S(\vec{k}, \vec{k}') = \frac{2\pi}{\hbar} \left|\langle\psi_{\vec{k}'}|\hat{H}_{int}|\psi_{\vec{k}}\rangle\right|^2 \delta\left[E(\vec{k}') - E(\vec{k})\right], \quad (2)$$

where the $\delta[.]$ term ensures energy conservation. The interaction Hamiltonian depends on the scattering type.

Due to the high doping in APAM devices, electron mobility is limited by ionized impurity scattering especially at low temperatures. Hence, we focus on the ionized impurity scattering only. By solving the Poisson equation for a n-type semiconductor, we obtain the screened electron-impurity interaction Hamiltonian as $\hat{H}_{int} = q^2 e^{-r/L_D}/(4\pi\varepsilon_0\varepsilon_s r)$, with L_D being the Debye length [10] and q being the elemental charge. L_D is given by $L_D = \sqrt{\varepsilon_0\varepsilon_s k_B T/(q^2 n\alpha)}$, where T is the lattice temperature and n is the electron density. Depending on the carrier statistics used in expressing the electron density when solving the Poisson equation, we obtain either $\alpha = 1$ for Boltzmann statistics, or $\alpha = \mathcal{F}_{-1/2}(\eta_F)/\mathcal{F}_{1/2}(\eta_F)$ for Fermi-Dirac (FD) statistics. $\mathcal{F}_l(\eta_F)$ is the Fermi-Dirac integral of the lth order [12], and $\eta_F = \mathcal{F}_{1/2}^{-1}(n/N_C)$, the inverse of the FD integral of the 1/2 order, with N_C being the conduction band effective density of states.

Assuming parabolic conduction band, empty final state (i.e., $1 - f(\vec{k}') \approx 1$), and weak energy dependence, we obtain the electron mobility due to impurity scattering as [10]

$$\mu = \frac{q}{m_c^*} \frac{128\sqrt{2\pi m_{dos}^*}\varepsilon_0^2\varepsilon_s^2(k_B T)^{3/2}}{q^4 N_d\left[\ln(1+\gamma^2) - \frac{\gamma^2}{1+\gamma^2}\right]}, \quad \gamma = \frac{2L_D}{\hbar}\sqrt{6k_B T m_{dos}^*}. \quad (3)$$

Here m_{dos}^* is the density-of-states effective mass, N_d is the doping density, ε_s is the static dielectric constant. To compute the electron mobility as a function of doping density and temperature, we assume $n = N_d$ and use the effective masses of bulk silicon [11], i.e., $m_{dos}^* = 1.08m_0$ and $m_c^* = 0.26m_0$. For the FD case, we also need to know the temperature dependence of N_C. For parabolic conduction band in silicon, $N_C(T)$ is given by

$$N_C(T) = 2.51 \times 10^{19} \left(\frac{m_{dos}^*}{m_0}\frac{T}{300}\right)^\beta = 2.82 \times 10^{19}\left(\frac{T}{300}\right)^\beta, \quad (4)$$

where the second equal sign assumes m_{dos}^* is independent of temperature. For silicon, $\beta = 1.5$ is often assumed. The electron mobility due to ionized impurity scattering in silicon is plotted in Fig. 2 as a function of the donor density at three different temperatures. Clearly, in the low doping regime, both Boltzmann and FD statistics produce the same mobility, and the mobility is higher for higher temperature since hot electrons are less likely scattered by impurities. As the doping density increases, the Boltzmann assumption becomes invalid, which happens at lower density for lower temperature.

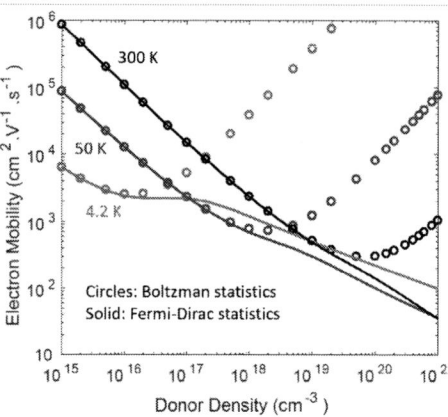

Fig. 2. Calculated electron mobility vs. donor density at three temperatures. The circles were obtained using the Boltzmann statistics in computing L_D (with $\alpha = 1$), while the solid lines were obtained using the FD statistics. The results were generated using $\beta = 1.5$ in computing $N_C(T)$.

It is noted that in the high doping regime, the electron mobility due to ionized impurity scattering alone indeed becomes small, on the order of 100 cm²/(V.s). Also note that the described mobility model is based on the 3D impurity scattering theory that is much simpler than the 2D model by Hwang [9]. A natural question is how the mobility values produced by this 3D scattering model compare with experimental data. In Fig. 3, we plotted the calculated mobility at high doping using three β values and compared with measured mobility in the literature. We found that by varying the β value slightly, the computed mobility values can reproduce the measured data surprisingly well. The variation of β value could be because the effective mass is temperature dependent [13].

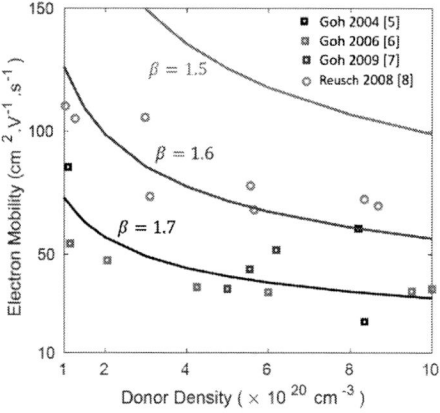

Fig. 3. Comparison of calculated electron mobility (solid lines) with experimentally extracted mobility (symbols) at 4 K. β is the exponent in (4). For comparison, the sheet doping density in experimental data was converted to 3D density assuming a 2-nm thickness.

To further evaluate semiclassical TCAD applicability to APAM devices, we simulated a simplified APAM wire shown in the inset of Fig. 4, using Sandia's open-source TCAD code, Charon [14-15], which is a 2D/3D, MPI-parallel, device simulation tool. The simulated sheet resistance versus the delta-layer doping density is shown in Fig. 4 and compared with cryogenic measurements by Goh et al. [6]. For comparison, the sheet doping density in Goh's paper was converted to 3D doping density assuming a 2-nanometer (nm) thick delta-layer as used in Charon's simulation. It is seen that, by using the extracted mobility values from [6], Charon reproduces the sheet resistance vs. doping density well when compared to the experimental data. The curves also follow the classical sheet resistance vs. doping density well, i.e., $R_s = 1/qN_d\mu_n t$, with t being the delta-layer thickness.

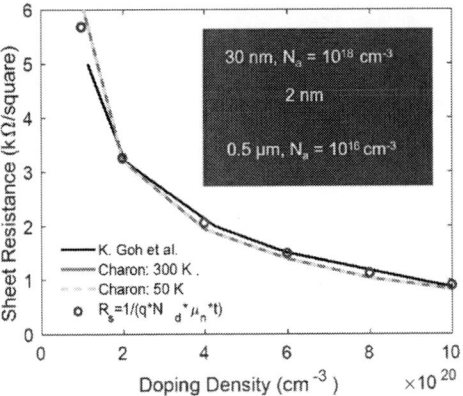

Fig. 4. Simulated sheet resistance vs. delta-layer doping density and comparison with measured data by Goh et al. [6]. Note that Charon results have no temperature dependence because a constant mobility at a given doping density was used. Inset: simulated APAM wire with contacts connecting the left and right edges of the red delta-layer.

From the above study, we see that the very low mobility in APAM wires can be explained using a simple 3D ionized impurity scattering model with the FD statistics. Furthermore, fast semiclassical TCAD simulation can reproduce sheet resistances of APAM wires, when proper electron mobility values are used, which can be obtained from either Hall measurements or a simple mobility model. While detailed quantum transport modeling may be intuitively more appropriate for cryogenic measurements, or given the tight confinement of electrons within the phosphorus layer, the above work indicates that semiclassical TCAD simulation provides a straightforward path to empirical understanding of APAM devices. In the following, we will apply semiclassical TCAD to realistic APAM devices to model current leakage and improve APAM device design for RT operation.

III. MODELING ASSISTED APAM RT DEVICE DESIGN

To design a structure to overcome the limitations of the device in Fig. 1, we apply Charon to model the current leakage through the p-type cap layer. The goal is to figure out what cap doping and thickness are needed to minimize leakage through the cap at RT. The simulated 2D structure is shown in Fig. 5, which represents a cross section of an APAM wire. Simulation results show hole current leakage in the cap layer is a trade-off between p-doping density and layer thickness. The current leakage for different cap doping and thickness is summarized in Fig. 6, where the leakage increases with increasing doping density and increasing thickness. The reason is that at a given thickness, the depletion width in the cap is smaller for a higher cap doping, and consequently more un-depleted holes lead to current leakage. For 10^{18}-cm^{-3} doping and 30-nm layer, the leakage is negligible because the cap is fully depleted at this condition. This result is also consistent with the PN junction depletion approximation, i.e., the depletion width, $W = \sqrt{2\varepsilon_0\varepsilon_s(N_d + N_a)V_{bi}/qN_dN_a} \approx 38$ nm for $N_d = 4 \times 10^{20}$ and $N_a = 10^{18}$ cm-3, which is comparable to the 30-nm layer thickness.

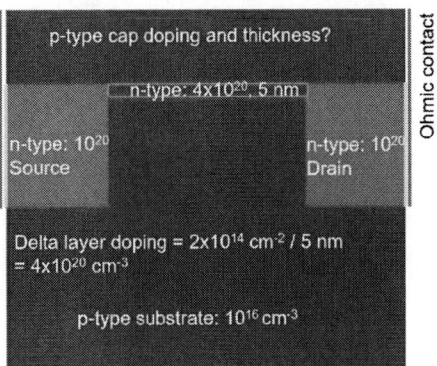

Fig. 5. Simulated 2D structure representing cross section of an APAM wire, where the delta-doped layer is approximated by a 5-nm doped region. Current leakage is simulated using the current difference between contacting and not contacting the cap layer.

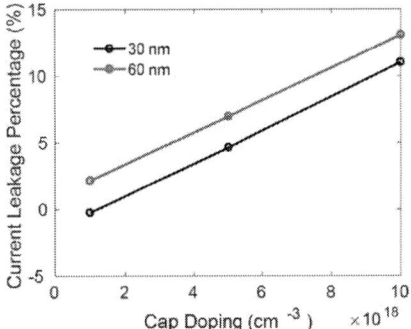

Fig. 6. Simulated current leakage as a function of the cap doping for cap layer thickness of 30 and 60 nm.

From the above study, we learned that by choosing appropriate doping and thickness for the cap layer, we could minimize the cap leakage. While depletion of the holes in the cap can suppress the leakage through the cap, additional care must be taken to suppress the leakage channel through the substrate. Therefore, we also isolate the contact regions from the cap and substrate regions using field oxide to further reduce possible hole leakage at RT. Fig. 7 shows a cross-section schematic of the modeling assisted new device design towards RT operation.

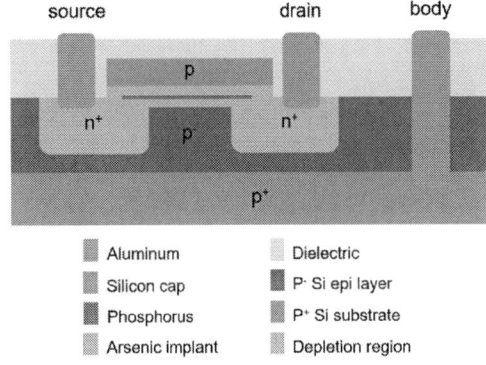

Fig. 7. Cross-section schematic of the modeling assisted new device design for RT operation. The red line denotes an APAM n-type wire. The wire contacts are isolated from the cap and substrate regions by field oxide.

Based on the device design in Fig. 7, we have fabricated Hall bar devices with and without the STM-doped delta layer.

279

A top view of the Hall bar is given in Fig. 8. Without the delta layer, the measured resistance at RT between the source and drain contacts (i.e., between contacts 4 and 8 in Fig. 8) is on the order of 300 MΩ, suggesting current leakage via the cap and substrate regions is minimal. With the delta layer, the measured wire resistance at RT is about 27 kΩ, close to the value of a similar APAM wire measured at 4 K. The delta-layer sheet electron density is calculated to be about 8×10^{13} cm^{-2} from four-probe measurement. This result also agrees with RT infrared measurements of the optical conductivity [16] of a similar APAM wire.

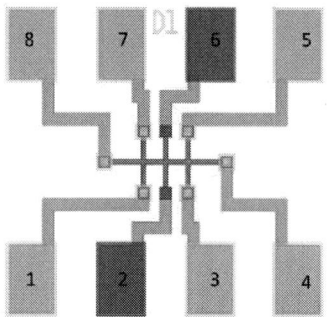

Fig. 8. Top view of APAM Hall bar where the main wire is between contacts 4 and 8 indicated by the purple line. The wire is 500-μm long and 20-μm wide.

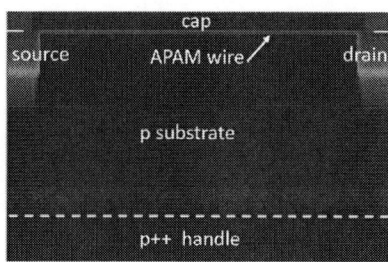

Fig. 9. Simulated 2D structure representing a 2-μm-long cross section of the RT Hall bar. The green solid lines indicate the source and drain Ohmic contacts. The isolation oxide between the cap and the Ohmic contacts is not included in the simulation since it does not affect current flow

To correlate simulation with experiment, we set up a simulation structure shown in Fig. 9. The APAM wire is 4-nm thick and its doping was set to 2×10^{20} cm^{-3} based on the measured sheet density of 8×10^{13} cm^{-2}. Doping profiles from secondary ion mass spectrometry (SIMS) of similar samples were used in the source/drain and substrate regions. The cap layer thickness and doping in similar samples can vary significantly due to unintentional variation in the STM process. For simulation purpose, the thickness of the cap layer was set to a designed value of 30 nm, while the cap doping was assumed to be a nominal value of 5×10^{18} cm^{-3}. Simulation was done using Charon [14]. Standard silicon models and parameters were used, e.g., the Slotboom model [17] for band gap narrowing, Shockley-Read-Hall (SRH) recombination, Fermi-Dirac statistics for high doping, and the Arora model [18] for carrier mobility. It turned out that the electron mobility computed using the Arora model is about 90 cm^2/(V.s) in the APAM wire, which is in the measured mobility range shown in Fig. 3. The simulated resistance is 105 Ω for the 2-μm-long wire, leading to an extrapolated value of 26 kΩ for the fabricated 500-μm-long wire. Without any calibration, the simulated resistance agrees surprisingly well with the measured data. There was some

uncertainty in the cap doping level in the fabricated device. However, our simulation results showed that, even for a range of two orders of magnitude difference in the cap doping, the simulated wire resistance was still on par with the measured value. This confirms that the current at RT is indeed mainly carried by the APAM wire and other leakage path is negligible.

IV. CONCLUSIONS

We have shown that the carrier transport in APAM devices can be modeled by semiclassical transport theory, especially at elevated temperatures, despite the nominally atomic monolayer doping. We showed that semiclassical TCAD simulation provided fast and instrumental insight to design APAM device for RT operation. With input from TCAD modeling, we have demonstrated RT operation of an APAM wire device for the very first time, which paves the path toward APAM enabled transistor for microelectronics.

REFERENCES

[1] F. J. Ruess, L. Oberbeck, M. Y. Simmons, K. E. J. Goh, A. R. Hamilton, T. Hallam, et al., "Toward atomic-scale device fabrication in silicon using scanning probe microscopy," Nano Letters, vol. 4, no. 10, pp. 1969–1973, 2004.

[2] D. R. Ward, S. W. Schmucker, E. M. Anderson, E. Bussmann, L. Tracy, T. Lu, et al., "Atomic precision advanced manufacturing for digital electronics," Elec. Dev. Fail. Anal. vol 22, pp. 4-10, 2020.

[3] Y. He, S. K. Gorman, D. Keith, L. Kranz, J. G. Keizer, and M. Y. Simmons, "A two-qubit gate between phosphorus donor electrons in silicon," Nature, vol. 571, pp. 371-375, 2019.

[4] D. Mamaluy, J. P. Mendez, X. Gao, and S. Misra, "Open-system quantum treatment of Si:P delta-layers and implications," unpublished.

[5] K. E. J. Goh, L. Oberbeck, M. Y. Simmons, A. R. Hamilton, and R. G. Clark, "Effect of encapsulation temperature on Si:P δ-doped layers," Appl. Phys. Lett., vol. 85, no. 21, pp. 4953-4955, 2004.

[6] K. E. J. Goh, L. Oberbeck, M. Y. Simmons, A. R. Hamilton, and M. J. Butcher, "Influence of doping density on electronic transport in degenerate Si:P δ-doped layers," Phys. Rev. B, vol. 73, 035401, 2006.

[7] K. E. J. Goh and M. Y. Simmons, "Impact of Si growth rate on coherent electron transport in Si:P δ-doped devices," Appl. Phys. Lett., vol. 95, art. 142104, 2009.

[8] T. C. G. Reusch, K. E. J. Goh, W. Pok, W.-C. N. Lo, S. R. McKibbin, and M. Y. Simmons, "Morphology and electrical conduction of Si:P δ-doped layers on vicinal Si(001)," J. of Appl. Phys., vol. 104, art. 066104, 2008.

[9] E. H. Hwang and S. Das Sarma, "Electronic transport in two-dimensional Si:P -doped layers," Phys. Rev. B, vol. 87, art. 125411, 2013.

[10] M. Lundstrom, Fundamentals of Carrier Transport, 2nd ed., Cambridge University Press, 2000.

[11] M. Lundstrom, "Notes on effective mass," nanohub.org, 2013.

[12] R. Kim and M. Lundstrom, "Notes on Fermi-Dirac integrals," 3rd ed., arXiv:0811.0116, August 2011.

[13] D. M. Riffe, "Temperature dependence of silicon carrier effective masses with application to femtosecond reflectivity measurements," J. Opt. Soc. Am. B, vol. 19, no. 5, pp. 1092-1100, 2002.

[14] https://charon.sandia.gov/.

[15] X. Gao, B. Kerr, and A. Huang, "Analytic band-to-trap tunneling model including band offset for heterojunction devices," J. Appl. Phys., vol. 125, art. 054503, 2019.

[16] A. Katzenmeyer, T. S. Luk, E. Bussmann, S. Young, E. Anderson, M. Marshall, et al., "Assessing atomically thin delta-doping of silicon using mid-infrared ellipsometry," J. Mat. Res., pp. 1-8, June 2020.

[17] J. W. Slotboom, "The pn production in silicon," Solid-State Electron., vol. 20, pp. 279-283, 1977.

[18] N. D. Arora, J. R. Hauser, D. J. Roulston, "Electron and hole mobilities in silicon as a function of concentration and temperature," IEEE. Trans. Electron Devices, vol. 29, no. 2, pp. 292-295, 1982.

Effects of the Dielectric Environment on Electronic Transport in Monolayer MoS₂: Screening and Remote Phonon Scattering

Maarten L. Van de Put*, Gautam Gaddemane†, Sanjay Gopalan, and Massimo V. Fischetti

Department of Materials Science and Technology, The University of Texas at Dallas, Richardson, Texas, USA
*Email: maarten.vandeput@utdallas.edu
†Now at *imec*, Leuven, Belgium

Abstract—We investigate theoretically the impact of the dielectric environment on electronic transport in monolayer MoS₂. In particular, we extend our first-principles Monte Carlo method to account for the screening of the electron-phonon interaction by the free carriers in the layer and the dielectric environment. In addition, we include the effect of remote-phonon scattering induced by the surrounding dielectrics. For monolayer MoS₂ on various dielectric substrates, we find that screening could improve the mobility significantly, but the inclusion of remote-phonon scattering degrades the mobility below its free-standing value. In our model, the introduction of gates in a dual-gate configuration does not appreciably decrease the remote-phonon interaction as it does in inversion layers or thicker films. However, for a double-gate field-effect transistor, we still obtain reasonable transport characteristics.

I. INTRODUCTION

Theoretical models are indispensable for the evaluation of 2D materials as potential channel materials in future field-effect transistors (FETs). Most modeling efforts focus on electronic transport in ideal free-standing layers, neglecting effects of the dielectric environment on the transport characteristics [1]–[9]. However, in experimental configurations and devices, the 2D layer is supported by oxides and often gated and, unlike in bulk materials, the dielectric response in 2D materials is dominated by the environment.

As we will demonstrate, the effects of the dielectric environment (substrates, oxides, gates) lead to significant changes in the mobility of supported 2D materials. Introducing a high-κ dielectric, or increasing the free-carrier concentration, screens the electron-phonon interaction, which improves the transport properties. Conversely, it is also well known that high-κ dielectrics can induce significant remote-phonon scattering that degrades performance. Due to their complex interplay, it is impossible to estimate, *a priori*, the combined effect of screening and remote-phonon scattering.

In this work, we extend our full-band Monte Carlo method for 2D materials [1], [2], to include the two major effects of the dielectric environment: screening and remote-phonon scattering. We briefly touch on the necessary theory and key approximations. We present results for two cases: a suspended monolayer of MoS₂, and a double-gate MoS₂ FET.

This work was supported in part by the Semiconductor Research Corporation (SRC).

II. THEORY

A. Screening of the electron-phonon interaction

We incorporate dynamic screening of the electron-phonon interaction by free-carriers and the dielectric environment. To achieve this, we apply a correction to the electron-phonon matrix elements obtained using density functional perturbation theory (DFPT). Neglecting the impact of the exchange and correlation functional, the interaction potential of a phonon mode η (*e.g.*, LA, LO, TA, or TO) for a free-standing layer is given by

$$\delta E_{\mathrm{ep}}(\mathbf{r}) \approx \int \mathrm{d}^3 r'\, G^{(0)}(\mathbf{r}, \mathbf{r}')\, \delta \rho^{(\eta)}(\mathbf{r}')\,, \tag{1}$$

where $\delta \rho^{(\eta)}$ is the response of the total charge to the phonon mode η, and $G^{(0)}(\mathbf{r}, \mathbf{r}')$ is the Green's function of the Poisson equation in vacuum.

For a 2D layer localized at $z = 0$ with an effective thickness h, we make the following assumptions:

- Only in-plane ionic displacements are considered, *i.e.*, no flexural modes (ZA / ZO).
- The charge-response to the ionic motion is localized at $z = 0$.
- The in-plane phonon wavevector $\mathbf{Q}$ is sufficiently small such that the Green's function is constant over the layer $|z| < h/2$.
- *Umklapp* processes can safely be neglected.

Using these assumptions, we find that the electron-phonon matrix elements are directly proportional to the Green's function of the 2D layer in vacuum $G^{(\mathrm{vac})}_{\mathbf{Q},\omega}(z, z')$. In particular, the matrix elements in vacuum, from a state in band n with in-plane wavevector $\mathbf{K}$ to band m with phonon mode η and in-plane wave-vector $\mathbf{Q}$, can be written as

$$g^{(\mathrm{vac})}_{nm\mathbf{K},\eta\mathbf{Q}} = \sum_{\mathbf{G}\mathbf{G}'} u^*_{m\mathbf{G},\mathbf{K}+\mathbf{Q}} u_{n\mathbf{G},\mathbf{K}} \sum_\alpha \mathrm{e}^{-\mathrm{i}\mathbf{Q}\cdot\boldsymbol{\tau}_\alpha}$$
$$\times \left(\frac{\hbar}{2M_\alpha \omega^{(\eta)}_{\mathbf{Q}}}\right)^{1/2} \mathbf{e}^{\eta,\alpha}_{\mathbf{Q}} \cdot \boldsymbol{\nabla}_{\boldsymbol{\tau}_\alpha} \rho(\mathbf{Q})\, G^{(\mathrm{vac})}_{\mathbf{Q},\omega^{(\eta)}_{\mathbf{Q}}}(0, 0)\,, \tag{2}$$

where $\hbar$ is the reduces Planck constant, $u_{n\mathbf{G},\mathbf{K}}$ are the Fourier components of the Bloch waves with reciprocal lattice vector

$\mathbf{G}$, α indicates the ion at position τ_α with mass M_α, and $\mathbf{e}_{\mathbf{Q}}^{\eta,\alpha}$ is the polarization vector, and $\omega_{\mathbf{Q}}^{(\eta)}$ is the phonon frequency.

The vacuum Green's function $G_{\mathbf{Q},\omega}^{(\text{vac})}(z,z')$ fully accounts for the dielectric response of the free-standing layer. To account for the dielectric response of the substrate and free-carriers in the layer, as well as boundary conditions imposed by metal gates, we replace $G_{\mathbf{Q},\omega}^{(\text{vac})}(z,z')$ by the complete environment Green's function $G_{\mathbf{Q},\omega}^{(\text{env})}(z,z')$. Since we only need to evaluate at $z=z'=0$, we can simply and efficiently rescale the free-standing electron-phonon matrix elements obtained from DFPT,

$$g_{nm\mathbf{K},\eta\mathbf{Q}}^{(\text{env})} = \frac{G_{\mathbf{Q},\omega}^{(\text{env})}(0,0)}{G_{\mathbf{Q},\omega}^{(\text{vac})}(0,0)} g_{nm\mathbf{K},\eta\mathbf{Q}}^{(\text{vac})} . \quad (3)$$

In practice, the Green's functions need to be approximated. To account for the dielectric response of the 2D layer, including the response of the free-carrier plasma, we refer the reader to Stern [10] and Maldague [11]. For small Q, the dielectric response of the oxides can be described using the generalized Lyddane-Sachs-Teller form as described in Refs. [12], [13].

B. Remote phonon scattering

When introducing polar substrates and oxides, we have to consider "remote phonon scattering", the interaction of coupled interface-plasmon phonon (IPP) modes with the charge carriers in the channel. The scattering potential is given by the usual Fröhlich-like term [12]. This evaluation requires the calculation of the dispersion of the hybrid IPP modes [13], [14]. Since the complete theory is too elaborate to present here, we refer to the Refs. [12]–[14] for a complete discussion.

In the results shown here, rather than calculating the dispersion of the IPP modes, we have approximated the dielectric response with two LO modes per oxide and have neglected the coupling with the plasmon mode. In this case, the interaction matrix element of mode ν (one of the LO modes) reduces to

$$\left| g_{nm\mathbf{K},\nu\mathbf{Q}}^{(\text{remote})} \right|^2 = \frac{e^2 \hbar \omega^{(\nu)}}{2\Omega_{\text{cell}} Q} \left| \frac{1}{\epsilon_{\text{high}}^{(\nu)}} - \frac{1}{\epsilon_{\text{low}}^{(\nu)}} \right| , \quad (4)$$

where $\epsilon_{\text{high}}^{(\nu)}$ is the total dielectric response of the 2D layer and surrounding dielectric environment when mode ν does not contribute, while $\epsilon_{\text{low}}^{(\nu)}$ is the total response where the mode fully responds.

III. RESULTS

We demonstrate the effects of screening and remote phonon interaction on transport through monolayer MoS_2 in two realistic configurations, a supported monolayer and a double-gate FET. The full-band Monte Carlo procedure we use to obtain mobility and FET characteristics has been detailed in Refs. [1], [3], [4]. The electronic band structure, phonon dispersion and free-standing electron-phonon matrix elements we start from have been published in Ref. [3]. The dielectric constant and layer thickness of MoS_2 have been obtained from Ref. [15].

A. Supported MoS2

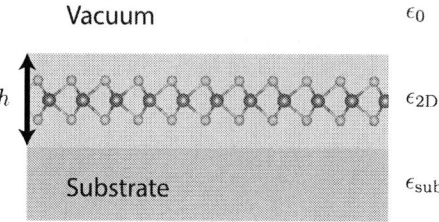

Fig. 1. A supported MoS_2 layer. The layer is supported by an infinitely thick substrate in vacuum. The substrate is either SiO_2, Al_2O_3, and HfO_3. The layer is assumed to have a thickness $h = 6.11\text{\AA}$ and uniform dielectric constant ϵ_{2D}.

We first simulate a supported monolayer of MoS_2 as depicted in Fig. 1. This system is representative of a common experimental setup used to probe the material's "intrinsic" transport characteristics. It is important to note that we consider an idealized, infinitely thick and defect-free substrate that does not change the electronic structure of the MoS_2 layer.

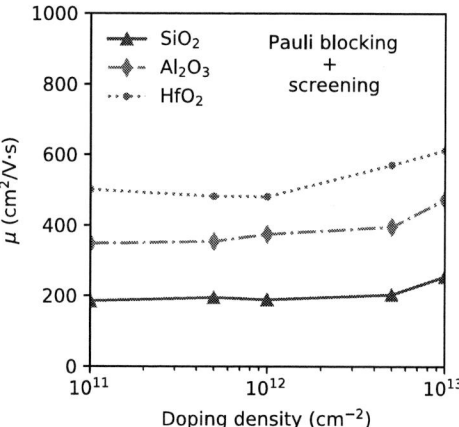

Fig. 2. The screened phonon-limited mobility of the supported MoS_2 layer shown in Fig. 1 with respect to the doping density for various substrates (SiO_2, Al_2O_3, and HfO_2). This includes the screening of the electron-phonon scattering by the substrate.

Figure 2 shows the effect of only free-carrier and dielectric screening on the mobility of a supported layer. For reference, the electron mobility of a free-standing layer, without screening is $127\,\text{cm}^2/(\text{Vs})$ [3]. As expected, the mobility scales with the dielectric constant of the substrate, with higher dielectric constant providing stronger screening of the electron-phonon interactions. Screening by a HfO_2 substrate, the highest κ dielectric we consider, increases the phonon-limited electron mobility of MoS_2 by up to four times, compared to the free-standing material ($> 500\,\text{cm}^2/(\text{Vs})$). Even with SiO_2 modest gains are shown, with an electron mobility around $200\,\text{cm}^2/(\text{Vs})$. Considering that our reference point is vacuum, the lowest-κ dielectric available, this should not be too surprising.

Looking at the variation of mobility with free-electron concentration (doping), we observe only a marginal improvement in mobility by increasing the doping from 10^{11} cm^{-2} to 10^{13} cm^{-2}. Since the free carriers are confined to the 2D plane of the layer, they cannot efficiently screen the long-wavelength electron-phonon scattering processes. Additionally, we see that degeneracy effects cause variations in the mobility. As the Fermi-level increases, Pauli blocking will result in decreased scattering to already occupied states, increasing mobility. However, with an increasing Fermi level, satellite valleys become available, which increases inter-valley scattering, reducing the mobility. For MoS$_2$, these degeneracy effects almost cancel each other and do not affect the mobility by appreciably.

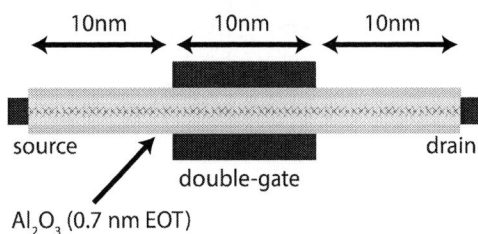

Fig. 4. A double gate MoS$_2$ FET. The oxide is Al$_2$O$_3$, with an equivalent oxide thickness (EOT) of 0.7 nm SiO$_2$. The lengths of the gate and source/drain extensions are 10 nm each. The channel is undoped, while the source/drain extensions are n-type doped with a concentration of 10^{13} cm^{-2}

with a equivalent oxide thickness (EOT) of 0.7 nm SiO$_2$.

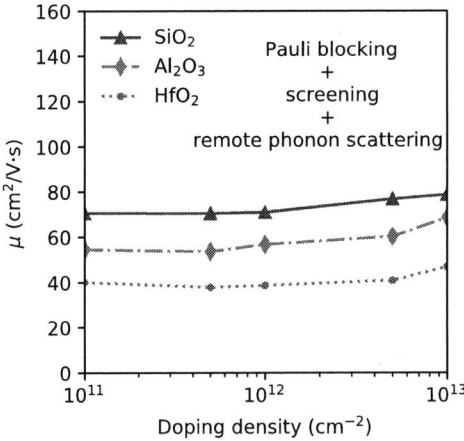

Fig. 3. The screened phonon-limited mobility of the supported MoS$_2$ layer shown in Fig. 1, including remote-phonon scattering induced by the dielectric with respect to the doping density for various substrates (SiO$_2$, Al$_2$O$_3$, and HfO$_3$).

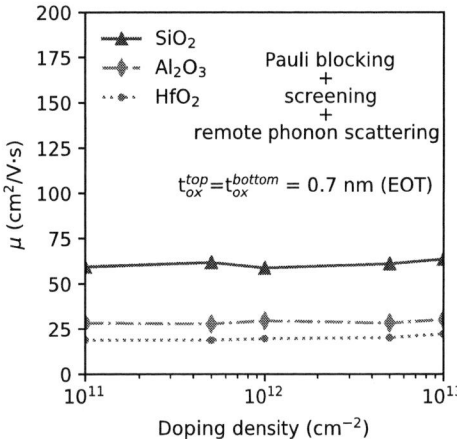

Fig. 5. The screened phonon-limited mobility of the supported MoS$_2$ layer shown in Fig. 1, including remote-phonon scattering induced by the dielectric with respect to the doping density for various substrates (SiO$_2$, Al$_2$O$_3$, and HfO$_3$).

While these results may seem promising, Fig. 3 shows that, once remote-phonon scattering is considered, the electron mobility of supported MoS$_2$ is degraded well below the free-standing mobility for all substrates we consider. Unsurprisingly, HfO$_2$ shows the largest degradation, reducing the mobility by more than a factor of three to 40 cm^2/(Vs). Perhaps more surprising, the electron mobility of MoS$_2$ supported on SiO$_2$ also drops to 70 cm^2/(Vs), almost two times lower than the intrinsic mobility of a free-standing layer. It is important to reiterate that this is a common experimental situation when measuring the intrinsic mobility of few-layer 2D materials. Our result suggests that, even under ideal circumstances (no defects or impurities and perfect contacts), the mobility of MoS$_2$, measured on a substrate, will not exceed 80 cm^2/(Vs) at room temperature.

B. Double-gate MoS$_2$ field-effect transistor

To estimate performance in a device, we simulate a MoS$_2$ FET, as shown in Fig. 4. We consider a double-gate structure

Looking at the channel region, Fig. 5 shows the mobility in the presence of various oxides in the double gate configuration. While creening is much stronger (not shown) in this configuration, adding an additional oxide doubles the number of remote-phonon branches and causes a net reduction in mobility in the channel for all oxides. We note that, unlike in bulk materials, screening by the metal gates only has a limited effect on suppressing the remote-phonon interaction. Indeed, in our model we consider the charges in the layer extremely confined, evaluating the remote-phonon interaction only at a single point in the center, which is not sensitive to screening by the gates. Whereas, in reality, the effect of the gates will be larger. Presumably, screening by the metal gates will be far less efficient in 2D materials than in (bulk) inversion layers or thick films due to the highly confined nature of the former.

For Al$_2$O$_3$, we perform fully self-consistent calculations on the FET structure shown in Fig. 4. Screening and remote phonon interaction are automatically updated to account for the local charge distribution and geometry. Figure 6 shows the

283

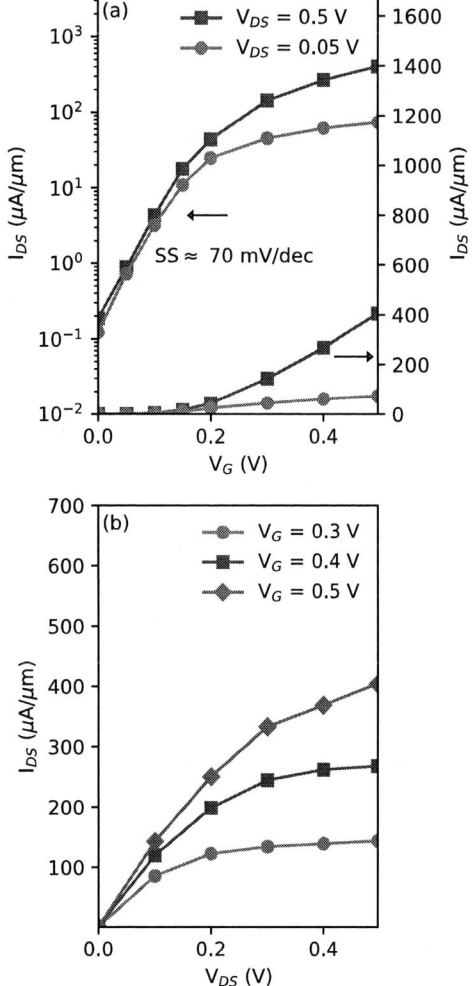

Fig. 6. Transfer (a) and output (b) characteristics of the device shown in Fig. 4, including all screening effects and remote phonon scattering.

transfer and output characteristics of the FET. When accounting for remote-phonon scattering, the transport characteristics compare favorably to state-of-the-art experimental results [16]. Compared to the mobility calculations of supported and gated MoS$_2$, the device characteristics are promising. Indeed, it is well known that in short devices that mobility is not a good measure for FET performance. Therefore, to design an optimized FET configuration, *e.g.*, materials selection and geometry, complete Monte-Carlo calculation including these effects are required.

IV. CONCLUSIONS

The effects of the dielectric environment on the mobility of 2D materials cannot be neglected. While dielectric screening can significantly improve the mobility, these positive effects are countered by the detrimental effect of remote-phonon scattering. For now, the only way to know to which side the balance will tip is to diligently study the transport properties using models that include the interactions with the dielectric environment. Since experimental measurements on supported layers and FETs are inherently affected by all of these physical processes, direct comparison to theoretical values in free-standing layers is perilous. From a device engineering perspective, the choice of the right material *combination* is essential, a task for which advanced modeling, as shown here, is crucial.

REFERENCES

[1] G. Gaddemane, W. G. Vandenberghe, M. L. Van de Put, S. Chen, S. Tiwari, E. Chen, and M. V. Fischetti, "Theoretical studies of electronic transport in monolayer and bilayer phosphorene: A critical overview," *Physical Review B*, vol. 98, p. 115416, Sep 2018.

[2] G. Gaddemane, W. G. Vandenberghe, M. L. Van de Put, E. Chen, and M. V. Fischetti, "Monte-Carlo study of electronic transport in non-σ_h-symmetric two-dimensional materials: silicene and germanene," *Journal of Applied Physics*, vol. 124, no. 4, p. 044306, 2018.

[3] G. Gaddemane, S. Gopalan, M. L. Van de Put, and M. V. Fischetti, "Limitations of ab initio methods to predict the electronic-transport properties of two-dimensional semiconductors: the computational example of 2H-phase transition metal dichalcogenides," *Journal of Computational Electronics*, vol. 1, p. 3, jun 2020.

[4] G. Gaddemane, M. L. V. de Put, W. G. Vandenberghe, E. Chen, and M. V. Fischetti, "Monte carlo analysis of phosphorene nanotransistors," *arXiv:2007.14940*, 2020.

[5] S. Gopalan, G. Gaddemane, M. L. Van de Put, and M. V. Fischetti, "Monte carlo study of electronic transport in monolayer InSe," *Materials*, vol. 12, no. 24, p. 4210, Dec 2019.

[6] T. Sohier, M. Calandra, and F. Mauri, "Two-dimensional fröhlich interaction in transition-metal dichalcogenide monolayers: theoretical modeling and first-principles calculations," *Phys Rev B*, vol. 94, p. 085415, Aug 2016.

[7] T. Sohier, D. Campi, N. Marzari, and M. Gibertini, "Mobility of two-dimensional materials from first principles in an accurate and automated framework," *Phys. Rev. Materials*, vol. 2, p. 114010, Nov 2018.

[8] Y. Lee, S. Fiore, and M. Luisier, "Ab initio mobility of single-layer MoS$_2$ and WS$_2$: comparison to experiments and impact on the device characteristics," in *2019 IEEE International Electron Devices Meeting (IEDM)*, 2019, pp. 24.4.1–24.4.4.

[9] C. Klinkert, Á. Szabó, C. Stieger, D. Campi, N. Marzari, and M. Luisier, "2-D materials for ultrascaled field-effect transistors: one hundred candidates under the ab initio microscope," *ACS Nano*, vol. 14, no. 7, pp. 8605–8615, 2020, pMID: 32530608.

[10] F. Stern, "Polarizability of a two-dimensional electron gas," *Physical Review Letters*, vol. 18, pp. 546–548, Apr 1967.

[11] P. F. Maldague, "Many-body corrections to the polarizability of the two-dimensional electron gas," *Surface Science*, vol. 73, pp. 296 – 302, 1978.

[12] M. V. Fischetti, D. A. Neumayer, and E. A. Cartier, "Effective electron mobility in Si inversion layers in metal-semiconductor systems with a high-κ insulator: The role of remote phonon scattering," *Journal of Applied Physics*, vol. 90, no. 9, pp. 4587–4608, 2001.

[13] Z.-Y. Ong and M. V. Fischetti, "Theory of remote phonon scattering in top-gated single-layer graphene," *Physical Review B*, vol. 88, p. 045405, Jul 2013.

[14] ——, "Theory of interfacial plasmon-phonon scattering in supported graphene," *Physical Review B*, vol. 86, no. 16, p. 165422, Oct 2012.

[15] A. Laturia, M. L. Van de Put, and W. G. Vandenberghe, "Dielectric properties of hexagonal boron nitride and transition metal dichalcogenides: from monolayer to bulk," *npj 2D Materials and Applications*, vol. 2, no. 1, p. 6, dec 2018.

[16] Q. Smets, G. Arutchelvan, J. Jussot, D. Verreck, I. Asselberghs, A. N. Mehta, A. Gaur, D. Lin, S. E. Kazzi, B. Groven, M. Caymax, and I. Radu, "Ultra-scaled MOCVD MoS$_2$ MOSFETs with 42nm contact pitch and 250μa/μm drain current," in *2019 IEEE International Electron Devices Meeting (IEDM)*, 2019, pp. 23.2.1–23.2.4.

Impact of Schottky Barrier on the Performance of Two-Dimensional Material Transistors

Sheng-Kai Su, Jin Cai, Edward Chen, Lain-Jong Li, and H.-S. Philip Wong
Corporate Research,
Taiwan Semiconductor Manufacturing Company (TSMC), Ltd.
No. 168, Park Ave. 2, Hsinchu Science Park, Hsinchu County, Taiwan
E-mail: sksua@tsmc.com

Abstract — Double-gated monolayer two-dimensional (2D) material transistor is expected to offer ideal (~60 mV/dec) subthreshold swing (SS) for gate lengths well below 10 nm. However, the ideal 2D transistor assumes Ohmic contacts whereas a realistic metal/2D Schottky contact can degrade SS. Transport simulations including scattering is necessary to correctly describe carrier thermalization and predict the SS degradation. Scaled 2D transistors with a Schottky barrier height (SBH) smaller than 100 meV and doping concentration in the extension region larger than 2×10^{13} cm^{-2} are required to achieve high performance.

Keywords- 2D van der Waals materials, WS$_2$, Schottky contact, sub-threshold swig, future CMOS technology

I. INTRODUCTION

For applications of abundant-data computing in the immediate future, high density transistors integrated with functional (e.g. memory) components are needed [1]. Unavoidable surface roughness of three-dimensional (3D) bulk materials (e.g. Si, Ge, and III-V compounds) limits the scaling of transistors [2]. Two-dimensional (2D) van der Waals materials with pristine surface in principle are promising for continuous transistor scaling and heterogeneous integration [2]. Owing to the appropriate transport effective mass and high phonon-limited mobility of both electrons and holes, monolayer (ML) tungsten disulfide (WS$_2$) is one of the potential channel candidates for future CMOS technology [2]. In addition to the intrinsic material property, Schottky contact between metal and 2D semiconductor is known as a challenge to achieve high performance [3, 4]. However, most of the studies treated intrinsic channel performance and Schottky contact resistance independently. The coupling effect between carrier injection from Schottky contact and carrier transport through the channel is ignored. In this work, we simulate the electrical characteristics of ML WS$_2$ nFET and pFET with integrated source/drain (S/D) Schottky contact considering phonon scatterings. The Schottky barrier degrades the performance of 2D transistors more than just increasing the contact resistance. Phonon scattering and the designs in the extension region could impact the sub-threshold swing of the extremely scaled transistors.

II. METHODOLOGY

The bandstructure calculation of ML WS$_2$ is based on density functional theory (DFT) and the generalized gradient approximation (GGA) in the form of Perdew-Burke-Ernzerhof (PBE) functional with spin-orbit coupling (SOC) applied [5]. The kinetic energy cutoff and k-point mesh were set to 110 Hartree and $10 \times 10 \times 1$, respectively. Figure 1a shows the calculated 2D k-space constant energy contours of the lowest energy conduction band of ML WS$_2$ with Brillouin-zone and high symmetry k-points indicated. Figure 1b shows the calculated bandstructure along high symmetry k-points. The top valence band with spin-orbit splitting is in good agreement with experimental results measured by angle-resolved photoemission spectroscopy (ARPES) [6]. The electronic band structure was then replicated by multi-valley effective mass approximation with non-parabolicity. The ML WS$_2$ is approximated as an ideal zero-thickness two-dimensional electron/hole gas (2DEG/2DHG). For tunneling simulations based on the WKB approximation, complex bandstructure inside the forbidden gap was computed perturbatively from the real-k states [7]. Electron and hole transports were simulated by self-consistently solving 2D Boltzmann transport and Poisson equations including acoustic and optical phonon scatterings [7, 8]. Schottky barrier height (SBH) as a parameter is defined as the difference between the contact metal workfunction and the band edge of ML WS$_2$ with Fermi energy fixed at the boundary.

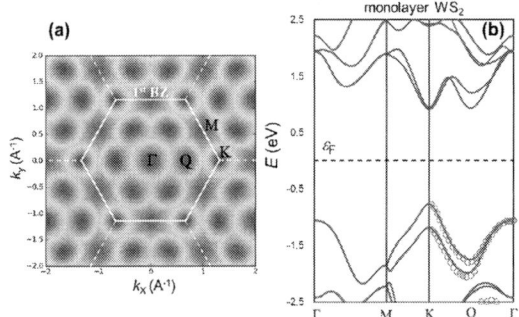

Fig. 1 (a) 2D *E-k* contour of the bottom conduction band of ML WS$_2$ with 1st Brillouin-zone (BZ) and high-symmetry k-points indicated. Red color for higher energy; blue color for lower energy. (b) Bandstruture of ML WS$_2$ along high-symmetry directions indicated in Fig. 1a with Fermi level ε_F at zero energy. Symbols are experimental results of ARPES from [6].

III. RESULT & DISCUSSION

Figure 2a shows the side view of the device with double-gated and S/D edge contact architecture for good electrostatic control and large effective width [2]. The gate length L_G and equivalent oxide thickness (EOT) used are 10 nm and 0.8 nm respectively. With minimal parasitic resistance, intrinsic electrical performance of n- and p-FETs ML WS_2 are nearly symmetric, as shown in Fig. 2b. Although the transport effective mass of hole ($\sim 0.4\ m_0$) is heavier than that of electron ($\sim 0.3\ m_0$), the large energy separation (> 300 meV) between K and Γ valleys of valence band (Fig. 1b) suppresses intervalley scattering resulting in higher phonon limited mobility than that of nFET [8]. Also, the large energy separation mitigates the velocity degradation from carrier populating into the satellite (Γ) valley with heavy mass at high V_G.

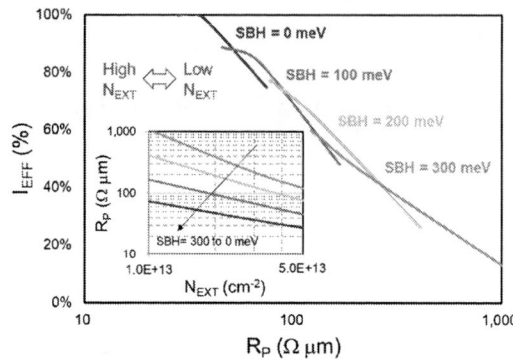

Fig. 3 Percentage of I_{EFF} [8] versus extracted parasitic resistance R_P (including contact and extension series resistances) at the same $I_{OFF} = 10^{-3}\ \mu A/\mu m$. Inset is R_P versus N_{EXT} for different Schottky barrier height (SBH) 0 (blue), 100 (green), 200 (yellow), and 300 (orange) meV.

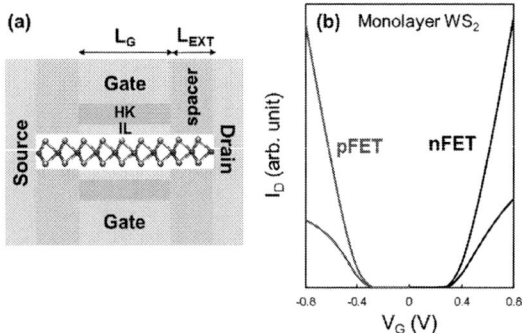

Fig. 2 (a) Simulated device structure with gate length $L_G = 10$ nm and EOT = 0.8 nm. Material of IL/HK and spacer are h-BN/HfO$_2$ and Si$_3$N$_4$ respectively. Under spacer is the extension region with extension length L_{EXT} and doping concentration N_{EXT}. (b) Simulated I_D-V_G of ML WS_2 nFET (black) and pFET (red) at $V_D = 0.05$ V and 0.75 V with $N_{EXT} = 5e13$ cm^{-2} and SBH = 0 eV.

We systematically studied 2D transistor performance as a function of SBH of the metal/2D contact and doping concentration in the extension region (N_{EXT}). A correlation between effective drive current I_{EFF} [9] and parasitic resistance R_P (including contact and extension series resistances) with the same I_{OFF} is shown in Fig. 3. As a design guideline, $\sim 70\%$ of the intrinsic 2D transistor drive current can be retained if the total R_p is within $\sim 100\ \Omega$-μm, requiring a SBH of <100 meV and a N_{EXT} of $\sim 2e13$ cm^{-2}. The lack of a universal correlation between I_{EFF} and R_P indicates that the impact from SBH is more than just the increase of parasitic resistance. Different extension doping could lead to different drain induced barrier lowing (DIBL). SBH could deplete the carrier in the short extension region inducing source starvation effect [10]. Moreover, the metal/2D Schottky contact can degrade the substhreshold swing (SS). Since maintaining good SS is important for future technology and is one of the motivation to choose layered 2D materials for continuous device scaling. We focus on discussing the SBH induced SS degradation in the following paragraphs.

Figure 4a shows the impact of the Schottky contact on SS of WS_2 pFET. For large SBH (e.g. > 300 meV), SS can degrade to > 70 mV/dec even for the double-gated architecture with extremely thin channel. The degradation is due to energy filtering effect by the Schottky barrier which provides high kinetic energy carriers that can not be effectively controlled by the gate. For a regular MOSFET with a thermalized heavily doped extension, most of the carriers have low energies following the Fermi-Dirac distribution. However, for Schottky barrier contacts, the energy of the carriers injected from the metal is filtered by the Schottky barrier. There could be more carriers with higher energies due to thermal injection over the barrier and tunneling near the top of the barrier. Take the case of SBH = 200 meV as an example, the SS is near ideal (~ 60 mV/dec) for $I_D < 10^{-2}\ \mu A/\mu m$ (channel barrier is still higher than the Schottky barrier), but becomes worse for larger I_D (V_G). As gate voltage increases and channel barrier reduces, those carriers filtered by the Schottky barrier can possess kinetic energies higher than the channel barrier which result in less effective gate control and degrade SS (Fig. 4b). Considering (especially optical phonon) scattering in the extension region is important and ballistic transport tends to be pessimistic in predicting SS (Fig. 4a). This is because optical phonon scattering can thermalize high kinetic energy carriers prescreened by the source-side Schottky barrier, and the carrier distribution at the source-side becomes closer to the Fermi-Dirac distribution at thermal equilibrium.

Figure 5a shows the impact of extension doping concentration N_{EXT} on SS for different SBH. For small SBH (e.g. 100 meV), SS degrades as doping concentration increases due to more serious short channel effect (SCE). For large SBH (e.g. 400 meV), however, SS becomes better as doping concentration increases. Considering the SB width thinning, higher doping can help recover some of the SBH induced SS degradation by increasing the tunneling probability of carriers with low energies through the SB (Fig. 4b). These two competing effects can be observed for medium SBH (e.g. 300 meV) at different current regions. A longer extension length L_{EXT} can do the same trick as higher N_{EXT} shown in Fig. 5b. With longer L_{EXT}, the scattering probability with phonon emission of high kinetic energy carriers increases, which lowers their energies. And it can be expected that the device behaves as a regular MOSFET as L_{EXT} becomes long enough with complete carrier thermalization. However, for future CMOS technology with small gate pitch, these knobs (L_{EXT} and N_{EXT}) have limited design window without adversely affecting either the SCE or extension series resistance.

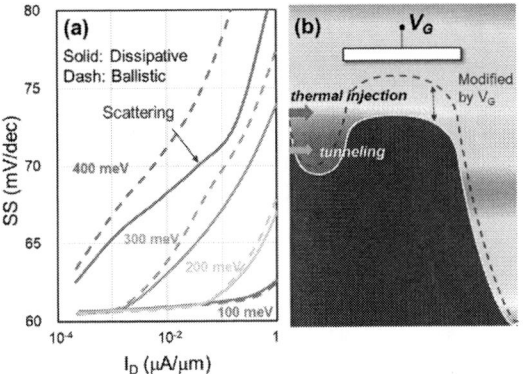

Fig. 4 (a) Sub-threshold swing SS versus drain current I_D for different SBH with L_{EXT} = 5 nm and N_{EXT} = 1e13 cm^{-2} assuming ballistic transport (dash line) and dissipative transport (solid line). (b) Distribution function (red for high and blue for low values) in energy space explains the result in (a). Horizontal axis is along the transport direction and vertical axis is energy. Grey dash line is at low V_G and white solid line is at high V_G.

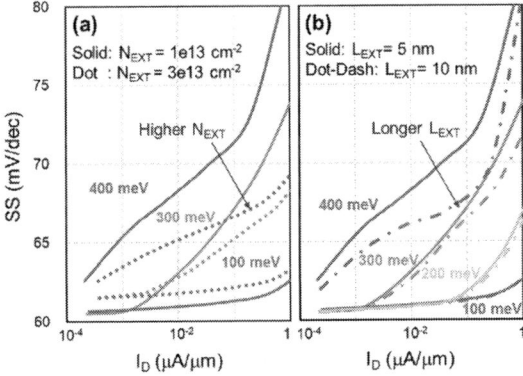

Fig. 5 SS versus I_D for different SBH. (a) Comparison of the effect of SBH on SS between low N_{EXT} = 1e13 cm^{-2} (solid line) and high N_{EXT} = 3e13 cm^{-2} (dot line) with the same L_{EXT} = 5 nm. (b) Comparison of the effect of SBH on SS between short L_{EXT} = 5 nm (solid line) and long L_{EXT} = 10 nm (dot-dash line) with the same N_{EXT} = 1e13 cm^{-2}.

IV. CONCLUSION

Monolayer WS$_2$ shows balanced nFET and pFET performances. Small Schottky barrier height for extremely scaled 2D FETs is more critical than that for Si technology nowadays. The impact of a high Schottky barrier is more than that of just the effect on contact resistance. Schottky contact can further degrade subthreshold swing and ballistic transport can over-estimate this effect. High extension doping concentration and long extension length can mitigate the swing degradation but need to be considered along with impact on short channel effect and

parasitic resistance.

REFERENCES

[1] A. M. M. Sabry et al., "Energy-efficient abundant-data computing: The N3XT 1,000x," Computer, vol. 48, pp. 24-33, 2015.

[2] S. K. Su, et al., "Layered semiconducting 2D materials for future transistor applications," (unpublished).

[3] D. Jena, K. Banerjee, and G. H. Xing, "Intimate contacts," Nat. Mater, vol. 13, pp. 1076-1078, 2014.

[4] D. S. Schulman, J. A. Andrew, and S. Das, "Contact engineering for 2D materials and devices," Chem. Soc. Rev., vol. 47.9, pp. 3037-3058, 2018.

[5] S. Smidstrup, et al., "QuantumATK: An integrated platform of electronic and atomic-scale modelling tools," J. Condens. Matter Phys., vol. 32.1, 015901, 2019.

[6] H. Henck, et al., "Electronic band structure of two-dimensional WS$_2$/graphene van der Waals heterostructures," Phys. Rev. B vol. 97, 155421, 2018.

[7] Z. Stanojević, et al., "Phase-space solution of the subband Boltzmann transport equation for nano-scale TCAD," SISPAD, pp. 65-68, 2016 & GTS Framework 2019 manual.

[8] Z. Jin, et al., "Intrinsic transport properties of electrons and holes in monolayer transition-metal dichalcogenides," Phys. Rev. B, vol. 90, 045422, 2014.

[9] M. H. Na, et al., "The effective drive current in CMOS inverters," IEDM, pp. 5.4.1-5.4.4, 2002.

[10] M. V. Fischetti, et al., "Scaling MOSFETs to 10 nm: Coulomb effects, source starvation, and virtual source model," J. Comput. Electron., vol. 8.2, pp. 60-77, 2009.

AC NEGF Simulation of Nanosheet MOSFETs

Sung-Min Hong and Phil-Hun Ahn

School of EECS, Gwangju Institute of Science and Technology, Gwangju, Republic of Korea, email: smhong@gist.ac.kr

Abstract—In this work, an AC nonequilibrium Green function (NEGF) simulation for nanosheet MOSFETs is presented. The AC NEGF equations are discretized using a decoupled mode-space approach for efficient implementation. The Poisson equation is solved self consistently to obtain the electrostatic potential. Our in-house device simulator, G-Device, is used to simulate the AC responses on nanosheet MOSFETs.

I. INTRODUCTION

During the last couple of decades, the NEGF (Nonequilibrium Green's function) formalism [1], [2], [3] has become the *de facto* standard method for the quantum transport modeling. It has been actively applied to the simulation of nanoscale transistors.

However, in most cases, the NEGF technique has been applied only to the DC (steady-state) simulation. It is difficult to find transient and AC simulation results for semiconductor devices. In contrast, TCAD tools are fully capable of DC, transient, and AC simulation. These days, even the BTE (Boltzmann transport equation) solvers [4], [5], [6], [7] can solve the transient Boltzmann equation.

Although recent theoretical advances have extended the method to the AC (small-signal) simulation [8], [9], it is difficult to find the AC NEGF simulation results of the electronic devices. It is expected that the frequency-domain analysis, which is fully consistent with the DC NEGF technique, can expand a new horizon of the quantum transport modeling. For example, the maximum limit of the intrinsic RF performance in the nanoscale MOSFET can be investigated by this analysis. Understanding the intrinsic limit of the RF performance is critically important to realize the terahertz (THz) electronics based on the MOSFET technology.

In this work, the AC NEGF code is implemented. Based upon the solution of the conventional DC NEGF simulation, the AC response to the voltage excitation with a given frequency is calculated by adopting the quantum transport theory. The AC response of the nanosheet MOSFET is evaluated. In Section II, the AC NEGF formalism is briefly explained. In Section III, the implementation of the AC NEGF formalism is discussed. A robust method to get the AC electrostatic potential is discussed. The simulation results for the nanosheet MOSFET are presented in Section IV. Finally, the conclusion is made in Section V.

II. METHODOLOGY

Detailed derivation procedure of the AC NEGF formalism can be found in [8], [9]. Only some important relations in [8] are selectively introduced below.

When the AC voltage excitation, whose amplitude and angular frequency are v_{amp} and ω, respectively, is applied to a contact, the retarded Green function is written as a sum of the DC retarded Green function and the AC one,

$$G^r(E) = G^r_{DC}(E) + G^r_{AC}(E), \qquad (1)$$

where E is the energy. It is noted that those quantities are matrices representing their position dependence, $\mathbf{r}$ and $\mathbf{r}'$. Here, the AC retarded Green function, $G^r_{AC}(E)$, is given by

$$G^r_{AC}(E) = \\ G^r_{DC}(E + \hbar\omega)\left[V_{AC} + \Sigma^r_{AC}(E + \hbar\omega, E)\right]G^r_{DC}(E), \quad (2)$$

where $\hbar$ is the reduced Planck constant, $V_{AC}(\omega)$ is the AC potential energy determined by the Poisson equation, and $\Sigma^r_{AC}(E + \hbar\omega, E)$ is the AC retarded self-energy. The AC retarded self-energy due to the voltage excitation at a contact is given by

$$\Sigma^r_{AC}(E + \hbar\omega, E) = -\frac{qv_{amp}}{\hbar\omega}\left[\Sigma^r_{DC}(E) - \Sigma^r_{DC}(E + \hbar\omega)\right], \tag{3}$$

where q is the absolute elementary charge.

Once the AC retarded Green function is obtained, the AC lesser Green function, $G^<_{AC}(E)$, is calculated. It is given by

$$\begin{aligned} G^<_{AC}(E) = {} & G^r_{DC}(E + \hbar\omega)\Sigma^<_{DC}(E + \hbar\omega)G^r_{AC}(E)^\dagger \\ & + G^r_{DC}(E + \hbar\omega)\Sigma^<_{AC}(E)G^r_{DC}(E)^\dagger \\ & + G^r_{AC}(E + \hbar\omega)\Sigma^<_{DC}(E)G^r_{DC}(E)^\dagger, \quad (4) \end{aligned}$$

where the AC lesser self-energy is also written as

$$\Sigma^<_{AC}(E) = -\frac{qv_{amp}}{\hbar\omega}\left[\Sigma^<_{DC}(E) - \Sigma^<_{DC}(E + \hbar\omega)\right]. \quad (5)$$

From the AC lesser Green function, the AC response of the electron density, $n_{AC}(\mathbf{r})$, can be easily obtained by an integration,

$$n_{AC}(\mathbf{r}) = -i\int\frac{dE}{2\pi}G^<_{AC}(\mathbf{r}, \mathbf{r}, E). \qquad (6)$$

As much as the contact self-energy is concerned, according to (3) and (5), the AC self-energies can be easily calculated from the DC ones. The DC retarded self-energy at the source (or drain) terminal is obtained by assuming the semi-infinite leads. [10] Also the DC lesser self-energy defined at the source terminal can be written as:

$$\Sigma^{<,S}_{DC}(E) = i(-2)\operatorname{Im}\left\{\Sigma^{r,S}_{DC}(E)\right\}f^s(E), \qquad (7)$$

where $f^s(E)$ is the Fermi-Dirac function evaluated at the source terminal. The expressions for the drain terminal are similar and not repeated here.

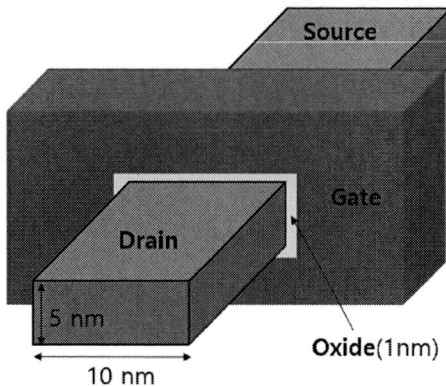

Fig. 1. Silicon nanosheet MOSFET with 5 nm × 10 nm cross section, 1 nm oxide thickness, and 7 nm channel length.

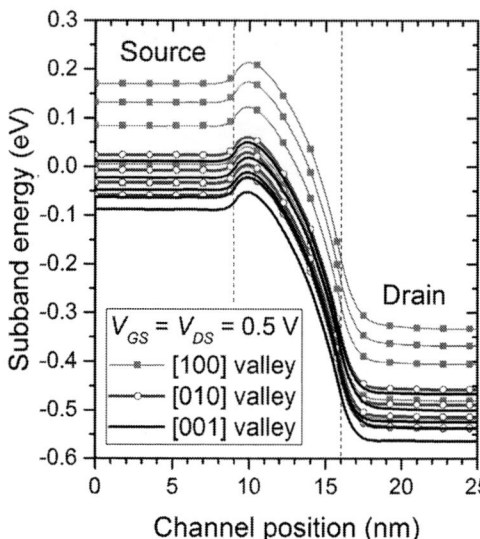

Fig. 2. Five lowest mode energies along the channel direction. Three elliptical valley pairs are considered.

A decoupled mode-space approach is used for an efficient implementation. Using this approach, real and imaginary parts are separated and obtained consistently with the mode-space approach. The size of the matrix is determined by the number of modes multiplied with the number of 1D real space grid points. The AC NEGF solver has been implemented in our in-house device simulator, G-Device. It was previously applied to the THz simulation [11] and the transient BTE simulation [12].

III. NUMERICAL RESULTS

In order to demonstrate the AC NEGF simulation, a silicon nanosheet transistor with 5 nm × 10 nm cross section, 1 nm oxide thickness, and 7 nm channel length is simulated, as illustrated in Fig. 1. A uniform yz cross section is assumed. The entire length of structure is 25 nm. Source/drain regions are 9-nm-long and the doping density is 10^{20} cm^{-3}, while the channel region is intrinsic. The gate contact is 7-nm-long and its workfunction is 4.3 eV. The effective mass Hamiltonian is employed to describe six elliptical valleys in the silicon conduction band. In addition, (001) surface and [100] channel are assumed. V_{DD} is set to be 0.5 V.

Before the AC NEGF simulation, the DC NEGF simulation is performed. Five lowest modes in each valley pair are considered in the simulation. An energy range of (-0.6 eV, 0.6 eV) is adopted. The ballistic transport is assumed. The self-consistent solution of the NEGF-Poisson system is obtained by using a conventional Gummel loop. The convergence criterion of the DC NEGF simulation is the potential error smaller than 10^{-5} times the thermal voltage. The five lowest mode energies for each valley pair is shown in Fig. 2. Since the z-directional length of the silicon channel (5 nm) is much smaller than the y-directional one (10 nm), the [001] valley whose z-directional mass is heavy has the lowest mode energy.

The DC input current-voltage curve of the transistor is shown in Fig. 3. The threshold voltage is about 0.23 V. The DC output current-voltage curve of the transistor is shown in Fig. 4. The ON current at $V_{GS} = V_{DC} = V_{DD}$ is about 50 μA. It is noted that curves in Figs. 3 and 4 are obtained

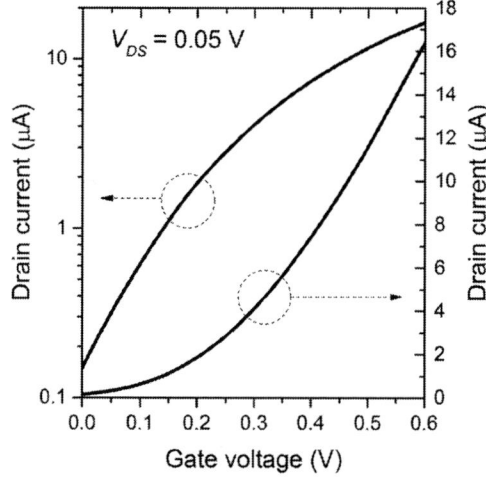

Fig. 3. DC I_D-V_G curve of the nanosheet transistor. $V_{DS} = 0.05$ V.

from the bias ramping procedure. The converged solution at the previous bias point is used as an initial solution of the Gummel loop.

When we recall (2) and (4), we need quantities like $G_{DC}^r(E + \hbar\omega)$, $G_{DC}^r(E)$, $G_{DC}^<(E + \hbar\omega)$, $G_{DC}^<(E)$, and so on. Therefore, the minimum frequency used in the AC NEGF simulation is determined by the energy spacing adopted in the DC NEGF simulation. For example, when the energy spacing of 0.2 meV corresponds to about 48 GHz. When we want to calculate the AC response at 0.2 THz, the maximum allowed energy spacing is 0.827 meV. In order to calculate the AC response at a very low frequency, a very fine energy grid should be introduced.

However, at the low frequency limit, such restriction can be relaxed. AC self-energies in (3) and (5) can be approximated

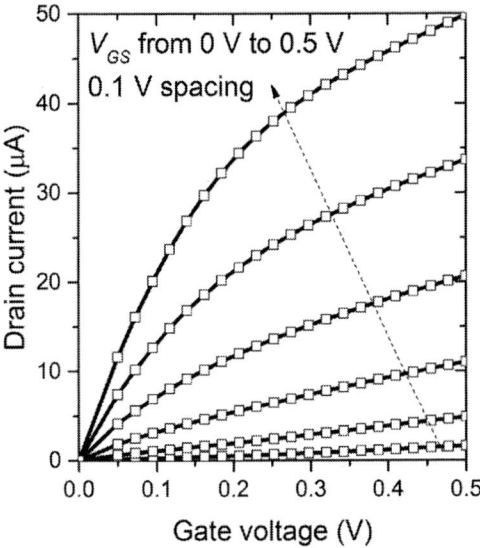

Fig. 4. DC I_D-V_D curve of the nanosheet transistor. Various V_{GS} values are considered.

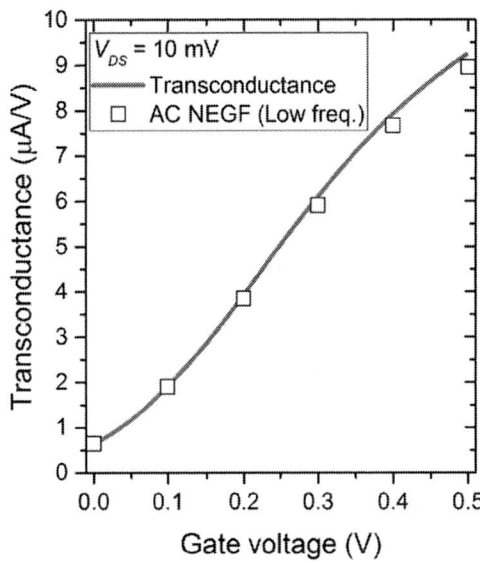

Fig. 5. Transconductance of the nanosheet transistor at the low frequency limit. Square symbols are results of the AC NEGF simulation. V_{DS} = 10 mV.

as

$$\lim_{\omega \to 0} \Sigma_{AC}^r(E + \hbar\omega, E) = qv_{amp}\frac{\partial \Sigma_{DC}^r}{\partial E}, \qquad (8)$$

$$\lim_{\omega \to 0} \Sigma_{AC}^<(E + \hbar\omega, E) = qv_{amp}\frac{\partial \Sigma_{DC}^<}{\partial E}, \qquad (9)$$

respectively. AC Green functions in (2) and (4) can be easily evaluated with the AC self-energies, $G_{DC}^r(E)$, and $G_{DC}^<(E)$.

At the low frequency limit, the admittance, $Y_{DG}(\omega)$, must converge to the transconductance. In Fig. 5, the transconductance of the nanosheet transistor, which is calculated from the DC simulation results with the finite difference, is shown. The low frequency admittance from the AC NEGF simulation shows a good agreement. Similarly, $Y_{DD}(\omega)$ at the low frequency limit should be an inverse of the output resistance. In Fig. 6, the output resistance of the nanosheet transistor is shown. Again, the AC NEGF simulation results show good agreement with the quasi-static results. In addition to the terminal currents, internal quantities such as the electron density should be checked. Electron density integrated over the yz-plane for a gate excitation or a drain excitation is shown in Fig. 7.

At nonzero frequencies, the wideband limit approximation is used for contacts, only because it greatly simplifies the implementation. A slightly modified version of the structure shown in Fig. 1 (long source/drain regions) is simulated. An isotropic effective mass is considered and only the lowest mode is calculated.

The AC total current is obtained by summing the particle current and the displacement current [8]. The frequency dependence of $Y_{GG}(\omega)$ is shown in Fig. 8. The real and imaginary parts are illustrated separately. $Y_{DG}(\omega)$ is also shown in Fig. 9. Its real part rapidly decreases at high frequencies. It is noted that the minimum frequency (48 GHz) is related with the

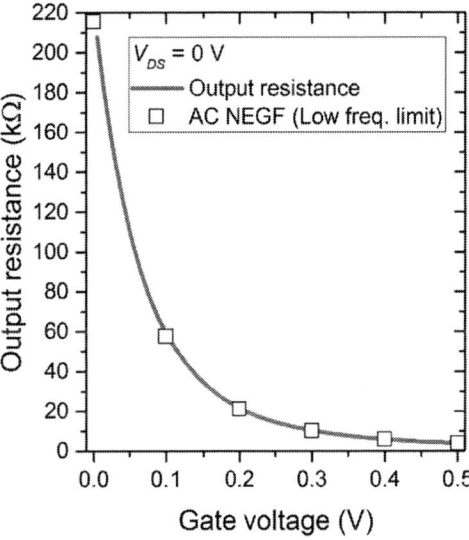

Fig. 6. Output resistance of the nanosheet transistor at the low frequency limit. Square symbols are results of the AC NEGF simulation. V_{DS} = 0 V.

energy spacing used in the DC NEGF simulation, as explained already.

IV. CONCLUSION

In summary, the AC NEGF simulation approach has been demonstrated for nanosheet MOSFETs. At the low frequency limit, the AC NEGF simulation results have been compared with the quasi-static results. AC terminal currents at nonzero frequencies have been evaluated under the wideband limit approximation. The result represents that the nanoscale device exhibits complicated frequency dependence on the external voltage excitation.

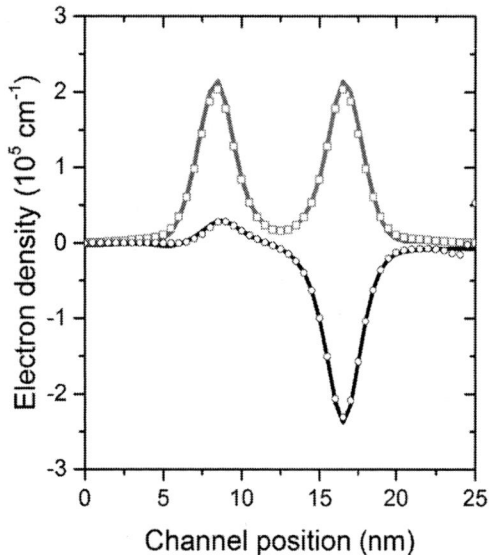

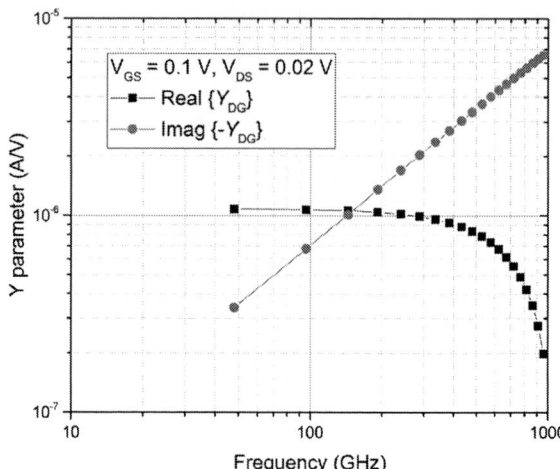

Fig. 7. Integrated electron density for a gate excitation (Red line) or a drain excitation (Black line) along the channel position. For the DC NEGF simulation, $V_{GS} = V_{DS} = 0.0$ V. The amplitude of a voltage excitation is 5 mV. Symbols represent the AC NEGF simulation results.

Fig. 9. $Y_{DG}(\omega)$ of the nanosheet transistor. The frequency varies from 48 GHz to 1 THz. $V_{GS} = 0.1$ V and $V_{DS} = 0.02$ V. The wideband limit approximation is applied.

REFERENCES

[1] S. Datta, "A simple kinetic equation for steady-state quantum transport," *Journal of Physics: Condensed Matter*, vol. 2, no. 40, pp. 8023–8052, oct 1990.

[2] ——, "Nanoscale device modeling: the greenâs function method," *Superlattices and microstructures*, vol. 28, no. 4, pp. 253–278, 2000.

[3] ——, *Quantum Transport: Atom to Transistor*. Cambridge University Press, 2005.

[4] S. Tiwari, M. V. Fischetti, and S. E. Laux, "Overshoot in transient and steady-state in GaAs, InP, Ga.47In.53As and InAs bipolar transistors," in *1990 International Electron Device Meeting*, 1990, pp. 435 438, doi: 10.1109/IEDM.1990.237 074.

[5] S. E. Laux and M. V. Fischetti, "Monte Carlo study of velocity overshoot in switching a 0.1-micron CMOS inverter," in *1997 International Electron Device Meeting*, 1997, pp. 877–890, doi: 10.1109/IEDM.1997.650 521.

[6] Y. Koseki, V. Rhzhii, T. Otsuji, V. V. Popov, and A. Satou, "Giant plasmon instability in a dual-grating-gate graphene field-effect transistor," *Physical Review B*, vol. 93, no. 24, p. 245408, June 2016, doi: 10.1103/PhysRevB.93.245408.

[7] S. Di, K. Zhao, T. Lu, G. Du, and X. Liu, "Investigation of transient responses of nanoscale transistors by deterministic solution of the time-dependent BTE," *Journal of Computational Electronics*, vol. 15, no. 3, pp. 770–777, September 2016, doi: 10.1007/s10825-016-0818-1.

[8] Y. Wei and J. Wang, "Current conserving nonequilibrium ac transport theory," *Phys. Rev. B*, vol. 79, p. 195315, May 2009. [Online]. Available: https://link.aps.org/doi/10.1103/PhysRevB.79.195315

[9] O. Shevtsov and X. Waintal, "Numerical toolkit for electronic quantum transport at finite frequency," *Phys. Rev. B*, vol. 87, p. 085304, Feb 2013. [Online]. Available: https://link.aps.org/doi/10.1103/PhysRevB.87.085304

[10] M. J. McLennan, Y. Lee, and S. Datta, "Voltage drop in mesoscopic systems: A numerical study using a quantum kinetic equation," *Phys. Rev. B*, vol. 43, pp. 13 846–13 884, Jun 1991. [Online]. Available: https://link.aps.org/doi/10.1103/PhysRevB.43.13846

[11] S.-M. Hong and J.-H. Jang, "Numerical simulation of plasma oscillation in 2-D electron gas using a periodic steady-state solver," *IEEE Transactions on Electron Devices*, vol. 62, no. 12, pp. 4192–4198, Dec 2015, 10.1109/TED.2015.2489220.

[12] ——, "Transient simulation of semiconductor devices using a deterministic Boltzmann equation solver," *IEEE Journal of the Electron Devices Society*, vol. 6, pp. 156–163, 2018, doi: 0.1109/JEDS.2017.2780837.

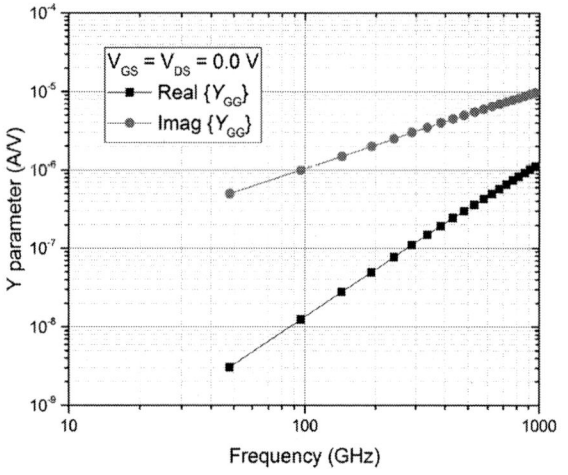

Fig. 8. $Y_{GG}(\omega)$ of the nanosheet transistor. The frequency varies from 48 GHz to 1 THz. $V_{GS} = V_{DS} = 0.0$ V. The wideband limit approximation is applied.

As a future work, the AC NEGF simulation will be extended to more general cases beyond the wideband limit approximation and the ballistic transport. Additionally, the proof of the current conservation under the conventional simulation set up would be an interesting future research topic.

ACKNOWLEDGEMENT

This work was supported by the National Research Foundation of Korea (NRF) grant funded by the Korea government (NRF-2019R1A2C1086656).

Enhanced Capabilities of the Nano-Electronic Simulation Software (NESS)

Cristina Medina-Bailon
Device Modelling Group,
School of Engineering,
University of Glasgow,
G12 8LT Glasgow, UK.
cristina.medinabailon@glasgow.ac.uk

Oves Badami
Device Modelling Group,
School of Engineering,
University of Glasgow,
G12 8LT Glasgow, UK.
qbadami@gmail.com

Hamilton Carrillo-Nuñez
Device Modelling Group,
School of Engineering,
University of Glasgow,
G12 8LT Glasgow, UK.
hacarrillo@gmail.com

Tapas Dutta
Device Modelling Group,
School of Engineering,
University of Glasgow,
G12 8LT Glasgow, UK.
tapas.dutta@glasgow.ac.uk

Daniel Nagy
Device Modelling Group,
School of Engineering,
University of Glasgow,
G12 8LT Glasgow, UK.
daniel.nagy@glasgow.ac.uk

Fikru Adamu-Lema
Device Modelling Group,
School of Engineering,
University of Glasgow,
G12 8LT Glasgow, UK.
fikru.adamu-lema@glasgow.ac.uk

Vihar P. Georgiev
Device Modelling Group,
School of Engineering,
University of Glasgow,
G12 8LT Glasgow, UK.
vihar.georgiev@glasgow.ac.uk

Asen Asenov
Device Modelling Group,
School of Engineering,
University of Glasgow,
G12 8LT Glasgow, UK.
asen.asenov@glasgow.ac.uk

Abstract—The aim of this paper is to present a flexible TCAD platform called Nano-Electronic Simulation Software (NESS) which enables the modelling of contemporary future electronic devices combining different simulation paradigms (with different degrees of complexity) in a unified simulation domain. NESS considers confinement-aware band structures, generates the main sources of variability, and can study their impact using different transport models. In particular, this work focuses on the new modules implemented: Kubo-Greenwood solver, Kinetic Monte Carlo solver, Gate Leakage calculation, and a full-band quantum transport solver in the presence of hole-phonon interactions using a mode-space $k \cdot p$ approach in combination with the existing NEGF module.

Index Terms—Integrated Simulation Environment, Drift-Diffusion, Kubo-Greenwood, Gate Leakage, Kinetic Monte Carlo, Non-Equilibrium Green's Function, Six band k.p, Effective mass, Variability

I. INTRODUCTION

As the 5 nm CMOS technology is being prototyped, the need for advanced nano-transistor transport simulation has increased. In particular, new device architectures and concepts rapidly evolve and add new simulation requirements. Since the quantum confinement and quantum transport phenomena are dominating the nano-scale transistor characteristics, they have to be incorporated in simulations. Moreover, the inclusion of statistical variability sources (such as charge and matter granularity and the complexity of the interface transitions) when the transistor dimensions are in the nanometer range are mandatory. Over the years, simulation tools with different

This project has received funding from EPSRC UKRI Innovation Fellowship scheme under grant agreement No. EP/S001131/1 (QSEE), No. EP/P009972/1 (QUANTDEVMOD) and No. EP/S000259/1(Variability PDK for design based research on FPGA/neuro computing); and from H2020-FETOPEN-2019 scheme under grant agreement No.862539-Electromed-FET OPEN.

levels of complexity and related computational cost have been developed by different software vendors and research groups, generally working in isolation: from classical (e.g. Drift-Diffusion (DD) [1]); through semi-classical (e.g. Monte Carlo (MC) [2] or direct Boltzmann Transport Equation (BTE) [3] solvers); to quantum transport solvers (e.g. Non-Equilibrium Green's Function (NEGF) [4]). In this work, we outline the development advances of the Nano-Electronic Simulation Software (NESS) [5] developed in the Device Modelling group at University of Glasgow, which is a flexible and modular platform combining different simulation paradigms in a unified simulation domain. Accordingly, it enables consistent simulations with techniques of varying complexity for a particular transistor structure highlighting and understanding the areas of applicability.

II. OVERVIEW OF NESS

Fig. 1 shows the five main modules of our simulation environment NESS: structure generator, effective mass extractor, material database, solvers, and outputs. First, the structure generator allows the creation and configuration of the 3D device structures [6], [7], including the following main variability sources: random discrete dopant (RDD), line edge roughness (LER), and metal gate granularity (MGG). Second, the effective mass extractor [8] can calculate the correct electron effective masses, in both confinement and transport directions, from the first principle simulations of the electronic bandstructure of nanowire transistors (NWTs) with technologically relevant cross-sectional area, shape, and transport orientations. Third, the material database provides the relevant material parameters for each solver, such as the work-function, affinity, or scattering parameters. Furthermore, the effective masses can be provided for each material from density functional theory (DFT), tight-binding (TB) methods, or directly from our effective mass extractor.

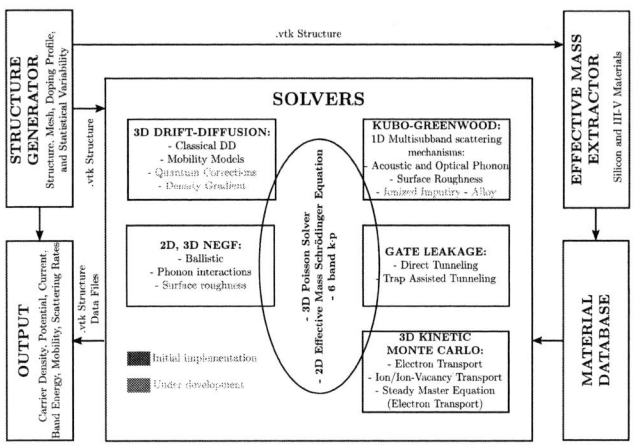

Fig. 1: Flowchart of NESS detailing its modular structure.

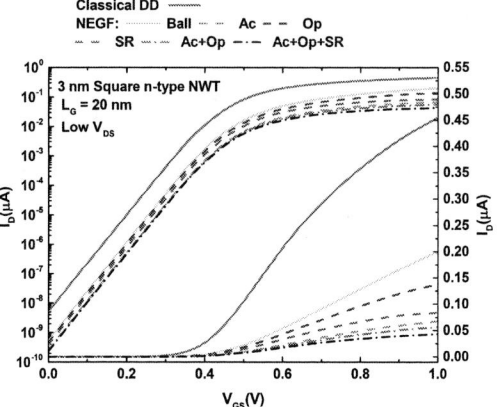

Fig. 2: I_D vs. V_{GS} characteristics for n-type square 3nm Si NWTs with L_G=20nm using the classical DD and the NEGF modules, assuming ballistic and dissipative transport with acoustic (Ac), g-type optical (Op) phonon and surface roughness (SR) scattering mechanisms

Fourth, different transport simulation solvers, with increasing computational complexity and cost, have been already implemented [5]. In general, each of them is self-consistently solved with the 3D Poisson and the 2D Schrödinger equations. The classical DD solver is based on the Sharfetter-Gummel discretisation of the semiconductor equation using Bernoulli functions. It contains different mobility models as well as quantum corrections based on Poisson-Schrödinger solutions [9], and density gradient. The coupled mode-space NEGF solver allows a quantum treatment of charge transport in nanodevices, naturally capturing quantum transport phenomena, such as tunnelling, coherence and particle-particle interactions. This solver can consider phonon scattering in addition to the transport in the ballistic limit. Furthermore, the NEGF solver implemented in NESS allows the calculation of band-to-band tunneling (BTBT) by using the Flietner model to compute the current in heterostructures with direct band gap [10] and the study of surface roughness (SR) scattering mechanism [11]. Finally, the simulation results are stored in text files (i.e. current, charge concentration) and also in vtk format for easy visualisation with freeware software, such as ParaView.

Fig. 2 shows the I_D vs. V_{GS} characteristics for n-type square 3nm Si NWTs with L_G=20nm considering both the classical DD module as well as the NEGF module assuming ballistic and dissipative transport with: *(i)* acoustic phonon (Ac), *(ii)* g-type optical phonon (Op); and *(iii)* SR scattering mechanisms. The new modules in NESS are described in detail in the following section.

III. NEW MODULES IN NESS

A. Kubo-Greenwood mobility module:

A 1D multisubband Kubo-Greenwood (KG) simulator [12] has been integrated in NESS in order to calculate the low-field electron mobility. This semi-classical approach combines the quantum effects based on the 1D multi-subband scattering rates of the most relevant scattering mechanisms in NWTs and

the semi-classical BTE by applying the KG formula within the relaxation time approximation. In Fig. 3, we present a comparison of the KG mobility with the mobility computed by the NEGF module considering the same scattering mechanisms as in Fig. 2: Ac, g-type Op (which corresponds to intravalley transitions) and SR. Apart from the scattering mechanisms included in Fig. 3, the f-type Op phonon (intervalley transitions) and the ionized impurity are also implemented in NESS.

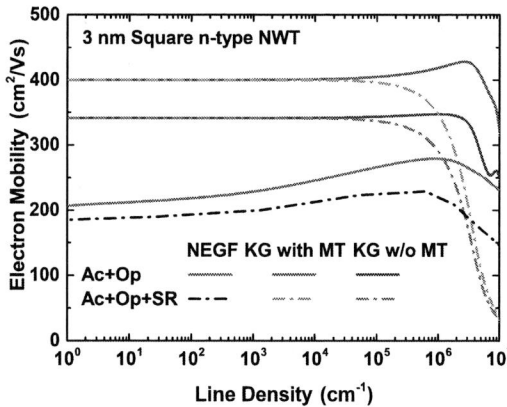

Fig. 3: Comparison of the electron mobility for a square 3nm Si NWT computed using NEGF and KG solvers combining acoustic (Ac) and g-type optical (Op) phonon scattering, and all the mechanisms (Ac + Op + surface roughness SR).

As this solver is based on the long-channel device approximation, the first step is to pre-calculate the required subband levels (E_l) and the corresponding wavefunctions (ξ_l) using a self-consistent Poisson-Schrödinger simulation in the presence of a low electric field in the transport direction. Currently, this pre-calculation is carried out using the NEGF module in NESS. The second step is to use these quantities to compute the scattering rates whose equations are directly derived from the Fermi golden rule [12]. Finally, the total mobility is estimated using either the Matthiessen rule, or by directly summing the scattering rates of each mechanism and the total mobility for each subband based on the KG formalism. In general, the advantage of both semi-classical alternatives to compute the total mobility in comparison to purely quantum transport simulations is that the rates are individually computed and then combined, hence reducing dramatically the computational cost.

B. Kinetic Monte Carlo module:

A kinetic Monte Carlo (kMC) module has been added to NESS, which is used to model memory devices with metal-oxide layers (e.g. Flash memory, RRAM, etc.). The kMC accounts for ionic and electronic transport, ion and ion-vacancy generation and recombination. Note that the diffusion barrier of the vacancies are relatively high, therefore we neglect the ion-vacancy transport [18]. In order to model the electronic transport, a comprehensive set of physics-based models are included to capture the Poole-Frenkel emission, Direct and Fowler-Nordheim (FN) tunneling, defect to defect tunneling, elastic and inelastic electrode to defect tunneling and vice versa tunneling models [16]. The electronic transport can also be modeled using the steady state master equation as in Ref. [17]. The electronic KMC (or steady state master equation), which is coupled to the ionic KMC, is used to model the RRAM devices [18] and the different aspects of the flash memory operation. An example for ion vacancy distributions for TiN/SiOx/TiN structure, during the forming, reset (overlapping of forming and reset ion vacancies) and set (overlapping of reset and set ion vacancies) processes are depicted in Fig. 4. It is important to highlight during the reset process that there are less vacancies near the bottom terminal. Fig. 5 shows the RRAM characteristic including forming, reset and set trajectories of a SiOx based RRAM. For these simulations, the steady state master equation was used to model the electronic transport while the ion dynamics were modeled using the kMC approach.

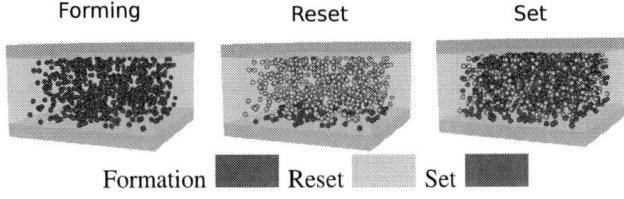

Fig. 4: Ion vacancy distributions for TiN/SiOx/TiN structure, during the forming, reset (overlapping of forming and reset ion vacancies) and set (overlapping of reset and set ion vacancies) processes.

C. Gate Leakage module:

We have implemented a module to calculate the gate leakage current in NESS. The gate leakage current is considered to be the sum of direct tunneling, FN tunneling, and trap assisted tunneling (TAT). The direct and FN tunneling are modelled using the Tsu-Esaki formulation [13], while the TAT is calculated using the methodology as proposed in Ref. [14]. This flexible module allows the user to set the

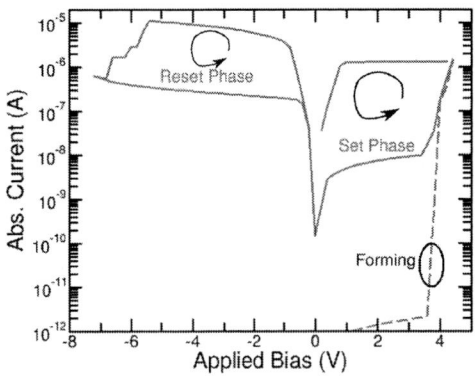

Fig. 5: Simulation characteristic of a bipolar SiOx based RRAM device. The compliance current of $1\mu A$ was enforced in the simulation.

trap parameters including trap density, energy and number of material dependent parameters. The module can also account for the impact of different sources of variability like oxide-semiconductor interface roughness, random discrete dopants, and random interface charges. Fig. 6 shows a comparison between the gate leakage calculated using the implemented module and experimental results.

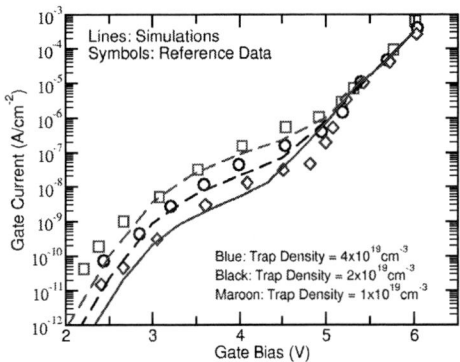

Fig. 6: Comparison between the reference gate current (symbols) and the simulated results (solid lines). The reference data is from [15].

D. Six Band $k \cdot p$ NEGF simulation capabilities:

We have implemented a mode-space full-band quantum transport solver with hole-phonon interactions by adding a six-band $k \cdot p$ Hamiltonian to the existing NEGF module in NESS [19]. The Hamiltonian is first unitary transformed from real-space to a reduced-order Hamiltonian discretized in the K-space which is constructed from sampling the Brillouin zone and obtaining the Bloch modes significant to the transport. The mode-space Hamiltonian is then generated by unitary transforming the reduced-order Hamiltonian in K-space [20], [21]. The expressions for the hole-phonon interactions self-energies within the mode-space representation are based on the self-consistent Born approximation and are given in Ref. [19]. Fig. 7 shows the efficient simulations of p-type square

295

3 nm Si NWT considering the most common crystallographic orientations ([100, 110, 111]) with coherent and dissipative transport. The energy transfer due to scattering in NEGF is shown in Fig. 8 where the LDOS is represented for the device considered in Fig. 7 with transport direction [100] and gate voltage of -0.6 V.

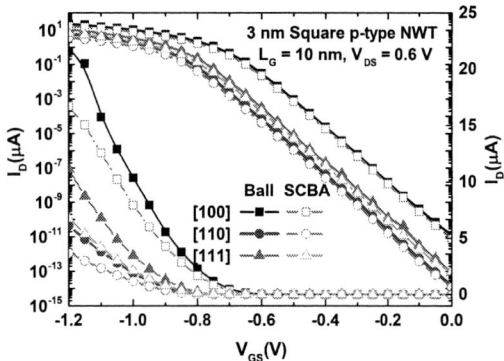

Fig. 7: I_D vs. V_{GS} characteristics for p-type square 3 nm Si NWTs considering the most common crystallographic orientations ([100, 110, 111]) with coherent and dissipative transport. The results have been simulated making use of the mode-space full-band quantum transport solver included in NESS by combining a six-band $k \cdot p$ Hamiltonian and the existing NEGF module.

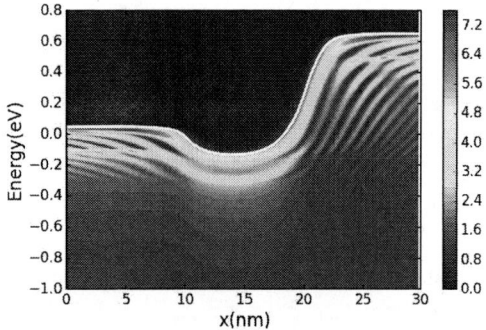

Fig. 8: LDOS for the device simulated in Fig. 7 with transport direction [100] and gate voltage -0.6 V. The white line corresponds to the conduction band.

IV. CONCLUSIONS

In this work, we have presented a general overview and the current stage of development of our flexible nanotransistor simulator NESS, with focus on the new modules implemented: Kubo-Greenwood solver, Kinetic Monte Carlo solver, Gate Leakage calculation, and a full-band quantum transport solver in presence of hole-phonon interactions using a mode-space $k \cdot p$ approach in combination with the existing NEGF module. The NESS modular architecture and unified simulation domain have allowed us to include solvers with different complexity and simulation techniques to study nanoelectronic structures in different areas of research and applications. The environment makes also easy its future expansion with new modules/solvers such as nano-interconnects and a direct Boltzmann solver.

REFERENCES

[1] G. Roy, A.R. Brown, F. Adamu-Lema, S. Roy, and A. Asenov, "Simulation study of individual and combined sources of intrinsic parameter fluctuations in conventional nano-MOSFETs", IEEE Transactions on Electron Devices, 53(12), pp. 3063-3070, 2006.

[2] L. Donetti, at al., "Multi-Subband Ensemble Monte Carlo Simulation of Si Nanowire MOSFETs", 2018 International Conference on Simulation of Semiconductor Processes and Devices (SISPAD), 2015.

[3] S. Jin, T.-W. Tang, and M. V. Fischetti, "Simulation of Silicon Nanowire Transistors Using Boltzmann Transport Equation Under Relaxation Time Approximation", IEEE Transactions on Electron Devices, 55(3), pp. 727-736, 2008.

[4] M. Luisier and G. Klimeck, "Atomistic full-band simulations of silicon nanowire transistors: Effects of electron-phonon scattering", Phys. Rev. B, 80, p. 155430, 2009.

[5] S. Berrada, et al, "Nano-Electronic Simulation Software (NESS): a flexible nano-device simulation platform", Journal of Computational Electronics, 19, pp. 1031—1046, 2020.

[6] J. Lee, et al, "Variability Predictions for the Next Technology Generations of n-type SixGe1-x Nanowire MOSFETs", Micromachines, 9(12), p. 643, 2018.

[7] Lee J., et al., "The Impact of Dopant Diffusion on Random Dopant Fluctuation in Si Nanowire FETs: A Quantum Transport Study", 2018 International Conference on Simulation of Semiconductor Processes and Devices (SISPAD), pp. 280–283, 2018.

[8] O. Badami, et al., "Comprehensive Study of Cross-Section Dependent Effective Masses for Silicon Based Gate-All-Around Transistors", Applied Sciences, 9(9), p. 1895, 2019.

[9] T. Dutta, C. Medina-Bailon, H. Carrillo-Nuñez, O. Badami, V.P. Georgiev, and A. Asenov, "Schrödinger Equation Based Quantum Corrections in Drift-Diffusion: A Multiscale Approach", IEEE Nanotechnology Materials and Devices Conference (NMDC), 2019.

[10] H. Carrillo-Nuñez, et al., "Random Dopant-Induced Variability in Si-InAs Nanowire Tunnel FETs: A Quantum Transport Simulation Study", IEEE Electron Device Lett. 39(9), pp. 1473–1476, 2018.

[11] O. Badami, S. Berrada, H. Carrillo-Nuñez, C. Medina-Bailon, V.P. Georviev, and A. Asenov, "Surface Roughness Scattering in NEGF using self-energy formulation", 2019 International Conference on Simulation of Semiconductor Processes and Devices (SISPAD), 2019.

[12] T. Sadi, et al., "Simulation of the Impact of Ionized Impurity Scattering on the Total Mobility in Si Nanowire Transistors", Materials, 12(1), p. 124, 2019.

[13] R. Tsu, and L. Esaki, "Tunneling in a finite superlattice", Applied Physics Letters, 22(11), pp. 562–564, 1973.

[14] M. Herrmann, A. and Schenk, "Field and high-temperature dependence of the long term charge loss in erasable programmable read only memories: Measurements and modeling", Journal of applied physics, 77(9), pp. 4522–4540, 1995.

[15] B. Ricco, G. Gozzi, M. and Lanzoni, "Modeling and simulation of stress-induced leakage current in ultrathin SiO/sub 2/films", IEEE Transactions on Electron Devices, 45(7), pp. 1554–1560, 1998.

[16] O. Badami, et al., "Multiscale modeling of charge trapping in molecule based flash memories". 2019 International Conference on Simulation of Semiconductor Processes and Devices (SISPAD), 2019.

[17] S. Yu, X. Guan, and H.S.P. Wong, "Understanding metal oxide RRAM current overshoot and reliability using kinetic Monte Carlo simulation", 2012 International Electron Devices Meeting (IEDM), pp. 26–1, 2012.

[18] T. Sadi, A. Mehonic, L. Montesi, M. Buckwell, A. Kenyon, and A. Asenov, "Investigation of resistance switching in SiO x RRAM cells using a 3D multi-scale kinetic Monte Carlo simulator, Journal of Physics: Condensed Matter, 30(8), p. 084005, 2018.

[19] H. Carrillo-Nuñez, C. Medina-Bailon, V.P. Georviev, and A. Asenov, "Full-band quantum transport simulation in presence of hole-phonon interactions using a mode-space k·p approach", Nanotechnology Journal, in press, 2020.

[20] M. Shin, "Full-quantum simulation of hole transport and band-to-band tunneling in nanowires using the $k \cdot p$ method", Journal of Applied Physics, 106(5), p. 054505, 2009.

[21] J.Z. Huang, W.C.J. Chew,J. Peng, C.Y. Yam, L.J. Jiang, and G.H. Chen, "Model order reduction for multiband quantum transport simulations and its application to p-type junctionless transistors", IEEE Transactions on electron devices, 60(7), pp. 2111–2119, 2013.

Electrostatic Potential Profile Generator for Two-Dimensional Semiconductor Devices

Seung-Cheol Han, Jonghyun Choi, and Sung-Min Hong

School of EECS, Gwangju Institute of Science and Technology, Gwangju, Republic of Korea, email: smhong@gist.ac.kr

Abstract—As efficiency is one of the bottlenecks of device simulation, we propose to employ deep neural networks to generate two-dimensional electrostatic potential profiles for efficiency. Supervising with previously obtained simulation results for various BJT devices, we train deep neural networks to generate an electrostatic potential profile as an initial guess for a non-equilibrium condition with estimating carrier densities by the frozen field simulation. With the generated potential profiles, we significantly reduce the number of Newton iterations without loss of accuracy.

I. INTRODUCTION

Deep neural networks are utilized in many fields including computer vision, natural language processing, and expert systems. In semiconductor research, deep neural networks are used in two different aspects. First, there are many efforts to develop a new hardware system to implement efficient deep neural networks [1], [2]. Second, the deep networks are used as an optimization tool in the technology development cycle [3], [4]. To the best of our knowledge, there is hardly an effort to improve efficiency of the device simulation. Since we have tremendous amount of solutions as results of multiple simulations, the deep neural network which learn the existing solutions can give us reasonably good estimation to the real solution. We propose to use deep neural network to quickly analyze the two-dimensional semiconductor devices.

In the device simulation, the computational cost depends on the number of bias points. Even when only the final bias point is of interest, the bias voltages should be gradually increased due to the severe nonlinearity of the semiconductor device equations. In Newton method, good initial solutions will expedite the convergence thus improve computational efficiency significantly. If the generated profiles are close enough to the numerical solutions, the bias ramping process can be skipped; the skip provides further computational gain. In our previous work [5], we showed that the deep neural networks could generate good initial profiles for the one-dimensional structures. As an extension to our previous work, we here propose deep neural networks for two-dimensional structures such as bipolar junction transistors (BJTs).

II. NEURAL NETWORKS

As discussed in our previous work [5], the electrostatic potential is a key quantity to be generated by deep neural networks. Under a fixed electrostatic potential profile, the electron and hole continuity equations become linear or at most locally nonlinear, depending on the adopted physical

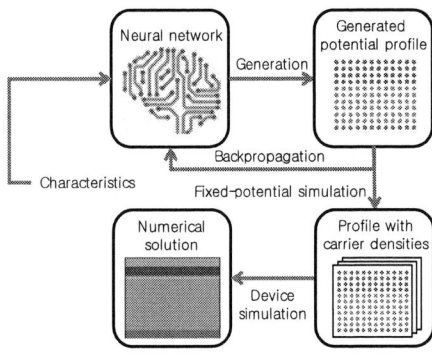

Fig. 1. A diagram for the proposed method. We consider two-dimensional structures in this work.

models. For this reason, a deep neural network is designed to generate electrostatic potential profiles.

Fig. 1 shows a diagram for the proposed method. We use some important device parameters (such as bias voltages and doping profiles) as input and use initial electrostatic profiles as output. As a loss function to train the deep network, we use mean squared error between generated and simulated electrostatic potential profiles [6]. To use the electrostatic profiles generated by the neural network as the initial solution of the device simulator, additional quantities such as the carrier densities are required. Fixed-potential simulations provide these quantities. To train the neural network structure, we use PyTorch library. For the device simulation, our in-house code written in C++ is used.

In our previous work [5], we use a deep neural network which is composed of multi-layer perceptrons (MLP) as the device structure is one-dimensional. To extend it to the two-dimensional device structure, we use the convolutional neural network (CNN). In particular, we use an auto encoder whose architecture is excerpted from the deep convolutional generative adversarial network structure (DCGAN) [7]. Fig. 2 shows the neural network structure employed in this work. The trained convolutional network can be applied to not only a single structure but also a set of several devices in a restricted parameter range, as demonstrated in Section III-B.

III. NUMERICAL RESULTS FOR BIPOLAR JUNCTION TRANSISTORS

We assume silicon devices at room temperature and consider both types of carriers (electrons and holes) in all of our simulations. In this preliminary study, constant carrier mobility

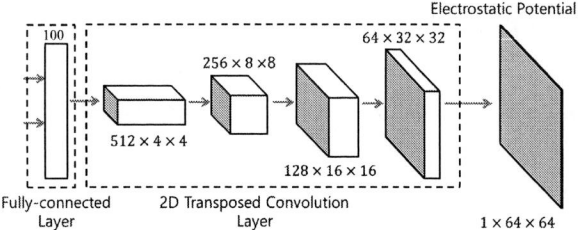

Fig. 2. Layer structure of the CNN adopted in the two-dimensional problem. The output layer generates a 64-by-64 matrix corresponding to the two-dimensional simulation domain. It is a modified version of the generator in the DCGAN in [6].

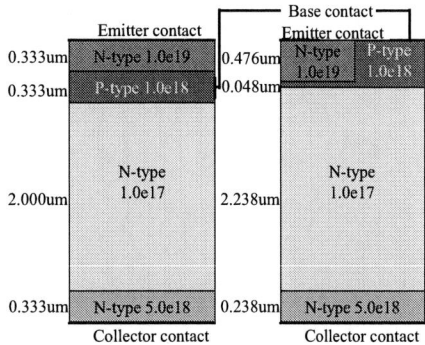

Fig. 3. Structures under consideration. A simple BJT (Left) and a more realistic BJT (Right). Numbers inside the device structures represent doping densities in cm^{-3}.

is adopted. But, more advanced mobility models can be readily used. No generation-recombination process is introduced.

A. Fixed device structures

Before applying our method to a set of several devices, two fixed device structures are tested in order to see the feasibility. The two BJT structures that we consider is shown in Fig. 3.

1) A simple BJT: First, we try our method on a simple BJT structure. It is a quasi-1D structure. The training data set contains 10,000 instances of 64-by-64 electrostatic potential profiles. Each instance is labeled with a pair of (V_B, V_C). The base voltage (V_B) varies from 0.0V to 0.5V. The collector voltage (V_C) is ranged from 0.0V to 0.4V. The training and validation errors are measured as functions of learning epoch as shown in Fig. 4.

After training the network with 100 epochs, we obtain a trained network to generate an approximate potential profile in the inference phase. Fig. 5 shows an example of the electrostatic potential profile generated by the trained neural network with its error. The maximum error of all test instances (not included in the training set) is lower than 21 mV. It implies that the generated potential profiles by the convolutional neural networks are close enough to the simulated profiles.

Fig. 6 shows the convergence behavior of the entire solution procedure to get the solution at the target bias condition. The target bias condition is $V_B = 0.5$V and $V_C = 0.4$V. When

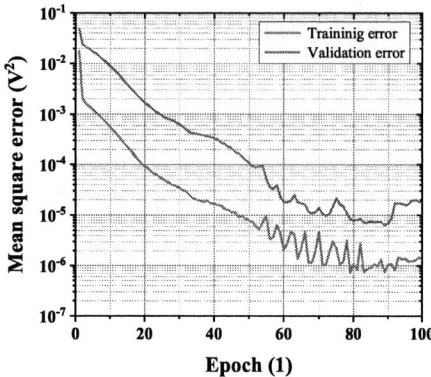

Fig. 4. Training and validation errors of a convolutional neural network, which is trained for the simple BJT structure.

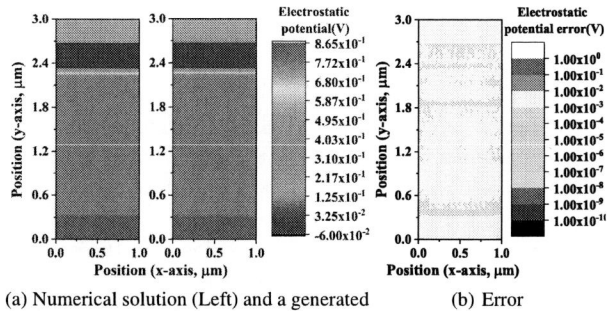

(a) Numerical solution (Left) and a generated potential profile(Right).

(b) Error

Fig. 5. (a) Generated potential profile and (b) its error of the simple BJT when $V_B = 0.4192$ V and $V_C = 0.2586$ V.

the maximum potential update is smaller than 10^{-10} V, the convergence criterion is satisfied. Starting with the generated potential profiles, the converged solution at the target bias condition is readily obtained in four Newton iterations with no bias ramping process. On the other hand, a conventional bias ramping with a uniform bias step ("Scratch condition" in the legend of Fig. 6) requires a larger number of Newton iterations. The number of Newton iterations can be reduced significantly with the generated potential profiles by comparing green curves and a red curve.

2) A more realistic BJT: Second, we try our method on a more realistic BJT structure. This structure has a base contact on the top surface. We configure the training data set for this structure following the same configuration of that of the first structure, and the same ranged input parameters to the first structure. The training and validation errors are shown in Fig. 7.

Fig. 8 shows an example of the generated electrostatic potential profile and its error. The maximum error of all test cases (not included in the training set) is lower than 38 mV, which is in a similar level with the simple BJT structure. It clearly demonstrates that two-dimensional doping profile does not introduce additional difficulties in the training phase. Also, its convergence behavior is shown in Fig. 9. The target bias

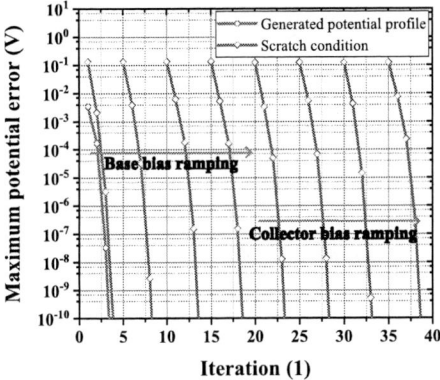

Fig. 6. Comparison of convergence behavior. The simulation with the generated initial profiles takes only four iterations for the converged solution. The target bias condition is $V_B = 0.5$ V and $V_C = 0.4$ V.

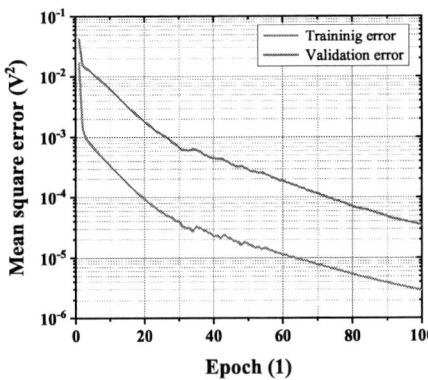

Fig. 7. Training and validation errors of a convolutional neural network, which is trained for the realistic BJT structure.

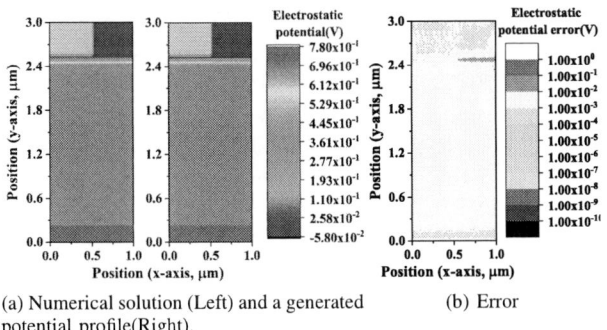

(a) Numerical solution (Left) and a generated potential profile(Right).

(b) Error

Fig. 8. (a) Generated potential profile and (b) its error of the more realistic BJT when $V_B = 0.4192$ V and $V_C = 0.2586$ V.

condition is $V_B = 0.5$ V and $V_C = 0.4$ V. Again, the number of iterations can be reduced significantly by comparing green curves and a red curve.

B. Various devices

We additionally present more results on various devices. Fig. 10 shows the template of device structures considered. Variable device parameters are as follow: X_1 is the length of

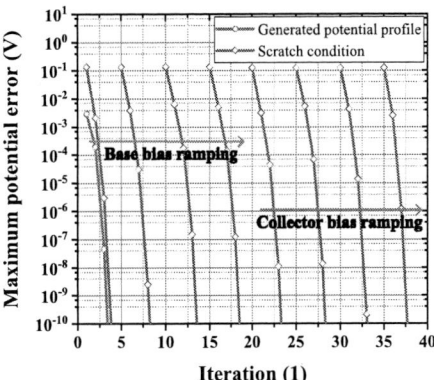

Fig. 9. Comparison of convergence behavior. The target bias condition is $V_B = 0.5$ V and $V_C = 0.4$ V.

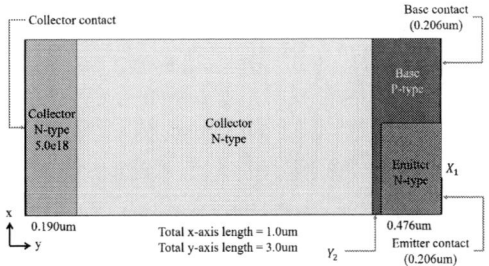

Fig. 10. Device structure under consideration. X_1 is the length of the emitter contact and Y_2 is the base thickness.

TABLE I
RANGES OF INPUT PARAMETERS

Parameter	Range
Collector voltage	$0.0\text{V} \sim 0.4\text{V}$
Base voltage	$0.0\text{V} \sim 0.5\text{V}$
X1 (Emitter length)	$0.254\mu m \sim 0.762\mu m$
Y2 (Base thickness)	$0.048\mu m \sim 0.019\mu m$
Collector doping density	$0.7 \times 10^{17}\text{cm}^{-3} \sim 1.3 \times 10^{17}\text{cm}^{-3}$
Base doping density	$0.7 \times 10^{18}\text{cm}^{-3} \sim 1.3 \times 10^{18}\text{cm}^{-3}$
Emitter doping density	$0.7 \times 10^{19}\text{cm}^{-3} \sim 1.3 \times 10^{19}\text{cm}^{-3}$

the emitter contact and Y_2 is the base thickness. In addition to these two geometrical parameters, the bias voltages (V_B and V_C) and the doping densities (in the base region, the collector region, and the emitter region) are also varied. These variable parameters are used as input parameters of the convolutional neural network.

Each input parameter in the training set is randomly picked within its range. The training data set consists of 25,000 instances of these input parameters and their corresponding potential profiles. Table I shows the ranges of input parameters.

The training and validation errors are shown in Fig. 11. Errors are successfully reduced after 100 epochs. Figs. 12 and 13 show two examples of the generated electrostatic potential profiles and their errors. These examples demonstrate that the convolutional neural network can generate the potential profile

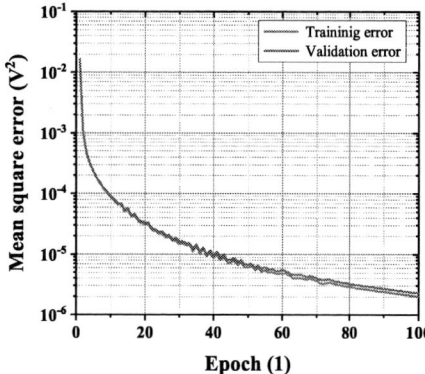

Fig. 11. Training and validation errors of a convolutional neural network which is trained for a set of various BJT structures.

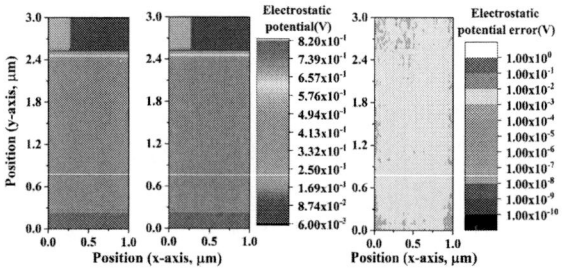

Fig. 12. Numerical solution (Left), a generated potential profile by the convolutional neural network (Center), and its error (Right). V_B = 0.5 V, V_C = 0.3 V, X_1 = 0.254 μm, Y_2 = 0.0476 μm, (Base doping)= 1.0×10^{18}cm^{-3}, (Emitter doping)= 1.1×10^{19}cm^{-3}, and (Collector doping)= 8.0×10^{16}cm^{-3}.

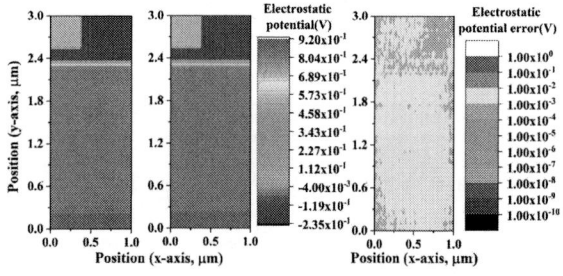

Fig. 13. Numerical solution (Left), a generated potential profile by the convolutional neural network (Center), and its error (Right). V_B = 0.25 V, V_C = 0.4 V, X_1 = 0.365 μm, Y_2 = 0.190 μm, (Base doping)= 1.3×10^{18}cm^{-3}, (Emitter doping)= 1.0×10^{19}cm^{-3}, and (Collector doping)= 1.3×10^{17}cm^{-3}.

TABLE II
COMPARISON OF THE NUMBER OF NEWTON ITERATIONS

Training structure	Maximum Newton iterations	
	Generated profiles	Conventional profiles
Simple BJT	4	38
More realistic BJT	4	39
Non-fixed BJT	5	36

even when the training set contains various structures with the length variation.

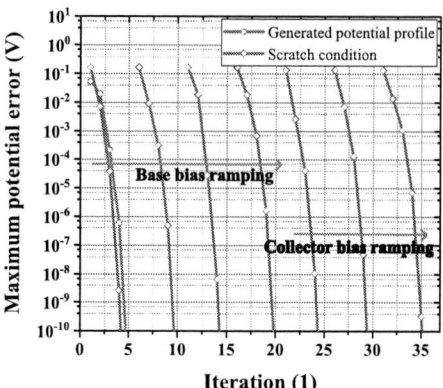

Fig. 14. Comparison of convergence behavior. The target bias condition is V_B = 0.5 V and V_C = 0.4 V. X_1 = 0.333μm, Y_2 = 0.190μm, (Base doping)= 7.7×10^{17}cm^{-3}, (Emitter doping)= 1.0×10^{19}cm^{-3}, and (Collector doping)= 1.2×10^{17}cm^{-3}.

Fig. 14 shows the convergence behavior for an arbitrarily chosen device. Table II compares the maximum number of Newton iterations for convergence between the generated potential profiles and the conventional initial profiles, shown in Figs. 6, 9, and 14.

IV. CONCLUSION

We show that the trained convolutional neural networks can generate the electrostatic potential profiles which are close to the solutions. These profiles can be used as good initial solutions to reduce the number of Newton iterations. It leads to significant computational gain in a semiconductor device simulation. Extensions to more general cases will be an interesting future research topic.

ACKNOWLEDGEMENT

This work was supported by the National Research Foundation of Korea (NRF) grant funded by the Korea government (NRF-2019R1A2C1086656). This work was also supported by Institute for Information & communications Technology Promotion(IITP) grant funded by the Korea government(MSIT) (No.2019-0-01351, Development of Ultra Low-Power Mobile Deep Learning Semiconductor With Compression/Decompression of Activation/Kernel Data).

REFERENCES

[1] W. Haensch, "Analog computing for deep learning: Algorithms, materials & architectures," in *IEDM*, 2018.
[2] J. Welser, J. Pitera, and C. Goldberg, "Future computing hardware for ai," in *IEDM*, 2018.
[3] B. Kim and M. Shin, "Machine-learning-based device optimization with tcad," in *KCS*, 2020.
[4] R. Orihara, R. Naraski, Y. Yoshinaga, Y. Morioka, and Y. Kokojima, "Approximation of time-consuming simulation based on generative adversarial network," in *ISCSA*, 2018.
[5] S.-C. Han and S.-M. Hong, "Deep neural network for generation of the initial electrostatic potential profile," in *SISPAD*, 2019.
[6] S. J. Russell and P. Norvig, *Artificial Intelligence-A Modern Approach*. Pearson Education London, 2010.
[7] A. Radford, L. Metz, and S. Chintala, "Unsupervised representation learning with deep convolutional generative adversarial networks," 2015, arXiv preprint, arXiv:1511.06434.

14-1 Agile Pathfinding Technology Prototyping: the Hunt for Directional Correctness

Daniel Chanemougame, Jeffrey Smith, Paul Gutwin, Brandon Byrns and Lars Liebmann

TEL Technology Center America, Albany, NY, USA

Abstract—New tools and methodologies are fused with conventional elements of the process-design-kit (PDK) and design enablement to introduce a rigorous yet fast and agile technology prototyping platform. This design technology co-optimization (DTCO) solution replaces the rigid, time and resources consuming PDK to enhance the core functions critical to evaluate power, performance area and cost (PPAC). Any technology definition with any device or process integration innovation can be evaluated at the standard cell level first, and then at the block level to explore and understand the requirements of different design applications. The flexibility and fast turn-around make it practical to imagine, test and compare many technology prototypes. From simple evolutions to innovative disruptions, the feasibility and value of the technology choices and hardware tools required can be identified early with great detail, significantly accelerating the development of future process tools. To illustrate the efficiency of the platform, complementary-FET (CFET) [1] technologies are compared to reference finFET technologies. As we approach the fundamental limits of dimensional scaling, with so many choices ahead of us including 3D constructs, we need efficient technology prototyping to navigate and steer in the right direction.

Keywords—Pathfinding DTCO, PDK, PPAC, LGAA, CFET, technology prototyping, compact modeling, parasitic extraction, timing characterization, synthesis place and route

I. INTRODUCTION

The bleeding edge of the semiconductor industry will likely transition to the lateral gate all around (LGAA) architecture. Beyond, the semiconductor scaling roadmap is uncertain. While the device architecture remains important, advanced technologies are susceptible to other critical components, which ultimately manifest themselves as parasitic resistance and capacitance (RC) loads. As technologies scale, conductors become smaller and closer to each other, resulting in the challenge of ever-increasing parasitic RC, wasting precious energy and performance during circuit operations. In the tremendous effort to limit waste and extend the roadmap, engineers are facing an explosion of critical PPAC technology options at all levels from device architecture and process integration to materials. As the challenges become harder than ever, more disruptive options must be considered as well. With so many combinations, the risk of implementing the wrong technology definition is significant with serious cost consequences.

Logic standard cells have long been the industry benchmark to model, predict and adjust the PPA value of a technology early in the definition. While ring oscillator (RO) circuit simulations are easy to run, major inputs specific to that technology are required which are typically part of a PDK. Building a PDK for advanced technologies, even a lighter evaluation version, is a multi-months affair, involving many experts from large design enablement teams (fig. 1). The concept of digital twins is common in the semiconductor industry. However imperfect, the PDK is a digital twin of a silicon technology: its ultimate goal is to provide the necessary inputs to predict the PPA so designers can fully leverage the technology. A collection of multiple software elements, the PDK is a remarkable and capable piece of modeling engineering, but extremely customized and rigid by definition. In the early stages of development, months before circuit simulations can reveal the PPA entitlement and potential roadblocks, major technology options have to be anchored to prevent development delays, making any significant corrections that much harder and leaving behind many options. These challenges have been quite severe and steadily growing since 20nm despite DTCO efforts.

Additionally, a classic PDK approach cannot practically address the speed, agility and variety of scope required in Pathfinding, as any process integration changes will necessitate multiple slow PDK updates. For this discussion, using CFET as a demonstration, we will present our work on an agile technology prototyping platform (fig. 2), built to increase the breadth of DTCO with reasonable resources, and accelerate the quantitative PPAC exploration of various technology definitions, ranging from evolutionary to disruptive, leaving no idea unexplored. To limit the scope and size of this paper, we will focus on the critical design enablement components leading up to synthesis place and route (SPnR) experiments for design applications exploration, and then discuss some of the results.

II. AGILE TECHNOLOGY PROTOTYPING

A modern production PDK has many components, but the transistor compact models and the parasitic RC extraction deck (fig. 1) represent the core functions essential to circuit simulations and therefore are the first steps in technology prototyping. The first two sections will review how we enabled these two critical capabilities without a PDK.

Ironically, as DTCO efforts in recent nodes managed to extend the scaling of standard cell height [2], cell routability started to suffer [3]. As a result, the scaling benefits achieved at the block level are reduced compared to what was initially observed at the standard cell level. This is a major problem for modern designs and technologies in their attempts to continue Moore's law. It is therefore necessary that the PPAC work extends to block level for a comprehensive analysis. Enabling block level SPnR experiments represents a significant step up in the design enablement efforts and requires a third critical component, characterization, which we will also describe in the third section.

By embracing a broad scope, ranging from process assumptions and critical design rules, to device models, parasitic extraction and characterization, and ultimately to block-level place and route, we can efficiently identify and quantify the complex PPAC tradeoffs associated with the process disruptions and innovations needed.

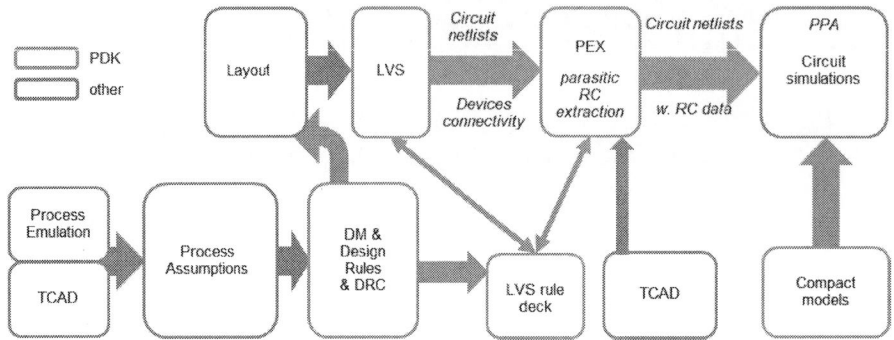

Fig. 1. High level description of the conventional PDK. PDK elements are represented in green. The goal is to enable circuit simulations to evaluate a technology's PPA. In addition to the time and resources involved, any significant change in the definition can easily break the flow, making this approach not suitable for the speed and agility required in Pathfinding.

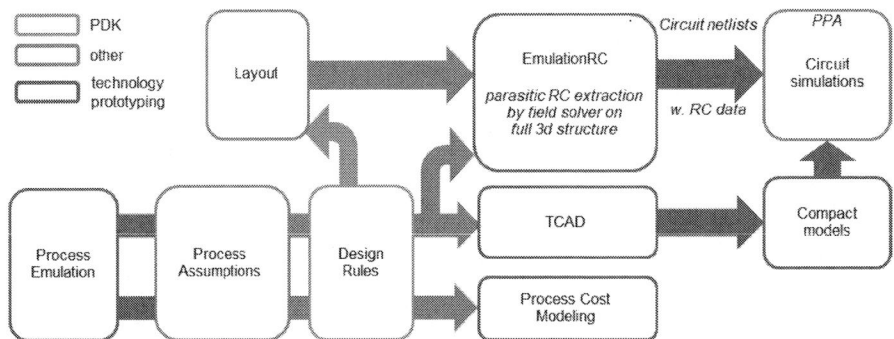

Fig. 2. High level description of the Agile Technology Prototyping platform developed and used in this study. The platform is designed to provide the core functions essential to circuit simulations without developing a PDK. This results in a consolidation of tools closer to the typical TCAD environment (red boxes). The tools involved are more rigorous and less dependent on expert user settings, therefore more flexible and robust.

A. Device compact models

Compact models such as BSIM-CMG [4] have come a long way to capture the behavior of advanced transistors by modeling more microscopic device phenomena [5]. An evaluation compact model still represents a significant effort, involving quality silicon and experienced modeling engineers. In Pathfinding, even if there is preliminary silicon device data, it is likely not good enough. Already used in early stages of development, the quality and accuracy of TCAD device simulations provide a great alternative to silicon measurements and a good starting point.

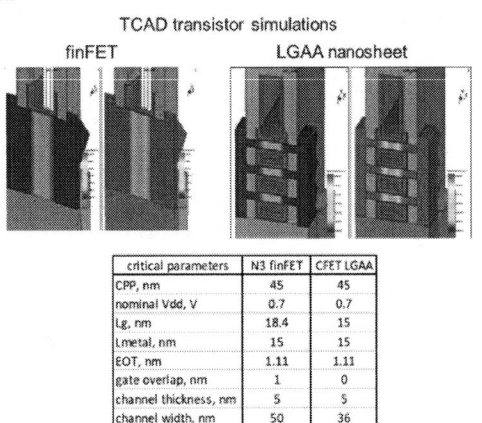

Fig. 3. Critical device geometries and parameters simulated for finFET and CFET LGAA.

Fig. 3 shows examples of the devices that were simulated for our N3 reference finFET and the CFET LGAA devices, as well as some critical dimensions and parameters. For the CFET, the NMOS and PMOS are simulated as individual LGAA devices, each with 3 stacked nanosheets, allowing more manageable computation time.

device		GPa	Idsat uA	Idsat µA/um	N/P Idsat	Idsat LGAA vs finFET	
finFET	P	-1.5	54	1070	1.3	/device	/um
	N	0	71	1418			
LGAA	P	0	95	881	2.0	1.8	-18%
	P	-0.75	141	1302	1.3	2.6	22%
	N	0	188	1739		2.6	23%

Fig. 4. FinFET Idsat vs CFET from fig. 3 devices, with different stress for CFET PMOS. The unstressed LGAA PMOS is 2x weaker than the NMOS. With PMOS stress, the N/P balance can be improved. The last 2 columns show the Idsat ratio between LGAA and finFET per device and then normalized by the total width of the device.

Additionally, a hybrid simulation approach was adopted to capture properly the electrostatics and transport effects in these advanced devices: the sub-threshold part of the IdVg curves is based on drift-diffusion (DD) to capture the leakage, while the above-threshold part is based on Monte-Carlo to capture transport effects and on-currents. IdVd curves are based on Monte-Carlo while CV curves are based on DD with calibrated density-gradient. An interesting point of note on Fig. 4 is that very much like planar devices on (100)/[110] with beta ratio of 2, the LGAA NMOS is twice as strong as the PMOS. If the PMOS can be strained, the N/P balance is significantly improved. Finally, Idsat is compared both per device and normalized per total width, between the single-fin finFET and the 3-nanosheet LGAA.

Fig. 5 shows the device geometries that were simulated to enable our CFET LGAA technology exploration. Only one geometry was done for the finFET reference. Different stress options were also applied for a total of 20 devices. This TCAD reference data provides solid foundational grounds. If a good model fit can be achieved, we will have a rigorous description of these advanced devices, and this is the primary goal: achieve accurate and pertinent Pathfinding circuit simulations.

LGAA	NMOS						PMOS										
stress Gpa	0						0						-0.75					
Lm nm	12			15			12			15			12			15		
NSW nm	10	24	36	10	24	36	10	24	36	10	24	36	10	24	36	10	24	36

Fig. 5. CFET LGAA devices simulated to extract compact models. Lm is Lmetal, and NSW is the nanosheet width. 2 reference finFET devices (N/PMOS) were simulated with Lm=15nm.

Once the reference TCAD data is generated for all the devices, the task of optimizing and fitting the compact model remains. We leveraged new tools and methodologies to automate the optimization strategies to best fit the TCAD reference curves. An initial BSIM-CMG model card is provided to get the optimization going. This new TCAD to compact model methodology is very effective in handling all the reference data and quickly generated more than 20 different model cards. Fig. 6 shows some example of fit achieved across different LGAA geometries. This methodology is the first significant design enablement improvement, as generating the many high quality compact models necessary for comprehensive advanced Pathfinding becomes practical. The TCAD becomes by far the slowest portion of the cycle, but Monte-Carlo tools are improving and make efficient use of parallel processing.

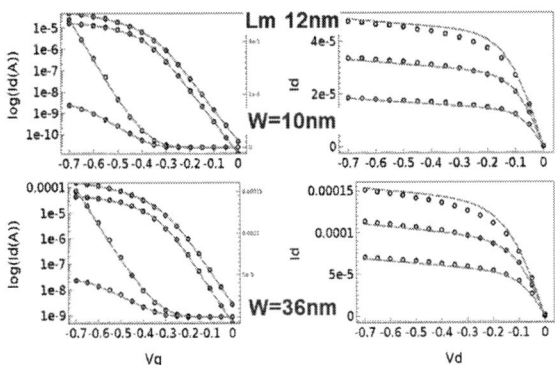

Fig. 6. Examples of TCAD to compact model extraction and fit on IdVg (both log and linear Id shown) and IdVd curves, shown for a PMOS LGAA (Lm=12nm, 2 nanosheet widths). Black circles: TCAD reference data, red curves: compact model. A similar quality of fit is achieved across all geometries.

B. Emulation-based parasitic extraction

Equipped with analytical compact models, we now need a circuit to perform SPICE simulations [6]. A standard cell is a circuit formed by transistors connected together to achieve its function through shared diffusions, gates and interconnects. To simulate this circuit, we therefore need a complete and accurate electrical description including all interconnects represented as parasitic resistance and capacitance elements, and all transistors. This description is known as a netlist and is a list of all components linked together by their associated electrical nodes. Each netlist is specific to a standard cell and its layout implementation.

All transistors must be identified and, despite the 2D representation of the layout, the 3D nature of interconnects must be captured. In a conventional PDK, the netlist generation is handled by the parasitic extraction deck (PEX), as shown in fig. 1. Building a PEX initially takes time and expert resources, but then running extraction is quite fast. The drawbacks are the lack of flexibility and accuracy which is continuously challenged by the complex 3D shapes of advanced technologies.

Highlighting the need for flexibility, the CFET architecture shown on fig. 7 represents a significant modeling challenge for the PEX. Process emulation is an excellent way to combine a cell layout with an integration flow to output a high fidelity, detailed 3D structure. Initially conceived for process integration development only, research very quickly focused on electrical RC extraction as well. This solution reached maturity after many years of development. The resulting netlist of emulation-based extraction is on the right of fig. 7.

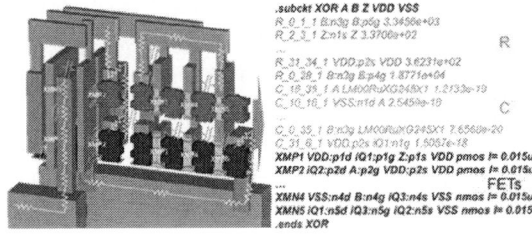

Fig. 7. On the left, CFET architecture with buried power rail. PMOS in blue at the bottom, NMOS in red on top. Parasitic RC elements originating from the interconnects are shown for illustration. On the right, an abbreviated example of emulation-based netlist extracted through rigorous field solver simulation.

The complex emulated 3D structure is directly meshed and exported to a rigorous field solver for extraction. The results depend very little on the level of tool setting expertise, and although running field solver simulations is much slower than running a traditional PEX, runtimes are largely manageable. The flexibility to accurately tackle any integration flow for any technology definition more than make up for longer runtime. With about 100 CPUs and 2TB of memory, a hundred cells can be extracted in about 24h, including cells with more than 30 transistors, making running multiple libraries or technology iterations possible within a week.

C. Logic library timing characterization

RO simulations provide meaningful PPA results with limited efforts on a few simple standard cells. Only a few layouts have to be created per technology definition, and correspondingly, only a few netlists need to be extracted. Block level experiments on the other hand require a much larger library of standard cells. Typical product libraries have hundreds, even thousands of standard cells, and each library is specific to a technology. Thankfully, for Pathfinding, the library size can be significantly reduced and still provide meaningful and pertinent results. For this work, we used about 80 cells. This is still a sizeable effort as the number of libraries or technology prototypes multiplies the total number of layouts that have to be created.

Logic designs at the block level are assembled by SPnR tools. The synthesis tool determines what standard cells are needed to achieve the block function. Through iterations and complex optimizations, all cells are physically placed and routed together to hit a combination of target density and speed. Evaluating the speed and power of a block-level circuit cannot use the circuit simulation methodology we described so far. Typical blocks have tens of thousands of standard cells, which means simulating hundreds of thousands of transistors together. SPICE simulations are not that efficient. Instead, the SPnR relies on a higher abstraction level: each standard cell is a black box with only input and output pins, with its complete power performance or timing profile attached to it. To generate this timing profile, each netlist of the library is run through many SPICE simulations where the output response is measured in terms of delay and power usage, while input signal conditions and output load capacitances are varied (fig. 8).

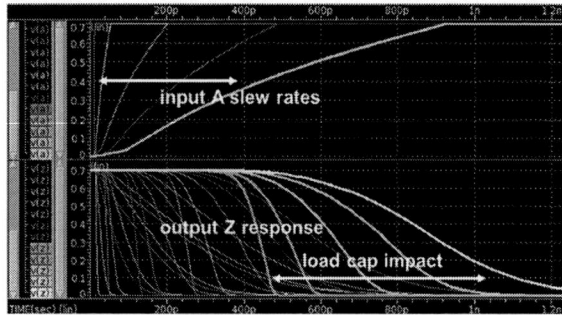

Fig. 8. SPICE simulations performed during timing characterization of a standard cell. As the input signal conditions and load capacitances are varied, the output response is measured.

The range of input signals is designed to represent the variety, shape and spread of signals that cells actually receive during operation in real designs. The SPnR tools will know precisely the cell's behavior and how and where to use it best in a design. The full library is characterized by running these simulations on all cells. The power and delay data collected on all cells is recorded and organized in a library timing file using the liberty format [7].

The challenge of characterization for Pathfinding is to strike the right balance between reasonable runtimes and sufficient accuracy and quality to achieve directional correctness in our results. In comparison, product characterization not only deals with much larger libraries, but must also cover the technology corners (multiple Vt flavors, process variability) and noise as well.

III. RESULTS AND DISCUSSION

Leveraging this platform to compare CFET to a reference finFET baseline, 18 technology prototypes were built out of different integration flows and design rules, each prototype containing a 80-cell library. 8 prototypes for finFET to provide a solid and relevant baseline, and 10 for CFET, all mixing driving strength, cell height, integration flow and power delivery options either conventionally through a M0 power rail (PR) or through a buried power rail (BPR).

A. Ring Oscillator simulations

The first step of PPA benchmarking of different technology prototypes is a 15-stage RO circuit with a fan-out of 3 to measure dynamic power usage and stage delay. For each prototype, 3 standard cells (INV, NAND2, NOR2) are run and their results averaged, resulting in more than 50 base RO simulations, multiplied by different compact model options.

We discuss here a small subset shown on fig. 9. On the left, the power-performance chart shows the impact of the LGAA PMOS device strength on circuit performance. Illustrated by 2 CFET libraries with different nanosheet widths, 2 PMOS stress conditions are compared: unstrained where the LGAA PMOS is 2x weaker than the NMOS, and strained where both are matched. When both devices are matched, we observe a 20% speed gain at iso-power, or 35% power reduction at iso-speed. In other words, if the CFET PMOS channel cannot be strained, significant power and performance are left on the table.

On the right of fig. 9, 5 CFET libraries with strained PMOS are compared to 3 reference finFET libraries (PMOS also strained, fig. 4), all mixing cell heights, drive strength and power rail options (M0 vs BPR). The first observation is that BPR offers only slightly better power-performance than M0 PR (black vs green circles). While BPR has a much lower resistance than M0 PR in these scaled libraries, the power vias associated with BPR are more resistive due to higher aspect ratios, limiting BPR advantage at least at the cell level. Next, we compare CFET to finFET as summarized in the table of fig. 9.

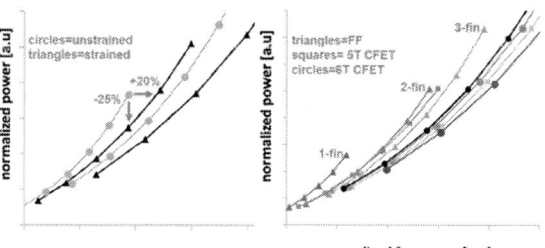

isoP/isoF		5T CFET low	6T CFET mid	5T CFET mid	6T CFET high
	freq	28%	47%	51%	56%
6.5T FF 1-fin	power	-35%	-47%	-50%	-53%
	area	-23%	-8%	-23%	-8%
	freq	5%	22%	26%	31%
6.5T FF 2-fin	power	-8%	-30%	-34%	-41%
	area	-23%	-8%	-23%	-8%
	freq	-5%	13%	17%	21%
8.5T FF 3-fin	power	10%	-20%	-25%	-29%
	area	-41%	-29%	-41%	-29%

Fig. 9. On the left: impact of the LGAA PMOS device strength illustrated on 2 CFET libraries with different nanosheet widths, with and without PMOS stress. On the right: 5 CFET libraries are compared to 3 reference finFET libraries. The table summarizes the PPA differences between CFET and finFET. Low, mid and high refer to the CFET drive strength, as 1-fin, 2-fin and 3-fin for finFET.

As expected, the CFET libraries with a stack of 3 nanosheets outperform the finFET in almost all cases while always being smaller. When comparing the 5T CFET libraries to the single-fin 6.5T finFET, the CFET ranges from 28 to 51% faster at iso-power, or 35 to 50% power savings at iso-speed, for the low and mid-drive strength respectively, while being 23% smaller (5 vs 6.5T). Compared to the 2-fin, this now ranges from 5 to 26% faster, or 8 to 34% power savings, for the low and mid-drive respectively, and 23% smaller. The 6T CFET with high-drive offers 31% more performance or 41% power savings, with an 8% area benefit (6 vs 6.5T). Finally, if compared to the 3-fin 8.5T finFET, the 5T CFET low-drive is slower by only 5%, or uses 10% more power, while the 5T mid-drive is 17% faster, or uses 25% less power, with a 41% area benefit (5 vs 8.5T). The 6T high-drive offers 21% more performance or 29% less power usage, with 29% area benefit (6 vs 8.5T).

CFET can provide area gains at the cell level and the potential to significantly outperform finFET. Nevertheless, if CFET is a roadmap solution, the industry will adopt it in 2 or 3 nodes after the transition from finFET to LGAA. The PPA gains must then reflect that. As shown here, the device still matters, finding ways to effectively balance N and PMOS drive strength will be critical. On the other hand, the capability to finely and broadly tune the drive strength within the same cell height is a very interesting and an important feature for designers. A library with high drive strength will be fast but tends to be big and burn a lot of power when speed is not required. Inversely, one with low drive strength will be power-efficient, but will fail to boost speed where needed. Designers typically push the technology through different libraries to provide the flexibility to cover a large PPA space, and achieve the best PPA tradeoffs in the various parts of their designs. CFET enables that design flexibility with finer granularity than finFET.

B. Place and route enabled

Very little difference is observed at the standard cell level between M0 and BPR, and that can be expected. Power delivery network (PDN) and other elements like the clock signal tree are evaluated properly only at block-level. The work presented here provides the necessary inputs to enable these experiments as illustrated in fig. 10. A CFET library with BPR and a finFET library with M0 are compared on a wire-dominated design made of around ten thousand cells. As the tool improves the density of the placement, violations of design rules occur, indicating the best area achieved. The results indicate a potential 29% density gain for CFET BPR compared to finFET M0. For scaled libraries, CFET allows for better integration of BPR than finFET, and BPR is critical for back-side PDN [8].

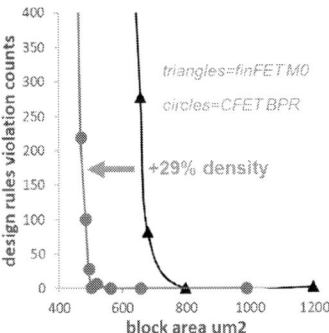

Fig. 10. Design rules violations as a function of the achieved block area after place and route. The block area achieved with CFET BPR is potentially 29% better than finFET M0.

C. Cost modeling

Wafer cost (including cycle time, depreciation, materials, facilities and labor) is just as important in Pathfinding as integration choices and PPA analyses. Our efforts have focused on automating the cost evaluation of any process integration flow to perform relative comparisons. Each process step of an emulation integration flow is linked to an industry-calibrated cost model, delivering accurate, fast and highly flexible relative cost comparisons. Any integration flows and variants can be compared on the fly, providing directional correctness for cost as well. A preliminary assessment of the front-end-of-line up through M1 indicates a 16% cost increase for CFET compared to the reference finFET. Accounting for the dilution of this added cost across a common back-end-of-line, and considering the higher transistor density achievable with CFET, the cost-per-transistor for CFET is anticipated to be competitive.

IV. SUMMARY

We developed a rigorous yet fast and agile technology prototyping platform dedicated to embrace the uncertainty of the scaling roadmap. Fueled by many new innovative and disruptive ideas, the platform efficiency makes it practical with limited resources to start tackling the explosion of technology choices by quantifying their PPAC impact at the standard cell level, but also by enabling block-level explorations for a more comprehensive holistic approach. SPnR is a complex field rich in expertise, and exploring this space extensively as well in Pathfinding will uncover many more technology innovations.

ACKNOWLEDGEMENTS

The authors acknowledge the valuable collaborations with Synopsys and IC Knowledge. This work would not have been possible without their support.

REFERENCES

[1] J. Ryckaert et al., "The Complementary FET (CFET) for CMOS scaling beyond N3" 2018 IEEE Symposium on VLSI Technology

[2] M. Garcia Bardon, Y. Sherazi, P. Schuddinck, D. Jang, D. Yakimets, P. Debacker et al, "Extreme scaling enabled by 5 tracks cells: Holistic design-device co-optimization for FinFETs and lateral nanowires", IEDM, 2016, pp. 28.2.1-28.2.4

[3] L. Liebmann, V. Gerousis, P. Gutwin, X. Zhuc, and J. Petykiewiczc, "Exploiting Regularity: Breakthroughs in Sub-7nm Place-and-Route" Proceeding SPIE, Design-Process-Technology Co-optimization for Manufacturability XI, vol. 10148, May 2017

[4] N. Paydavosi, S. Venugopalan, Y. S. Chauhan, J. Duarte, S. Jandhyala, A. Niknejad, et al., "Bsim-spice models enable finfet and utb ic designs", IEEE Access, vol. 1, pp. 201, 2013

[5] M. Miura-Mattausch, "Compact Modeling Perspective – Bridge to Industrial Applications", International Conference on Simulation of Semiconductor Processes and Devices (SISPAD), 2019, pp. 319-322

[6] Nagel, L. W, and Pederson, D. O., "SPICE (Simulation Program with Integrated Circuit Emphasis)", Memorandum No. ERL-M382, University of California, Berkeley, Apr. 1973

[7] Open Source Liberty. http://www.opensourceliberty.org

[8] D. Prasad et al., "Buried Power Rails and Back-side Power Grids: Arm® CPU Power Delivery Network Design Beyond 5nm," 2019 IEEE International Electron Devices Meeting (IEDM), San Francisco, CA, USA, 2019, pp. 19.1.1-19.1.4

14-2

Self-Aligned Single Diffusion Break Technology Optimization Through Material Engineering for Advanced CMOS Nodes

Ashish Pal, El Mehdi Bazizi, Liu Jiang, Mehdi Saremi, Blessy Alexander and Buvna Ayyagari-Sangamalli

Applied Materials, Santa Clara, USA; Email: ashish_pal@amat.com

Abstract— **Though single diffusion break (SDB) acts as an efficient area-scaling enabler for current CMOS technology nodes, it degrades devices' variability performance, which can be mitigated by enabling self-aligned SDB (SA-SDB) technology. Unfortunately, SA-SDB causes PMOS performance degradation due to channel stress relaxation. To solve this issue, we propose material engineering of SA-SDB technology to improve PMOS performance. Using 3D-TCAD simulations, we show that by using stressed oxide for the SA-SDB cavity fill, both PMOS and NMOS device performance can be improved. Furthermore, using ring-oscillator as a representative circuit for CMOS technology evaluation, we showed that the circuit performance can be improved by 13-21% for 2-3 GPa stress in the oxide, thus enabling simultaneous area-scaling and circuit and variability performance improvement with SA-SDB technology for advanced CMOS nodes.**

Keywords— Self-aligned Single Diffusion Break, FinFET, Stress relaxation, Ring-oscillator

I. INTRODUCTION

Single diffusion break (SDB) as an area scaling enabler is replacing double diffusion break (DDB) in recent CMOS technology nodes [1-3]. Along with area-scaling advantage, SDB comes with the problem of mask placement error leading to epi-volume reduction and drive current variability. To eliminate the mask placement error issue, self-aligned single diffusion break (SA-SDB) technology has recently been proposed [4]. However, SA-SDB impacts the PMOS device performance heavily by relaxing the compressive channel stress induced by the source/drain SiGe source/drain epitaxy. In this paper, we propose a material engineering approach, consisting of stressed oxide deposition in the SA-SDB cavity to recover and further boost the PMOS channel stress and hence the device and circuit performance.

Figure 2. Typical inverter layout for 3nm node showing important layers – fin, gate, dummy gate and SA-SDB. 2 fin structures are assumed for both NMOS and PMOS.

II. DESCRIPTION OF APPROACH

3D-TCAD FinFET process (Fig. 1a) and device simulations are performed to analyze the impact of SA-SDB on device performance for 3nm node (Fig. 1b). In SA-SDB module, first a lithography step is performed using the SA-SDB patterns (Fig. 2). Next the dummy gate and fin are etched creating the SA-SDB cavity, which is then filled with oxide, followed by CMP for planarization. The 3 nm (Fig 3) baseline device does not go through the SA-SDB module. The number of devices (NOD) between the SA-SDB patterns can vary depending on the complexity of the standard cell. Fig. 4 shows a cross-section of device with SA-SDB for NOD=3.

For PMOS with NOD=1, the device has SA-SDB pattern on its both source and drain sides. As a result, after the dummy-gate & fin etch in SA-SDB module, both source & drain SiGe epitaxy can expand, which leads to complete relaxation of the compressive stress in the channel (Fig. 5b).

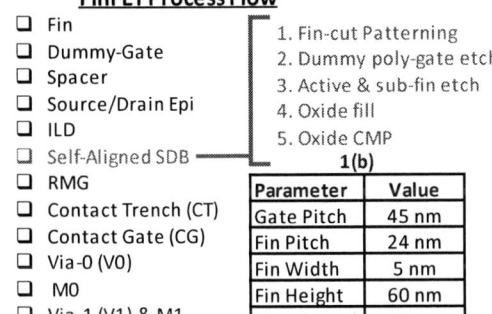

1(a)

FinFET Process Flow

- ❑ Fin
- ❑ Dummy-Gate
- ❑ Spacer
- ❑ Source/Drain Epi
- ❑ ILD
- ❑ Self-Aligned SDB
- ❑ RMG
- ❑ Contact Trench (CT)
- ❑ Contact Gate (CG)
- ❑ Via-0 (V0)
- ❑ M0
- ❑ Via-1 (V1) & M1

1. Fin-cut Patterning
2. Dummy poly-gate etch
3. Active & sub-fin etch
4. Oxide fill
5. Oxide CMP

1(b)

Parameter	Value
Gate Pitch	45 nm
Fin Pitch	24 nm
Fin Width	5 nm
Fin Height	60 nm
Gate Length	14 nm

Figure 1. (a) FinFET process flow used in TCAD process simulation with self-aligned single diffusion break module (SA-SDB). (b) Typical 3nm node parameters used in the TCAD simulations.

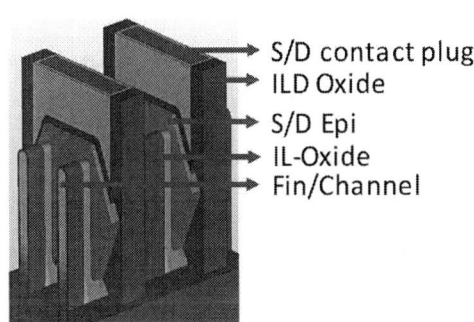

S/D contact plug
ILD Oxide
S/D Epi
IL-Oxide
Fin/Channel

Figure 3. 3D-TCAD 2-fin FinFET structure generated by process simulation using the process flow depicted in figure 1

307

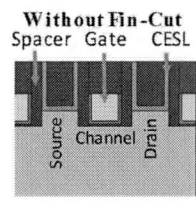

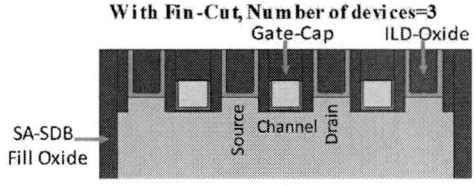

Figure 4. Cross-section of simulated devices with and without SA-SDB fin-cut showing different parts of the FinFET

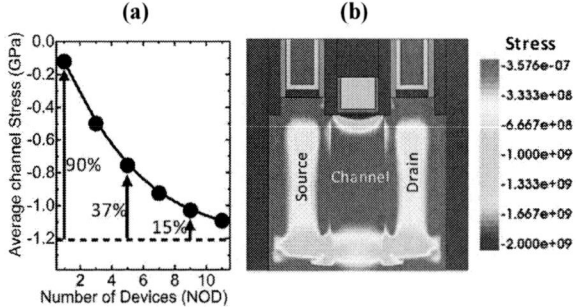

Figure 5. (a) Degradation in channel stress as a function of number of devices (NOD) between 2 fin-cuts (b) FinFET stress distribution for NOD=1, showing very small amount of stress inside channel.

Thus, for the PMOS device with NOD=1, the stress relaxation is close to 90% (Fig. 5a), in line with [4]. The channel stress can be recovered by depositing a stressed oxide during the oxide fill step in SA-SDB module.

III. RESULTS AND DISCUSSION

We will focus on PMOS device with NOD=1 to analyze for stress recovery as this device suffers from maximum stress loss. When a stressed oxide is used for SA-SDB cavity fill, the stress gets transferred from oxide to the channel, indicating the effectiveness of this method (Fig. 6a). A compressive stress in

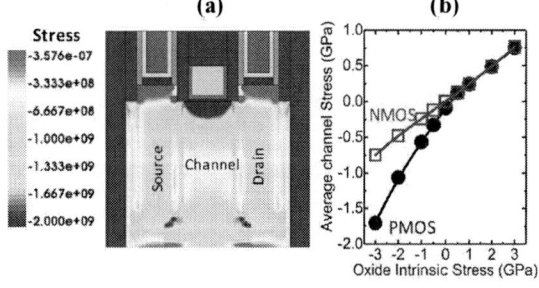

Figure 6. (a) FinFET stress distribution for NOD=1, when 2GPa compressive stress is introduced in the SA-SDB fill oxide, showing recovery of channel stress. (b) Average channel stress as a function of SA-SDB fill oxide intrinsic compressive and tensile stress for NOD=1

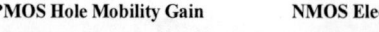

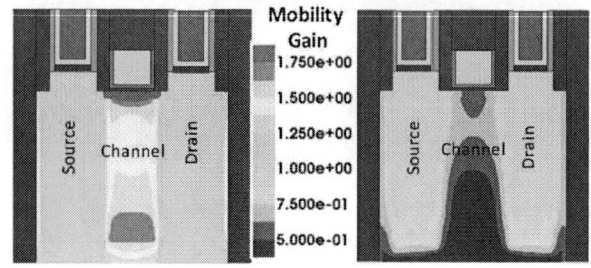

Figure 7. PMOS and NMOS mobility gain when 2GPa compressive stress is used in SA-SDB fill oxide with NOD=1.

the oxide helps to recover compressive channel stress for the PMOS device (Fig. 6b).

A tensile stressed oxide may be preferred for NMOS, leading to tensile channel stress. Having differently stressed oxide fill for NMOS and PMOS calls for additional lithography and process steps. Hence there is a trade-off between performance, cost of fabrication and process complexity.

For a 2 GPa compressively stressed oxide, PMOS channel mobility is boosted by a factor of 1.5 for most part of the channel (Fig. 7). The top of the fin has relatively higher compressive stress, leading to higher mobility boost in this region. Opposite to PMOS, the mobility in NMOS channel is reduced by a factor of 3 when an oxide with 2 GPa intrinsic compressive stress is used for SA-SDB cavity fill (Fig. 7). However, for 2 GPa intrinsic tensile stress in oxide, PMOS mobility decreases to ~0.6X compared to the case without stress in oxide (Fig. 8). The NMOS mobility increases to ~2.4X under same scenario.

The off-currents increases for both tensile and compressive stress and for both NMOS and PMOS due to bandgap and barrier height reduction (Fig. 9). Like mobility, NMOS and PMOS on-current increases with tensile and compressive oxide stress respectively.

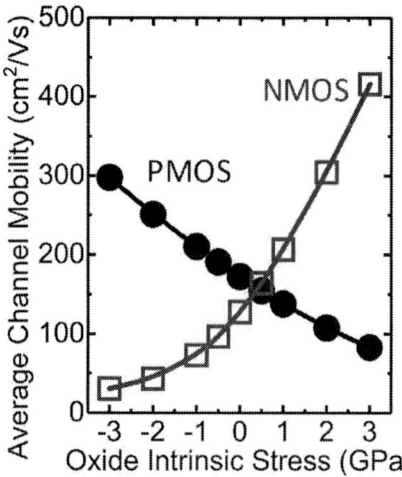

Figure 8. Average NMOS and PMOS channel mobility as a function of SA-SDB fill oxide intrinsic stress for NOD=1

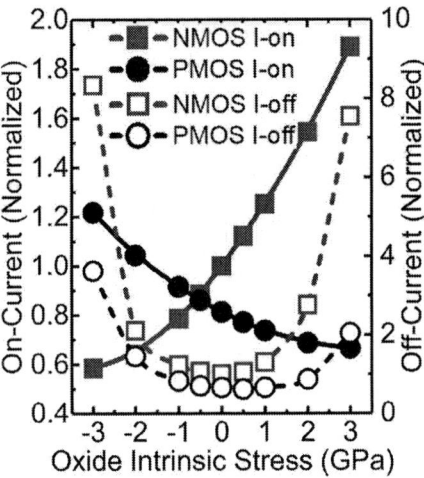

Figure 9. NMOS and PMOS on-current and off-current as a function of SA-SDB fill oxide intrinsic stress for NOD=1

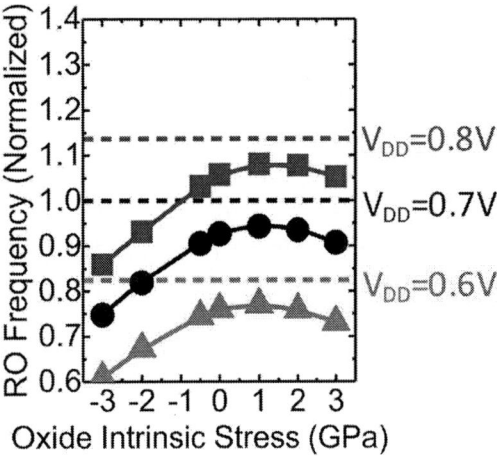

Figure 11. 31-stage ring-oscillator frequency when same type of stress (either tensile or compressive) is used for both NMOS and PMOS in SA-SDB fill oxide.

At constant off-current (achieved by work-function tuning), the PMOS device on-current degrades by 16% when the SA-SDB fin-cut is performed and its cavity fill oxide is deposited without any intrinsic stress (Fig. 10). The on-current for PMOS can be fully recovered with 2GPa intrinsic compressive stress in SA-SDB oxide fill. With 2 GPa tensile stress in oxide, the PMOS on-current degrades by 31% compared to the baseline case. For NMOS, the baseline device does not have any stress in channel. Hence performing SA-SDB fin-cut does not impact its drive-current. With 2GPa compressive stress in oxide fill, its drive current degrades by 40% (Fig. 10). However, with 2 GPa tensile stress in oxide, the NMOS drive current can be boosted by 42%.

To analyze the impact of stress recovery in SA-SDB technology on circuit performance, 31-stage ring-oscillator (RO) simulations are performed using the framework described in [5]. Two scenarios are studied. In the first one, same oxide fill is performed for both NMOS and PMOS. Thus both NMOS and PMOS have same type of stress. For compressive stress, NMOS drive-current degrades. So overall RO performance (frequency) degrades for compressive stress (Fig. 11). For tensile stress, PMOS drive-current degrades. Thus, the RO frequency degrades as PMOS becomes the bottle-neck. Thus for either tensile and compressive stress, the

RO frequency is lower than the baseline case (without SA-SDB fin-cut). In this scenario, the RO performance (frequency) is optimum with ~1 GPa tensile stress and is ~ 5% lower than the baseline case.

In the second scenario, a tensile and compressively stressed oxide fill with same magnitude of stress are assumed for NMOS and PMOS respectively. Since both NMOS and PMOS drive current improve in this scenario, it is possible to achieve same RO performance as of baseline with 1GPa oxide stress (Fig. 5b). Using higher amount of oxide stress - such as 2 or 3 GPa, RO performance can be enhanced upto 13% and 21% respectively, thus highlighting the effectiveness of stressed oxide deposition.

IV. CONCLUSION

Using 3D-TCAD simulations, we demonstrated that the stress loss in PMOS with self-aligned single diffusion break technology can be fully recovered using compressively stressed oxide fill. We also showed that using oxide fill with

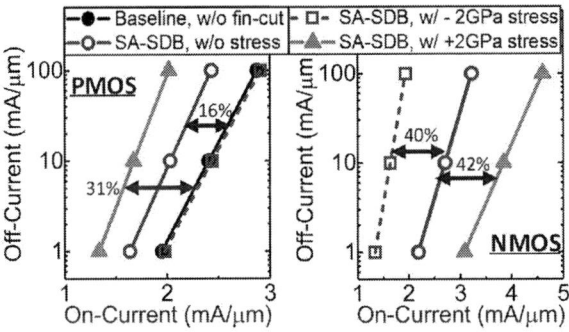

Figure 10. I-on vs I-off characteristics of NMOS and PMOS in presence of SA-SDB, with and without stress in fill oxide

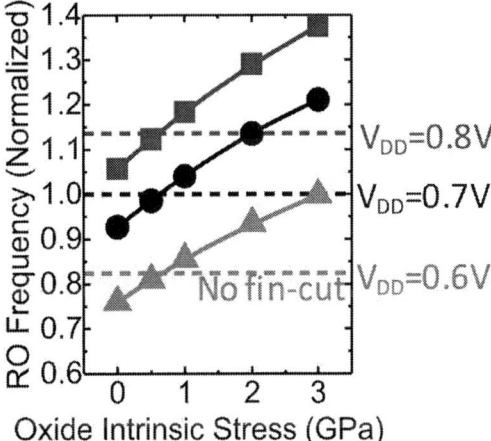

Figure 12. 31-stage ring-oscillator frequency when opposite types of stress are used for both NMOS and PMOS in SA-SDB fill oxide. Tensile stress is used for NMOS, whereas compressive stress is used for PMOS

compressive stress oxide for PMOS and tensile stress for NMOS can improve both device performance. In this scenario, the ring-oscillator performance can be improved by 13-21% for 2 - 3 GPa intrinsic oxide stress.

REFERENCES

[1] G. Yeap et al, "5nm CMOS Production Technology Platform featuring full-fledged EUV, and High Mobility Channel FinFETs with densest 0.021μm2 SRAM cells for Mobile SoC and High Performance Computing Applications," IEEE Internaional Electron Device Meeting (IEDM), 2019, pp. 36.7.1-36.7.4

[2] W. C. Jeong et al, "True 7nm Platform Technology featuring Smallest FinFET and Smallest SRAM cell by EUV, Special Constructs and 3rd Generation Single Diffusion Break," IEEE Symposium on VLSI Technology, 2018, pp. 59-60

[3] C. Auth et al, "A 10nm high performance and low-power CMOS technology featuring 3rd generation FinFET transistors, Self-Aligned Quad Patterning, contact over active gate and cobalt local interconnects," IEEE Internaional Electron Device Meeting (IEDM), 2017, pp. 29.1.1-29.1.4

[4] K. Miyaguchi et al, "Single and Double Diffusion Breaks in 14nm FinFET and Beyond," International Conference on Solid State Devices and Materials (SSDM), 2017, pp. 219-220

[5] A. Pal et al, "Impact of MOL/BEOL Air-Spacer on Parasitic Capacitance and Circuit Performance at 3 nm Node," International Conference on Simulation of Semiconductor Processes and Devices (SISPAD), 2019, pp. 1-4

L-UTSOI: A compact model for low-power analog and digital applications in FDSOI technology

Sébastien Martinie
CEA, LETI, MINATEC Campus,
Univ. Grenoble Alpes
Grenoble, France
sebastien.martinie@cea.fr

Olivier Rozeau
CEA, LETI, MINATEC Campus,
Univ. Grenoble Alpes
Grenoble, France
olivier.rozeau@cea.fr

Thierry Poiroux
CEA, LETI, MINATEC Campus,
Univ. Grenoble Alpes
Grenoble, France
thierry.poiroux@cea.fr

Patrick Scheer
STMicroelectronics
Crolles, France
patrick.scheer@st.com

Salim El Ghouli
STMicroelectronics
Crolles, France
salim.elghouli@st.com

Mihyun Kang
Samsung
South Korea
mihyun.kang@samsung.com

André Juge
STMicroelectronics
Crolles, France
andre.juge@st.com

Harrison Lee
Samsung
South Korea
lhw114@samsung.com

Abstract— **With the maturity of CMOS technologies and their use for low power various analog and digital applications, some additional effects must be modeled or enhanced to improve the accuracy of SPICE models. Indeed, with the decrease of supply voltages/currents and the use of the back bias in Fully-Depleted Silicon On Insulator (FDSOI) technologies, the devices operate close to the weak-moderate inversion, where gm/Id figure is impacted by effects like the depletion of source/drain electrodes and the parasitic currents such as Impact ionization current in moderate inver-sion and Gate Leakage current in weak inversion, can have a significant impact on the model accuracy. This paper describes the latest significant improvements of L-UTSOI model (formerly Leti-UTSOI) related to version 102.4. These model extensions are validated against Silicon experimental data.**

Keywords— *L-UTSOI, compact model, SPICE, FDSOI.*

I. INTRODUCTION

L-UTSOI [1] is a compact model dedicated to Fully-Depleted on Silicon-On-Insulator (FDSOI) Metal-Oxide-Semiconductor Field Effect Transistors (MOSFETs) technologies with low doped channel, developed at CEA-LETI and previously named Leti-UTSOI [2]. This SPICE model, used in semiconductor industry, is a standard model [1] available in most commercial simulators. L-UTSOI is a surface-potential based model that accounts for the creation of a strong inversion layer at the both interfaces and contains all relevant physical effects such as mobility degradation at high field, velocity saturation, short channel effects, gate currents, backplane depletion, gate induced source/drain leakage, STI induced stress effects, self-heating, etc [3]. L-UTSOI model gives an accurate description of currents, charges and their derivatives (i.e. transconductance, conductance and capacitances and higher order derivatives). The latter is especially important for analog and RF circuit designs.

In this paper, we present the new physical insights included in L-UTSOI model version 102.4 [3]. In Section II, we focus on the gm/Id dedicated to short channel transistor and, in section III, we describe the impact ionization evidence and modeling,

in section IV, we detail accurate model for overlap gate currents. And last section illustrates L-UTSOI model extraction using 28nm FDSOI technology.

II. FOCUS ON GM/ID: INTRODUCTION OF EFFECTIVE DOPING

L-UTSOI model is used for analog designs and gm/Id is one of the key figures of merit. To address this point we have improve initial short channel model [2] in moderate inversion regime. Figure 1.a and 1.b illustrates the discrepancy on short channel gm/Id figures of merit. The origin of this inaccuracy comes from the Source and Drain (S/D) depletion, which affects the transistor DC behavior in the subthreshold and moderate inversion regimes. In L-UTSOI, this effect of S/D depletion is accounted by modifying the boundaries conditions at the source/drain sides in the calculation of the 2D electrostatic effect (i.e. as a correction to the surface potential calculation that modulates the Short Channel Effect (SCE) and the Drain Induced Barrier Lowering (DIBL)).

The starting point of this improvement for the short channel devices is the calculation of the longitudinal profile (x direction) of the surface electrostatic potential accounting for short channel effects [4]:

$$\psi_{2D(x)} = \psi_{1D} + (\psi_{edge} - \psi_{1D})e^{-x/\lambda_{2D}} + (\psi_{edge} - \psi_{1D})e^{-(L-x)/\lambda_{2D}} \quad (1)$$

where ψ_{1D} is a smooth minimum function between the long channel surface potential at the center and its threshold value (to extend the validity in moderate inversion), ψ_{edge} is the electrostatic potential at the edge of S/D regions and λ_{2D} is the bi-dimensional characteristic length of the device.

This expression is valid in the subthreshold regime, where the longitudinal electric field at the edge of the S/D region is, respectively: $E_{x,edge} = \pm(\psi_{edge} - \psi_{1D})/\lambda_{2D}$. In the depleted region of the S/D electrodes, we simplify the bi-dimensional Poisson's equation by considering that a fraction α of the depletion charge acts on the longitudinal electric field. Writing x_{dep} the depleted region, we have:

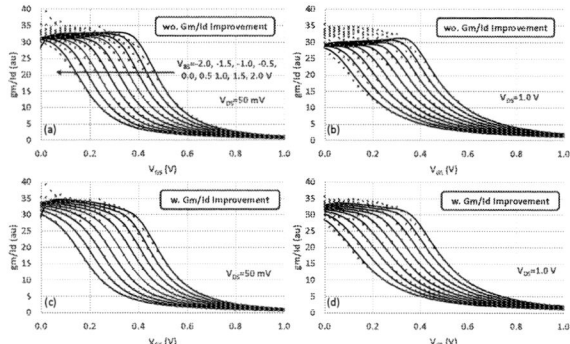

Fig. 1: Comparison between experiments from 28nm technology (symbol) and model (line): For short & wide device gm/Id versus VGS for different VBS without (a) & (b) and with (c) & (d) model improvement (extraction of dedicated NSDDC parameter). (note current has been normalized).

$$\varepsilon_{Si} E_{x,edge} = q\alpha N_{sd} x_{dep} \qquad (2.a)$$

$$\psi_{electrode} - \psi_{edge} = E_{x,edge} x_{dep}/2 \qquad (2.b)$$

where ε_{Si} is the dielectric permittivity of silicon, q is the electron charge, N_{sd} is the S/D doping level and $\psi_{electrode}$ is the electrostatic potential in the neutral part of the considered electrode (S or D). Together with the previously electric field expression, we obtain the electrostatic potential at electrode edge as:

$$\psi_{edge} = \psi_{1D} + \psi_{dep} \left[\sqrt{1 + 2(\psi_{electrode} - \psi_{1D})/\psi_{dep}} - 1 \right] \qquad (3)$$

where $\psi_{dep} = q\alpha N_{SD} \lambda_{2D}{}^2 / \varepsilon_{Si}$ is the depletion potential parameter and where the key parameter to modulate the depletion and, consequently gm/Id, is the effective doping αN_{SD} (named NSDDC as model parameter). The implementation of this effect is detailed in the user's manual [3].

Figure 1.c and 1.d shows the modeling results obtained on short and wide channel transistor gm/Id for a wide range of back biases from -2V to +2V in linear and saturation regimes. An accurate description of the device 2D electrostatics is achieved over this wide back bias range, including the moderate inversion regime thanks to implementation of S/D depletion.

III. IMPACT IONIZATION EVIDENCE AND MODELING: LOW-LEVEL AVALANCHE MULTIPLICATION CURRENT

For a transistor operating in saturation, the longitudinal electric field at the drain side may reach very high values. In this case, electrons travelling through the channel from S to D are accelerated and energy gain can create extra electron-hole pairs by exciting electrons from the valence band into the conduction band. This phenomenon is commonly referred to Impact Ionization (II). In this way, an avalanche of free carriers may arise and the initial flux of carriers is multiplied. In FDSOI transistor, this effect is more significant for thicker oxide transistors, featuring higher supply voltages, and only weak avalanche occurs. This component increases the source/drain

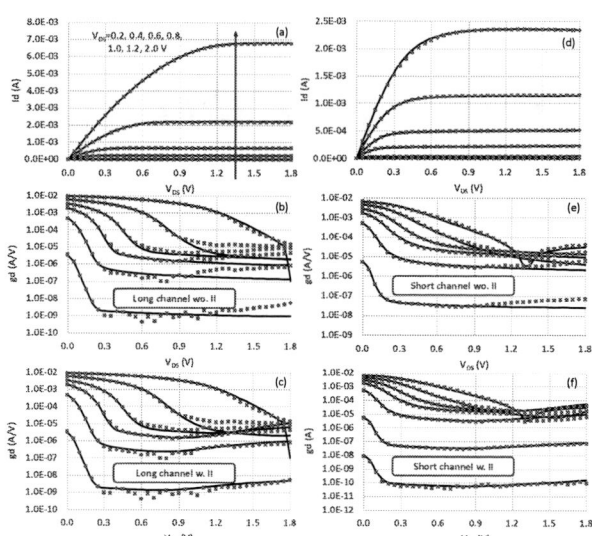

Fig. 3: Comparison between experiments from 28nm technology (symbol) and model (line): For long & wide device (respectively short & wide) (a) drain current versus V_{DS} (respectively (d)), (b) first derivative of drain current (gd) versus V_{DS} without Impact Ionization (respectively (e)) and (c) gd versus V_{DS} with Impact Ionization (respectively (f)) for different V_{DS}. (note current has been normalized).

current (in opposition to bulk MOSFET where the impact ionization results in a bulk/drain current).

Formally, the inclusion of this model in term of physics and code implantation is close to that of PSP model [5]. For low-level avalanche multiplication, the avalanche current is [6]:

$$I_{aval} = I_{DS}. \int_0^L \alpha_n. dx \sim A1. \Delta V_{sat}. I_{DS}. e^{-\frac{a_2}{\Delta V_{sat}}} \qquad (4)$$

where A1 is the II pre-factor, $a_2 = A2.(T_{KC}/T_{KR})^{STA2}$, A2 is the II exponent, STA2 is the temperature parameter, T_{KC} & T_{KR} are the channel temperature & reference temperature, $\Delta V_{sat} = V_{DS} - A3. V_{DSsat}$ where V_{DSsat} is the drain to source saturation voltage internally calculated using the surface potential calculation, A3 is the saturation dependence II parameter and I_{DS} is the drain to source current. These 3 model parameters (A1, A2 and A3) are extracted similarly as done with PSP model [5].

Figure 2 shows the comparison between experimental data and simulation including or not this impact ionization current on thick oxide transistor. The agreement is very good using the II current and as expected, the avalanche is small enough to be accurately modeled with the weak avalanche equation previously described. Note also that this extraction has been realized with the self-heating effect activated in the model to capture as accurately as possible the negative output conductance for high drain current levels.

IV. ACCURATE MODEL FOR OVERLAP GATE CURRENTS: FOWLER-NORDHEIM CURRENT

In this section, we detail the improvements brought to the gate current model in L-UTSOI. In fact, the gate current model is built similarly as in PSP model [7]. Initially this model comes

312

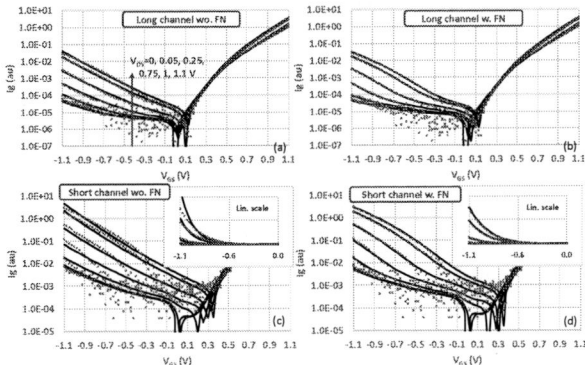

Fig. 3: Comparison between experiments from 28nm technology (symbol) and model (line): gate current versus V_{GS} for long & wide device (respectively short & wide (c-d)) without Fowler-Nordheim (a) (respectively (c)) and with Fowler-Nordheim (b) (respectively (d)) for different V_{DS}. (note current has been normalized).

from [8], where the integration over energy is skipped to obtain explicit function by assuming that all electrons have the same kinetic energy. So, the gate current density at point x in the channel is given by [7-8]:

$$J_{g_{(x)}} = J_0 . \int_0^\infty D_{(E,x)} . Fs_{(E,x)} . dy \sim J_0 . D_{(x)} . Fs_{(x)} \qquad (5)$$

where $D_{(E,x)}$ is the transmission coefficient calculated in the WKB approximation, $Fs_{(E,x)}$ is the supply function and x is the position along the channel. This transmission formulation requires further simplification for circuit design. The Taylor expansion is used to obtain [7]:

$$D_{(x)} = e^{-B0.f_{(zg)}} \qquad (6.a)$$

$$f_{(zg)} = \frac{1 - (1 - zg)^{1.5}}{zg} \sim -\frac{3}{2} + G2 . zg + G3 . zg \qquad (6.b)$$

where zg is the voltage drop through the gate oxide normalized to the built-in voltage, J_0 and B_0 are constant related to the tunneling current and G2 and G3 are parameters related to Taylor expansion. These two last parameters are the well-known GC2 and GC3 parameters in PSP [5] or GC2OVINV and GC3OVINV in L-UTSOI for overlap region [3]. Reference [7]: clearly explains that these parameters are tunable to compensate the huge assumptions made previously on the integral over energy. In fact, these parameters directly affect the electric field effect on the current shape.

As illustrated on Figure 3.a and 3.c, the extraction of the gate current parameters on long & wide and short & wide channel NMOS transistors evidences that the model is well aligned with experimental data. Nevertheless, it appears to be difficult to perfectly reproduce the gate current for high drain bias in overlap region. GIDL is not the root cause of such effect because this component is smaller than the gate current. The discrepancy observed on different gate current curves is evident for large negative V_{GS} and positive V_{DS} (in NMOS convention); in fact, we observe 2 different behaviors at low and high V_{GD} values.

If we deeply investigate our actual approach (equation 5 and 6), the transmission coefficient is close to the Direct Tunneling

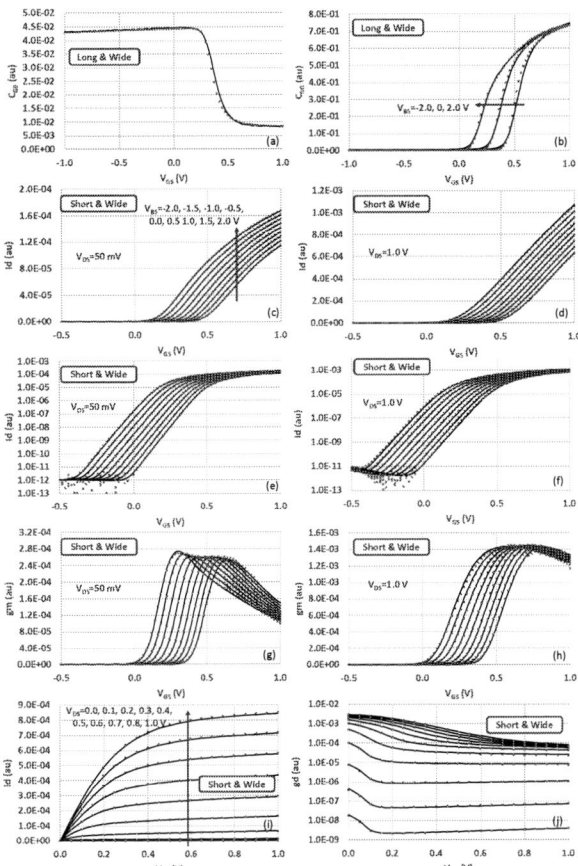

Fig. 1: Comparison between experiments from 28nm technology (symbol) and the model (line). C_{GB} (a) & C_{GG} (b) versus V_{GS} for long & wide device. For short and wide device: current versus V_{GS} in linear regime in lin (c) and log (e) scale (respectively (d) and (f) in saturation regime); first derivative of drain current (gm) versus V_{GS} in linear (g) and saturation (h) regime and Id versus V_{DS} (i) and first derivative of current versus V_{DS} (j). (note current and capacitance has been normalized).

formulation [9]. Historically [10], Fowler-Nordheim tunneling was sufficient to reproduce the gate current in technology generations with oxide thickness above 3-4nm. Indeed, the gate current is dominated by the tunneling of electrons through a triangular dielectric barrier in that case. However, for thinner oxide and with the reduction of the supply voltage the barrier is close to trapezoidal shape and Direct Tunneling is no longer negligible. Nevertheless, the both effects of negative gate voltage and positive drain voltage strongly increase the potential drop across the oxide. Following the definition introduced in [10], we propose to add an additional term related to high electric field close to Fowler-Nordheim equation. The modified gate current equation is:

$$J_g \sim J_0 . Fs_{(x)} . \left[e^{-B0.\left(f_{(zg)}\right)} + FNOV . e^{-B0.\frac{GCOVFN}{zg}} \right] \qquad (7)$$

where FNOV and GCOVFN are the new parameters that govern this Fowler-Nordheim term. Figure 3.b and 3.d, illustrates the comparison between the new model and experimental data, the behavior has been enhanced on the full range of bias conditions.

313

V. L-UTSOI MODEL EXTRACTION USING 28NM FDSOI TECHNOLOGY

Following the model extraction strategy [3], we have realized a parameter extraction using 28nm FDSOI technology. L-UTSOI is able to model all the features of FDSOI transistor characteristics for any bias configuration, including the case of the strong forward back bias, where two channels take place at the front and back interfaces of the thin silicon body.

Figure 4 illustrates the very good agreement on different electrical characteristics: capacitances (4.a & 4.b C_{GB} and C_{GG} versus V_{GS}) and currents with their associated derivatives (4.c to 4.i Id, gm, gd versus V_{GS} and V_{DS}) in different back bias ranges for short and wide devices. This accuracy is obtained thanks to the activation of backplane depletion, quantum confinement, mobility distinction between front & back channels and physical effects detailed in previous parts.

VI. CONCLUSION

In this paper, we have presented the recent improvements brought to L-UTSOI model to improve its accuracy and predictability in moderate and weak inversion regions. By comparison with experimental data, we have illustrated the enhanced accuracy obtained on gm/Id thanks to the introduction of source/drain depletion effect, on the output conductance thanks to the implementation of the impact ionization effect, and on gate tunneling current, thanks to the adding of a Fowler-Nordheim component. The L-UTSOI model is available in all major commercial SPICE simulators. Those three enhancements significantly contribute to demonstrate L-UTSOI readiness for Low Power circuit design applications

REFERENCES

[1] http://www.si2.org/2020/03/10/lutsol/

[2] T. Poiroux, O. Rozeau, P. Scheer, S. Martinie, M-A. Jaud, M. Minondo, A. Juge, J.C. Barbé, M. Vinet, "Leti-UTSOI2.1: A Compact Model for UTBB-FDSOI Technologies—Part II: DC and AC Model Description", Transaction of Electron Device, vol 62, no. 9, september 2015.

[3] L-UTSOI 102.4.0 user's manual.

[4] K.K. Young, "Short-Channel Effect in Fully Depleted SO1 MOSFET's", Transaction of Electron Device, vol. 36, no. 2, februrary 1989.

[5] PSP 103.6 Verilog-A and documentation. http://www.cea.fr/cea-tech/leti/pspsupport.

[6] R. van Langevelde, A.J. Scholten and D.B.M. Klaassen, Physical Background of MOS Model 11, Level 1101, 2003.

[7] Xin (Ben) Gu, Ten-Lon Chen, Gennady Gildenblat et al, "A Surface Potential-Based Compact Model of n-MOSFET Gate-Tunneling Current", IEEE TRANSACTIONS ON ELECTRON DEVICES, VOL. 51, NO. 1, JANUARY 2004.

[8] R. Tsu and L. Esaki, "Tunneling in a finite superlattice," Appl. Phys. Lett., vol. 22, pp. 562–564, 1973.

[9] M. Depas, B. Vermeire, P.W. Mertens, R.L. Van Meirhaeghe, M.M. Heyns, "Determination of tunneling parameters in ultra-thin oxide layer poly-Si/SiO2/Si Structures", Solid-State Electronics, Vol. 38, No. 8. pp, 1465-1471. 1995.

[10] Juan C. Ranuarez et al, "A review of gate tunneling current in MOS devices", Microelectronics Reliability 2006.

[11] TCAD Sentaurus Device Manual, Synopsys, Inc.: 0-2018.06.

14-4

Electromigration Model for Platinum Hotplates

Lado Filipovic

Institute for Microelectronics, Technische Universität Wien, 1040 Vienna, Austria
Email: filipovic@iue.tuwien.ac.at, ORCID: 0000-0003-1687-5058

Abstract—Microheaters are frequently applied in the design of semiconductor metal oxide gas sensors in order to heat the sensing layer and induce the surface chemical reactions which promote molecular adsorption. One of the most common materials used for the microheater layer is platinum. In this manuscript, a model for electro-migration is developed and implemented to study vacancy dynamics and the thereby-induced stress in platinum-based microheaters for gas sensor applications. The model is then applied to study the impact of the individual components which contribute to vacancy transport, including electro-migration, thermo-migration, and stress-migration. We find that these structures have very high thermal gradients, making the impact of thermo-migration component higher than the electro-migration component in the early stages of vacancy transport, unlike in copper-based interconnects. Therefore, improving the temperature uniformity of the microheater design should lead to a longer operating time before failure.

Index Terms—Electromigration, platinum, microheater, Semiconductor metal oxide gas sensor

I. INTRODUCTION

There is a plethora of different types of sensors which are based on micro-elecro-mechanical systems (MEMS) microheater technology and rely on a an elevated temperature in order to enable the sensor's primary functionality. These high temperatures are commonly provided by an integrated micro-heater or hotplate. Typical examples of such sensors are chemoresistive gas sensors based on semiconductor metal oxide (SMO) films [1] and nondispersive infrared (NDIR) optical gas sensors [2]. In SMO sensors, the MEMS heater is applied to elevate the sensing film's temperature to values between 200°C and 500°C in order to enable the surface chemical reaction to take place, while NDIR optical gas sensors require temperatures above 500°C to provide the infrared emission source [2]. A typical schematic of the layers composing the membrane of the hotplate is given in Figure 1(a). Here, we note the presence of a heat spreading plate, which is not electrically connected, but serves to help reduce the temperature gradient in the sensing layer [3]. A typical top view of the structure is shown in Figure 1(b), where we note the suspension beams, which hold the heated components above the silicon wafer, creating the air cavity. At the required high temperatures and current densities, electro-migration (EM) may lead to a failure in the hotplate's metallization lines [4], [5]. More importantly, the inherent non-uniformity in the temperature distribution may exacerbate the thermo-migration (TM) effect, as the migration of vacancies increases with increasing temperature gradients [6].

In order to satisfy the industry's push towards More-than-Moore integration [7] and reduce the cost of sensor fabrica-

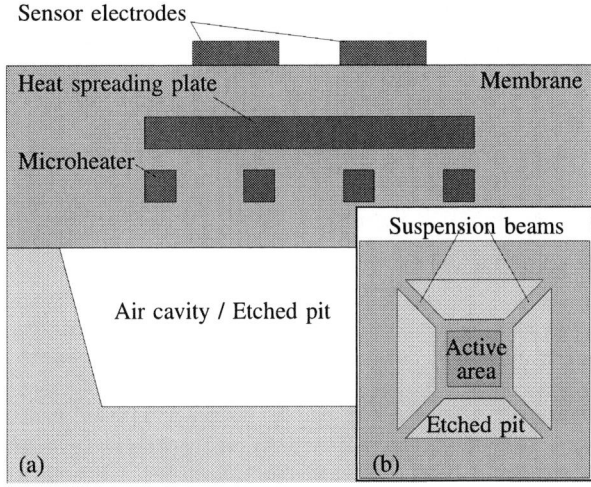

Fig. 1. (a) Two-dimensional cross-cut schematic of the layers composing the membrane of the hotplate and (b) the top view of the suspended membrane used in SMO gas sensors, showing the suspension beams, used to suspend the heated components in air above the silicon wafer.

tion, the entire sensor, including the microheater element and membrane which houses it, should be fabricated using complementary metal oxide semiconductor (CMOS) technology, which means that the heater material should be available in a CMOS fabrication facility. While initially polysilicon and aluminum were used, these materials were quickly shown to be prone to oxidation and early EM failure. Therefore, the most common materials used in CMOS-integrated hotplates today are platinum [1] and tungsten [8]. Tungsten, while being practically EM resistant, tends to form an oxide when exposed to air at high temperatures [9]. Platinum, on the other hand, is very stable, has a linear temperature response, and is resistant to oxidation at a broad range of operating temperatures [1]. Nevertheless, there is some evidence of degrading EM behavior in thin platinum lines [5], which is very difficult to quantify experimentally since there are always more effects influencing the different material stacks at very high operating temperatures, including diffusion of atoms from the adhesion layer, thermal expansion, and cracking and delamination [10]. Using published material parameters for atom and vacancy diffusivity in platinum, we have devised an EM model for microheater applications. Using this model, we are able to study the impact of EM, including thermo-migration and stress-migration, on microheaters and micro-

315

heater arrays for SMO sensor applications. Furthermore, we quantify the importance temperature uniformity on the EM reliability. Previously, we have already shown that improving the uniformity results in a faster return to the sensor baseline and improved sensitivity [11].

II. Electromigration Model

A physical description of EM is provided in [12], where the driving force for EM failure is shown to stem from the accumulation of vacancies, leading to increased tensile stress and the generation of a void which can cause immediate cracking or it can grow under further EM stress to induce open-circuit failure. At the same time, the reduction of vacancies, or accumulation of atoms, at the other end of the metal line forms hillocks, resulting in a compressive stress. To model the EM phenomena, three components must be solved at each simulation time step, as shown by Algorithm 1 in Figure 2: The electro-thermal problem, vacancy dynamics, and solid mechanics. The electro-thermal problem determines the electric field and induced temperature gradient in the structure, the vacancy dynamics problem calculates the movement of vacancies through the line, and the solid mechanics components calculates the stress and strain induced by the accumulation of vacancies. Usually, EM simulations are performed under accelerated conditions, achieved by applying a very high current density and high ambient temperature. Thereafter, using Black's equation, the behavior can be extrapolated back to operating conditions in order to find the time to failure under normal operating conditions [5], [13], [14].

To simulate EM in microheaters, only an increased current density is applied and the ambient temperature is kept at room temperature. The high current density and electric field in the platinum layer create a high temperature by Joule heating. Not prescribing a static temperature allows to form natural temperature gradients in the simulated microheater, which is commonly ignored when performing EM simulations in, e.g., copper interconnects [13], [15], but is essential in metal lines which potentially have a non-uniform temperature distribution. Since EM simulations are typically very time consuming, even for simple interconnect lines, in order to enable EM modeling for complex micro-hotplate geometries and membranes, we propose an accelerated approach. For this approach, first the steady state solution to the electro-thermal problem is found and the result is used when solving vacancy dynamics and solid mechanics in the time dependent simulation, depicted by Algorithm 2 in Figure 2. This is appropriate in the case of EM modeling since the final current density and temperature distribution in the platinum film is usually reached in under one ms [3], while EM phenomena is observed at times many orders of magnitude higher [5].

The electro-thermal problem solves the heat equation to find the temperature induced by the applied current; then the vacancy migration dynamics is solved to find the vacancy concentration inside the platinum layer; finally, the mechanical problem is solved to find the stress induced due to the accumulation of vacancies [13]. Calculating the vacancy dynamics

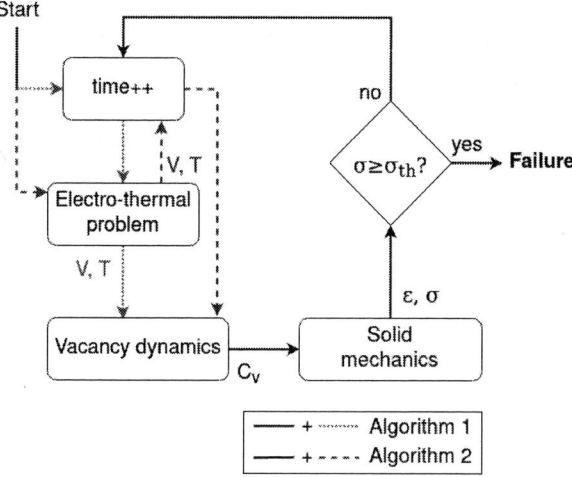

Fig. 2. Flowchart for the electro-migration model showing the paths for Algorithm 1 (dotted red): The electro-thermal problem is solved at every time step and Algorithm 2 (dashed blue): Steady-state solution for the electro-thermal problem is found prior to time discretization. The black arrows apply to both algorithms

is at the core of EM modeling. The change in the vacancy concentration C_v in time t is found by summing the effects of the vacancy flux $\vec{J}_v$ and vacancy generation/annihilation G. The equation which governs this mechanism is given by

$$\frac{\delta C_v}{\delta t} = -\nabla \cdot \vec{J}_v + G, \tag{1}$$

where the vacancy flux is determined by the diffusivity D_v and the sum of the individual EM, TM, and stress-migration (SM) components, represented by $eZ^*\nabla\varphi$, $Q^*\nabla T/T$, and $f\Omega\nabla\sigma$, respectively, by

$$\vec{J}_v = -D_v \left[\nabla C_v + \frac{C_v}{kT} \left(eZ^*\nabla\varphi - \frac{Q^*}{T}\nabla T + f\Omega\nabla\sigma \right) \right], \tag{2}$$

where k is the Boltzmann constant, T is temperature, e is the elementary charge, φ is the electrical potential, and σ is the induced stress. The other parameters used in the model and their values in the case of platinum are given in Table I. The vacancy diffusivity also depends on the induced stress and is calculated using

$$D_v = D_{v0} \exp\left(\frac{\Omega\sigma - E_a}{kT}\right). \tag{3}$$

The vacancy generation/annihilation term G from Eq. (1) is calculated using the Rosenberg-Ohring function [16]

$$G = \frac{1}{\tau}\left(C_v - C_{v0}\right), \tag{4}$$

which is applied at the grain boundaries and material interfaces. When an atom is exchanged by a vacancy, the effective volume change leads to the build-up of strain [17], and ultimately in a stress σ. The stress is determined by Hooke's law, assuming elastic material properties [18]. Once the induced stress reaches a critical threshold σ_{th}, the metal layer will either crack or a void will form, which then grows, increasing the line resistance, until an effective open circuit failure is induced.

TABLE I
PUBLISHED PARAMETERS WHICH WERE USED FOR THE EM SIMULATION
OF PLATINUM MICROHEATERS.

Sym.	Description	Value	Ref.
C_{v0}	Equilibrium vacancy concentration	$1.07 \times 10^{16} \mathrm{cm}^{-3}$	[19]
D_{v0}	Pre-exponential factor for vacancy diffusivity	$0.22 \mathrm{cm}^2 \mathrm{s}^{-1}$	[20]
E_a	Activation energy for diffusion	2.89eV	[20]
f	Vacancy relaxation factor	0.5	[21]
Ω	Atomic volume of Pt	$1.51 \times 10^{-29} \mathrm{m}^3$	–
Q^*	Heat of transport	0.68eV	[6]
τ	Vacancy relaxation time	200ms	[22]
Z^*	Effective valence	0.3	[23]

III. RESULTS AND DISCUSSION

To assess the reliability of the microheater, only the membrane is simulated, while the silicon wafer below is ignored, as it does not contribute to the EM phenomena. The membrane is suspended using four arms, which are attached to the silicon wafer and are used as fixed points for the mechanical simulation, as shown in Figure 1. The presented model is applied to study EM in a typical meander microheater on a $1.8\mu\mathrm{m} \times 1.8\mu\mathrm{m}$ membrane with platinum lines with a cross section of $100\mathrm{nm} \times 100\mathrm{nm}$ and a 200nm pitch, embedded in a 300nm thick SiO_2 membrane, shown in Figure 3. A current density of $11.3\mathrm{MA/cm}^2$, corresponding to a power dissipation of 0.77mW, was applied in order to reach a temperature of $1660°\mathrm{C}$ at the center of the membrane, which drops to about $1600°\mathrm{C}$ at the membrane edges. Heat conduction through the suspension beams and air, as well as air convection were applied in order to obtain a realistic temperature profile for the applied power.

After performing the time-dependent EM simulation on this structure for a period of 10^6 seconds, with the top and bottom of the membrane exposed to air, the resulting maximum EM-induced stress is calculated, as shown in Figure 4. We note that Algorithm 2 provides no loss of accuracy while reducing the simulation time by 40% compared to Algorithm 1. Furthermore, the obtained stress is near $1\mathrm{GN/m}^2$, which is significant enough to suggest that it needs to be considered when designing microheaters. Currently, most studies only look at the intrinsic and thermal stresses when evaluating the mechanical stability of the proposed designs, while EM is largely ignored [10].

The implementation of this model also allows to analyze the individual contributions to vacancy dynamics which can be from EM ($eZ^*\nabla\varphi$), TM ($Q^*\nabla T/T$), and SM ($f\Omega\nabla\sigma$) components from Eq. (2). These are shown in Figure 5, where it is clear that, as expected, SM becomes the main driving force in the later stages. It is noteworthy that the temperature non-uniformity, which results in an increased

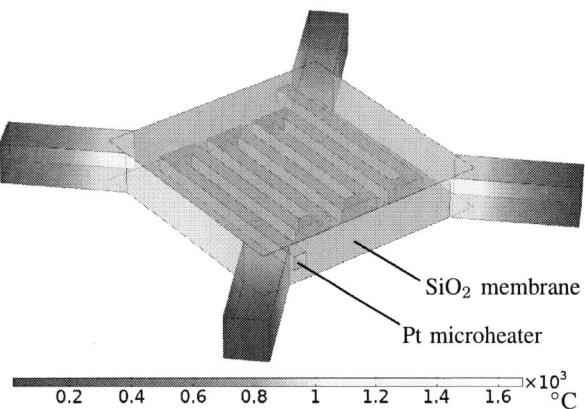

Fig. 3. Temperature ($°\mathrm{C}$) and geometry of a typical meander platinum hotplate in an SiO_2 membrane after applying a current density of $11.3\mathrm{MA/cm}^2$, reaching a maximum temperature of $1600°\mathrm{C}$. The pitch of the microheater lines is 200nm with a cross-section of $100\mathrm{nm} \times 100\mathrm{nm}$.

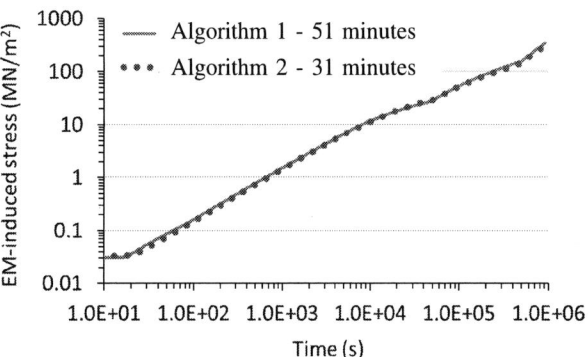

Fig. 4. EM-induced stress when applying $11.3\mathrm{MA/cm}^2$ to the microheater from Figure 3 using Algorithm 1 and Algorithm 2 from Figure 2.

temperature gradient and thereby in TM, has a significant impact in the early stages of vacancy transport. In fact, for this hotplate design, the ∇T component has a slightly greater impact than the $\nabla\varphi$ component, which is not the case for copper (Cu) interconnects. The differences in the impact of EM and TM components in the meander hotplate is visually shown in Figure 6. When we analyze the individual terms in previous Cu studies [13], it can be observed that the $\nabla\varphi$ component is three orders of magnitude larger than the ∇T component. From this, we can conclude that the vacancy-induced reliability for platinum-based hotplates is significantly impacted by temperature gradients in the metal lines and that the lifetime of these devices can be improved by improving the temperature uniformity.

IV. CONCLUSION

A model for vacancy dynamics under the influence of EM and TM in platinum micro-hotplates is presented, along with two algorithms for its solution. The first algorithm solves all required steps, which includes the electro-thermal, the vacancy

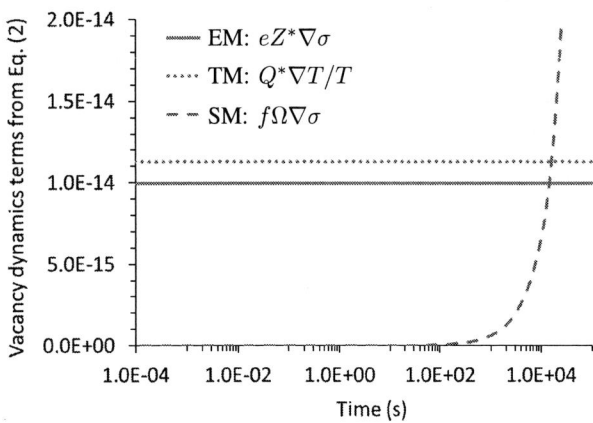

Fig. 5. Maximum contributions from different components from Eq. (2) to the vacancy dynamics in platinum microheaters. It is evident that the temperature non-uniformity (∇T) plays a significant role in EM.

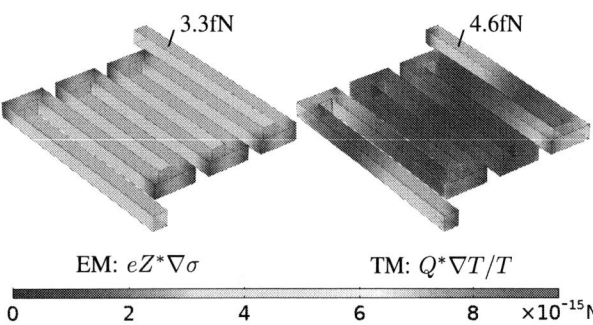

Fig. 6. Distribution of the EM and TM components of the vacancy dynamics model in a typical platinum meander microheater, shown in Figure 3.

dynamics, and the mechanical stress components, at every time step, while the second accelerated algorithm calculates the static electro-thermal component once and then applies this solution to the time-discretized vacancy dynamics calculation. This ultimately allowed for a 40% speedup in the simulation of a platinum-based meander microheater with no loss in accuracy. This is possible due to the fact that the potential and temperature distribution in the microheater reach a steady-state in less than one millisecond, while EM requires a much larger time-scale. The accelerated algorithm was then applied to study the impact of temperature non-uniformity on the lifetime of the aforementioned microheater. We found that the thermal component is larger than the electrical component in these devices, meaning that high temperature gradients in such microheaters could induce premature failure.

REFERENCES

[1] A. Lahlalia, L. Filipovic, and S. Selberherr, "Modeling and simulation of novel semiconducting metal oxide gas sensors for wearable devices," *IEEE Sensors Journal*, vol. 18, no. 5, pp. 1960–1970, 2018, doi: 10.1109/JSEN.2018.2790001.

[2] D. Popa and F. Udrea, "Towards integrated mid-infrared gas sensors," *Sensors*, vol. 19, no. 9, pp. 2076(1–15), 2019, doi: 10.3390/s19092076.

[3] L. Filipovic and S. Selberherr, "Thermo-electro-mechanical simulation of semiconductor metal oxide gas sensors," *Materials*, vol. 12, no. 15, pp. 2410–1–2410–37, 2019, doi: 10.3390/ma12152410.

[4] J. Spannhake, A. Helwig, G. Müller, G. Faglia, G. Sberveglieri, T. Doll, T. Wassner, and M. Eickhoff, "SnO2:Sb – A new material for high-temperature MEMS heater applications: Performance and limitations," *Sensors and Actuators B: Chemical*, vol. 124, no. 2, pp. 421–428, 2007, doi: 10.1016/j.snb.2007.01.004.

[5] R. Rusanov, H. Rank, T. Fuchs, R. Mueller-Fiedler, and O. Kraft, "Reliability characterization of a soot particle sensor in terms of stress- and electromigration in thin-film platinum," *Microsystem Technologies*, vol. 22, no. 3, pp. 481–493, 2015, doi: 10.1007/s00542-015-2576-6.

[6] S. Ho, T. Hehenkamp, and H. Huntington, "Thermal diffusion in platinum," *Journal of Physics and Chemistry of Solids*, vol. 26, no. 2, pp. 251–258, 1965, doi: 10.1016/0022-3697(65)90152-6.

[7] W. Arden, M. Brillouët, P. Cogez, M. Graef, B. Huizing, and R. Mahnkopf, "More-than-Moore white paper," 2010, online: http://itrs2.net/uploads/4/9/7/7/49775221/irc-itrs-mtm-v2_3.pdf.

[8] S. Z. Ali, A. D. Luca, R. Hopper, S. Boual, J. Gardner, and F. Udrea, "A low-power, low-cost infra-red emitter in CMOS technology," *IEEE Sensors Journal*, vol. 15, no. 12, pp. 6775–6782, 2015, doi: 10.1109/jsen.2015.2464693.

[9] A. Warren, A. Nylund, and I. Olefjord, "Oxidation of tungsten and tungsten carbide in dry and humid atmospheres," *International Journal of Refractory Metals and Hard Materials*, vol. 14, no. 5-6, pp. 345–353, 1996, doi: 10.1016/s0263-4368(96)00027-3.

[10] R. Coppeta, A. Lahlalia, D. Kozic, R. Hammer, J. Riedler, G. Toschkoff, A. Singulani, Z. Ali, M. Sagmeister, S. Carniello, S. Selberherr, and L. Filipovic, "Electro-thermal-mechanical modeling of gas sensor hotplates," in *Sensor Systems Simulations*. Springer International Publishing, 2019, pp. 17–72, doi: 10.1007/978-3-030-16577-2_2.

[11] A. Lahlalia, O. L. Neel, R. Shankar, S. Selberherr, and L. Filipovic, "Improved sensing capability of integrated semiconducting metal oxide gas sensor devices," *Sensors*, vol. 19, no. 2, pp. 374(1–14), 2019, doi: 10.3390/s19020374.

[12] H. Ceric, H. Zahedmanesh, and K. Croes, "Analysis of electromigration failure of nano-interconnects through a combination of modeling and experimental methods," *Microelectronics Reliability*, vol. 100-101, pp. 113 362(1–6), 2019, doi: 10.1016/j.microrel.2019.06.054.

[13] L. Filipovic, "A method for simulating the influence of grain boundaries and material interfaces on electromigration," *Microelectronics Reliability*, vol. 97, pp. 38–52, 2019, doi: 10.1016/j.microrel.2019.04.005.

[14] J. Black, "Electromigration failure modes in aluminum metallization for semiconductor devices," *Proceedings of the IEEE*, vol. 57, no. 9, pp. 1587–1594, 1969, doi: 10.1109/proc.1969.7340.

[15] R. de Orio, H. Ceric, and S. Selberherr, "Physically based models of electromigration: From Black's equation to modern TCAD models," *Microelectronics Reliability*, vol. 50, no. 6, pp. 775–789, 2010, doi: 10.1016/j.microrel.2010.01.007.

[16] R. Rosenberg and M. Ohring, "Void formation and growth during electromigration in thin films," *Journal of Applied Physics*, vol. 42, no. 13, pp. 5671–5679, 1971, doi: 10.1063/1.1659998.

[17] H. Ceric, R. Heinzl, C. Hollauer, T. Grasser, and S. Selberherr, "Microstructure and stress aspects of electromigration modeling," in *AIP Conference Proceedings*, vol. 817, 2006, pp. 262–269, doi: 10.1063/1.2173558.

[18] R. Kirchheim, "Stress and electromigration in Al-lines of integrated circuits," *Acta Metallurgica et Materialia*, vol. 40, no. 2, pp. 309–323, 1992, doi: 10.1016/0956-7151(92)90305-x.

[19] A. Seville, "Effects of vacancies on the physical properties of platinum," *Platinum Metals Review*, vol. 19, no. 3, pp. 96–99, 1975, online: https://www.technology.matthey.com/pdf/pmr-v19-i3-096-099.pdf.

[20] F. Cattaneo, E. Germagnoli, and F. Grasso, "Self-diffusion in platinum," *Philosophical Magazine*, vol. 7, no. 80, pp. 1373–1383, 1962, doi: 10.1080/14786436208213170.

[21] A. Čížek and V. Doláková, "A dilatometric study of platinum during repeated quenching," *Czechoslovak Journal of Physics*, vol. 22, no. 4, pp. 302–310, 1972, doi: 10.1007/bf01689616.

[22] A. H. Seville, "Studies of the specific heat of platinum by modulation methods," *Physica Status Solidi (a)*, vol. 21, no. 2, pp. 649–658, 1974, doi: 10.1002/pssa.2210210230.

[23] J. P. Dekker, A. Lodder, and J. van Ek, "Theory for the electromigration wind force in dilute alloys," *Physical Review B*, vol. 56, no. 19, pp. 12 167–12 177, 1997, doi: 10.1103/physrevb.56.12167.

Compact Modeling of Radiation Effects in Thin-Layer SOI-MOSFETs

Mitiko Miura-Mattausch
HiSIM Research Center
Hiroshima University
Higashi-Hiroshima, Japan
mmm@hiroshima-u.ac.jp

Hideyuki Kikuchihara
HiSIM Research Center
Hiroshima University
Higashi-Hiroshima, Japan
Kikuchihara532@hiroshima-u.ac.jp

Shunsuke Baba
1st Research Unit
JAXA
Tsukuba, Japan
baba.shunsuke@jaxa.jp

Dondee Navarro
HiSIM Research Center
Hiroshima University
Higashi-Hiroshima, Japan
dondee-navarro@hiroshima-u.ac.jp

Takahiro Iizuka
HiSIM Research Center
Hiroshima University
Higashi-Hiroshima, Japan
iizuka@hiroshima-u.ac.jp

Keita Sakamoto
1st Research Unit
JAXA
Tsukuba, Japan
sakamoto.keita@jaxa.jp

Hans Juergen Mattausch
HiSIM Research Center
Hiroshima University
Higashi-Hiroshima, Japan
hjm@hiroshima-u.ac.jp

Abstract— **Radiation can generate huge amounts of carriers in thin-layer SOI-MOSFETs, which change the device-internal potential distribution, known as an origin for of malfunction of circuits. 2D-numerical device-simulation analysis shows that the radiation-generated electrons initially flow-out from the SOI layer to both source and drain electrodes, which moderates the radiation-effect magnitude on device currents in this beginning stage. Subsequent enhancement of the current flow is due to accumulated holes caused by the potential barrier at source/channel junction. Compact modeling of the carrier movements during the initial radiation stage and of the hole-accumulation dynamics is based on the dynamically generated carrier densities. The developed compact model has been implemented into SPICE and model evaluation has been done by comparison to 2D-numerical device-simulation results. Under the off-state, it is shown that circuits can be easily switched to operation condition. Under the on-state, it is demonstrated that circuits can easily malfunction by operating differently from the designed circuit function. Though the radiation itself happens only for a short time, the radiation-induced effects continue for a rather time long, which causes serious effects in the circuits and is explained by the capacitor features of the SOI-MOSFET**

Keywords—radiation effect, ETSOI, SOTB, compact model, HiSIM, circuit simulation, malfunction

I. INTRODUCTION

SOI-MOSFET technology is often used to improve device reliability under radiation, thus suppressing malfunction of circuits in radiative environments [1, 2]. The key for predicting radiation-induced circuitry instabilities is an accurate compact model, which includes the microscopic radiation effects on the device, such as the potential redistribution within the device. Previously, phenomenological modeling has been done with existing models without explicit carrier-generation modeling due to the deficit of a fundamental modeling technique. Here, our focus is given on the carrier-generation modeling based on the radiation features and characterized by physical quantities. The dynamics of the generated carriers are modeled in conjunction with the potential redistribution by developing

two key model components. One is the movement of carriers caused by the radiation, which changes the MOSFET reaction to the operating condition. The second is the carrier-accumulation delay, which causes a long-term radiation effect. The developed radiation-effect model is implemented in the compact model HiSIM_SOI series [3] for the SOI MOSFET.

II. MODELING AND IMPLEMENTATION OF RADIATION EFFECT

A. Single MOSFET Operation

Transient characteristics of a single nMOS under radiation are investigated with 2D numerical device simulation (2D-Sim) [4]. A thin SOI-MOSFET device is studied, which represents one of the important devices utilized in space activities [5, 6]. The investigation condition with a radiation event is depicted in Fig. 1. The body-potential node V_{body} is introduced to monitor the substrate condition and no current flow is possible, because the gate voltage V_{gs} is set to zero so that the device is in the off-state. The radiation particle is assumed to hit through the device middle with a diameter of $0.1\mu m$ and a linear energy transfer of $25MeV$-cm^2/mg. The generation rate is assumed to have a Gaussian profile with a characteristic time of 10ps. Fig. 2 shows the potential distribution from 2D-Sim along the channel at different time points after the radiation event. A short time of 5ps after the radiation impact, the potential in the drain contact has already reduced to zero and then recovers gradually to the initial value of 1.2V after 30ps. This behavior is explained by the huge flow of generated electrons to the drain contact, where they disappear.

2D-Sim results of the immediate separation of generated electrons and holes and their flow-out from the channel are depicted in Fig. 3. Nodes of all observed current flows are additionally depicted in Fig. 1. Most of the time, the drain current I_{drain} (electron flow) and the source current I_{source} (hole flow) are identical with opposite signs. However, it can be seen that electrons flow out from both electrodes during the initial stage (see Fig. 3b), which is due to the extremely high electron/hole densities induced within the SOI layer. Since the gate is biased to zero, such huge carrier amounts cannot be sustained within the SOI layer. Therefore, electrons flow out immediately to the drain as well as to the source. However, the electron flow to the source is diminished shortly after the radiation event. Since the induced source-channel field due to the built-in potential prevents holes from flowing out, they are accumulated within the SOI layer for a relatively long time, as can be seen in Fig. 4. The resulting positively-charged SOI layer reduces the built-in potential and thus causes the hole flow-out through the source [7].

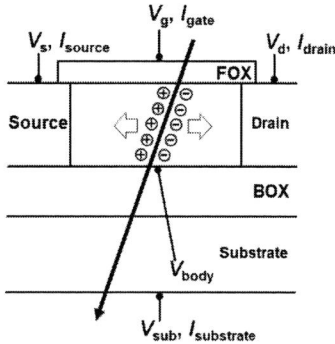

Figure 1: Schematic of studied radiation event in the SOI-MSOFET, where V_{body} is introduced to monitor the charging condition in the SOI layer.

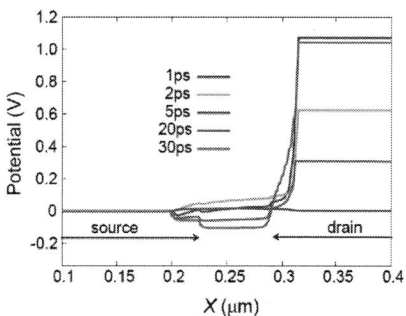

Figure 2: 2D device simulation results of the potential distribution along the device (X) at different time steps during radiation. The gate voltage V_{gs} is set to zero and the drain voltage V_{ds} to 1.2V. t=0 refers to the radiation starting and t=5ps is the end (see SEU in Fig. 4).

Consequently, the main modeling tasks are accurate descriptions of generated charge G and remaining accumulated hole charge Q_h in the SOI layer [8]. Since G is simultaneously induced during the radiation event, the model includes several parameters (see Table 1), which are extracted from measurable currents [8]. Q_h is obtained by integrating G over time, as shown in Fig. 5. Here, the generation occurs multiplicatively and takes time until completion. Therefore, Q_h is the integrated G including a delay. Additionally, holes also flow out from the SOI layer continuously through the diffusion mechanism after the generation is over. The relationship between the current and the generated carriers is modeled consistently with the basic device equations (see Table 2). Since the linear energy transfer of the radiation is extremely high and their crossing track is very thin, the energy attenuation can be neglected.

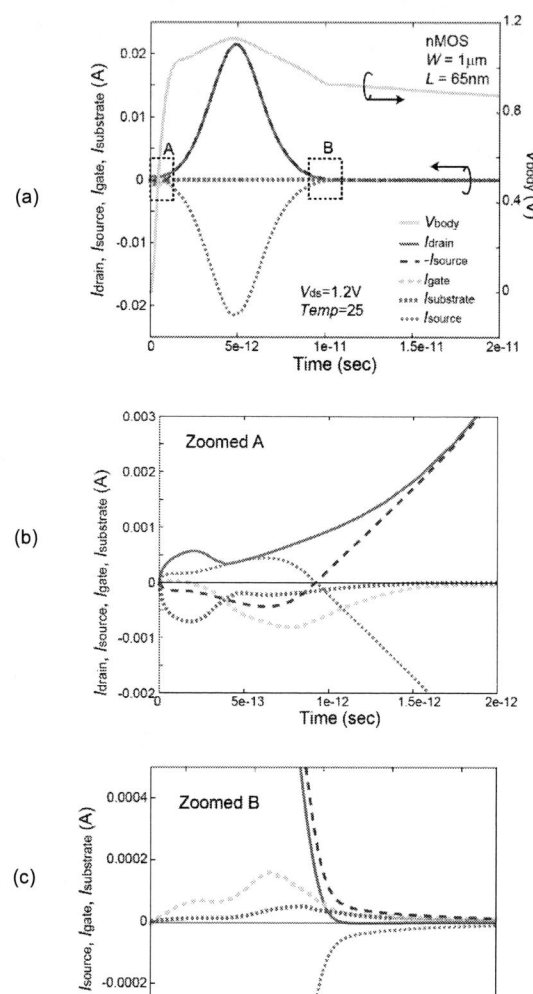

Figure 3: (a) 2D device simulation currents flowing to different electrodes during radiation event. The potential value at the bottom of the SOI layer in the channel middle V_{body} is depicted together. Zoomed features of (b) the A square and (c) the B square are also shown together. The radiation starts at t=0 and ends at t=5ps (see Fig. 4).

B. Implementation into SPICE

Circuit simulation is performed with a SPICE simulator, where the developed compact radiation model is implemented, using the industry standard compact model HiSIM_SOI [3] as its platform. The radiation model is implemented with consideration of two important characteristics. One is the generated electron flow-out from both electrodes until the MOSFET is turned on, namely, until Q_h becomes large enough so that more current can flow due to the increased substrate potential. The other is the Q_h inclusion in the Poisson equation to obtain self-consistent potential distributions. By solving the Poisson equation simultaneously, the carrier dynamics according to the device reaction is automatically modeled in a correct way.

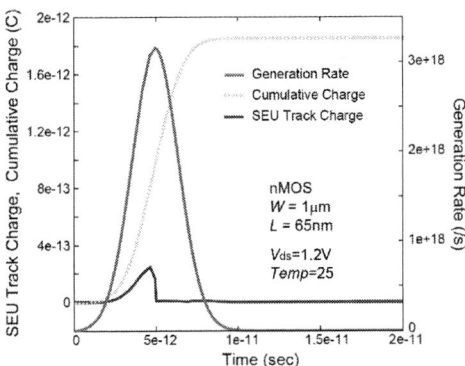

Figure 4: 2D-simulated generation rate and cumulated hole charge Q_h, which refers to the integration of the generated charge G over time. The applied radiation is also depicted in the graph. The SEU (Single Event Upset) particle streak ends at t=5ps.

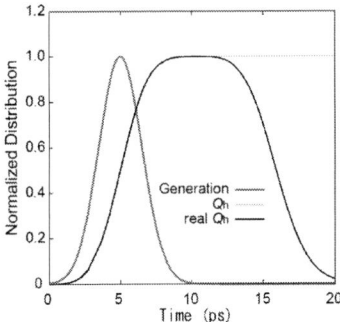

Figure 5: Normalized G and Q_h of model-calculation results, where the gradual hole flow-out is also considered in the quantity "real Q_h". Model parameters are depicted in Table 1.

Table 1. Model parameters and their notations.

Model Parameters	
Generation current delay:	RADIPDA0
Q_h charging delay:	RADQHA0
Q_h release delay:	RADQHDLY
Characteristic time for G:	RADPSIG

Table 2. Important device equations.

I_{RD}: generation current

$$I_{RD,aft}(t) = I_{RD,aft}(t - \Delta t) + \frac{\Delta t}{\tau_{qd} + \Delta t}\left(I_{RD,aft}(t) - I_{RD,aft}(t - \Delta t)\right)$$

$$I_{RD,aft}(t) = delay\left(RAD \_ I_{RD}, RADIPDA0\right)$$

$$\tau_{qd} = TAUFAC \cdot RADPSIG$$

Q_h: accumulated hole charge

$$Q_{h,aft}(t) = Q_{h,aft}(t - \Delta t) + \frac{\Delta t}{\tau_{qh} + \Delta t}\left(Q_{h,aft}(t) - Q_{h,aft}(t - \Delta t)\right)$$

$$Q_{h,aft}(t) = RADQHA0 \cdot \int_0^t Q_{RD}(t)\,dt$$

$$\tau_{qh} = TAUFACQ \cdot RADPSIG$$

$$Q_{h,out,aft}(t) = Q_{h,out,aft}(t - \Delta t) + \frac{\Delta t}{\tau_{qh} + \Delta t}\left(Q_{h,out,aft}(t) - Q_{h,out,aft}(t - \Delta t)\right)$$

$$Q_{h,out,aft}(t) = delay\left(Q_{h}, RADQHDLY\right)$$

$$\tau_{qh} = TAUFACQO \cdot RADPSIG$$

G: generation rate

$$G(y,t) = \alpha e^{-\alpha y}\phi(t)$$

$$\phi(t) = \frac{RADETA \cdot RADFLUX \cdot RADLAMBDA}{\left(hbar = 1.0546e^{-34}\right)\cdot\left(c = 3e^8\right)} P(t)$$

$$P(t) = e^{\frac{(t - RADPT1 \cdot 3 \cdot RADPSIG)^2}{2 \cdot RADPSIG^2}}$$

ϕ: photon flux
α: attenuation coefficient

III. RESULTS AND VERIFICATION

A. Inverter-Circuit Operation during Off-State

The radiation effects are investigated with an inverter circuit under V_{gs}=0 and V_{dd}=1.2V as depicted in Fig. 6a. Since V_{gs}=0, no current flow occurs and thus V_{out} is kept at 1.2V. A radiation is assumed to hit on the n-MOSFET at Time=0 so that electrons and holes are generated. The electrons start to flow out as observed in the electron current, and V_{out} is temporarily reduced to zero. The holes are accumulated within the substrate, which increases the substrate potential V_{body}. The calculated switching performance with the developed model is compared with 2D-Sim results in Fig. 6b. The calculated V_{body} is also depicted together. Fig. 6c compares model-simulation results for V_{out} with consideration of the initial electron flow and without. It can be seen that the initial current flow-out from both source- and drain contacts prevents the drastic change of the carrier distribution within the device and realizes smooth adjustment of the circuit reaction to the radiation event.

B. Inverter-Circuit Operation during On-State

Fig. 7 illustrates a malfunction event of the studied inverter circuit depicted in Fig. 6a. The simulation is performed under the assumption that the radiation event happens before the inverter's regular gate-controlled switching-off operation, which is scheduled to start at t=50ps and to completed at t=60ps. It can be seen, that an additional switching-off and switching-on glitch event occurs due to the radiation influence before the intentional gate-controlled switching-off. Because of the statistical nature of radiation events, precise prediction of a malfunction is not possible. However, it is shown here, that serious malfunction cases can be predicted with a certain statistical probability by using the developed compact model.

REFERENCES

[1] J. R. Schwank, V. Ferlet-Cavrois, M. R. Shaneyfelt, P. Paillet, and P. E. Dodd, "Radiation effects in SOI technologies," IEEE Trans. Nucl. Sci., vol. 50, no. 3, pp. 522-538, Jun. 2003.

[2] D. Kobayashi, K. Hirose, V. Ferlet-Cavrois, D. McMorrow, T. Makino, H. Ikeda, Y. Arai, and M. Ohno, "Device-physics-based analytical model for single-event transients in SOI CMOS Logic," IEEE Trans. Nuclear Science, vol. 56, pp. 3043-3049, Dec. 2009.

[3] N. Sadachika, D. Kitamaru, Y. Uetsuji, D. Navarro, M. M. Yusoff, T. Ezaki, H. J. Mattausch, M. Miura-Mattausch, "Completely Surface-Potential-Based Compact Model of the Fully Depleted SOI-MOSFET Including Short-Channel Effects," IEEE Trans. Electron Devices, vol. 53, pp. 2017-2024, Sept. 2006; M. Miura-Mattausch, U. Feldmann, Y. Fukunaga, M. Miyake, H. Kikuchihara, F. Ueno, H. J. Mattausch, T. Nakagawa, and N. Sugii, "Compact modeling of SOI MOSFETs With Ultrathin Silicon and BOX Layers," IEEE Trans. Electron. Devices, vol. 61, pp. 255-265, Feb. 2014; HiSIM_SOI User's Manual, Hiroshima University, 2018; HiSIM_SOTB User's Manual, Hiroshima University.

[4] ATLAS User's Manual, Silvaco Inc., 2015.

[5] G. E. Davis, L. R. Hite, T. G. W. Blake, C. –E. Chen, H. W. Lam, and J. R. DeMoyer, "Transient radiation effects in SOI memories," IEEE Trans. Nucl. Scie., vol. NS-32, pp. 4432-4437, Dec. 1985.

[6] D. Kobayashi, K. Hirose, T. Oto, Y. Kakehashi, O. Kawasaki, T. Makino, T. Ohshima, D. Matsuura, T. Narita, M. Kato, S. Ishii, and K. Masukawa, "Heavy-ion soft errors in back-biased thin-BOX SOI SRAMs: Hundredfold sensitivity due to line-type multicell upsets," IEEE Trans. Nuclear Science, vol. 65, pp. 523-532, Jan. 2018.

[7] G. Suzuki, K. Konno, D. Navarro, N. Sadachika, Y. Mizukane, O. Matsushima and M. Miura-Mattausch, "Time-domain-based modeling

of carrier transport in lateral p-i-n photodiode," International Conference on Simulation of Semiconductor Processes and Devices, pp. 107-110, Tokyo, Sept. 2005; T. Ezaki, G. Suzuki, K. Konno, O. Matsushima, Y. Mizukane, D. Navarro, M. Miyake, N. Sadachika, H. J. Mattausch, M. Miura-Mattausch, "Physics-Based Photodiode Model Enabling Consistent Opto-Electronic Circuit Simulation", Digest Int. Electron Device Meeting, pp. 187-190, San Francisco, Dec. 2006.

[8] K. Konno, O. Matsushima, D. Navarro, and M. Miura-Mattausch, "High frequency response of *p-i-n* photodiodes analyzed by an analytical model in Fourier space," J. Appl. Phys. vol. 96. pp. 3839-3844, Oct. 2004.

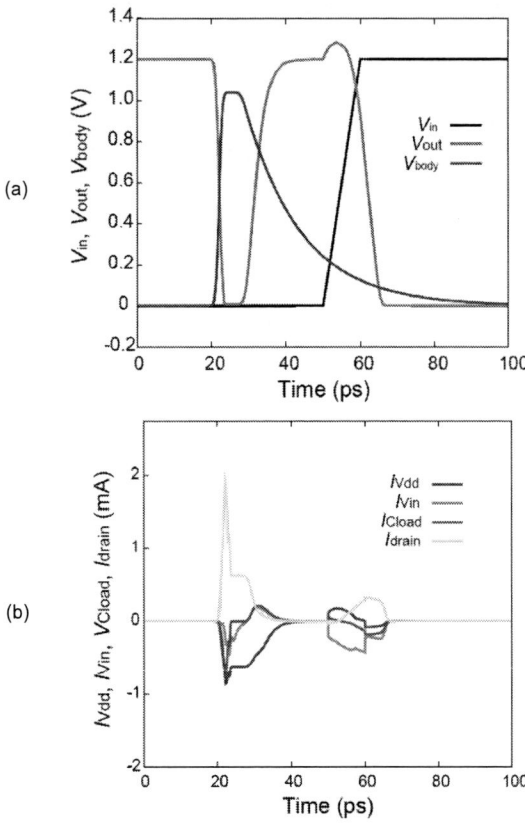

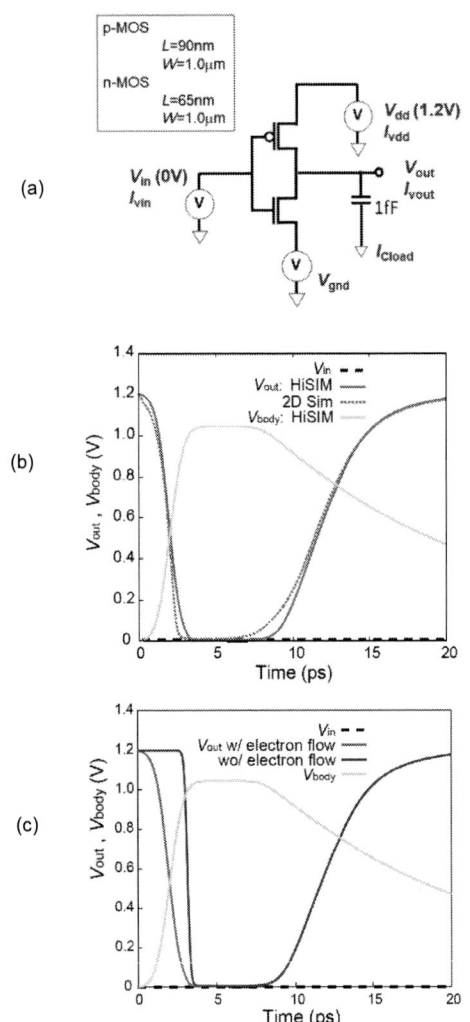

Figure 7: (a) Switching-performance malfunction of the studied inverter circuit (see Fig. 6a) due to a radiation event at the n-MOSFET before an intentional gate-controlled switching-on of the n-MOSFET. Without the radiation V_{out} stays 1.2V until switching-on starts at t=50ps. Due to the accumulated charge Q_h induced by the radiation V_{body} increases, which switches on the nMOSFET, causing the V_{out} inversion. (b) Transient current flows to different directions are summarized. I_{drain} is the current flowing between the nMOSFET and pMOSFET connection and V_{DD}, which is approximately equal to $I_{Vdd}+I_{Cload}$. It is seen that the first current flows between 20ps and 40ps by the radiation could be much larger than the expected ones.

Figure 6: (a) Studied inverter circuit; (b) comparison of switching performance from model calculation results with those from 2D-Sim results under n-MOSFET off condition, and (c) switching performance with (red line) and without (blue line) consideration of the electron flow-out during the initial stage of the radiation.

14-6

Complementary FET Device and Circuit Level Evaluation Using Fin-Based and Sheet-Based Configurations Targeting 3nm Node and Beyond

Liu Jiang, Ashish Pal, El Mehdi Bazizi, Mehdi Saremi, He Ren, Blessy Alexander and Buvna Ayyagari-Sangamalli

Applied Materials, Santa Clara, USA; Email: liu_Jiang@amat.com

Abstract— **Complementary FET (CFET), implemented by stacking NMOS and PMOS on top of each other, is considered as an emerging option to continue logic scaling beyond 3nm node. It can be configured with a fin-on-fin (fin-based CFET) or sheet-on-sheet (sheet-based CFET) structures. In this paper, we use 3D-TCAD simulation to compare those two configurations at both device and circuit levels. For accurate comparison between these two CFET configurations, we deploy a drift-diffusion simulation framework, calibrated to semi-classical sub-band BTE (Boltzmann Transport Equation). We show that for the same effective channel width, nMOS of sheet-based CFET has 10% higher drive-current compared to fin-based CFET. For pMOS, sheet-based CFET shows 5% lower drive-current compared to fin-based CFET. When compared for the same device footprint with increased nanosheet width, nMOS and pMOS sheet-based CFET shows 73% and 47% higher drive current respectively compared to fin-based CFET. Using 31-stage ring-oscillator as a representative circuit, we show that for the same electrical channel width, the circuit performance of the sheet-based CFET is 2.6% higher than the fin-based CFET at Vdd of 0.7V. When compared for the same device footprint, sheet-based CFET shows 9% higher circuit performance compared to the fin-based CFET.**

Keywords— *Complementary FET, FinFET, Nano-sheet FET, DTCO, Ring Oscillator, Circuit Modeling*

I. INTRODUCTION

In current CMOS technology, feature size scaling of FEOL and BEOL are reaching their limit due to weaker electrostatic controllability and increased portion of parasitic. Complementary FET (CFET) is an attractive option to continue logic area scaling. First, CFET stacks nMOS and pMOS vertically, thus the layout area previously utilized for nMOS devices becomes available. Second, this vertical stacking of nMOS and pMOS offers simplified access of the transistor terminals [1], such as by shorting n/p drain contacts internally. This allows the design of the standard cell library with lower number of M0 tracks, further reducing the standard cells height and area. Overall, it has been shown that using CFET, 25% area scaling can be achieved for both logic and SRAM [2]. Currently FinFET as well as Nanosheet are considered as main candidates for advanced CMOS technology. Similarly, CFET can also be realized using a fin-based configuration (FBC) or a sheet-based configuration (SBC). In this paper, we perform both device and circuit level evaluation for those two configurations at 3nm node dimensions.

II. DESCRIPTION OF APPROACH

First, we performed 3D-TCAD process simulations using a CFET process flow (Fig. 1) and 3nm CFET dimensions (Fig. 2). CFET inverter layout (Fig. 3) assumes two fins for FBC and one N/P stack for SBC shown in Fig. 4. The N/P stack in SBC is assumed to consist of two nanosheets for each nMOS and pMOS to achieve the same electrical channel width as in FBC. In both CFET structures, nMOS is stacked on top of pMOS (with 30nm separation) so that pMOS can benefit from substrate induced stress for mobility improvement. The nMOS device is assumed un-stressed while a 500MPa compressive stress is assumed for pMOS.

CFET Process Flow

❏ Si/SiGe Fin Formation
❏ Dummy Gate
❏ Spacer
❏ Cavity Etch
❏ Inner Spacer
❏ Bottom Epi & Contact
❏ NP Isolation
❏ Top Epi
❏ ILD
❏ RMG
❏ Top Contact

Fig. 1. CFET process flow

Parameter	Value
Gate Pitch	45 nm
Gate Length	15 nm
N/P Space	30nm
Fin Pitch	26nm
Fin Height	18nm
Fin Width	5nm
Nanosheet Pitch	39nm
Nanosheet Width	18nm
Sheet Thickness	5nm

Fig. 2. 3nm CFET parameters used in 3D-TCAD simulations

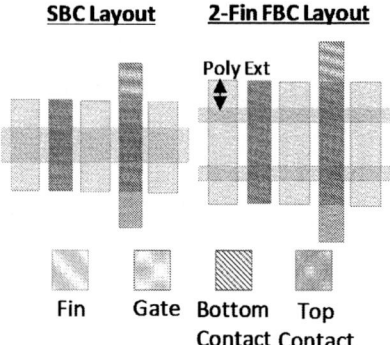

SBC Layout **2-Fin FBC Layout**

Poly Ext

Fin Gate Bottom Top
Contact Contact

Fig. 3. Layout for SBC and 2-Fin FBC

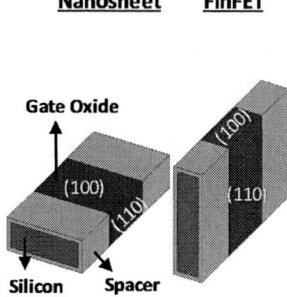

Nanosheet **FinFET**

Gate Oxide

(100) (100)

(110) (110)

Silicon Spacer

Fig. 5. Simplified Nanosheet and FinFET structures used for Sub-band BTE (Boltzmann Transport Equation) simulation.

For proper comparison, both FBC and SBC are assumed to have the same size of silicon channels, with a cross-section of 18nm by 5nm. Both devices are assumed to have the same poly extension, set to half of the fin-to-fin spacing (Fig. 3).

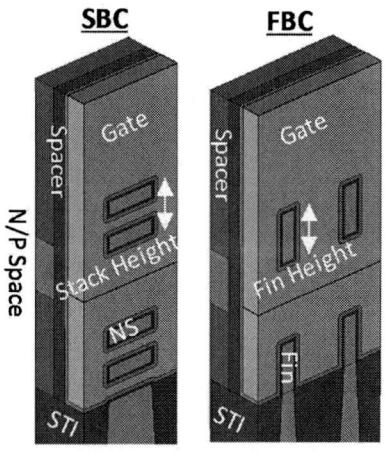

SBC **FBC**

Spacer Gate Spacer Gate

N/P Space Stack Height Fin Height

NS Fin

STI STI

Fig. 4. Nanosheet-based CFET and Fin-based CFET, showing only half gate pitch from the middle of channel to the middle of the Drain. 18nm of fin height by 5nm of fin width, Fin height are used for nMOS and pMOS in FBC which is equal to Si/SiGe/Si stack height in SBC.

Under these assumptions, SBC offers a 25% footprint reduction of the active area. Additionally, for both nMOS and pMOS, SBC and FBC are assumed to have the same doping profile and gate work function. This leads to very similar electrostatics (subthreshold slope and DIBL) for FBC and SBC. To accurately model the nature of electron and hole transport at these ultra-short 3nm node dimensions, first we solved the semi-classical sub-band Boltzmann transport equations (BTE), self-consistently with the Poisson and Schrodinger equations for both FinFET and nanosheet structures (Fig.5). The current-voltage characteristics are then used for calibration of the simplified drift-diffusion modeling framework. The density gradient, low field mobility and high-field saturation model parameters in the drift-diffusion framework are tuned for this calibration.

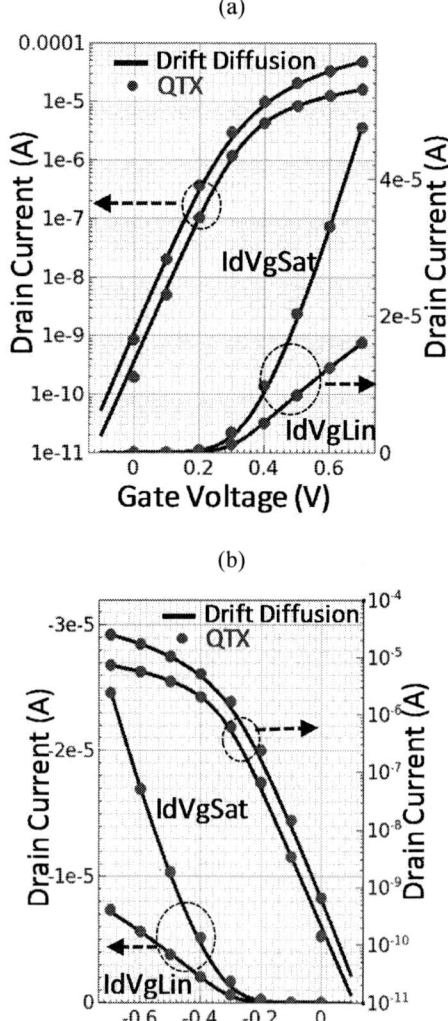

Fig. 6. Calibration of Drift-diffusion models to Sub-band BTE simulations: IDVG characteristics of (a) nMOS, (b) pMOS

324

A good matching is achieved in the current-voltage characteristics for both nMOS (Fig. 6a) and pMOS (Fig. 6b) in both linear and saturation conditions. Simultaneously, similar carrier density profile (Fig. 7) is also achieved between the semi-classical sub-band BTE and simplified drift-diffusion framework.

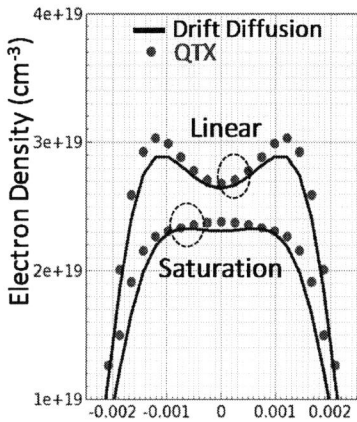

Fig. 7. Calibration of electron density distribution for Drift-diffusion models to Sub-band BTE simulation

Using this calibrated drift-diffusion framework, the performances of FBC and SBC devices are further analyzed. To evaluate FBC and SBC at circuit level, we perform 31-stage ring-oscillator (RO) simulation considering only the FEOL part, using the simulation framework described in [4].

III. RESULTS AND DISCUSSIONS

Fig. 8 shows the electron mobility distribution along gate pitch direction in nMOS for SBC and FBC. Mobility profile along fin pitch direction in the middle of channel is shown for nMOS (Fig. 9a) and pMOS (Fig. 9b).

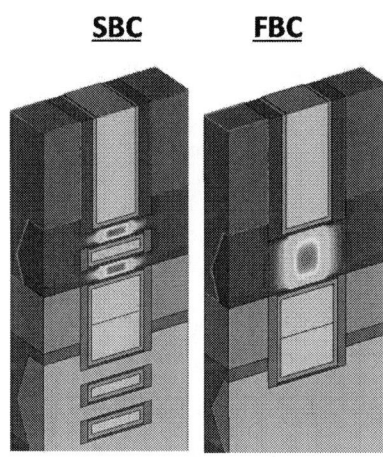

Fig. 8. 3D eMobility distribution in nMOS.

SBC channel is dominated by (100) surface while FBC channel is dominated by (110) surface. For silicon, (100) surface provides higher electron mobility than (110) surface.

Hence, higher electron mobility is seen in SBC nMOS. On the other hand, hole mobility for (100) surface is lower than the one of (110) surface, thus SBC pMOS shows lower hole mobility.

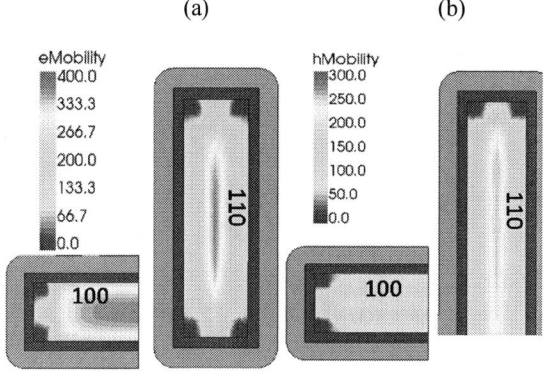

Fig. 9. Mobility distribution for Sheet and Fin in the middle of channel for (a) nMOS and (b) pMOS. Electron mobility is higher for (100) surface orientation compared to (110); hole mobility is lower for (100) surface orientation compared to (110).

Fig. 10 shows Ron and Cinv comparison for FBC and SBC. With the same channel volume, SBC nMOS shows 16% lower Ron (normalized series resistance) compared to FBC while pMOS shows 10% higher Ron. Due to 25% smaller footprint for SBC, parallel capacitance from gate to drain is lower, leading to 9% lower Cinv (inversion gate capacitance) compared to FBC.

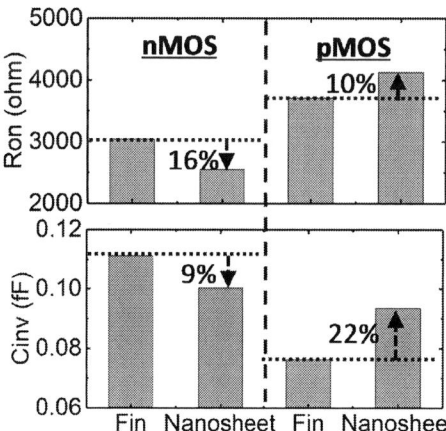

Fig. 10. Ron and Cinv comparison for FBC and SBC.

On the other hand, pMOS in SBC structure shows 22% higher Cinv that could be explained by two main reasons: (i) a parasitic bottom device is formed when the nanosheet structure is fabricated on bulk Si substrate. This parasitic device contributes more to capacitance than to current. (ii) bottom gate height is increased due to an additional sacrificial SiGe layer for SBC pMOS, and adds capacitance from gate to drain.

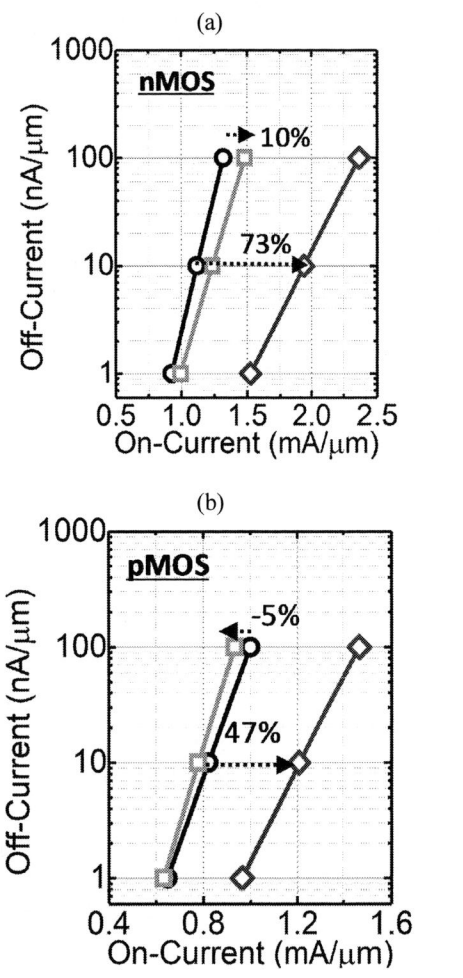

Fig. 11. Ion-Ioff performance comparison for (a) nMOS and (b) pMOS. SBC nMOS shows 10% Ion-Ioff improvement over FBC; SBC pMOS shows 5% Ion-Ioff degradation compared to FBC. For both nMOS and pMOS, increasing Nanosheet width to 31nm (to match the same active device footprint as CFET with 2 Fin design) significantly improved performance.

When comparing the on-state current at the same off-state current, SBC nMOS shows 10% Ion improvement over FBC (Fig.11a) while SBC pMOS shows 5% Ion degradation compared to FBC (Fig.11b). To match the same active device footprint as of FBC CFET with 2 fin design, nanosheet width in SBC is increased to 31nm. With increased nanosheet width, nMOS and pMOS drivability can be higher than FBC by 73% and 47% respectively due to higher effective channel width.

To evaluate the combined impact of drive-current and parasitic capacitance on circuit performance, a 31-stage ring oscillator is simulated as a representative circuit. SBC overall RO performance (frequency) is 2.6% better compared to FBC

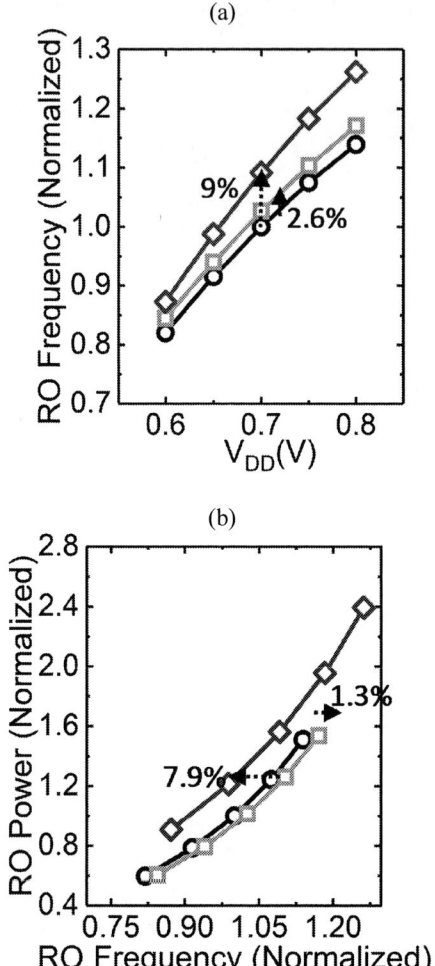

Fig. 12. Ring-oscillator Performance comparison of 3 CFET structures

(Fig.12a). With increased nanosheet width, RO performance can be 9% better than FBS due to higher drivability. For the same power consumption, SBC with 18nm nanosheet width shows 1.3% higher performance (Fig.12b). With increased nanosheet width, the capacitance in SBC becomes higher, which leads to lower performance.

Reference

[1] J. Ryckaert et al., "The Complementary FET (CFET) for CMOS scaling beyond N3", Sym. VLSI Tech, vol. T13.3, pp. 141, 2018.

[2] P. Schuddinck, et al., "Device- circuit- & block-level evaluation of CFET in a 4 track library", Proc. Symp. VLSI Technol., pp. T204-T205, Jun. 2019.

[3] Synopsys, Sentaurus, www.synopsys.com

[4] A. Pal et al, "Impact of MOL/BEOL Air-Spacer on Parasitic Capacitance and Circuit Performance at 3 nm Node," International Conference on Simulation of Semiconductor Processes and Devices (SISPAD), 2019, pp. 1-4.

Via Size Optimization for Optimum Circuit Performance at 3 nm node

Sushant Mittal[1], Ashish Pal, Mehdi Saremi, El Mehdi Bazizi*, Blessy Alexander and Buvna Ayyagari

[1]Applied Materials, Bangalore, India; Applied Materials, Santa Clara, USA
* ElMehdi_Bazizi@amat.com

Abstract— **Via size and placement for layer-to-layer connection needs careful assessment. Small via size offers compact pitch and denser connections between metal layers, while larger via size offers reduced resistance for better performance. In this paper, an optimization scheme for via size is presented, without changing the density of via allocation. We show that increasing via CD reduces resistance, resulting in enhanced performance. However, this also results in increased capacitance between different circuit nodes, which causes degradation in performance. These two opposite effects result in an optimum via CD, which offers best performance. We also show that this optimum via CD depends on the resistivity of the via material and the dielectric constant of inter-layer dielectric (ILD) surrounding the via. Via design guidelines for TiN/Co via material and for a futuristic barrier-less metal with equivalent resistivity 1/10th of cobalt via, is presented for different dielectric constants of surrounding dielectrics.**

Keywords—Via optimization, DTCO, Ring Oscillator, Circuit Modeling, FinFET, Nano-sheet FET, RC delay, Process Optimization.

I. INTRODUCTION

Vias are used for connecting two metal lines at different level. The size of the via is usually designed keeping in consideration either higher packing density at advanced nodes leading to smaller size, or higher current density for a smaller voltage drop to support power grids, resulting in a larger size [1]. The metal used for the via or the dielectric surrounding the via, are usually not considered while designing a layout. An increment in the size of the via leads to reduced via resistance, which favors improved performance.

If the size of the via is increased keeping the pitch constant, an increment in size of the via results in reduced gap between two adjacent via. Thus, the capacitance between two via increases with increment of via CD, which results in reduced performance. Therefore, an optimum size of the via exists which results in best performance, and would depend on the resistivity of the via metal and dielectric surrounding the via. In this paper, using detailed front-end of line (FEOL), back-end of line (BEOL) and circuit simulations, we show such optimum size of the via resulting in optimum ring-oscillator (RO) performance, as a function of via metal and dielectric parameters.

II. DESCRIPTION OF APPROACH

The ring-oscillator (RO) circuit simulation framework (Fig. 1) is used to see the impact of via CD modulation on RO stage delay. Firstly, FinFET based FEOL TCAD deck is established at 3 nm node for both NMOS and PMOS transistors. Next, the BSIM-CMG compact model is calibrated to the TCAD generated IV/CV characteristics. The 3nm inverter layout for middle-of-line (MOL) & BEOL layers is drawn next (Fig. 2 (a)), using which the BEOL 3D structure is generated (Fig. 2 (b)). The parasitic resistance and capacitance of MOL & BEOL is extracted from this 3D structure. Using the FEOL compact model and MOL/BEOL parasitic for inverter and interconnect, the 31-stage RO simulations are performed. This flow is discussed in detail in [2] and [3]. The parameters used in this study are shown in Fig. 3. Via-0 (Fig. 2(b)) is used for CD optimization.

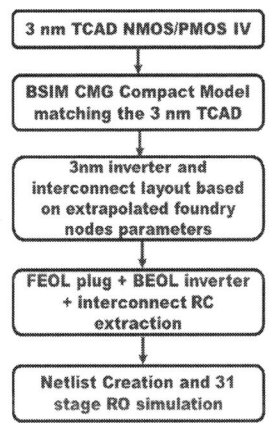

Fig. 1. Flow chart for 31-stage ring-oscillator (RO) performance evaluation, already presented in [2] [3].

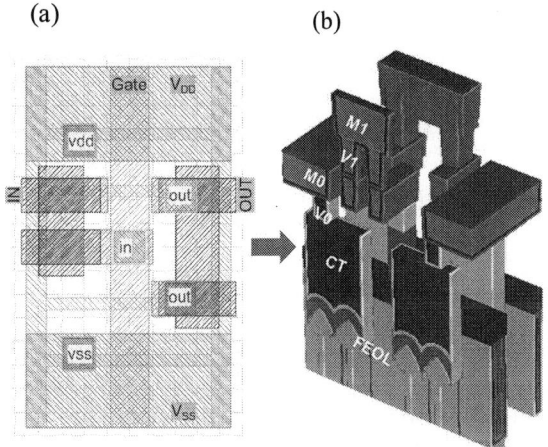

Fig. 2. (a) Inverter Layout at 3 nm node. Via-0 connections are highlighted (b) Schematic of the BEOL inverter structure of an inverter, used for RC extraction.

Parameter	Value (nm)
Gate Pitch	45
Fin Pitch	24
W_{FIN}	5
H_{FIN}	60
Spacer W	5
L_G	14
Contact CD	13
CT/CG CD	13
Via 0 CD	13

Fig. 3. List of FEOL and BEOL parameters used in the study.

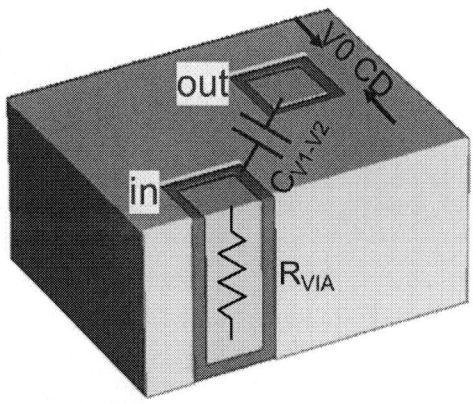

Fig. 4. Via-0 at in and out node from (b). The resistance of the via and capacitance between them is highlighted. Goal of this paper is to optimize Via-0 CD for best circuit performance.

III. RESULTS AND DISCUSSIONS

To study the impact of via CD modulation, the via connected to IN and OUT port of the inverter are considered (Fig. 2 (b) and Fig. 4). Fig. 4 shows the resistance of a via, and the capacitance between two via, which in turn decides the capacitance between IN and OUT port of an inverter.

Fig. 5 (a) shows the normalized resistance of via vs. via CD, for two via metals: Co and a futuristic barrier-less metal with equivalent resistivity $1/10^{th}$ of cobalt via. The normalization is done with respect to Co via resistance at 13 nm CD. Co based via has a TiN liner of 2 nm thickness. Fig. 5 (b) shows the relative capacitance between the two via vs. via CD, for different dielectric constant of ILD surrounding the via. For a via CD increment of 8 nm, resistance reduces ~70 % while the capacitance increases by 2× (Fig. 5(a) & (b)).

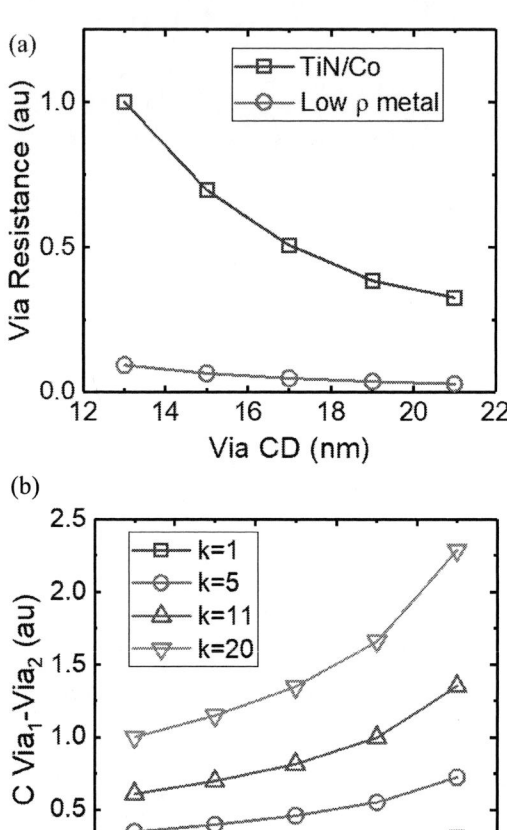

Fig. 5. (a) Via-0 resistance vs. Via CD for different Via-0 material systems. Via resistance improves as per the expectation. (b) Capacitance between two vias vs. Via-0 CD, for different ILD dielectric constants. Coupling capacitance increases as a function of Via CD.

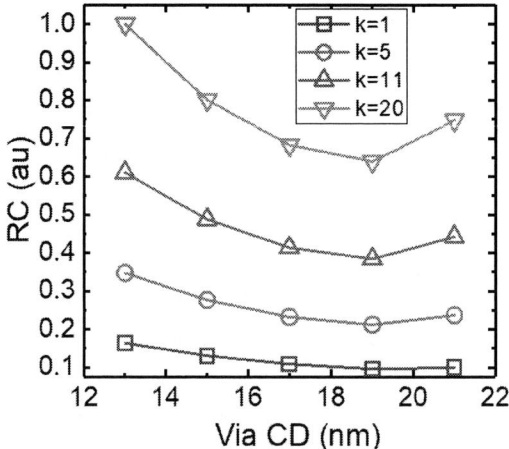

Fig. 6. RC time constant as a function of Via CD for TiN/Co via material, for different ILD k. At lower k, increasing Via-0 CD only helps in reducing time constant. At higher k values (>= 5), an optimum via CD exists for minimum RC time constant.

328

The collective effect of resistance reduction and capacitance increment with via CD is shown in Fig. 6, where in the RC time constant (multiplication of via R and C) is shown for TiN/Co, for various ILD dielectric constant. It is observed that for ILD k=1 (for air spacers [3]), the via CD increment only results in RC time constant reduction, as the capacitance does not increase as significantly with increment in via CD. However, for ILD k = 5 and above, the RC vs. via CD is a U-shape curve, depicting an optimum via CD point with lowest RC.

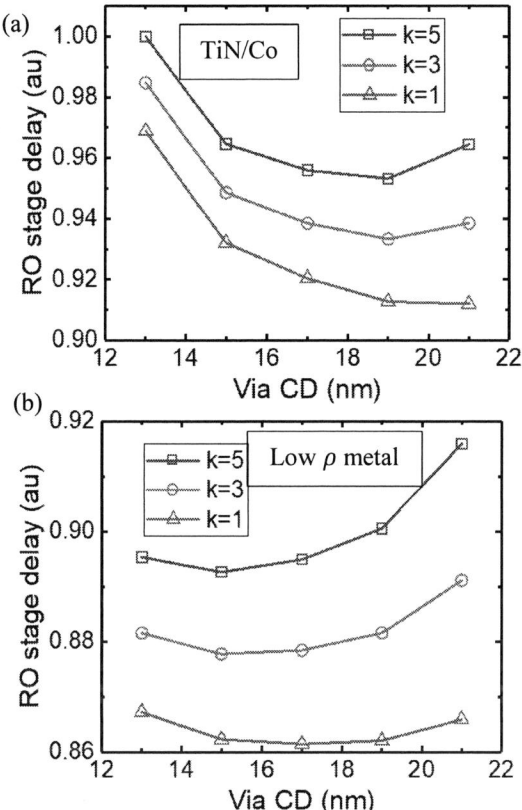

Fig. 7. (a) RO stage delay vs. Via CD for TiN/Co Via-0 material, for ILD k of 1,3 and 5. We observe that RO delay first decreases with reduction in Via-0 CD and then increases. Reduction is attributed to reduction in via resistance and increment is attributed to increased capacitive coupling. Thus, an optimum performance point exists. At low k (k=1), the optimum point is not seen because capacitance is very low. (b) same as (a) but for lower ρ Via-0 material. For lower ρ, the resistance is very low to begin with. Thus, increasing CD does not give much performance benefit as capacitance degradation starts to degrade the performance earlier

The entire circuit RO delay with such via CD change is shown in Fig. 7 (a) for TiN/Co and (b) for low resistivity metal (ρ). The plots are normalized with respect to TiN/Co metal at nominal via CD of 13 nm. Analogous to individual layer RC time constant, RO delay also suggests that for ILD k=1, the via CD increment will only help in improving performance. However, for both k = 3 and 5, a via CD of 19 nm shows best performance with a RO stage delay reduction of ~ 5 %. For the lower ρ metal, increasing via CD does not yield much performance boost as resistance of via is already low to begin

with. At ILD k = 1, there is an optimum point at CD = 17 nm. However, for k=3 and 5, an optimum point exists at lower via CD.

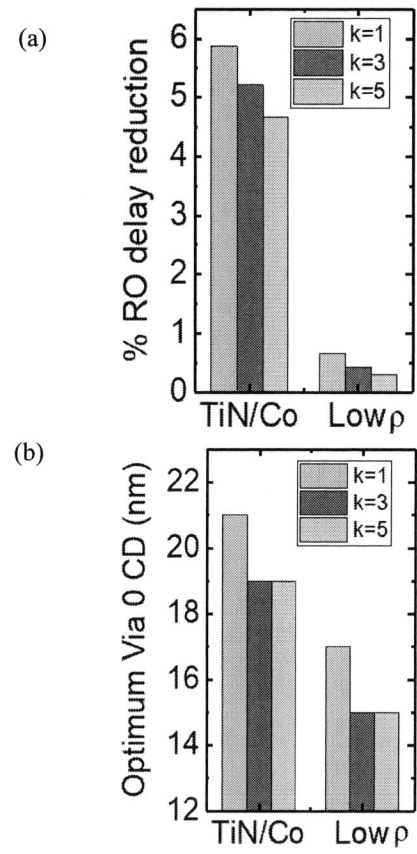

Fig. 8. (a) Maximum reduction in RO delay for TiN/Co and lower ρ Via-0 material system, at ILD k = 1, 3 and 5. For Co, ~ 5% reduction is observed, however for lower ρ material, benefit is less than 1 %. (b) Via-0 CD for optimum performance at ILD k = 1, 3 and 5, for Co and lower ρ metal. For Co material system, a larger Via-0 CD of 19-21 nm would be beneficial, but for a low resistance material system, a smaller Via-0 CD of 15-17 nm would be optimum.

Fig. 8 (a) summarizes the performance boost which can be obtained for both TiN/Co and low ρ via material. Using proper via CD, TiN/Co can enable performance boost about 5 %. The optimum via CD is shown in Fig. 8 (b), for both metals, and for different ILD k's. For air-spacer, larger via CD can give better performance, however for ILD k=3 and 5, a smaller via CD can give an optimum performance point. One way to achieve larger via CD is application of suitable mask bias. For example, a 3 nm mask bias would result in 19 nm via CD at a nominal via CD of 13 nm.

IV. CONCLUSION

We propose via material system (via material resistivity & surrounding ILD's dielectric constant) based via CD design at 3 nm node to enable up to 5 % performance gain. Such a method can also be used in congruence with via CD design for other BEOL layers to enable even larger performance gains.

REFERENCES

[1] Mikhail R. Baklanov, Paul S. Ho and Ehrenfried Zschech, Advanced Interconnects for ULSI Technology, Wiley Publishers 2012

[2] S. Mittal et al., "Highly-Doped Through-Contact Silicon Epi Design at 3 nm node," 2019 Device Research Conference (DRC), Ann Arbor, MI, USA, 2019, pp. 55-56, doi: 10.1109/DRC46940.2019.9046479.

[3] A. Pal et al., "Impact of MOL/BEOL Air-Spacer on Parasitic Capacitance and Circuit Performance at 3 nm Node," 2019 International Conference on Simulation of Semiconductor Processes and Devices (SISPAD), Udine, Italy, 2019, pp. 1-4, doi: 10.1109/SISPAD.2019.8870410.

Time-Resolved Mode Space based Quantum-Liouville type Equations applied onto DGFETs

1st Lukas Schulz
High Frequency Institute
TU Dortmund
Dortmund, Germany
lukas.schulz@tu-dortmund.de

2nd Dirk Schulz
High Frequency Institute
TU Dortmund
Dortmund, Germany
dirk2.schulz@tu-dortmund.de

Abstract—**The investigation of a time-resolved quantum transport analysis is a major issue for the future progress in engineering tailored nanoelectronic devices. In this contribution, the time dependence is addressed along with the single-time formulation of quantum mechanics based on the von-Neumann equation in center-mass coordinates. This equation is investigated utilizing a distinct set of basis functions leading to so-called Quantum-Liouville type equations, which are combined with the mode space approximation to investigate the time-resolved behavior of double gate field effect transistors including the self-consistent Hartree potential.**

Index Terms—**double gate field effect transistor, mode space expansion, numerical methods, time-resolved quantum transport, von-Neumann equation, Wigner equation**

I. INTRODUCTION

Switching processes in today's sophisticated transistors are the heart of modern information technology. Unfortunately, they are responsible for the essential dissipation of power. To account for these effects, quantum kinetic numerical methods are desirable. In principle, the corresponding quantum kinetics can be subdivided into two major approaches, as there are double-time methods, such as the non-equilibrium Green's function (NEGF) formalism [1], as well as single-time methods based on the statistical density matrix, i.e. the Wigner formalism [2]. Time-resolved simulations based on the NEGF formalism are numerically challenging [3]. To overcome these restrictions, methods based on the wavefunction have been introduced [4], [5]. The Wigner formalism provides a quantum description of the system in terms of a phase space formulation [6]. In contrast to the NEGF formalism, the time-resolved quantum transport can be readily determined in both, the coherent as well as the incoherent regime. Indeed, the latter can be addressed along with the simple inclusion of a problem dependent collision operator. However, the adequate choice of the collision operator is a vivid field of discussion [7].

Recently, efforts have been made to numerically solve the von-Neumann equation in center-mass coordinates [8]–[10]. Indeed, this approach seems to be highly versatile, since numerical advantages of a real-space approach can be combined with those of a phase space representation due to simple algebraic basis transformations [8]. Consequently, these different

The authors gratefully acknowledge the financial support by the German Research Funding Association Deutsche Forschungsgemeinschaft under grant SCHU-2016/8-1.

algebraic basis transformations lead to the introduction of the so-called Quantum-Liouville type Equations (QLTE).

Following this approach, the QLTE are combined with the mode space approximation (MSA) [11], [12]. to investigate the time-resolved behavior of a double gate field effect transistor (DGFET). The self-consistent Hartree potential is considered, whereas mode coupling effects and scattering mechanisms are neglected at this stage for a straightforward comparison along with the NEGF formalism within the stationary regime.

II. THEORY

In the following, the major relations with regard to the presented approach are briefly discussed.

A. Time-Resolved Mode Space Approximation

When quantum transport is considered in multiple spatial directions, the numerical effort of the solution procedure is in general quiet expensive. As a consequence, accurate methods reducing the overall numerical costs are needed. Right here, the MSA comes into play [11], [12]. Along with the MSA the Hamiltonian with regard to the lateral direction is expanded in terms of basis functions leading to a set of effective one-dimensional equations. When the carriers are confined with regard to the lateral direction, the corresponding lateral eigenvectors are employed to expand the Hamiltonian. These eigenvectors are also known as the modes and, typically, only a few eigenvectors with the lowest energy eigenvalues characterize the transport behavior, which can be estimated from the value of the Fermi-level. Considering the structure

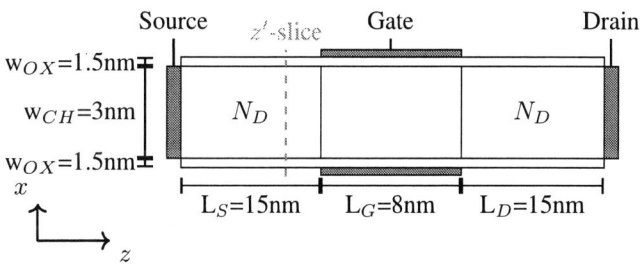

Fig. 1. Schematic diagram of the investigated DGFET including the structural parameters. The dashed line conceptually reflects the slice formalism.

of the analyzed DGFET, as shown in Fig. 1, the corresponding

Schrödinger equation within the effective mass approximation reads

$$\left\{ -j\hbar\frac{\partial}{\partial t} - \frac{\hbar^2}{2m_z^\nu}\frac{\partial^2}{\partial z^2} - \underbrace{\frac{\hbar^2}{2m_x^\nu}\frac{\partial^2}{\partial x^2} + V(x,z,t)}_{\mathcal{H}_C} \right\}\Psi^\nu(x,z,t) = 0,$$
(1)

where x is the confinement direction and z represents the transport direction. The effective masses m_x^ν and m_z^ν depend on the ν-th valley. The potential V includes the conduction band potential, the effect from transient external fields as well as the self-consistent Hartree-Fock potential V_H. The Hamiltonian $\mathcal{H}_C$ captures the confinement and can be diagonalized for each slice z, as depicted in Fig. 1, and timestep t, which enter the equation parametrically, according to

$$\mathcal{H}_C\Phi_m^\nu(x;z,t) = E_m^\nu(z,t)\Phi_m^\nu(x;z,t) = E_m^\nu(z,t)\,|m,\nu\rangle$$
(2)

with $E_m^\nu(z,t)$ being the eigenenergy and $|m,\nu\rangle$ representing the spatial distribution of the m-th mode. Distinct modes are orthonormal $\langle \nu,n|\nu,m\rangle = \delta_{n,m}$. Introducing the expansion coefficient $\varphi_m^\nu(z,t)$, the wavefunction can be expanded as

$$\Psi^\nu(x,z,t) = \sum_m \varphi_m^\nu(z,t)\,|m,\nu\rangle,$$
(3)

and inserted in (1). Exploiting the orthogonality condition as well as (2), an equation of motion for the expansion coefficient $\varphi_m^\nu(z,t)$ can be derived

$$j\hbar\frac{\partial}{\partial t}\varphi_m^\nu = \left\{ -\frac{\hbar^2}{2m_z^\nu}\frac{\partial^2}{\partial z^2} + E_m^\nu(z,t) \right\}\varphi_m^\nu(z,t) = \mathcal{H}_m^\nu\varphi_m^\nu,$$
(4)

where due to the parametric dependence the relations $\frac{\partial}{\partial t}|m,v\rangle = 0$ and $\frac{\partial}{\partial z}|m,v\rangle = 0$ are assumed leading to the uncoupled MSA. The limitations of the latter spatial approximation are discussed elsewhere [11].

B. *Quantum-Liouville type Equations*

To solve the transport governed by (4), the equation of motion for the statistical density matrix is employed [13]

$$j\hbar\frac{\partial}{\partial t}\rho = [\mathcal{H},\rho] + j\hbar\Gamma(\rho),$$
(5)

where the operator Γ contains all interaction mechanisms, such as for instance electron-phonon scattering. Here, only the open system related to in and out scattering at the device boundaries is considered.

Introducing the center-mass coordinates $z = \chi + \frac{1}{2}\xi$ and $z' = \chi - \frac{1}{2}\xi$ as part of the Weyl transform, the von-Neumann equation can be rewritten accordingly as

$$\frac{\partial}{\partial t}\rho_m^\nu(\chi,\xi,t) = \frac{j\hbar}{m_z^\nu}\frac{\partial^2}{\partial\chi\partial\xi}\rho_m^\nu +$$
$$\frac{1}{j\hbar}\left\{ E_m^\nu\left(\chi+\frac{1}{2}\xi,t\right) - E_m^\nu\left(\chi-\frac{1}{2}\xi,t\right) - jW(\xi) \right\}\rho_m^\nu,$$
(6)

where the complex absorbing potential [9] is introduced to account for appropriate boundary conditions in the ξ-space. Here, the statistical density matrix is defined as $\rho_m^\nu(\chi,\xi,t) =$

$\varphi_m^\nu(\chi+\frac{1}{2}\xi)\varphi_m^{\nu\dagger}(\chi-\frac{1}{2}\xi)$. From the statistical density matrix, the current density j and the carrier density n can be obtained as

$$n_m^\nu(\chi,t) = \rho_m^\nu\big|_{\xi=0} \quad \text{and} \quad j_m^\nu(\chi,t) = \frac{\hbar}{m_z^\nu}\Im\left\{ \frac{\partial}{\partial\xi}\rho_m^\nu\big|_{\xi=0} \right\}.$$
(7)

The von-Neumann equation (6) is discretized within the real space applying a finite volume scheme with regard to the ξ-direction [9] leading to a discretized statistical density matrix $\rho_m^\nu(\chi,t)$. With regard to the in- and out-scattering at the device contacts, the concept of the inflow boundary conditions is employed [14]. In this manner, the statistical density matrix must be transformed into a presentation, which allows the distinction with respect to the flow behavior. Introducing the transformation matrix $[T]$

$$\rho_m^\nu(\chi,t) = [T]\cdot a_m^\nu(\chi,t),$$
(8)

the vector of expansion coefficients $a_m^\nu(\chi,t)$ can be assigned to the flow behavior. For the numerical discretization of the remaining transformed coupled transport equations, the introduction of the phase space exponential operator [9] leads to the formulation

$$\frac{\mathrm{d}}{\mathrm{d}t}A_m^\nu(t) = [\Gamma_m^\nu(t)]\cdot A_m^\nu(t) + B_m^\nu,$$
(9)

where the supervector $A_m^\nu(t)$ contains the vectors for all expansion coefficients $a_m^\nu(\chi,t)$ for each slice of the discretized χ-direction. The supervector B represents the inflow boundary conditions. The system (9) can be readily solved in the stationary case ($\frac{\mathrm{d}}{\mathrm{d}t}A = 0$) as well as the transient case applying standard approaches, i.e. Crank-Nicolson schemes [15] or Low-Storage-Runge-Kutta methods [9].

C. *Inclusion of Self-consistency*

Assuming an instantaneous adaption of the field to the carrier-density it is sufficient to solve the Poisson equation for the Hartree potential instead of a wave equation in both, the transient and in the stationary case [16]

$$\nabla\cdot(\epsilon(x,z)\nabla V_H) = q^2\left(n(x,z,t) - N_D(x,z) \right).$$
(10)

The total carrier density $n(x,z,t)$ is calculated from the corresponding modal carrier density in each valley by

$$n(x,z,t) = \sum_{\nu,m} n_m^\nu(z,t)\cdot|\Phi_m^\nu(x;z,t)|^2,$$
(11)

where $n_m^\nu(\chi,t) = n_m^\nu(z,t)$ is utilized. The function $N_D(x,z)$ represents the spatial doping profile. The current density is obtained by accumulating the modal current-densities in each valley. As usual, the Newton-Raphson method is employed along with Dirichlet boundary conditions at the gate contact and von-Neumann boundary conditions at all remaining boundaries (Fig. 1) in the stationary regime

$$V_H(x,z)\big|_{x,z\in\text{gate}} = E_f + q\phi_\mathrm{m} - q\chi_\mathrm{ch} - qU_\mathrm{gate},$$
$$\frac{\partial V_H(x,z)}{\partial\eta}\bigg|_{x,z\notin\text{gate}} = 0$$
(12)

332

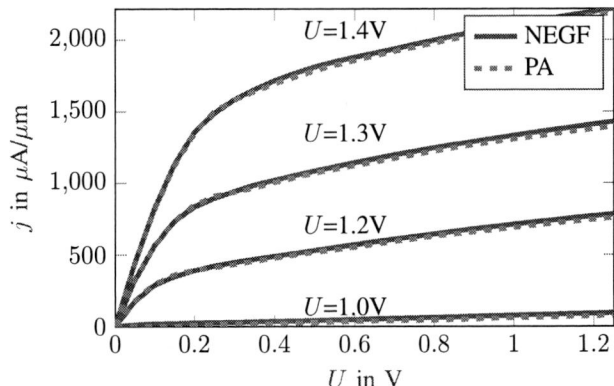

Fig. 2. Comparison of the stationary current densities.

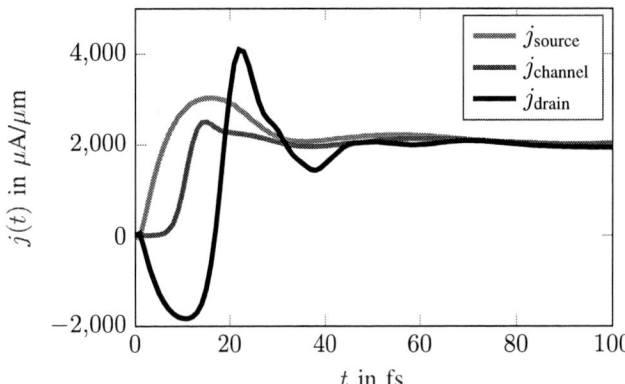

Fig. 3. Current density in dependence of the time within different locations of the bias due to a switch of the gate bias $U_{\text{gate}} = 0.8\text{V}$ to 1.4V at $t = 0\text{fs}$. A constant source-drain bias $U = 1.0\text{V}$ is applied.

with η being a coordinate belonging to a direction perpendicular to the boundary. The parameters E_f, $q\phi_{\text{m}}$ and $q\chi_{\text{ch}}$ define the Fermi-level, the work function of the metal and the affinity of the channel, respectively. The voltage U_{gate} is applied at the gate contacts. For the evaluation of the expression $\frac{\mathrm{d}}{\mathrm{d}V_H} n$ as part of the Jacobian matrix appearing in the Newton-Raphson method, a Maxwell distribution is presumed for the carrier density [16], so that the relation $\frac{\mathrm{d}}{\mathrm{d}V_H} n = \frac{n}{k_B T}$ holds.

For the transient solution, the Newton-Raphson method cannot be utilized as discussed in [16]. As a consequence, a direct solution procedure is employed, where the Poisson equation is solved once for each time step applying Dirichlet boundary conditions at the drain and source contacts utilizing the relations

$$V_H(x, z, t)\Big|_{x,z \in \text{gate}} = E_f + q\phi_{\text{m}} - q\chi_{\text{ch}} - qU_{\text{gate}}(t)$$

$$V_H(x, z, t)\Big|_{x,z \in \text{source/drain}} = -qU_{\text{source/drain}}(t), \quad (13)$$

$$\frac{\partial V_H(x, z, t)}{\partial \eta}\Big|_{x,z \notin \text{gate,source,drain}} = 0$$

since the use of von-Neumann boundary conditions at the contacts implies charge neutrality near the contacts, which may not hold for the time-resolved behavior.

III. STATIONARY REGIME: COMPARISON TO THE NEGF APPROACH

For the evaluation of the proposed approach, the DGFET with the structural parameters as defined in Fig. 1 is analyzed. The material of the gate-contact is Ag, for which the work function $q\phi_{\text{m}} = 4.74\text{eV}$ is given. The insulator material is SiO_2 ($\epsilon = 3.9\epsilon_0$) and the electron affinity of the channel material $Ga_{0.47}In_{0.53}As$ ($\epsilon = 13.9\epsilon_0$) is assumed to be $q\chi_{\text{ch}} = 4.5\text{eV}$. A constant doping concentration $N_D = 2 \cdot 10^{25}\text{m}^{-3}$ is assumed within source and drain regions. The ambient temperature is $T = 300K$ and the isotropical effective masses of the oxid and channel are given by $0.5m_0$ and $0.041m_0$, respectively. Up to three modes have been taken into account. From the results obtained for these cases, it can

be concluded that the device behavior is dominated by the first mode. Hence, the inclusion of a larger modal basis does not provide significant more information in the considered case and is, therefore, neglected.

To assess the accuracy of our proposed method, a comparison to the results obtained from the NEGF approach [12] is performed in the stationary regime. The discretization widths $\Delta \chi (= \Delta z) = 0.1\text{nm}$ and $\Delta y = 0.25\text{nm}$ are employed, whereas the ξ-direction is discretized within the interval $[-60, 60]\text{nm}$ utilizing $N_\xi = 400$ discretization points. Here, the results are shown for an expansion of the statistical density matrix in terms of a $N = 200$ orthogonal plane wave basis. Further bases have been investigated, such as sinusoidal-like basis functions given by the eigenvectors of the discretized diffusion matrix. Since they do not provide a deeper insight, the results are not shown. However, for the cases considered, a slightly better convergence rate is observed for the plane wave basis.

In Fig. 2, the current densities obtained from the NEGF approach are compared to the current densities obtained from the proposed approach (PA) applying different gate biases U_{gate} as well as different drain-source biases U.

For all considered cases, a fairly good agreement can be observed from Fig. 2. Due to this agreement, the proposed approach can be validated, predestinating this approach for the time-resolved device analysis.

IV. TRANSIENT RESPONSE OF A DGFET

Essentially, two major switching mechanisms can be investigated: the source-drain bias U is constant and the gate voltage U_{gate} is time-dependend and vice versa. The initial statistical density matrix at $t = 0\text{fs}$ is obtained from the stationary solution of (9) ($\frac{\mathrm{d}}{\mathrm{d}t} = 0$) in the self-consistent case for the corresponding applied bias situation. Then, for $t > 0\text{fs}$ the device is driven towards non-equilibrium by either switching the gate bias or the source-drain bias. The transport equation (9) is solved mutually along with the Poisson equation for each discrete timestep applying (13).

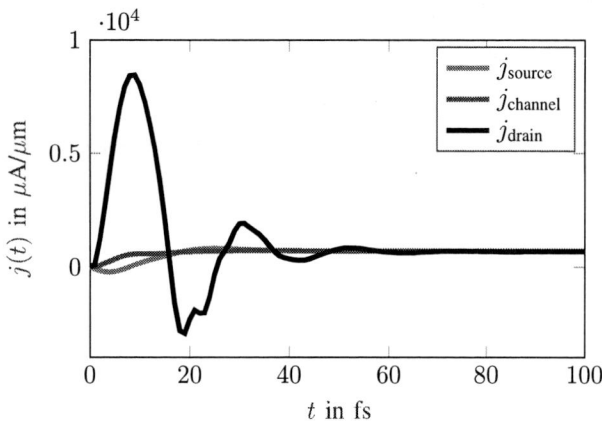

Fig. 4. Transient evolution of the current density within different locations for a constant gate bias $U_{gate} = 1.2V$ and a switch of the source-drain bias from $U = 0.0$V to $U = 1.0$V.

A. Switching the Gate Voltage

To start with, the "OFF" to "ON"-state switching process is analyzed. Initially, the DGFET is in a steady state at $t = 0$fs applying a constant voltage between source and drain of $U = 1.0$V and a gate bias $U_{gate} = 0.8$V. At $t > 0$fs the gate voltage is switched to a constant value of $U_{gate} = 1.4$V. The corresponding time-resolved evolution of the current density at different locations within the DGFET, as there are the source, the middle of the channel and the drain, are depicted in Fig. 3. As it can be seen from the spatially non-constant current densities, the device is in a strong non-equilibrium condition. After approximately 50fs, the major transient effects are decayed and the current densities convergence towards the steady state current density. From Fig. 2 the corresponding stationary current can be determined to a value of approximately $2000\mu A/\mu m$, which coincides along with the steady state result obtained from the time-resolved approach.

B. Switching the Source-Drain Voltage

Now, the gate bias is kept constant $U_{gate} = 1.2$V and the voltage between source and drain is switched at $t > 0$fs from 0.0V to 1.0V. As before, the time-resolved evolution of the current density is depicted for the same spatial location in Fig. 4. Remarkably, the transient current density at the drain j_{drain} takes extremely large values in comparison to the steady state current of approximately $700\mu A/\mu m$, which again coincides with the stationary current as can be observed from Fig. 2. In practice, these large temporal spikes of the current are undesirable since they lead to a large power consumption as well. However, when interaction mechanisms are considered, the current density is effectively reduced. Nonetheless, when phonon-emission processes are considered, the increasing device temperature plays a decisive role.

V. SUMMARY AND CONCLUSION

A self-consistent, time-resolved mode space approach based on Quantum-Liouville type equations has been analyzed,

demonstrated and validated. For the stationary regime, an excellent coincidence with the non-equilibrium Green's function approach has been obtained. The approach has been extended to the transient case enabling a deeper insight into the device performance. Since the time-resolved Green's function formalism is computationally extremely demanding, the proposed approach represents a beneficial alternative in this particular case.

REFERENCES

[1] P. Vogl and T. Kubis, "The non-equilibrium green's function method: An introduction," *J. Comput. Electron*, vol. 9, no. 3-4, pp. 237–242, 2010.

[2] J. Weinbub and D. Ferry, "Recent advances in wigner function approaches," *Appl. Phys. Rev.*, vol. 5, no. 4, p. 041 104, 2018.

[3] E. P. et al., "Time-dependent transport in graphene nanoribbons," *Phys. Rev. B*, vol. 82, p. 035 446, 3 Jul. 2010. DOI: 10.1103/PhysRevB.82.035446.

[4] B. Novakovic and G. Klimeck, "Atomistic quantum transport approach to time-resolved device simulations," in *Proc. of SISPAD*, 2015, pp. 8–11.

[5] B. G. et al., "Numerical simulations of time-resolved quantum electronics," *Physics Reports*, vol. 534, no. 1, pp. 1–37, 2014. DOI: https://doi.org/10.1016/j.physrep.2013.09.001.

[6] E. P. Wigner, "On the quantum correction for thermodynamic equilibrium," in *Part I: Physical Chemistry. Part II: Solid State Physics*, Springer, 1997, pp. 110–120.

[7] R. I. et al., "Wigner-function formalism applied to semiconductor quantum devices: Need for nonlocal scattering models," *Phys. Rev. B*, vol. 96, p. 115 420, 11 Sep. 2017. DOI: 10.1103/PhysRevB.96.115420.

[8] L. Schulz and D. Schulz, "Subdomain algorithm for the numerical solution of the liouville-von-neumann equation," *Proc. of IWCN*, 2019.

[9] ——, "Numerical analysis of the transient behavior of the non-equilibrium quantum liouville equation," *IEEE Trans Nanotechnol*, vol. 17, no. 6, pp. 1197–1205, 2018.

[10] R. K. et al., "A revised wigner function approach for stationary quantum transport," in *Proc. of LSSC*, Springer, 2019, pp. 403–410.

[11] R. V. et al., "Simulating quantum transport in nanoscale transistors: Real versus mode-space approaches," *J. Appl. Phys.*, vol. 92, no. 7, pp. 3730–3739, 2002.

[12] Z. R. et al., "Nanomos 2.5: A two-dimensional simulator for quantum transport in double-gate mosfets," *IEEE Trans. Electron Devices*, vol. 50, no. 9, pp. 1914–1925, 2003.

[13] R. Rosati and F. Rossi, "Scattering nonlocality in quantum charge transport: Application to semiconductor nanostructures," *Phys. Rev. B*, vol. 89, p. 205 415, 20 May 2014. DOI: 10.1103/PhysRevB.89.205415.

[14] W. R. Frensley, "Wigner-function model of a resonant-tunneling semiconductor device," *Phys. Rev. B*, vol. 36, no. 3, p. 1570, 1987.

[15] L. Schulz and D. Schulz, "Formulation of a phase space exponential operator for the wigner transport equation accounting for the spatial variation of the effective mass," *J Comput. Electron*, 2020.

[16] B. A. Biegel and J. D. Plummer, "Comparison of self-consistency iteration options for the wigner function method of quantum device simulation," *Phys. Rev. B*, vol. 54, pp. 8070–8082, 11 Sep. 1996. DOI: 10.1103/PhysRevB.54.8070.

Power Device Degradation Estimation by Machine Learning of Gate Waveforms

Hiromu Yamasaki, Koutaro Miyazaki, Yang Lo, A. K. M. Mahfuzul Islam, Katsuhiro Hata,
Takayasu Sakurai, and Makoto Takamiya

The University of Tokyo, Tokyo, Japan

Abstract— The emitter resistance (R_E), the junction temperature (T_J), the collector current (I_C), and the threshold voltage (V_{TH}) of power devices are key parameters that determine the reliability of power devices. Adding dedicated sensors to measure the key parameters, however, will increase the cost of the power converters. To solve the problem, power device degradation estimation methods by the machine learning of gate waveforms are proposed. Two methods are shown in this paper. First, in order to detect the bond wire lift-off of power devices, the estimation of the number of the connected bond wires using the linear regression of two feature points extracted from the gate waveforms of a SiC MOSFET is shown using SPICE simulations. Then, in order to detect the power device degradation, the estimation of R_E, T_J, I_C, and V_{TH} using the convolutional neural network (CNN) with the gate waveforms of an IGBT for input is shown using both simulations and measurements.

Keywords- CNN, Neural Network, IGBT, power device, degradation, reliability, gate

I. INTRODUCTION

Highly reliable power devices are required, because the power devices are important components of power electronics systems that support society. As examples of the power devices, an insulated gate bipolar transistor (IGBT) and a SiC MOSFET are discussed in this paper. The most frequent cause of the long-term degradation of the IGBTs is the bond wire lift-off due to the thermal cycles [1,2], where emitter bond wires are peeled-off from the pad. Fig. 1 (a) shows five important parameters related to the bond wire lift-off in IGBTs, the number of the connected bond wires (n), the emitter resistance (R_E), the junction temperature (T_J), the collector current (I_C), and the threshold voltage (V_{TH}). In order to measure the values of n, R_E, T_J, I_C, and V_{TH}, however, adding dedicated sensors for current, voltage and temperature will increase the cost of the power converters. Several previous papers tried to reduce the number of sensors. The estimation of I_C for an IGBT using the collector-emitter voltage (V_{CE}), gate-emitter voltage (V_{GE}), T_J, and V_{TH} as input has been investigated based on a neural network [3]. It is, however, very costly to know these voltages and T_J beforehand using sensors and only I_C can be estimated. The estimation of T_J by measuring the peak voltage of the gate drive voltage has been proposed [4]. The estimation, however, is only for T_J and since it does not use an AI-based approach, the estimation needs very high precision of the peak voltage measurement without noise, which is practically very difficult.

To solve the problems, two power device degradation estimation methods by the machine learning of the gate voltage (V_G) waveforms are proposed in this paper. Fig. 1 (b) shows a schematic of the V_G waveforms and two feature points ($V_{G,BENT}$ and $V_{G,MIN}$), where the definitions will be shown later. In Section II, as shown in Fig. 1 (c), in order to detect the bond wire lift-off of power devices, the estimation of n using the

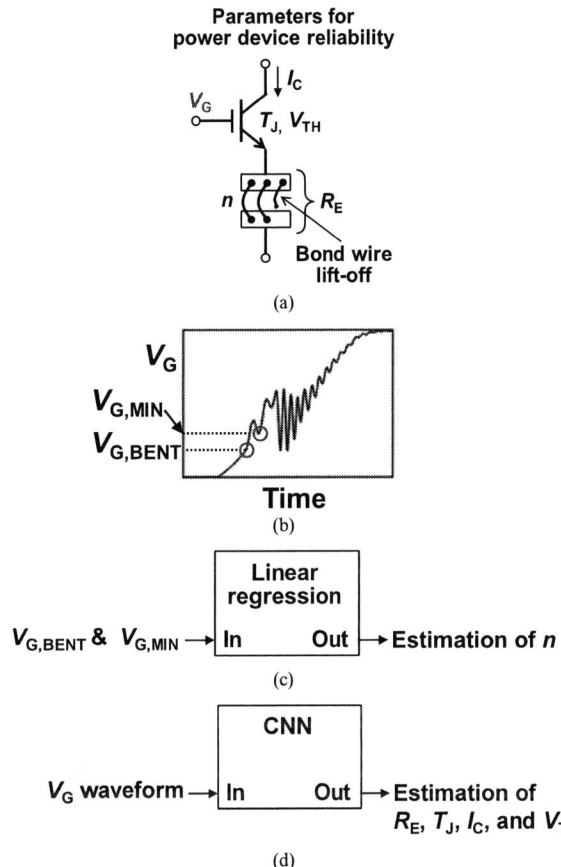

Fig. 1. Overview of this paper. (a) Parameters related to bond wire lift-off in IGBTs. (b) Schematic of V_G waveforms and two feature points ($V_{G,BENT}$ and $V_{G,MIN}$). (c) Estimation of n using linear regression of two feature points. (d) Estimation of R_E, T_J, I_C, and V_{TH} using CNN with V_G waveforms.

linear regression of two feature points extracted from V_G waveforms of a SiC MOSFET is shown using simulations. In Section III, as shown in Fig. 1 (d), in order to detect the power device degradation, the estimation of R_E, T_J, I_C, and V_{TH} using the convolutional neural network (CNN) with V_G waveforms of an IGBT for input is shown using both simulations and measurements.

II. DETECTION OF BOND WIRE LIFT-OFF USING LINEAR REGRESSION OF TWO FEATURE POINTS

In order to detect the bond wire lift-off of power devices, in this chapter, the estimation of n using the linear regression of two feature points ($V_{G,BENT}$ and $V_{G,MIN}$) extracted from V_G waveforms of a SiC MOSFET is shown. As shown in Fig. 1 (b), $V_{G,BENT}$ is defined the inflection point of V_G around V_{TH}, and $V_{G,MIN}$ is defined the first minima of V_G.

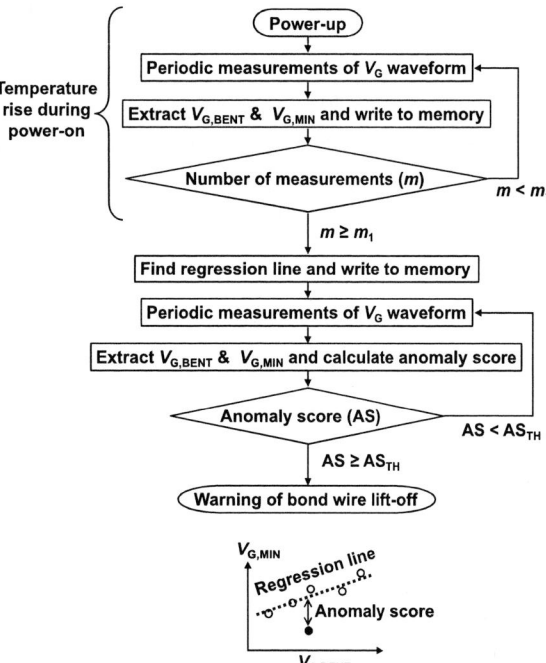

Fig. 2. Flowchart of proposed detection method of bond wire lift-off.

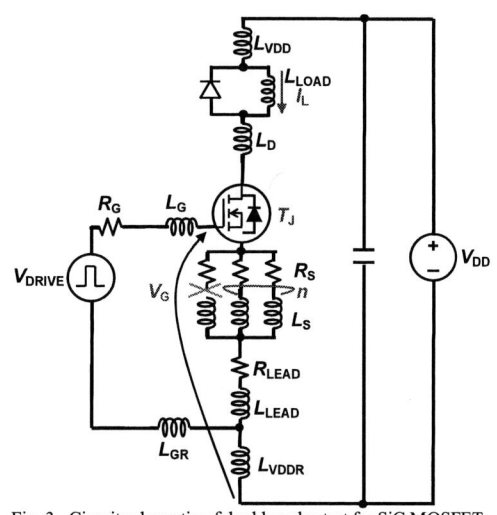

Fig. 3. Circuit schematic of double pulse test for SiC MOSFET.

Table I Parameters for SPICE simulation.

	Name	Value
V_{DD}	Supply voltage	300 V
L_{VDD}	Parasitic inductance between diode and capacitor	6 nH
L_{LOAD}	Load inductance	1.5 mH
L_D	Parasitic inductance between diode and drain	3 nH
R_G	External gate resistance	2.5 Ω
L_G	Parasitic inductance between gate driver and gate	6 nH
n	Number of connected bond wires	3, 2, 1
R_S	Parasitic resistance of bond wires (n = 3, 2, 1)	0.270 mΩ, 0.405 mΩ, 0.810 mΩ
L_S	Parasitic inductance of bond wires (n = 3, 2, 1)	1.15 nH, 1.72 nH, 3.44 nH
R_{LEAD}	Parasitic resistance of lead of source	7.41 mΩ
L_{LEAD}	Parasitic inductance of lead of source	3.55 nH
L_{GR}	Parasitic inductance between ground of gate driver and lead of source	3 nH
L_{VDDR}	Parasitic inductance between lead and capacitor	6 nH
V_{DRIVE}	Gate driver voltage source	0 V / 15 V

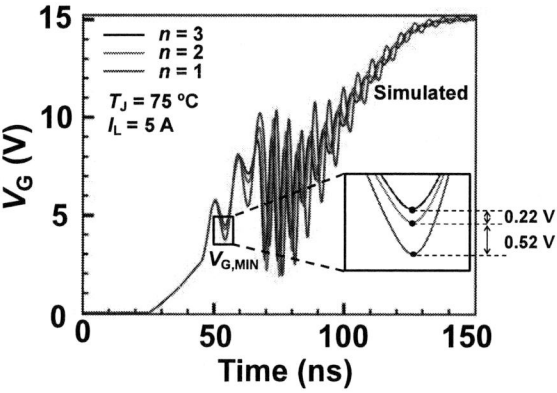

Fig. 4. Simulated V_G waveforms with varied n at T_J = 75 ℃ and I_L = 5 A.

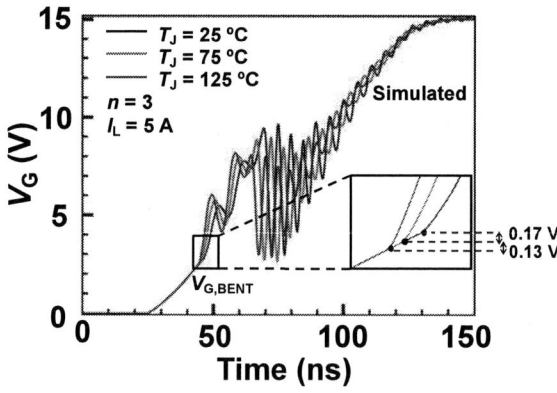

Fig. 5. Simulated V_G waveforms with varied T_J at n = 3 and I_L = 5 A.

Fig. 2 shows a flowchart of the proposed detection method of the bond wire lift-off. During the initial power-on of a power converter, in order to collect two feature points with different temperatures, V_G waveforms are periodically measured m_1 times, $V_{G,BENT}$ and $V_{G,MIN}$ are extracted, and they are written to memory. Then, a regression line is found using the data of $V_{G,BENT}$ and $V_{G,MIN}$, and the regression line is written to memory. After that, during the normal operation of the power converter, V_G waveforms are periodically measured, $V_{G,BENT}$ and $V_{G,MIN}$ are extracted, and an anomaly score is calculated. The anomaly score is the distance between the two feature points and the regression line. When the anomaly score is larger than the predetermined value (AS_{TH}), the bond wire lift-off is detected.

Fig. 3 shows a circuit schematic of the double pulse test for the SiC MOSFET (C3M0060065D, 650V, 37A) to demonstrate the estimation of n using the linear regression of $V_{G,BENT}$ and $V_{G,MIN}$. Table I shows the parameters for the

SPICE simulation of the circuit. In the circuit simulations, n is varied to emulate the bond wire lift-off, and T_J and the load current (I_L) are varied, because T_J and I_L are key parameters to determine V_G during the operation of the power converters.

Fig. 4 shows the simulated V_G waveforms with varied n at T_J = 75 ℃ and I_L = 5 A. $V_{G,MIN}$ clearly changes with n, while $V_{G,BENT}$ does not depend on n, because the R_S and L_S change with n as shown in Table I and $V_{G,BENT}$ is determined by V_{TH}.

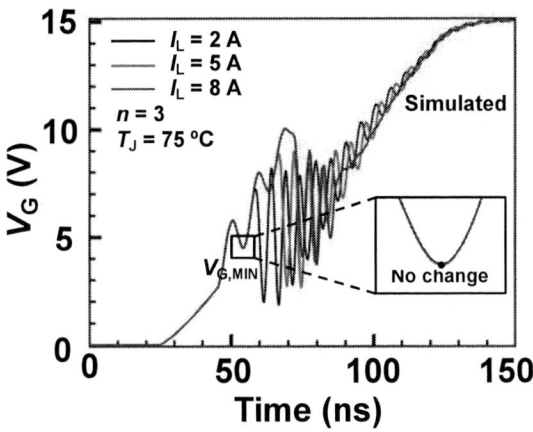

Fig. 6. Simulated V_G waveforms with varied I_L at $n = 3$ and $T_J = 75$ °C.

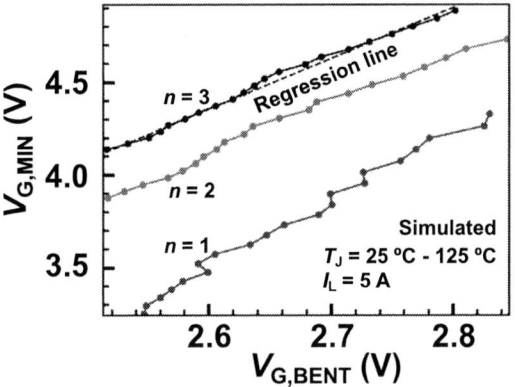

Fig. 7. Simulated relationship between $V_{G,MIN}$ and $V_{G,BENT}$ with varied T_J and n at $I_L = 5$ A.

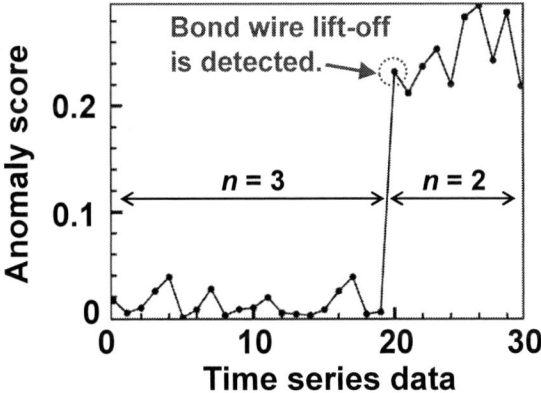

Fig. 8. Simulated time series data of anomaly score at $I_L = 5$ A.

Fig. 5 shows the simulated V_G waveforms with varied T_J at $n = 3$ and $I_L = 5$ A. Both $V_{G,BENT}$ and $V_{G,MIN}$ change with T_J, because T_J changes V_{TH}. Fig. 6 shows the simulated V_G waveforms with varied I_L at $n = 3$ and $T_J = 75$ °C. Both $V_{G,BENT}$ and $V_{G,MIN}$ do not depend on I_L. The above results show that n can be estimated by a regression analysis of $V_{G,BENT}$ and $V_{G,MIN}$, because $V_{G,BENT}$ depends on T_J, and $V_{G,MIN}$ depends on both n and T_J.

Fig. 7 shows the simulated relationship between $V_{G,MIN}$ and $V_{G,BENT}$ with varied T_J and n at $I_L = 5$ A. When T_J increases, both $V_{G,BENT}$ and $V_{G,MIN}$ decrease, because V_{TH} is reduced.

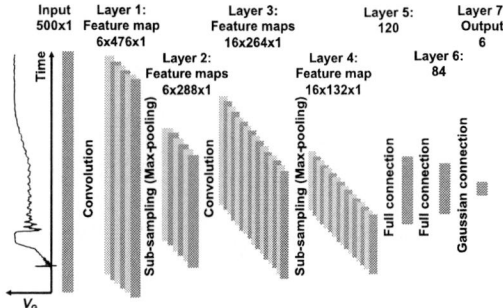

Fig. 9. CNN network adopted for measurement.

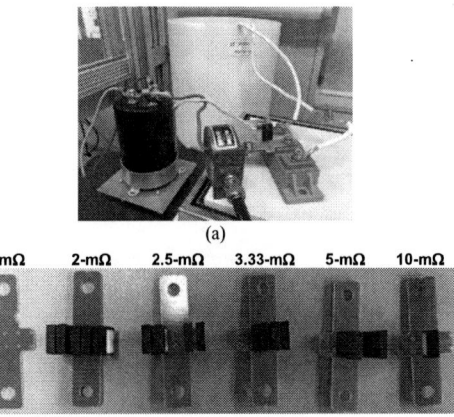

Fig. 10. (a) Measurement setup of IGBT. (b) Varied R_E to emulate the bond wire lift-off.

When n increases, only $V_{G,MIN}$ decreases as shown in Fig. 4. The regression line for $n = 3$ to calculate the anomaly score is also shown. When the bond wire lift-off occurs, n changes from 3 to 2, the distance between the data and the regression line increases, and the anomaly score increases, thereby detecting the bond wire lift-off. Fig. 8 shows the simulated time series data of the anomaly score at $I_L = 5$ A. In each data point, T_J is randomly changed between 25 °C and 125 °C. n is 3 in the first 20 data points, while n is 2 in the last 10 data points. When n changes from 3 to 2, the anomaly score suddenly increases from sub-0.04 to above-0.2, thereby detecting the bond wire lift-off.

III. ESTIMATION OF R_E, T_J, I_C, AND V_{TH} USING CONVOLUTIONAL NEURAL NETWORK

In order to detect the power device degradation, in this chapter, the estimation of R_E, T_J, I_C, and V_{TH} using CNN with V_G waveforms of an IGBT for input is shown using both simulations and measurements [5]. In [5], the estimation of R_E, T_J, I_C, and V_{TH} using simulations and the estimation of R_E and I_C using measurements are shown. In this paper, the measurements are shown.

Fig. 9 shows the CNN network architecture adopted for a measurement. The input is a voltage waveform of the gate driver output during the turn-on of IGBT. The input vector length is chosen to be 500, which corresponds to 6-ns cycle sampling for the entire turn-on process of 3-μs. The output is the categorized in 6 classes.

Fig. 10 shows the measurement setup with an IGBT (2MBI100TA-060-50, 600 V, 100 A) and varied R_E to emulate the bond wire lift-off. R_E is varied by changing the number of resistors in parallel. The measured waveforms are shown in Fig. 11. It is seen from the measured waveforms that it is

337

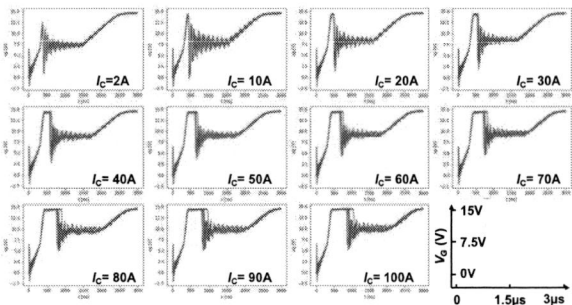

Fig. 11. Measured V_G waveforms with varied I_C.

Table II Six classes of R_E and I_C.

(a) R_E: Emitter resistance

Class #	0	1	2	3	4	5	Total
Label: R_E=	0mΩ	2mΩ	2.5mΩ	3.33mΩ	5mΩ	10mΩ	
# of waveforms	1100	1100	1100	1100	1100	1100	6600

(b) I_C: Collector current

Class #	0	1	2	3	4	5	Total
Label: I_C range	0A	15A	30A	45A	60A	75A	100A
# of waveforms	1200	600	1200	600	1200	1800	6600

Table III Estimation results using CNN.

(a) R_E: Emitter resistance

Success rate: 0.995

		Correct class						Ratio
		0	1	2	3	4	5	
Guessed class	0	275	3	0	0	0	0	0.989
	1	0	265	0	0	0	0	1
	2	0	0	264	0	0	0	1
	3	0	0	0	262	4	0	0.985
	4	0	0	0	1	279	0	0.996
	5	0	0	0	0	0	297	1
	Ratio	1	0.989	1	0.996	0.986	1	

# of epochs: 74

(b) I_C: Collector current

Success rate: 1.000

		Correct class						Ratio
		0	1	2	3	4	5	
Guessed class	0	315	0	0	0	0	0	1
	1	0	144	0	0	0	0	1
	2	0	0	272	0	0	0	1
	3	0	0	0	155	0	0	1
	4	0	0	0	0	298	0	1
	5	0	0	0	0	0	466	1
	Ratio	1	1	1	1	1	1	

# of epochs: 213

difficult to estimate R_E and I_C using human eyes. The machine recognition is needed for estimating the parameters. R_E and I_C are categorized into 6 classes this time as shown in Table II. The results of estimation using the CNN approach are tabulated in Table III. The success rate is very high amounting up to 99.5%. The number of epochs in the table shows the number of deep learning optimization loops needed for the entire learning process. Using i7 Intel processor with the clock rate of 2.5-GHz, one epoch needs approximately 1-s, which is sufficiently fast to be practical.

IV. CONCLUSIONS

Two power device degradation estimation methods by the machine learning of gate waveforms are proposed and demonstrated with simulations and measurements. Introducing the machine learning into the power electronics will be a key to increase the reliability of the power devices and to reduce the cost for various sensors.

REFERENCES

[1] V. Smet, F. Forest, J.-J. Huselstein, F. Richardeau, Z. Khatir, S. Lefebvre and Berkani, "Ageing and failure modes of IGBT modules in high-temperature power cycling," IEEE Trans. Ind. Electron. vol. 58, no. 10, pp. 4931-4941, Oct. 2011.

[2] Z. Wang, B. Tian, W. Qiao and L. Qu, "Real-time aging monitoring for IGBT modules using case temperature," IEEE Trans. Ind. Electron. vol. 63, no. 2, pp. 1168-1178, Feb. 2016.

[3] X. Zeng, Z. Li, W. Gao, M. Ren, J. Zhang, Z. Li and B. Zhang. "A novel virtual sensing with artificial neural network and k-means clustering for IGBT current measuring," IEEE Trans. Ind. Electron. vol. 65, no. 9, pp. 7343-7352, Sep. 2018.

[4] N. Baker, S. M. Nielsen, F. Iannuzzo and M. Liserre, "Online junction temperature measurement using peak gate current," IEEE Applied Power Electronics Conference and Exposition, pp. 1270-1275, 2015.

[5] K. Miyazaki, Y. Lo, A. K. M. M. Islam, K. Hata, M. Takamiya, and T. Sakurai, "CNN-based Approach for Estimating Degradation of Power Devices by Gate Waveform Monitoring," IEEE International Conference on IC Design and Technology, pp. 104-107, June 2019.

Machine Learning Prediction of Defect Formation Energies in a-SiO$_2$

Diego Milardovich[1], Markus Jech[1], Dominic Waldhoer[1,2], Michael Waltl[1,2], and Tibor Grasser[1]

[1]Institute for Microelectronics, Technische Universität Wien,
Gußhausstraße 27–29, 1040 Vienna, Austria
[2]Christian Doppler Laboratory for Single-Defect Spectroscopy in Semiconductor Devices,
Gußhausstraße 27–29, 1040 Vienna, Austria
E-mail: [milardovich | jech | waldhoer | waltl | grasser]@iue.tuwien.ac.at

Abstract—Due to its stochastic nature, the calculation of defect formation energies in amorphous structures is a CPU-intensive task. We demonstrate the use of machine learning to predict defect formation energies to significantly minimize the number of required calculations. Different combinations of *descriptors* and machine learning algorithms are used to predict the formation energies of hydroxyl E' center defects in amorphous silicon dioxide structures. The performance of each combination is analyzed and compared to results obtained from direct *ab initio* calculations.

I. INTRODUCTION

Ab initio methods, particularly those based on density functional theory (DFT), are routinely used in material science to calculate electronic and structural properties. However, these methods have the disadvantage of being highly computationally expensive. This disadvantage limits the use of ab initio methods to small systems (on the order of few hundred atoms) and short time scales (on the order of tens of ps).

The design of modern electronic devices heavily relies on reliability considerations of new fabrication processes and emerging materials (e.g. 2D materials). In this regard, DFT studies can provide valuable information on underlying defects and physical mechanisms impacting the device reliability. Even so, the high computational costs often prohibit a direct employment of DFT within process or device simulations, particularly for amorphous structures like gate dielectrics, where statistical data has to be gathered.

A promising solution is offered by using machine learning (ML) to reduce the computational demands required to study certain aspects of device reliability. In this work, we study the possibility of using ML models to calculate the formation energies of hydroxyl E' center defects in amorphous silicon dioxide (a-SiO$_2$) structures [1]. Studying the behavior of these defects, especially their formation during device processing, is of great importance for the development of modern microelectronics, since they are suspected to be responsible for bias temperature instability (BTI) and random telegraph noise (RTN) in MOS transistors [2–4].

Since the formation of hydroxyl E' centers depends on the availability of hydrogen, the concentration of these defects is not directly accessible in DFT or ML models. However, it could be derived from a kinetic Monte Carlo (KMC) process model [5], coupled with ML-based on-the-fly prediction of formation energies as a function of the local environment.

II. METHODOLOGY

In order to build a ML model which is able to predict certain electronic properties, it is first necessary to represent the atomistic structure in a way which is compatible with the selected ML algorithm. Such mathematical representation of the structure is called a *descriptor*. The conceptual workflow from structures to the final predictions is shown in Fig. 1.

In this study, we prepared 16 different a-SiO$_2$ structures, each with a total of 216 atoms. These structures were created with LAMMPS [6] by using the ReaxFF [7] force field and the melt-quench technique, as described in [8]. An example of these structures is shown in Fig. 2. Within these structures, 1271 hydroxyl E' centers were created, as shown in Fig. 3. Their formation energies were extracted using DFT; the calculations were performed using the PBE functional [9] in the CP2K software package [10]. The resulting distribution of formation energies can be seen in Fig. 4.

Three popular ML models were used to predict the formation energy of the hydroxyl E' centers: neural network (NN), kernel ridge regression (KRR) and decision tree (DT), as implemented in the scikit-learn package [11]. These models were combined with 2 local descriptors common in the literature: *smooth overlap of atomic positions* (SOAP) [12] and *atom-centered symmetry functions* (ACSF) [13], implemented in the Python package DScribe [14].

It is clear from Fig. 5 that there is an inverse correlation between the formation energy of an hydroxyl E' center defect and the length of the Si-O bond which has to be broken in order to form the defect. In other words, the formation energy is influenced by directly accessible geometric quantities (e.g., bond-lengths and bond-angles). Based on this observation, we propose a simple geometry-based descriptor: *bond-lengths and bond-angles* (BLBA).

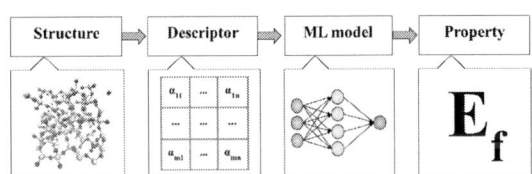

Fig. 1: Workflow to predict a property from an atomistic structure. The structure must be represented by a descriptor. Then, a ML model can be trained and subsequently be used to predict the desired property in new atomistic structures.

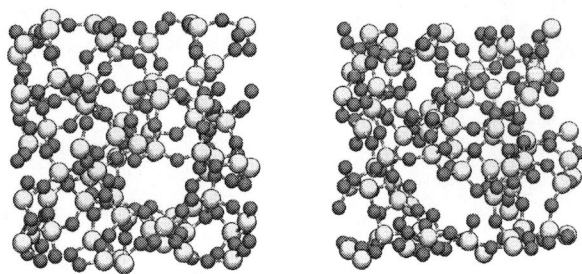

Fig. 2: Example of two amorphous a-SiO$_2$ structures used in this work to train and test the ML models. Each structure contains 216 atoms and a total of 16 such structures are used in this work.

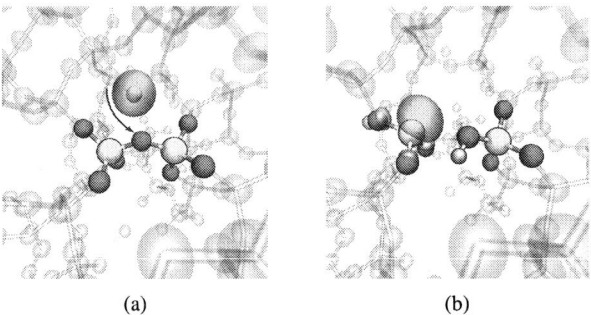

(a)   (b)

Fig. 3: Hydrogen interaction with the a-SiO$_2$ matrix (a) can lead to formation of a hydroxyl group, as well as breakage of a strained Si-O bond, resulting in a hydroxyl E' center defect (b). The blue bubbles show the spin-density associated with interstitial hydrogen and the hydroxyl E' center defect, respectively.

Our BLBA descriptor is constructed according to the following steps:

1) Create the neighbor list of Nth order. The starting point of this list is the Oxygen atom for which we would like to predict the formation energy of an hydroxyl E' center defect. The first step is to list the atoms bonded to this Oxygen atom. Then, every step consists in listing the atoms bonded to the atoms found in the previous step, until a maximum number of iterations, N, is reached (in our study, N = 3).

2) Extract the bond-lengths for every pair of bonded atoms in the neighbor list created in step 1. Use these values to create a bond-lengths vector. The bond lengths within one iteration are ordered in descending order. Note that the values from different iteration steps are kept separated to retain information about the distance to the defect site.

3) Extract the bond-angles between triplets in the neighbor list created in step 1. Use these values to create a bond-angles vector. As for the bond-lengths, the angles are arranged in descending order.

4) The BLBA descriptor is created by merging the bond-lengths vector and bond-angles vector, in this order. Therefore, this descriptor is a vector which consists of the properly ordered bond-lengths and bond-angles of the neighbourhood of the possible defect site.

Although our BLBA descriptor cannot describe the local environment in the same level of detail of SOAP and ACSF descriptors, it displays very valuable properties: since it contains only physical values which are relative to the possible defect site, our descriptor is invariant to spatial translations and rotations of the coordinate system. Moreover, since its components are ordered, it is also invariant with respect to permutation of indexes. Finally, it is highly compact, i.e. it contains sufficient information to be used for the prediction of the formation energies of hydroxyl E' center defects, while keeping its size and complexity to a minimum.

The performance of our BLBA descriptor will also be considered, as a demonstration of the potential of simpler geometry-based descriptors. The 1271 hydroxyl E' center defects were randomly divided into a training and a testing data set, with a ratio of 4:1. Every permutation of descriptor and ML model was trained with the training set and used to predict the values for the testing set. In every case, the mean absolute error (MAE) was calculated between the predictions and the targets.

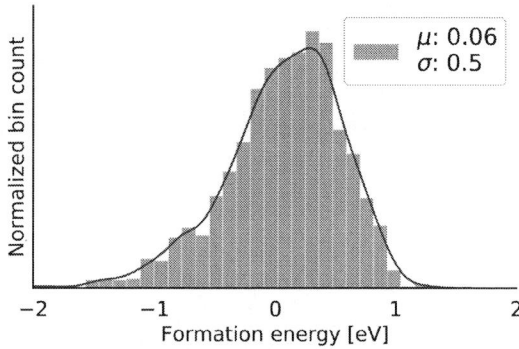

Fig. 4: Formation energies of hydroxyl E' center defects in a-SiO$_2$, obtained with DFT calculations, together with the mean (μ) and standard deviation (σ). The broad distribution is due to the amorphous nature of the structures.

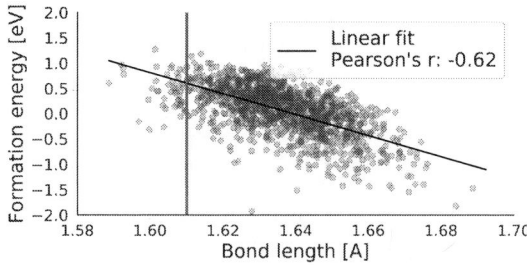

Fig. 5: Correlation between formation energies and bond-lengths. The formation energy decreases for larger bond lengths, so defects preferably form at strained Si-O bonds. The red line indicates the Si-O bond-length in alphaquartz [15].

340

III. RESULTS AND DISCUSSION

The results of this study are summarized in Fig. 6, where the prediction errors for the formation energies in the testing set are presented, together with its MAE, for every combination of descriptor (SOAP, ACSF and BLBA) and ML model (NN, KRR and DT) studied in this work.

Ultimately, the worst results were obtained with the ACSF descriptor in conjunction with a DT model, which presents a MAE of $0.39\,eV$. On the other hand, the best results were obtained with the SOAP descriptor in conjunction with a NN model, which presents a MAE of $0.26\,eV$. This was expected, since SOAP describes the local environment in a higher level of detail. In our study, SOAP required 108 parameters to describe the local environment of the atom of interest, while ACSF required 36 and BLBA 17. A logical explanations is that the information contained in these extra parameters allows the SOAP descriptor to achieve more accurate predictions of formation energies of defects in the atomistic structures.

Overall, BLBA performed similarly to SOAP and ACSF. However, it is important to note the following differences:

- BLBA is more compact than SOAP and ACSF, since it is able to properly describe the relevant local environment with a significantly lower amount of information.

- SOAP and ACSF have several additional parameters particularly designed to represent the chemical surroundings. This makes them more accurate, but it also means that the user requires a deeper understanding of the underlying atomistic nature. Moreover, such necessity to adjust parameters which depend on the specific atomistic structure makes them less accurate when this information is not available. On the contrary, BLBA does not require to adjust any parameter (with the exception of N, the number of iterations, but this adjustment does not require any previous knowledge of the atomistic structure). Therefore, BLBA might perform better in atomistic structures of which there is a limited previous knowledge.

- Every element of BLBA represents a physical property of the atomistic structure (since each element of the descriptor represents a particular bond-length or bond-angle). This does not only make it simpler than SOAP and ACSF, it also gives BLBA a much higher interpretability.

- Our BLBA descriptor performed slightly better than the SOAP and ACSF descriptors when combined with a KRR or DT model. We believe that this could be explained given the fact that our BLBA descriptor is able to describe the local atomistic environments with a lower amount of parameters. Therefore, it performs better with reduced data-sets, by avoiding overfitting. However, more complex ML models, such as our NN model, might make use of the extra information provided by the SOAP descriptor to increase the formation energy predictions accuracy.

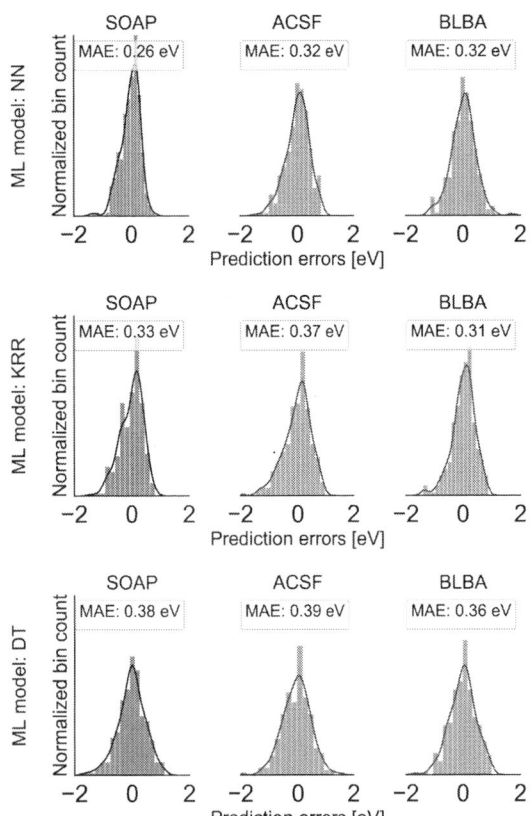

Fig. 6: Error distributions and MAE in the prediction of formation energy of hydroxyl E' center defects in a-SiO$_2$ for the different combinations of descriptors (SOAP, ACSF and BLBA) and ML models (NN, KRR and DT).

Even though there are noticeable differences in the accuracy of the defect formation energy predictions, it is important to remember that the predictions of all the solutions studied in this work are considered to be accurate enough for our practical application.

In our particular case, we are interested in using a combination of a descriptor and a ML model to predict the formation energies of defects in a-SiO$_2$. Such predictions will be combined with an KMC method in future works, in order to assess the likelihood of defects formation in specific conditions. In other words, this work can be considered as a first step towards a more sophisticated ultimate goal: to predict reaction barriers by using a combination of a KMC method and a ML model. It is also important to remember that the accuracy of the predictions is highly dependent on the atomistic structures we consider and the electronic property we want to extract or predict from them. Therefore, the results obtained in this work do not necessarily imply that the prediction accuracy of these descriptors would be unchanged if we apply them to different materials or different electronic properties from the one considered here. In case of applying a descriptor-based ML solution to predict a certain electronic property, it is crucial to consider different combinations of descriptors and ML models and to analyze the results obtained from them before choosing the best combination.

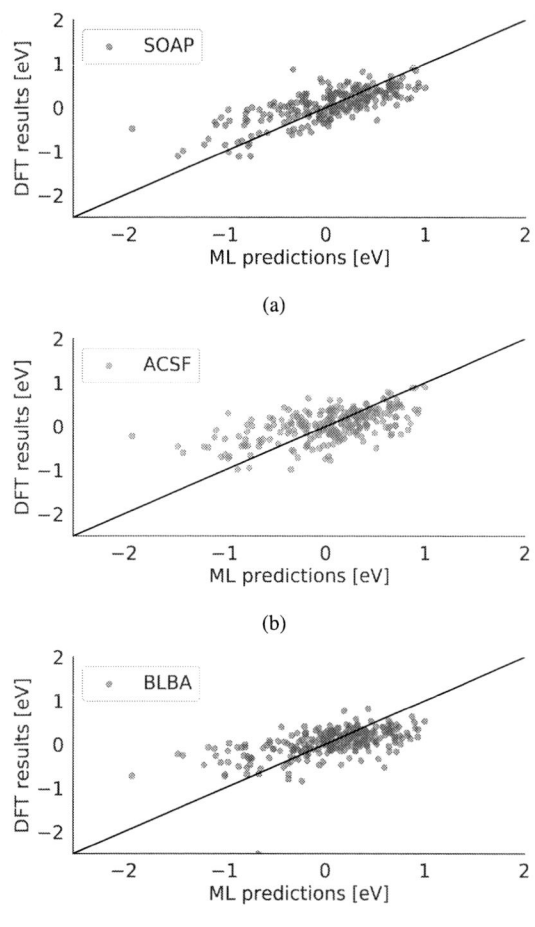

(a)

(b)

(c)

Fig. 7: Correlation between the DFT calculation results and the ML-based predictions using (a) SOAP, (b) ACSF and (c) BLBA descriptors (in all cases, combined with a neural network). All descriptors perform similarly. However, SOAP shows a slightly better correlation in general, particularly for negative values.

In order to allow a clear comparison between our BLBA descriptor and the well-established SOAP and ACSF descriptors, a correlation between the DFT results and the predictions made by using each of these descriptors is shown in Fig. 7. In all cases, the descriptors were combined with a NN model, since it was the ML model which showed the best results in the previous section.

It can be noted that there is a clear linear correlation between the DFT results and the predictions made by all the descriptors. The fact that all distributions are roughly centered around the identity line shows there is no considerable systematic error in the predictions and no strong overfitting of the ML model.

The SOAP descriptor yields slightly better results than ACSF and BLBA, particularly for negative formation energies. However, all descriptors are accurate enough for our practical application.

IV. CONCLUSIONS

The study of reliability in modern electronic devices requires the use of simulation techniques in order to assess the effect of new materials and changes in the fabrication processes.

The direct study of these reliability issues posed by new materials and changes in the fabrication processes using DFT is currently a computational challenge due to the high costs. Therefore, computationally cheaper solutions must be found in order to substitute the direct application of DFT and other ab initio methods.

In this context, there are several practical applications in which such DFT calculations could be aided or even replaced by computationally inexpensive ML models combined with well-established descriptors, as shown in this work and in other examples in the literature [16–18].

Novel descriptors can be developed for specific applications, in the same way in which our BLBA descriptor was developed to be used in the ML-based prediction of formation energies of hydroxyl E' center defects in a-SiO$_2$. This approach could prove particularly useful in the study of amorphous materials, where large statistics are needed. Moreover, this solution could be applied to other defects, as well as to new materials, providing an enormous potential to aid the development of novel electronic devices.

Finally, apart from providing computationally inexpensive solutions to practical problems in the study of modern electronic devices reliability, descriptor-based ML solutions show an enormous potential to provide a deeper understanding of the mechanisms by which defects are formed and of the relationships between atomistic structure properties and the formation of such defects.

V. ACKNOWLEDGMENTS

This project has received funding from the European Union's Horizon 2020 research and innovation programme under grant agreement No. 871813, within the framework of the project Modeling Unconventional Nanoscaled Device FABrication (MUNDFAB).

REFERENCES

[1] Al-Moatasem El-Sayed et al. *Phys. Rev. Lett.*, 114, 2015.
[2] Tibor Grasser et al. *IEEE International Electron Devices Meeting*, pages 21.1.1–21.1.4, 2014.
[3] Yannick Wimmer et al. *Proc. R. Soc*, 472: 20160009, 2016.
[4] Wolfgang Goes et al. *Microelectron. Reliab.*, 87:286–320, 2018.
[5] Antonino La Magna et al. *Phys. Status Solidi*, 216, 2018.
[6] Steve Plimpton. *J. of Comp. Phys.*, 117:1–19, 1995.
[7] Adri C. T. van Duin et al. *J. Phys. Chem.*, 105:9396–9409, 2001.
[8] Al-Moatasem El-Sayed et al. *Solid State Electron.*, 109:68–71, 2013.
[9] John P. Perdew et al. *Phys. Rev. Lett.*, 77, 1996.
[10] Jürg Hutter et al. *WIREs Comput Mol Sci*, 4:15–25, 2013.
[11] Fabian Pedregosa et al. *Journal of ML Res.*, 12:2825–2830, 2011.
[12] Albert P. Bartók et al. *Phys. Rev. B*, 87, 2013.
[13] Jörg Behler. *Journal of Chem. Phys.*, 134, 2011.
[14] Lauri Himanen et al. *Comp. Phys. Com.*, 106949, 2019.
[15] T. Demuth et al. *J. Phys.: Condens. Matter*, 11:3833–3874, 1999.
[16] Felix Faber et al. *Int. Journal of Quantum Chem.*, 115:1094–1101, 2015.
[17] Shin Kiyohara et al. *Science Advances*, 2(11), 2016.
[18] Atsuto Seko et al. *Phys. Rev. B*, 95, 2017.

15-3

Novel Optimization Method using Machine-learning for Device and Process Competitiveness of BCD Process

Junhyeok Kim[1], Jae-Hyun Yoo[1], Jaehyun Jung[2], Kwangtea Kim[2], Jaehyun Bae[1], Yoon-suk Kim[1],
OhKyum Kwon[2], UiHui Kwon[1], DaeSin Kim[1]

[1] Data and Information Technology Center, [2] 8inch PA Group, Foundry Manufacturing Technology Center
Device Solution Business, Samsung Electronics Co., Ltd.,
Hwasung-si, Gyeonggi-do, Republic of Korea.
* Email Address:j_hyeok.kim@samsung.com (J.Kim).

Abstract— The novel optimization method for BCD(Bipolar-CMOS-DMOS) process development based on Machine Learning(ML) and statistical process modeling considering the entire wafer variation is proposed to improve the device and process competitiveness. The self-align PBODY process is used for high-performance N-type Lateral Diffused Metal Oxide Semiconductor(NLDMOS) in BCD process and it also is related to stability in PMIC operation. The process modeling embracing the performance and the stability of LDMOS is performed with TCAD using inline data. For the development of BCD process, the PBODY process parameters are optimized through the ML algorithms and the condition is verified with TCAD and silicon test. Finally, we can secure new low voltage NLDMOS with the improved performance and stability respectively for without any degradation in the new 0.13μm BCD process.

Keywords—BCD process, optimization, machine-learning, LDMOS, statistical modelling

I. INTRODUCTION

Recently, the PMIC becomes more important as the market of portable electronic products growing up and expanding. PMIC performs a role of supplying stable power to a number of blocks and converting an input voltage to fit each block. The BCD process providing various LDMOS for 8V to 40V applications is well-known for a suitable process for PMIC [1,2]. However, the BCD process has some underlying difficulties owing to the limited number of layers relating to the cost of production [3]. Diverse types of LDMOS and hundreds of devices are manufactured simultaneously with the same IIP layer and different pattern sizes in BCD process, which is a root cause of increased process instability. Therefore, the simulation of not only the performance of the device but also the variation of the process is requested to develop the BCD process.

In this paper, the novel simulation flow using machine-learning and statistical process modelling that suggested in Fig.1 is proposed to improving both performance and process stability of new low voltage NLDMOS with 0.13μm BCD process, which is a suitable of a device for mobile application using 3.3 ~5V battery in Fig. 2 [4].

II. METHOD

A. Statistical process modeling

LDMOS is made with the Self-align BODY IIP process that creates the stable short channel regions and forms the gate poly pattern simultaneously shown in Fig. 3(a) [5]. In order to decrease the substrate resistance (Rsub) for the reliability of LDMOS, higher energy and more dosage are

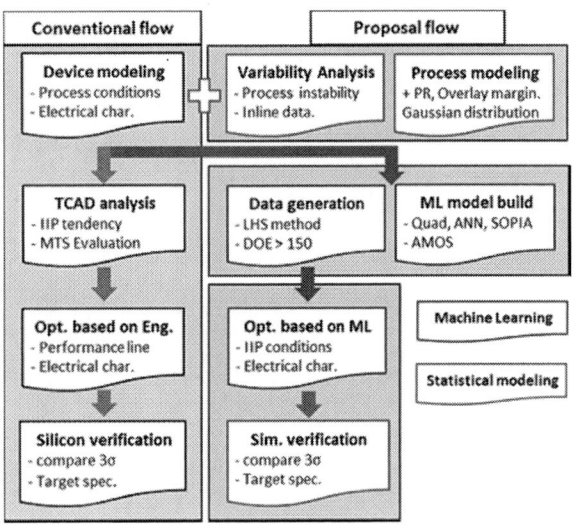

Fig. 1. New simulation methodology using machine learning and statistical process modeling

used in self-align BODY accompanied with the thicker photoresist (PR) [6,7]. However, the excessive PR etching is inevitable to form the precise pattern with thick PR and it causes the PR slope variation across the entire wafer. Therefore, the shapes of PR patterns are bound to be different depending on the location for the reason that a variety type of LDMOS are formed at the same process step. Particularly, the difference between inner and outer PR shapes of PBODY process is the outstanding issue of LDMOS, mainly used as a form of multi-finger structure. Figure 3(b) shows the Vtgm/Idsat distribution by such variation, especially between inner-outer transistors due to PR pattern density [8].

The relation between the different PR slopes and the Vtgm/Idsat distribution is verified with TCAD simulation. The channel doping profile determining Vtgm of LDMOS varies with PR slope, due to the difference of boron concentration penetrated through PR shown in Fig. 4 and 5.

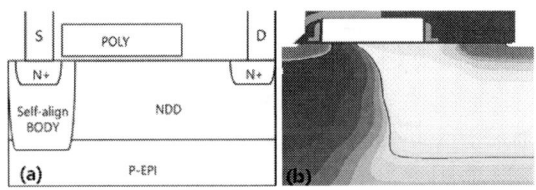

Fig. 2. NLDMOS: cross-section (a) and simulation structure (b).

For co-optimization of process variation and performance, the statistical process modeling for the variation of PR slope is implemented based on inline Scanning Electron Microscope (SEM) image and Critical Dimension(CD) data. The angles of PR slope are calculated from the difference between the Top CD and the bottom CD of PR. The mean value and the standard deviation of PR slope are obtained built on the inline data, as shown in Fig. 6. The PR slope distribution is assumed that comes from Gaussian distribution. Then, Vtgm/Idsat distribution of LDMOS is calibrated with the PR distribution on TCAD. The other variation factors like the overlay and the open area of IIP mask are ignored because the self-align scheme has the advantage that makes such variations smaller.

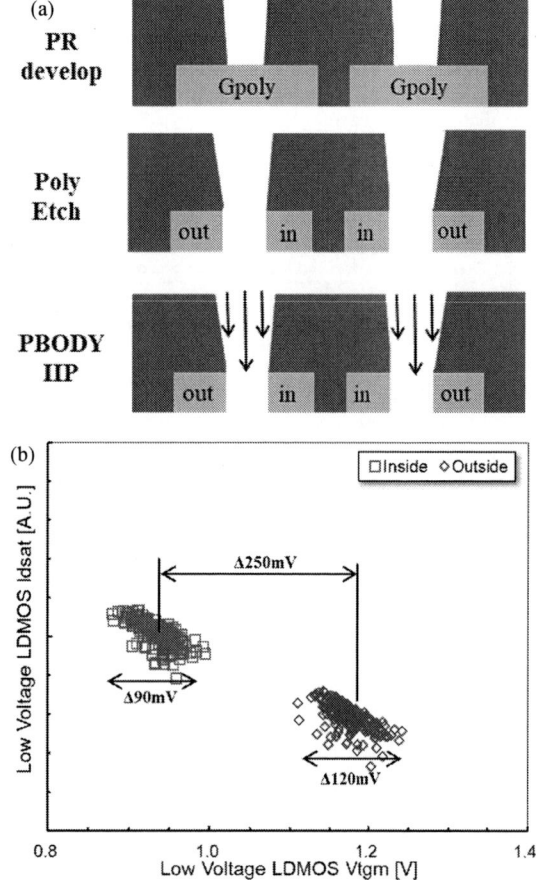

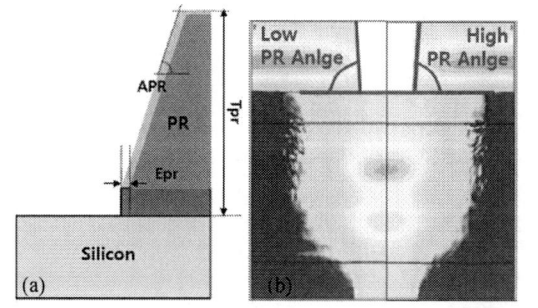

Fig. 3. PR variation of self-align BODY after etch in inner-outer pattern (a) and Vtgm/Idsat distribution including layout effect of LDMOS (b).

Fig. 4. Self-align BODY PR half-structure (a) and TCAD modeling IIP profile including PR variation (b).

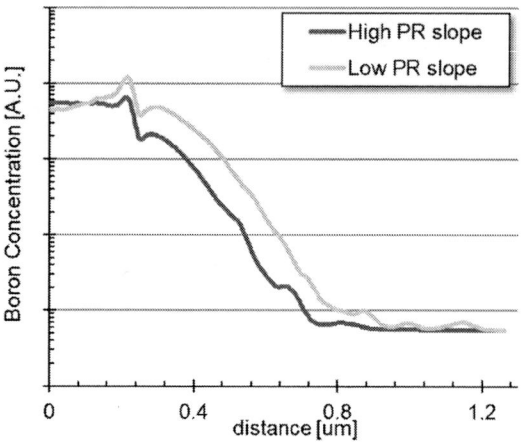

Fig. 5. Lateral Boron Concentration after anneal from TCAD simulation form the channel of the low voltage LDMOS

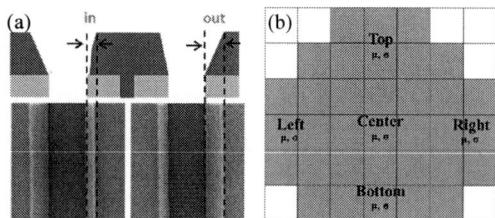

Fig. 6. Top view of self-align BODY PR after implant (a) and entire wafer CD data map example after PR developing (b)

B. Machine learning

In optimization for the electrical characteristics and robustness of the process, Automatic Multi-objective Optimization solution(AMOS) is used with PBODY IIP parameters for the new NLDMOS, because PBODY IIP dosage not affect other devices' characteristics. AMOS is an in-house tool that performs automatically data generations, data model learning, and optimization at once, based on ML.

Firstly, 150 conditions that consist of IIP parameters were selected for Latin Hypercube Sampling (LHS) method [9]. And, 5 electrical parameters of both types of low and high voltage LDMOS related to performance and reliability are trained through AMOS with TCAD data shown in Fig. 7. Then, several regression models like quadratic, Artificial neural network (ANN), Samsung open platform for Intelligence Application auto regression model (SOPIA) were evaluated. Finally, SOPIA model which totally records the high test R^2 was chosen among those for development for new NLDMOS [10]. As shown in Fig. 8, the regression result is attained R^2 more than 0.98 and has a good match with the TCAD data.

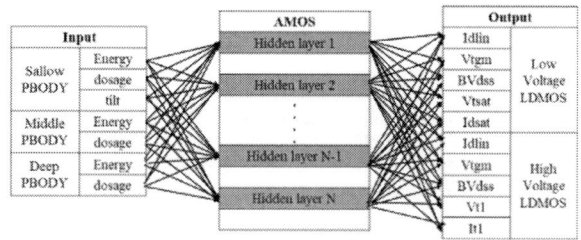

Fig. 7. AMOS regression diagram between PBODY IIP parameters and electric parmaters of two types of LDMOS.

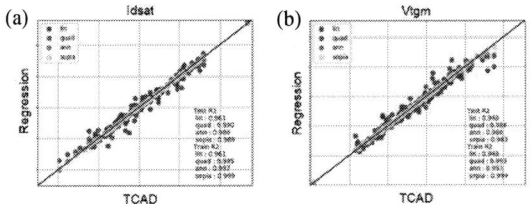

Fig. 8. AMOS regression consistency: Idsat (a) and Vtgm (b) of LDMOS

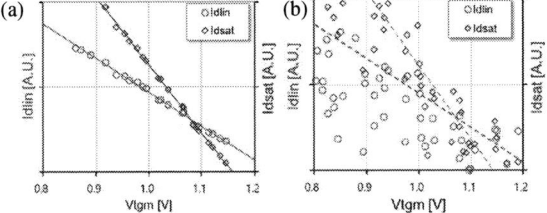

Fig. 9. The simulated data extracted from conventional DOE (a) and the simulated data obtained through LHS method (b).

III. SIMULATION

The new simulated data obtained through our method shows the improved conditions compare to the current performance line while the new simulated data obtained through our method shows the simulated data extracted from conventional DOE is placed on the same performance line with the current silicon data shown in Fig 9. It means there is a key to develop the device under conditions that changed the existing our perspective. The condition optimized with ML indicated that lower energy of deep PBODY is enough to satisfy reliability spec. Also, Shallow PBODY dosage increases to compensate for lowered deep PBODY IIP in the view of PBODY Vtgm. Through the machine learning model, the shallower and stronger IIP conditions are selected in terms of reliability and performance.

And the ΔVtgm distribution was evaluated simultaneously with the performance of the condition using the statistical process modeling. ΔVtgm is compared correspondingly with standard deviations of Vtgm distribution derived from PR slope variation. It is founded that ΔVtgm of low voltage LDMOS is more affected by weakening deep PBODY IIP than increasing shallow PBODY IIP dosage. Because NDD IIP dosage is much lower than PBODY IIP, a small portion of deep PBODY IIP penetrated PR change the length of the channel, causing a variation in Vtgm.

Finally, the proposal condition shows the performance is improved 5% and ΔVtgm decreases 30% compared to current NLDMOS shown in Fig. 10 and 11. Using the new simulation method, the optimized condition that enhances the stability and the performance is obtained at once, whereas the conventional method might need a few steps of iteration to optimize both of them.

IV. EXPERIMNET RESULT

Figure 12 and Figure 13 show the silicon data for two test conditions of new low voltage NLDMOS, slightly changed from the proposal condition in the light of productivity. These results also show a similar trend with the simulated one, although there is a little difference.

The difference between inner and outer transistor's Vtgm is reduced a level lower than 1/5 to existing silicon data in both of two cases as shown in Fig. 12. In the view of performance,

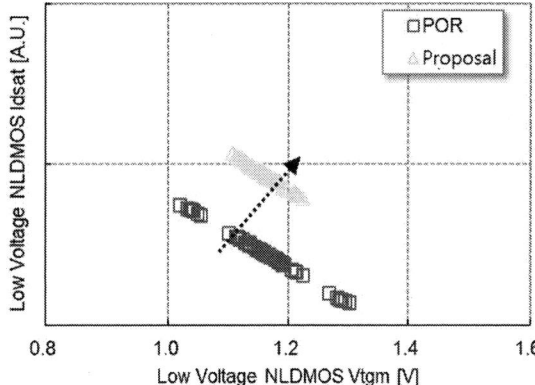

Fig. 10. ML optimization results of NLDMOS performance .

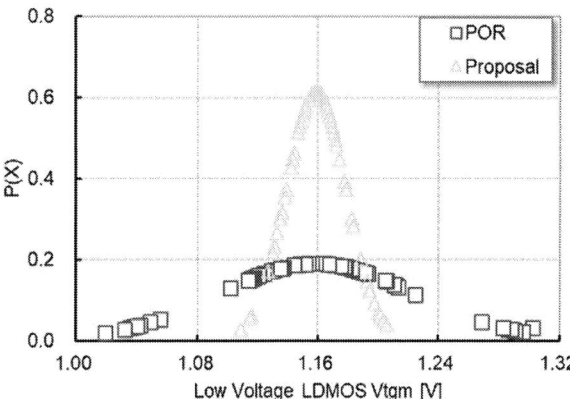

Fig. 11. ML optimization results of NLDMOS Vtgm distribution.

Idsat of case #2 is higher than POR and case #1 at the same Vtgm as shown in Fig. 13. Because the low voltage LDMOS operates at high gate bias, Idsat characteristic is also important not only Ron.sp, representing the operation at a low gate bias.

From this work, the new low voltage NLDMOS has improved the performance of 5% and process stability (ΔVtgm) from 270mV to 44mV.

V. CONCLUSION

The new optimization method that can be satisfied with the device performance and the process stability is proposed. The PBODY process, which is the key process in manufacturing LDMOS, cause dispersion of device electrical characteristic, and it is statically modeled with TCAD simulation.

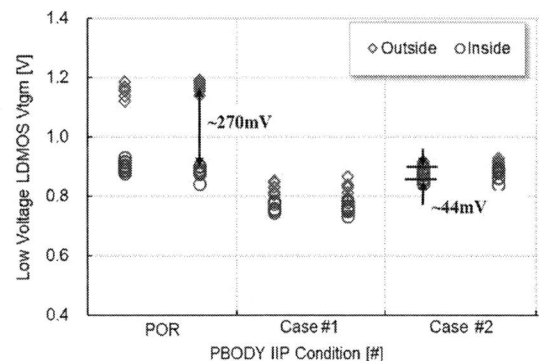

Fig. 12. Vtgm distribution of ML optimized LDMOS IIP condition

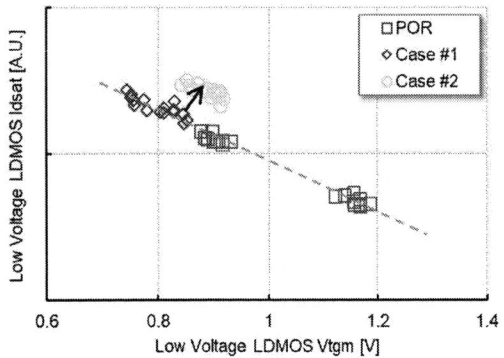

Fig. 13. Vtgm and Idsat graph of ML optimized LDMOS IIP condition related to performanc.

Especially, the relation between PR slope variation and Vtgm distribution is verified using the entire wafer's in-line data and TCAD analysis. For development of LDMOS in BCD process, PBODY IIP parameters are trained and optimized with AMOS, which is an in-house tool for ML. Also, the condition selected by Machine-learning is tested in BCD process and shows a good match with simulation data.

Machine-leaning and statistical process modeling are applied to the new method and the effectiveness of the method is verified through TCAD and silicon analysis. Finally, we obtain the improved result for performance and ΔVtgm of new LDMOS and enhanced the competitiveness of the BCD process.

REFERENCES

[1] KO, Kwang-Young, et al. BD180LV-0.18 μm BCD Technology with Best-in-Class LDMOS from 7V to 30V. In: 2010 22nd International Symposium on Power Semiconductor Devices & IC's (ISPSD). IEEE, 2010. p. 71-74.

[2] KO, Kwang-Sik, et al. HB1340-Advanced 0.13 um BCDMOS technology of complimentary LDMOS including fully isolated transistors. In: 2013 25th International Symposium on Power Semiconductor Devices & IC's (ISPSD). IEEE, 2013. p. 159-162.

[3] Rose M, Bergveld HJ. Integration trends in monolithic power ICs: Application and technology challenges. IEEE Journal of Solid-State Circuits. 2016 May 30;51(9):1965-74.

[4] Chang H, et al. Advanced 0.13 um smart power technology from 7V to 70V. In2012 24th International Symposium on Power Semiconductor Devices and ICs 2012 Jun 3 (pp. 217-220). IEEE.

[5] MAO, Kun, et al. A 0.18-um LDMOS With Excellent Ronsp and Uniformity by Optimized Manufacture Process. IEEE Transactions on Semiconductor Manufacturing, 2018, 32.1: 129-133.

[6] HOWER, P., et al. A rugged LDMOS for LBC5 technology. In: Proceedings. ISPSD'05. The 17th International Symposium on Power Semiconductor Devices and ICs, 2005. IEEE, 2005. p. 327-330.

[7] HOWER, Philip L. Safe operating area-a new frontier in Ldmos design. In: Proceedings of the 14th International Symposium on Power Semiconductor Devices and Ics. IEEE, 2002. p. 1-8.

[8] KIM, Hyun-Woo, et al. Comprehensive analysis of sources of total CD variation in ArF resist Perspective. In: Advances in Resist Technology and Processing XXI. International Society for Optics and Photonics, 2004. p. 254-265.

[9] HELTON, Jon C.; DAVIS, Freddie Joe. Latin hypercube sampling and the propagation of uncertainty in analyses of complex systems. Reliability Engineering & System Safety, 2003, 81.1: 23-69.

[10] Cameron AC, Windmeijer FA. An R-squared measure of goodness of fit for some common nonlinear regression models. Journal of econometrics. 1997 Apr 1;77(2):329-42.

15-4

Real-Time TCAD: a new paradigm for TCAD in the artificial intelligence era

Sanghoon Myung*, Jinwoo Kim, Yongwoo Jeon, Wonik Jang, In Huh, Jaemin Kim, Songyi Han,
Kang-hyun Baek, Jisu Ryu, Yoon-Suk Kim, Jiseong Doh, Jae-ho Kim, Changwook Jeong, Dae Sin Kim

Computational Science and Engineering Team, Data and Information Technology Center,
Device Solution Business, Samsung Electronics Co., Ltd., Gyeonggi-do 18448, Korea.
* Email Address: shoon.myung@samsung.com

Abstract- **This paper presents a novel approach to enable real-time device simulation and optimization. State-of-the-art algorithms which can describe semiconductor domain are adopted to train deep learning models whose input and output are process condition and doping profile / electrical characteristic, respectively. Our framework enables to update automatically deep learning models by estimating the uncertainty of the model prediction. Our Real-Time TCAD framework is validated on 130nm processes for display driver integration circuit (DDI), and 1) prediction time was 530,000 times faster than conventional TCAD, and time spent for process optimization was reduced by 300,000 times compared to human expert, 2) the model achieved average accuracy of 99% compared to TCAD simulation results, and thus, 3) process development time for DDI was reduced by 8 weeks.**

Keywords—TCAD, neural network (NN), convolutional neural network (CNN), recurrent neural network (RNN), Real-time prediction, Real-time optimization, active learning.

I. INTRODUCTION

Over the past few decades, the number of transistors in an integrated circuit (IC) has doubled every two years, as predicted by the Moore`s law. As the size of transistors scales down, predicting non-ideal phenomena, e.g., short-channel effects, becomes important. Generally, engineers rely on the technology computer-aided design (TCAD) simulators to predict and solve these complex physical phenomena [1]. Although TCAD simulators require proper calibration of the model parameters to match simulation results with measurements, they have been successful so far especially in terms of predicting the performance of transistors. However, due to several hours of simulation time, the role of TCAD has been limited to predict transistor performance in the research and development phase, not in the mass production phase. Recently, to address this problem, adopting artificial intelligence (AI) technologies to decrease TCAD simulation turn-around-time (TAT) has been investigated. One of the advantages of employing AI in the manufacturing industry is that it opens a possibility of fast decision making without human intervention, which in turn, can significantly shorten the development cycle. However, current AI studies for manufacturing are not mature enough to realize fully-automated decision making due to the following reasons: First, low model prediction power, most of previous studies[2-4] use some shallow learning (SL) such as artificial neural network, random forest, Gaussian process and so on. One of the limitation of SL models is that they are difficult to interpret the relationship of output values. For instance, SL cannot accurately predict the current-versus-voltage characteristics curves, whose current levels (outputs) are closely correlated with each other, according to semiconductor device theory. Concrete example will be presented in Section III. Second,

trustworthiness problem of the model, it is important to avoid potential risks of blindly trusting the result of the models by checking predictive uncertainty. However, previous studies cannot catch this problem because they focused on improving the performance of the model.

In this paper, we present a novel framework, called Real-Time TCAD (RTT), to enable the automated training and decision making. The contribution of this paper is mainly threefold: First, we proposed the novel algorithm combining recurrent neural network (RNN) / convolutional neural network (CNN) architecture in semiconductor domain. Second, the novel approach to enable automatic update of DL models and to avoid the risks by using calculated uncertainty. Last, we experimentally showed that our RTT model achieves the average accuracy 99% compared to TCAD simulation, and prediction and optimization time is ~ 530,000 times faster and 300,000 times faster than conventional TCAD and human expert, respectively. Process development time for DDI was reduced by 8 weeks.

II. BACKGROUND

A. TCAD Setup

Before illustrating our RTT framework, TCAD simulation 130nm process will be briefly introduced. This process is well-calibrated so that TCAD results show great agreement with measurement data. Simulation dimension is shown in Fig. 1 and the square indicated by dashed line is RTT-process model's output dimension. The input variable consists of 10 variables associated with structure and 24 variables of ion implantation. The process simulator receives input variables as zero-dimensional (0D) scalar values and solve dopant implantation, diffusion, and activation phenomena to calculate final doping profile of MOSFET as a two-dimensional (2D) image. The device simulator receives same input values as the process simulator, and solves electron transport to calculate *I-V* curves as one-dimensional (1D) arrays, and 0D scalar values such as threshold voltage (V_T), breakdown voltage (BV) and etc.

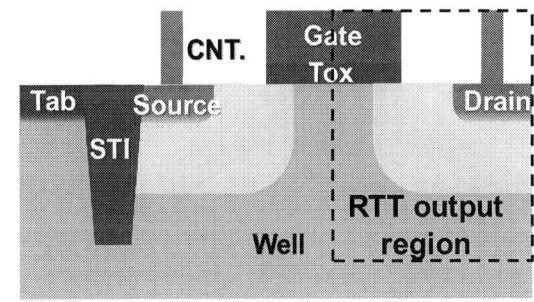

Fig. 1. Simulation structure. Dashed line area is a simulation

B. Pefromance Metric for RTT Process Model

To assess the performance of RTT-process model, Structure SIMilarity (SSIM) and intersection over union (IOU) are used. SSIM is a widely used metric for image quality assessment in the computer vision field [5]. By examining doping profile images with different SSIM values, minimum SSIM required for the RTT-process model is set to 0.94, although for images of large luminance and contrast, >0.96 is required to ensure good image quality [6]. Since SSIM alone cannot capture important regions of MOSFET doping profiles, such as n/p-type junction areas, IOU of important regions (Fig. 2 (a)) is also used. The minimum IOU required is set to 0.9 since the area should match at least 95% (Fig. 2 (b)). Finally, average of SSIM and IOU are used as a score of RTT-process models and the target score is ≥0.92.

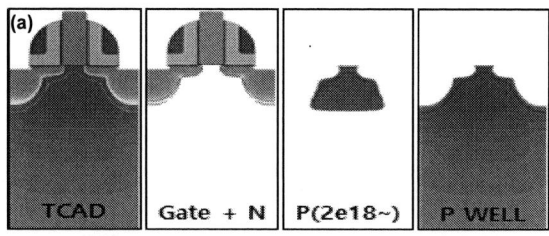

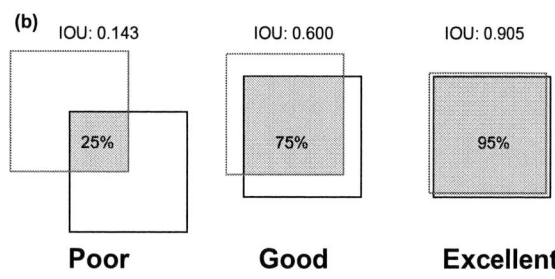

Fig. 2. (a) Classification of regions by doping type. (b) An example of IOU for square boxes.

C. Pefromance Metric for RTT Device Model

For RTT-device model assessment, we introduce mean-absolute-percentage-accuracy (MAPA) as follows:

$$\text{MAPA} = \frac{1}{N} \sum_{i=1}^{N} \left(1 - \left| \frac{Y_i - \bar{Y}_i}{Y_i} \right| \right) \tag{1}$$

where Y and $\bar{Y}$ are true and prediction value, respectively and N is the number of samples. However, MAPA alone cannot capture sensitivity discrepancy between true and prediction values. Hence, we additionally introduce the coefficient of determination (R-squared), which measures how much the model predicted values statistically matches with the true values. To take advantage of both metrics, average of MAPA and R-squared is used as a model performance score of RTT-device models and the target score is set to ≥0.98.

III. METHODOLOGY

A. Preliminary

Generally, TCAD-based semiconductor process optimization is done in a heuristic and iterative manner. For the optimization, firstly TCAD simulations for the process are set up and then iterative adjustment of process conditions until target device characteristics are met. Since a single simulation takes several hours, usually the performance optimization takes several days to a few weeks. This TAT limits the use of TCAD-based optimization results in a mass production phase. To be useful in mass production phase, simulation TAT and error should be within a few seconds and less than 1%, respectively. By adopting deep ensemble approach [7], continual addition of TCAD data and re-training of the models can be fully automated in our framework (Fig. 3).

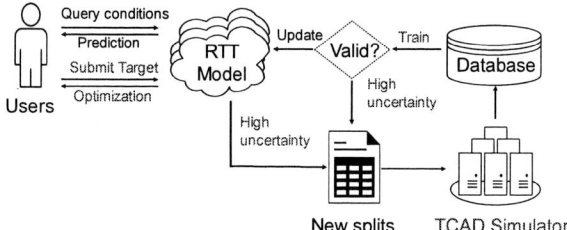

Fig. 3. Illustraion of RTT framework.

B. RTT for Process Simulation

TCAD process simulator takes process conditions as inputs, solves continuum physical models, and outputs distribution of doping concentration as images. Since the output of RTT process model is an image, it is natural to adopt CNN structure for the model [8]. From a baseline CNN structure, several algorithm engineering techniques are applied to enhance the model performance. First of all, the residual block, introduced in ResNet [9], is added to prevent vanishing gradients and ensure efficient training of the model. Secondly, swish activation is used in order to enhance training accuracy [10]. Last but not least, Group Normalization (GN) [11] is also applied. Normalization techniques make training loss landscape smoother and thus enable fast and stable training of deep learning models [12]. Although batch normalization [13] is the one of the most famous for normalization technique, GN is used for the model to achieve stable training for small batch size. In semiconductor process data, it is important to capture critical regions and dimensions, such as oxide thickness and effective channel length, but conventional CNN structures are not good at processing positional information of images [14]. Thus, to boost the model's ability to capture important regions by embedding positional information of data, we added coordinate convolution which adds Cartesian coordinate for every convolution. As shown Fig. 4, the final score of the model is 0.96, which is above our target score, 0.92.

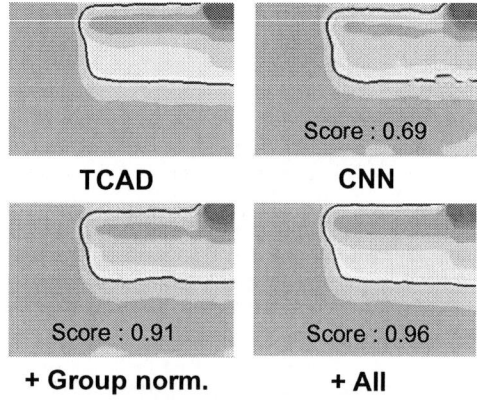

Fig. 4. An example RTT process model output image as model engineering techniques are applied. The score is an average of SSIM and IOU.

C. RTT for Device Simulation

Device simulator takes process condition as inputs, and calculates electron transport and then, generates electrical characteristics such as I-V curves, BV, VT as outputs. Most preliminary work to predict such characteristics with AI has been developed using SL, which cannot catch relationship between outputs. For example, let's say one of test data is out of training range (Fig. 5 (a)). For SL, the model needs to extrapolate to predict out-of-training-range test data, while when relationships of the point with its neighbors are considered, the point may not be regarded as out-of-training range. To capture relationships between neighboring points, we adopted a RNN structure, which is specially designed to learn sequential data, to RTT device models. In a RNN structure, the current is predicted from zero to max gate (drain) voltage sequentially for a fixed drain (gate) voltage (Fig. 5 (b)). As applying the RNN structures with Layer Normalization (LN) [15], the model score increases from 0.62 to 0.89. In addition, from domain knowledge on device physics, current values of neighboring drain and gate voltages are highly correlated. Hence, we adopted a 2D CNN structure to capture relations between neighboring gate bias and drain bias at the same time. For the precise prediction, an accurate calculation of derivatives of current with respect to voltage, is essential. Thus, in order to ensure RTT device model learns both current and trans-conductance values simultaneously, we used average of mean-squared-error and derivative losses. The device model cannot be predict proper G_M without the derivative loss even when it can accurately predict I-V curves (Fig. 6 (a)). The final score of device model is 0.99 (Fig. 6 (b)).

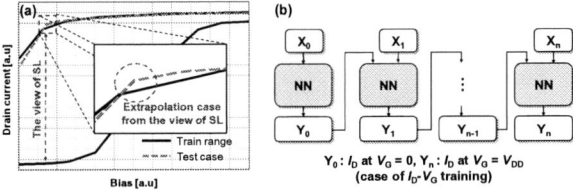

Fig. 5. (a) Example test data to illustrate the weakness of SL. (b) RNN structure for RTT device model.

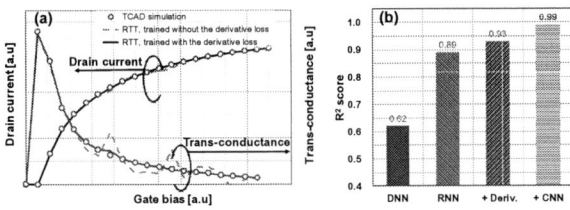

Fig. 6. (a) I_D and G_M-versus-V_G curve whether applying derivative loss or not. (b) RTT device model score as model structure engineering is applied.

D. Optimization

Once RTT models are accurately trained, they can be directly used to find the optimal process condition to achieve target device performances. For this task, optimization algorithms such as genetic algorithm (GA), simulation annealing (SA), gradient descent (GD) are commonly used. In general, GA and SA are better at finding global optimum but requires more iterations than GD. Since RTT model is based on neural network, which is easily differentiable, 10,000 GD from different Latin hypercube sample points [16] are executed to find global optimum in real-time. By comparing the optimization results with the results obtained by GA, it is

confirmed that the optimal point found is global optimum. Regarding TAT, our model found the global optimum within 60 seconds, which is 280 times faster than GA.

E. Confidence Interval Prediction

One of the well-known problem of deep learning models is that it is hard to tell whether the model prediction is trustable or not. When the model is asked to predict electrical characteristics of unseen process conditions, due to not enough train data or queried data out-of-training range, it is desired that the model outputs prediction results with low confidence rather than completely wrong prediction results with high confidence. To do that, we changed loss function which learns confidence interval, assumed by multivariate Gaussian process [17]. The loss function is negative-log-likelihood (NLL):

$$NLL = \frac{(y-\mu(x))^2}{2\sigma^2(x)} + \frac{\log\sigma^2(x)}{2}, \qquad (2)$$

where $\mu(x)$ and $\sigma^2(x)$ are mean and variance of model outputs. The numerator of the first term of eqn. (2) is introduced to reduce the gap between the mean of predicted values and the truth, i.e. it is same as mean-squared-error. If the mean-squared-error is relatively large, the model is likely to have large variance values to decrease the first term of the loss. But since the increase of the variance increases the second term of the loss, the model learns an appropriate variance values balancing between two terms. However, a single model tends to over-fit to the train samples, so to overcome overfitting issue, ensemble method is applied [7]. In training phase, each model is trained with different samples generated by bootstrap algorithm (Fig. 7 (a)). In the testing phase, the predicted outputs of the models are the mean and variance of the ensemble mixture (Fig. 7 (b)). When queried data point is in the training range, confidence interval of I-V curve is negligible so engineers can trust the prediction and performs experiments based on the prediction results (Fig. 8 (a)). For queried data point out-of-training range, since the model can tell the queried point is unseen during the training, predicted confidence interval is significantly large, as expected (Fig. 8 (b)). After RTT model is automatically retrained with previously unseen data points, which is obtained by TCAD simulations, the RTT model accurately predicts I-V curve with low confidence interval (Fig. 8 (c)).

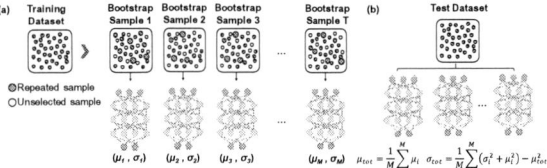

Fig. 7. (a) The models are independently trained using bootstrap techniques in training phase. (b) The ensemble prediction as Gaussian whose mean and variance are the mean and variance of the mixture, respectively in test phase.

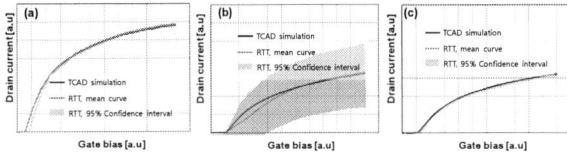

Fig. 8. Prediction uncertainty plot (a) in training range (b) out-of-training range (c) after retraining out-of-training range sample.

IV. EXPERIMENT RESULTS

A. RTT Models's Performance

TCAD simulator which is calibrated with 130 nm DDI process are used to generate TCAD data to train and evaluate the proposed RTT model. We generated 1,050 TCAD simulation samples with changing input variables. Among them, 1,000 samples are used in the model training phase, and the remaining 50 samples are used to evaluate the performance of trained model. The evaluated test score of the process model and the device model are equal to 0.96 and 0.99, respectively. Prediction times of the RTT process and device models are 60 and 0.1 seconds, respectively, while them of TCAD process and device simulations are ~1,200 and ~530,000 times longer, respectively.

B. Process Optimization Results

Next, the RTT device model is used to optimize actual semiconductor process. Our targets are listed in Table I: while keeping V_T as close to the target value as possible, I_{ON}, BV_{OFF}, and BV_{ON} are (I_{OFF} is) required to be maximized (minimized). With the RTT model, optimal process condition can be searched within 50 seconds, which is 300,000 times faster than the human expert (Fig. 9 (a)). Also, it is confirmed that at the searched optimal condition, the results of RTT device model agree with them of the TCAD simulation results with 99% accuracy (Table VI and (Fig. 9 (b))). Finally, our optimal condition has resulted in 11.4% gain of performance on average compared to the reference condition of the process. With the assist of RTT model optimization, process development time for DDI is reduced by 8 weeks.

TABLE I. OPTIZIATION RESULTS.

	V_T	I_{ON}	I_{OFF}[b]	BV_{OFF}	BV_{ON}
Target[a]	1	> 1	< 1	> 1	> 1
Exp.[c]	0.98	0.99	0.33	1.06	N/A
Opt.	1.04	1.07	0.56	1.09	1
TCAD	1.02	1.08	0.6	1.09	1
Acc.	98.4%	99.7%	98.1%	100%	99.6%

[a.] Numbers are normalized by target values for confidential reasons.

[b.] For I_{OFF}, $\log(I_{OFF})$ is used for accuracy calculation.

[c.] Exp. are experimental results on 130nm DDI process.

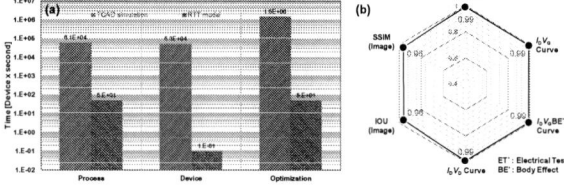

Fig. 9. (a) Prediction time for each task. (b) Performance score of RTT models on TCAD simulation results at optimal process condition.

V. CONCOLUSION

We propose a new approach to realize real-time TCAD prediction and optimization for semiconductor device designs. For 130nm DDI, simulation time is 530,000 times faster than conventional TCAD simulators, and the process development time is reduced by 8 week.

The whole procedure described here, from TCAD simulation data generation to trained model deployment and process optimization, can be automated to realize fully automated technology design.

REFERENCES

[1] Dennard, Robert H., et al. "Design of ion-implanted MOSFET's with very small physical dimensions." IEEE Journal of Solid-State Circuits 9.5 (1974): 256-268.

[2] Shi, Mengchao, Pinghui Mo, and Jie Liu. "Deep neural network for accurate and efficient atomistic modeling of phase change memory." IEEE Electron Device Letters 41.3 (2020): 365-368.

[3] Chen, Jing, et al. "Powernet: SOI Lateral Power Device Breakdown Prediction With Deep Neural Networks." IEEE Access 8 (2020): 25372-25382.

[4] Carrillo-Nuñez, Hamilton, et al. "Machine learning approach for predicting the effect of statistical variability in Si junctionless nanowire transistors." IEEE Electron Device Letters 40.9 (2019): 1366-1369.

[5] Wang, Zhou, et al. "Image quality assessment: from error visibility to structural similarity." IEEE transactions on image processing 13.4 (2004): 600-612.

[6] Bartolini, Andrea, Martino Ruggiero, and Luca Benini. "Visual quality analysis for dynamic backlight scaling in LCD systems." 2009 Design, Automation & Test in Europe Conference & Exhibition. IEEE, 2009.

[7] Lakshminarayanan, Balaji, Alexander Pritzel, and Charles Blundell. "Simple and scalable predictive uncertainty estimation using deep ensembles." Advances in neural information processing systems. 2017.

[8] Ulyanov, Dmitry, Andrea Vedaldi, and Victor Lempitsky. "Deep image prior." Proceedings of the IEEE Conference on Computer Vision and Pattern Recognition. 2018.

[9] He, Kaiming, et al. "Deep residual learning for image recognition." Proceedings of the IEEE conference on computer vision and pattern recognition. 2016.

[10] Ramachandran, Prajit, Barret Zoph, and Quoc V. Le. "Searching for activation functions." arXiv preprint arXiv:1710.05941 (2017).

[11] Wu, Yuxin, and Kaiming He. "Group normalization." Proceedings of the European conference on computer vision (ECCV). 2018.

[12] Santurkar, Shibani, et al. "How does batch normalization help optimization?." Advances in Neural Information Processing Systems. 2018.

[13] Ioffe, Sergey, and Christian Szegedy. "Batch normalization: Accelerating deep network training by reducing internal covariate shift." arXiv preprint arXiv:1502.03167 (2015).

[14] Liu, Rosanne, et al. "An intriguing failing of convolutional neural networks and the coordconv solution." Advances in Neural Information Processing Systems. 2018.

[15] Ba, Jimmy Lei, Jamie Ryan Kiros, and Geoffrey E. Hinton. "Layer normalization." arXiv preprint arXiv:1607.06450 (2016).

[16] Tang, Boxin. "Orthogonal array-based Latin hypercubes." Journal of the American statistical association 88.424 (1993): 1392-1397.

[17] Nix, David A., and Andreas S. Weigend. "Estimating the mean and variance of the target probability distribution." Proceedings of 1994 ieee international conference on neural networks (ICNN'94). Vol. 1. IEEE, 1994.

Application of Noise to Avoid Overfitting in TCAD Augmented Machine Learning

Sophia Susan Raju
Electrical Engineering
San Jose State University
San Jose, USA
sophiasusan.raju@sjsu.edu

Boyan Wang
Center for Power Electronics Systems
Virginia Polytechnic Institute and State University
Blacksburg, VA, U.S.A
wangboyan@vt.edu

Kashyap Mehta
Electrical Engineering
San Jose State University
San Jose, USA
kashyap.mehta@sjsu.edu

Ming Xiao
Center for Power Electronics Systems
Virginia Polytechnic Institute and State University
Blacksburg, VA, U.S.A
mxiao@vt.edu

Yuhao Zhang
Center for Power Electronics Systems
Virginia Polytechnic Institute and State University
Blacksburg, VA, U.S.A
yhzhang@vt.edu

Hiu-Yung Wong*
Electrical Engineering
San Jose State University
San Jose, CA, U.S.A
hiuyung.wong@sjsu.edu

Abstract— **In this paper, we propose and study the use of noise to avoid the overfitting issue in Technology Computer-Aided Design-augmented machine learning (TCAD-ML). TCAD-ML uses TCAD to generate sufficient data to train ML models for defect detection and reverse engineering by taking electrical characteristics such as Current-Voltage, *IV*, and Capacitance-Voltage, *CV*, curves as inputs. For example, the model can be used to deduce the epitaxial thicknesses of a *p-i-n* diode or the ambient temperature of a Schottky diode being measured, based on a given *IV* curve. The models developed by TCAD-ML usually have overfitting issues when it is applied to experimental *IV* curves or *IV* curves generated with different TCAD setup. To avoid this issue, white Gaussian noise is added to the TCAD generated curves before ML. We show that by choosing the noise level properly, overfitting can be avoided. This is demonstrated successfully by using the TCAD-ML model to predict 1) the epitaxial thicknesses of a set of TCAD silicon diode *IV*'s generated with different settings (extra doping variations) than the settings in the training data and 2) the ambient temperature of experimental *IV*'s of Ga_2O_3 Schottky diode. Moreover, domain expertise is not required in the ML process.**

Keywords—Gallium Oxide, Machine Learning, Noise, Schottky Barrier Diode, TCAD

I. INTRODUCTION

TCAD augmented machine learning (TCAD-ML) uses TCAD to generate sufficient data to train the ML model for various purposes [1][2]. For example, the model can be used to deduce the physical parameters, such as the epitaxial thicknesses or the ambient temperature of a diode, based on a given *IV* curve. However, models developed by TCAD-ML are usually overfitted and cannot be used to predict the physical parameters of TCAD *IV* curves generated with different settings [3] or experimental *IV* curves [4]. This is because the machine cannot understand that the variation in the *IV* curves can be caused by factors it has not seen in the training.

It is expected that, with domain expertise (i.e. extracting V_{bi}, I_{on}, I_{off}, subthreshold swing from the IV curves before applying ML[5][6]), overfitting problem may be reduced. However, this tremendously reduces the benefit of using ML as domain expertise is required to pre-process the data before ML. Thus, the algorithms developed and the experience

garnered cannot be transferred directly to other problems. The overfitting issue can also be understood as the fact that the machine trained by TCAD *IV* is "over-confident" and tries to attribute any variations in the *IV* only to the parameter variations it learned in the training dataset. As a result, when there are new variation sources (such as *IV* generated with different settings in TCAD or experimental data with noise and other unintentional variations), the machine fails to predict the correct physical parameters. To avoid this problem, Principal Components Analysis and Autoencoder are used in [4] and [3], respectively, while domain expertise is not required.

In this paper, we further propose to apply noise to the *IV* generated by TCAD to avoid overfitting without using domain expertise. By adding noise to the TCAD generated IV curves, the machine trained will understand that variations in *IV* curves can be caused by some hidden variations and will not be over-fitted. This is demonstrated successfully by using the TCAD-ML model to predict 1) the epitaxial thicknesses of the given TCAD *IV*'s of Silicon diode and 2) the ambient temperature of experimental *IV*'s of Ga_2O_3 Schottky diode in which a better performance in predicting experimental temperature than PCA in [4] is achieved.

II. PREDICTION OF SI *P-I-N* DIODE EPITAXIAL THICKNESSES

Three sets of Si *p-i-n* diode *IV*'s were generated using TCAD simulations. Sentaurus Process [7] simulation is used to create the structures and Sentaurus Device [8] is used for electrical simulations. Essential physical models are turned on, including Fermi-Dirac statistics, doping dependent, and high field saturation models for carrier mobilities, Schottky-Reed-Hall Recombination (SRH), and non-local Band to Band tunneling (BTBT). 80-bit ExtendedPrecision is used to avoid noisy reversed curves. Poisson, electron, and hole continuity equations are solved self-consistently.

Set1 is for ML model training and has ~2000 1-D *p-i-n* diodes with layer thicknesses being varied independently and uniformly but with fixed doping concentrations (Table I). The ML model is trained to deduce the intrinsic layer (t_i) and total n- and p-layer thicknesses ($t_p + t_n$) based on any given *IV* curve. *Set2* and *Set3* have constant thicknesses but the doping concentrations are varied randomly. They are used for testing the ability of the *Set1* trained ML model of deducing the

*Corresponding Author: hiuyung.wong@sjsu.edu

TABLE I. *P-I-N* DIODES SIMULATION SETUP FOR SET1, SET2 AND SET3.

Data	Thicknesses (nm)			Doping (cm⁻³)		
	n	i	p	n	i	p
Set1	150 -250	5 -16	150 -250	10^{20}	10^{17}	10^{20}
Set2	200	10	200	5×10^{18} -5×10^{20}	1×10^{13} -1×10^{17}	5×10^{18} -5×10^{20}
Set3	350	15	250	5×10^{18} -5×10^{20}	1×10^{13} -1×10^{17}	5×10^{18} -5×10^{20}

thicknesses when unseen variations (in this case, doping concentrations) are introduced. This is to emulate the experimental data which typically contains multiple variation sources. Note that *Set2* has thicknesses equal to the means of *Set1*. If a machine trained by *Set1* can predict the thicknesses in *Set2* well, there is a possibility that the machine just uses the means it learned in *Set1* instead of correctly predicting the thicknesses in *Set2*. Therefore, *Set3* which has thicknesses different from the means in *Set1* is created to confirm the machine's prediction ability.

80% of *Set1* data is used for training (using linear regression) and 20% is used for validation to develop *Model1*. Fig. 1 shows that it deduces the thicknesses of the validation data very well. However, when this model is used to deduce the thicknesses of the *IV*'s from *Set2* (with t_i = 10nm and $t_n + t_p$ = 400nm), it produces spurious results (e.g. negative thicknesses or t_i > 2000nm). This is because of overfitting as the machine "thinks" that all variations in the *IV* curves are caused by the thickness variations alone. To explain the *IV* variation caused by doping in *Set2*, it deduces the spurious results. To avoid overfitting, we propose to add noise to the *IV* curves in *Set1* for training, so that the machine will be able to isolate the features that are affected by the thicknesses by "understanding" that there are variations (due to noise) that cannot be explained by using thicknesses alone. White Gaussian noise is thus added to the *IV* curves in *Set1* with a certain signal-to-noise ratio (SNR) in dB. Fig. 2 shows an example of the final *IV* after noises are added. When SNR is larger, the noise signal is smaller.

Fig. 3 shows that with noise (SNR = 12.5dB, *Model2*), the validation result in *Set1* becomes worse in the prediction of $t_n + t_p$ (the machine is less confident or less overfitted). However, spurious results in predicting the thicknesses in *Set2* and *Set3* can be completely avoided. The prediction of *Set2*

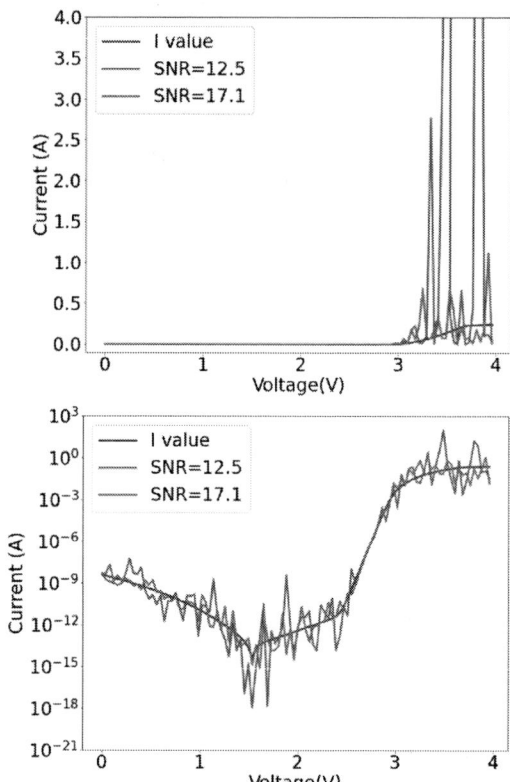

Fig. 2: *IV* curve from Dataset1 with (red and green) and without (blue) noise. Top: Current in linear scale. Bottom: Current in logarithmic scale.

and *Set3* t_i is fairly good that the prediction average is only about 2nm different from the actual values. However, the prediction of $t_n + t_p$ is not as good as desired. Prediction of *Set2* and *Set3* are 107nm and 236nm of difference from the actual values, respectively.

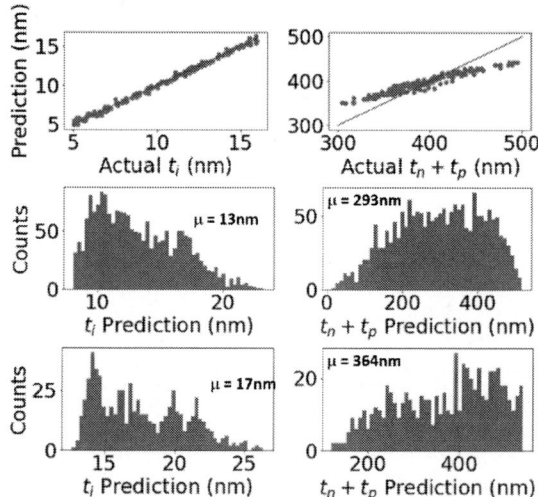

Fig. 3: *Model1* with SNR = 12.5dB performance. Top: validation result in *Set1* (scatter plots). Middle: *Set2* prediction histograms. Bottom: *Set3* prediction histograms

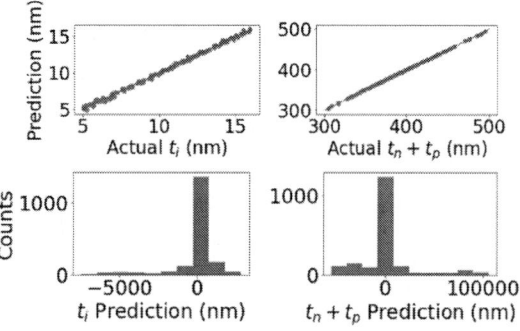

Fig. 1: *Model1* without noise. Top: validation result in *Set1* (scatter plots). Bottom: Prediction of *Set2* thicknesses (histograms).

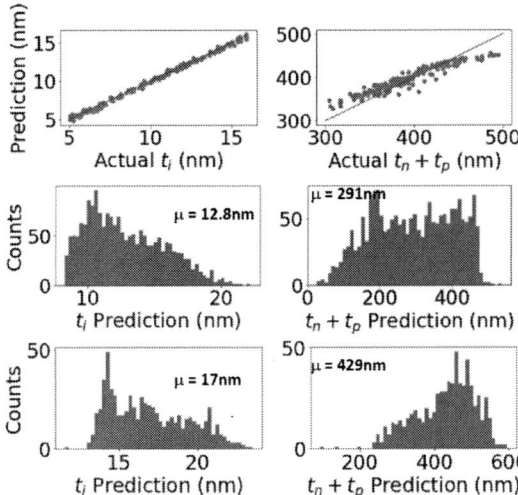

Fig. 4: *Model3* with SNR = 17.1 dB performance. Top: validation result in *Set1* (scatter plots). Middle: *Set2* prediction histograms. Bottom: *Set3* prediction histograms.

To achieve a better prediction, smaller noise (SNR = 17.1dB) is applied but with linear regression of 3rd order polynomial (*Model3*). Fig. 4 shows that it gives better validation results in *Set1* than *Model2* (therefore, it is more confident). It gives similar results in t_i prediction as *Model2*. But it improves the prediction of *Set3* $t_n + t_p$ by reducing the error to 171nm. Therefore, it is important to choose the noise level carefully to achieve the best performance. More importantly, the prediction means of t_i and $t_n + t_p$ changes by about 5nm and 140nm, respectively, from *Set2* and *Set3*, which is very similar to the actual changes (5nm for t_i and 200nm for $t_n + t_p$). This confirms that the trained machine is able to predict the thicknesses well and is not just using *Set1*'s means.

III. PREDICTION OF GA2O3 DIODE AMBIENT TEMPERATURE

Ga2O3 is an emerging ultra-wide bandgap semiconductor and have been attracting more attention recently [9]-[11]. Potentially, it can also be used as a high-temperature sensor. Fig. 5 shows the structure. The drift region of the Schottky diode has large variations in drift layer thickness (t_D) and doping concentration (N_D) due to the immature technology of Ga2O3. Moreover, special chemical treatment was applied before anode metal deposition to achieve different effective workfunction (*WF*). The experimental *IV*'s are measured at 3

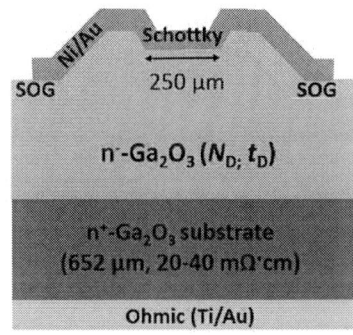

Fig. 5: Structure of the Ga2O3 Schottky Diode fabricated.

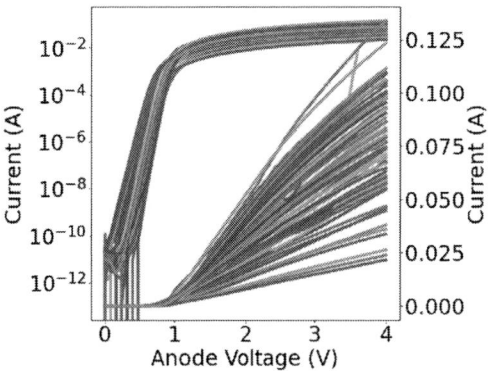

Fig. 6: 67 experimental *IV's* of Ga2O3 Schottky Diodes with variations in N_D, t_D, and effective workfunction measured at 3 different ambient temperatures. Left: log scale. Right: linear scale

different ambient temperatures. Fig. 6 shows 67 measured *IV*'s. Note that the experimental *IV*'s have a lot of non-idealities including the measurement noise and contact resistance variations.

TCAD models and parameters are calibrated to the experimental Ga2O3 Schottky diodes in [4]. Essential models such as dopants incomplete ionization and Philips Unified Mobility Model (PhuMob) [12] are calibrated to the literature [13]. A few normal experimental *IV*'s are then used to calibrate TCAD *IV*'s from 300K to 510K. Self-heating simulation is turned on and solved self-consistently with Poisson equation and electron and hole continuity equations. The calibration details can be found in [4].

Afterward, 10,000 *IV*'s are generated in TCAD with variations in ambient temperature, *WF*, t_D, and N_D using a server with 2 Fourteen-Core Intel Xeon Processor E5-2690 v4 2.60GHz with hyperthreading. A machine is trained using similar settings in *Model3* to predict these variations with the TCAD *IV*'s as inputs. The machine is then used to deduce the ambient temperature of experimental *IV*'s.

As showed in Fig. 7, the model deduces spurious ambient temperature on the experimental *IV*'s and the predicted

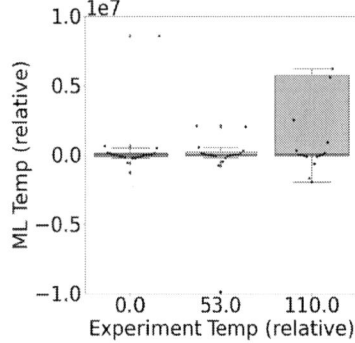

Fig. 7: Prediction of measurement ambient temperature using the machine trained with TCAD *IV*'s (*no noise*).

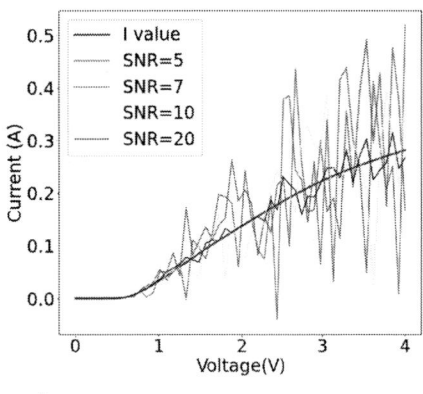

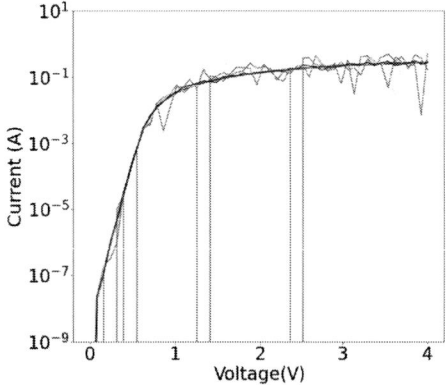

Fig. 8: TCAD Ga₂O₃ Schottky Barrier Diode *IV* curves with and without noise. Top: Current in linear scale. Bottom: Current in logarithmic scale.

temperatures can reach $\pm10^6$K when noise is not added to the TCAD *IV*s.

Therefore, noise (SNR = 7dB) is added to the TCAD *IV* curves before ML as shown in Fig. 8. With the noise and using *Model3*, the machine can predict the ambient temperature variation range very well (Fig. 9), despite the non-idealities in the experimental *IV*s and the presence of unseen variations. The result is also better than the range predicted by using PCA in [4].

CONCLUSION

Noise addition to TCAD *IV*s is proposed to avoid overfitting in TCAD-augmented machine learning. This is demonstrated successfully in two cases. One is to deduce the layer thicknesses of Si *p-i-n* diode *IV*s generated by TCAD with known thicknesses and extra variation in doping. It is statistically shown that noise addition can minimize overfitting. The second case is to deduce the ambient temperature based on experimental Ga₂O₃ Schottky diode *IV*s. This demonstrates the practical application of TCAD-augmented ML without domain expertise by adding white Gaussian noise.

ACKNOWLEDGMENT

This project was supported by San Jose State University College of Engineering New Faculty Start-up Fund. Hiu Yung Wong thanks Synopsys, Inc. for the donation of TCAD licenses.

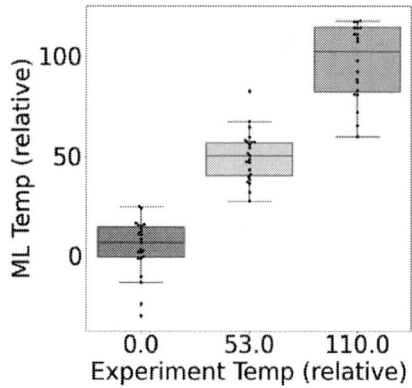

Fig. 9: Prediction of measurement ambient temperature using the machine trained with TCAD *IV*s (*SNR = 7dB*).

REFERENCES

[1] Y. S. Bankapalli and H. Y. Wong, "TCAD Augmented Machine Learning for Semiconductor Device Failure Troubleshooting and Reverse Engineering," 2019 International Conference on Simulation of Semiconductor Processes and Devices (SISPAD), Udine, Italy, 2019, pp. 21-24, doi: 10.1109/SISPAD.2019.8870467.

[2] C. Teo, K. L. Low, V. Narang and A. V. Thean, "TCAD-Enabled Machine Learning Defect Prediction to Accelerate Advanced Semiconductor Device Failure Analysis," 2019 International Conference on Simulation of Semiconductor Processes and Devices (SISPAD), Udine, Italy, 2019, pp. 17-20, doi: 10.1109/SISPAD.2019.8870440.

[3] K. Mehta, S. S. Raju, M. Xiao, B. Wang, Y. Zhang and H. Y. Wong, "Improvement of TCAD Augmented Machine Learning Using Autoencoder for Semiconductor Variation Identification and Inverse Design," in IEEE Access, vol. 8, pp. 143519-143529, 2020, doi: 10.1109/ACCESS.2020.3014470.

[4] Hiu Yung Wong, Ming Xiao, Boyan Wang, Yan Ka Chiu, Xiaodong Yan, Jiahui Ma, Kohei Sasaki, Han Wang, and Yuhao Zhang, "TCAD-Machine Learning Framework for Device Variation and Operating Temperature Analysis with Experimental Demonstration", submitted to IEEE Journal of Electron Devices Society (J-EDS).

[5] H. Carrillo-Nuñez, N. Dimitrova, A. Asenov and V. Georgiev, "Machine Learning Approach for Predicting the Effect of Statistical Variability in Si Junctionless Nanowire Transistors," in IEEE Electron Device Letters, vol. 40, no. 9, pp. 1366-1369, Sept. 2019, doi: 10.1109/LED.2019.2931839.

[6] J. Chen et al., "Powernet: SOI Lateral Power Device Breakdown Prediction With Deep Neural Networks," in IEEE Access, vol. 8, pp. 25372-25382, 2020, doi: 10.1109/ACCESS.2020.2970966.

[7] Sentaurus™ Process User Guide Version O-2018.06, June 2018.

[8] Sentaurus™ Device User Guide Version O-2018.06, June 2018.

[9] M. Higashiwaki, and G. H. Jessen, "Guest Editorial: The dawn of gallium oxide microelectronics," *Appl. Phys. Lett.*, vol. 112, no. 6, p. 060401, Feb 2018. doi: 10.1063/1.5017845.

[10] S. J. Pearton, Jiancheng Yang, Patrick H. Cary IV, F. Ren, Jihyun Kim, Marko J. Tadjer, and Michael A. Mastro, "A review of Ga₂O₃ materials, processing, and devices," *Appl. Phys. Rev.*, vol. 5, no. 1, p. 011301, Jan. 2018. doi: 10.1063/1.5006941.

[11] B. Wang, M. Xiao, X. Yan, H. Y. Wong, J. Ma, K. Sasaki, H. Wang, and Y. Zhang, "High-voltage vertical Ga₂O₃ power rectifiers operational at high temperatures up to 600K," *Appl. Phys. Lett.* 115, 263503 (2019); https://doi.org/10.1063/1.5132818.

[12] D. B. M. Klaassen, "A unified mobility model for device simulation—I. Model equations and concentration dependence," *Solid-State Electronics*, Vol. 35, No. 7, pp.953-959, 1992. doi: 10.1016/0038-1101(92)90325-7.

[13] K. Goto, K. Konishi, H. Murakami, Y. Kumagai, B. Monemar, M. Higashiwaki, A. Kuramata, and S. Yamakoshi, "Halide vapor phase epitaxy of Si doped β-Ga2O3 and its electrical properties," *Thin Solid Films* 666 (2018) 182–184. doi: 10.1016/j.tsf.2018.09.006.

Automatic Device Model Parameter Extractions via Hybrid Intelligent Methodology

Cheng-Che Liu[1,2], Yiming Li[1,2,3,*], Ya-Shu Yang[1,2], Chieh-Yang Chen[1,2], and Min-Hui Chuang[1,2]

[1]Parallel and Scientific Computing Laboratory; [2]Institute of Communications Engineering;
[3]Department of Electrical and Computer Engineering, National Chiao Tung University, 1001 Ta-Hsueh Rd., Hsinchu 30010, Taiwan;
*Tel: +886-3-5712121 ext. 52974; Fax: +886-3-5726639; Email: ymli@faculty.nctu.edu.tw

Abstract—We report an advanced hybrid intelligent methodology for device model parameter extractions combining multiobjective evolutionary algorithms, numerical optimization methods, and unsupervised learning neural networks on a unified optimization framework. The results between experimentally measured data and the calculation from industrial standard compact models are accurate, stable and convergent rapidly for all I-V curves. Verifications from diodes, bipolar transistors, MOSFETs, FinFETs, to nanowire MOSFETs confirm the robustness of the developed prototype, where the extraction is within 5% of accuracy.

Keywords—Automatic model parameter extraction; Hybrid intelligent methodology; Multiobjective evolutionary algorithms.

I. INTRODUCTION

Compact models for various semiconductor devices have been indispensable bridges between foundries and integrated circuit (IC) design companies [1-3]. For a specified compact mode, IC design companies design products according to model cards that characterize the electrical behaviors of transistors fabricated by foundries. The CMOS scaling have reached nanometer sizes following the Moore's law [4]. The conventional planner MOSFETs face physical challenges to further reducing channel length. The multiple-gate MOSFETs, bulk FinFETs and nanowire MOSFETs, are able to suppress short-channel effects enabling further device scaling [5]. Nowadays, commercial available extraction tools BSIMpro+® and MBP® are empirically relying on manual adjustment. The device-engineers-intensive work is very time consuming. We had proposed a single-objective genetic algorithm (GA)-based [6-7] hybrid intelligent methodology for MOSFETs model parameter extraction running on a unified optimization framework (UOF) [8]. However, model parameter extraction considers many sets of I-V and C-V curves simultaneously; thus, it is a multiobjective optimization problem (MOP) in nature. It will be more efficient and useful in semiconductor industry if we can advance the hybrid intelligent methodology combining with a multiobjective evolutionary algorithm (MOEA) [9-12] to improve the extraction efficiency for bulk FinFETs and nanowire MOSFETs.

In this work, we advance the aforementioned hybrid intelligent methodology for automatic model parameter extraction of bulk FinFETs and nanowire MOSFETs. The new approach combines the MOEA with an objective-function-decomposition technique, numerically robust Levenberg-Marquardt (LM) method, and unsupervised learning neural network (NN). We have successfully implemented a prototype of the extraction on UOF.

II. THE HYBRID INTELLIGENT METHODOLOGY

Figure 1 shows the newly MOEA-based hybrid intelligent methodology and the implemented prototype. The system has modules: file, solver, calculator, and problem. First, we through file class to read the mask, input, parameter, and output files required by UOF. Second, the imported data from the file class will be sent to the solver class, which controls the problem and calculator classes. Third, the problem class defines various optimization problems and return to the solver class. Final, we use different model calculators to calculate I-V curves of devices. Notably, curves of devices can be calculated via external circuit simulators (e.g. HSPICE® or PSPICE®) or from the built-in analytical equations of compact models.

We apply a nonlinear least square technique to formulate the optimization problem of parameter extractions, as listed in Eq. (1), where a penalty-boundary intersection method is used to decompose the dimension of MOP. Both the MOEA and LM are applied to optimize all parameters. The NN traces the trend of errors of original curves and their first derivatives. To minimize errors between measurement and model, the MOEA will compute the corresponding score of fitness. When the MOEA obtains an updated solution, the LM is activated to search the nearby local optima, and, simultaneously, the NN suggests proper searching directions based on an overall fitness. The prototype automatically extracts improved solutions via the evolutionary process which will terminate when a given root-means-square error reaches for all curves.

III. RESULTS AND DISCUSSION

We implement a prototype of model parameter extraction to verify the automatic extraction for various devices, such as p-n diodes, BJTs, MOSFETs, bulk FinFETs, and nanowire MOSFETs, respectively; Fig. 2 shows a system architecture and the logo of the implemented tool. Figure 3(a) shows the extracted (lines) and measured (dots) I-V and transconductance (g_m) curves of a p-n diode. Figure 3(b) shows the Gummel plot of the extracted Gummel-Poon model of a npn BJT. All errors between measured data and extracted current are within 5%. Adopted compact models and final extracted parameters are partially listed in Table 1.

For MOSFETs, we extract model parameters 10 times independently for 130, 85, and 75 nm, respectively, where all devices have the same width of 10 μm, oxide thickness of 3.3 nm, and doping concentration of 3×10^{18} cm^{-3} at 300 K. Figures 4(a)-(d) and (e)-(h) show the extracted I_D-V_{GS}, g_m, I_D-V_{DS}, and outputconductance (g_d) curves of the 130 and 85 nm n-MOSFET, respectively. The results between the model and measured show the accuracy of the extraction and the extracted 10-times curves show its robustness. Figures 5(a)-(d) show the extracted I_D-V_{GS}, g_m, I_D-V_{DS}, and g_d curves of the 75 nm n-MOSFET, respectively. Figures 6(a)-(b) show 10-time plots of the CPU time verse the number of evolution for 85- and 75-nm n-MOSFET, respectively. It costs about 10 minutes to complete an extraction within 5% error when the BSIM 4.21 model is applied. The average errors of I_D-V_{GS}, g_m, I_D-V_{DS}, and g_d for various dimensions n-MOSFET are listed in Table 2. Partially extracted parameters for an n-MOSFET with W/L = 10 μm/75 nm are listed in Table 3. The 10-time extracted threshold voltage V_{th} and $g_{m,\ max}$ for various dimensions are listed in Table 4. Above tests clearly confirm the accuracy and robustness of the automatic extraction.

We do further apply the tool to extract bulk FinFETs and n-/p-type nanowire MOSFETs, as shown in Figs. 7 and 8. For the 26-nm drawn gate length bulk n-type bulk FinFET with a

7-nm channel thickness and a 42-nm channel height, at least five objectives are considered simultaneously to extract a set of BSIM-CMG model. Figures 7(a)-(d) show the extracted I_D-V_{GS}, g_m, I_D-V_{DS}, and g_d curves, respectively. The average errors of five curves and times of bulk n-FinFET are listed in Table 5. The extracted threshold voltage $g_{m,\,max}$, $V_{t,\,lin}$, $V_{t,\,sat}$, DIBL, I_{off}, I_{on}, and SS are listed in Tables 6 and 7. Partially extracted parameters are listed in Table 8. Figures 8(a)-(h) show the extracted I_D-V_{GS}, g_m, I_D-V_{DS}, and g_d curves, respectively, of 10-nm n-/p-type nanowire MOSFETs with a 31-nm channel width. Not shown here, we have also applied the work to extract C-V curves for various devices before model parameter extractions from I-V curves.

IV. CONCLUSIONS

In summary, we have successfully advanced a hybrid intelligent methodology for model parameter extraction. It bases on (i) the multi-objective evolutionary algorithm with a decomposition technique; (ii) the Levenberg–Marquardt method; and (iii) unsupervised learning neural network on the unified optimization framework. Various extractions have been tested which imply its practicality in automatic model parameter extraction. This work can benefit device engineers to perform parameter extraction with engineering acceptable model accuracy for emerging CMOS devices.

ACKNOWLEDGMENT

This work was supported in part by the Ministry of Science and Technology, Taiwan, under Grant MOST 109-2221-E-009-033, Grant MOST 108-2221-E-009-008, and Grant MOST 108-3017-F-009-001, and in part by the "Center for mmWave Smart Radar Systems and Technologies" under the Featured Areas Research Center Program within the framework of the Higher Education Sprout Project by the Ministry of Education in Taiwan.

REFERENCES

[1] M. V. Dunga et al., "Modeling Advanced FET Technology in a Compact Model," *IEEE Trans. Electron. Devices*, vol. 53, pp. 1971-1978, 2006.

[2] G. Crupi et al., "A comprehensive review on microwave FinFET modeling for progressing beyond the state of art," *Solid-State Electron.*, vol. 80, pp. 81-95, 2013.

[3] J. P. Duarte et al., "Unified FinFET compact model: Modelling trapezoidal triple-gate FinFETs," in *SISPAD*, pp. 135-138, 2013.

[4] R. R. Schaller, "Moore's law: past, present and future," *IEEE Spectrum*, vol. 34, pp. 52-59, 1997.

[5] Y. Taur et al., "CMOS scaling into the nanometer regime," *IEEE Proc.*, vol. 85, pp. 486-504, 1997.

[6] Y. Li and Y.-Y. Cho., "Intelligent BSIM4 Model Parameter Extraction for Sub-100 nm MOSFET era," *Jpn. J. Appl. Phys.*, vol. 43, pp. 1717-1722, 2004.

[7] Y. Li, "Parallel Genetic Algorithm for Intelligent Model Parameter Extraction of Metal-Oxide- Semiconductor Field Effect Transistors," *Mater. Manuf. Process.*, vol. 24, pp. 243-249, 2009.

[8] Y. Li, S.-M. Yu, and Y.-L., "Electronic Design Automation Using a Unified Optimization Framework," *Math Comput Simul*, vol. 79, pp. 1137-1152, 2008.

[9] S.-C. Hung et al., "Circuit-Simulation-Based Multi-objective Evolutionary Algorithm for Design Optimization of a-Si:H TFTs Gate Driver Circuits under Multi-Level Clock Driving," *IEEE/OSA J. Display Technol.*, vol. 11, pp. 640-645, 2015.

[10] Y. Li and C.-Y. Chen, "Capacitance Characteristic Optimization of Germanium MOSFETs with Aluminum Oxide by using Semiconductor-Device-Simulation-Based Multi-Objective Evolutionary Algorithm Method," *Mater. Manuf. Process.*, vol. 30, pp. 520-528, 2015.

[11] K. Deb et al., "A fast and elitist multiobjective genetic algorithm: NSGA-II," *IEEE Trans. Evol. Comput.*, vol. 6, pp. 182-197, April 2002.

[12] Q. Zhang and H. Li, "MOEA/D: A Multiobjective Evolutionary Algorithm Based on Decomposition," *IEEE Trans. Evol. Comput.*, vol. 11, pp. 712-731, 2007.

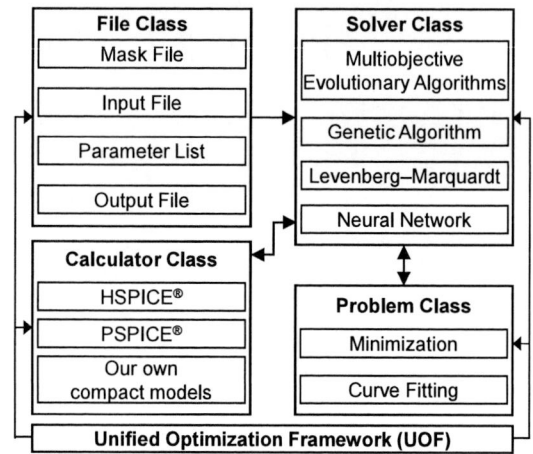

Fig. 1. The unified optimization framework (UOF) of the hybrid intelligent methodology.

$$Minimize \left\{ \frac{1}{N} \sum_{j=1}^{N} \frac{\sqrt{\left(G_{i,j} - M_{i,j}\left(\bar{V}_{i,j}, \bar{P}_{i,j}\right)\right)^2}}{\left|Max\left(G_{i,j}\right)\right|} \right\} \quad for\ each\ i, \quad (1)$$

where for each set of I-V or C-V curves (i.e., for each i), the model of MOP to be solved, where G, M, N, V, P are measured data, calculated results, number of total points, applied voltage, and model parameters, respectively.

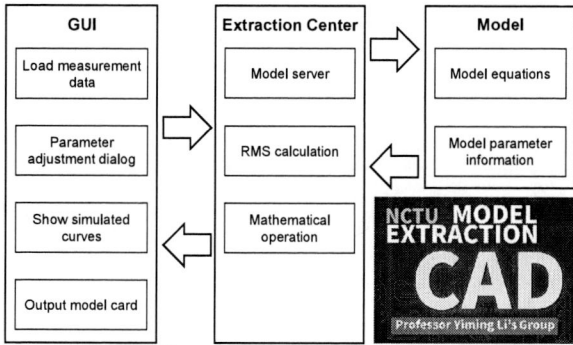

Fig. 2. A system architecture and the logo of the implemented tool.

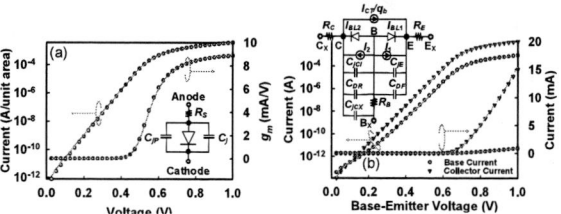

Fig. 3. For a *p-n* diode, (a) the extracted (lines) and measured (dots) of the normalized I-V curves in both the linear and log scales. For an *npn* BJT, (b) the extracted (lines) and measured (dots) Gummel plot. The inset in (a) and (b) are the equivalent circuit model of the *p-n* diode and Gummel-Poon model of *npn* BJT.

Table 1. List of the extracted parameter of the *p-n* diode and *npn* BJT.

List of the extracted parameter of diode		
Parameter Names	Initial Settings	Extracted Results
I_S (A/unit area)	1.0×10^{-14}	8.196×10^{-13}
N	1.0	1.07107
$R_S (\Omega)$	0.0	104.053
List of the extracted parameter of BJT		
I_S (A)	1.0×10^{-16}	3.32×10^{-15}
N_F	1.0	0.98
$R_E (\Omega)$	0.0	16.53
$R_C (\Omega)$	0.0	32.49
I_{SE} (A)	0.0	1.216×10^{-14}

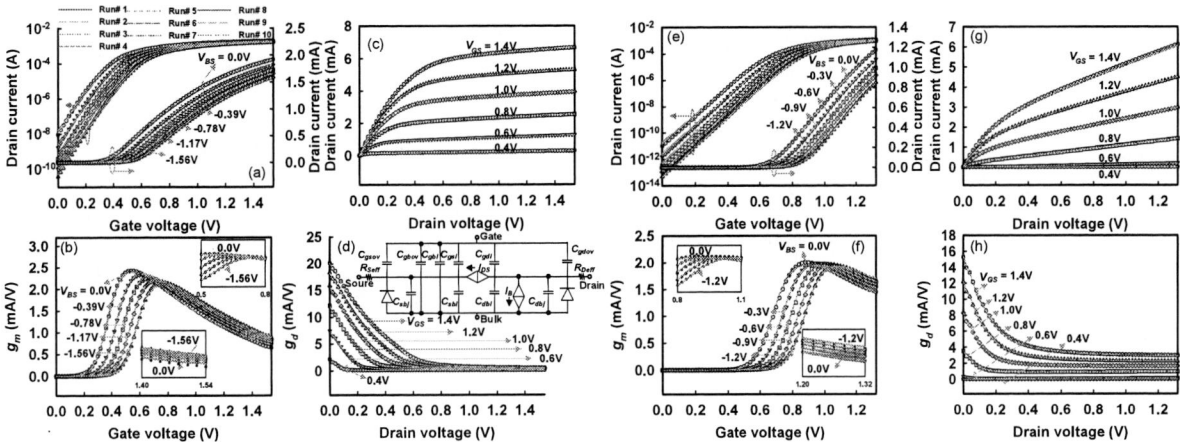

Fig. 4. Width/Length (W/L; μm/nm) = (a)-(d) 10 /130 and (e)-(h) 10/85 n-type MOSFETs; 10-times extracted (lines with colors) and measured (dots) data for (a) and (e) I_D-V_{GS}, (b) and (f) g_m, (c) and (g) I_D-V_{DS}, and (d) and (h) g_d curves, respectively. The inset in (d) is the equivalent circuit model of the MOSFET.

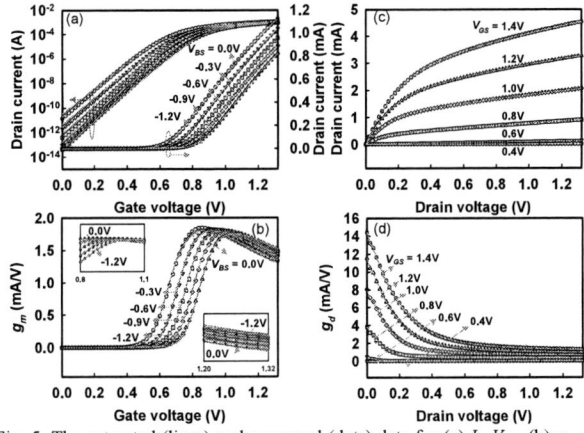

Fig. 5. The extracted (lines) and measured (dots) data for (a) I_D-V_{GS}, (b) g_m, (c) I_D-V_{DS}, and (d) g_d curves for an n-MOSFET with W/L = 10 μm/75 nm.

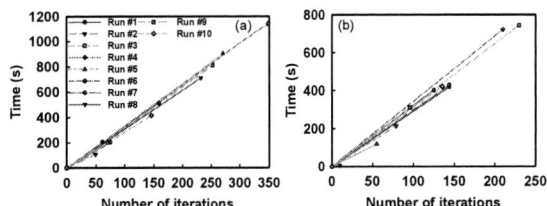

Fig. 6. 10-time plots of the CPU time verse the number of evolutions of n-MOSFETs with (a) L = 85 nm and (b) L = 75 nm, respectively.

Table. 2. List of the average errors among I-V sets of the extracted devices.

Length	Average Errors (%)		
	130 nm	85 nm	75 nm
$\log(I_D)$-V_{GS} (%)	2.876922	2.935092	3.160965
I_D-V_{GS} (%)	1.487478	1.116974	1.172822
g_m (%)	4.969339	4.801862	4.821634
I_D-V_{DS} (%)	2.638325	3.797678	3.199965
g_d (%)	2.200646	3.778339	1.989134

Table 3. Partial list of the extracted parameters for an n-MOSFET with width/length = 10 μm/75 nm at one time.

Parameter Names	Initial Settings	Extracted Results	Parameter Names	Initial Settings	Extracted Results
V_{TH0} (V)	0.7	0.359248	N_{factor}	1.0	1.06441
K_1 (V$^{1/2}$)	0.5	0.858372	U_0 (m^2/(Vs))	0.067	0.0216208
K_2	0.0	-0.146184	U_A (m/V)	1.0×10^{-9}	-5.07149×10^{-10}
D_{VT0}	2.2	2.69279	U_B (m^2/V^2)	1.0×10^{-19}	3.1358×10^{-19}
D_{VT1}	0.53	1.08794	V_{SAT} (m/s)	80000	56119.9
C_{IT} (F/m^2)	0.0	0.000410095	P_{CLM}	1.3	0.584354
V_{OFF} (V)	-0.08	-0.0797791	D_{ROUT}	0.56	1.46222

Table 4. List of the 10-time extracted physical quantities V_{th} and $g_{m,max}$ for the devices with three different channel length = 130, 85, and 75 nm, respectively.

Length	130 nm				85 nm				75 nm			
Measured V_t, $g_{m,max}$	V_t (V)		$g_{m,max}$ (mA/V)		V_t (V)		$g_{m,max}$ (mA/V)		V_t (V)		$g_{m,max}$ (mA/V)	
	0.243689		2.4265		0.565841		2.00023		0.557628		1.84322	
Run #	V_t	Error (%)	$g_{m,max}$	Error (%)	V_t	Error (%)	$g_{m,max}$	Error (%)	V_t	Error (%)	$g_{m,max}$	Error (%)
1	0.2418	0.7645	2.442	0.6429	0.5711	0.9400	1.9672	1.7529	0.5619	0.7726	1.8118	1.7046
2	0.2417	0.7796	2.4429	0.6758	0.5693	0.6123	1.9612	1.9512	0.5607	0.5555	1.8100	1.8022
3	0.2406	1.2331	2.4486	0.9107	0.5687	0.5176	1.9609	1.9662	0.5606	0.5329	1.8049	2.0789
4	0.2548	4.5858	2.4020	1.0063	0.5702	0.7732	1.9634	1.8412	0.5606	0.5472	1.8084	1.8890
5	0.2485	0.8662	2.4432	0.6882	0.5688	0.5396	1.9695	1.5363	0.5603	0.4841	1.8111	1.7420
6	0.2466	1.2337	2.4469	0.8407	0.5690	0.5669	1.9764	1.1913	0.5607	0.5583	1.8005	2.3177
7	0.2543	4.3800	2.3949	1.3022	0.5704	0.8177	1.9665	1.6863	0.5610	0.6186	1.8026	2.2037
8	0.2562	5.1584	2.4081	0.7582	0.5690	0.5595	1.9681	1.6063	0.5600	0.4336	1.8326	0.5761
9	0.2414	0.9355	2.4432	0.6882	0.5700	0.7511	1.9597	2.0262	0.561	0.6047	1.8065	1.9921
10	0.2541	4.2764	2.3940	1.3393	0.5681	0.4132	1.9582	2.1012	0.5607	0.5611	1.8034	2.1603

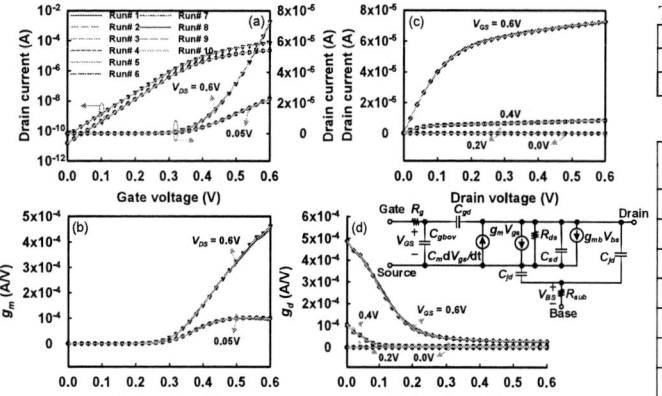

Table 5. List of average errors and CPU time of the extracted device.

$\log(I_D)$-V_{GS} (%)	0.1919993	I_D-V_{DS} (%)	4.767431
I_D-V_{GS} (%)	0.499764	g_d (%)	3.57325
g_m (%)	1.603667	CPU Time (s)	582.726

Table 6. List of the 10-time extracted physical quantities $g_{m,\,max}$.

Measured	$g_{m,\,max}$ (10^{-4}A/V)	
	4.6281	
Run #	$g_{m,\,max}$	Error (%)
1	4.6252	0.0626
2	4.6359	0.1685
3	4.5879	0.8686
4	4.5714	1.2251
5	4.6432	0.3262
6	4.6066	0.4645
7	4.6162	0.2571
8	4.5062	2.6339
9	4.6458	0.3824
10	4.6121	0.3457

Fig. 7. For the 26 nm n-type bulk-FinFET with a 7-nm channel width and a 42-nm channel height, the extracted (lines) results from the BSIM-CMG model and the measured (dots) data are shown for (a) I_D-V_{GS}, (b) g_m, (c) I_D-V_{DS}, and (d) g_d. The inset in (d) is the equivalent circuit model of the bulk FinFET. Notably, at least 51 parameters are optimized in the BSIM-CMG model with the 108 version.

Table 7. List of the 10-time extracted physical quantities $V_{t,\,lin}$, $V_{t,\,sat}$, DIBL, I_{off}, I_{on}, and SS, respectively.

Measured	$V_{t,\,lin}$ (V)		$V_{t,\,sat}$ (V)		DIBL		I_{off} (10^{-11}A)		I_{on} (10^{-5}A)		SS (mV/dec)	
	0.282628		0.252134		0.0554436		5.4031		7.2658		70.4044	
Run #	$V_{t,\,lin}$	Error (%)	$V_{t,\,sat}$	Error (%)	DIBL	Error (%)	I_{off}	Error (%)	I_{on}	Error (%)	SS	Error (%)
1	0.2827	0.0536	0.2546	0.9822	0.0512	7.6243	5.4690	1.2209	7.2838	0.2478	71.2009	1.13127
2	0.2823	0.0872	0.2538	0.6788	0.0518	6.4241	5.5807	3.2883	7.2865	0.2859	71.1617	1.0756
3	0.2832	0.2335	0.2472	1.9230	0.0654	18.065	5.5085	1.9518	7.2707	0.0686	69.2127	1.69271
4	0.2829	0.1011	0.2543	0.8615	0.0520	6.1847	5.4632	1.1138	7.2758	0.1376	71.1064	0.99713
5	0.2817	0.3256	0.2533	0.4775	0.0515	6.9652	5.5788	3.2531	7.2555	0.1412	71.0163	0.86914
6	0.2819	0.2403	0.2552	1.2385	0.0485	12.468	5.4814	1.4502	7.2945	0.3959	71.4011	1.41573
7	0.2832	0.231	0.2545	0.9576	0.0522	5.7781	5.5410	2.5537	7.2798	0.1937	71.2968	1.26752
8	0.2817	0.3172	0.2533	0.4872	0.0515	6.9685	5.4317	0.5302	7.2102	0.7639	70.7927	0.55155
9	0.2811	0.5183	0.2539	0.7010	0.0495	10.602	5.5430	2.5909	7.2757	0.1365	71.1187	1.01458
10	0.2833	0.2620	0.2536	0.5953	0.0540	2.4922	5.4956	1.7121	7.3131	0.6518	70.9697	0.80287

Table 8. Partial list of the extracted parameters for the 26-nm n-type bulk FinFET.

Parameter Names	Initial Settings	Extracted Results	Parameter Names	Initial Settings	Extracted Results
C_{IT} (F/m^2)	0.0	5.146×10^{-7}	V_{SAT} (m/s)	85000	99998.7
C_{DSC} (F/m^2)	0.007	0.0219993	U_0 (m^2/Vs)	0.03	0.0650832
C_{DSCD} (F/m^2)	0.007	0.0142987	U_{CS}	1.0	8.403×10^{-5}
D_{ITO}	0.0	0.99285	P_{DIBL1}	1.30	0.199277
D_{IT1}	0.6	0.376897	D_{ROUT}	1.06	0.271762
K_1 (V$^{1/2}$)	0.0	3.249×10^{-7}	P_{VAG}	1.0	0.599929
E_{TA0}	0.6	4.917×10^{-6}	P_{CLM}	0.013	1.211×10^{-6}
D_{SUB}	1.06	0.977978	P_{CLMG}	0.0	0.522503

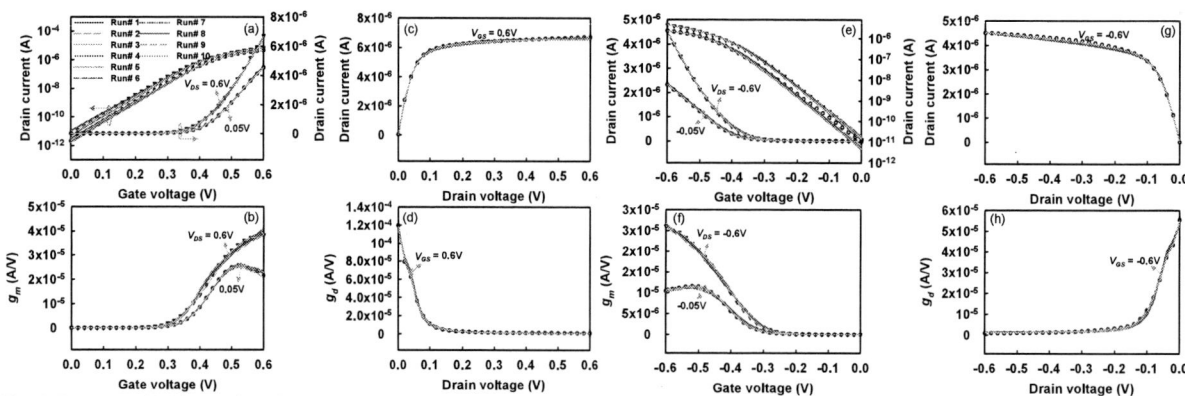

Fig. 8. For 10 nm (a)-(d) n- and (e)-(h) p-type nanowire MOSFETs with a 31-nm channel width, the extracted (lines) and measured (dots) data are shown for (a) and (e) I_D-V_{GS}, (b) and (f) g_m, (c) and (g) I_D-V_{DS}, and (d) and (h) g_d curves, respectively. Similar to the bulk FinFET, the BSIM-CMG model with the 108 version has been adopted for parameters are optimized among the five objectives, where errors are within 5% for all sets of I-V curves.

358

Physics-Informed Graph Neural Network for Circuit Compact Model Development

Xujiao Gao
Electrical Models & Simulation
Sandia National Laboratories
Albuquerque, USA
xngao@sandia.gov

Andy Huang
Electrical Models & Simulation
Sandia National Laboratories
Albuquerque, USA
ahuang@sandia.gov

Nathaniel Trask
Computational Mathematics
Sandia National Laboratories
Albuquerque, USA
natrask@sandia.gov

Shahed Reza
Component & Systems Analysis
Sandia National Laboratories
Albuquerque, USA
sreza@sandia.gov

Abstract—We present a Physics-Informed Graph Neural Network (pigNN) methodology for rapid and automated compact model development. It brings together the inherent strengths of data-driven machine learning, high-fidelity physics in TCAD simulations, and knowledge contained in existing compact models. In this work, we focus on developing a neural network (NN) based compact model for a non-ideal PN diode that represents one nonlinear edge in a pigNN graph. This model accurately captures the smooth transition between the exponential and quasi-linear response regions. By learning voltage dependent non-ideality factor using NN and employing an inverse response function in the NN loss function, the model also accurately captures the voltage dependent recombination effect. This NN compact model serves as basis model for a PN diode that can be a single device or represent an isolated diode in a complex device determined by topological data analysis (TDA) methods. The pigNN methodology is also applicable to derive reduced order models in other engineering areas.

Keywords—*physics informed graph neural network (pigNN), compact model development, topological data analysis (TDA), TCAD, non-ideality factor, Shockley-Read-Hall*

I. Introduction

To rapidly reduce the often multi-year cycle of reliable circuit compact model development (CMD), data-driven machine learning (ML) methods have recently been employed to derive compact models [1-3] and even optimize design at the circuit level [4-5]. However, as clearly recognized in [6], compact models generated from data-driven ML can lead to unphysical results due to the nature of a physics-agnostic method. To address this issue, researchers have tried to analytically formulate the most significant physics for a given device as constraints in the ML training [6]. The obvious drawback with this remedy is that only a few components of the entire complex device physics are considered, so the resulting compact model (CM) is still not predictive outside the training data range. Most recently, data-driven NN models with ties to PDEs are being investigated [7] for CMD, and

inspired the ML training used in this work. However, their models rely on device terminal response data and ignore the internal physical information offered by technology aided computer design (TCAD) simulation. To incorporate and facilitate TCAD in CMD, we propose a Physics-Informed Graph Neural Network (pigNN) methodology for rapid and automated CMD, which brings together the inherent strengths of data-driven ML, high-fidelity physics in TCAD simulation, and knowledge contained in existing compact models.

II. Methodology

The motivating principle grounding our methodology lies in a fundamental shift in the order of compact model construction. Traditionally, a circuit CM is obtained by piecing together individual geometry-independent electronic components, informed by adjacent and correlated physical phenomena within the device. Our proposed pigNN methodology reverses the order of CM assembly: first, we use TCAD device simulation results to identify the graph topology physically intrinsic to a given device accounting for important physical effects; then, we deduce the circuit components required to populate the graph in order to form the circuit CM.

As depicted in Fig. 1, our pigNN workflow contains four stages. First, during the Physics Priming (PP) stage, loosely calibrated TCAD device models are used to simulate a given device throughout its operating regime. We use Sandia's open-source TCAD Charon [8] code which allows us to have full access to physical quantities. Second, physically significant regions are identified (RR stage). From the concert of physical fields, we determine physically important regions as they evolve through a sweep of bias/time conditions. ML classification and Topological Data Analysis (TDA) for processing the fields are the primary tools in this stage. Third, we process the data from stage RR using TDA and graph cut techniques to seed a ML process to determine the intrinsic device topology (TT stage). Finally, electronic components are selected to functionally represent the physical interactions along the edges of a device topology (II stage). In this stage, established compact models (ECM) will be utilized to guide the interaction identification; Sandia's Xyce simulator [9] is used for circuit simulation, allowing us to streamline pigNN-based compact models directly into Xyce. Stages RR, TT, and II are trained and adapted using available experimental data until a robust physics-informed CM is achieved.

This work is funded under Laboratory Directed Research and Development (LDRD) program, Project No. 218474, at Sandia National Laboratories. Sandia National Laboratories is a multimission laboratory managed and operated by National Technology and Engineering Solutions of Sandia, LLC., a wholly owned subsidiary of Honeywell International, Inc. for the U.S. Department of Energy's National Nuclear Security Administration under contract DE-NA-0003525.

This paper describes objective technical results and analysis. Any subjective views or opinions that might be expressed in the paper do not necessarily represent the views of the U.S. Department of Energy or the United States Government.

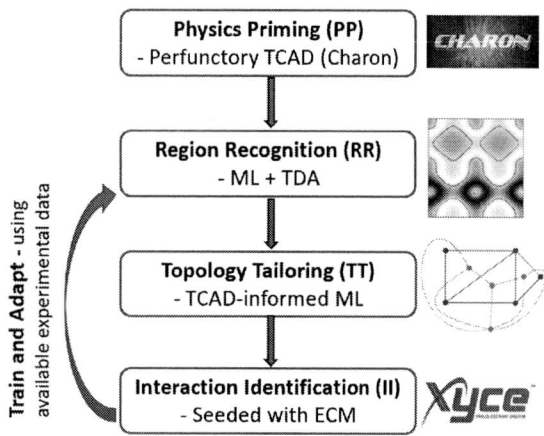

Fig. 1. Workflow of the pigNN-CMD methodology. The second and third graphs in the right column are taken from Wikipedia for illustration purpose only.

III. APPLICATION EXAMPLE

Details of the pigNN methodology are described in [10] where pigNN is demonstrated on a 2D resistor network and a nonlinear non-CMD problem [10]. For the 2D resistor problem, a spectral partitioning scheme (a TDA method) [10] was used to determine important physical regions based on TCAD solutions, which map to a linear resistor network. The TensorFlow based NN was used to achieve parameter-free learning by enforcing Kirchhoff's current law at each node via partial different equation (PDE) constrained optimization.

To move beyond linear elements, the most important non-linear device for CMD is a PN diode. Multiple PN diodes are widely used in compact models of semiconductor devices. Before we can model multiple PN diodes or mixed electronic components in a graph using pigNN, we need to first build an accurate compact model for a non-ideal diode that represents one edge in a pigNN graph. In this work, we focus on developing NN based compact model for a non-ideal PN diode under forward bias. A single PN diode represents an isolated PN junction determined by TDA based on spatial physical quantities such as electric potential and/or space charge profiles in a real device. The obtained NN compact model will serve as PN diode model basis in a complex device.

We started by simulating an arbitrarily selected silicon 1D PN diode in Charon. It is 0.5-μm long in both P and N sides and has a symmetric, uniform doping of 10^{16} cm^{-3}. Our baseline TCAD simulation used the most basic physical models such as constant mobility and absence of recombination/generation processes. For this baseline case, the simulated results from Charon are shown as solid lines in Fig. 2. It is well-known that the current-voltage (I-V) curve for a non-ideal PN diode shows an exponential region and a quasi-linear region. TCAD simulation naturally captures the smooth transition from exponential to quasi-linear response. However, popular diode compact models [11] cannot model a smooth transition switching between the two regimes. Even in the quasi-linear regime, the diode conductance is not constant, but varies nonlinearly with voltage as seen in the inset of Fig. 2. Bandwell and Jayakumar [12] proposed using the Lambert W-function to smoothly interpolate the exponential and the quasi-linear regions without a switch. However, the model does not incorporate a voltage dependent non-ideality factor, which can account for the voltage dependent recombination

process governed by the carrier transport drift diffusion continuity equations.

To develop a compact model that captures the diode's smooth I-V response determined by TCAD physics, we propose the following CM for the edge current in a pigNN graph:

$$I_D = [1 - w(V_D)]I_0 \left(e^{\frac{qV_D}{\eta k_B T}} - 1.0\right) + w(V_D)(V_D - V_{tr})G_0$$

$$+ w(V_D)(V_D - V_{tr})G_{NN}(V_D). \tag{1}$$

Here $w(V_D)$ is a voltage dependent weighting function, i.e.,

$$w(V_D) = 0.5[\tanh(10^5(V_D - V_{tr})) + 1.0], \tag{2}$$

which is a step-like but smooth function between 0 and 1 with the transition occurring at V_{tr}. The factor of 10^5 controls how steep the transition occurs, which is a heuristic value at this point. We will explore initializing the factor from physical fields contained in TCAD results. $G_{NN}(V_D)$ is a NN component to model the voltage dependent conductance in the quasi-linear regime. Other variables learned by NN are I_0, η, V_{tr}, and G_0. We used the TensorFlow library [13] to train the neural network and model parameters. Equation (1) can be interpreted as a diode, a constant conductance, and a voltage dependent conductance in parallel. However, the diode is turned on when $V_D < V_{tr}$ and the conductance branches are turned on when $V_D \geq V_{tr}$. The transition is determined by the smooth function in (2). To compare the NN learned conductance with Charon simulation result, we rewrite (1) as

$$I_D = [1 - w(V_D)]I_0 \left(e^{\frac{qV_D}{\eta k_B T}} - 1.0\right) + V_D G_D(V_D), \tag{3}$$

where $G_D(V_D)$ is the diode conductance in the quasi-linear region and given by

$$G_D(V_D) = w(V_D)\left(1 - \frac{V_{tr}}{V_D}\right)G_0 + w(V_D)\left(1 - \frac{V_{tr}}{V_D}\right)G_{NN}. \tag{4}$$

Using initial values obtained from simulated I-V results for the learning variables, we achieved excellent agreement between the NN based CM (dots in Fig. 2) and TCAD results (solid lines in Fig. 2) for our baseline diode.

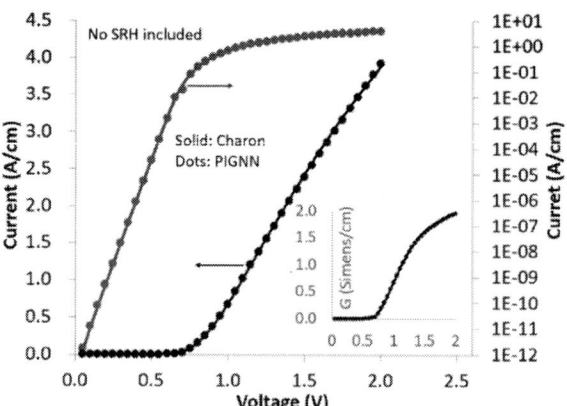

Fig. 2. Diode current-voltage comparison between Charon (solid lines) and pigNN learned results (dots) in both linear (left axis) and log (right axis) scales for the case of no SRH. Inset shows the diode conductance, $G_D(V_D)$, vs. voltage comparison between Charon (solid) and pigNN (dots).

It is well-known that the non-ideality factor η in (1) may model the Shockley-Read-Hall (SRH) recombination effect in the low forward bias region. As a next step, we enabled the SRH recombination in our Charon simulation of the diode with four different carrier lifetimes. Charon results are shown as solid lines in Fig. 3. Using the compact model in (1), the NN learned I-V curves (dots in Fig. 3) match Charon results quite well except in the exponential region. The learned diode conductance, $G_D(V_D)$, in the quasi-linear region, are in excellent agreement with Charon simulated conductance as shown in Fig. 4. Note the diode conductance has a strong non-linear voltage dependence.

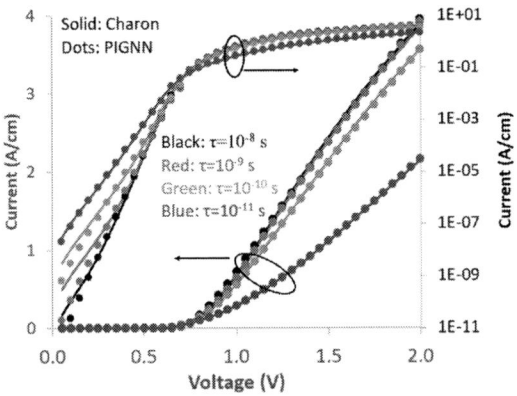

Fig. 3. Diode current-voltage comparison between Charon (solid lines) and pigNN results (dots) in both linear (left axis) and log (right axis) scales for TCAD simulations including SRH with 4 different carrier lifetimes.

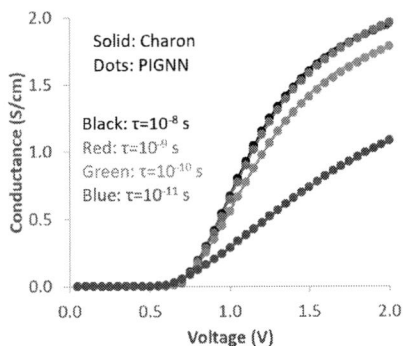

Fig. 4. Diode conductance, $G_D(V_D)$, vs. voltage comparison between Charon (solid) and pigNN (dots) for the same simulations as in Fig. 3.

We see in Fig. 3 that the NN learned I-V responses did not accurately reproduce the Charon results in the exponential region. This is because our NN compact model above used a voltage independent non-ideality factor, η, for a given carrier lifetime, while the Charon simulation naturally captures the voltage dependent SRH recombination effect. The pigNN learned variables for the results shown in Figs. 3-4 are listed in Table I. Clearly, the leakage current (I_0) and the non-ideality factor (η) both increase with decreasing lifetime (τ), i.e., increased SRH recombination, as expected, since they are both directly associated with recombination processes. The transition voltage (V_{tr}) is somewhat related to the built-in potential of a diode, so its value does not change much until at a very short lifetime. The constant conductance (G_0) is more a fitting parameter than related to any physical quantity, while the important conductance quantity is the diode conductance, $G_D(V_D)$, defined in (4).

TABLE I.   pigNN Learned Variables for Results in Fig. 3

τ (s)	I_0 (A/cm)	η	V_{tr}	G_0
No SRH	1.71×10^{-13}	1.0	0.68	2.82
10^{-8}	1.35×10^{-12}	1.13	0.71	1.79
10^{-9}	5.14×10^{-12}	1.21	0.68	2.30
10^{-10}	2.37×10^{-10}	1.46	0.67	1.89
10^{-11}	1.02×10^{-8}	1.75	0.57	0.37

It is clear that a voltage independent non-ideality factor in a diode compact model cannot correctly capture the SRH recombination effect. To include the recombination effect in our compact model (1), we apply NN to learn the voltage dependent non-ideality factor. Since we know $\eta = 1$ for the case of no recombination, we rewrite $\eta(V_D) = 1 + \eta_{NN}(V_D)$ with $\eta_{NN}(V_D)$ being the NN-learned factor accounting for the recombination effect. With two NN-learned components in the model (1), we find the choice of the NN loss function becomes very important. When η is voltage independent, we can achieve decent results shown in Figs. 3-4 by defining the NN loss function as $\mathcal{L} = (I_{NN} - I_t)^2 / I_t^2$. Here I_t is the target current obtained from either experiment or TCAD simulation, and I_{NN} is the NN learned current. When using NN to learn $\eta_{NN}(V_D)$, we find this simple loss function is insufficient due to the twelve order of magnitude difference in the current response. A good loss function seems to be a mixture of a logarithmic function (for the exponential I-V region) and a linear function (for the quasi-linear region). However, it is difficult heuristically figuring out a proper weighting of the two functions. To address this challenge, we propose to incorporate an approximate inverse (i.e., voltage-current) function of the target I-V response in the NN loss function. Specifically, given a I-V response, we can apply a regression method to obtain a function that approximates the voltage as a function of the current. We stress that an approximation is sufficient since it is only used in the loss function. Fig. 5 shows a comparison between the approximate V-I curve (red line) and the Charon result for the case of $\tau = 10^{-9}$ s. There exist many possible functions to approximate a diode V-I response. We somewhat arbitrarily chose this form: $v(I) = (aI)^b \times \tanh(dI - e) + fI$, where v and I represent voltage and current, respectively, a, b, d, e, f are parameters that can be determined using a regression method. We used the Python sci.optimize module to obtain these parameters. Then we define the loss function as $\mathcal{L} = [v(I_{NN}) - v(I_t)]^2 / [v(I_t)]^2$. The $v(I)$ function maps the logarithmically separated current range (twelve orders of magnitude) to a linearly separated voltage range, which allows us to effectively minimize the error between learned and target currents.

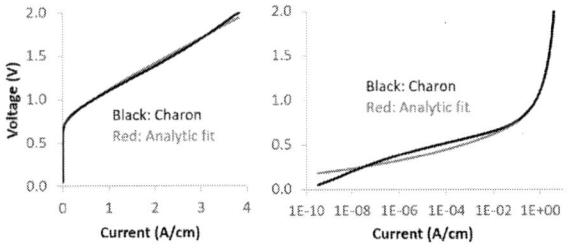

Fig. 5. Approximate voltage-current function (red line) of the Charon result for the case of $\tau = 10^{-9}$ seconds (black line) on the linear current scale (left panel) and the log scale (right panel).

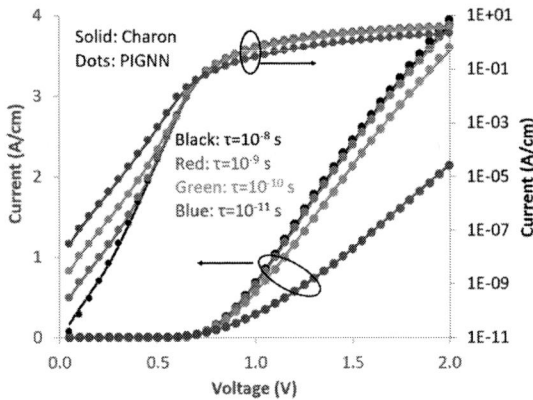

Fig. 6. Diode current-voltage comparison between Charon (solid lines) and pigNN results (dots) obtained using voltage-dependent non-ideality factors in both linear (left axis) and log (right axis) scales for TCAD simulations including SRH with four different carrier lifetimes.

Using the compact model in (1) with voltage dependent NN-learned non-ideality factor $\eta(V_D)$ and the proposed loss function, we achieve excellent agreement between pigNN and Charon results as shown in Fig. 6 in both the exponential and linear regions for the four different carrier lifetimes. The NN-learned $\eta(V_D)$ for the pigNN results in Fig. 6 are plotted in Fig. 7 as a function of voltage for different lifetimes. Except for the largest lifetime of 10^{-8} seconds, the non-ideality factors for other three cases start around 2 at the lowest voltage point and decrease with increasing voltages. We know from semiconductor physics that a pure recombination current yields a non-ideality of 2. Therefore, our NN-learned non-ideality factor is consistent with the physics that the SRH recombination dominates the current at low voltages and its effect decreases with increasing voltages. As the forward voltage increases, the PN junction barrier is reduced, so the diffusion current becomes more dominant, until the drift current takes over in the quasi-linear current-voltage regime. For the case of 10^{-8}-second lifetime, the non-ideality factor shows an opposite trend in the low voltages. This is more numerical than physical since the recombination effect is very small at this lifetime. Note that the non-ideality factor values have no significance for voltages above the transition voltages V_{tr}, which are listed in Table II for the pigNN results shown in Fig. 6. This is because our compact model in (1) contains a weighting function that smoothly masks out the diode response for voltages above V_{tr}.

TABLE II. PIGNN LEARNED PARAMETERS FOR RESULTS IN FIG. 6

τ (s)	I_0 (A/cm)	V_{tr} (V)	G_0 (S/cm)
10^{-8}	2.22×10^{-12}	0.62	2.50
10^{-9}	1.85×10^{-10}	0.63	1.63
10^{-10}	1.87×10^{-9}	0.57	1.46
10^{-11}	1.96×10^{-8}	0.52	0.83

IV. CONCLUSIONS

We have presented a Physics-Informed Graph Neural Network (pigNN) methodology for rapid and automated compact model development, which brings together the inherent strengths of data-driven ML, high-fidelity physics in TCAD simulation, and knowledge contained in existing compact models. Using data-driven neural network, we have developed an accurate compact model of a non-ideal PN diode that represents one nonlinear edge in a pigNN graph. A PN diode can be a single device or one isolated diode in a complex device determined by TDA methods. The NN compact model accurately captures the smooth transition between the exponential and quasi-linear response regions and the voltage-dependent recombination effect in a non-ideal PN diode under forward bias. We are applying the full pigNN methodology which enforces Kirchhoff's current law at each node in a pigNN graph to develop compact models for bipolar transistors. The methodology is also being explored to derive reduced order models in other engineering areas such as mechanics and electromagnetics.

REFERENCES

[1] H. B. Hammouda, M. Mhiri, Z. Gafsi, and K. Besbes, "Neural-based models of semiconductor devices for SPICE simulator," Amer. J. Appl. Sci., vol. 5, no. 4, pp. 785-791, 2008.

[2] H. Yu, M. Swaminathan, C. Ji, D. White, "A method for creating behavioral models of oscillators using augmented neural networks," Proc. IEEE 26th Conf. Elect. Perform. Electron. Packag. Syst. (EPEPS), pp. 1-3, Oct. 2017.

[3] R. Trinchero, M. Larbi, H. M. Torun, F. G. Canavero and M. Swaminathan, "Machine learning and uncertainty quantification for surrogate models of integrated devices with a large number of parameters," in IEEE Access, vol. 7, pp. 4056-4066, 2019.

[4] S. Koziel, A. Bekasiewicz and P. Kurgan, "Rapid EM-driven design of compact RF circuits by means of nested space mapping," IEEE Microw. Wireless Compon. Lett., vol. 24, no. 6, pp. 364-366, 2014.

[5] H. Yu, H. Chalamalasetty and M. Swaminathan, "Modeling of voltage-controlled oscillators including I/O behavior using augmented neural networks," IEEE Access, vol. 7, pp. 38973-38982, 2019.

[6] M. Li, O. İrsoy, C. Cardie and H. G. Xing, "Physics-inspired neural networks for efficient device compact modeling," IEEE J. Explor. Solid-State Computat. vol. 2, pp. 44-49, Dec. 2016.

[7] K. Aadithya, P. Kuberry, B. Paskaleva, P. Bochev, K. Leeson, A. Mar, T. Mei, and E. Keiter, "Development, demonstration, and validation of data-driven compact diode models for circuit simulation and analysis," arXiv:2001.01699, 2020.

[8] https://charon.sandia.gov, May 2020.

[9] https://xyce.sandia.gov, v7.1, 2020.

[10] N. Trask, A. Huang, and X. Hu, "Exact physics for scientific machine learning: learning a discrete exterior calculus on graphs," unpublished, August, 2020.

[11] https://www.pspice.com/discrete/diodes.

[12] T. C. Bandwell and A. Jayakumar, "Exact analytic solution for current flow through diode with series resistance," Electron. Lett., vol. 36, no. 4, pp. 291-292, 2000.

[13] https://www.tensorflow.org/about/bib.

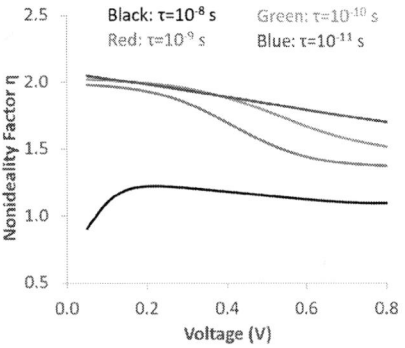

Fig. 7. NN learned non-ideality factor $\eta(V_D)$ as a function of voltage for the pigNN results shown in Fig. 6.

Theoretical study of electronic transport in monolayer SnSe

Sanjay Gopalan*, Gautam Gaddemane[†][1], Maarten L. Van de Put[‡], and Massimo V. Fischetti[§]

Department of Materials Science and Engineering
The University of Texas at Dallas, Richardson, Texas 75080
* Email: sanjay.gopalan@utdallas.edu
[†] Email: gautamg88@gmail.com
[‡] Email: maarten.vandeput@utdallas.edu
[§] Email: max.fischetti@utdallas.edu

Abstract—**Monolayer SnSe is a two-dimensional (2D) material with an indirect band gap ($\sim$ 0.92 eV) that can be obtained relatively easily by exfoliating bulk SnSe crystals. Like most 2D van der Waals monolayers, its layered nature reduces or eliminates the defects found in bulk materials, such as surface interface roughness and dangling bonds. Here, we show promising results of first-principle calculations of the low-field mobility and high-field characteristics of monolayer SnSe by implementing the full-band Monte Carlo approach.**

Keywords—**Monolayer SnSe, DFT, Monte Carlo, transport, mobility**

I. INTRODUCTION

The discovery of graphene has stimulated interest in two-dimensional materials, such as silicene [1]- [3], germanene [1]-[4], phosphorene [5]- [9] and transition metal dichalcogenides (TMDs) [10]- [16]. Monolayer SnSe has gained attention in VLSI applications, thanks to recent progress in growth techniques [17]- [18] and resulting in a low defect density [19] and in a promising measured carrier mobility [20]- [21].

Unfortunately, there are discrepancies in the intrinsic electron mobility calculated for monolayer SnSe. Shi *et al.* [22] performed a study of the transport properties of monolayer SnSe calculating the mobility within the deformation potential theory. They predicted the electron mobility in monolayer SnSe to be 757 cm^2 V^{-1} s^{-1}. However, the use of constant deformation potentials fails to account for the anisotropy of the matrix elements, leading to an overestimation of the mobility. Gaddemane *et al.* [5] have shown the importance of selecting the correct physical models and numerical approximations. For example, the lack of an accurate treatment often leads to an overestimation of mobility, as we suspect is the case in Ref. [22].

In order to clarify the situation, in this paper, we present a theoretical study of electronic-transport properties of monolayer SnSe. The band structure is calculated using density functional theory (DFT) as implemented in Quantum ESPRESSO (QE) [23] with the Perdew-Burke-Enzerhoff generalized-gradient approximation (GGA-PBE) [24] for the

This work has been supported in part by the Taiwan Semiconductor Manufacturing Company, Ltd (TSMC) and the Semiconductor Research Corporation (SRC-nCORE)

exchange-correlation functional, and the Optimized Norm-Conserving Vanderbilt (ONCV) [25] pseudopotential. The phonon spectrum is obtained using QE which employs density functional perturbation theory (DFPT). The electron-phonon scattering rates are calculated to first order using Fermi's golden rule. The electron-phonon Wannier (EPW) [26] code of the QE suite is used to calculate the electron-phonon matrix elements on a fine mesh in the first Brillouin Zone (BZ) by an interpolation strategy based on maximally-localized Wannier functions. Having obtained the electron-phonon scattering rates, we solve the Boltzmann Transport Equation (BTE) using a full-band Monte Carlo method. The Monte Carlo calculations for the low-field mobility calculations are performed assuming a zero electric field and estimating the electron mobility from the diffusion constant, using Einstein's relation. A uniform electric field is assumed when calculating the high-field behavior. A detailed description of the theoretical method is given in Ref. [1]

II. RESULTS AND DISCUSSION

A. Band Structure and Phonon Spectrum

In Fig. 1, we show the calculated band structure of monolayer SnSe along the high-symmetry directions in the first BZ. Monolayer SnSe is an indirect band gap material with a band gap of 0.92 eV, with the conduction band minimum along the X-Γ (C_1 valley) direction and the valence band maximum along Y-Γ (C_2 valley) direction. We observe 5 satellite valleys ($C_1 - C_5$) in the first conduction band (see Fig.1) and the energy difference between each valley to conduction band minimum is given in Table 1. At present, there is disagreement about whether monolayer SnSe is a direct band gap or indirect band gap material as experimental investigations focusing on monolayer SnSe are scarce. Previous theoretical calculation done by Gomes *et al.* [27] found that SnSe is a direct band gap material. On the other hand, Guo et al. [28] have determined that it is an indirect band gap material. This difference stems from the discrepancy caused when using different 'flavors of DFT' which can be seen in 2D TMDs as studied by Gaddemane *et al.* [15]. From our calculations, we observe

[1] Now at imec, Kapeldreef 75, 3001 Heverlee, Belgium

an energy difference between the direct and indirect band gap of 23 meV. Such a small energy difference between the conduction band minima along the Γ-X and Γ-Y directions in monolayer SnSe indicates that the semiconductor properties can be tuned from direct to indirect or vice versa by the application of external controls (*e.g.*, strain).

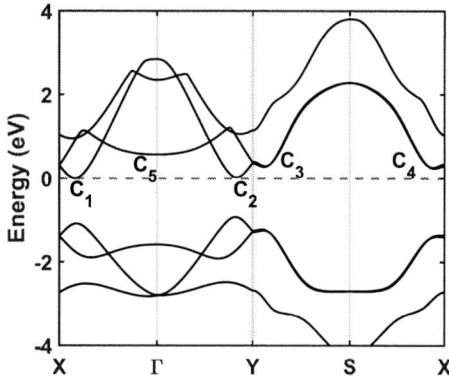

Fig. 1. Calculated band sturcture for monolayer SnSe along the high-symmetry directions of the first Brillouin Zone using the GGA-PBE exchange co-relational functionals

TABLE I
ENERGY DIFFERENCE BETWEEN CONDUCTION MINIMUM VALLEYS

Valleys	ΔE (meV)
$C_1 - C_2$	23
$C_1 - C_3$	272
$C_1 - C_4$	239
$C_1 - C_5$	567

Figure 2 illustrates the phonon dispersion of the 12 phonon branches that result from the presence of 4 atoms in the unit cell. The optical branches corresponding to the longitudinal optical phonons have a maximum frequency away from the Γ point in the BZ. This is a well-known phenomenon known as 'overbending' or 'Kohn anomaly' [29], which has been previously discussed in the case of monolayer h-BN [30] and monolayer InSe [16].

B. Scattering Rates

The density of states (Fig. 3), as well as the electron scattering rates (Fig. 4), are calculated and plotted as a function of electron kinetic energy (averaged over equi-energy surfaces). One can see a step-like increase in the density of states and electron scattering rates at about 25 meV and 560 meV. These steps are due to the onset of intervalley scattering between valleys C_1 and C_2 (at 25 meV) and between valleys C_1 and C_5 (at 560 meV). The intervalley scattering between other valleys cannot be distinguished as the interactions between them are weak.

C. Mobility and Velocity-field Characteristics

From the calculated diffusion constant, we obtain a low-field electron mobility of 72 cm^2 V^{-1} s^{-1}. This mobility is

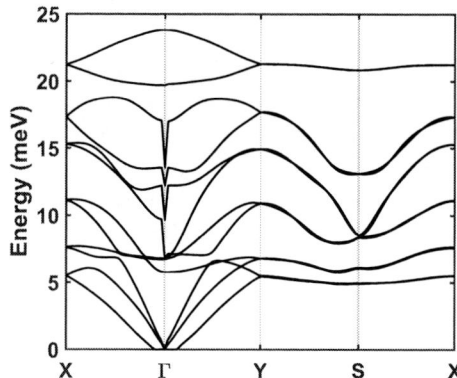

Fig. 2. Calculated phonon spectrum for monolayer SnSe along the high-symmetry directions of the first Brillouin Zone using the GGA-PBE exchange co-relational functionals

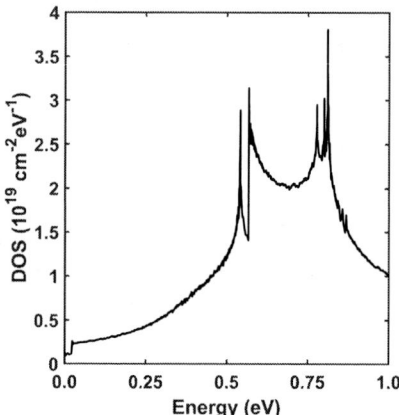

Fig. 3. Density of states plotted as a function of the electron energy with DFT calculations performed using the GGA-PBE functionals

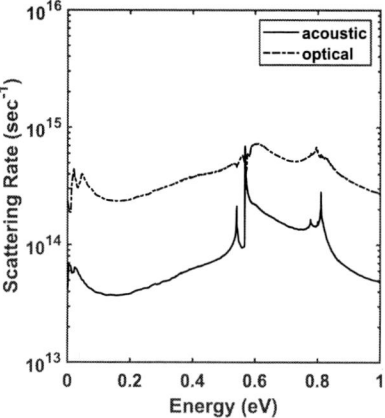

Fig. 4. The room-temperature (T = 300K) electron scattering rates plotted as a function of the initial electron energy (average over equi-energy surfaces) with DFT calculations performed using the GGA-PBE functionals

promising when compared to other widely studied 2D materials, such as phosphorene, silicene, and germanene. Finally, we have studied the velocity- and energy-field characteristics by assuming a homogeneous applied electric field along the zigzag direction [31]. The zigzag direction corresponds to Γ-X direction in k-space. We show, in Fig. 5a, with the dashed line, the drift velocity obtained from the mobility calculated from the diffusion constant.

Figure 6 shows the electron occupation in the first BZ. At low fields, the electrons populate only the regions near the conduction band minima at C_1 and C_2, as seen in Fig. 6a. However, at high fields, above 4×10^4 V/cm, the electrons become hot reaching an energy of 150 meV. However, due to low mobility when compared to bulk silicon, we do not observe the saturation of electron velocity at an electric field of 10^5 V/cm. Due to a significant increase of the average energy, the electrons in the C_1-valley gain enough energy to scatter to other valleys.

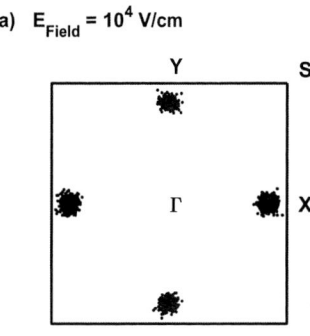

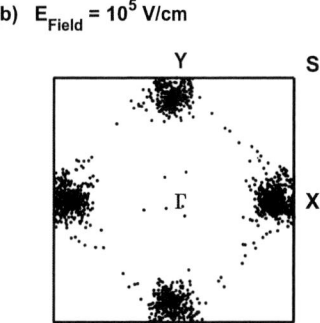

Fig. 6. Room temperature distribution of electrons in the first Brillouin Zone under the acceleration caused by a homogeneous electric field of strength.

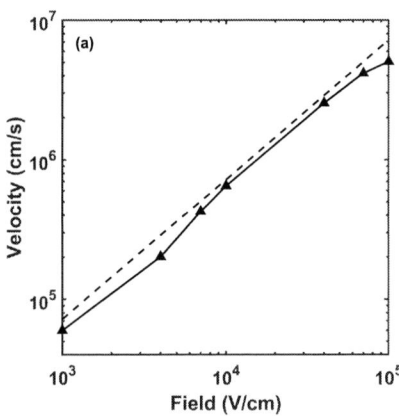

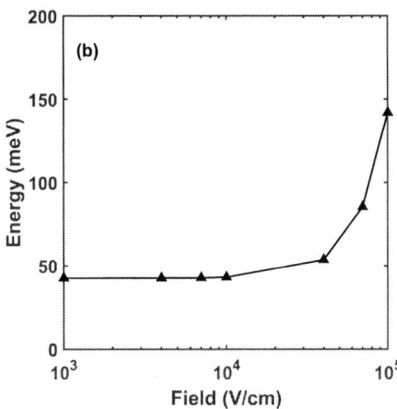

Fig. 5. (a) Velocity-field characteristics of monolayer SnSe and their corresponding (b) average energy-field plot at room temperature (T = 300K).

III. CONCLUSION

In an attempt to clarify the confusing situation regarding experimental and theoretical results regarding electron transport

in SnSe, we have performed DFT calculations as accurate as possible [1]. From these calculations we find a band gap of 0.92 eV, making this material suitable for complementary-logic applications. Using the full-band Monte Carlo method, we have calculated the low-field electron mobility and velocity-field characteristics of monolayer SnSe. We found an electron mobility of 72 $cm^2\,V^{-1}\,s^{-1}$.

In order to assess the potential of monolayer SnSe as a candidate for a channel material in high-performance in field-effect devices, it is interesting to compare the intrinsic mobility of monolayer SnSe with the electron mobility of 1nm-thick Si slab. The calculated electron mobility of monolayer SnSe (72 $cm^2\,V^{-1}\,s^{-1}$) exceeds the electron mobility of 1nm-thick Si (50 $cm^2\,V^{-1}\,s^{-1}$), when accounting for surface roughness [32] under a transverse gate-field of 10^6 V/cm.

Considering the relatively large electron mobility we have obtained here and the recent progress in its growth techniques, monolayer SnSe should be viewed with interest for VLSI applications [33].

ACKNOWLEDGMENT

We acknowledge Dr. Edward Chen for his idea to study monolayer SnSe and also for his support.

REFERENCES

[1] G. Gaddemane, W. G. Vandenberghe, M. L. Van De Put, E. Chen, M. V. Fischetti, "Monte-Carlo study of electronic transport in non-σ_h-symmetric two-dimensional materials: Silicene and germanene," J. Appl. Phys., vol. 124, p. 044306, 2018.

[2] M. Houssa, E. Scalise, K. Sankaran, G. Pourtois, V. V. Afanas'ev, A. Stesmans, "Electronic properties of hydrogenated silicene and germanene," Appl. Phys. Lett., vol. 99, p. 223107, 2011.

[3] N. J. Roome and J. D. Carey, "Beyond graphene: Stable elemental monolayers of silicene and germanene," ACS Appl. Mater. Interfaces, vol. 6, pp. 7743-7750, 2014.

[4] M. E. Davila, L. Xian, S. Cahangirov, A. Rubio, G. L. Lay, "Germanene: A novel two-dimensional germanium allotrope akin to graphene and silicene," New J. Phys., vol. 16, p. 095002, 2014.

[5] G. Gaddemane, W. G. Vandenberghe, M. L. Van De Put, S. Chen, S. Tiwari, M. V. Fischetti, "Theoretical studies of electronic transport in monolayer and bilayer phosphorene: A critical overview," Phys. Rev. B., vol. 98, p. 115416, 2018.

[6] A. Castellanos-Gomez, L. Vicarelli, E. Prada, J. O. Island, K.L. Narasimha-Acharya, S. I. Blanter, D. J. Groenendijk, M. Buscema, G. A. Steele, J. V. Alvarez, et al. "Isolation and characterization of few-layer black phosphorus," 2D Mater., vol. 1, p. 025001, 2014.

[7] L. Li, Y. Yu, G. J. Ye, Q. Ge, X. Ou, H. Wu, D. Feng, X. H. Chen, Y. Zhang, "Black phosphorus field-effect transistors," Nat. Nanotechnol., vol. 9, pp. 372-377, 2014.

[8] H. Liu, A. T. Neal, Z. Zhu, Z. Lou, X. Xu, D. Tomanek, P. D. Ye, "Phosphorene: An unexplored 2D semiconductor with high hole mobility," ACS Nano., vol. 8, pp. 4033-4041, 2014.

[9] J. Qiao, X. Kong, Z. -H. Hu, F. Yang, W. Ji, "High-mobility transport anisotropy and linear dichroism in few-layer black phosphorus," Nat. Commun., vol. 5, p. 4475, 2014.

[10] A. Rawat, N. Jena, A. De Sarkar, "A comprehensive study on carrier mobility and artificial photosynthetic properties in group VI B transistion metal dichalcogenide monolayers," J. Mater. Chem., vol. A 6, pp. 8693-8704, 2018.

[11] T. Sohier, D. Campi, N. Marzari, M. Gibertini, "Mobility of two-dimensional materials from first principles in an accurate and automated framework," Phy. Rev. Mater., vol. 2, p. 114010, 2018.

[12] K. Kaasbjerg, K. S. Thygesen, K. W. Jacobsen, "Phonon-limited mobility in n-type single-layer MoS2 from first principles," Phy. Rev. B, vol. 85, p. 115317, 2012.

[13] Z. Jin, X. Li, J. T. Mullen, K. W. Kim, "Intrinsic transport properties of electrons and holes in monolayer transistion-metal dichalcogenides," Phy. Rev. B, vol. 90, p. 045422, 2014.

[14] W. Li, "Electrical transportt limited by electron-phonon coupling from Boltzmann transport equation: An ab initio study of Si, Al, and MoS2," Phy. Rev. B, vol. 92, p. 075405, 2015.

[15] G. Gaddemane, S. Gopalan, M. L. Van de Put, M. V. Fischetti, "Limitations of ab inito methods to predict the electronic-transport properties of two-dimensional semiconductors: the computational example of 2H-phase transistion metal dichalcogenides," J. Comput. Electron, 2020. https://doi.org/10.1007/s10825-020-01526-1

[16] S. Gopalan, G. Gaddemane, M. L. Van de Put, M. V. Fischetti, "Monte Carlo study of electronic transport in monolayer InSe," Materials, vol. 12, no. 24, p. 4210, 2019.

[17] W. Shi, M. Gao, J. Wei, J. Gao, C. Fan, E. Ashalley, H. Li, Z. Wang, "Tin Selenide (SnSe): Growth, Properties, and Applications," Adv. Sci. vol. 5, no. 4, p. 1700602, 2018.

[18] T. Inoue, H. Hiramatsu, H. Hosono, T. Kamiya, "Heteroepitaxial growth of SnSe films by pulsed laser deposition using Se-rich targets," J. Appl. Phys., vol. 118, p. 205302, 2015.

[19] B. D. Tracy, X. Li, X. Liu, J. Furdyna, M. Dobrowolska, D. J. Smith, "Characterization of structural defects in SnSe2 thin films grown by molecular beam epitaxy on GaAs (111)B substrates," J. Cryst. Growth, vol. 453, p. 58, 2016.

[20] L.-C. Zhang, G. Qin, W. -Z. Fang, H. -J. Cui, Q. -R. Zheng, Q. -B. Yan, G. Su, "Tinselenidene: a Two-dimensional auxetic material with ultralow lattice thermal conductivity and ultrahigh hole mobility," Sci. Rep., vol. 6, p. 19830, 2016.

[21] L. B. Shi, M. Yang, S. Cao, Q. You, Y. Y. Niu, Y. Z. Wang, "Elastic behavior and intrinsic carrier mobility for monolayer SnS and SnSe: First-principles calculations," Appl. Surf. Sci., vol. 492, p. 435, 2019.

[22] L. -B. Shi, M. Yang, S. Cao, Q. You, Y. -Y. Niu, Y. -Z. Wang, "Elastic behavior and intrinsic carrier mobility for monolayer SnS and SnSe: First principles calculation," Appl. Surf. Sci., vol. 492, pp. 435-448, 2019.

[23] P. Giannozzi et al. "Quantum ESPRESSO: A modular and open-source software project for quantum simulations of materials," J. Phys. Condens. Matter, vol. 21, p. 395502, 2009.

[24] J. P. Perdew, Y. K. Burke, M. Ernzerhof, "Generalized Gradient Approximation made simple," Phy. Rev. B, vol. 88, p. 085117, 2013.

[25] D. R. Hamann, "Optimized norm-conserving Vanderbilt pseudopotentials," Phy. Rev. B, vol. 88, p. 085117, 2013.

[26] S. Ponce, E. R. Margine, C. Verdi, F. Guistino, "EPW: Electron-phonon coupling, transport and superconducting properties using maximally localized Wannier functions," Comput. Phys. Commun., vol. 209, pp. 116-133, 2016.

[27] L. C. Gomes and A. Carvalho, "Phosphorene analogues: Isoelectronic two-dimensional group-IV monochalcogenides with orthorhombic structure," Phys. Rev. B, vol. 92, p. 085406, 2015.

[28] S. -D. Guo and Y. H. Wang, "Thermoelectric properties of orthorhombic group IV-VI monolayers from the first-principles calculations," J. Appl. Phys., vol. 121, p. 034302, 2017.

[29] W. Kohn, "Image of the fermi surface in the vibration spectrum of a metal," Phys. Rev. Lett., vol. 2, p. 393, 1959.

[30] K. H. Michel, B. Verberck, "Theory of elastic and piezoelectric effects in two-dimensional hexagonal boron nitride," Phy. Rev. B, vol. 80, p. 224301, 2009.

[31] A. Shafique and Y. -H. Shin, "Thermoelectric and phonon transport properties of two-dimensional IV-VI compounds," Scientific Reports, vol. 7, p. 506, 2017.

[32] F. Gamiz and M. V. Fischetti, "Monte Carlo simulation of double-gate silicon-on-insulator inversion layers: The role of volume inversion," J. Appl. Phys., vol. 89, p. 5478, 2001.

[33] L. Li, Z. Chen, Y. Hu, X. Wang, T. Zhang, W. Chen, Q. Wang, "Single-Layer single-crystalline SnSe nanosheets," J. Am. Chem. Soc., vol. 135, no. 4, p. 1213, 2013.

Transient simulation of graphene FET gated by electrolyte medium

Koki Arihori, Matsuto Ogawa and Satofumi Souma[†]
Department of Electrical and Electronic Engineering,
Kobe University, Kobe 657-8501, Japan
[†]email: ssouma@harbor.kobe-u.ac.jp

Junko Sato-Iwanaga and Masa-aki Suzuki
Innovation Promotion Sector, Panasonic Corporation,
Osaka 571-8508, Japan

Abstract—We present a numerical study on the electrical conduction characteristics of the graphene channel FET with electrolyte medium for gate control. By using the tight-binding formalism to calculate the electronic band structure and the Nernst-Planck-Poisson (NPP) equation to calculate the formation of the electric double layer at the interface of the ionic liquid, we found that the drain current after the EDL is formed is almost independent of the IL thickness, while the transient behavior is greatly influenced by the thickness of ionic liquid. In addition, we present our simulation results for the case of solid electrolyte gate, where the effect of finite ion concentration in the solid electrolyte has been successfully taken into account appropriately by using the extended NPP equation.

I. INTRODUCTION

Graphene has been considered as an attractive material for the nanoelectronics applications [1]. Here the wide range of controllability in the electric field applied to graphene is an important issue in realizing wide range of controllability in carrier concentration and electronic current.

Among various approaches to apply electric field to graphene, including traditional top gating via gate insulator, the use of electrolyte medium [i.e., ionic liquid (IL) or ionic solid (IS)] is beneficial in obtaining high carrier density in graphene, where the formation of an electric double layer (EDL) with the thickness ~ 1 nm at the electrolyte medium/graphene interface is essential in inducing high electric field and consequently a high density of carriers [2], [3], [4], [5]. However, it has not been explored theoretically enough in how the thicknesses, ion concentrations, and the diffusion coefficients of IL and IS play the roles in the resulting electronic current flowing through graphene. With such motivation, we present a numerical study on the electrical conduction characteristics of the graphene channel FET using electrolyte medium for gate control.

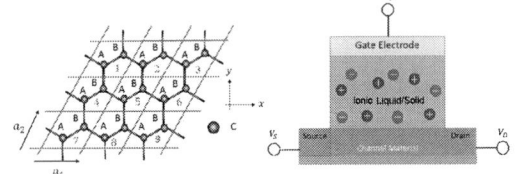

Fig. 1. (Left): Crystal structure of graphene. (Right): Schematic illustration of IL gated graphene FET.

II. THEORETICAL FORMALISM

A. Tight-binding (TB) formalism

In Fig. 1 we show the schematic illustration of the device structure considered in this study, where the graphene channel part is described by the TB method and the electrolyte medium (IL or IS) part is described by the Nernst-Planck-Poisson equation. Their coupling is assumed to be electrostatic and is also described by the Poisson's equation. Electronic properties of graphene can be calculated by solving the eigenvalue problem $H(\boldsymbol{k})|\psi_{lk}\rangle = E_l(\boldsymbol{k})|\psi_{lk}\rangle$ [$l = 1$ (2) corresponds to the valence (conduction) band] with the Hamiltonian

$$H(\boldsymbol{k}) = \begin{pmatrix} 0 & H_{\mathrm{BA}}^*(\boldsymbol{k}) \\ H_{\mathrm{BA}}(\boldsymbol{k}) & 0 \end{pmatrix}, \qquad (1)$$

$$H_{\mathrm{BA}}(\boldsymbol{k}) = -t\left(2e^{ik_y a_0/2}\cos(k_x\sqrt{3}a_0/2) + e^{-ik_y a_0}\right), \quad (2)$$

where a_0 is the spacing between nearest-neighbor atoms and $t = 2.7$ eV is the nearest neighbor hopping energy [6].

B. Current calculation method under the ballistic condition

Once the band structure $E_l(\boldsymbol{k})$ is calculated, the electronic current flowing through graphene attached at both ends to ideal electrodes can be calculated as

$$I = \frac{q}{\hbar}\frac{1}{N_{\mathrm{kp}}S_{\mathrm{UC}}}\sum_{l}\sum_{\boldsymbol{k}\in\mathrm{BZ}}^{N_{\mathrm{kp}}}|v_l(\mathbf{k})|$$
$$\times\left[f\left(E_l(\mathbf{k}) + U_{\mathrm{C}} - E_{\mathrm{FL}}\right) - f\left(E_l(\mathbf{k}) + U_{\mathrm{C}} - E_{\mathrm{FR}}\right)\right]. \quad (3)$$

Here N_{kp} is the total number of $\mathbf{k}$ sampling points within the 1st Brillouin zone (BZ), $q = -e$ is the charge of an electron, $E_l(\mathbf{k})$ and $v_l(\mathbf{k})$ are the energy and the group velocity for the energy eigenstate $|\psi_{lk}\rangle$, and $f(E)$ is the Fermi distribution function, where $E_{\mathrm{FL/FR}}$ is the Fermi energy in the left/right electrode so that $V_{\mathrm{D}} = (E_{\mathrm{FL}} - E_{\mathrm{FR}})/e$ is the drain voltage, $U_{\mathrm{C}} = -e\varphi(z_{\mathrm{gra}})$ with $\varphi(z_{\mathrm{gra}})$ being the electrostatic potential at the graphene layer.

C. Nernst-Planck-Poisson equation

The electrostatic potential $\varphi(z_{\mathrm{gra}})$ at the graphene layer, required in Eq. (3) is determined by the coupling with the

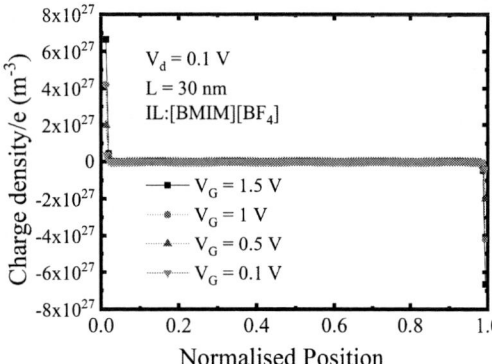

Fig. 2. Charge density in units of e in the IL region of IL-gated FET for various gate voltages at the time 50 ns.

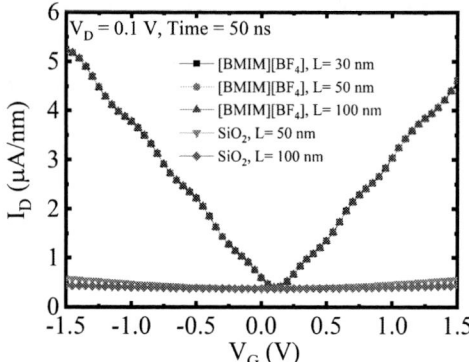

Fig. 3. I_D-V_G characteristics for various thicknesses of IL as indicated in the figure at the time 50 ns for IL-gated FET. Results for the conventional FET with SiO_2 gate insulator with the thickness 50 nm and 100 nm are also shown for comparison.

electrolyte (IL or IS) region above graphene. The electrolyte region is described by the Nernst-Planck-Poisson equation:

$$\frac{dc_j(z,t)}{dt} = -\frac{d}{dz}F_j(z,t) + [G_j(z,t) - R_j(z,t)], \quad (4)$$

$$F_j(z,t) = -D_j\frac{dc_j(z,t)}{dz} + c_j(z,t)q_j M_j\left(-\frac{d\varphi(z,t)}{dz}\right), \quad (5)$$

$$-\frac{d}{dz}\varepsilon(z)\frac{d\varphi(z,t)}{dz} = q_+ c_+(z,t) + q_- c_-(z,t) + \rho_{\mathrm{gra}}(z,t). \quad (6)$$

Here $c_j(z,t)$ is the ion density for cation/anion ($j = +/-$), $F_j(z,t)$ is the flux density along the z-direction, $G_j(z,t)$ and $R_j(z,t)$ are the generation and recombination rate, $\varphi(z)$ is the electrostatic potential in whole region including electrolyte (IL or IS) and graphene, D_j is the diffusion coefficient, $M_j = D_j/k_B T$ is the mobility, $q_{\pm} = \pm e$, and $\rho_{\mathrm{gra}}(z,t)$ is the charge density within the graphene layer. We used parameter values for representative IL and IS, which are [BMIM][BF4] [7] and LiPO4 [8], respectively. In the actual simulations, the ion density $c_j(z,t)$ is initially assumed to be homogeneous with the values $c_{\pm} = 1.6 \times 10^{27}$ m^{-3} and $c_{\pm} = 6.51 \times 10^{27}$ m^{-3} in IL ([BMIM][BF4]) and IS (LiPO4), respectively. We also note that, in the case of IS (LiPO4) only the cations (Li$^+$ ions) are mobile ions [8], and they move through a limited number of vacancies, so that the diffusion coefficient D has to be generalized to take into account the maximum density ν as $D'_j = D_j/[1 - c_j(z,t)/\nu]$ [8], [9], [10], [11]. We assume $\nu = 7.0 \times 10^{27}$ m^{-3} for IS throughout this paper.

III. RESULTS AND DISCUSSIONS

We first consider the case of IL-gated FET. Figure 2 shows the cation and anion density distributions in IL ([BMIM][BF4]) region for various gate voltages. Here it is seen that the EDL has been formed near the IL/graphene and IL/gate electrode interfaces by applying a gate voltage, and the ion densities near the interfaces increased as the gate voltage increased.

In Fig. 3 we show the comparison of I_D-V_G characteristics for various thicknesses of ionic liquid as indicated in the figure

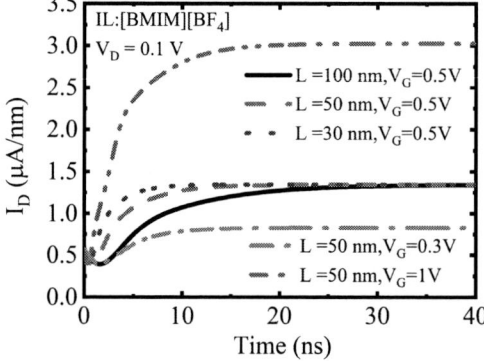

Fig. 4. Drain current I_D as a function of time for various IL thicknesses and applied gate voltages.

at the time 50 ns when the EDL formation is completed. Results for the conventional FET with SiO_2 gate insulator with the thickness 50 nm and 100 nm are also shown for comparison. Here it was found that the drain current when the EDL formation was completed is almost independent of the IL thickness L. This is a reasonable result since the thickness of the EDL is around 1 nm and the amount of charge accumulated in graphene is determined by EDL. It should be noted that the use of IL enables much larger I_D than the conventional SiO_2 gate insulator with the same thickness.

In Fig. 4 we compare how the transient characteristics until the EDL is formed (until the current converges) differs depending on the gate voltage and the thickness of the IL. We first observe that the initial current is independent of L and V_G. This is because the surface potential at the graphene is initially assumed to be zero. Importantly, the time required for the current to be saturated is longer for thicker IL thickness L, but is almost independent of V_G. This is interpreted to be due to the longer time required for ion to move toward the interface for thicker IL.

Next we consider the case of IS-gated FET. Figure 5 shows

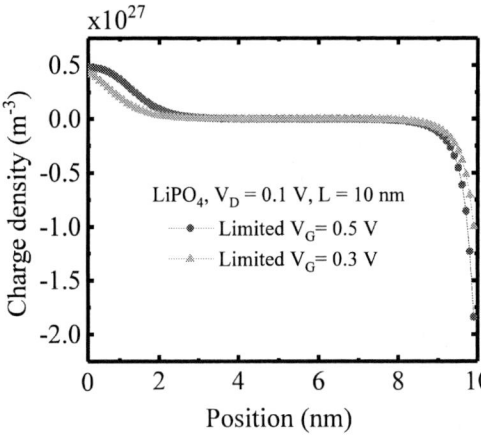

Fig. 5. Charge density distribution in units of e in IS region with thickness 10 nm in IS-gate FET for various gate voltages at the time 700 μs. "Limited" stands for the limited Li$^+$ density situation imposed by the maximum density ν.

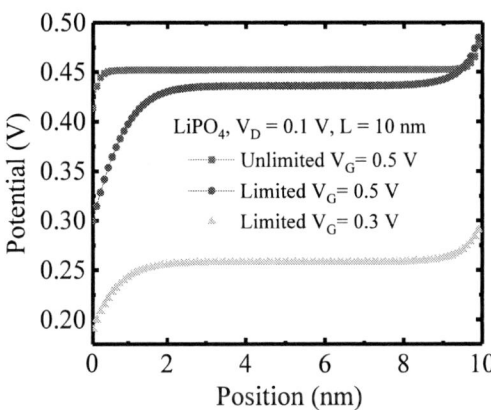

Fig. 6. Electrostatic potential distributions in IS for various gate voltages at the time 700 μs. "Limited" and "unlimited" stand for the limited and unlimited Li$^+$ density situations, where the former case the density is imposed by the maximum density ν.

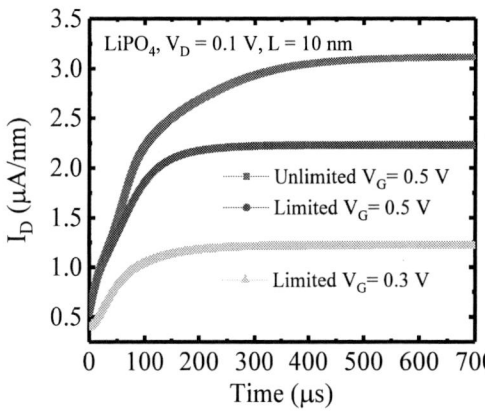

Fig. 7. Drain current I_D as a function of time for various applied gate voltages for IS-gated FET.

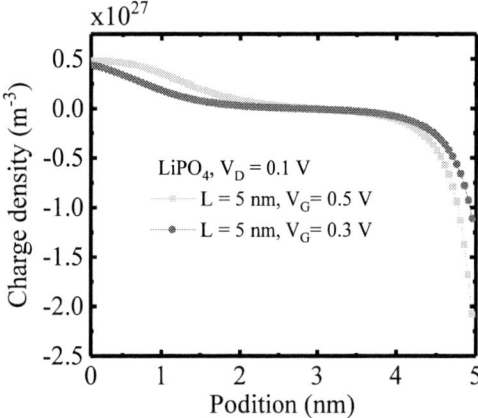

Fig. 8. Charge density in units of e in IS with thickness 5 nm for various gate voltages at the time 700 μs.

the cation and anion density distributions in IS (LiPO$_4$) region for various gate voltages. Here, by applying a positive gate voltage to the IS, Li$^+$ ions start to fill up subsequently from the gate electrode side to the graphene side, and the flat charge density region (fully occupied Li$^+$ region) starts to appear near the graphene side, where the charge density at the fully occupied Li$^+$ region is determined by the imposed maximum cation density $\nu = 7.0 \times 10^{27}$ m^{-3} subtracted by the homogeneous anion density $c_- = 6.51 \times 10^{27}$ m^{-3}. In addition, the length of the fully occupied region increases with the gate voltage. This is interpreted to be due to the movement of Li$^+$ ions through the limited numbers of vacancies.

In Fig. 6 we show the internal electrostatic potential in the IS region for various gate voltages. By placing an upper limit on the cation density (maximum density ν), the potential drop in the fully occupied region (of Li$^+$ ions) becomes large compared to unlimited case, indicating that the gate electric

field is screened within the fully occupied region by Li$^+$ ions and the graphene surface is not effectively influenced by the gate voltage. In addition, the potential drop in the fully occupied region becomes larger as the gate voltage increases. This is due to the fact that the length of the fully occupied region of Li$^+$ ions becomes longer as the gate voltage increases.

Figure 7 is to compare how the transient characteristics of I_D are influenced by the upper limit of the ion density and value of the gate voltage through the EDL formation in IS. Here we observe that the saturated values of the drain current at around 700 μs is diminished by placing the upper limit of the ion density. This is interpreted to be due to the lower surface potential at the graphene in the limited case shown in Fig. 6. On the other hand, we can see that the current saturation time (time required for the current to be saturated) becomes shorter by placing the upper limit of the cation density. This is because in the unlimited case the Li$^+$ ions are required to reach the graphene surface to establish the steady state, while in the limited case they are allowed to move only until

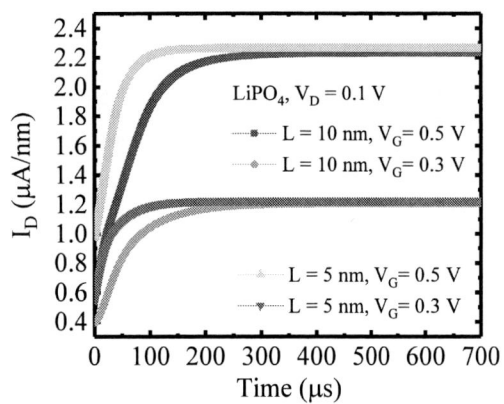

Fig. 9. Drain current I_D as a function of time for various IS thicknesses and applied gate voltages.

around the edge of the fully occupied region ($\sim$ 2 nm in Fig. 6), meaning the shorter time required to establish the steady state in the limited case. Therefore the value of the saturation current and the current saturation time are in the relationship of trade-off. Moreover, we can also see that the saturation time in IS case is much longer than the IL case. This is because of the four orders of magnitude smaller values of the diffusion coefficient in IS than in IL.

In Fig. 8 we plotted the spatial distributions of charge densities in the IS region of thinner IS-gate FET with the thickness of 5 nm for two different gate voltages. Here it can be seen that the spatial extent of the region occupied fully by Li$^+$ ions is almost the same ($\sim$ 1 nm for $V_G = 0.5$ V) as for the IS thickness of 10 nm shown in Fig. 5. This observation suggests that the spatial extent of the region fully occupied by the Li$^+$ ions is determined mainly by the gate voltage. Another important observation here is that the spatial extent of the charge neutral region at around the center of IS becomes narrower as the thickness of IS decreases and the gate voltage increases. More detailed analyses for such situations including the case of extremely thinner IS will be presented elsewhere.

Finally in Fig. 9 we compare how the transient characteristics of the drain current in IS-gated FET vary with the gate voltage and the IS thickness. Important observations here are that the current saturation time is longer for thicker IS case and the saturation current value is almost independent of the IS thickness. These are the similar characteristics as in the IL-gated FET case.

IV. CONCLUSION

We presented a numerical study on the electrical conduction characteristics of the graphene channel FET using electrolyte medium for gate control. By using the tight-binding formalism to calculate the electronic band structure and the Nernst-Planck-Poisson (NPP) equation to calculate the formation of the electric double layer of the ionic liquid, we found that the drain current after the EDL is formed is almost independent of the IL thickness, while the transient behavior is greatly influenced by the IL thickness L. Moreover, the electrical conduction characteristics of the graphene channel FET with the ionic solid gate have also been successfully studied by using the extended NPP equation, where the finite ion concentration in the solid electrolyte has been taken into account appropriately.

ACKNOWLEDGMENT

This work was partially supported by JSPS KAKENHI Grant No. 19H04546.

REFERENCES

[1] K. S. Novoselov, V. I. Fal'ko, L. Colombo, P. R. Gellert, M. G. Schwab, and K. Kim, Nature, **490**, 192 (2012).
[2] X. He, N. Tang, L. Gao, J. Duan, Y. Zhang, F. Lu, F. Xu, X. Wang, X. Yang, W. Ge, and B. Shen, Appl. Phys. Lett. **104** 143102 (2014).
[3] T. Tsuchiya, K. Terabe, and M. Aono, Appl. Phys. Lett. **105**, 183101 (2014)
[4] M. Philippi, I. Gutiérrez-Lezama, N. Ubrig, and A. F. Morpurgo, Appl. Phys. Lett. **113**, 033502 (2018).
[5] A. Kumar, P. B. Pillai, X. Song, and M. M. De Souza, ACS Appl. Mater. Interfaces **10**, 19812 (2018).
[6] S. Souma, M. Ueyama and M. Ogawa, Appl. Phys. Lett. **104** 213505 (2014).
[7] S. Tsuzuki, W. Shinoda, H. Saito, M. Mikami, H. Tokuda, and M. Watanabe, J. Phys. Chem. B **113**, 10641 (2009).
[8] D. Danilov, R. A. H. Niessen, and P. H. L. Notten, J. Electorochem. Soc. **158**, A215 (2011).
[9] M. Landstorfer, S. Funken and T. Jacob, Phys. Chem. Chem. Phys. **13**, 12817 (2011).
[10] M. S. Kilic, M. Z. Bazant, A. Ajdari, Phys. Rev. E **75**, 021503 (2007).
[11] S. Braun, C. Yada, and A. Latz, J. Phys. Chem. C **119**, 22281 (2015).

Quantum Transport Simulations of Phosphorene Nanoribbon MOSFETs: Effects of Metal Contacts, Ballisticity and Series Resistance

Mirko Poljak[*] and Mislav Matić
Computational Nanoelectronics Group, Micro and Nano Electronics Laboratory
Faculty of Electrical Engineering and Computing, University of Zagreb, Zagreb, Croatia
[*]E-mail: mirko.poljak@fer.hr

Abstract—**Performance of phosphorene nanoribbon (PNR) MOSFETs at "3 nm" logic technology node is studied using atomistic quantum transport simulations, with an emphasis on the impact of metal contacts, series resistance and transport ballisticity. We find that realistic metal contacts decrease drain current by up to 70%, which corresponds to more than 1400 $\Omega\mu$m in contact resistance (R_{SD}). On the other hand, setting R_{SD} to 270 $\Omega\mu$m, as foreseen by the International Roadmap for Devices and Systems (IRDS), PNR MOSFETs would need to operate at 50% to 70% of their ballistic limit, depending on PNR width, in order to meet IRDS targets.**

Keywords—*ballisticity, FET, metal contacts, nanoribbon, NEGF, phosphorene, series resistance, quantum transport*

I. INTRODUCTION

Phosphorene is a promising 2D material as an alternative to graphene or transition metal dichalcogenides for end-of-roadmap logic devices due to its layered structure, acceptable bandgap, appropriate carrier effective masses and mobility, and a higher immunity to crystal defects than e.g. graphene [1]–[5]. Large-area phosphorene MOSFETs have been reported experimentally, demonstrating promising performance and contact resistance values [6]–[8]. In terms of modeling and simulation, phosphorene MOSFETs have been studied mostly in the ballistic limit, whereas in cases where dissipative transport is covered the contacts have been assumed to be ideal i.e. having identical structure as the channel [3], [9], [10]. Quantum confinement by width engineering in phosphorene nanoribbons (PNRs) provides an additional way of tuning the electronic and transport features relevant for logic FET applications, such as the bandgap and charge carrier effective masses [2], [4], [11], [12]. In this work, we investigate the performance of PNR MOSFETs using atomistic quantum transport modeling and simulations coupled with top-of-the-barrier (ToB) model. We study the impact of metal contacts, parasitic series resistance, transport ballisticity and PNR width scaling on several FET figures-of-merit (FOMs), and find conditions under which PNR FETs could fulfill industry requirements at the "3 nm" logic CMOS technology node.

II. METHODOLOGY

The 15 nm-long nanoribbon Hamiltonians are expressed in an atomistic tight-binding (TB) basis [13]. Quantum transport based on non-equilibrium Green's function (NEGF) formalism is employed to find geometry-dependent material

properties such as density of states (DOS) and transmission. In terms of source/drain (S/D) contacts attached to the channel within NEGF simulations we study two cases. First, ideal contacts assume that S/D contact regions are semi-infinite semiconducting PNRs identical to the channel, and here the surface Green's functions (SGFs) are solved by the Sancho-Rubio method [14]. Second, metal contacts are modeled using the wide-band limit (WBL) approximation in which we set the contact-channel coupling strength to $t = 3$ eV and metal DOS at the Fermi level of $g(E = E_F) = 0.2$ eV^{-1} (while we do not assume any specific metal and ignore possible Schottky barriers, the metal DOS value corresponds to that of Au(111) [15]). The WBL replaces iterative procedures for calculating the SGFs, and these two parameters (t and $g(E = E_F)$) lead to constant imaginary elements in the surface Green's function of metal contacts equal to $\Sigma^R_{S/D} = -0.9i$ eV.

Ballistic current-voltage (I-V) characteristics are obtained using top-of-the-barrier (ToB) model that self-consistently solves electrostatics, i.e. Poisson equation, coupled to DOS and quantum transmission results from NEGF [16], [17]. Since transport is ballistic, the drain current is found from the Landauer formula, after self-consistence is achieved for ToB potential and electron density. A major drawback of the ToB model is that tunneling is not included, which makes it reliable for predicting FET performance only for gate lengths above approximately 15 nm where direct S-D tunneling should be negligible given the phosphorene effective masses [16]. In addition, bandgap values for the examined PNRs are larger than ≈ 1.5 eV, which should be high enough to suppress band-to-band tunneling (BTBT) at 0.7 V supply voltage (V_{DD}).

For our 15 nm-long PNR MOSFETs we analyze the impact of nanoribbon width (W) scaling in the ≈ 0.5 nm to ≈ 5.5 nm range, assume a gate oxide with equivalent oxide thickness (EOT) of 1 nm, undoped channel, and S/D doping of $m = 0.001$, where m is the molar fraction of areal density of phosphorus atoms. Quasi-Fermi levels in S/D regions are determined to maintain charge neutrality. In order to set the threshold voltage (V_{th}) as projected in the International Roadmap for Devices and Systems (IRDS) at "3 nm" logic node (N3) [18] and provide meaningful results, we set a significantly tighter requirement for the OFF-state current (I_{OFF}) equal to 0.87 nA/μm for all devices in this work. The IRDS sets this value at 10 nA/μm, however, ToB model results in perfect subthreshold slope (SS) of 60 mV/dec in all cases, whereas the projected SS in IRDS at N3 equals 82 mV/dec.

This work was sponsored by the Croatian Science Foundation (HRZZ) under the project CONAN2D (UIP-2019-04-3493).

III. RESULTS AND DISCUSSION

Transfer characteristics reported in Fig. 1 show normal FET operation with V_{th} of about 0.25 V. In case of ideal contacts (Fig. 1a), maximum achievable normalized drain current (ON-state current, I_{ON}) is 1.85 mA/μm. We note that *I-V* curves are closely spaced, indicating a weak modification of the normalized I_{ON} with nanoribbon width. However, drain current is significantly lower in PNR FETs with metal contacts (Fig. 1b), reaching up to 0.52 mA/μm, i.e. 28% of the maximum obtained for ideal contacts. In this case the *I-V* characteristics are again matched except for the narrowest PNR with $W = 0.49$ nm that exhibits significantly lower I_{ON}.

Influence of nanoribbon width scaling on I_{ON}, shown in Fig. 2, reveals a monotonic decrease of the absolute I_{ON} (Fig. 2a) and a non-monotonic behavior of the width-normalized I_{ON} (Fig. 2b). For the absolute ON-state current values, when the width increases, I_{ON} increases linearly in narrower PNR FETs, while it rises sub-linearly in wider PNRs. The shape of the I_{ON} - W curve in Fig. 2a, in the case of ideal contacts, results in a non-trivial width-dependence of the normalized I_{ON} in Fig. 2b that exhibits a local minimum for $W = 3.43$ nm. In the case of metal contacts, normalized I_{ON} seems to saturate at about 0.5 mA/μm for PNR widths above 1.5 nm. This behavior is expected since the absolute I_{ON} in Fig. 2a increases sub-linearly when the PNR width increases.

These results show that the inclusion of metal contacts reduces the current in comparison to ideal contact regions, meaning that the usually assumed nanoribbon S/D regions significantly over-estimate FET current-driving capabilities. Metal contacts described with WBL indeed have a wide energy-band, allowing electron injection from contacts for the entire energy-band of the channel. However, the bandstructure of S/D regions is not identical to that of the channel region, which causes qualitative and quantitative changes. Namely, the shape of the allowed energy states changes considerably, and a non-unitary transmission through the channel is observed due to quantum interference that is possible in coherent transport. These effects are seen as oscillations in transmission and DOS and, moreover, metal contacts lead to increased DOS and transmission within the bandgap. This topic is not analyzed further in this work since the main conclusions are known from previous studies on graphene nanoribbons (GNRs) with non-ideal contacts [19]. Therefore, we can conclude that PNR FETs with metal contacts generally exhibit lower I_{ON}, higher I_{OFF}, and a significantly lower ON-OFF ratio as discussed in the next paragraph.

Figure 3 reports I_{ON} vs. ON-OFF current ratio for PNR FETs with ideal and metal contacts. Generally, PNR devices with ideal nanoribbon S/D regions exhibit greater I_{ON} and ON-OFF current ratio, with the latter being in excess of 10^6. The inserted literature data for other contending 2D-material-based MOSFETs [9], [20]–[22] shows that PNR FETs with ideal contact regions exhibit similar performance as germanene FETs or GNR FETs. Moreover, PNR FETs with ideal contacts show higher ON-OFF current ratio than arsenene, antimonene and silicene FETs. Even metal-contact PNR devices fulfill IRDS requirement for the ON-OFF ratio ($> 2 \cdot 10^5$), however, neither group of the examined PNR FETs satisfies the I_{ON} requirement (1.95 mA/μm) since the maximum obtained equals 1.85 mA/μm for the device with 4.41 nm-wide PNR.

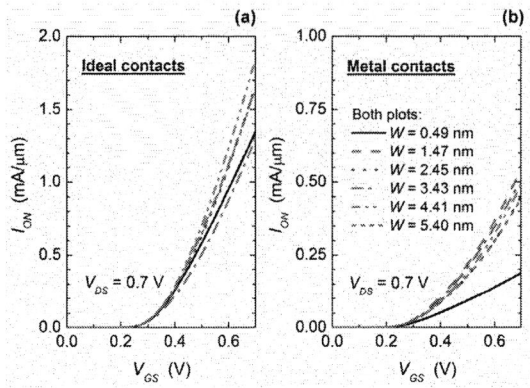

Fig. 1. Transfer characteristics of PNR MOSFETs at $V_{DS} = 0.7$ V for (a) ideal nanoribbon contacts and (b) metal contacts in the wide-band limit. Including metal contacts reduces I_{ON} by up to 72% (equiv. to $R_{SD} = 1470$ Ωμm).

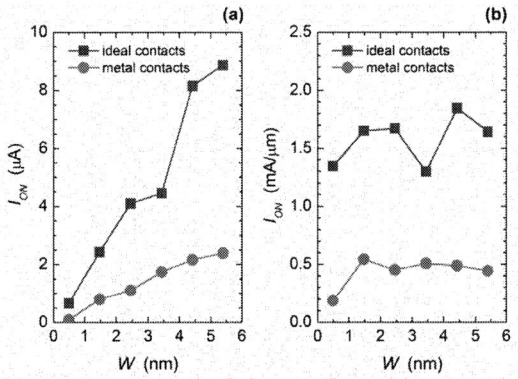

Fig. 2. ON-state current versus PNR width with current (a) expressed in absolute units and (b) normalized by PNR width. Absolute I_{ON} decreases monotonically while normalized I_{ON} exhibits local minima/maxima.

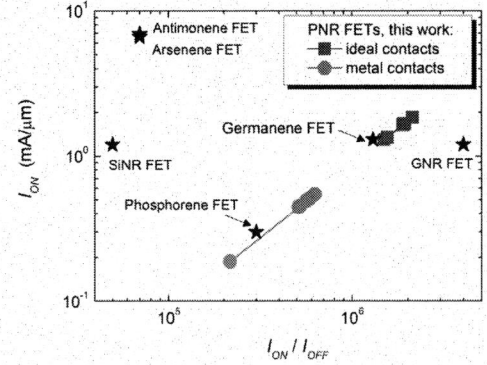

Fig. 3. ON-state current vs. ON-OFF current ratio for PNR MOSFETs with ideal and metal contacts. Data for other 2D material FETs from literature is inserted for comparison.

In addition to current-related FOMs, we investigate the intrinsic delay time (τ), as an intrinsic limit to switching speed, and power-delay product (PDP), which gives the limit of dynamic power dissipation. The PDP vs. delay plot is reported in Fig. 4 for PNRs with ideal and metal contacts, with inserted literature data for other 2D material FETs. We observe that PDP generally increases with increasing delay, and while both PNR FET groups exhibit similar switching energies of up to 0.04 fJ/μm, delay time differs significantly. In the case of ideal contacts, delay reaches up to only 31 fs, while the lower driving currents in devices with metal contacts lead to much higher delay of up to 129 fs. This increase means that significantly lower switching speeds are achievable when more realistic metal contacts are considered. In terms of PDP (switching energy), both PNR FET structures are comparable to germanene FETs and outperform MoS_2 FETs and even antimonene/arsenene FETs. As for the delay (switching speed), PNR FET with metal contacts are comparable to all the examined alternatives, while ideal-contact PNR devices outperform all others.

As discussed previously, PNR FETs with metal contacts that are modeled with the WBL approximation exhibit I_{ON} values that on average reach 28% of the ballistic limit in devices with ideal contacts. While the contact parameters presented in the methodology section seem reasonable, the observed I_{ON} decrease corresponds to total contact/series resistance (R_{SD}) of 1470 Ωμm. While this value is comparable to experimentally measured resistance in phosphorene FETs [8], it is way above IRDS requirements. Therefore, we apply the required maximum $R_{SD} = 275$ Ωμm according to IRDS at N3 on device characteristics obtained numerically for PNR FETs with ideal contacts. Figure 5 reports I_{ON} for various PNR widths when $R_{SD} = 275$ Ωμm is assumed, and moreover reports I_{ON} dependence on transport ballisticity. This plot allows us to determine ballisticity values needed to fulfill the IRDS targets for the ON-state current that equals 0.91 mA/μm with non-zero R_{SD}. We find that ballisticity levels needed range from 50% to 70%, depending on the nanoribbon width, in order to attain the IRDS goal.

The obtained ballisticity limits seem reasonable and achievable in nanodevices. For example, recent theoretical/simulation work on 10.5 nm-long and wide-phosphorene (not nanoribbons) MOSFETs demonstrated 88-93% ballisticity when full-band electron-phonon scattering is included in the simulations. In order to provide an initial assessment of the ballisticity limit in the case of PNR FETs, we perform dissipative quantum transport simulations for the ON-state with optical phonons (OPs) included. Here, 4.4 nm-wide PNR is described with a single-band effective-mass Hamiltonian, for the sake of simplicity and numerical efficiency. Electron-phonon scattering was included using the self-consistent Born approximation [23], [24], real part of the scattering self-energy was neglected and only diagonal elements are retained. Optical phonon energy of $E_{OP} = 32$ meV and OP coupling strength of $D_{OP} = 170$ eV/nm is used in the simulations [10]. Figure 6 shows electron current density in the ON-state for the 4.4 nm-wide PNR FET in the case of coherent ballistic (Fig. 6a) and non-coherent dissipative (Fig. 6b) transport. The inclusion of OPs causes a small change in the current spectrum, with the distribution on the drain side being somewhat wider and with a lower average carrier energy, and no change in the electrostatic profile. For this device, I_{ON} in the dissipative case reaches 82% of the coherent ballistic limit, which means that PNR FETs are expected to fulfill IRDS goals for all the examined PNR widths if R_{SD} can be lowered to 275 Ωμm

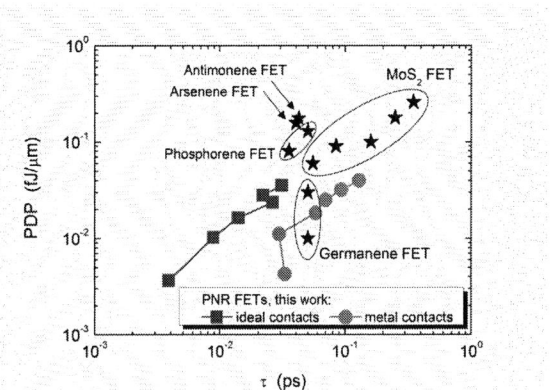

Fig. 4. Power-delay product vs. intrinsic delay time in PNR MOSFETs compared to other reported results from the literature. PNR devices outperform MoS_2, antimonene, arsenene and phosphorene FETs.

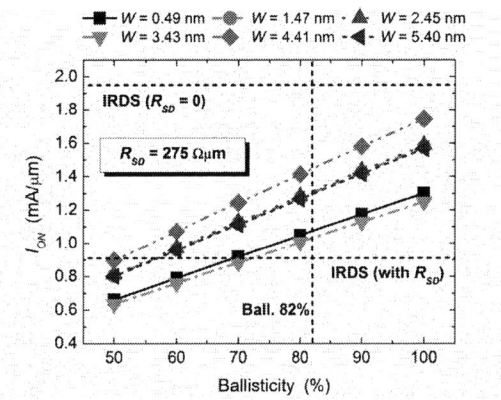

Fig. 5. Impact of ballisticity on the ON-state current in PNR MOSFETs with included parasitic series resistance of S/D regions ($R_{SD} = 275$ Ωμm).

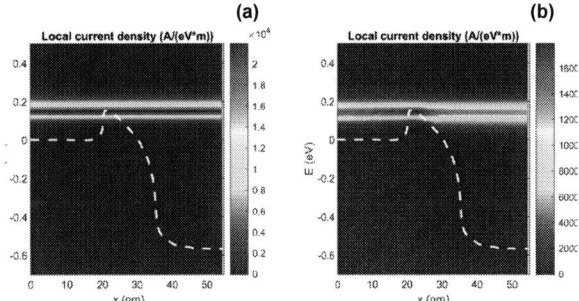

Fig. 6. Position and energy-resolved ON-state current density in 4.4 nm-wide PNR MOSFETs under (a) coherent ballistic transport and (b) dissipative transport with optical phonon scattering ($E_{OP} = 32$ meV, $D_{OP} = 170$ eV/nm). In the dissipative case, the ON-state current reaches 82% of the ballistic limit.

(see discussion related to Fig. 5). We note that including more subbands, or introducing additional scattering mechanisms such as acoustic phonon or lattice defect scattering, would reduce the ballisticity additionally, but the extent of that decrease is beyond the scope of this work.

IV. CONCLUSION

Several important figures-of-merit of PNR MOSFETs with 15 nm-long channel are analyzed using atomistic NEGF simulations coupled with ToB model. The focus is on exploring the impact of metal contacts in WBL approximation, series resistance, ballisticity and nanoribbon width scaling on the performance of PNR transistors. We find that including realistically described metal contacts in NEGF simulations decreases I_{ON} by up to 72%, which is equivalent to R_{SD} of 1470 $\Omega\mu$m. While this R_{SD} value agrees with experiments, it is more than 5× larger than the resistance limit set by IRDS. By including the IRDS-required series resistance, we find that PNR FETs should operate at 50% to 70% of the ballistic limit in order to attain IRDS goal for I_{ON}. Dissipative simulation with optical phonons reveals ballisticity of 82% in the ON-state. If good metal-phosphorene contacts are engineered such that R_{SD} can be lowered to about 270 $\Omega\mu$m, PNR FETs with sub-5 nm-wide nanoribbon channels would be promising to fulfill IRDS-HP requirements at "3 nm" logic technology node.

REFERENCES

[1] H. Liu *et al.*, "Phosphorene: An Unexplored 2D Semiconductor with a High Hole Mobility," *ACS Nano*, vol. 8, no. 4, pp. 4033–4041, Apr. 2014, doi: 10.1021/nn501226z.

[2] S. Das, W. Zhang, M. Demarteau, A. Hoffmann, M. Dubey, and A. Roelofs, "Tunable Transport Gap in Phosphorene," *Nano Letters*, vol. 14, no. 10, pp. 5733–5739, Oct. 2014, doi: 10.1021/nl5025535.

[3] A. Szabo, R. Rhyner, H. Carrillo-Nunez, and M. Luisier, "Phonon-limited performance of single-layer, single-gate black phosphorus n- and p-type field-effect transistors," in *2015 IEEE International Electron Devices Meeting (IEDM)*, Dec. 2015, pp. 297–300, doi: 10.1109/IEDM.2015.7409680.

[4] M. Poljak and T. Suligoj, "Immunity of electronic and transport properties of phosphorene nanoribbons to edge defects," *Nano Research*, vol. 9, no. 6, pp. 1723–1734, Jun. 2016, doi: 10.1007/s12274-016-1066-1.

[5] M. Poljak, "Electron Mobility in Defective Nanoribbons of Monoelemental 2D Materials," *IEEE Electron Device Letters*, vol. 41, no. 1, pp. 151–154, Jan. 2020, doi: 10.1109/LED.2019.2952661.

[6] S. Das, M. Demarteau, and A. Roelofs, "Ambipolar Phosphorene Field Effect Transistor," *ACS Nano*, vol. 8, no. 11, pp. 11730–11738, Nov. 2014, doi: 10.1021/nn505868h.

[7] Y. Du, H. Liu, Y. Deng, and P. D. Ye, "Device Perspective for Black Phosphorus Field-Effect Transistors: Contact Resistance, Ambipolar Behavior, and Scaling," *ACS Nano*, vol. 8, no. 10, pp. 10035–10042, Oct. 2014, doi: 10.1021/nn502553m.

[8] N. Haratipour, M. C. Robbins, and S. J. Koester, "Black Phosphorus p-MOSFETs With 7-nm HfO2 Gate Dielectric and Low Contact Resistance," *IEEE Electron Device Letters*, vol. 36, no. 4, pp. 411–413, Apr. 2015, doi: 10.1109/LED.2015.2407195.

[9] F. Liu, Y. Wang, X. Liu, J. Wang, and H. Guo, "Ballistic Transport in Monolayer Black Phosphorus Transistors," *IEEE Transactions on Electron Devices*, vol. 61, no. 11, pp. 3871–3876, Nov. 2014, doi: 10.1109/TED.2014.2353213.

[10] A. Afzalian and G. Pourtois, "ATOMOS: An ATOmistic MOdelling Solver for Dissipative DFT Transport in Ultra-Scaled HfS2 and Black Phosphorus MOSFETs," in *Proc. SISPAD*, 2019, pp. 199–202.

[11] E. Taghizadeh Sisakht, M. H. Zare, and F. Fazileh, "Scaling laws of band gaps of phosphorene nanoribbons: A tight-binding calculation," *Phys. Rev. B*, vol. 91, no. 8, p. 085409, Feb. 2015, doi: 10.1103/PhysRevB.91.085409.

[12] M. Poljak and T. Suligoj, "The Potential of Phosphorene Nanoribbons as Channel Material for Ultrascaled Transistors," *IEEE Trans. Electron Devices*, vol. 65, no. 1, pp. 290–294, Jan. 2018, doi: 10.1109/TED.2017.2771345.

[13] A. N. Rudenko and M. I. Katsnelson, "Quasiparticle band structure and tight-binding model for single- and bilayer black phosphorus," *Phys. Rev. B*, vol. 89, no. 20, p. 201408, May 2014, doi: 10.1103/PhysRevB.89.201408.

[14] M. P. L. Sancho, J. M. L. Sancho, J. M. L. Sancho, and J. Rubio, "Highly convergent schemes for the calculation of bulk and surface Green functions," *J. Phys. F: Met. Phys.*, vol. 15, no. 4, p. 851, 1985, doi: 10.1088/0305-4608/15/4/009.

[15] D. Kienle and A. W. Ghosh, "Atomistic Modeling of Metal-Nanotube Contacts," *J Comput Electron*, vol. 4, no. 1, pp. 97–100, Apr. 2005, doi: 10.1007/s10825-005-7116-7.

[16] A. Rahman, Jing Guo, S. Datta, and M. S. Lundstrom, "Theory of ballistic nanotransistors," *IEEE Transactions on Electron Devices*, vol. 50, no. 9, pp. 1853–1864, Sep. 2003, doi: 10.1109/TED.2003.815366.

[17] S. Kaneko, H. Tsuchiya, Y. Kamakura, N. Mori, and M. Ogawa, "Theoretical performance estimation of silicene, germanene, and graphene nanoribbon field-effect transistors under ballistic transport," *Appl. Phys. Express*, vol. 7, no. 3, p. 035102, Mar. 2014, doi: 10.7567/APEX.7.035102.

[18] "IEEE International Roadmap for Devices and Systems (IRDS), 2019 Update." https://irds.ieee.org/.

[19] G. Liang, N. Neophytou, M. S. Lundstrom, and D. E. Nikonov, "Ballistic graphene nanoribbon metal-oxide-semiconductor field-effect transistors: A full real-space quantum transport simulation," *Journal of Applied Physics*, vol. 102, no. 5, p. 054307, Sep. 2007, doi: 10.1063/1.2775917.

[20] G. Fiori and G. Iannaccone, "Simulation of Graphene Nanoribbon Field-Effect Transistors," *IEEE Electron Device Letters*, vol. 28, no. 8, pp. 760–762, Aug. 2007, doi: 10.1109/LED.2007.901680.

[21] G. Pizzi, M. Gibertini, E. Dib, N. Marzari, G. Iannaccone, and G. Fiori, "Performance of arsenene and antimonene double-gate MOSFETs from first principles," *Nature Communications*, vol. 7, p. 12585, Aug. 2016.

[22] Y. Zhao, A. AlMutairi, and Y. Yoon, "Assessment of Germanane Field-Effect Transistors for CMOS Technology," *IEEE Electron Device Letters*, vol. 38, no. 12, pp. 1743–1746, Dec. 2017, doi: 10.1109/LED.2017.2763120.

[23] M. Luisier, "Quantum transport beyond the effective mass approximation," Doctoral Thesis, ETH Zurich, 2007.

[24] M. Pourfath, *The Non-Equilibrium Green's Function Method for Nanoscale Device Simulation*. Wien: Springer-Verlag, 2014.

High-Performance Metal-Ferroeletric-Semiconductor Nanosheet Line Tunneling Field Effect Transistors with Strained SiGe

Narasimhulu Thoti[1,2], Yiming Li[1-4,*], Sekhar Reddy Kola[1,2], and Seiji Samukawa[5]

[1]Parallel and Scientific Computing Laboratory; [2]EECS International Graduate Program; [3]Institute of Communications Engineering; [4]Department of Electrical and Computer Engineering, National Chiao Tung University, 1001 Ta-Hsueh Rd., Hsinchu 30010, Taiwan; [5]Institute of Fluid Science, Tohoku University, Sendai 980-8577, Japan
*Tel: +886-3-5712121 ext. 52974; Fax: +886-3-5726639; Email: ymli@faculty.nctu.edu.tw (Y. Li)

Abstract—Nanosheet line tunnel-field effect transistors (NLTFETs) are for the first time proposed by utilizing the advantages of ferroelectricity through HZO materials. Three ferroelectric line TFETs have been proposed and investigated. Among these, the metal-ferroelectric-semiconductor (MFS) structure has shown superior performance than the other two variants. The factors of electric field and electron barrier tunneling have been addressed to govern the performance of these structures. In addition, the effects of the ferroelectric ($Hf_{0.5}Zr_{0.5}O_2$) thickness (t_{FE}) and the dielectric constant have been discussed. The MFS NLTFETs can effectively utilize the advantages of ferroelectric than the other variants. High on-current of 175.6 µA/µm and low off-current of 38.4 aA/µm are achieved at t_{FE} of 4 nm through proper utilization of gate-overlap on to the drain side. Furthermore, the proposed MFS structure successfully delivers low average and minimum subthreshold swings even at very thin t_{FE}.

Keywords—*Nanosheet line TFETs, MFS, $Hf_{0.5}Zr_{0.5}O_2$, and SiGe.*

I. INTRODUCTION

The tunnel-field effect transistors (TFETs) are promising aspirants for future energy-efficient electronics. In this prospective, band-to-band tunneling (BTBT) is the principal mechanism that governs DC characteristics of TFETs. Nevertheless, TFETs suffer from low tunneling probability due to its limitations in tunneling length or width (λ) and barrier height ($\Delta\Phi$). Hence, major investigation is carried out by considering various semiconductors (e.g., direct bandgap and 2D materials) to increase the tunneling probability [1]–[3]. Alternatively, this limitation is also overwhelmed to some extent through the excess field generations with the concept of line tunneling [4]. Even though various options of using line tunneling concepts have been proposed [5]–[8] the completely scaled n-epitaxial layer geometry has shown most promising choice in improving the tunneling probability [9], [10].

The ferroelectric utilization on TFETs is proved to be an alternative and a proficient option in further improving the limitations [11]–[14]. This is mainly due to the enhancement of internal voltage at source-channel junction that provides excess source to enhance the tunneling probability at low applied potentials. Thus, ferroelectric not only improves the on-current (I_{on}) but also enhances the steepness of the slope under sub-thermionic emission. Hence, the line TFETs

This work was supported in part by the Ministry of Science and Technology, Taiwan, under Grant MOST 109-2221-E-009-033, Grant MOST 108-2221-E-009-008, and Grant MOST 108-3017-F-009-001, and in part by the "Center for mmWave Smart Radar Systems and Technologies" under the Featured Areas Research Center Program within the framework of the Higher Education Sprout Project by the Ministry of Education in Taiwan".

TABLE I. DEVICE SPECIFICATIONS AND MATERIALS USED

Parameter	Material	Value
Gate length (L_G)	TiN	15 nm
Channel length (L_{ch})	Silicon	10 nm
Channel thickness (t_{Ch})	Silicon	5 nm
Channel width (W)	Silicon	10 nm
Source-overlap length (S_{Ov})	$Si_{1-x}Ge_x$	5 nm
Gate-overlap on drain (D_{Or})	Silicon	5 nm
Gate oxide thickness (t_{ox})	HfO_2	3 nm
Ferroelectric layer thickness (t_{FE})	$Hf_{0.5}Zr_{0.5}O_2$	4 nm
Gate-metal thickness (t_G)	TiN	3 nm
Gate-electrode work-function	TiN	4.4 eV
Source doping concentration (p^{++})	Boron	5×10^{20}
Channel doping concentration (p)	Boron	1×10^{16}
Drain doping concentration (n^+)	Arsenic	1×10^{19}
Epitaxial doping concentration (n)	Arsenic	5×10^{18}

(efficient than point-TFETs) with utilization of ferroelectric is the effective choice and need to be addressed [14].

The metal-ferroelectric-insulator-semiconductor (MFIS) metal-ferroelectric-metal-insulator-semiconductor (MFMIS), and the metal-ferroelectric-semiconductor (MFS) are the possible options that have been realized as negative capacitance (NC) FETs [15]–[17]. The basic advantage of the MFMIS is eliminating the leakage current by controlling the equipotential surface through the extra metal layer. However, the MFMIS structures are experimentally failed in utilization of polarization effect. Considering MFIS structures that utilizes the benefits of NC but not fully by the semiconducting channel due to the existence of dielectric layer. Apart from MFIS and MFMIS, the MFS structures have been shown exceptionally well in utilizing the benefits of ferroelectricity to improve the device characteristics by the use of 2D materials in delivering ultra-low power consumptions and reducing the switching time of various applications [17].

On the other hand, FETs with nanosheet structures are promising in emerging devices owing to their controllability in sinking scattering of electrons/holes and in generating sufficient fields for TFETs. In addition, nanosheet devices can be more beneficial for high-packed density with its reduced effective width of the device [18]. Notably, devices with nanosheet structures are also considered as effective choices in ultrascaled technology nodes, such as sub-5-nm [19]–[21]. Thus, we first demonstrate the nanosheet line TFET (NLTFET) with a MFS structure and compare with other variants of MFMIS and MFIS. The rest of the paper is organized as follows: design and simulation of the explored NLTFET are shown in Sec. II. Results and discussion on the specified structures are shown in Sec. III. Conclusions are listed in Section IV.

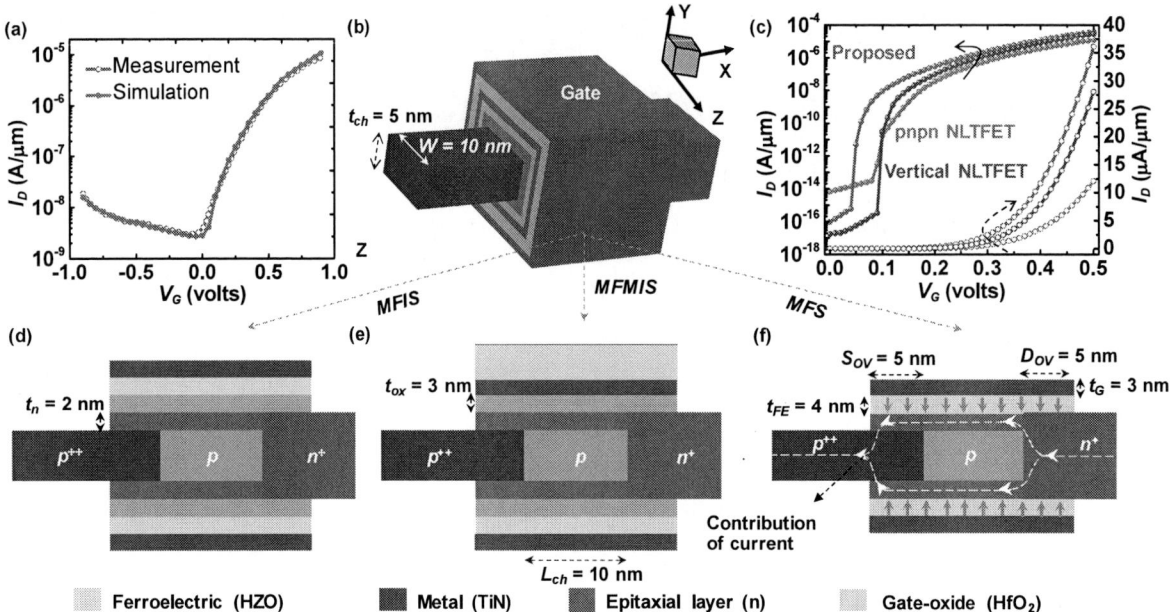

Fig. 1. (a) Validation of simulated data with experimental data [24]. (b) Proposed structures of NLTFETs with (d), (e) and (f) as MFIS, MFMIS, and MFS (with internal line TFET specifications), respectively. The doping concentrations of regions source (p^{++}), channel (p), epitaxial layer (n), and drain (n^+) are considered as 5×10^{20}, 1×10^{16}, 5×10^{18}, and 1×10^{19} cm^{-3}, along with the work-function as 4.4 eV. (c) Comparative analyses on the proposed structure (MFIS) with existed simulations [22], [28]. Note that the ferroelectric pnpn NLTFET and Vertical NLTFET in are designed and modelled as MFIS NLTFETs, based on the internal structure shown in [22], [28].

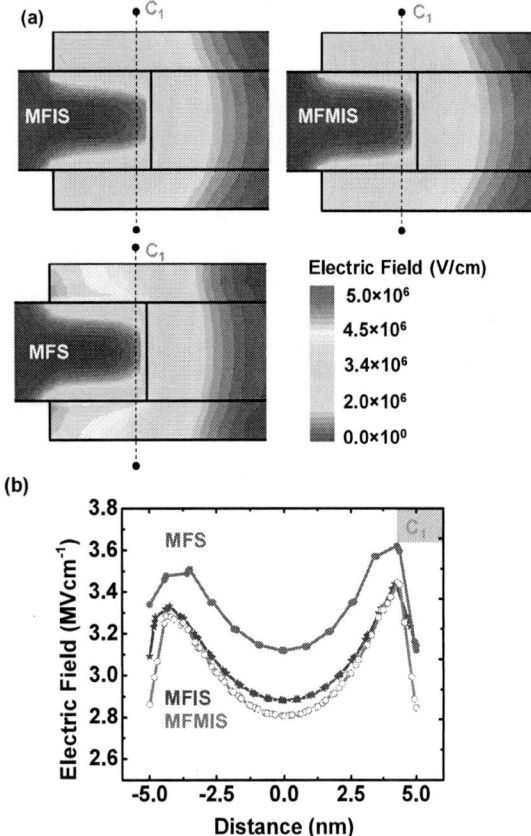

Fig. 2. (a) The electric field profile distributions and (b) its variation through n-epitaxy-source regions (along the cut-line C_1) for the proposed structures of MFIS, MFMIS, and MFS based NLTFETs.

II. DEVICE MODELLING AND SIMULATION

To design and analyze the TFETs, 3D device simulations are performed, where the band-to-band tunneling models as dynamic nonlocal tunneling and Hurkx trap-assisted tunneling (TAT) are calibrated with respect to the material considerations ($Si_{1-x}Ge_x$, where x as the Ge fraction and HfO_2 as dielectric) [9]. The simulations are experimentally validated for ferroelectric TFET, as depicted in Fig. 1a [24]. In addition, the proposed device performance is estimated through reasonable comparison with recently explored architecture, can be found in Fig. 1c [22]. The material properties of $Hf_{1-x}Zr_xO_2$ with Zr fraction (x) of 0.5 is a suitable option in delivering the significant polarization and electric fields [23], [24]. Furthermore, $Hf_{0.5}Zr_{0.5}O_2$ has the superior band alignment with Si and $Si_{0.6}Ge_{0.4}$ respectively, therefore the devices with these material choices can be able to achieve low leakage currents even at highly scaled ferroelectric thickness (t_{FE}) [25]. The designed and simulated structures naming ferroelectric NLTFETs as MFIS, MFMIS and MFS are illustrated in Figs. 1b, d-f based on the specifications as listed in Table I.

III. RESULTS AND DISCUSSIONS

Distributions of the electric field in the proposed structures along the cut-lines C_1 are depicted in Fig. 2. It can be observed that the MFS NLTFETs can able to utilize the ferroelectric polarization effectively and there by producing significant field compared to MFIS and MFMIS structures. In addition, the surface potential can be further improved through the reduction of oxide capacitance (C_{ox}), thus achieving more internal voltage (V_{int}) through ferroelectric capacitance (C_{FE}) [26]. Hence, the expression for internal voltage generated through MFS structure can be modified to (by neglecting C_{ox})

376

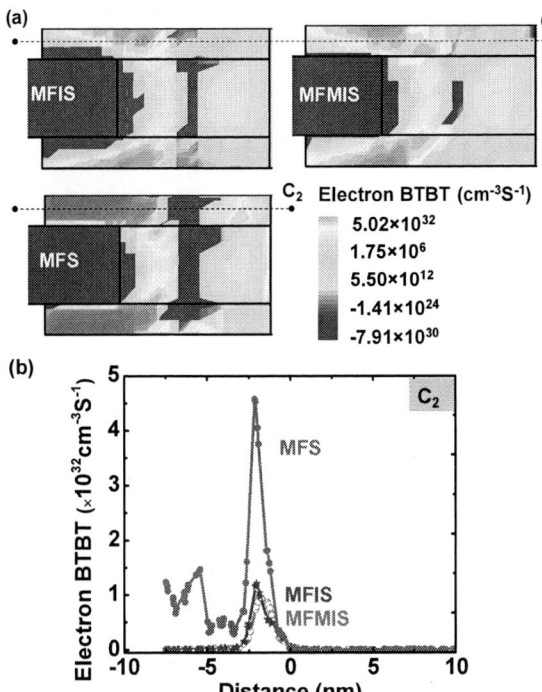

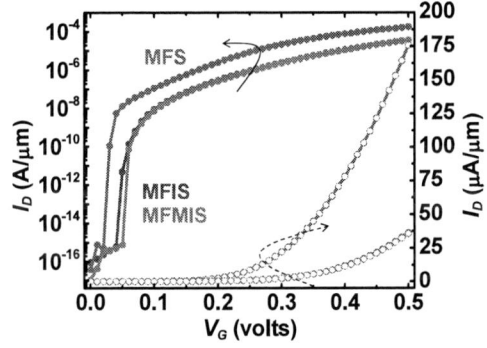

Fig. 3. (a) The distributions of electron BTBT on the proposed structures (MFIS, MFMIS, and MFS), (b) including their tunneling rate along the cut-line C_2 (contributor of line tunneling).

Fig. 4. The I_D-V_G characteristics of MFS, MFIS and MFMIS structures at the bias level of $V_D = V_G = 0.5$ V and at t_{FE} of 4 nm.

$$V_{int} = \frac{V_{gate}}{1 + \frac{C_{NS}}{C_{NS}}\left(\frac{1}{C_{FE}}\right)}, \qquad (1)$$

where, V_{gate} as the applied voltage and C_{NS} as the nanosheet capacitance. Therefore, it is understood that the negative capacitance effect is fully utilized by reducing the series capacitance (or C_{ox}) [17]. Eventually, the electron BTBT can be replicated as stronger (line/vertical tunneling) for MFS structure than other structures. This is because the reduction in oxide thickness reduces the screening tunneling length as well, based on the expression

$$\lambda = \sqrt{\frac{\varepsilon_{NS}}{\varepsilon_{ox}}t_{ox}t_{NS}}, \qquad (2)$$

TABLE II. EXTRACTED DC RESULTS OF MFS AND MFIS NLTFETs

Structure	t_{FE} (nm)	I_{off} (aA)	I_{on} (μA)	V_t (V)
MFS FeLTFET	5	34.63	211.46	0.215
	4	38.36	175.6	0.210
	3	38.96	211.56	0.209
	2	36.01	188.56	0.210
	1	34.06	187.7	0.213
MFIS FeLTFET	5	91.6	37.3	0.315
	4	94.3	36.1	0.318
	3	94.5	35.5	0.321
	2	127.23	32.6	0.327
	1	138.1	34.2	0.321

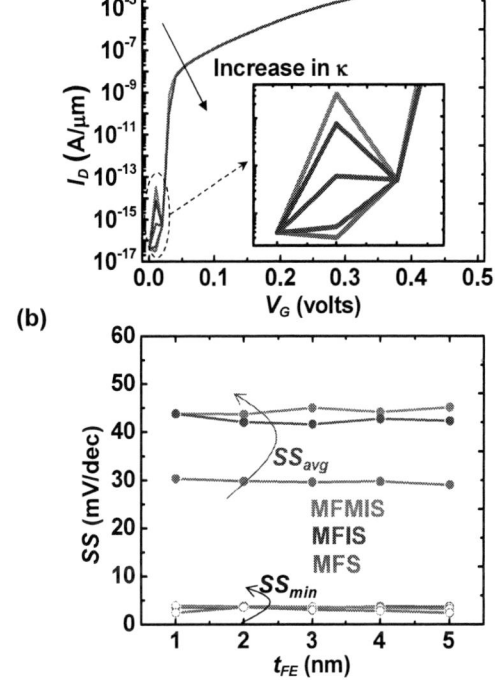

Fig. 5. (a) The extracted I_D-V_G for variation in dielectric constant (κ) of the $Hf_{0.5}Zr_{0.5}O_2$ on MFS NLTFET structure. Inset shows the off-state current fluctuation for increased κ (red to green) at t_{FE} of 4 nm. (b) The observed SS_{min}, SS_{avg} for the proposed structures by scaling t_{FE}.

where ε_{NS}, ε_{ox}, t_{NS}, and t_{ox} are the dielectric constant and thicknesses of the nanosheet (semiconductor) and oxide materials. Hence, the tunneling probability as specified in Fig. 3 can be significantly improved to two orders of electron BTBT rate for the MFS NLTFETs. Since, the coercive field (E_c) strongly correlated on tunneling mechanism and thus corresponds to an increased the tunneling rate. The performance of proposed structures (three variants) at t_{FE} of 4 nm is depicted in Fig. 4. It is understood that the on-current boost is approximately 5 times higher than the MFIS structure (understood from Table II). Interestingly, good agreement with I_{off} is achieved, even for reduced t_{FE} due to wrapped around gate and gate-overlap on to drain (D_{OV}). However, the controllability of negative-capacitance regime in MFS structure's is difficult than the MFIS/MFMIS structures [17]. Thus the I_{off} at low bias levels is observed to be fluctuated upon scaled t_{FE} (can be seen from Fig. 4). The extracted I_{on}

377

state that the controllability of negative capacitance is the crucial factor in the absence of dielectric. Thus, the fluctuated I_{on} can be seen for the scaled t_{FE} as listed in Table II. In addition, based on the matched C_{FE} with that of C_{NW}, the high I_{on} can be seen at the t_{FE} of 3 nm with reasonable I_{off}. However, the fluctuation in threshold voltage is insignificant. It means, the higher gate-bias and very low bias levels are critical in controlling the MFS structures. Furthermore, the scalability of dielectric constant (κ) in $Hf_{0.5}Zr_{0.5}O_2$ is analyzed in MFS structures and is depicted in Fig. 5a. As like MFIS structures, the MFS structures are also invariant for variation in κ values, except at the off-state condition, can be seen in Fig. 5a. Because, the variation in capacitance is crucial for varied dielectric thickness and which is negotiate at high potential levels [27]. The performance of average and min subthreshold swing swings (SS_{min} and SS_{avg}) are extracted and depicted in Fig. 5b. The results state that the factor of average subthreshold swing is greatly improved through major impact of ferroelectric polarization below the subthreshold regime. This is due to enhanced transport mechanism MFS structures is made steeper SS_{avg} than either MFIS or MFMIS structures. In addition, the results state that these values are consistent upon highly scaled t_{FE} values (viewed from Fig. 5b). Therefore, we state that the MFS based NLTFETs are delivering superior performance, provided that the controllability should be properly maintained. Nevertheless, the proposed structure has been made for the needs of high-drive current and low I_{off} with very good average and minimum subthreshold swings.

IV. CONCLUSIONS

We have designed and studied three types of ferroelectric NLTFETs with $Si_{0.6}Ge_{0.4}$ as source. The relevant discussions on the device performance has been addressed with physical constraints of electric field and BTBT on the proposed structures (MFS, MFIS, and MFMIS). Among all, the MFS based NLTFETs has been delivering superior performance in terms of I_{on}, I_{off} and well maintained V_t. Furthermore, the low and steeper subthreshold swings have been achieved on the proposed MFS structure. Thus, we conclude that the proposed structure has been made in delivering superior DC performance with the needs of high-drive current and low I_{off} with very good average and minimum subthreshold swings.

REFERENCES

[1] G. Alymov, V. Vyurkov, V. Ryzhii, and D. Svintsov, "Abrupt current switching in graphene bilayer tunnel transistors enabled by van Hove singularities," *Nat. Publ. Gr.*, vol. 6, no. April, pp. 1–8, 2016.

[2] G. V Resta *et al.*, "Devices and Circuits Using Novel 2-D Materials : A Perspective for Future VLSI Systems," *IEEE Trans. Very Large Scale Integr. Syst.*, vol. 27, no. 7, pp. 1486–1503, 2019.

[3] T. A. Ameen, H. Ilatikhameneh, P. Fay, A. Seabaugh, R. Rahman, and G. Klimeck, "Alloy Engineered Nitride Tunneling Field-Effect Transistor: A Solution for the Challenge of Heterojunction TFETs," *IEEE Trans. Electron Devices*, vol. 66, no. 1, pp. 736–742, 2019.

[4] A. S. Verhulst, D. Leonelli, R. Rooyackers, and G. Groeseneken, "Drain voltage dependent analytical model of tunnel field-effect transistors," *J. Appl. Phys.*, vol. 110, no. 2, pp. 024510-1–10, 2011.

[5] Z. Jiang *et al.*, "Quantum Transport in AlGaSb/InAs TFETs With Gate Field In-Line With Tunneling Direction," *IEEE Trans. Electron Devices*, vol. 62, no. 8, pp. 2445–2449, 2015.

[6] W. Cheng *et al.*, "Fabrication and Characterization of a Novel Si Line Tunneling TFET with High Drive Current," *IEEE J. Electron Devices Soc.*, vol. 8, no. March, pp. 336–340, 2020.

[7] Ashita, S. A. Loan, and M. Rafat, "A High-Performance Inverted-C Tunnel Junction FET with Source-Channel Overlap Pockets," *IEEE*

Trans. Electron Devices, vol. 65, no. 2, pp. 763–768, 2018.

[8] S. Blaeser *et al.*, "Novel SiGe/Si line tunneling TFET with high Ion at low Vdd and constant SS," in *Technical Digest - International Electron Devices Meeting, IEDM*, 2015, pp. 22.3.1-22.3.4.

[9] N. Thoti, Y. Li, S. R. Kola, and S. Samukawa, "Optimal Inter-Gate Separation and Overlapped Source of Multi-Channel Line Tunnel FETs," *IEEE Open J. Nanotechnol.*, vol. 1, no. May, pp. 1–1, 2020.

[10] N. Thoti, Y. Li, S. R. Kola, and S. Samukawa, "New Proficient Ferroelectric Nanosheet Line Tunneling FETs with Strained SiGe through Scaled n - epitaxial Layer," in *IEEE 20th International Conference on Nanotechnology (IEEE-NANO)*, 2020, pp. 319–323.

[11] M. Kobayashi, "A perspective on steep-subthreshold-slope negative-capacitance field-effect transistor," *Appl. Phys. Express*, vol. 11, pp. 110101-1–20, 2018.

[12] H. Li, P. Xu, L. Xu, Z. Zhang, and J. Lu, "Negative capacitance tunneling field effect transistors based on monolayer arsenene, antimonene, and bismuthene," *Semicond. Sci. Technol.*, vol. 34, no. 8, pp. 085006-1–8, 2019.

[13] M. Kobayashi, K. Jang, N. Ueyama, and T. Hiramoto, "Negative Capacitance for Boosting Tunnel FET performance," *IEEE Trans. Nanotechnol.*, vol. 16, no. 2, pp. 253–258, 2017.

[14] A. M. Ionescu, L. Lattanzio, G. A. Salvatore, L. De Michielis, K. Boucart, and D. Bouvet, "The hysteretic ferroelectric tunnel FET," *IEEE Trans. Electron Devices*, vol. 57, no. 12, pp. 3518–3524, 2010.

[15] J. Jo and C. Shin, "Negative capacitance field effect transistor with hysteresis-free sub-60-mV/decade switching," *IEEE Electron Device Lett.*, vol. 37, no. 3, pp. 245–248, 2016.

[16] S. Salahuddin and S. Datta, "Use of negative capacitance to provide voltage amplification for low power nanoscale devices," *Nano Lett.*, vol. 8, no. 2, pp. 405–410, 2008.

[17] X. Wang *et al.*, "Two-dimensional negative capacitance transistor with polyvinylidene fluoride-based ferroelectric polymer gating," *npj 2D Mater. Appl.*, vol. 1, no. 1, pp. 1–6, 2017.

[18] N. Loubet *et al.*, "Stacked nanosheet gate-all-around transistor to enable scaling beyond FinFET," *Dig. Tech. Pap. - Symp. VLSI Technol.*, vol. 5, no. 1, pp. T230–T231, 2017.

[19] "The International Roadmap for Devices and Systems," *IEEE*, 2018. [Online]. Available: https://irds.ieee.org/editions/2018.

[20] J. Jeong, J. S. Yoon, S. Lee, and R. H. Baek, "Comprehensive Analysis of Source and Drain Recess Depth Variations on Silicon Nanosheet FETs for Sub 5-nm Node SoC Application," *IEEE Access*, vol. 8, no. February, pp. 35873–35881, 2020.

[21] J. S. Yoon, S. Lee, J. Lee, J. Jeong, H. Yun, and R. H. Baek, "Reduction of Process Variations for Sub-5-nm Node Fin and Nanosheet FETs Using Novel Process Scheme," *IEEE Trans. Electron Devices*, vol. 67, no. 7, pp. 2732–2737, 2020.

[22] V. P. H. Hu, H. H. Lin, Y. K. Lin, and C. Hu, "Optimization of Negative-Capacitance Vertical-Tunnel FET (NCVT-FET)," *IEEE Trans. Electron Devices*, vol. 67, no. 6, pp. 2593–2599, 2020.

[23] S. J. Kim, J. Mohan, S. R. Summerfelt, and J. Kim, "Ferroelectric $Hf_{0.5}Zr_{0.5}O_2$ Thin Films: A Review of Recent Advances," *Miner. Met. Mater. Soc.*, vol. 71, no. 1, pp. 246–255, 2019.

[24] A. Saeidi *et al.*, "Effect of hysteretic and non-hysteretic negative capacitance on tunnel FETs DC performance," *Nanotechnology*, vol. 29, no. 9, p. 095202, 2018.

[25] Y. Peng *et al.*, "Band alignments at $Hf_{1-x}Zr_xO_2$/Si and $Hf_{0.52}Zr_{0.48}O_2$/$Si_{0.55}Ge_{0.45}$ interfaces," *Superlattices Microstruct.*, vol. 130, no. 2, pp. 519–527, 2019.

[26] A. Saeidi *et al.*, "Nanowire Tunnel FET with Simultaneously Reduced Subthermionic Subthreshold Swing and Off-Current due to Negative Capacitance and Voltage Pinning Effects," *Nano Lett.*, vol. 20, no. 5, pp. 3255–3262, 2020.

[27] S. Mookerjea, R. Krishnan, S. Datta, and V. Narayanan, "Effective Capacitance and Drive Current for Tunnel FET (TFET) CV/I Estimation," *IEEE Trans. Electron Devices*, vol. 56, no. 9, pp. 2092–2098, 2009.

[28] J. K. Mamidala, R. Vishnoi, and P. Pandey, *Tunnel Field-Effect Transistors (TFET): Modelling and Simulation*, 2017th ed. UK: John Wiley & Sons, Ltd, 2017.

16-5

A First-principles Study on the Strain-induced Localized Electronic Properties of Dumbbell-shape Graphene Nanoribbon for Highly Sensitive Strain Sensors

1st Qinqiang Zhang
Department of Finemechanics,
Graduate School of Engineering
Tohoku University
Sendai, Japan
zhang.qinqiang@rift.mech.tohoku.ac.jp

2nd Ken Suzuki
Fracture and Reliability Research
Institute
Tohoku University
Sendai, Japan
kn@rift.mech.tohoku.ac.jp

3rd Hideo Miura
Fracture and Reliability Research
Institute
Tohoku University
Sendai, Japan
hmiura@rift.mech.tohoku.ac.jp

Abstract—The electronic properties of graphene nanoribbons (GNRs) have a function of the ribbon width. It can vary from metallic-like ones to semiconductive-like ones when the width of single GNR is changed. Therefore, the novel structure of GNRs called dumbbell-shape GNR (DS-GNR) was proposed to achieve the development of highly sensitive, reliable, and deformable strain sensors. The DS-GNR consists of one long narrow GNR coalesced by two wide segments of GNRs at its both ends. The wide segments of the original DS-GNR possess the metallic-like electronic properties and the narrow segment of the original DS-GNR has the semiconductive-like electronic properties. In this study, the strain-induced change of the electronic band structure of DS-GNR was analyzed by using the first-principles calculations. The range of the applied uniaxial tensile strain on DS-GNR was from 0% to 10%. When the length of the narrow segment of DS-GNR is longer than 4.3 nm, the effective bandgap located in the narrow segment changes obviously with the change of applied strain. The result indicates that the piezoresistive effect appears in the narrow segment of DS-GNR, and thus high strain sensitivity of its resistivity can be applied to strain sensors.

Keywords—*graphene nanoribbon, dumbbell-shape, strain-induced, first-principles, local electronic properties*

I. INTRODUCTION

The further aging society requires 24-hour and high-quality nursing cares despite the shortage of caregivers. It is, therefore, indispensable to develop wearable and smart self-health-monitoring systems. Hitherto, conventional strain sensors cannot fulfill the requirements, such as the full range of the large deformation of joints with human body and the detection of fine change of vibration of the heartbeat's pulse. From this point of view, graphene, a two-dimensional (2D) monolayer material is one of the most attractive candidate materials to substitute the conventional components such as silicon utilized in electronic devices. Graphene exhibits unique super-conductive properties, high intrinsic strength, and large deformability [1]. Moreover, narrow armchair graphene nanoribbons (aGNRs) are the candidates for a sensing component of highly sensitive strain sensors owing to its appreciable bandgaps, and thus, piezo-resistance effect [2]. The aGNR manifests semiconductive-like properties with its large bandgaps. While aGNR has relatively small bandgaps exhibiting metallic-like properties. The bandgap monotonically increases when the width of aGNR decreases in all cases [3].

To fabricate highly sensitive, reliable, and deformable strain sensors with low-cost and mass-productivity, the authors have proposed a novel structure, a dumbbell-shape GNR (DS-GNR) structure, as depicted in Fig. 1 [4]. The wide segment of DS-GNR shows metallic-like properties and should have ohmic contact with external metal electrodes. The narrow segment as the sensing element in the DS-GNR shows wide variation of bandgap as a strong function of the total number of carbon atoms along its width direction. Since the DS-GNR consists of three single GNRs with different widths but only carbon atoms, it is readily to be fabricated by a dry-etching process from a large graphene sheet. In our previous study, a smooth-electron-flow existing in the junction area between the wide and narrow segment of DS-GNR has been confirmed [5]. In this study, the strain-induced change of the electronic band structure of DS-GNRs were analyzed by using the first-principles calculations based on density functional theory (DFT) to clarify the strain-induced change of electronic properties under uniaxial tensile strain.

II. SIMULATION MODEL

The simulation model of DS-GNR has wide and narrow segments as depicted in Fig. 2. The wide segment was defined to possess metallic-like properties in all cases to simulate wide GNRs or graphene in practical conditions. The narrow

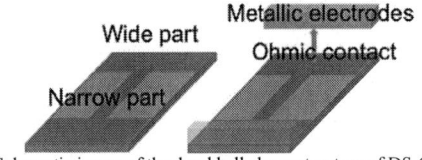

Fig. 1. Schematic image of the dumbbell-shape structure of DS-GNR for the use of strain sensors

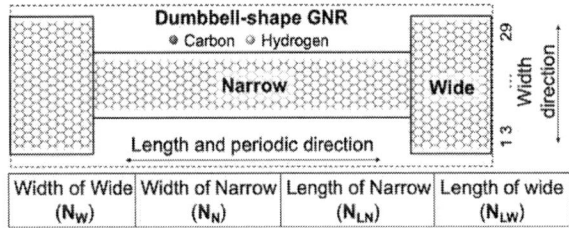

Fig. 2. Simulation model of DS-GNR. One group of six-membered ring of carbon atoms along the length direction is defined as one length of narrow and wide segment. Total number of carbon atoms connected as a dimer line along the width direction was defined as N_W and N_N.

segment was selected to have both large bandgaps and small bandgaps. The dashed line indicates supercell and it had periodic condition along the length direction, the horizontal direction in the inset. All the dangling bonds of carbon atoms on the outer frame of the structure were terminated by hydrogen atoms. Four parameters are defined to describe the structure of the DS-GNR. The width of the wide segment, N_W, the width of the narrow segment of N_N, the length of the narrow segment N_{LN}, and the length of the wide segment, N_{LW}. N_W and N_N indicate the number of carbon atoms connected as a dimer line along the width direction. N_{LN} and N_{LW} indicate two dimer lines as one unit along the length direction. The length of each segment was modelled to be longer than 10 six-membered rings because it was found that the length of the interaction area around the junction between the narrow and wide segments was about 5 six-membered rings [6]. To estimate the effect of strain, the position of each atom along the length direction was stretched uniformly from 0% to 10%.

The VASP for GPU code was utilized for fully optimizing the large dumbbell-shape structure by the RMM-DIIS method with the maximum force between two atoms less than 0.1 eV Å⁻¹. The SIESTA package was used for the calculations of electronic band structure and the local density of states by utilizing the conjugate gradient (CG) method. The generalized gradient approximation (GGA) in the Perdew-Burke-Ernzerhof (PBE) form was used in both codes. In this study only the structure with 0% strain was fully optimized. The structures with applied strain were calculated by single-point CG method.

III. STRAIN-INDUCED CHANGE OF ELECTRLNIC BAND STURCTURE

Fig. 3 shows the example change of the electronic band structure of DS-NGR ($N_W = 17$, $N_N = 07$, $N_{LN} = 20$, $N_{LW} = 05$), the local density of states (LDOS), under the application of uniaxial tensile strain. The change of the LDOS in the representative two regions, line 1 (ln 1) and line 15 (ln 15), which were the center regions of the wide and narrow segments, respectively. The upper graph of LDOS in Fig. 3 describes the change of electronic band structure in the wide segment. The bottom graph of LDOS indicates that in the narrow segment. In the center region of the wide segment, the bandgap between the most inner peak of the LDOS increased with strain. It indicates that the electronic properties in the wide segment changed from metallic-like ones to semiconductive-like ones with the increment of the applied strain. On the other hand, the electronic band structure in the narrow segment changed from semiconductive-like ones to metallic-like ones with strain. The change rate of LDOS in the narrow segment was larger than that in the wide segment as indicated by the color arrows.

Similar change appeared in the DS-GNR ($N_W = 29$, $N_N = 11$, $N_{LN} = 20$, $N_{LW} = 05$) as shown in Fig. 4. In this structure, however, the effective bandgap in both segments increased monotonically with strain. Thus, the electronic properties in both regions changed from metallic-like ones to semiconductive-like ones.

These results clearly indicate that piezoresistive effect appears in the narrow segment as was expected, and therefore, high strain sensitivity of its resistivity can be applied to strain sensors. To assure the reliable operation, however, it is important to minimize the strain in the wide segment, which should show metallic properties.

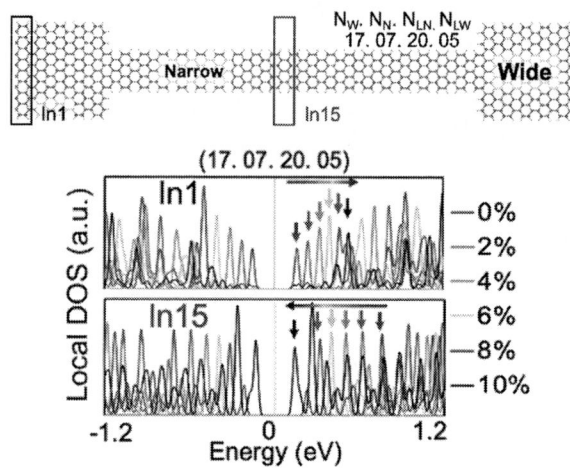

Fig. 3. Change of the electronic band structure in the center regions of the wide (ln 1) and narrow (ln 15) segments of DS-GNR (17, 07, 20, 05) under strain

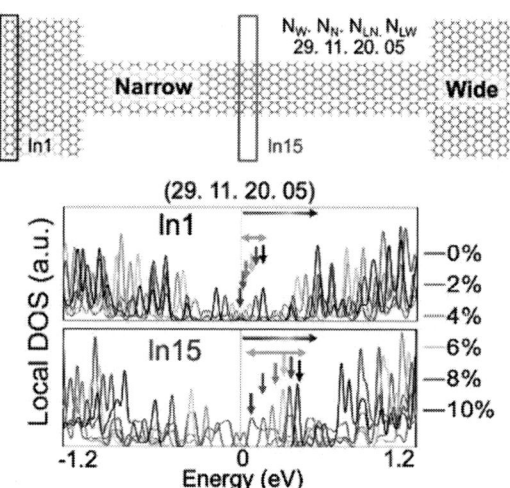

Fig. 4. Change of the electronic band structure in the center regions of the wide (ln 1) and narrow (ln 15) segments of DS-GNR (29, 11, 20, 05) under strain

IV. STRAIN-INDUCED CHANGE OF ELECTRON ORBITAL DISTRIBUTION

Figs. 5(a) and 5(b) show the electron orbital distribution of DS-GNR (29, 11, 20, 05) on the highest occupied molecular orbitals (HOMO). Yellow region indicates positive phase of the movement of electrons and green region indicates that of negative phase. As shown in the Figs. 5(a) and 5(b), the colored region is located in the wide segments of DS-GNR with 0% and 10% strain. The results reveal that the possibility of the movement of electrons in the wide segment of DS-GNR is higher than that of in the narrow segment under strain, and thus the strain sensitivity of the narrow segment of DS-GNR is higher than that of in the wide segment owing to its large resistivity. The figures of electron orbital distribution of DS-GNR with strain from 0% to 10% have identical pattern at the HOMO 1, thus, those are not shown here.

In particular, the electron orbital distribution of DS-GNR with 10% strain exhibits a gradient transition region around the junction between the wide and the narrow segment. It

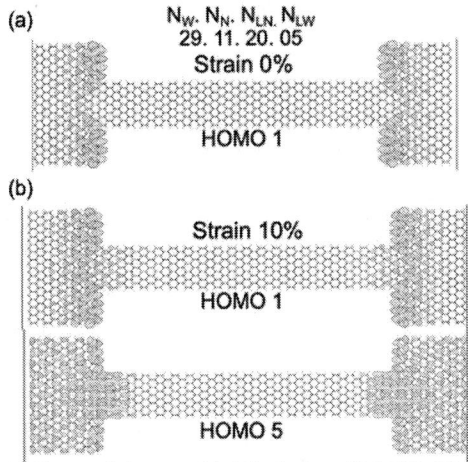

(a) $N_W. N_N. N_{LN}. N_{LW}$
29. 11. 20. 05
Strain 0%

HOMO 1

(b)

Strain 10%

HOMO 1

HOMO 5

Fig. 5. (a) Change of electron orbital distribution of DS-GNR (29, 11, 20, 05) at the first HOMO (HOMO 1) with 0% strain (b) Change of electron orbital distribution of DS-GNR (29, 11, 20, 05) at the HOMO 1 and the fifth HOMO (HOMO 5) with 10% strain

indicates that electrons transfer smoothly from the wide segment to the narrow segment of DS-GNR with 10% strain. In addition, the electron orbital distribution of DS-GNR with 10% strain exhibits localized pattern on the fifth HOMO (HOMO 5). This result indicates that piezoresistive effect has more impact in the narrow segment with 10% strain than that of 0% strain, and thus, higher strain sensitivity of its resistivity can be achieved for use in strain sensors with applied strain.

V. STRAIN-INDUCED CHANGE OF ELECTRLNIC BAND STURCTURE WITH FIXED BONDING LENGTH IN THE WIDE SEGMENT OF DS-GNR

In this sub-section, another condition is hypothesized that the DS-GNR is suspended for use in highly deformable strain sensors. Since the widths between the wide and the narrow segment are relatively different, the stress appeared in two segments should be different with the same force applied in DS-GNR under uniaxial tensile strain. Therefore, the strain deformation in the wide and the narrow segments is different in this condition. When the width of wide segment becomes much wider than that of narrow segment or the wide segment becomes large area graphene, the carbon-carbon bonding length in the wide segment of DS-GNR should show negligible change during the applied strain, in other words, no strain appears in the wide segment. Therefore, the strain-induced change of electronic band structure with the fixed carbon-carbon bonding length in the wide segment of DS-GNR was also analyzed by the first-principles calculations here. It is, likewise, applied to the condition which the DS-GNR has suspended GNR in the narrow segment but with deposited metal electrodes on the wide segment (no strain in the wide segment).

The bonding length between the wide and the narrow segment changed abruptly in this study as the pre-calculation in order to simplify the cumbersome simulation factors. Thus, only the position of each atom along the length direction in the narrow segment of DS-GNR was stretched uniformly from 0% to 10% in this condition.

Fig. 6 shows the example change of the electronic band structure of DS-GNR ($N_W = 29$, $N_N = 11$, $N_{LN} = 20$, $N_{LW} = 05$), the LDOS, under the application of uniaxial tensile strain with

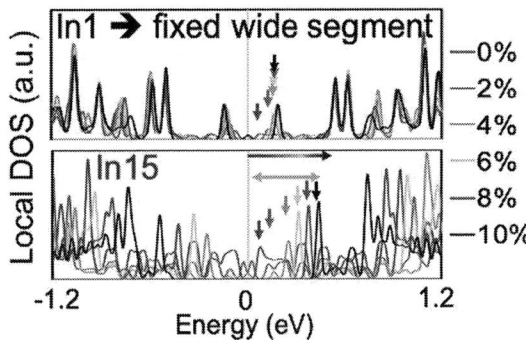

Fig. 6. Change of the electronic band structure in the center regions of the wide (ln 1) and narrow (ln 15) segments of DS-GNR (29, 11, 20, 05) with fixed carbon-carbon bonding length in the wide segment under strain

fixed carbon-carbon bonding length in the wide segment. The upper graph of LDOS in Fig. 6 describes the change of electronic band structure of DS-GNR with the fixed wide segment. In the center region of the fixed wide segment, the peaks of bandgap are superposed practically under strain. It indicates that the electronic properties in the wide segment has nearly no change with the increment of the applied strain which was metallic-like properties. The bottom graph of LDOS in Fig. 6 presents the change of electronic band structure of DS-GNR with uniaxial strain in the narrow segment. However, the electronic band structure in the narrow segment changed from metallic-like ones to semiconductive-like ones. The change is identical to that of DS-GNR without fixed carbon-carbon bonding length in the wide segment which already indicated above as shown in Fig. 4.

The results clearly indicate that piezoresistive effect appears in the narrow segment even there is no piezoresistive effect in the wide segment, and therefore, high strain sensitivity of its resistivity can be applied to strain sensors by utilizing DS-GNR.

Fig. 7 expresses the change of effective bandgap between single GNRs and DS-GNRs with fixed carbon-carbon bonding length in the wide segment under strain. The solid line with square points (ln 1) indicates the change of effective bandgap in the wide segment of DS-GNR. The slope of it is nearly flat with strain close to 10%. It indicates that the strain sensitivity of wide segment is low which should show metallic properties continuously under strain. Whereas, the strain sensitivity is still high in the narrow segment (solid line with triangle points, ln 15) regardless the low strain sensitivity in the wide segment. It has great potential for enhancing the reliability of strain sensors utilizing DS-GNRs with multi-GNRs in the narrow segment due to its high strain sensitivity of its resistivity when the length of the narrow segment is longer than 10 N_{LN} (4.3 nm).

The dumbbell-shape structure has stable ohmic contact to the external metal electrodes owing to its metallic-like properties in the wide segment under strain in this condition. In addition, the DS-GNR consists of only carbon atoms, hence, it has the smooth-electron-flow between the wide segment and the narrow segment.

VI. CONCLUSIONS

The change of the local electronic band structure of DS-GNRs under uniaxial strain was analysed by applying the first-principles calculation. High strain sensitivity was validated in

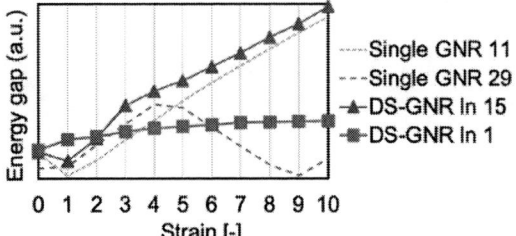

Fig. 7. Change of the effective bandgap in Single GNRs and in the center region of the wide (ln 1) and the narrow (ln 15) segments of DS-GNR with fixed carbon-carbon bonding length in the wide segment under strain

the GNR narrower than 70 nm. The sensitivity was found to increase with the decreases in the width of GNR. Therefore, the DS-GNR exhibits great potential to be applied to highly sensitive, reliable and deformable strain sensors.

ACKNOWLEDGMENT

This research activity has been supported partially by Japanese special coordination funds for promoting science and technology, Japanese Grants-in-aid for Scientific Research, and Tohoku University. This research was supported partially by Murata Science Foundation and JSPS KAKENHI Grant Number JP16H06357. The authors would like to express their sincere thanks to the crew of Center for Computational Materials Science of the Institute for Materials Research, Tohoku University for their continuous support of the supercomputing facilities.

REFERENCES

[1] M. Young, The Technical Writer's Handbook. Mill Valley, CA: University Science, 1989. C. Lee, X. Wei, J. W. Kysar, and J. Hone, "Measurement of the Elastic Properties and Intrinsic Strength of Monolayer Graphene", SCIENCE, 321, (2008), pp. 385-388.

[2] Nakagawa, R., Wang, Z., & Suzuki, K. (2018, November). Area-Arrayed Graphene Nano-Ribbon-Base Strain Sensor. In *ASME 2018 International Mechanical Engineering Congress and Exposition* (pp. V010T13A008-V010T13A008). American Society of Mechanical Engineers.

[3] Jowesh Goundar, Takuya Kudo, Qinqiang Zhang, Ken Suzuki, Hideo Miura, "Strain and Photovoltaic Sensitivities of Dumbbell-Shape GNR-Base Sensors", Proc. of IMECE2019, IMECE2019-11076, (2019), pp. 1-6.

[4] Q. Zhang, T. Kudo, K. Suzuki and H. Miura, 'Theoretical study of electronic band structure of dumbbell-shape graphene nanoribbons for highly-sensitive strain sensors', Proc. of ASME International Mechanical Engineering Congress and Exposition, No. IMECE2018-88431. (2018), pp. 1-6.

[5] Zhang, Q., Kudo, T., Gounder, J., Chen, Y., Suzuki, K., & Miura, H. (2019, September). Theoretical Study of the Edge Effect of Dumbbellshape Graphene Nanoribbon with a Dual Electronic Properties by First-principle Calculations. In *2019 International Conference on Simulation of Semiconductor Processes and Devices (SISPAD)* (pp. 1-4). IEEE.

[6] Takuya Kudo, Qinqiang Zhang, Ken Suzuki, Hideo Miura, "First-Principle Analysis of the Effect of Strain on Electronic Transport Properties of Dumbbell-Shape Graphene Nanoribbons", Proc. of IMECE2019, No. IMECE2019-11107, (2019), pp. 1-6.

LN

Multiband Phase Space Operator for Narrow Bandgap Semiconductor Devices

1st Lukas Schulz
High Frequency Institute
TU Dortmund
Dortmund, Germany
lukas.schulz@tu-dortmund.de

2nd Dirk Schulz
High Frequency Institute
TU Dortmund
Dortmund, Germany
dirk2.schulz@tu-dortmund.de

Abstract—The analysis of the charge carrier transport within modern device concepts of nanoelectronics and nanophotonics as well as THz technology requires the inclusion of multiband Hamiltonians. These can then be used to consider not only intraband transitions but also interband transitions as well as effects based on the existence and interaction of light and heavy holes. For this purpose appropriate multiband Hamiltonians must be applied for a suitable numerical analysis. On the basis of the quantum Liouville equation, a formalism is derived how multiband Hamiltonians can be integrated into advanced and recently developed Wigner transport based algorithms utilizing a phase space operator and which multiband models are appropriate. The presented formalism is demonstrated by its application onto resonant tunnel diodes that take advantage of interband effects within narrow band gap semiconductor devices.

Index Terms—Interband tunneling, multiband Hamiltonians, phase space operator, Wigner equation.

I. INTRODUCTION

The quantum mechanical description of nanoelectronic devices in terms of a phase space formulation provided by the Wigner formalism is of considerable interest as this formalism offers the possibility to analyze transient effects and to incorporate interaction mechanisms effectively. So far, Wigner based formalisms are predominantly based on single band approximations considering the conduction band but are lacking of the inclusion of multiband models. Typically, existing formalisms are applied onto large bandgap double barrier resonant tunneling diodes [1]–[3] and field effect transistors [4]–[6], so that intraband dynamics essentially affect the device performance. However, when the bandgap shrinks and band-coupling effects shall be considered, the corresponding interband kinetics must be adequately included.

A recently proposed Wigner based formalism is extended, which is based on a phase space operator [7] considering non-local quantum effects and on the complex absorbing potential (CAP) formalism [8] in order to adequately account for the finiteness of the computational domain.

Two distinct multiband approaches are suitable for the application within the Wigner formalism, namely the two band Kane-model [9] as well as the multiband-envelope function (MEF)-model [10], [11]. The latter model leads to a similar

The authors gratefully acknowledge the financial support by the German Research Funding Association Deutsche Forschungsgemeinschaft under grant SCHU-2016/8-1.

formulation as in the Luttinger-Kohn-model, however, in comparison the interband kinetics are not neglected. The MEF-model is preferable due to the direct physical interpretation of the electron and hole wave functions [12]. In addition, the interband coupling is only present in the case of a non constant external potential. This fact leads to a simplification of the formulation of boundary conditions as intermixing states are avoided at the boundaries. Because of the enumerated reasons, the MEF-model is particularly distinctive and, therefore, is applied onto the Wigner equation.

II. MEF-WIGNER TRANSPORT EQUATION

To demonstrate the approach, the two band Schrödinger equation within the MEF framework [12] is chosen and given by

$$i\hbar \frac{\partial}{\partial t} \Psi(z,t) = \mathcal{H}\Psi(z,t) = \begin{bmatrix} H_c(z) & H_i(z) \\ H_i(z) & H_v(z) \end{bmatrix} \Psi(z,t), \quad (1)$$

where the vector $\Psi = (\Psi_c(z), \Psi_v(z))^T$ contains the wavefunctions for the conduction band Ψ_c as well as the valence band Ψ_v. The diagonal terms of the Hamiltonian $\mathcal{H}$ describe the intraband dynamics, whereas the off-diagonal elements account for the interband dynamics. The Hamiltonian $\mathcal{H}$ is defined according to

$$
\begin{aligned}
H_{c,v}(z) &= -\frac{\hbar^2}{2m_{c,v}} \frac{\partial^2}{\partial z^2} + E_{c,v} + V(z) \\
H_i(z) &= -\frac{\hbar^2 P}{m_0 E_g} \frac{\partial}{\partial z} V(z),
\end{aligned}
\quad (2)
$$

where m_c and m_v are the effective masses of the conduction band and the valence band, respectively, and m_0 is the free electron mass. $V(z)$ includes the external applied biases, the self-consistent Hartree-Fock potential as well as the bandgap discontinuities due to heterostructures [12]. The parameter P is related to Kane's matrix momentum element and $E_g = E_c - E_v$ is the bandgap determined by the conduction band edge energy E_c and valence band edge energy E_v.

The elements of the real space statistical density matrix ρ_{ij} with $i, j = c, v$ are defined as

$$\rho_{ij} = \Psi_i \left(\chi + \frac{1}{2}\xi, t \right) \Psi_j^\dagger \left(\chi - \frac{1}{2}\xi, t \right) \quad (3)$$

and after the application of the Wigner-Weyl transform the following multiband Wigner functions f_{ij} result:

$$f_{ij}(\chi, k, t) = \int d\xi \, \exp(-\imath k\xi) \cdot \rho_{ij}\left(\chi + \frac{1}{2}\xi, \chi - \frac{1}{2}\xi, t\right). \tag{4}$$

Finally, the application of the Wigner-Weyl transform onto the Liouville von Neumann equation leads to a system of coupled Wigner equations

$$\frac{\partial}{\partial t}f_{cc} = +\mathcal{A}\frac{\partial}{\partial \chi}f_{cc} + \Theta[\hat{V}]f_{cc} + \Theta[\hat{V}^-]f_{cv} - \Theta[\hat{V}^+]f_{vc}$$

$$\frac{\partial}{\partial t}f_{cv} = +\left[\mathcal{D}\frac{\partial^2}{\partial \chi^2} + \mathcal{G}\right]f_{cv} + \Theta[\hat{V}]f_{cv} +$$

$$\Theta[\hat{V}^-]f_{cc} - \Theta[\hat{V}^+]f_{vv},$$

$$\frac{\partial}{\partial t}f_{vc} = -\left[\mathcal{D}\frac{\partial^2}{\partial \chi^2} + \mathcal{G}\right]f_{vc} + \Theta[\hat{V}]f_{vc} +$$

$$\Theta[\hat{V}^-]f_{vv} - \Theta[\hat{V}^+]f_{cc}$$

$$\frac{\partial}{\partial t}f_{vv} = -\mathcal{A}\frac{\partial}{\partial \chi}f_{vv} + \Theta[\hat{V}]f_{vv} + \Theta[\hat{V}^-]f_{vc} - \Theta[\hat{V}^+]f_{cv} \tag{5}$$

where the occuring operators are defined as follows:

$$\mathcal{A} = -\frac{\hbar k}{m}, \quad \mathcal{D} = -\frac{\hbar}{4\imath m}, \quad \mathcal{G} = \frac{1}{\imath \hbar}\left\{\frac{\hbar^2 k^2}{m} + E_g\right\}, \tag{6}$$

$$\Theta[\hat{V}]f_{ij} = \frac{1}{\imath \hbar}\int \frac{dk'}{2\pi}\hat{V}(\chi, k - k')f_{ij}(\chi, k', t) \tag{7}$$

and

$$\Theta[\hat{V}^{\pm}]f_{ij} = \frac{\hbar P}{\imath m_0 E_g}\int \frac{dk'}{2\pi}\hat{V}^{\pm}(\chi, k - k')f_{ij}(\chi, k', t). \tag{8}$$

Additionally, for the demonstration of the multiband formalism, the effective masses within the conduction band are assumed to be equal with $m = m_c = -m_v$, so that a comparison with results obtained by existing multiband approaches could be possible, i.e. [11]. However, the extension is straightforward and does not affect the multiband formalism as well as the discretization due to the phase space operator presented here. The terms including the potential distributions are given by

$$\hat{V}(\chi, k) = \int d\xi \, \exp(-\imath k\xi)$$

$$\times \left\{V\left(\chi + \frac{1}{2}\xi\right) - V\left(\chi - \frac{1}{2}\xi\right) - \imath W(\xi)\right\}, \tag{9}$$

where the CAP has been introduced by an addend $\imath W(\xi)$. The terms related to the interband dynamics are defined by

$$\hat{V}^{\pm}(\chi, k) = \int d\xi \, \exp(-\imath k\xi)\mathcal{E}\left(\chi \pm \frac{1}{2}\xi\right) \tag{10}$$

introducing the electrical field $\mathcal{E}(z) = -\frac{d}{dz}V(z)$. As can be observed from these coupled Wigner equations, the standard single band Wigner equation for the conduction band arises from (5), when no external potentials are present or the Kane's matrix momentum is set to zero.

A. Numerical Discretization

For the numerical solution of the system of coupled Wigner equations (5), the formalism proposed in [7], [13] for the single band case is extended towards the multiband case. The momentum variable is discretized utilizing a standard finite difference scheme leading to a semi-discrete Wigner equation, which then is solved along with the phase space operator as in [13]. Due to the discretization utilizing N_k discretization points, only a finite range of momentum related values k within the interval $[-k_{max}, k_{max}]$ can be taken into account. Conceptually, the discretization of (5) with regard to the momentum variable k results in a system of coupled partial differential equations, which can be written as

$$\frac{\partial}{\partial t}\boldsymbol{f}_{cc} = +\boldsymbol{A}\frac{\partial}{\partial \chi}\boldsymbol{f}_{cc} + \boldsymbol{B}(\chi)\boldsymbol{f}_{cc} + \boldsymbol{B}^-(\chi)\boldsymbol{f}_{cv} - \boldsymbol{B}^+(\chi)\boldsymbol{f}_{vc}$$

$$\frac{\partial}{\partial t}\boldsymbol{f}_{cv} = +\left\{\boldsymbol{D}\frac{\partial^2}{\partial \chi^2} + \boldsymbol{G}\right\}\boldsymbol{f}_{cv} + \boldsymbol{B}(\chi)\boldsymbol{f}_{cv} +$$

$$\boldsymbol{B}^-(\chi)\boldsymbol{f}_{cc} - \boldsymbol{B}^+(\chi)\boldsymbol{f}_{vv}.$$

$$\frac{\partial}{\partial t}\boldsymbol{f}_{vc} = -\left\{\boldsymbol{D}\frac{\partial^2}{\partial \chi^2} + \boldsymbol{G}\right\}\boldsymbol{f}_{vc} + \boldsymbol{B}(\chi)\boldsymbol{f}_{vc} +$$

$$\boldsymbol{B}^-(\chi)\boldsymbol{f}_{vv} - \boldsymbol{B}^+(\chi)\boldsymbol{f}_{cc}$$

$$\frac{\partial}{\partial t}\boldsymbol{f}_{vv} = -\boldsymbol{A}\frac{\partial}{\partial \chi}\boldsymbol{f}_{vv} + \boldsymbol{B}(\chi)\boldsymbol{f}_{vv} + \boldsymbol{B}^-(\chi)\boldsymbol{f}_{vc} - \boldsymbol{B}^+(\chi)\boldsymbol{f}_{cv} \tag{11}$$

Due to the discretization, the diffusion process related operators $\mathcal{A} \to \boldsymbol{A}, \mathcal{D} \to \boldsymbol{D}$ and the intraband dynamics related drift operators $\mathcal{G} \to \boldsymbol{G}, \Theta[V] \to \boldsymbol{B}$ as well as the interband dynamics related drift operator $\Theta[V^{\pm}] \to \boldsymbol{B}^{\pm}$ are introduced. The elements of the corresponding matrix representation (11) are indicated by the subscripts $j, j' \in 1, \ldots, N_k$ and given by

$$\boldsymbol{A}_{j,j'} = \frac{-\hbar k_j}{m}\delta_{j,j'} \qquad \boldsymbol{D}_{j,j'} = \frac{-\hbar}{4\imath m}\delta_{j,j'}$$

$$\boldsymbol{G}_{j,j'} = \left\{\frac{\hbar^2 k^2}{m} + E_g\right\}\frac{\delta_{j,j'}}{\imath \hbar} \qquad \boldsymbol{B}_{j,j'} = \frac{\Delta k}{2\pi}\frac{\hat{V}(\chi, k_j - k_j')}{\imath \hbar}$$

$$\boldsymbol{B}^{\pm}_{j,j'} = \frac{\hbar P}{\imath m_0 E_g}\frac{\Delta k}{2\pi}\hat{V}^{\pm}(\chi, k_j - k_j') \tag{12}$$

with $\delta_{j,j'}$ being the Kronecker-delta. The approximation of the integral kernels, as apparent in the drift related matrices $\boldsymbol{G}$, $\boldsymbol{B}$ and $\boldsymbol{B}^{\pm}$, is carried out applying a standard midpoint rule leading to the relations

$$\hat{V}(\chi, k_j - k_j') = \sum_{i=1}^{N_k}\Delta\xi \exp(-\imath(k_j - k_j')\xi_i) \cdot$$

$$\left\{V\left(\chi + \frac{\xi_i}{2}\right) - V\left(\chi - \frac{\xi_i}{2}\right) - \imath W(\xi_i)\right\} \, .$$

$$\hat{V}^{\pm}(\chi, k_j - k_j') = \sum_{i=1}^{N_k}\Delta\xi \exp(-\imath(k_j - k_j')\xi_i) \cdot \mathcal{E}\left(\chi \pm \frac{\xi_i}{2}\right) \tag{13}$$

The system of coupled discretized Wigner equations is reformulated in order to derive a phase space operator addressing the spatial approximation with regard to the variable χ and following the concept as described in [13]. Furthermore, a supervector $f = (f_{\mathrm{cc}}, f_{\mathrm{cv}}, f_{\mathrm{vc}}, f_{\mathrm{vv}})^T$ is introduced containing all the discretized Wigner functions (4). Along with these abbreviations, the coupled Wigner equations (11) can be rewritten as

$$\left[D_2 \frac{\mathrm{d}^2}{\mathrm{d}\chi^2} + D_1 \frac{\mathrm{d}}{\mathrm{d}\chi} + D_0(\chi) \right] f(\chi) = 0, \qquad (14)$$

which is the standard form for the derivation of the phase space operators containing a derivative up to the 2nd order in the χ-direction [13]. The matrices $[D_2], [D_1], [D_0]$ of the order $4N_k \times 4N_k$ are given by

$$D_2 = \mathrm{diag}\,(0, D, -D, 0)$$
$$D_1 = \mathrm{diag}\,(A, 0, 0, -A)$$
$$D_0(\chi) = \begin{bmatrix} B(\chi) & B^-(\chi) & -B^+(\chi) & 0 \\ B^-(\chi) & B(\chi) + G & 0 & -B^+(\chi) \\ -B^+ & 0 & B(\chi) - G & B^-(\chi) \\ 0 & -B^+(\chi) & B^-(\chi) & B(\chi) \end{bmatrix} \tag{15}$$

with 0 being a $N_k \times N_k$ zero-valued matrix. As can be observed, only the matrix D_0 contains non zero-valued off-(block-)diagonal elements including the interband coupling effects. Following the procedure as described in detail in [13], the discretized Wigner functions f_{ij} are obtained.

B. Inflow Boundary Conditions for Quantum Transport

Similar to the conventional single band case [1]–[3], a quantum statistical distribution function has to be defined at the boundaries. The injection of electrons is then related to the sign of the corresponding momentum k. The inflow of electrons is provided by the distribution function $f_{l,r}^e$ for the left contact considering positive values and for the right contact with negative values given by

$$f_{\mathrm{cc}}(\chi_{1,N_\chi}, k \gtrless 0, t) = f_{l,r}^e(k \gtrless 0). \tag{16}$$

For the hole transport, a relationship must be chosen in a way so that the holes are injected where the electrons are flowing out [14] leading to

$$f_{\mathrm{vv}}(\chi_{1,N_\chi}, k \gtrless 0, t) = f_{l,r}^h(k \gtrless 0) \tag{17}$$

with $f_{l,r}^h$ being the hole distribution function at the left and right contact. Obviously, the components f_{cv} and f_{vc} of the discretized Wigner function can be assumed to be zero-valued functions at the boundaries of the computational domain. In addition, von-Neumann boundary conditions are provided, namely

$$\frac{\mathrm{d}}{\mathrm{d}\chi} f_{\mathrm{cv}} = \frac{\mathrm{d}}{\mathrm{d}\chi} f_{\mathrm{vc}} = 0. \tag{18}$$

In a similar manner as for the single band case described in [13], the boundary conditions are incorporated into the system matrix.

	contact	spacer	barrier	well	barrier	spacer	contact
L in nm	30	15	3	5	3	15	30
V in eV	0	0	0.11	0.51	0.11	0	0

TABLE I: Parameters of the resonant interband tunneling diode.

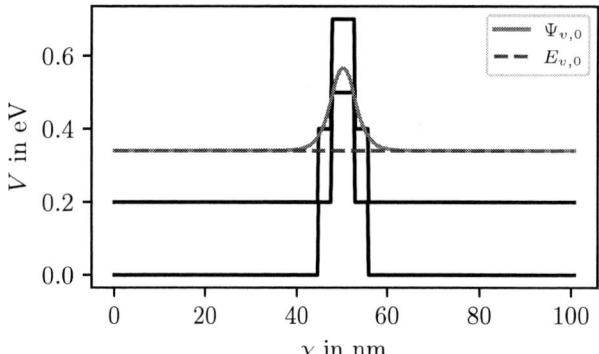

Fig. 1: Band diagram of the resonant interband tunneling diode along with the wavefuntion $\Psi_{v,0}$ (with $E_{v,0}$) of the bounded state within the valence band is depicted.

III. NUMERICAL EVALUATION

To demonstrate the capability of the proposed formalism, a resonant structure is analyzed, in which interband tunneling effects are assumed to be prominent leading to a negative differential resistance. Here, a similar structural and material parameters are adopted from [11]. The relevant structural parameters of the resonant interband tunneling diode are provided in Table I and the band diagram is depicted in Fig. 1. The bounded valence band state $\Psi_{v,0}$ located at approximately $E_{v,0} = 0.34\mathrm{eV}$ is depicted, too, as the results obtained later on are discussed with regard to the bounded valence band state. This bounded state serves as the resonant state and its value $E_{v,0}$ is obtained by solving the Schrödinger equation. For the discretization of the computational domain, a step width size $\Delta\chi = 0.25\mathrm{nm}$ has been chosen. The momentum grid is discretized along with $N_k = 100$ discretization points within the interval $[-1.5\mathrm{nm}^{-1}, 1.5\mathrm{nm}^{-1}]$. The band gap is assumed to be $E_g = 0.2\mathrm{eV}$ and the momentum matrix element is set to $P = 5\mathrm{nm}^{-1}$. The effective mass is set to a value of $m = 0.027m_0$ and the doping concentration at the contacts is $1 \cdot 10^{24}\mathrm{m}^{-3}$.

The current-voltage characteristic for the device is depicted in Fig. 2 presuming a linear voltage drop in between the left and right contact regions. The bias sweep within 0V and 0.55V is discretized with 56 points and the current density obtained for each discrete bias point is obtained by summing up both current densities stemming from the conduction as well as the valence band [13]. As expected, the negative differential resistance behavior can be observed and coincides with the result given in [12]. In addition, when analyzing the location the current peak at approximately $U_{\mathrm{peak}} \approx 0.13\mathrm{V}$, a fairly good agreement with the determined eigenenergy $E_{v,0}$ of the resonant state can be established. This resonant state is about 0.14eV above the conduction band energy, additionally justifying the qualitative behavior of the current

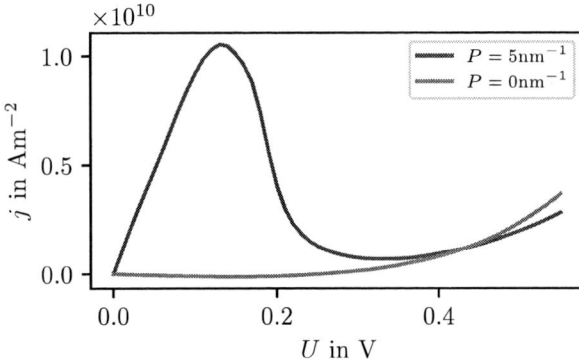

Fig. 2: Current-voltage characteristic for the resonant interband tunneling diode.

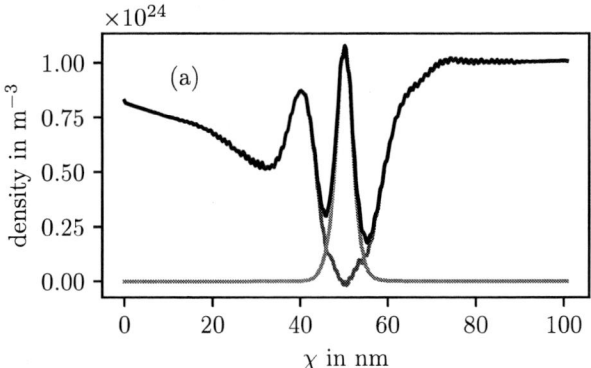

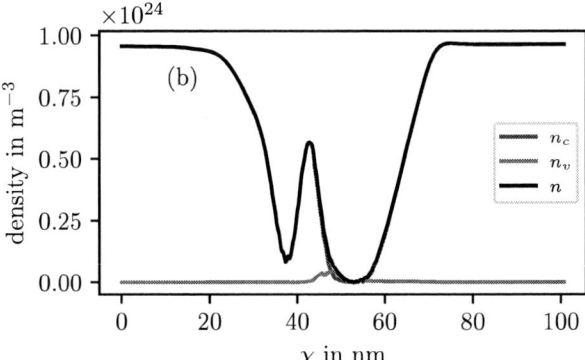

Fig. 3: Carrier density depicted for the peak voltage $U_{\text{peak}} = 0.13$V (a) and $U_{\text{valley}} = 0.35$V (b).

density. Furthermore, when the momentum matrix element P is zero valued, no negative differential resistance can be observed from Fig. 2 as expected. Further on, the carrier densities for the valence band n_v, the conduction band n_c and the corresponding sum $n = n_c + n_v$ are shown in Fig. 3 for the peak voltage at about $U_{\text{peak}} = 0.13$V as well as for the valley voltage at about $U_{\text{valley}} = 0.35$V, respectively. These densities are determined by taking the zeroth order moment of the corresponding Wigner function. The carrier density for the peak voltage case U_{peak} exhibits large values within the well region due to the strong interaction with the resonant state, which assists the tunneling process. However, when the carrier density for the valley voltage U_{valley} is considered, the resonant state does not contribute to the transport process leading to an extremely small density within the well regions.

IV. CONCLUSION

The proposed formalism forms the basis for further device oriented analysis, for which different band edge discontinuities, different effective masses e. g. related to the Luttinger-Kohn parameters, multiband models including strain effects as well as different momentum matrix elements and their spatial dependence can be considered revealing the full potential of the Wigner formalism.

REFERENCES

[1] W. R. Frensley, "Wigner-function model of a resonant-tunneling semiconductor device," *Phys. Rev. B*, vol. 36, pp. 1570–1580, 3 Jul. 1987.

[2] ——, "Boundary conditions for open quantum systems driven far from equilibrium," *Rev. Mod. Phys.*, vol. 62, pp. 745–791, 3 Jul. 1990.

[3] H. Tsuchiya, M. Ogawa, and T. Miyoshi, "Simulation of quantum transport in quantum devices with spatially varying effective mass," vol. 38, no. 6, pp. 1246–1252, Jun. 1991.

[4] D. Querlioz, J. Saint-Martin, V. Do, A. Bournel, and P. Dollfus, "A study of quantum transport in end-of-roadmap dg-mosfets using a fully self-consistent wigner monte carlo approach," vol. 5, no. 6, pp. 737–744, Nov. 2006.

[5] H. Kosina, "Wigner function approach to nano device simulation," *Int. J. Comput. Eng. Sci.*, vol. 2, no. 3-4, pp. 100–118, 2006.

[6] H. Jiang and W. Cai, "Effect of boundary treatments on quantum transport current in the green's function and wigner distribution methods for a nano-scale dg-mosfet," *J. Comput. Phys.*, vol. 229, no. 12, pp. 4461–4475, 2010.

[7] D. Schulz and A. Mahmood, "Approximation of a phase space operator for the numerical solution of the wigner equation," vol. 52, no. 2, pp. 1–9, 2016.

[8] L. Schulz and D. Schulz, "Complex absorbing potential formalism accounting for open boundary conditions within the wigner transport equation," vol. 18, pp. 830–838, 2019.

[9] G. Borgioli, G. Frosali, and P. F. Zweifel, "Wigner approach to the two-band kane model for a tunneling diode," *Transp. Theory Stat. Phys.*, vol. 32, no. 3-4, pp. 347–366, 2003.

[10] O. Morandi, "Multiband wigner-function formalism applied to the zener band transition in a semiconductor," *Phys. Rev. B*, vol. 80, p. 024 301, 2 Jul. 2009.

[11] G. Frosali and O. Morandi, "A quantum kinetic approach for modeling a two-band resonant tunneling diode," *Transp. Theory Stat. Phys.*, vol. 36, no. 1-3, pp. 159–177, 2007.

[12] O. Morandi and M. Modugno, "Multiband envelope function model for quantum transport in a tunneling diode," *Phys. Rev. B*, vol. 71, no. 23, Jun. 2005.

[13] L. Schulz and D. Schulz, "Formulation of a phase space operator accounting for the spatial variation of the effective mass within the wigner equation," *J. Comput. Electron*, 2020.

[14] P. Weetman and M. S. Wartak, "Wigner function modeling of quantum well semiconductor lasers using classical electromagnetic field coupling," *J. Appl. Phys.*, vol. 93, no. 12, pp. 9562–9575, 2003.

AUTHOR INDEX

Abe K 75
Acosta-Alba P 89
Adamu-Lema F 237, 241, 273, 293
Agnew M 145
Ahn H 19
Ahn PH 289
Alexander B 307, 323, 327
Amoroso SM 35
Anderson E 181, 277
Antonelli M 133
Arai F 161
Arfelli F 133
Arihori K 367
Asai H 55
Asai H 93
Asenov A 237, 241, 273, 293
Asenov P 35
Atmaca G 233
Avila Herrera F 109
Ayyagari-Sangamalli B ... 307, 323, 327

Badami O 241, 273, 293
Bae J 343
Baek KH 347
Baek S 47
Balasingam A 197
Barnes JP 89
Baumgartner O 63
Bayle R 173
Bazizi EM 307, 323, 327
Bejenari I 67, 71
Bhagdikar S 117
Biasiol G 133

Blaise P 249
Blonkowski S 173
Bournel A 27
Brown AR 35
Buckley J 233
Burenkov A 67, 71
Byun J 205

Cai J 285
Campbell D 181, 277
Carrato S 133
Carrillo-Nunez H 293
Cautero G 133
Cervenka J 189
Chan K 47
Chanemougame D 301
Charles J 249
Chen CY 355
Chen E 285
Chen YD 101
Cheng CY 149
Cho K 47
Choe G 165
Choi J 297
Chowdhury N 113
Chuang MH 23, 355
Collard E 233
Cristiano F 89
Cuesta J 269
Cueto O 173
Curvers B 71

Dagault L 89

Davier B	27	Guichard E	249
De Angelis D	133	Gwoziecki R	233
Deretzis I	67, 71		
Doh J	347	Hamano T	125
Dollfus P	27, 145	Han S	347
Driussi F	133	Han SC	297
Duenkel S	169	Hattori J	55, 93
Dumas P	43	He R	323
Dutta T	237, 241, 273, 293	Hemeryck A	43
		Henry H	173
Ender J	209, 213	Herrmann T	169
Eriguchi K	125	Hilario F	43
Escoffier R	221	Hong SM	217, 289, 297
Esseni D	133	Hosotani K	161
		Hossinger A	39
Fauziah K	15	Huang A	359
Filipovic L	59, 315	Huang J	101
Fiorentini S	209, 213	Huet H	89
Fischetti MV	281, 363	Huet K	71
Fujita Y	15	Hur J	165
Fujiwara M	161		
Fukuda K	55, 93	Iizuka K	137
Fukuda M	265	Iizuka T	109, 157, 319
		Ikeda H	15
Gaddemane G	281, 363	Ikegami T	93
Gandus G	177	Ishikawa Y	265
Gao X	181, 277, 359	Ishimabuchi H	51
Georgiev V	241, 273, 293	Itoh KM	1
Georgiev VP	31	Ivie J	181, 277
Ghouli SE	311		
Godoy A	269	Jain R	169
Goes W	209, 213	Jang I	19
Gopalan S	281, 363	Jang JH	217
Grasser T	339	Jang W	347
Grebot J	145	Jaud MA	221, 233
Grimaldi MG	71	Jech M	339

Jeon J 19
Jeon Y 347
Jeong C 253, 347
Jeong J 47
Jiang L 307, 323
Joseph E 101
Juge A 311
Julliard PL 43
Jung J 343

Kamakura Y 15
Kamino T 137
Kang D 205
Kang G 19
Kang JH 261
Kang M 47, 311
Kapoor U 249
Kariya N 161
Karner M 63
Kato T 161
Kayama Y 51
Kerdiles S 89
Khan AI 165
Kikuchihara H 109, 157, 319
Kim D 19, 47, 347
Kim DS 51, 253, 343
Kim J 19, 343, 347
Kim K 343
Kim MJ 217
Kim S 47
Kim Y 253
Kim YS 343, 347
Klemenschits X 59
Kobayashi M 83
Kola SR 79, 375
Kondo M 161

Koshimoto H 51
Kosik R 189
Kosina H 185, 189
Kotani J 55
Kubis T 249
Kunikiyo T 137
Kuroda T 93
Kurusu T 161
Kwon OK 343
Kwon U 51
Kwon UH 343

La Magna A 67, 71
Lacombe E 145
Lapham P 273
Le ND 27
Lee H 311
Lee J 35
Lee K 47
Lee Y 177
Lemus DA 249
Lespinasse B 71
Levy A 197
Li H 197
Li K 225
Li LJ 285
Li Y 23, 79, 355, 375
Lin XW 35
Liu CC 355
Liu W 101
Lombardo SF 71
Lopez A 145
Lu TM 181, 277
Luisier M 177

Ma T 7

Mabuchi T	105	Nebesnyi V	237
Maciazek P	241	Nestor DR	157
Mahapatra S	113, 117, 121	Nguyen A	201
Maki D	153	Nguyen H	201
Mamaluy D	181, 277	Nichetti C	133
Mamdy B	145	Nishitani K	161
Manouvrier JR	145	Nogita T	15
Manstetten P	39		
Marin EG	269	Oda K	75
Martinie S	221, 311	Ogawa M	153, 367
Matic M	371	Ohuchi K	161
Matsukawa T	75	Onoue S	161
Matsuki T	75	Orio R	209, 213
Mattausch HJ	109, 158, 319	Oshiyama A	11
McGhee J	31	Oussaiti Y	145
Mediana-Rull A	269		
Medina-Bailon C	241	Pal A	307, 323, 327
Medina-Bailon C	273, 293	Pala M	3, 27, 145
Mehta K	351	Palestri P	133
Mendez JP	181	Parent D	201
Menk RH	133	Parihar N	113
Milardovich D	339	Park J	217
Misra S	181, 277, 327	Pasadas F	269
Miura M	379	Passerone D	177
Miura-Mattausch M	109, 157, 319	Philippe T	173
Mohamedou M	209, 213	Pichler P	67, 71
Mohiyaddin FA	253	Pilotto A	133
Monsieur F	43	Plappa M	173
Mori N	93	Plissonnier M	233
Moroz V	35	Poiroux T	221, 233, 311
Mugny G	145	Poljak M	371
Mukhopadhyay S	113		
Myung S	253, 347	Quenette V	145
Nagy D	293	Raina P	197
Navarro D	157, 319	Raju SS	351

Reyntjens PD 97
Reza S 359
Richter R 169
Rideau D 43, 145
Royet AS 89
Rozeau O 221, 311
Ruiz FG...269
Ryu 261
Ryu J 253, 347
Rzepa G 63

Saint-Martin J 27
Saito H 225
Samadder T 113
Samukawa S......................................375
Saremi M 307, 323, 327
Sato H 137
Sato-Iwanaga J 367
Scharinger A 39
Scheer P 311
Schmucker S 181, 277
Schulz D 331, 383
Schulz L 331, 383
Sciuto A 71
Selberherr S 59, 209, 213
Selmi L 133
Sengupta A..241
Seo J 229
Sharma U 121
Shigyo N 129
Shim W 165
Shimada Y.. 161
Shimokawa J......................................161
Shin M 205, 299
Shin Y 47
Sonoda K 137

Sor´ee B 97
Souma S............................... 153, 367
Stanojevic Z 63
Steiner K 63
Steinhartova T................................. 133
Strof G 63
Su SK 285
Suwa T... 141
Suzuki K 379
Suzuki M 367
Suzuki Y 15
Sverdlov V 185, 209, 213

Takamiya M 335
Takatsuka H 109
Tanimoto H 161
Teo KH 225
Thoti N 79, 375
Tiwari R 113
Tiwari S 97
Tokuhira H.. 161
Tokumasu T 105
Tomita M 75
Toral-Lopez A 269
Townsend M 249
Tracy L 181, 277
Trask N 359
Triozon F 221
Tsuda M.. 161

Uchiyama Y 161
Uene N 105
Urabe K 125

Van de Put ML...................97, 281, 363
Vandenberghe WG 97

Vasileska D 149

Vaysset A 221

Venimadhavan S........................... 201

Venkattraman A............................ 201

Verstraete A 71

Waldhoer D 339

Waltl M 339

Wang B 351

Wang Q 101

Watanabe T 15, 75

Wehbe-Alause H 145

Weinbub J 39

Widjaja Y 237

Wong H....................................... 193

Wong HSP................................... 285

Wong HY 201, 351

Xiao M....................................... 351

Yagyu E 225

Yaita J 55

Yamada S 51

Yamashita T 137

Yang T 35

Yang YS...................................... 355

Yasuhara S 105

Yokota Y 161

Yoo J... 51

Yoo JH 343

Yu S.. 165

Yu TH 245

Yvon A 233

Zaitsu M...................................... 105

Zaka A 169

Zhang Q...................................... 379

Zhang Y 351

IEEE
445 Hoes Lane
Piscataway, NJ 08854-4141

ISBN 978-1-7281-7354-2